A LIE GROUP: ROTATIONS IN QUANTUM MECHANICS

A LIE GROUP:
ROTATIONS IN QUANTUM MECHANICS

Jean-Marie NORMAND

Commissariat à l'Energie Atomique
Centre d'Etudes Nucléaires de Saclay

1980

NORTH-HOLLAND PUBLISHING COMPANY
AMSTERDAM • NEW YORK • OXFORD

ISBN: 0 444 86125 4

Published by:
NORTH-HOLLAND PUBLISHING COMPANY—AMSTERDAM • NEW YORK • OXFORD

Sole distributors for the U.S.A. and Canada
ELSEVIER NORTH-HOLLAND, INC.
52 Vanderbilt Avenue
New York, N.Y. 10017

Library of Congress Cataloging in Publication Data

Normand, Jean Marie, 1947-
A Lie group, rotations in quantum mechanics.

Includes index.
1. Rotation groups. 2. Lie groups. 3. Quantum theory. I. Title.
QC174.17.R65N67 530.1'2 80-22977
ISBN 0-444-86125-4

PRINTED IN THE NETHERLANDS

To my Father

PREFACE

The three-dimensional rotation group and the quantum theory of angular momenta are widely studied subjects and many works have already been devoted to them. Therefore, it comes natural to ask why one more book concerning these subjects. The main reason for this, emphasized by the choice of the title, is that this volume is devoted not only to the rotation group in Quantum Mechanics. Actually, our attempt is based on the following remarks. The rotations form a symmetry group which clearly plays an important role in many fields of fundamental physics such as for instance the study of molecules, nuclei, and elementary particles. Furthermore, from a pedagogical point of view this group and the associated one of unitary and unimodular two-dimensional complex matrices are at the same time sufficiently simple so that intuition can play a role, yet sufficiently complex to exhibit most features of compact Lie groups. My aim has thus been twofold : (1) To bring out a set of general methods and results concerning the representations of compact Lie groups, and mainly the linear ones. (2) To derive many specific properties and computational methods for the representations of the three-dimensional rotation group in Quantum Mechanics.

Needless to say, this monograph is not intended to be, in any sense, an exhaustive treatise of group theory in Quantum Mechanics. Among the numerous main omissions let us quote for instance the classification of simple Lie algebras and also the study of the representations of the symmetric group. This volume obviously owes a great deal to many excellent textbooks, review articles, and lecture notes ; especially works of V. Bargmann and E.P. Wigner. The list of references is not meant to be an exhaustive bibliography about group theory and rotations. It quotes only those sources which I know and find helpful. This is why I apologize in advance to all of those who will not find their work properly quoted.

The book is roughly divided into four parts. The first five chapters are devoted to general remarks concerning the rotation group and its representations. The unitary representations of the two-dimensional special unitary group are studied in Chaps. 6,7. The important mathematical tool of tensor algebra is developed in detail in Chaps. 8,9. The last two chapters are devoted to the representations in commonly used functional vector spaces and to the applications to Quantum Mechanics. The Appendices A–D which are relative to vector spaces, algebras, and groups, are set out in the form of mathematical definitions and theorems available for an understanding of the text. The other appendices are concerned with specific results, tables, and examples in connection with the rotation group.

The reader is assumed to have a general knowledge of Quantum Mechanics and Mathematics. The book is intended for both experimentalists and theorists, but also for students from the graduate level. The main formulae concerning the rotation group are collected at the end of most of the chapters. The attempt at presenting a pedagogical survey of the subject sometimes leads to the introduction of cumbersome notations which, of course, can be omitted by those who are familiar with the underlying difficulties.

I wish to thank all friends and colleagues in the Service de Physique Théorique in Saclay who helped me. I especially have a great debt towards Gilbert Mahoux and Madan Lal Mehta who read the manuscript and made criticisms and suggestions. I also thank Jean-Michel Drouffe, Pierre Moussa, and Jacques Raynal for stimulating discussions. I am grateful to Neil Sullivan who did his best to palliate my insufficient mastery of the English language.

It would have been impossible to carry through this task without the help of my laboratory. I shall never thank sufficiently Dany Bunel who typed the manuscript with as much talent as prettiness. I am greatful to Madeleine Porneuf who did the unpleasant task of re-reading the proofs and Geneviève Gaujour (D.I.N.T.) who carefully drew the figures. I wish to thank the Commissariat à l'Energie Atomique and its staff who supported the work.

Why not admit in the end a last reason for having written this book : the author took pleasure to do it.

Saclay
July, 1980

Jean-Marie Normand

TABLE OF CONTENTS

A LIE GROUP:
ROTATIONS IN QUANTUM MECHANICS

CHAPTER 1

PRELIMINARY REMARKS ABOUT LINEAR ALGEBRA AND NOTATIONS

We are going to deal with vector spaces and we suppose this is a well known subject, at least when the dimension is finite (Appendix A.1). Nevertheless, as an introduction we want to point out the importance of the intrinsic character of the objects we use (vectors, tensors,...). By intrinsic we mean independent of the choice of basis. This is precisely the aim of tensor algebra which, once a basis is chosen, introduces a one-to-one correspondence between intrinsic objects and sets of objects (components, matrix elements, metric tensor, structure constants,...). The intrinsic character implies that these sets of objects are characterized by their laws of transformation under a change of basis. Later on (Chap. 8) we will even forget about vector spaces and only keep the idea of sets of objects of any kind (components of vectors, tensorial operators,...) which are defined with respect to a system of reference, and which are subject to a special linear transformation when this system is changed.

At the same time, we specify the notations for tensor algebra which are used throughout these lectures. This is done only for vector and matrix representatives of linear operators, but can easily be generalized to tensors of any order.

Finally, we recall the two interpretations of a bijective linear application, namely the active and the passive points of view. The so-called equivalence between them is pointed out.

1.1. SOME ELEMENTARY RESULTS OF LINEAR ALGEBRA

1.1.1. VECTOR SPACES

The definition of a vector space is recalled in the Appendix A.1. Only finite-dimensional vector spaces are considered here. For the generalities about linear algebra the Einstein convention is used, i.e. :

(1) Each monomial with indices occurring twice, in upper position in one factor and lower position in another has to be summed over the values 1,2,...,n, if n is the dimension of the considered vector space.

(2) Any equality, where there exist free indices (not summed over) at the same level in both sides of the equality, has to be considered for all values of the free indices.

Let us consider an n-dimensional vector space E_n over the field K ; K will always be commutative and, except in this introductory chapter, K will be the real or complex field (R or C). A set of n linearly independent vectors e_i denoted by (e_i), is a basis of E_n, and any vector v can be expressed as a linear combination of them :

$$v = e_i \, v^i \quad . \tag{1-1}$$

A change of basis is defined by a nonsingular matrix t :

$$e'_i = e_j \, t^j_i \quad , \quad e_j = e'_i (t^{-1})^i_j \quad . \tag{1-2}$$

This induces the following change for the components :

$$v = e_j v^j = e'_i \, v'^i = e_j \, t^j_i \, v'^i \quad , \quad v^j = t^j_i \, v'^i \quad . \tag{1-3}$$

The set (v^i) is a tensorial set of components of the intrinsic object v. This set is characterized by the law of transformation given in Eq. (1–3), namely any set of elements of the field which experiences such a law of transformation under a change of basis is the set of components of a vector in E_n.

Unlike the e'_i, the v'^i are given in terms of the v^i through the matrix t^{-1} and not the matrix t :

$$v'^i = (t^{-1})^i_j \, v^j \quad . \tag{1-4}$$

By definition (e_i) is called a covariant set and (v^i) a contravariant one. The difference of behaviour of both sets (e_i) and (v^i) under a change of basis is underlined by the up–and–down index notation. More generally, the level of the indices of an object, independently of its nature (basis vector, components,...), characterizes its law of transformation under a change of basis, i.e. its tensorial character :

The upper (resp. lower) index of t (resp. t^{-1}) always refers to objects defined with respect to the first basis.

Nevertheless this up–and–down index notation is unusual for the rotation matrices $(\mathcal{D}^j_{mm'})$ defined in Subsec. 6.1.2. There all the indices are written on the same level, i.e. :

$$v = e_i \, v_i \quad , \quad e'_i = e_j \, t_{ji} \quad , \quad v_j = v'_i \, t_{ij} \quad , \tag{1-5}$$

where, as always the summation is over indices which occur twice. This notation is related to the previous one by $t^i_j = t_{ij}$. Thus :

The left (resp. right) index of t (resp. t^{-1}) refers to objects defined with respect to the first basis.

Although less powerful, the customary notation in Eq. (1–5) will be used for the rotation matrices ($\mathcal{D}^j_{mm'}$).

1.1.2. DUAL VECTOR SPACES

The set of linear mappings of E_n into the field K is the dual vector space E^*_n of E_n. A vector space structure over the same field K can easily be given to E^*_n. This space has the same dimension as E_n. A particular choice of basis of $\overset{*}{E}_n$ is the associated dual basis (e^{*i}) defined by :

$$\forall\, v = e_i\, v^i \quad , \qquad e^{*i}(v) = v^i \quad . \tag{1-6}$$

Clearly (e^{*i}) is a contravariant set, as denoted by the position of the index, since under a change of basis its law of transformation reads :

$$e'^{*i}(v) = v'^i = (t^{-1})^i_j\, v^j = (t^{-1})^i_j\, e^{*j}(v) \quad , \quad e'^{*i} = (t^{-1})^i_j\, e^{*j} \quad . \tag{1-7}$$

The correspondence :

$$v = v^i e_i \in E_n \leftrightarrow v^* = \sum_{i=1}^{n} v^i\, e^{*i} \in E_n^* \tag{1-8}$$

defines an isomorphism between E_n and E_n^* which depends upon the choice of the basis (e_i) of E_n. Whereas, iterating the duality yields the dual vector space E_n^{**} of E_n^*, and there exists an intrinsic isomorphism between E_n and E_n^{**}. Indeed, let (e_i^{**}) be the associated dual basis of E_n^{**} defined by :

$$\forall\, v^* = v_i\, e^{*i} \in E_n^* \quad , \quad e_i^{**}(v^*) = v_i \quad .$$

Then the correspondence :

$$v = v^i e_i \in E_n \leftrightarrow v^{**} = v^i e_i^{**} \in E_n^{**} \tag{1-9}$$

is an isomorphism between E_n and E_n^{**} which is independent of the choice of the basis (e_i), since both bases (e_i) and (e_i^{**}) transform in the same way under a change of basis, i.e. both sets of basis vectors have the same tensorial character.

1.1.3. THE SET OF LINEAR APPLICATIONS

The set of linear mappings of E_n into itself is called $L(E_n)$. The following considerations can be extended in a straightforward way when images are taken in another vector space E' over the same field K. One can easily introduce a vector space structure on the set $L(E_n)$ over the field K.

The main consequence of linearity is that any element A of $L(E_n)$ is defined by specifying the images of any basis vector of E_n. This can be done by a set of elements in K, i.e. the matrix elements of A in the basis (e_i) :

$$A\, e_i = e_j\, A^j_i \quad . \tag{1-10}$$

With the notations of Eq. (1–5), by definition the summation is over the left index for the image of a basis vector :

$$A\, e_i = e_j\, A_{ji} \quad . \tag{1-11}$$

For any vector v one has :

$$A\,v = A(e_i v^i) = A(e_i)v^i = e_j A^j_i v^i = e_j A^j_i e^{*i}(v) \quad .$$

This implies the following relation between the intrinsic linear application A and the basis dependent set (A^j_i) of elements in K :

$$A = e_j\, A^j_i\, e^{*i} \quad . \qquad (1\text{-}12)$$

One can show that the set of linear applications $(e_j e^{*i})$ is a basis of $L(E_n)$. In agreement with the general rule specified previously, the law of transformation of matrix elements of A under a change of basis is, using Eqs. (1–2,7) :

$$A = e_j A^j_i e^{*i} = e'_k A'^k_\ell e'^{*\ell} = e_j t^j_k A'^k_\ell (t^{-1})^\ell_i e^{*i} \;\Rightarrow\; A^j_i = t^j_k A'^k_\ell (t^{-1})^\ell_i \quad , \qquad (1\text{-}13)$$

or in matrix notation :

$$A = t\,A'\,t^{-1} \quad . \qquad (1\text{-}14)$$

From this equation one deduces that the determinant of a linear operator is an intrinsic quantity, i.e. independent of its matrix representative :

$$\det A = (\det t)(\det A')(\det t)^{-1} = \det A' \quad .$$

1–The choice of the summation over the left index with notations of Eqs. (1–5,11) implies for the transformed–vector components a summation over the right index :

$$(A\,v)_j = A_{ji}\, v_i \quad . \qquad (1\text{-}15)$$

2–If one adopts the column–matrix notation for the vector components, this leads to the following row and column interpretation of A :

$$A^j_i \quad \text{jth row} \;,\; \text{ith column} \quad . \qquad (1\text{-}16)$$

Among the linear applications, let us consider the subset of regular applications A, i.e. such that A^{-1} exists or equivalently det A $\neq 0$. The set of these regular mappings form a group (Def. C.1), which is isomorphic (Def. C.22) to the linear group over the field K : GL(n,K), i.e. the set of $n\times n$ regular matrices the elements of which belong to K. One subgroup (Def. C.9) of this group is the subset of $n \times n$ regular and unimodular matrices, namely the determinant of which is one. This is the special linear group SL(n,K) which is isomorphic to the subgroup of regular and unimodular linear applications. Moreover SL(n,K) is an invariant subgroup (Def. C.11) of GL(n,K) since for any regular application A and any unimodular one B, ABA^{-1} is still unimodular.

1.1.4. HERMITIAN VECTOR SPACES AND UNITARY GROUPS

i) A vector space E_n over the complex field (K = C) is said to be Hermitian if an inner product of any two vectors v and w is defined by a Hermitian sesquilinear form g, such that the associated quadratic form is positive definite :

$$(v,w) = g(v,w) \quad . \tag{1-17}$$

Let E_n be a vector space over the complex field. A sesquilinear form is a mapping of $E_n \times E_n$ into the complex field which is linear for the right argument and antilinear for the left one :

$$g(\lambda_1 v_1 + \lambda_2 v_2, \lambda_3 v_3) = \lambda_1^* \lambda_3 \, g(v_1,v_3) + \lambda_2^* \lambda_3 \, g(v_2,v_3) \quad .$$

This form is said to be :
(1) Hermitian if for all v and w, $g(v,w) = [g(w,v)]^*$;
(2) nondegenerate if for every v, $g(v,w) = 0$ implies $w = 0$; and otherwise degenerate.
The associated quadratic form is defined for every v by : $q(v) = g(v,v)$.
This quadratic form is said to be :
(1) definite if $q(v) = 0$ if and only if $v = 0$,
this implies the nondegenerate character of the sesquilinear form g ;
(2) positive if for any nonzero v, q(v) is real positive.

Two vectors are orthogonal if their inner product vanishes. One can prove the Schwarz and Minkowski inequalities :

$$|g(v,w)| \leqslant [q(v)q(w)]^{1/2} \quad , \quad q(v+w)^{1/2} \leqslant q(v)^{1/2} + q(w)^{1/2} \quad . \tag{1-18}$$

These relations allow us to define a Hermitian norm which is a real non-negative number :

$$\|v\| = q(v)^{1/2} \quad . \tag{1-19}$$

A norm is a mapping of E_n into R^+ which fulfils the conditions :

$$\|v\| = 0 \Leftrightarrow v = 0 \quad , \quad \|\lambda v\| = |\lambda| \; \|v\| \quad , \quad \|v+w\| \leqslant \|v\| + \|w\| \quad .$$

ii) In defining an inner product no reference is made to a basis, so that inner product and norm are intrinsic objects on E_n. Once again the linear character of g implies that the inner product is determined by assigning values to inner products of the basis vectors, this defines the metric tensor :

$$g_{ij} = (e_i,e_j) = g_{ji}^* \quad . \tag{1-20}$$

In terms of the components, the inner product of any two vectors, $v = e_i v^i$ and $w = e_j w^j$, takes the form :

$$(v,w) = v^{i*} \, g_{ij} \, w^j \quad . \tag{1-21}$$

Under a change of basis defined by a matrix t, Eq. (1–2), the law of transformation of the metric tensor is, in agreement with the general rule given in Subsec. 1.1.1 :

$$g'_{ij} = (e'_i, e'_j) = t_i^{k*} t_j^{\ell} (e_k, e_\ell) = t_i^{k*} t_j^{\ell} g_{k\ell} \quad , \tag{1-22}$$

or in matrix notation :

$$g' = t^+ g \; t \quad . \tag{1-23}$$

One proves that the Hermitian tensor (g_{ij}) can be diagonalized by a unitary change of basis, i.e. with a unitary matrix t in Eq. (1–2) : $t^+t = tt^+ = \mathbb{1}$ (Theorem 4.2, Subsec. 4.1.2,i)). More generally let us now show that :

Theorem 1.1. The nondegenerate character of the sesquilinear form implies that the determinant of the metric tensor is nonzero.

Proof. In terms of the components, the nondegenerate character of the form g can be stated in the following way :

$$\forall\, (w^i) \quad w^{i*} g_{ij} v^j = 0 \;\Leftrightarrow\; g_{ij} v^j = 0 \;\Rightarrow\; (v^j = 0) \quad .$$

Therefore, the above system of linear equations having only the trivial solution, $\det(g_{ij})$ is nonzero.

Given this result, for a nondegenerate Hermitian sesquilinear form one can construct an orthonormal basis (ε_i) :

$$(\varepsilon_i, \varepsilon_j) = \delta_{ij} \quad . \tag{1-24}$$

On such a basis, the inner product takes the simple form :

$$(v,w) = v^{i*} \, \delta_{ij} \, w^j = \sum_i v^{i*} \, w^i \quad . \tag{1-25}$$

The necessary and sufficient condition for a change of basis defined by a matrix t to preserve the orthonormal character of a basis is that this matrix t be unitary. The proof is obvious from Eq. (1–23) :

$$g' = g = \mathbb{1} \;\Rightarrow\; t^+ t = \mathbb{1} \quad . \tag{1-26}$$

iii) Let us now give other important consequences of the nondegenerate character of the Hermitian form based on Theorem 1.1.

1–Any vector v determined by its <u>contravariant components</u> (v^i) with respect to a given basis (e_i) can be defined as well by its <u>covariant components</u> (v_i) such that :

$$v_i = (e_i, v) = g_{ij} \, v^j \quad . \tag{1-27}$$

Since det g is nonzero there exists a one–to–one correspondence between both sets of components :

$$v^i = \gamma^{ij} v_j \quad , \quad \gamma^{ij} g_{jk} = g_{kj} \gamma^{ji} = \delta^i_k \quad . \tag{1-28}$$

Actually, using the above procedure to stepwise raise both indices of g_{ij}, one gets :

$$g^i_j = \gamma^{ik} g_{kj} = \delta^i_j \quad , \quad g^{ij} = \gamma^{jk} g^i_k = \gamma^{jk} \delta^i_k = \gamma^{ji} \quad . \tag{1-29}$$

Then Eq. (1–28) reads :

$$v^i = v_j g^{ji} \quad , \quad g^{ji} g_{jk} = g_{kj} g^{ij} = \delta^i_k \quad . \tag{1-30}$$

One has to take care that in the first above equation the summation is over the left index of g^{ji}, while it is over the right one of g_{ij} in Eq. (1–27). This plays a role when the metric tensor is not symmetrical, cf. Subsec. 8.3.3.2. For an orthonormal basis :

$$g_{ij} = \delta_{ij} \quad , \tag{1-31}$$

therefore the covariant and contravariant components are numerically equal although their tensorial characters are different.

2–Now one may ask : does there exist a basis (e^j) such that the covariant components (v_j) of v are the components of v with respect to this basis ? This leads to the definition of a <u>biorthogonal basis</u> (e^j) of a given basis (e_i) according to :

$$(e_i, e^j) = \delta^j_i \quad . \tag{1-32}$$

The level of the index of e^j (i.e. the tensorial character) is justified below. Since the inner product is defined by a nondegenerate form the basis (e^j) is uniquely defined by :

$$e^i = g^{ij} e_j \quad , \quad e_i = e^j g_{ji} \quad . \tag{1-33}$$

If (e_i) is orthonormal, it is identical to its biorthogonal basis. The metric tensor with respect to the basis (e^j) is :

$$(e^i, e^j) = g^{ij*} \quad . \tag{1-34}$$

Under a change of basis, the biorthogonal basis (e'^i) of the basis (e'_i) is related to the biorthogonal basis (e^i) of the basis (e_i) by :

$$e'_i = e_j t^j_i \quad \Leftrightarrow \quad e'^i = (t^{-1})^{i*}_j e^j \quad .$$

In agreement with the level of the indices, this is the law of transformation of a contravariant set (apart from the conjugate complex, due to the sesquilinear character of the form g). The components (v'_j) of a vector $v = e_i v^i$ with respect to the basis (e^j) are the covariant components (v_j), indeed :

$$v = e_i v^i = e^j g_{ji} v^i = e^j v_j \quad .$$

3–Let us illustrate the previous remarks in an Euclidean (Subsec. 2.1.1) two–dimensional plane. The basis (e_i) is nonorthogonal, but for simplicity assumed to be normalized.

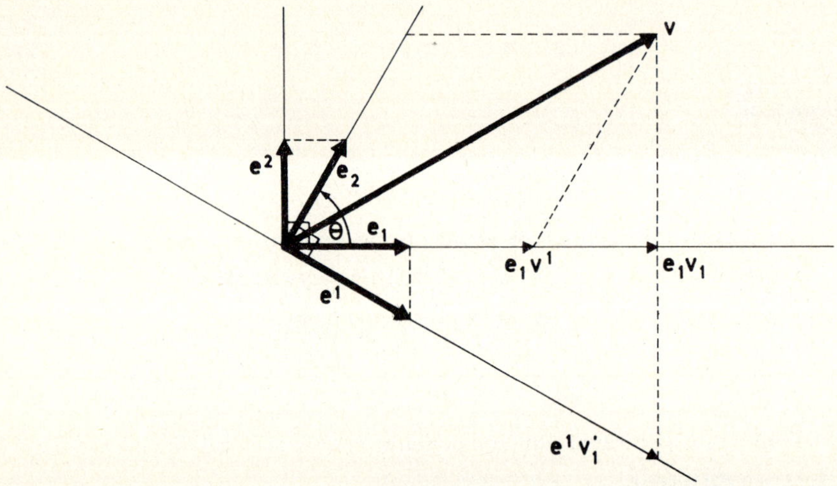

Fig. 1.1. Contravariant and covariant components of v, respectively, (v^i) and (v_i).

According to Eq. (2–2), one has (see Fig. 1.1) :

$$g = \begin{pmatrix} 1 & \cos\theta \\ \cos\theta & 1 \end{pmatrix} \quad , \quad g^{-1} = \frac{1}{\sin^2\theta} \begin{pmatrix} 1 & -\cos\theta \\ -\cos\theta & 1 \end{pmatrix} \quad .$$

Then the contravariant and covariant components correspond, respectively, to oblique and orthogonal projections :

$$v_1 = v^1 + v^2 \cos\theta \quad .$$

The biorthogonal basis to (e_i) is constructed on Fig. 1.1, for instance :

$$e^1 = \frac{1}{\sin^2\theta} e_1 - \frac{\cos\theta}{\sin^2\theta} e_2 \quad .$$

From the geometrical relation $\dfrac{\|e_1 v_1\|}{\|e_1\|} = \dfrac{\|e^1 v'_1\|}{\|e^1\|}$, one obtains $v'_1 = v_1$.

4–We finally want to point out that the definition of components of a vector in Quantum Mechanics (Sec. 3.1) corresponds to the covariant components, Eq. (1–27). Moreover for a linear application A the matrix elements are defined in Quantum Mechanics by :

$$A_{ij} = (e_i, Ae_j) = g_{ik} A^k_j \quad . \tag{1-35}$$

This definition yields a different characteristic equation when one wants to diagonalize a matrix A given in terms of its elements A_{ij} :

$$Av = \lambda v \Rightarrow (e_i, Av) = \lambda(e_i, v) \Rightarrow (A_{ij} - \lambda g_{ij})\, v^j = 0 \quad . \tag{1-36}$$

iv) A special type of mappings of interest of a Hermitian space into itself is that of the isometric applications. By definition they preserve the inner product :

$$\forall\ v,w \quad (v,w) = (Av,Aw) \quad . \tag{1-37}$$

This implies the conservation of the norm. Conversely, any application which preserves the norm is an isometric application.

The knowledge of the quadratic form q uniquely defines the sesquilinear form g by :

$$\forall\ v,w \qquad g(v,w) = \frac{1}{4}[q(v+w) - q(v-w)] - \frac{i}{4}[q(v+iw) - q(v-iw)] \quad .$$

Therefore, if for any v $q(Av) = q(v)$, one has :

$$\forall\ v,w \qquad g(Av,Aw) = g(v,w) \quad .$$

We now prove the following theorem (Wigner's theorem given in Subsec. 3.2.2,i) is a generalization of it) :

Theorem 1.2. Let A be an isometric surjective application which maps a Hermitian space E_n onto itself, then :
(1) A is a one-to-one application and its inverse, denoted A^{-1} is also an isometric application,
(2) A is a linear application.

To prove the first statement, let us assume that there exist v_1 and v_2 such that $Av_1 = Av_2$. Then for any w, as A is an isometric application :

$$(Aw,Av_1) = (Aw,Av_2) = (w,v_1) = (w,v_2) \quad .$$

Due to the sesquilinear character of the Hermitian form (w,v_1-v_2) vanishes, and the form being by definition nondegenerate, v_1 and v_2 are equal. Thus A is a bijective application, and its inverse A^{-1} is obviously an isometric application :

$$(A^{-1}v,\ A^{-1}w) = (AA^{-1}v,\ AA^{-1}w) = (v,w) \quad .$$

To prove the second statement, one gets from above :

$$(v,Aw) = (AA^{-1}v,Aw) = (A^{-1}v,w) \quad ,$$

and by definition $(A^{-1}v,w)$ is linear in w. Thus if w is a linear combination of vectors w_i :

$$(v,A\lambda^i w_i) = \lambda^i(v,Aw_i) = (v,\lambda^i Aw_i) \Rightarrow (v,A\lambda^i w_i - \lambda^i Aw_i) = 0 \quad .$$

As the Hermitian form is nondegenerate, one has :

$$A\lambda^i w_i = \lambda^i Aw_i \quad .$$

This ends the proof of the theorem.

In all bases the matrix representative of an isometric application A fulfils the relations :

$$(e_i,e_j) = g_{ij} = (Ae_i,Ae_j) = A_i^{k*} A_j^{\ell} g_{k\ell} \quad , \tag{1-38}$$

or in matrix notation :

$$g = A^+ g A \quad . \tag{1-39}$$

Thus the determinant of A satisfies the intrinsic condition :

$$|\det A|^2 = 1 \quad . \tag{1-40}$$

Since the matrix A fulfils Eq. (1–39), it depends upon less parameters than n^2 complex numbers. The metric tensor (g_{ij}) being a Hermitian matrix, Eq. (1–39) is equivalent to the conservation under the mapping A of the n real norms g_{ii} and of the n(n–1)/2 complex inner products g_{ij} for i smaller than j. Thus any matrix representative of an isometric application depends at most upon :

$$2n^2 - [n+2 \frac{n(n-1)}{2}] = n^2 \text{ real parameters} \quad .$$

In the particular case of an orthonormal basis, the matrix representative of A is a unitary matrix. The set of the $n\times n$ complex unitary matrices is the unitary group U(n). The elements of U(n) depend upon n^2 real parameters. The set of isometric applications is a subgroup of the linear group GL(n,C). This isometric group $I(E_n)$ is isomorphic to the unitary group U(n). A subgroup of $I(E_n)$ and also of the special linear group SL(n,C) is the set of isometric and unimodular applications. This subgroup is isomorphic to the special unitary group SU(n), namely the set of $n\times n$ complex unitary and unimodular matrices. The elements of SU(n) depend upon n^2-1 real parameters. It is shown in Appendix C.5.4.3,i) that U(n) and SU(n) are compact Lie groups (Defs. C.47,60, cf. also Subsec. 4.1.4).

1.1.5. POINT SPACES

i) **Definitions.** A point space $E_n^{\boldsymbol{\cdot}}$ is a set of elements called points, such that there exists a mapping of $E_n^{\boldsymbol{\cdot}} \times E_n^{\boldsymbol{\cdot}}$ into a vector space E_n :

$$\forall\, M,N \in E_n^{\boldsymbol{\cdot}} \times E_n^{\boldsymbol{\cdot}} \rightarrow v = MN \in E_n \quad . \tag{1-41}$$

This application has to fulfil the following conditions :

(1) for any point M and any vector v, there exists a unique point N such that v = MN,

(2) for any points M, N, and O , one has :

$$OM + MN = ON \quad . \tag{1-42}$$

In Eq. (1–41) one has to distinguish between v and MN : v is called a free vector, and MN a linked vector for MN is defined by two given points.

One calls subspace of $E^{\bullet}_n$, associated to a point M and a subspace P of E_n, the set P_M of points N such that MN belongs to P. The dimension of P is by definition the dimension of P_M. One can show that any point M' of P_M defines the same subspace $P_{M'}$. In particular a line of $E^{\bullet}_n$ is by definition any one-dimensional subspace of $E^{\bullet}_n$. Such a set is defined by a point O and a vector v, such that for any M of this line :

$$OM = \lambda v \qquad \lambda \in K \quad .$$

ii) **Cartesian coordinates.** From the definition of $E^{\bullet}_n$, once a particular point O of $E^{\bullet}_n$ is chosen, there exists a one-to-one correspondence between $E^{\bullet}_n$ and E_n :

$$\forall\, M \in E^{\bullet}_n \leftrightarrow v_M = OM \in E_n \quad . \tag{1-43}$$

The vector v_M can be determined by its components (v^i) on a given basis (e_i) of E_n :

$$v_M = OM = e_i\, v^i \quad . \tag{1-44}$$

A point O and a basis (e_i) of E_n is called a reference frame (O,e_i) of $E^{\bullet}_n$, (v^i) being the Cartesian coordinates of M in this reference frame.

A change of reference frame is defined by a nonsingular matrix t for the change of the basis in E_n and by a vector τ for the change of the reference point of $E^{\bullet}_n$. From Eqs. (1-2,42) one has :

$$OM = e_j\, x^j \quad , \quad O'M = e'_i\, x'^i$$

$$e'_i = e_j\, t^j_i \quad , \quad \tau = OO' = e_j \tau^j$$

$$OM = OO' + O'M = e_j \tau^j + e_j\, t^j_i\, x'^i$$

$$x^j = \tau^j + t^j_i\, x'^i \quad , \quad x'^i = (t^{-1})^i_j (x^j - \tau^j) \quad . \tag{1-45}$$

iii) **Linear point transformations.** A mapping $\mathcal{A}$ of $E^{\bullet}_n$ into itself is called a linear point transformation if the relation OM = λ ON is preserved for the images by $\mathcal{A}$, i.e. :

$$\forall \lambda \in K \qquad OM = \lambda ON \xrightarrow{\mathcal{A}} O'M' = \lambda O'N' \quad . \tag{1-46}$$

As a consequence, to each linear point transformation $\mathcal{A}$ on $E^{\bullet}_n$ corresponds a linear application A on E_n such that :

$$A(OM) = O'M' \quad . \tag{1-47}$$

The converse is not true, but it is sufficient to give A and the image O' = $\mathcal{A}$(O) of any point O to determine $\mathcal{A}$ by :

$$\forall\, M \to M' = \mathcal{A}(M) \quad , \quad O'M' = A(OM) \quad . \tag{1-48}$$

We denote this application by :

$$\mathcal{A} = (OO',A) \quad , \tag{1-49}$$

where the vector OO' may be replaced by any vector MM', M and M' being in correspondence by $\mathcal{A}$:

$$MM' = MO + OO' + O'M' = OO' + (A - \mathbf{1})(OM) \quad . \tag{1-50}$$

Let us now classify the set of linear point transformations in the following way :
(1) If A is the identity mapping **1**, because of Eq. (1–50), for any M one has :

$$MM' = OO' = \tau \in E_n \quad . \tag{1-51}$$

Then $\mathcal{A}$ is called a <u>translation</u> and we denote it :

$$\mathcal{A} = (\tau,\mathbf{1}) = T(\tau) \quad . \tag{1-52}$$

(2) If A is different from the identity mapping, then we have to distinguish two cases :
–There exists at least one fixed point. Then, for a given fixed point, there is an isomorphism between the subset of these linear point transformations and $L(E_n)$. We need to specify a fixed point O in order to define $\mathcal{A}$ by :

$$\mathcal{A} = (O,A) \quad . \tag{1-53}$$

There may exist p fixed points : $O_i = \mathcal{A}(O_i)$ for i=1,..,p, this set of p points defines a subspace $P_{O_1 \ldots O_p}$ of $E_n^{\cdot}$ of dimension less or equal to min(p–1,n). Clearly any point of $P_{O_1 \ldots O_p}$ is also a fixed point. Let us consider for example the case of two distinct fixed points O_1 and O_2 which define the line $P_{O_1 O_2}$, then for any point M of this line one has :

$$\lambda \neq 1 \qquad O_1M = \lambda O_2M \xrightarrow{\mathcal{A}} O_1M' = \lambda O_2M' \quad ,$$

and thus MM'(1–λ) is equal to zero, i.e. M and M' are the same. With the notation of Eq. (1–53), the existence of fixed points O_i different from O correspond to the existence of eigenvectors OO_i of A with eigenvalue 1.
–There does not exist any fixed point. Then using the composition of applications, one gets :

$$\mathcal{A} = (OO',A) = (OO',\mathbf{1})(O,A) = T(OO')(O,A) \quad , \tag{1-54}$$

namely a noncommutative product of a translation and a transformation of type given in Eq. (1–53).

No reference frame is specified in the definition of a linear point transformation. Once a reference frame (O,e_i) is chosen, Eq. (1–48) can be written in terms of the coordinates as :

$$v'^i = \tau^i + A^i_j \, v^j \quad , \tag{1-55}$$

the A^i_j being the matrix elements of A, and the τ^i the components of OO' with respect to the basis (e_i).

1.1.6. HERMITIAN POINT SPACES

i) Let E_n be a Hermitian vector space, then $E^{\bullet}_n$ is said to be a Hermitian point space. The norm introduced on E_n, Eq. (1–19), allows us to define a distance on $E^{\bullet}_n$ by :

$$\forall\ M,N \in E^{\bullet}_n \qquad d(M,N) = ||MN|| \quad . \qquad (1\text{-}56)$$

A distance on $E^{\bullet}_n$ is a mapping of $E^{\bullet}_n \times E^{\bullet}_n$ into R_+ which fulfils the conditions :

$$d(M,N) = 0 \Leftrightarrow M = N \quad , \quad d(M,N) = d(N,M) \quad , \quad d(O,M) \leqslant d(O,N) + d(N,M) \quad .$$

A space with a distance is said to be a metric space.

A reference frame (O,e_i) of $E^{\bullet}_n$ is said to be orthonormal if the basis (e_i) of E_n is orthonormal.

ii) An isometric point transformation $\mathcal{A}$ is by definition a mapping of $E^{\bullet}_n$ into itself which preserves the distances :

$$\forall\ M,N \qquad d(M,N) = d(\mathcal{A}M,\mathcal{A}N) \quad . \qquad (1\text{-}57)$$

From the definition of the distance and Eq. (1–47), it follows that to any isometric transformation $\mathcal{A}$ corresponds an isometric application A on E_n.

Let $I(E^{\bullet}_n)$ be the set of surjective isometric point applications in the point space $E^{\bullet}_n$. From Theorem 1.2 (Subsec. 1.1.4), these applications are linear and bijective. $I(E^{\bullet}_n)$ is the isometric group defined in $E^{\bullet}_n$, with the notation of Eq. (1–49) the multiplication law for this group reads :

$$\mathcal{A}_1\mathcal{A}_2 = (OO_1,A_1)(OO_2,A_2) = (OO_1 + A_1(OO_2), A_1A_2) \quad , \qquad (1\text{-}58)$$

(we have seen, Eq. (1–50), that $\mathcal{A}_1$ and $\mathcal{A}_2$ can be both expressed using the same point O).

The set of translations $T(E^{\bullet}_n)$, Eq. (1–52), is a subgroup of $I(E^{\bullet}_n)$ for which the multiplication law reads :

$$T(\tau_1)T(\tau_2) = T(\tau_1+\tau_2) \quad . \qquad (1\text{-}59)$$

This subgroup is commutative or Abelian (Def. C.4), because of the commutative character of the addition in E_n. Moreover $T(E^{\bullet}_n)$ is an invariant subgroup of $I(E^{\bullet}_n)$, since for any translation T and any isometric point application $\mathcal{A}$, $\mathcal{A}\ T\ \mathcal{A}^{-1}$ is still a translation, the linear operator in E_n associated to this product being from Eq. (1–58) $A\,\mathbf{1}\ A^{-1} = \mathbf{1}$.

Another important subgroup of $I(E^{\bullet}_n)$ is the set of applications of the type (O,A), Eq. (1–53), for a given fixed point O. This subgroup is isomorphic to the group $I(E_n)$ of isometric mappings of E_n onto itself, and thus also to the unitary group U(n). The multiplication law for this subgroup is :

$$(0,A_1)(0,A_2) = (0,A_1A_2) \quad . \tag{1-60}$$

The isometric group $I(E^{\cdot}_n)$ is the semi-direct product (Def. C.25) of the group of translation $T(E^{\cdot}_n)$ by the group of applications of the type (O,A), with the multiplication law given in Eq. (1–58).

In terms of the coordinates in a given orthonormal reference frame (O,e_i), any isometric point transformation takes the form of Eq. (1–55), where $(A^i_{\ j})$ is a unitary matrix :

$$v'^i = \tau^i + A^i_j v^j \ , \quad A^i_k A^{k*}_j = \delta^i_j \quad . \tag{1-61}$$

The isometric group $I(E^{\cdot}_n)$ is thus isomorphic to the group of elements of the type (τ, U) where τ denotes the components of a vector and U a unitary matrix of U(n). The multiplication law for this group is :

$$(\tau_1,U_1)(\tau_2,U_2) = (\tau_1+U_1\tau_2,U_1U_2) \quad , \tag{1-62}$$

and its elements depend upon $n+n^2$ real parameters.

1.2. THE ACTIVE AND THE PASSIVE POINTS OF VIEW OF A LINEAR APPLICATION, THE EQUIVALENCE BETWEEN THEM

1.2.1. THE CASE OF A VECTOR SPACE

Let A be a bijective linear application, its matrix representative belongs to the group GL(n,K). To any vector v, A associates a new vector Av. This is the active point of view of the transformation. It is fully intrinsic, viz. independent of the choice of a basis. On a particular basis (e_i), this application takes the form :

$$Av = A(e_i v^i) = e_j(A^j_i v^i) = e_j(Av)^j \quad . \tag{1-63}$$

Since A is a bijective application, A^{-1} exists, and from any basis (e_i) one gets a new one $(A^{-1}e_i)$, where the new basis vectors are the images by A^{-1} of the previous ones. This change of basis is the passive interpretation of the application A and, unlike the active point of view, it is not intrinsic. The new components of v on this new basis $(A^{-1}e_i)$ are using Eq. (1–63) :

$$v = A^{-1}Av = (A^{-1}e_j)(Av)^j \quad . \tag{1-64}$$

By comparing Eqs. (1–63,64) we see that the components of Av in a basis (e_j) and of v in the basis $(A^{-1}e_j)$ are the same. This is the meaning of the equivalence between the active and the passive points of view of a bijective linear application.

1.2.2. THE CASE OF A POINT SPACE

Let A be a bijective linear point transformation on $E^{\bullet}_n$. For this, it is necessary and sufficient that the associate linear application A, Eq. (1–47), be bijective. The active point of view of the transformation A corresponds to the intrinsic mapping of $E^{\bullet}_n$ into itself defined by Eq. (1–46). Once a reference frame (O,e_j) is chosen, in terms of the coordinates the transformation takes the form of Eq. (1–55) :

$$O' = A(O) \quad , \quad M' = A(M) \quad , \quad OM' = OO' + A(OM) = e_j(\tau^j + A^j_i x^i) \ . \qquad (1\text{-}65)$$

Since A is bijective, using the previous results of Subsec. 1.2.1, one gets a new reference frame $(A^{-1}O, A^{-1}e_j)$ from the first one (O,e_j). This is the passive interpretation of A, namely a change of reference frame. The new coordinates of M with respect to this new reference frame are from Eqs. (1–45,65) :

$$O'' = A^{-1}(O) \quad , \quad O'M = O''O + OM = A^{-1}(OO' + O'M') = (A^{-1}e_j)(\tau^j + A^j_i x^i) \quad . \qquad (1\text{-}66)$$

As it has been done previously, we see comparing Eqs. (1–65,66) that the coordinates of $A(M)$ in the reference frame (O,e_j) and of M in the reference frame $(A^{-1}O, A^{-1}e_j)$ are the same. This is the equivalence between the active and the passive interpretations of a linear bijective point application.

These two points of view of a linear mapping are extended to one–to–one applications of a manifold onto itself in Subsec. 10.2.1,iii).

CHAPTER 2

DEFINITION OF THE ROTATION GROUP

Following the same scheme as for the Hermitian case in Subsects. 1.1.4,5, the Euclidean vector space and point space are defined in Sec. 2.1. Sec. 2.2. deals with the orthogonal and special orthogonal groups, and discusses essentially the problem of the diagonalization of an orthogonal matrix. One then concentrates upon the three-dimensional case in Sec. 2.3, where in particular two parametrizations of the rotations are described.

2.1. THE REAL EUCLIDEAN CASE

2.1.1. EUCLIDEAN VECTOR SPACES

2.1.1.1. Definition

A vector space E_n over the real field (K = R) is called an Euclidean vector space if an inner product of any two vectors v and w is defined by a symmetrical bilinear form g :

$$(v,w) = g(v,w) = g(w,v) \quad , \tag{2-1}$$

such that the associated quadratic form is positive definite. Most of what was said previously in the case of a Hermitian metric applies in the Euclidean case. The Schwarz inequality, Eq. (1-18), allows us to define to within a sign an angle θ between vectors by :

$$(v,w) = \|v\| \, \|w\| \cos\theta \quad . \tag{2-2}$$

In a given basis (e_i) the metric tensor is real symmetrical :

$$(e_i,e_j) = g_{ij} = g_{ji} \quad . \tag{2-3}$$

One proves (Theorem 4.2, Subsec. 4.1.2) that it can be diagonalized by a real orthogonal change of basis : the matrix t in Eq. (1-2) being orthogonal ($tt^t = 1$). In an orthonormal basis the inner product takes the simple form :

$$(v,w) = v^i\delta_{ij}w^j = \sum_i v^iw^i \quad , \tag{2-4}$$

or v^tw in matrix notation. The transformation from one orthonormal basis into another is accomplished by an orthogonal matrix t.

2.1.1.2. The isometric group

As in the Hermitian case the isometric applications A preserve by definition the Euclidean inner product :

$$\forall\ v,w \qquad (Av,Aw) = (v,w) \quad . \tag{2-5}$$

The set of isometric surjective applications which map an Euclidean space onto itself is the isometric group $I(E_n)$. The matrix representative of these applications, with respect to any basis, fulfils the condition equivalent to Eq. (1–39), i.e. in matrix notation :

$$g = A^t\ g\ A \quad . \tag{2-6}$$

Therefore the matrix (A^i_j) depends upon less than n^2 real parameters. The same type of calculation done for the isometric applications in the Hermitian case (Subsec. 1.1.4,iv)) yields a dependence upon $n(n-1)/2$ real parameters. From Eq. (2–6) one gets for any isometric application A the intrinsic condition :

$$\det A = \pm 1 \quad . \tag{2-7}$$

The application A is called a proper or an improper isometric application according to whether the determinant of A is +1 or -1.

The subset of proper isometric applications is an invariant subgroup of $I(E_n)$, indeed we have seen in Subsec. 1.1.3 that the special linear group SL(n,R) is an invariant subgroup of the linear group GL(n,R).

By definition the proper isometric linear applications on the Euclidean space E_n are the rotations, and their set is the rotation group.

The subset of improper isometric applications can be generated from the rotation group and any improper isometric application A. Indeed the inverse of A is an improper application and the product of two improper applications is a proper one. Therefore any improper isometric application B can be written as :

$$B = A(A^{-1}B) = (B\ A^{-1})A \quad ,$$

namely as the product of a given improper application A by the proper one $A^{-1}B$ (or BA^{-1}). With respect to a given orthonormal basis, let us consider the set of improper applications S with a diagonal matrix representative of elements plus one or minus one, the number of minus signs being odd. The square of these applications is the identity and their number is 2^{n-1}. These applications are called the discrete symmetries.

2.1.1.3. The n-fold product and the vector product

In an Euclidean vector space E_n the n-fold product of n vectors is defined by the value of the determinant of their components with respect to any orthonormal basis (e_i) :

$$v_i = e_j\, v_i^j \quad , \quad (v_1 \dots v_n) = \det(v_i^j) \quad . \tag{2-8}$$

The value of this product is preserved under a proper isometric application A and has its sign changed under an improper one :

$$\det A = \varepsilon = \pm 1 \quad , \quad (Av_1 \dots Av_n) = \varepsilon(v_1 \dots v_n) \quad . \tag{2-9}$$

In the passive point of view of a linear application (Subsec. 1.2.1) this means that the n-fold product is not intrinsic since, although it is invariant under any proper orthonormal change of basis, its sign is changed under an improper one. The orthonormal bases can then be classified into two sets : bases of the same set being related by proper isometric applications and bases of different sets being related by improper ones. If we consider only one set of orthonormal bases the space is said to be oriented. Therefore in an oriented space, the n-fold product is an intrinsic quantity. By definition the n-fold product is an alternating n-linear form on E_n, consequently :

$$(v_1 \dots v_i \dots v_j \dots v_n) = -(v_1 \dots v_j \dots v_i \dots v_n) \quad . \tag{2-10}$$

It is zero if and only if the n vectors are linearly dependent.

Let us now introduce the vector product of n–1 vectors $w = v_1 \times \dots \times v_{n-1}$ by the relation :

$$\forall v \quad (w,v) = (v_1 \dots v_{n-1}, v) \quad . \tag{2-11}$$

This equation defines w up to a sign, but if the space is oriented, the n-fold product is then intrinsic and the vector product w is uniquely defined (the Euclidean form being nondegenerate, $(w,v) = (w',v)$ for every v implies $w = w'$). From its definition, one obtains some obvious properties of the vector product :

(1) It is zero if any subset of the n–1 vectors is composed of linearly dependent vectors.

(2) It changes sign under a transposition of any two vectors.

(3) It is orthogonal to each of the vectors v_i, $i=1,\dots,n-1$.

(4) The vector product is stable under a proper isometric application A and changes sign under an improper one :

$$\det A = \varepsilon = \pm 1 \quad , \quad A(v_1 \times \dots \times v_{n-1}) = \varepsilon\, Av_1 \times \dots \times Av_{n-1} \quad . \tag{2-12}$$

Actually, there exists a more general tensorial frame in which both the n-fold product and the vector product of n–1 vectors are defined. The external algebra of order p on the vector space E_n is the set $E^{(p)}$ of completely contravariant and antisymmetrical tensors of order p on E_n. One shows that this external algebra is a vector space of dimension $\binom{n}{p}$ over the same field as E_n.

The n–fold product defined by :

$$\det (v_i^j) = \varepsilon_{i_1 \ldots i_n} v_1^{i_1} \ldots v_n^{i_n} \quad ,$$

where :

$$\varepsilon_{i_1 \ldots i_n} = \begin{cases} 0 & \text{if two indices are equal} \\ (-1)^{\pi(p)} & \pi(p) \text{ being the parity of the substitution } p = \begin{pmatrix} 1 & \ldots n \\ i_1 & \ldots i_n \end{pmatrix} \end{cases} ,$$

is an element of the external algebra $E_n^{(n)}$ of order n on E_n, which is a one–dimensional vector space.

The vector product of n–1 vectors is actually a completely contravariant and antisymmetrical tensor of order n–1 on E_n. The set of these tensors is a $\binom{n}{n-1}$ = n–dimensional vector space which is thus isomorphic to E_n. This is why, according to custom the vector product of n–1 vectors is identified with a vector in Eq. (2–11).

2.1.2. EUCLIDEAN POINT SPACES

i) As in the Hermitian case one defines the real Euclidean point space $E_n^{\cdot}$ associated with the Euclidean vector space E_n. In this point space, the isometric point applications are the linear mappings of $E_n^{\cdot}$ onto itself which preserve the Euclidean distance defined as in Eq. (1–56). The set of these applications is the isometric group $\mathcal{I}(E_n^{\cdot})$.

ii) The isometric group $\mathcal{I}(E_n^{\cdot})$ may be divided into subsets as follows :

1–The subset of translations, Eq. (1–52), $\mathcal{T}(E_n^{\cdot})$ is still an Abelian invariant subgroup, isomorphic to the additive group of E_n.

2–Likewise the set of applications of the type (O,A), Eq. (1–53), for a given fixed point O is a subgroup of $\mathcal{I}(E_n^{\cdot})$ which is isomorphic to the isometric group $I(E_n)$ on the Euclidean vector space E_n. These mappings $\mathcal{A}$ = (O,A) are also called proper or improper applications according to whether A is proper or improper. One can distinguish :

(1) By definition the proper applications $\mathcal{A}$ = (O,A) with one or more fixed points are called rotations. They form an invariant subgroup of $\mathcal{I}(E_n^{\cdot})$, called the rotation group on the Euclidean point space $E_n^{\cdot}$.

These rotations can be classified according to the dimension of the invariant point subspace (Subsec. 1.1.5,iii)). We have seen that the existence of fixed points other than O corresponds to the existence of eigenvectors of A with eigenvalue +1. For a proper application A, different from the identity, the order of the eigenvalue +1 is at most n–2. This discussion will be continued in Subsec. 2.2.1,ii).

(2) For a given fixed point O, the set of improper isometric applications can be generated from the rotation group and any improper application (O,A). Indeed as in the Euclidean vector space any improper application (O,B) can be written as :

$$(O,B) = (O,A)\,(O,A^{-1}B) = (O,B\,A^{-1})\;\;(O,A) \quad ,$$

namely as the product of a given improper isometric application (O,A) by a proper one $(O,A^{-1}B)$ (or $(O,B\,A^{-1})$). Once a reference frame is chosen, one can adopt one of the 2^{n-1}

improper applications $S = (O,S)$, where S are the discrete symmetries defined in Subsec. 2.1.1.2, to generate all the other improper isometric applications. These mappings S, explicitly defined with respect to a given orthonormal reference frame (O,e_i), are also called discrete symmetries. For example in the three-dimensional case the four discrete symmetries correspond, respectively, to the matrix representatives of S :

$$S_1 = \begin{pmatrix} -1 & 0 & 0 \\ 0 & 1 & 0 \\ 0 & 0 & 1 \end{pmatrix} , \quad S_2 = \begin{pmatrix} 1 & 0 & 0 \\ 0 & -1 & 0 \\ 0 & 0 & 1 \end{pmatrix} , \quad S_3 = \begin{pmatrix} 1 & 0 & 0 \\ 0 & 1 & 0 \\ 0 & 0 & -1 \end{pmatrix} , \quad S_4 = \begin{pmatrix} -1 & 0 & 0 \\ 0 & -1 & 0 \\ 0 & 0 & -1 \end{pmatrix} , \qquad (2\text{-}13)$$

namely : $S_i = (O,S_i), i=1,2,3$ are the discrete symmetries with respect to the two-dimensional point spaces (planes) defined by the point O and the couples of vectors, respectively, e_2e_3, e_3e_1, and e_1e_2 ; S_4 is the discrete symmetry with respect to the point O or the space inversion. Actually in this three-dimensional case, and more generally when the dimension is odd, it is convenient to choose the space inversion $(O,-\mathbb{1})$ to generate the subset of improper isometric applications.

2.2. THE ORTHOGONAL GROUPS

2.2.1. THE ORTHOGONAL GROUP O(n)

i) In an n-dimensional Euclidean vector space E_n the matrix representative of an isometric application A with respect to an orthonormal basis (e_i) is an orthogonal matrix, i.e. :

$$A^t A = \mathbb{1} \quad .$$

The set of real n × n orthogonal matrices is the orthogonal group O(n).

This group is isomorphic to the isometric group $I(E_n)$ on the Euclidean vector space E_n (Subsec. 2.1.1.2) or to the group of isometric applications (O,A) for a given fixed point O on the Euclidean point space $E_n^{\bullet}$ (Subsec. 2.1.2,ii)). We have already seen that these matrices depend upon $n(n-1)/2$ real parameters.

ii) Let us now consider the problem of the diagonalization of such an orthogonal matrix A of O(n). The eigenvalues λ are the solutions of the characteristic equation :

$$\det(A-\lambda\, \mathbb{1}) = 0 \quad ,$$

which is a polynomial in λ of degree n with real coefficients. It is convenient to solve this equation on the complex field. Let denote, respectively, by $\{E_n\}$ and $\{A\}$ the complex extensions of the Euclidean space E_n and of the isometric application A (Defs. A.8,9). From Prop. A.4,A and $\{A\}$ have the same matrix representative, respectively, in the basis (e_i) of E_n and $(\{e_i,0\})$ of $\{E_n\}$. Moreover, $\{E_n\}$ is a Hermitian vector space, and from Prop. A.7

$\{A\}$ is an isometric application on $\{E_n\}$. The conservation of the Hermitian norm on $\{E_n\}$ implies that the modulus of the eigenvalues λ is one :

$$\forall\, W \in \{E_n\} \quad \|W\| = \|\{A\}\, W\| = |\lambda|\ \|W\| \quad .$$

Therefore, the eigenvalues are either real namely +1 or -1, or complex and hence occur in complex conjugate pairs $e^{-i\theta}$ and $e^{i\theta}$ ($\theta \neq k\pi$). Furthermore the product of the eigenvalues is equal to the determinant of A, i.e. ± 1. Thus if the dimension is odd, there exists an odd number of real eigenvalues, and for a proper matrix +1 is necessarily one of the eigenvalues. The trace, also called in group theory the character (Def. C.31), takes for an orthogonal matrix the form :

$$\mathrm{Tr}\, A = 2\, \Sigma \cos\theta + n_+ - n_- \quad , \tag{2-14}$$

where n_+ and n_- are, respectively, the order of the eigenvalues +1 and -1, the sum Σ runs over the different values of θ.

In the Hermitian vector space $\{E_n\}$, the eigenvectors W and W^*, corresponding, respectively, to the eigenvalues $e^{-i\theta}$ and $e^{i\theta}$, are orthogonal since θ is different from an integer multiple of π and :

$$(W,W^*) = (AW,AW^*) = e^{2i\theta}(W,W^*) \quad .$$

These vectors are said to be conjugate complex vectors since their components on the basis $(\{e_i,0\})$ are complex conjugate.

They are also called isotropic vectors because their Euclidean norm is zero :

$$\theta \neq k\pi \quad , \qquad (W,W) = (AW,AW) = e^{-2i\theta}(W,W) \quad ,$$

where (,) denotes the inner product on E_n.

To return to the Euclidean space, let us perform the following change of basis in each two-dimensional subspace of $\{E_n\}$ spanned by the orthonormal vectors W and W^* :

$$W_1 = \frac{1}{\sqrt{2}}(W + W^*) \quad , \quad W_2 = \frac{1}{i\sqrt{2}}(W - W^*) \quad . \tag{2-15}$$

By construction the vectors W_1 and W_2 are orthonormal on $\{E_n\}$ and moreover have real components on the basis $(\{e_i,0\})$. Therefore they are of the type $\{W_1,0\}$ and $\{W_2,0\}$ and thus can be identified (Prop. A.2) with the vectors W_1 and W_2 of E_n which are also orthonormal vectors on this space. On the particular basis W_1,W_2, the diagonal matrix :

$$\begin{pmatrix} e^{-i\theta} & 0 \\ 0 & e^{i\theta} \end{pmatrix} \text{ takes the form } \begin{pmatrix} \cos\theta & -\sin\theta \\ \sin\theta & \cos\theta \end{pmatrix} \quad .$$

The eigenvectors corresponding to the eigenvalues ± 1 are also of the type $\{W,0\}$. Finally, by a suitable change of basis any orthogonal matrix can be put into the form :

$$A = \begin{pmatrix} \cos\theta & -\sin\theta & & & \\ \sin\theta & \cos\theta & & & 0 \\ & & \ddots & & \\ & & & \ddots & \\ & 0 & & & \pm 1 \end{pmatrix} \quad . \tag{2-16}$$

2.2.2. THE SPECIAL ORTHOGONAL GROUP SO(n)

The set of real $n \times n$ proper orthogonal matrices is the special orthogonal group SO(n) :

$$A^t A = 1\!\!1 \quad , \quad \det A = 1 \quad . \tag{2-17}$$

This invariant subgroup of O(n) is isomorphic to the rotation group either on the Euclidean vector space (Subsec. 2.1.1.2) or on the Euclidean point space (Subsec. 2.1.2,ii)). These matrices depend upon n(n-1)/2 real parameters. The most general two-dimensional proper orthogonal matrix has the form $\begin{pmatrix} \cos\theta & -\sin\theta \\ \sin\theta & \cos\theta \end{pmatrix}$. Hence using Eq. (2–16) for a proper orthogonal matrix the order of the eigenvalue -1 is even and thus the order of the eigenvalue +1 has the same parity as the dimension n. As U(n) and SU(n), both groups O(n) and SO(n) are compact Lie groups (Appendix C.5.4.3,i), Subsec. 4.1.4).

Many details about the unitary and rotation groups and their representations are given for instance by Murnaghan[42,43]. Let us now focus our attention on the three-dimensional case.

2.3. THE THREE-DIMENSIONAL CASE

2.3.1. THE ISOMETRIC GROUP ON THE THREE-DIMENSIONAL EUCLIDEAN POINT SPACE

i) The point space of nonrelativistic physics is the three-dimensional Euclidean point space $E_3^{\cdot}$. Let us recall the different coordinate systems which are commonly used to define a point with respect to a given orthonormal reference frame (O,e_i) :

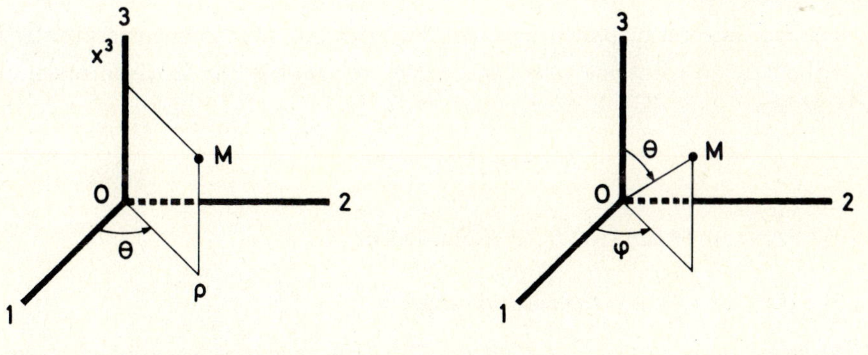

Fig. 2.1. Cylindrical polar coordinates and spherical polar coordinates.

(1) Cartesian coordinates : (x^1,x^2,x^3)

$$OM = x^i e_i \quad . \tag{2-18}$$

(2) Cylindrical polar coordinates : (x^3, ρ, θ)

$$x^1 = \rho\cos\theta \quad , \quad x^2 = \rho\sin\theta \quad , \quad 0 \leqslant \theta \leqslant 2\pi \quad . \tag{2-19}$$

(3) Spherical polar coordinates : (r, θ, φ)

$$x^1 = r\sin\theta\cos\varphi \ , \ x^2 = r\sin\theta\sin\varphi \ , \ x^3 = r\cos\theta \ , \ 0 \leqslant \theta \leqslant \pi \ , \ 0 \leqslant \varphi \leqslant 2\pi \quad . \tag{2-20}$$

The orientation of the angles is specified later on by Eq. (2–27).

The spherical polar coordinates can be generalized for n greater than three. One considers the orthogonal projection of the n-dimensional vector $v = r\,\hat{v}_n$, where $\hat{v}_n$ is a unitary vector, on the axis e_1 and the orthogonal n-1-dimensional subspace :

$$\hat{v}_n = \cos\theta_1 e_1 + \sin\theta_1 \hat{v}_{n-1} \quad .$$

One then iterates this procedure for the unitary vector $\hat{v}_{n-1}$ in the previous subspace generated by $e_2, e_3, \ldots, e_n$. This yields the following coordinates :

$$\begin{aligned}
x^1 &= r\cos\theta_1 \\
x^2 &= r\sin\theta_1\cos\theta_2 \\
x^3 &= r\sin\theta_1\sin\theta_2\cos\theta_3 \\
&\ldots\ldots\ldots\ldots\ldots\ldots\ldots \\
x^{n-1} &= r\sin\theta_1\sin\theta_2\sin\theta_3\ldots\cos\theta_{n-1} \\
x^n &= r\sin\theta_1\sin\theta_2\sin\theta_3\ldots\sin\theta_{n-1} \quad .
\end{aligned} \tag{2-21}$$

ii) The isometric group $I(E_3^{\cdot})$ on the Euclidean point space $E_3^{\cdot}$ is the semi-direct product, Eq. (1–58), of the Abelian invariant subgroup of translations $T(E_3^{\cdot})$ by the subgroup of isometric applications of the type (O,A) for a given fixed point O. This last group has been divided into :

(1) The proper isometric applications (det A = +1) namely the rotations, the set which is the rotation group on $E_3^{\cdot}$.

(2) The improper isometric applications (det A = -1), their set can be generated by the product of one of them by all rotations. One can choose for example the space inversion Π such that to any M :

$$M \xrightarrow{\Pi} M' \quad , \quad OM' = -\,OM \quad . \tag{2-22}$$

2.3.2. TWO PARAMETRIZATIONS OF ROTATIONS

2.3.2.1. The rotation axis and the rotation angle

i) We have already noticed (Subsec. 2.1.2,ii)) that the rotations can be classified according to the dimension of the invariant point subspace, the dimension of this subspace being equal to the order of the possible eigenvalue +1. For a three-dimensional rotation different from the identity, +1 is an eigenvalue of order one, cf. Eq. (2–16), let $\hat{n}$ be a corresponding eigenvector of norm one. The invariant subspace is thus the line :

$$M \quad , \quad OM = \lambda\, \hat{n} \quad . \qquad (2\text{-}23)$$

This is the <u>rotation axis</u> defined for example by the point O and the vector $\hat{n}$.

We assume that the Euclidean vector space E_3 is oriented, namely we only consider orthonormal bases called <u>right-handed bases</u>, generated from one of them e_1, e_2, e_3, by proper orthogonal matrices. The three-fold product, also called mixed product, Eq. (2–8), and the vector product, Eq. (2–11), are then uniquely defined. For any right-handed basis e_i, e_j, e_k one has :

$$(e_i,e_j,e_k)=1 \;;\; e_i \times e_j = \sum_k \varepsilon_{ijk} e_k \;,\; \varepsilon_{ijk} = \begin{cases} \text{parity of the substitution } \begin{pmatrix} 1 & 2 & 3 \\ i & j & k \end{pmatrix} \\ 0 \text{ in the other cases} \end{cases} \quad . \qquad (2\text{-}24)$$

The <u>Levi-Civita symbol</u> ε_{ijk} is the totally antisymmetrical tensor of order three. One establishes the Lagrange and Gibbs formulae, respectively :

$$(\|v_1\|\,\|v_2\|)^2 = (v_1.v_2)^2 + \|v_1 \times v_2\|^2 \quad , \qquad (2\text{-}25)$$

$$v_1 \times (v_2 \times v_3) = (v_1.v_3)v_2 - (v_1.v_2)v_3 \quad , \qquad (2\text{-}26)$$

where the dot denotes as usual the Euclidean inner product. Then from the definition, Eq. (2–2), of the angle between two vectors, one gets :

$$\| v_1 \times v_2 \| = \| v_1 \| \, \| v_2 \| \; \sin\theta \quad .$$

The angle θ is uniquely specified (modulo 2π) in the following way ; let e_i, e_j, e_k be a right-handed orthonormal basis such that both vectors v_1 and v_2 belong to the plane spanned by e_1 and e_2, then :

$$v_1 \cdot v_2 = \cos\theta \,\|v_1\|\,\|v_2\| \quad , \quad v_1 \times v_2 = \sin\theta \,\|v_1\|\,\|v_2\|\, e_3 \quad . \qquad (2\text{-}27)$$

The corresponding three-dimensional Euclidean point space is also said to be <u>oriented</u>. We choose to represent the <u>right-handed orthonormal reference frame</u> (O,e_1,e_2,e_3), associated to the right-handed basis e_1,e_2,e_3, as in Fig. 2.2.

Let us now consider a point M which does not belong to the rotation axis. The plane containing the point and orthogonal to the rotation axis intersects it at the point H (see Fig. 2.3). The image M' of M by the rotation $R = (O,R)$ is in this plane since the inner product is preserved :

$$\hat{n}.HM = \hat{n}.HM' = 0 \quad .$$

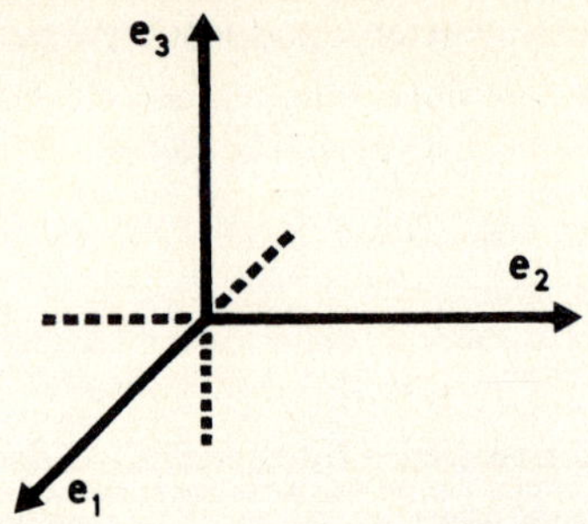

Fig. 2.2. Right-handed basis.

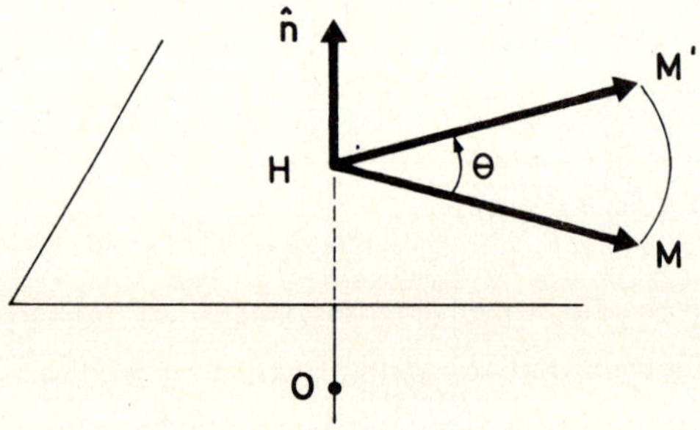

Fig. 2.3. Rotation axis and rotation angle.

By definition the three vectors $v = HM$, $\hat{n}\times v$, and $\hat{n}$ form a right-handed orthogonal basis which can be normalized. We have shown, Eq. (2–16), that in such a basis the matrix representative of R takes the form :

$$R = \begin{pmatrix} \cos\theta & -\sin\theta & 0 \\ \sin\theta & \cos\theta & 0 \\ 0 & 0 & 1 \end{pmatrix} \quad . \tag{2-28}$$

The angle θ , which is independent of the point M, is called the rotation angle. For any v orthogonal to the rotation axis we get :

$$v' = Rv = \cos\theta \; v + \sin\theta \; \hat{n}\times v \quad . \tag{2-29}$$

The angle θ can be obtained to within a sign from the trace, Eq. (2–14) :

$$\mathrm{Tr}\; R = 2\cos\theta + 1 \quad . \tag{2-30}$$

For an infinitesimal rotation, viz. with an infinitesimal angle ε :

$$R = \mathbf{1} + \varepsilon\,\hat{n}\times + 0(\varepsilon^2) \quad . \tag{2-31}$$

A rotation $\mathcal{R} = (O,R)$ is then defined by a fixed point O, a unitary vector $\hat{n}$, and a rotation angle θ oriented by the choice of Eq. (2–27) ; we denote it as $R(O;\theta,\hat{n})$. One also introduces the vector :

$$\Theta = \theta\,\hat{n} \quad , \tag{2-32}$$

which, with the fixed point O, uniquely defines a rotation. Conversely, a rotation does not uniquely define a vector Θ, since instead of rotating by θ about $\hat{n}$ one can equally rotate by $2\pi-\theta$ about $-\hat{n}$. We may remove this indetermination by the restriction :

$$0 \leqslant \theta \leqslant \pi \quad . \tag{2-33}$$

Then, there exists a one–to–one correspondence between the points in the three-dimensional ball D^3 of radius π (Def. C.49) and the rotations for a given fixed point O, except for the points on the surface (rotations of angle π) for which opposite points on the surface refer to the same rotation (see Fig. 2.4). Opposite points not on the surface correspond to rotations inverse of one another :

$$R(O\,;\,\theta,\hat{n}) \rightarrow R^{-1} = R(O\,;\,\theta,-\hat{n}) \quad . \tag{2-34}$$

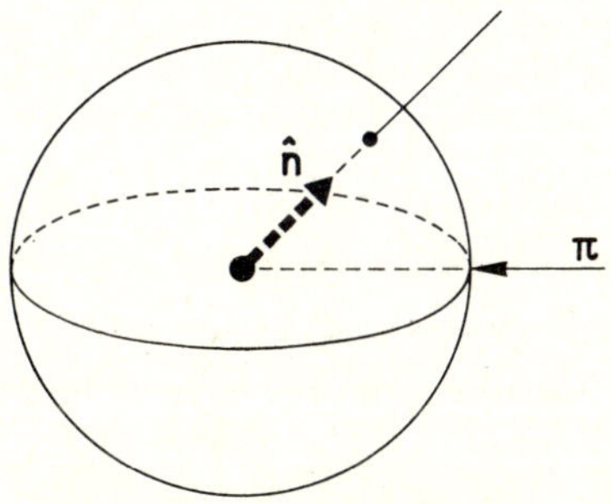

Fig. 2.4. The space of parameters θ and $\hat{n}$ for SO(3). Compare with Fig. 4.1 for SU(2).

Later on (Subsec. 7.1.3.2) a metric is defined on this space of parameters such that it becomes a Riemannian manifold.

We only introduce rotations with the same fixed point O, the set of which is a group isomorphic to the rotation group on the Euclidean vector space E_3. Therefore, from now on one only considers the rotations on E_3, denoted by :

$$R(\theta,\hat{n}) = R(\Theta) \quad . \tag{2-35}$$

ii) It is emphasized that the parametrization of a rotation $R(\theta,\hat{n})$ by an angle and a unitary vector is intrinsic, namely independent of any choice of basis.

Let ρ be any rotation, one has the important multiplication law :

$$\rho R(\theta,\hat{n})\rho^{-1} = R(\theta,\rho\hat{n}) \quad . \tag{2-36}$$

Indeed, the rotation axis of $\rho R(\theta,\hat{n})\rho^{-1}$ is $\rho\hat{n}$ since this unitary vector is invariant under this rotation :

$$\rho R(\theta,\hat{n})\rho^{-1}\rho\hat{n} = \rho R(\theta,\hat{n})\hat{n} = \rho\hat{n} \quad .$$

Moreover, a rotation preserves the oriented angle between vectors defined by Eq. (2–27) and their norms, consequently the rotation angle of $\rho R(\theta,\hat{n})\rho^{-1}$ is also θ since, let v be any vector orthogonal to $\rho\hat{n}$, one has :

$$\rho R(\theta,\hat{n})\rho^{-1}v.v = R(\theta,\hat{n})\rho^{-1}v.\rho^{-1}v = \cos\theta\ \|v\|^2 \quad ,$$

$$\rho R(\theta,\hat{n})\ \rho^{-1}\ v\times v = R(\theta,\hat{n})\rho^{-1}v\times\rho^{-1}v = \sin\theta\|v\|^2\hat{n} \quad .$$

Let us now give some consequences of the above multiplication law :

(1) The classes of conjugate elements of the rotation group (Def. C.20) are the subsets of rotations having the same rotation angle. Thus the classes of the rotation group depend upon the continuous variable θ , $0 \leqslant \theta \leqslant \pi$, (instead of a discrete variable for finite groups). With the previous parametrization of rotations, the elements of given classes lie on concentric spheres.

(2) The rotation R_r which brings e_3 on the vector $r(r,\theta,\varphi)$, Eq. (2–20), the rotation axis $\hat{n}$ being orthogonal to the plane defined by e_3 and r (see Fig. 2.5), is given by :

$$R_r = R(\varphi,e_3)R(\theta,e_2)\,R(\varphi,e_3)^{-1} \quad . \tag{2-37}$$

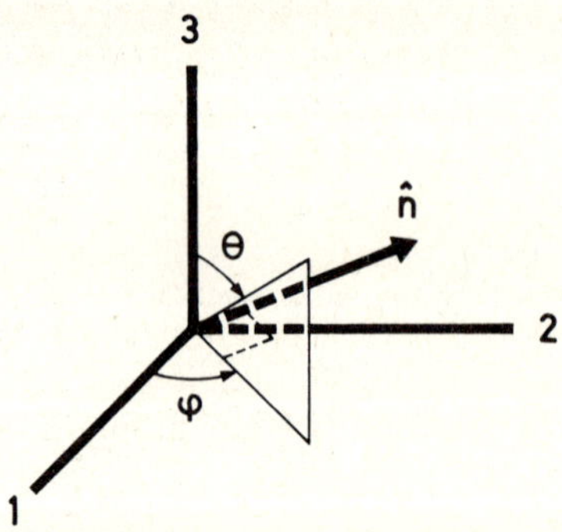

Fig. 2.5. The rotation R_r.

(3) A rotation of angle θ about the axis e_1 can be written in terms of rotations about the axes e_2 and e_3 as :

$$R(\theta,e_1) = R(\tfrac{\pi}{2}, -e_3)R(\theta,e_2)\,R(\tfrac{\pi}{2}, -e_3)^{-1} \quad . \tag{2-38}$$

iii) In Appendix E.1 one can find :

(1) The matrix representative, with respect to an orthonormal basis, of the rotation $R(\theta,\hat{n})$.

(2) The expression of the rotation angle and the components of the rotation axis in terms of

the matrix elements of R, with respect to an orthonormal basis.

(3) The variations of the Cartesian and spherical polar coordinates under the infinitesimal rotations $R(\varepsilon, e_i)$.

2.3.2.2. The Euler angles

i) The description of the displacement of a rigid body due to a rotation about a fixed point (Subsec. 11.4.1) leads to the characterization of any rotation by three angles which define the relative position of two right-handed orthogonal reference frames with the same origin. We adopt the following definitions (also used by many authors, cf. Appendix F.1,i)) : a rotation φ about the axis e_3 (precession), then a rotation θ about the axis x'_2 (nutation), and finally a rotation ψ about the axis X_3 (proper rotation), see Fig. 2.6.

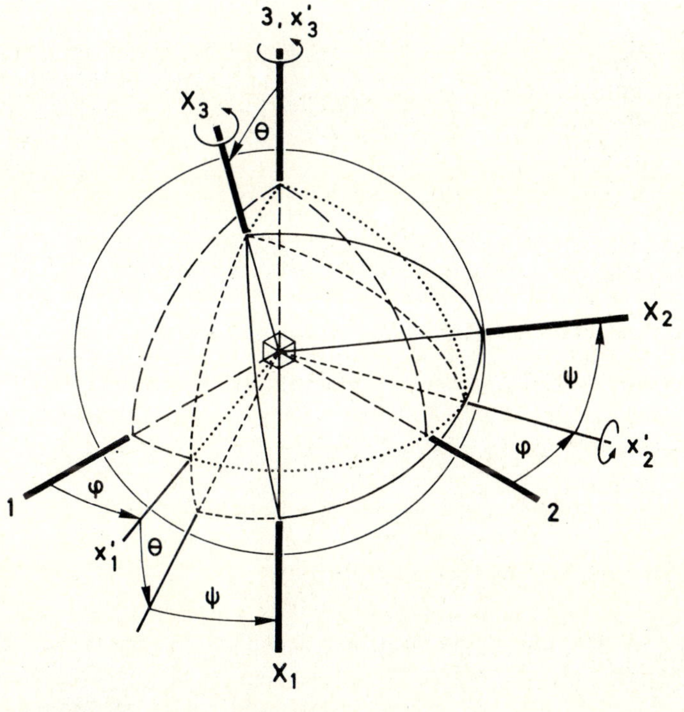

Fig. 2.6. Definition of the Euler angles.

The possible values of φ, θ, and ψ are restricted so as to preserve a one-to-one correspondence between parameters and rotations :

$$0 \leqslant \varphi \leqslant 2\pi \quad , \quad 0 \leqslant \theta \leqslant \pi \quad , \quad 0 \leqslant \psi \leqslant 2\pi \quad . \tag{2-39}$$

Nevertheless the following surface points (φ, θ, ψ) of the parameter space have to be identified :

$$(1) \begin{cases} 0,\theta,\psi \\ 2\pi,\theta,\psi \end{cases} \quad (2) \begin{cases} \varphi,\theta,0 \\ \varphi,\theta,2\pi \end{cases} \quad (3)\ \{\theta=0,\ \varphi+\psi=\text{cst}\} \quad (4)\ \{\theta=\pi,\ \varphi-\psi=\text{cst}\}\ . \tag{2-40}$$

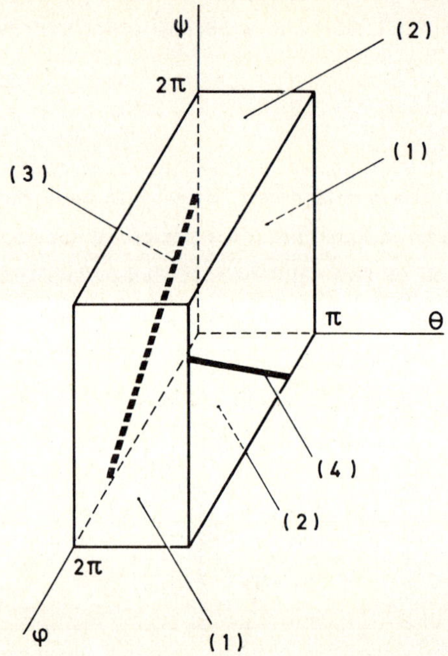

Fig. 2.7. The Euler angle parameter space for SO(3). Compare with Fig. 4.3 for SU(2).

It should be noted that the spherical polar coordinates Eq. (2–20) (φ,θ) of the axis X_3, with respect to the original reference frame, are identical to the Euler angles φ and θ[3]. Actually, this is a particular case of a general property of the group O(n). Let us consider the set O(n)/O(n-1), namely the set of classes defined by the equivalence relation (Prop. C.21) :

$$g,g'\in O(n) \quad , \quad g\sim g' \Leftrightarrow \exists h\in O(n-1) \ , \ g' = gh \quad .$$

The $n\times n$ matrices of O(n) can be written by blocks as :

$g=\left(\begin{array}{c|c} g_1 & g_2 \\ \hline g_3 & g_4 \end{array}\right)$, where g_4 is a number and $\left(\frac{g_2}{g_4}\right)$ the components of an n–dimensional unitary vector. Then any element of the class of g is of the type, assuming a particular form for h as an element of O(n) :

$$g' = gh = \left(\begin{array}{c|c} g_1 & g_2 \\ \hline g_3 & g_4 \end{array}\right)\left(\begin{array}{c|c} h & 0 \\ \hline 0 & 1 \end{array}\right) = \left(\begin{array}{c|c} g_1h & g_2 \\ \hline g_3h & g_4 \end{array}\right) \quad ,$$

namely they all have in common the unitary vector $\left(\frac{g_2}{g_4}\right)$. Conversely, let us give a geometrical interpretation : for any n-dimensional unitary vector one considers its orthogonal n-1–dimensional subspace, and in this subspace n-1 orthonormal vectors which are defined to within a n-1-dimensional orthogonal application. Thus the set O(n)/O(n-1) is isomorphic to the set of unitary n-dimensional vectors. Therefore a matrix of O(n) is defined by an n-dimensional unitary vector and a matrix of O(n-1). In the three-dimensional case, having given the unitary vector by its polar coordinates (φ,θ), it remains to define a matrix of O(2) by an angle ψ (the proper rotation).

ii) We point out that the parametrization in terms of the Euler angles, unlike the previous one $R(\theta,\hat{n})$, is not intrinsic, viz. one has to specify the reference frame with respect to which the Euler angles are defined. This is why we put a subscript a to recall this dependence on the reference axes : $R(\varphi,\theta,\psi;a)$.

From the group structure and using the first parametrization $R(\theta,\hat{n})$, the rotation $R(\varphi,\theta,\psi;a)$ is the ordered product of three rotations :

$$R(\varphi,\theta,\psi;a) = R(\psi,\hat{X}_3)R(\theta,\hat{x}_2')R(\varphi,e_3) \quad . \tag{2-41}$$

Using the multiplication law of the group, Eq. (2–36), one now proves two relations :

(1) The first one :

$$R(\varphi,\theta,\psi;a) = R(\varphi,e_3)R(\theta,e_2)R(\psi,e_3) \quad , \tag{2-42}$$

allows us to express the rotation $R(\varphi,\theta,\psi;a)$ in terms of rotations about the original axes e_2 and e_3. This result is obtained by using two times the law of Eq. (2–36) :

$$R(\psi,\hat{X}_3) = R(\theta,\hat{x}_2')R(\psi,e_3)R^{-1}(\theta,\hat{x}_2') \quad ,$$

$$R(\theta,\hat{x}_2') = R(\varphi,e_3)R(\theta,e_2)R^{-1}(\varphi,e_3) \quad .$$

(2) The second one is for any rotation ρ :

$$\rho R(\varphi,\theta,\psi;a)\rho^{-1} = R(\varphi,\theta,\psi;\rho a) \quad , \tag{2-43}$$

which emphasizes the nonintrinsic character of the parametrization by the Euler angles. Using the first relation, Eq. (2–42), and the multiplication law of Eq. (2–36) the proof is obvious :

$$\rho R(\varphi,\theta,\psi;a)\rho^{-1} = \rho R(\varphi,e_3)\rho^{-1}\,\rho R(\theta,e_2)\rho^{-1}\,\rho R(\psi,e_3)\rho^{-1} = R(\varphi,\rho e_3)R(\theta,\rho e_2)R(\psi,\rho e_3) .$$

From the first relation, Eq. (2–42), one gets the inverse of a rotation parametrized by its Euler's angles :

$$R^{-1}(\varphi,\theta,\psi;a) = R(\psi,-e_3)R(\theta,-e_2)R(\varphi,-e_3) \quad . \tag{2-44}$$

The relation of Eq. (2–43) allows us to go back to the original axes e_1, e_2, and e_3, since one has :

$$R(\psi,-e_3) = R(\pi,e_1)R(\psi,e_3)R(\pi,e_1)^{-1} \quad ,$$

$$R(\theta,-e_2) = R(\pi,e_1)R(\theta,e_2)R(\pi,e_1)^{-1} \quad ,$$

and thus :

$$R(\varphi,\theta,\psi;a)^{-1} = R(\pi,e_1)R(\psi,\theta,\varphi;a)R(\pi,e_1)^{-1} \quad . \tag{2-45}$$

If one does not take care of the restriction of Eq. (2–39), R^{-1} reads :

$$R(\varphi,\theta,\psi;a)^{-1} = R(-\psi,-\theta,-\varphi;a) \quad . \tag{2-46}$$

iii) In Appendix E.2. one can find :

(1) The matrix representative of $R(\varphi,\theta,\psi;a)$ with respect to the orthonormal basis a.

(2) The rotation angle and the rotation axis in terms of φ, θ, and ψ.

(3) The variations of the Euler angles under the infinitesimal rotations $R(\varepsilon, e_i)$.

2.3.2.3. Remarks about the manifold structure of the rotation group

It is shown in Appendix C.5.4.3,i) that SO(3) is a connected three-parameter Lie group, and thus in particular an analytic three-manifold. The previous discussion looked only for a faithful parametrization, i.e. a one–to–one correspondence between the rotation group and a set of three real parameters, without taking care of the topology of the group. Actually, in the manifold language (Def. C.52), to the two parametrizations defined previously, correspond only local charts (U = open set in the group, f = one–to–one correspondence between U and f(U) open set in R^3), which do not cover the whole group, i.e. which do not define by themself an atlas (the subset U is not equal to the group) :

(1) To the parametrization by means of the vector $\Theta = \theta\hat{n}$ corresponds the local chart (U_1,f_1) where :

– U_1 is the open set in the rotation group of rotations parametrized by $\theta\hat{n}$ with θ in $[0,\pi[$,

– f_1 is the one–to–one mapping such that :

$$\forall\, R \in U_1 \qquad f(R) = \Theta = \theta\hat{n} \quad ,$$

– $f_1(U_1)$ is the open three-dimensional ball D^3 of radius π (surface points excluded).

Thus this local chart is not defined for all rotations of angle $\pi(\Theta = \pi\hat{n},\ \forall\hat{n})$.

(2) Similarly, to the parametrization by the Euler angles corresponds the local chart (U_2,f_2) where :

– U_2 is the open set in the rotation group of rotations parametrized by (φ,θ,ψ) which do not fulfil one of the conditions given by Eq. (2–40),

– f_2 is the one–to–one mapping such that :

$$\forall\, R \in U_2 \qquad f_2(R) = (\varphi,\theta,\psi) \quad ,$$

– $f_2(U_2)$ is the open set in R^3 defined by :

$$0 < \varphi < 2\pi \quad , \quad 0 < \theta < \pi \quad , \quad 0 < \psi < 2\pi \quad .$$

Thus this local chart is not defined for all rotations parametrized by (φ,θ,ψ) which fulfil one of the conditions given by Eq. (2–40). In particular, this local chart is not defined for the unity of the group.

The previous remarks are useful for the study of the invariant integration on the rotation group, discussed in Sec. 7.1 (cf. Subsecs. 4.1.4,i) and 4.2.4,i) for a similar discussion for the group SU(2)).

Let us remark that there exists a one–to–one correspondence between the rotation group and for instance the points of the three–dimensional ball of radius π , the surface points of one hemisphere being excluded. But with respect to the topology of SO(3) induced by this one–to–one correspondence from the usual topology of R^3, the group is not a topological one. Indeed the application $(R,R') \rightarrow RR'$ is not continuous in for instance $R = R' = R(\pi, \hat{n})$, cf. the same discussion for the group U(1) parametrized by θ in $[0,2\pi[$ given in Appendix C.5.4.3,i). This emphasizes the difference between a simple parametrization and a local chart.

FORMULAE FOR CHAPTER 2

Euclidean vector space E_3

Right-handed orthonormal basis :

$$a = (e_i, i=1,2,3) \quad \forall x \in E_3 \quad x = x^i e_i$$

$$e_i . e_j = g_{ij} = \delta_{ij} \quad , \quad e_i \times e_j = \varepsilon_{ijk} e_k$$

$$\| x \|^2 = x.x = \sum_{i=1}^{3} (x^i)^2 = r^2 \quad \begin{cases} x^1 = r \sin\theta\cos\varphi & 0 \leqslant \theta \leqslant \pi \\ x^2 = r \sin\theta\sin\varphi & 0 \leqslant \varphi \leqslant 2\pi \\ x^3 = r \cos\theta \end{cases}$$

$$v_1 . v_2 = \cos\theta \, \| v_1 \| \, \| v_2 \|$$

v_1 and $v_2 \in$ plane $e_1, e_2 \Rightarrow v_1 \times v_2 = \sin\theta \, \| v_1 \| \, \| v_2 \| \, e_3$

$$(\| v_1 \| \, \| v_2 \|)^2 = (v_1 . v_2)^2 + \| v_1 \times v_2 \|^2$$

$$(v_1, v_2, v_3) = (v_1 \times v_2) . v_3 \quad , \quad v_1 \times (v_2 \times v_3) = (v_1 . v_3) v_2 - (v_1 . v_2) v_3 \quad .$$

Rotation angle and rotation axis

$\forall R \in$ rotation group on E_3 :

$$R = R(\theta, \hat{n}) \quad \| \hat{n} \| = 1 \quad 0 \leqslant \theta \leqslant \pi$$

$$R(\theta, \hat{n})\hat{n} = \hat{n} \quad , \quad \forall v, v.\hat{n} = 0 \quad R(\theta, \hat{n}) v = \cos\theta \, v + \sin\theta \, \hat{n} \times v$$

$$R(\varepsilon, \hat{n}) = \mathbf{1} + \varepsilon \hat{n} \times + 0(\varepsilon^2)$$

$$R(\theta, \hat{n})^{-1} = R(\theta, -\hat{n}) \quad , \quad \mathrm{Tr}\, R(\theta, \hat{n}) = 2\cos\theta + 1 \quad .$$

$\forall \rho \in$ rotation group on E_3 :

$$\rho R(\theta, \hat{n}) \rho^{-1} = R(\theta, \rho\hat{n})$$

$$\hat{q} = \hat{q}(\theta, \varphi) \quad , \quad \| \hat{q} \| = 1 \quad , \quad R_{\hat{q}} e_3 = \hat{q} \quad , \quad \hat{n}.e_3 = \hat{n}.\hat{q} = 0$$

$$R_{\hat{q}} = R(\theta, \hat{n}) = R(\varphi, e_3) R(\theta, e_2) R(\varphi, -e_3)$$

$$R(\theta, e_1) = R(\tfrac{\pi}{2}, -e_3) R(\theta, e_2) R(\tfrac{\pi}{2}, -e_3)^{-1} \quad .$$

Euler's angles

$\forall$ R $\in$ rotation group on E_3 and a right-handed orthonormal basis $a = (e_i)$

$$R = R(\varphi,\theta,\psi;a) \qquad 0 \leqslant \varphi \leqslant 2\pi \quad , \quad 0 \leqslant \theta \leqslant \pi \quad , \quad 0 \leqslant \psi \leqslant 2\pi$$

$$R(\varphi,\theta,\psi;a) = R(\varphi,e_3)R(\theta,e_2)R(\psi,e_3)$$

$$R(\varphi,\theta,\psi;a)^{-1} = R(\pi,e_1)R(\psi,\theta,\varphi;a)R(\pi,e_1)^{-1} = R(-\psi,-\theta,-\varphi;a)$$

$\forall \rho \in$ rotation group on E_3 :

$$\rho R(\varphi,\theta,\psi;a)\rho^{-1} = R(\varphi,\theta,\psi;\rho a) \quad , \quad \rho a = (\rho e_i) \quad .$$

CHAPTER 3

GENERALITIES ABOUT SYMMETRY LAWS IN QUANTUM MECHANICS

The first section is a brief summary of the formalism of usual Quantum Mechanics. Sec. 3.2 deals with the symmetry laws in Quantum Mechanics. They are defined in Subsec. 3.2.1. The basic Wigner theorem with its proof is given in Subsec. 3.2.2. Subsec. 3.2.3 gives a brief account of the consequences of this theorem. When there exists a symmetry group, one is led to consider the representations up to a phase of this group. Sec. 3.3 is devoted to the invariance and conservation laws. Finally, in Sec. 3.4 all these general considerations are applied to the three-dimensional rotation group.

3.1. DESCRIPTION OF PHYSICAL SYSTEMS IN QUANTUM MECHANICS

Let us only sketch the general formalism.

i) In order to describe a physical system in Quantum Mechanics one introduces a Hilbert space $\mathcal{H}$. The measurable quantities i.e. the observables are associated with self-adjoint operators acting on $\mathcal{H}$. These operators belong to an algebra (Def. B.1) $\mathcal{A}$ over the complex field. Actually when for certain physical systems there exist superselection rules[41], then among the operators of $\mathcal{A}$ only those which commute with all the superselection operators Q (like the electric charge, the baryonic charge...) are observables. Then, $\mathcal{H}$ is the direct sum $\mathcal{H} = \sum_i \mathcal{H}_i$ of Hilbert spaces $\mathcal{H}_i$ called superselection sectors, and each observable A has a representative A_i in $\mathcal{H}_i$: $A = \sum_i P_i A_i P_i$, where P_i is the projector on $\mathcal{H}_i$. The index i labels the spectrum of the superselection operators.

For instance, to describe a one-dimensional spinless particle, the Hilbert space of functions is $L^2(R, dr)$, and the algebra $\mathcal{A}$ is spanned by the operators $\mathbf{r}^m \mathbf{p}^n$ (m and n being non-negative integers) where $\mathbf{r}$ and $\mathbf{p}$ are, respectively, the position and momentum operators such that :

$$[\mathbf{r},\mathbf{p}] = i\hbar\,\mathbf{1} \quad . \tag{3-1}$$

A quantum mechanical state is a linear form L defined on $\mathcal{A}$, i.e. it is an element of the dual space $\mathcal{A}^*$ of $\mathcal{A}$. The image by L of any operator A in $\mathcal{A}$ is a complex number L(A) which is the expectation value of the operator A in the state L :

$$\langle A\rangle_L = L(A) \quad . \tag{3-2}$$

A linear form L in A^* describes a physical state provided L satisfies the following conditions :

(1) For the unit operator $\mathbb{1}$ of A^*, $L(\mathbb{1}) = 1$.

(2) For any self-adjoint operator A, L(A) is real.

(3) L is non-negative, namely for any positive self-adjoint operator A, $L(A) \geqslant 0$.

(4) For identical particles, L has to obey a definite symmetry principle.

ii) We limit ourselves to the usual case (i.e. countable degrees of freedom) for which the linear form L can be realized by an operator ρ . This density operator ρ defined on the Hilbert space H is such that for any operator A one has :

$$\langle A\rangle_L = L(A) = \mathrm{Tr}(\rho A) \quad , \tag{3-3}$$

where the trace is taken on H . The previous conditions for L to be a physical state imply respectively for ρ :

(1') $\mathrm{Tr}\ \rho = 1$. (3-4)

(2') ρ is a self-adjoint operator $\rho = \rho^+$. (3-5)

(3') ρ is non-negative, for any $|\varphi\rangle$ in H , $\langle\varphi|\rho|\varphi\rangle \geqslant 0$. (3-6)

(4') For identical particles ρ commutes with the symmetrization or antisymmetrization operator, according to the statistics of the particles.

The density operator ρ also represents an observable since its expectation value in a given state ρ', gives the probability for the state ρ' to occur in ρ , namely the transition probability for the state ρ' to the state ρ .

Among the possible states there are the pure states defined by a density operator equal to a projector $|\varphi\rangle\langle\varphi|$ on a ray $|\varphi\rangle$ (in Weyl's terminology[62]) in H , namely the subset of normalized vectors $e^{i\theta}|\varphi\rangle$ where θ is any real number. Pure states are necessarily eigenstates of superselection operators and thus belong to a given superselection sector. The transition probability from the pure state $|\psi\rangle\langle\psi|$ to the pure state $|\varphi\rangle\langle\varphi|$ is given by $|\langle\varphi|\psi\rangle|^2$ for normalized states. As any vector of a vector space, a pure state $|\varphi\rangle$ can be described by its components in a given complete orthonormal basis ($|x\rangle$) :

$$|\varphi\rangle = \int d\mu(x)\,|x\rangle\langle x|\varphi\rangle \quad , \tag{3-7}$$

with the measure $d\mu\,(x)$. The covariant components , Eq. (1–27), are :

$$\varphi(x) = \langle x|\varphi\rangle \quad , \tag{3-8}$$

where φ is the wave function of the state $|\varphi\rangle$ with respect to the basis ($|x\rangle$). This wave function is defined up to a multiplicative phase factor.

iii) Only the relation, Eq. (3–3) : $\langle A\rangle = \mathrm{Tr}(\rho A)$ between the density operator ρ and the observable A is physically meaningful. Therefore the time evolution equations can take different forms according to how the time dependence is divided between the density operator and the observable A. Let us recall two commonly used representations :

1–In the Schrödinger representation (denoted by the subscript S) the algebra of operators is generated by time independent operators ($A_{S,i}$). Nevertheless one can imagine time dependent operators in the Schrödinger representation, equal to linear combinations of the $A_{S,i}$ with time dependent complex coefficients :

$$A_S(t) = \sum_i a_i(t)\, A_{S,i} \quad . \tag{3-9}$$

The dynamics is entirely described by the time evolution of the density operator :

$$\rho_S(t) = U(t,t_o)\rho_S(t_o)U(t,t_o)^{-1} \quad , \tag{3-10}$$

where the time evolution operator fulfils the Schrödinger equation which reads in a Galilean reference frame as :

$$i\hbar \frac{\partial}{\partial t} U(t,t_o) = H_S(t)U(t,t_o) \quad , \quad U(t_o,t_o) = \mathbf{1} \quad , \tag{3-11}$$

$H_S(t)$ being the Hamiltonian operator in the Schrödinger representation. In a conservative system $H_S(t)$ is time independent.

2–On the other hand in the Heisenberg representation (denoted by the subscript H) the dynamics is described by the evolution in time of the operators :

$$A_H(t,t_o) = U(t,t_o)^{-1}A_S(t)U(t,t_o) \quad , \tag{3-12}$$

while the density operator :

$$\rho_H(t,t_o) = U(t,t_o)^{-1}\rho_S(t)\ U(t,t_o) = \rho_S(t_o) \quad , \tag{3-13}$$

is independent of t. It has to be noticed that the Heisenberg representation is defined with respect to a reference time t_o.

3.2. SYMMETRY LAWS IN QUANTUM MECHANICS

3.2.1. DEFINITION

Let us first give an intuitive idea of what is a symmetry law. For simplicity a geometrical example is used, but all the following remarks are valid for any type of symmetry. We consider a physical system symbolized by the letter S (associated for instance with its geometrical shape). Some physical observable of this system is measured by means of an apparatus symbolized by an arrow (for instance a telescope to observe the shape). The intuitive idea of a symmetry law is then the following :

(1) (2)

A measure is performed in the situation (1). Next the physical system S and the apparatus → are transformed in the same way by an application S (a geometrical displacement in our example) and a measure is performed in the new situation (2). The transformation S is a symmetry law if the measurements give the same values in both situations (1) and (2) (i.e. if one observes the same shape in (1) and (2)). Actually, the definition of a symmetry law is more restrictive. Indeed, let us assume in our previous example that the physical system S is confined in a box. Then only the displacements which leave the physical system S inside the box are allowed.

Let us emphasize that the existence of a symmetry law as previously sketched does not imply anything :

(1) neither about the symmetry of the physical systems themselves (for instance only perfectly spherical objects should be spherically symmetrical with respect to rotations),

(2) nor about the evolution in time, namely about the dynamics of the systems (Subsec. 3.3,iii)).

In the quantum mechanical formalism, a state is described by a density operator ρ, an observable by a self-adjoint operator A, the possible box by the choice of the Hilbert space and what can be measured is the expectation value $\mathrm{Tr}(\rho A)$ (or for a pure state $|\varphi\rangle$, the matrix element $\langle\varphi|A|\varphi\rangle$). This leads us to the definition[41] :

A symmetry law S holds if there exist two surjective applications $A \to {}^{S}A$ of the set of observables onto itself, and $\rho \to {}^{S}\rho$ of the set of states onto itself, such that :

$$\mathrm{Tr}(\rho A) = \mathrm{Tr}({}^{S}\rho\, {}^{S}A) \quad . \tag{3-14}$$

A pure state is transformed by S into a pure state since $\mathrm{Tr}\,\rho^2 = 1$ is characteristic of a pure state, and ρ being an observable setting $A = \rho$ in Eq. (3–14) one gets :

$$\mathrm{Tr}\rho^2 = 1 = \mathrm{Tr}({}^{S}\rho)^2 \quad . \tag{3-15}$$

The condition of Eq. (3–14) implies that for pure states there exists a surjective mapping between rays $|\underline{\varphi}\rangle \to |{}^{S}\underline{\varphi}\rangle$ which preserves the transition probabilities :

$$|\langle\underline{\varphi}|\underline{\psi}\rangle|^2 = |\langle{}^{S}\underline{\varphi}|{}^{S}\underline{\psi}\rangle|^2 \quad . \tag{3-16}$$

In the quantum mechanical formalism as well as in the definition of the symmetry law given above the observer never occurs. Only physical systems and their observables are introduced. One does not consider the problem of a global quantum mechanical theory where both physical systems and observers would be described in the same frame.

3.2.2. THE WIGNER THEOREM

i) The question then arises to know which laws of transformation are possible on the density operators ρ (or on the pure states $|\varphi\rangle$ in the Hilbert space H) and which are compatible with the constraint of Eq. (3–16) ? The answer to this question is given by :

Theorem 3.1. Wigner's theorem. To any symmetry law, associated with the surjective mapping $|\underline{\varphi}\rangle \rightarrow |{}^{S}\underline{\varphi}\rangle$ such that Eq. (3–16) is fulfilled, there exists an operator $U(S)$ defined on the Hilbert space, which is either unitary (linear) or antiunitary (antilinear), namely : $UU^+ = U^+U = \mathbf{1}$, where U^+ is defined by $\langle\varphi|U^+|\psi\rangle = \langle U\varphi|\psi\rangle$ in the unitary case, and $\langle\varphi|(U^+|\psi\rangle) = \langle U\varphi|\psi\rangle^*$ in the antiunitary one, and such that $U(S)|\varphi\rangle$ belongs to the ray $|{}^{S}\underline{\varphi}\rangle$ whenever $|\varphi\rangle$ belongs to the ray $|\underline{\varphi}\rangle$. This operator $U(S)$ is uniquely defined up to a multiplicative phase factor.

Wigner's theorem is a generalization of Theorem 1.2 given in Subsec. 1.1.4,iv). For these two theorems one should note the necessity of the surjective character of the mapping. Let us be more precise in the statement of Wigner's theorem :

One considers a Hermitian vector space E with a finite dimension, or an infinite dimension and such that there exists a complete orthonormal basis. The bracket notation is used for vectors and inner products. Let U be a surjective mapping on E which preserves the modulus of the inner product :

$$\forall\, |\varphi\rangle , |\psi\rangle \in E \quad |\varphi'\rangle = U|\varphi\rangle, |\psi'\rangle = U|\psi\rangle , \quad |\langle\varphi'|\psi'\rangle| = |\langle\varphi|\psi\rangle| \quad . \tag{3-17}$$

Then there exists a choice of phases such that U is either linear unitary or antilinear unitary (antiunitary). The mapping U is then uniquely defined up to an overall multiplicative phase factor.

Proof. (Wigner[63], Appendix to Chap. 20 ; Messiah[39], Chap. XV, § 2). Clearly the image by U of any vector in E is defined up to an arbitrary phase factor :

$$U|\varphi\rangle = e^{i\alpha_\varphi}|\varphi'\rangle \quad , \quad \forall\, \alpha_\varphi \in R \quad .$$

The proof below consists in showing that these phases can be properly chosen so that for every pair of vectors $|\varphi\rangle$ and $|\psi\rangle$, either

$$\langle U\varphi|U\psi\rangle = \langle\varphi|\psi\rangle \qquad \text{or} \qquad \langle U\varphi|U\psi\rangle = \langle\varphi|\psi\rangle^* \quad .$$

1–Although this is not necessary, let us assume for simplicity that the complete orthonormal basis is countable and written as ($|i\rangle$, i=1,2,...). Then, for any choice of phases the images $|i'\rangle$ by U of the basis vectors $|i\rangle$ also form a complete orthonormal basis of E . Indeed, using Eq. (3–17), the $|i'\rangle$ are orthonormal :

$$\forall\, i,j \qquad |\langle i'|j'\rangle| = |\langle i|j\rangle| = \delta_{ij} \Rightarrow \langle i'|j'\rangle = \delta_{ij} \quad .$$

Consequently the $|i'\rangle$ are linearly independent, since if there exists a linear combination $\lambda^i|i'\rangle$ equal to the zero vector, then for all j :

$$\lambda^i \langle j'|i'\rangle = \lambda^j = 0 \quad .$$

Moreover the set ($|i'\rangle$) is complete since if there exists a vector $|\varphi'\rangle$ orthogonal to all vectors $|i'\rangle$, then U being surjective there exists at least a vector $|\varphi\rangle$ such that $|\varphi'\rangle = U|\varphi\rangle$, and using Eq. (3–17) $|\varphi\rangle$ is orthogonal to all $|i\rangle$, which is absurd since the set ($|i\rangle$) is assumed to be complete.

2–Let us denote by $|e_i\rangle$ the vectors defined by :

$$i=2,3,\ldots \qquad |e_i\rangle = |1\rangle + |i\rangle \quad .$$

Then using Eq. (3–17) :

$$|\langle j'|e_i'\rangle| = |\langle j|e_i\rangle| = \begin{cases} 1 & \text{if } j=1 \text{ or } i \\ 0 & \text{otherwise} \end{cases} ,$$

therefore the expansion of $|e_i'\rangle$ on the complete orthonormal basis $(|i'\rangle)$ reads :

$$|e_i'\rangle = x_i(|1'\rangle + y_i|i'\rangle) \quad ,$$

where $|x_i| = |y_i| = 1$.
A new choice of phases, denoted by the subscript 1 is defined as follows :

$$|1'\rangle_1 = |1'\rangle \quad ; \quad i=2,3,\ldots \quad |i'\rangle_1 = y_i|i'\rangle \quad , \quad |e_i'\rangle_1 = \frac{1}{x_i}|e_i'\rangle \quad .$$

One gets :

$$i=2,3,\ldots \qquad |e_i'\rangle_1 = |1'\rangle_1 + |i'\rangle_1 \quad .$$

Let $|\varphi\rangle$ be any vector in E having a nonzero component on $|1\rangle$, and $|\varphi'\rangle$ its image by U :

$$|\varphi\rangle = \varphi^i|i\rangle \quad , \quad \varphi^1 \neq 0 \quad \xrightarrow{U} |\varphi'\rangle = \varphi'^i|i'\rangle_1 \quad .$$

Since Eq. (3–17) implies for all i=1,2... :

$$|\varphi'^i| = |{}_1\langle i'|\varphi'\rangle| = |\langle i|\varphi\rangle| = |\varphi^i| \quad , \tag{3-18}$$

a new choice of phase for $|\varphi'\rangle$ can be defined such that :

$$|\varphi'\rangle_1 = \varphi^1/\varphi'^1|\varphi'\rangle = \varphi^1|1'\rangle_1 + \sum_{i\geq 2} \varphi_1'^i|i'\rangle_1 \quad .$$

Using Eq. (3–17) for $|e_i\rangle$ and $|\varphi\rangle$ yields :

$$i=2,3,\ldots \quad |{}_1\langle e_i'|\varphi'\rangle_1| = |\varphi^1 + \varphi_1'^i| = |\langle e_i|\varphi\rangle| = |\varphi^1 + \varphi^i| \quad .$$

Combining the above equations with Eq. (3–18) implies :

$$\varphi^1\varphi_1'^{i*} + \varphi^{1*}\varphi_1'^i = \varphi^1\varphi^{i*} + \varphi^{1*}\varphi^i \quad .$$

In this equation $\varphi_1'^{i*}$ can be eliminated by using $\varphi_1'^i\varphi_1'^{i*} = \varphi^i\varphi^{i*}$, providing $\varphi^i \neq 0$, otherwise $\varphi_1'^i$ is also equal to zero. Then one gets a quadratic equation for $\varphi_1'^i$:

$$i=2,3,\ldots \quad \varphi^{1*}(\varphi_1'^i)^2 - (\varphi^1\varphi^{i*} + \varphi^{1*}\varphi^i)\varphi_1'^i + \varphi^1|\varphi^i|^2 = 0 \quad ,$$

with the solutions :

$$i=2,3,\ldots \text{ (a) } \varphi_1'^i = \varphi^i \quad , \quad \text{(b) } \varphi_1'^i = (\varphi^1/\varphi^{1*})\varphi^{i*} \quad .$$

One notes that these are still solutions when $\varphi^i = 0$. In the case (a) one has :

$$i=1,2,\ldots \qquad \varphi_1'^i = \varphi^i \quad . \tag{3-19}$$

In the case (b) :

$$\varphi_1'^1 = \varphi^1 \quad ; \quad i=2,3,\ldots \quad \varphi_1'^i = (\varphi^1/\varphi^{1*})\varphi^{i*} \quad .$$

Then the new choice of phase for $|\varphi'\rangle$:

$$|\varphi_2'\rangle = \varphi^{1*}/\varphi^1|\varphi'\rangle_1 \quad ,$$

leads to :

$$|\varphi'>_2 = \varphi'^{i}_2 \, |i'>_1 \quad ; \quad i=1,2,\ldots \quad \varphi'^{i}_2 = \varphi^{i*} \quad . \qquad (3\text{-}20)$$

In the above calculation the vector $|1>$ plays a special role in the definition of $|e_i>$ and in assuming that $|\varphi>$ has a nonzero first component. Actually, if $|\varphi> = 0$ then also $|\varphi'> = 0$, and for any nonzero vector $|\varphi>$ there exists at least a nonzero component on the basis $(|i>)$ which can be taken as the first one by a change of notation.

Thus for any vector $|\varphi>$ in E there exists a choice of phases such that $|\varphi'> = U|\varphi>$ is either in case (a), Eq. (3–19), or in case (b), Eq. (3–20). We now drop the subscripts relative to the choice of phases and assume that the choice which leads to Eqs. (3–19,20) has been made.

3–It still remains to prove that all the vectors in E are simultaneously either in the case (a) or in (b). Consider the vector :

$$|\alpha> = \alpha^{i}|i> \quad , \quad \alpha_i \in C \quad .$$

As any vector in E , $|\alpha>$ belongs either to the case (a) or (b). Let us assume for instance that it belongs to the case (a), i.e. :

$$|\alpha'> = U|\alpha> = \alpha^{i}|i'> \quad .$$

We have to show that any other vector $|\varphi> = \varphi^{i}|i>$ in E is in the same case. From Eq. (3–17) one has :

$$|<\alpha'|\varphi'>| = |\alpha^{i*}\varphi'^{i}| = |<\alpha|\varphi>| = |\alpha^{i*}\varphi^{i}| \quad .$$

Now $|\varphi>$ belongs either to the case (a) or (b). It belongs to case (a), because :

$$|\alpha^{i*}\varphi^{i*}| \neq |\alpha^{i*}\varphi^{i}| \quad ,$$

provided that the components φ^{j} are not of the form :

$$j=1,2,\ldots \qquad \varphi^{j} = |\varphi^{j}|e^{i\theta} \quad , \quad \theta \in R \quad .$$

If it is so, the vector $e^{-i\theta}|\varphi>$, which has all its components real, belongs to the case (a) (as well as to the case (b)).

A similar argument can be given if one assumes that $|\alpha>$ belongs to the case (b).

4–Finally in the case (a), for all vectors $|\varphi>$ in E there exists a choice of phases such that :

$$i=1,2,\ldots \qquad \varphi'^{i} = \varphi^{i} \quad .$$

The operator U is linear since for all vectors $|\varphi> = \varphi^{i}|i>$, $|\psi> = \psi^{i}|i>$, and complex numbers λ and μ :

$$U(\lambda|\varphi> + \mu|\psi>) = U(\lambda\varphi^{i} + \mu\psi^{i})|i> = (\lambda\varphi^{i} + \mu\psi^{i})|i'> = \lambda|\varphi'> + \mu|\psi'> \quad .$$

Moreover for all $|\varphi>$ and $|\psi>$ in E one has :

$$<\varphi'|\psi'> = <\varphi|\psi> \quad ,$$

i.e. U is a linear unitary operator.

In the case (b), for all $|\varphi>$ in E there exists a choice of phases such that :

$$i=1,2,\ldots \qquad \varphi'^{i} = \varphi^{i*} \quad .$$

The operator U is antilinear since, with the previous notations :

$$U(\lambda|\varphi> + \mu|\psi>) = (\lambda^{*}\varphi^{i*} + \mu^{*}\psi^{i*})|i'> = \lambda^{*}|\varphi'> + \mu^{*}|\psi'> \quad .$$

Furthermore,U is an antilinear unitary (or antiunitary) operator since for all $|\varphi>$ and $|\psi>$ in

E one has :

$$<\varphi'|\psi'> = <\varphi|\psi>^* \quad .$$

Clearly the previous conclusions in both cases (a) and (b) are not affected by an overall change of phase of all image vectors by U, i.e. :

$$\forall\ |\varphi> \in E \quad , \qquad \alpha \text{ real independent of } |\varphi> \quad ,$$

$$\text{(a)} \qquad \varphi'^j = e^{i\alpha}\ \varphi^j \quad , \qquad \text{(b)} \qquad \varphi'^j = e^{i\alpha}\ \varphi^{j*} \quad .$$

Namely U is uniquely defined up to a multiplicative phase factor.

Let us remark that in an one–dimensional vector space, the two cases (a) and (b) are then related by a change of phase.

ii) Let us now consider the transformation of any observable A. For a pure state Eq. (3–14) reads :

$$<\varphi|A|\varphi> = <^S\varphi|^S A|^S\varphi> \quad . \tag{3-21}$$

From Wigner's theorem $|^S\varphi> = U(S)\ |\varphi>$, and since the matrix element $<\varphi|A|\varphi>$ is real one gets for $U(S)$ unitary as well as antiunitary :

$$<\varphi|A|\varphi> = [<\varphi|U(S)^+]\ ^S A\ [U(S)|\varphi>] = <\varphi|[U(S)^{+\,S}A\ U(S)]\,|\varphi> \quad .$$

Since this result is true for all $|\varphi>$, one obtains the law of transformation of a self–adjoint operator A :

$$^S A = U(S)\ A\ U(S)^{-1} \quad . \tag{3-22}$$

3.2.3. SYMMETRY GROUPS AND REPRESENTATIONS UP TO A PHASE

i) There exists a symmetry group G if to any element g of the group G there corresponds a symmetry law as defined above in Subsec. 3.2.1.

Then, Wigner's theorem implies that to any g of G there corresponds a unitary or antiunitary operator U(g) acting on the Hilbert space and uniquely defined up to a multiplicative phase factor. Namely, to any g of G there corresponds, following Bargmann[4)] , a ray operator which is the set of all operators c U(g), for a fixed U(g) and c being any complex number of modulus one. Any operator contained in this set is called a representative of the ray operator.

Let us denote by e the unity of G, then any representative of the ray operator which corresponds to e, reads :

$$U(e) = c\,\mathbf{1} \quad , \quad |c| = 1 \quad ,$$

since **1** is one of the representatives. By a general change of phase of all representatives :

$$\forall\ g \in G \qquad U(g) \to c^*\ U(g) \quad ,$$

it is possible to choose c = 1. In the following we will assume that this choice has been made :

$$U(e) = \mathbf{1} \quad . \tag{3-23}$$

Let us now prove that there exists a system of phase factors $\omega(g_1,g_2)$ such that for all g_1 and g_2 in G :

$$U(g_1)U(g_2) = \omega(g_1,g_2)U(g_1g_2) \quad , \quad |\omega(g_1,g_2)| = 1 \quad . \tag{3-24}$$

Indeed if $|\varphi\rangle$ belongs to the ray $|\underline{\varphi}\rangle$, then from Wigner's theorem, both $U(g_1g_2)|\varphi\rangle$ and $U(g_1)U(g_2)|\varphi\rangle$ belong to the ray $|^{g_1g_2}\underline{\varphi}\rangle$. Therefore one can write :

$$U(g_1)U(g_2)|\varphi\rangle = \omega(g_1,g_2,|\varphi\rangle)U(g_1g_2)|\varphi\rangle \quad , \quad |\omega(g_1,g_2,|\varphi\rangle)| = 1 \quad .$$

Using the linearity or antilinearity of the operators U, one has :

$$\begin{aligned} U(g_1)U(g_2)(|\varphi\rangle + |\psi\rangle) &= \omega(g_1,g_2,|\varphi\rangle + |\psi\rangle)U(g_1g_2)(|\varphi\rangle + |\psi\rangle) \\ &= \omega(g_1,g_2,|\varphi\rangle)U(g_1g_2)|\varphi\rangle + \omega(g_1,g_2,|\psi\rangle)U(g_1g_2)|\psi\rangle \quad . \end{aligned}$$

If $|\varphi\rangle$ and $|\psi\rangle$ are taken linearly independent, one obtains :

$$\omega(g_1,g_2,|\varphi\rangle + |\psi\rangle) = \omega(g_1,g_2,|\varphi\rangle) = \omega(g_1,g_2,|\psi\rangle) \quad ,$$

i.e. $\omega(g_1,g_2,|\varphi\rangle)$ does not depend upon $|\varphi\rangle$, let us denote it by $\omega(g_1,g_2)$, q.e.d. .

This phase factor has to fulfil the strong condition implied by the associativity relation for the product of operators, for all g_1,g_2, and g_3 in G :

$$[U(g_1)U(g_2)]U(g_3) = U(g_1)[U(g_2)U(g_3)] \quad . \tag{3-25}$$

The operators U(g) fulfilling Eq. (3–24) for all g in G are said to form a representation of G (Def. C.27) up to a phase or a ray representation.

The phase factor $\omega(g_1,g_2)$ depends upon the selection of representatives $U(g_1)$, $U(g_2)$, and $U(g_1g_2)$ for the ray operators which correspond respectively to g_1,g_2, and g_1g_2. Indeed, consider a different choice, say :

$$\forall\, g \in G \qquad U'(g) = \theta(g)U(g) \quad , \quad |\theta(g)| = 1 \quad ,$$

and let $\varepsilon(g)$ be equal to plus one or minus one according to whether U(g) is unitary or antiunitary. Then one has :

$$U(g_2) = \theta(g_2)^{-1}\, U'(g_2) \quad , \quad U(g_1) = \theta(g_1)^{-1}U'(g_1) \quad ,$$

$$U(g_1)U(g_2) = \theta(g_1)^{-1}\theta(g_2)^{-\varepsilon(g_1)}U'(g_1)U'(g_2) \quad ,$$

$$U'(g_1g_2) = \theta(g_1g_2)U(g_1g_2) = \theta(g_1g_2)\omega(g_1,g_2)U(g_1)U(g_2) \quad .$$

Therefore the new phase factor is related to the old one by :

$$U'(g_1g_2) = \omega'(g_1,g_2)U'(g_1)U'(g_2) \quad ,$$
$$\omega'(g_1,g_2) = \theta(g_1g_2)\omega(g_1,g_2)\theta(g_1)^{-1}\theta(g_2)^{-\varepsilon(g_1)} \quad . \qquad (3\text{-}26)$$

Two phase factors ω' and ω related in this fashion are called equivalent or of the same type. Indeed, to be of the same type defines an equivalence relation among the set of representations U(g) of G up to a phase.

Finally, the existence of a symmetry group leads to a classification of physical systems : these systems are described in the Hilbert space of representation up to a phase of the symmetry group. The representations are characterized by :

(1) $\varepsilon(g) = \pm 1$, i.e. the unitary or antiunitary character of U(g).

(2) A type of phase factor $\omega(g_1,g_2)$, i.e. a class of equivalent phase factors in the sense of Eq. (3–26).

(3) For given $\varepsilon(g)$ and type of phase factor, a class of representation up to an unitary equivalence. Namely for given $\varepsilon(g)$ and $\omega(g_1,g_2)$, U(g) and U'(g) define unitary equivalent representations (Defs. C.29,30) if there exists a regular linear operator A such that :

$$\forall\, g \in G \qquad U'(g) = A\, U(g)\, A^{-1} \quad .$$

ii) In a general way one can be more specific for the three previous points which characterize the representations, i.e. respectively :

1–Unitary or antiunitary operators. If G is a connected Lie group (Defs. C.45,60), U(g) is unitary for all g, i.e. $\varepsilon(g) = +1$. Indeed, in a neighbourhood of the unity of G one can always write $g = g_1g_1$, and the corresponding operator U(g), equal to $U(g_1)U(g_1)$ up to a multiplicative phase factor, is necessarily linear. Moreover any g is the product of a finite number of elements of G in the vicinity of the unity, and thus for all g, U(g) is unitary as the product of a finite number of unitary operators.

2–The type of phase factor. This is the main and difficult point.

First a remark about the phase factor in a special case ; if G contains an element k the square of which is the identity, and if U(k) is antiunitary, one then has :

$$U(k)U(k) = \omega(k,k)\mathbf{1} \quad \text{with} \quad \omega(k,k) = \pm 1 \quad . \qquad (3\text{-}27)$$

Indeed, using the choice of phase defined by Eq. (3–23), and setting $g_1 = g_2 = g_3 = k$ in Eq. (3–25) one obtains :

$$U(k)U(k)U(k) = U(k)\omega(k,k)\mathbf{1} = \omega(k,k)^*U(k) = \omega(k,k)\mathbb{1}\; U(k) = \omega(k,k)U(k) \quad ,$$

therefore $\omega(k,k)$ is real and thus equal to plus one or minus one.

The representations up to a phase, or ray representations, of a connected Lie group have been studied in detail by Bargmann[4]. It is out of the scope of these lectures to reproduce this study. Let us only quote the main result for nearly all the groups usually considered in physics with the exception of the Galilean group :

For a large class of Lie groups G, the continuous representations up to a phase of G are true representations (i.e. representations of the same type as those for which the phase factor is identical to one) of its universal covering group G^*.

The universal covering group G^* is the smallest connected group, simply-connected (Def. C.46) such that G is the image of G^* in a homomorphism and G^* is locally isomorphic to G (Subsec. 4.1.4,ii), cf. Theorem 4.3).

As a consequence, when the above result applies, the phase factor ω can be ignored providing one considers G^* instead of G as the true symmetry group.

3–The classes of unitary equivalent representations. This is a problem of true group representations, i.e. to find all the inequivalent irreducible representations (Def. C.32) of a group. Some results about this problem are given in Appendix C.4 for finite groups and in Sec. 7.2 for compact Lie groups.

3.3. INVARIANCE AND CONSERVATION LAWS

i) If S is a symmetry law, the observable A is said to be invariant under this symmetry law, or there exists an invariance law of A by S, if the self-adjoint operator A_S in the Schrödinger representation is such that :

$$^{S}A_S = A_S \quad . \tag{3-28}$$

We emphasize that this definition is explicitly given in terms of a particular representation of the operator, namely the Schrödinger representation. Thus, the above equation does not imply the same equality for the operators in the Heisenberg representation for instance, except, as we shall see below, if the dynamics itself is invariant under the symmetry law.

ii) Let us give the main consequences of the above assumption of invariance for the observable A (the subscript S always refers to objects in the Schrödinger representation) :

1–From the law of transformation of an observable, Eq. (3–22), one obtains :

$$[A_S, U(S)] = 0 \quad . \tag{3-29}$$

Later (Subsec.11.2.2) we shall see that if A is invariant under a continuous symmetry group, the above equation written for the infinitesimal transformations implies the commutation of A_S with the Schrödinger representation of the generators of the group representation.

2–If $|\varphi\rangle_S$ is an eigenvector of A_S with the eigenvalue a, then the state $|^{S}\varphi\rangle_S$ is also an eigenvector of A_S with the same eigenvalue. Indeed, since a is real, for unitary or antiunitary $U(S)$, one has :

$$A_S|{}^{\mathcal{S}}\varphi\rangle_S = U(\mathcal{S})U(\mathcal{S})^{-1}A_S U(\mathcal{S})\,|\varphi\rangle_S = U(\mathcal{S})A_S|\varphi\rangle_S = a|{}^{\mathcal{S}}\varphi\rangle_S \quad . \tag{3-30}$$

Thus if A is invariant under a symmetry law the eigenvalues of A_S are degenerate.

iii) Two special cases are important :

1–If A_S is taken as the density operator $\rho_S(t)$, then the physical state is said to be invariant under the symmetry law at the given time t :

$$[\rho_S(t), U(\mathcal{S})] = 0 \quad . \tag{3-31}$$

For a pure state $|\varphi t\rangle_S$ this implies :

$$|{}^{\mathcal{S}}\varphi t\rangle_S = \omega|\varphi t\rangle_S \quad , \tag{3-32}$$

where ω is a phase factor. These properties are preserved during the time evolution only if the dynamics is invariant by $\mathcal{S}$ (cf. below).

2–If the Hamiltonian H is invariant under a symmetry law $\mathcal{S}$, the above consequences Eqs. (3–28–30) hold, especially for the degeneracy of the spectrum of H_S. Moreover the evolution operator $U(t,t_o)$, Eq. (3–11), which fulfils the integral equation :

$$U(t,t_o) = \mathbf{1} + \frac{1}{i\hbar}\int_{t_o}^{t} dt' H_S(t')U(t',t_o) \quad , \tag{3-33}$$

commutes with the operator $U(\mathcal{S})$:

$$\forall t \qquad [U(t,t_o), U(\mathcal{S})] = 0 \quad . \tag{3-34}$$

This implies that the mapping generated by the symmetry law $\mathcal{S}$, either on the states or on the observables, is stable under the time evolution. This result may be schematized as follows for a density operator for instance :

$$\begin{array}{ccc} \rho_S(t_o) & \xrightarrow{\;U(t,t_o)\;} & \rho_S(t) \\ \downarrow \mathcal{S} & & \downarrow \mathcal{S} \\ {}^{\mathcal{S}}\rho_S(t_o) & \xrightarrow{\;U(t,t_o)\;} & {}^{\mathcal{S}}\rho_S(t) \end{array} \quad . \tag{3-35}$$

The commutation relation of Eq. (3–34) yields conservation laws for the operators $U(\mathcal{S})$, or if there exists a continuous symmetry group for the generators of the group representation, and thus for all functions of these generators.

If both the Hamiltonian H and a given observable A are invariant under a symmetry law $\mathcal{S}$, then the Heisenberg representation of A is also unchanged by $\mathcal{S}$,i.e. :

$${}^{\mathcal{S}}A_H(t,t_o) = U(\mathcal{S})U(t,t_o)^{-1}A_S(t)U(t,t_o)U(\mathcal{S})^{-1} = A_H(t,t_o) \quad . \tag{3-36}$$

Similarly if both H and ρ are invariant under $\mathcal{S}$, then the physical state is invariant under the symmetry law S at any time, i.e. Eq. (3–31) is valid for all t.

3.4. APPLICATION TO THE THREE–DIMENSIONAL ROTATION GROUP

Assuming that the three–dimensional rotation group is a symmetry group, one has to define in Quantum Mechanics the representations up to a phase of the group SO(3).

Let us now make precise the three points, defined in Subsec. 3.2.3,ii), which characterize these representations :

(1) Since SO(3) is a connected Lie group the representations are unitary.

(2) The representations up to a phase of SO(3) are true representations of its universal covering group, namely the group SU(2) : the set of 2×2 complex unitary and unimodular matrices, cf. Chap. 4. The consideration of SU(2) instead of SO(3) is important, especially it allows us to understand the existence of half–odd–integer spin particles.

(3) It is shown in Subsec. 7.2.2.3 that, up to an unitary equivalence, the representations $\mathcal{D}^j$[SU(2)] are the only unitary and irreducible representations of the group SU(2) (completeness of the representations $\mathcal{D}^j$[SU(2)]).

It is worthwhile to emphasize that, throughout the remainder of these notes, we only postulate the existence of a symmetry group but never refer to the possible invariance of the Hamiltonian and more generally of the dynamics.

CHAPTER 4

THE GROUP SU(2) AND ITS LIE ALGEBRA

This chapter is devoted to the study of the group SU(2) in Secs. 4.1,2 and its Lie algebra $\mathcal{W}(2)$ in Sec. 4.3. The relation between the group and its Lie algebra is mainly contained in the Baker–Campbell–Hausdorff formula given in Sec. 4.4 and Appendix D. This study for SU(2) gives us the opportunity to discuss general methods which can be applied to all Lie groups with a semi-simple Lie algebra. In particular, whenever it can easily be done the more general case of the group SU(n) is discussed.

In order to make the best use of this chapter it is recommended to first go through Appendices B, C, and D, where several definitions and general properties of algebras and groups are given.

4.1. THE GROUP, ITS PARAMETERS, AND SOME REMARKS ABOUT ITS TOPOLOGY

4.1.1. DEFINITION OF THE GROUP SU(n) AND OF ITS CENTRE

The group SU(n) is the set of complex $n \times n$ unitary and unimodular matrices :

$$UU^+ = U^+U = \mathbf{1} \quad , \quad \det U = 1 \quad . \tag{4-1}$$

The <u>centre</u> (Def. C.3) of SU(n) is the subset of matrices of SU(n) which commute with all the elements of this group. This subset is an Abelian invariant subgroup of SU(n). From Prop. C.46, which is a direct consequence of Schur's lemma (Prop. C.45), the elements of the centre of SU(n) are multiples of the unit matrix. Since they are unitary and unimodular, they read :

$$U_p = e^{-i\varphi_p}\mathbf{1} \quad , \quad \varphi_p = \frac{p}{n}2\pi + 2k_p\pi \quad , \; p=1,\ldots,n \quad , \quad k_p \in Z \quad , \tag{4-2}$$

i.e. $e^{-i\varphi_p}$ are the nth roots of unity. The centre of SU(n) is thus isomorphic to the cyclic group (Def. C.8) Z_n of order n, or also to the additive group of positive, negative, or zero integers modulo n.

As for n greater than or equal to two Z_n is a proper Abelian invariant subgroup of SU(n), this group is therefore not semi-simple and also not simple (Def. C.12).

Since the centre is an invariant subgroup, one can define the factor group (Prop. C.25) $SU(n)/Z_n$, the elements of which are the subsets of matrices of the type UU_p, with U in SU(n) and all U_p in Z_n. For instance in the case n=2 the two elements of the centre are **1** and − **1**, and the factor group $SU(2)/Z_2$ is the set of matrices of SU(2) defined up to a sign.

4.1.2. THE EXPONENTIAL FORM

i) **Theorem 4.1.** Any $n\times n$ unitary matrix U (an element of the unitary group U(n)) can be written as :

$$U = e^{-iW} \quad , \tag{4-3}$$

where W is an n-dimensional Hermitian matrix. This matrix W is not uniquely defined since :

$$U = e^{-iW} = e^{-iW'} \Leftrightarrow W' = W + 2\pi V(k_j)\, V^+ \quad , \tag{4-4}$$

where V is a unitary matrix which diagonalizes U, and (k_j) denotes the diagonal matrix, with elements k_j belonging to Z. Moreover if U is unimodular, i.e. belongs to SU(n), then :

$$\mathrm{Tr}\, W = 2k\pi \quad , \quad k\in Z \quad . \tag{4-5}$$

Proof :

1–**Theorem 4.2.** A necessary and sufficient condition that a complex $n\times n$ matrix A can be diagonalized by a unitary matrix is that A be a normal matrix, i.e. fulfils the condition :

$$AA^+ = A^+A \quad . \tag{4-6}$$

We quote the proof of this theorem as given for example by Mehta[38] (Subsec. 4.3, Theorem 4.3.6). Let λ_1 be an eigenvalue and x a corresponding eigenvector with Hermitian norm unity. It is always possible to construct an orthonormal basis with first basis vector x. Then the matrix $U_1 = (x,Y)$, where the first column is x and the columns of the $n\times(n-1)$ matrix Y are the other basis vectors, is unitary, therefore :

$$x^+x = 1 \quad , \quad Y^+x = 0 \quad .$$

Calculating $U_1^+ A U_1$, we get :

$$U_1^+AU_1 = \begin{pmatrix} x^+ \\ Y^+ \end{pmatrix} A(xY) = \begin{pmatrix} x^+Ax & x^+AY \\ Y^+Ax & Y^+AY \end{pmatrix} = \begin{pmatrix} \lambda_1 & x^+AY \\ 0 & AY \end{pmatrix} \quad .$$

If A is normal, so is $U_1^+ A U_1$, giving us two relations :

(1) $x^+AYY^+A^+x = 0$, i.e. the Hermitian norm of the row vector x^+AY is zero, implying that $x^+AY = 0$,

(2) Y^+AY is normal.

Thus :

$$U_1^+AU_1 = \begin{pmatrix} \lambda_1 & 0 \\ 0 & A_1 \end{pmatrix} \quad ,$$

where A_1 is a normal $(n-1)\times(n-1)$ matrix. The process can be repeated on A_1. Let U'_2 be a unitary $(n-1)\times(n-1)$ matrix such that :

$$U_2'^{+} A_1 U_2' = \begin{pmatrix} \lambda_2 & 0 \\ 0 & A_2 \end{pmatrix} \quad ,$$

where A_2 is an $(n-2) \times (n-2)$ normal matrix. Putting :

$$U_2 = U_1 \begin{pmatrix} 1 & 0 \\ 0 & U_2' \end{pmatrix} \quad ,$$

U_2 is a unitary matrix such that :

$$U_2^{+} A U_2 = \begin{pmatrix} \lambda_1 & & 0 \\ & \lambda_2 & \\ 0 & & A_2 \end{pmatrix} \quad .$$

The construction of U can thus be carried out step by step. Hence $A^+A = AA^+$ is a sufficient condition to ensure that A can be diagonalized by a unitary matrix.

Let us remark that this result can also be obtained as follows. Any complex matrix A can be uniquely written as :

$$A = A' + iA'' \quad ,$$

where A' and A" are Hermitian matrices. If A is normal, then :

$$AA^+ = A^+A \Leftrightarrow [A', A''] = 0 \quad .$$

Now referring to the theorem : two commuting Hermitian matrices can be simultaneously diagonalized by a unitary matrix, one gets the result.

Conversely, if there exists a unitary matrix U which diagonalizes A :

$$A = U^+DU \quad ,$$

where D is a diagonal matrix, then clearly A is normal since :

$$AA^+ = U^+DU\, U^+D^+U = U^+DD^+U = U^+D^+DU = A^+A \quad .$$

2–The unitary (as also the Hermitian and antihermitian) matrices are normal matrices. Therefore, for any unitary matrix U there exists a unitary matrix V which diagonalizes it :

$$U = V D V^+ \quad , \qquad V^+ = V^{-1} \quad , \tag{4-7}$$

where D is a diagonal matrix. Both matrices V and D are not uniquely defined since they correspond to a choice of the order of the eigenvalues, if there exist degenerate eigenvalues unitary transformations within the corresponding eigenspaces can be performed, and moreover V is defined up to a multiplicative phase factor. The matrix U being unitary its eigenvalues are of modulus one :

$$D = \left(e^{-id_j}\right) \quad , \tag{4-8}$$

where d_j are real numbers defined modulo 2π :

$$d_j = \lambda_j + 2k_j\pi \qquad k_j \in Z \quad . \tag{4-9}$$

Let us denote by d the diagonal matrix (d_j). Then one has $D = e^{-id}$ and expanding the exponential yields :

$$U = e^{-iVdV^+} \quad ,$$

where $W = VdV^+$ is a Hermitian matrix.

This exponential series is well defined since for any n–dimensional matrix A with bounded elements :

$$A = (a^i_j) \quad , \quad \forall\, i,j \quad |a^i_j| \leqslant c \qquad c \in R \quad ,$$

the exponential series $\sum\limits_{p=0}^{\infty} \frac{1}{p!} A^p$ is convergent. Indeed the i,j matrix element is the sum of the series with the general term $u_p = \frac{1}{p!} a^{i\,(p)}_j$, where $a^{i\,(p)}_j$ denotes the i,j matrix element of A^p. From the condition on $|a^i_j|$ one deduces by induction that :

$$|a^{i\,(p)}_j| \leqslant n^{p-1}\, c^p \quad .$$

Therefore, one has :

$$p=0,1,\ldots \qquad |u_p| \leqslant \frac{n^{p-1}}{p!}\, c^p \quad ,$$

and thus the series with general term u_p is absolutely convergent.

3–In the product VdV^+ there only remains the indetermination of the eigenvalues given by Eq. (4–9) :

$$W = V(\lambda_j)V^+ + 2\pi V(k_j)V^+ \quad , \quad k_j \in Z \quad .$$

Moreover if U is unimodular, then one has :

$$\det U = 1 = \exp\left[-i \sum_{j=1}^{n} (\lambda_j + 2k_j\pi)\right] \quad ,$$

which is equivalent to the condition :

$$\mathrm{Tr}\, W = \sum_{j=1}^{n} (\lambda_j + 2k_j\pi) = \sum_{j=1}^{n} \lambda_j + 2k\pi = 2k'\pi \quad , \quad k,k' \in Z \quad .$$

Without changing U which belongs to SU(n), it is possible to choose W with a trace equal to zero :

$$W = W^+ \quad , \quad \mathrm{Tr}\, W = 0 \quad . \tag{4-10}$$

To show how to take care of the above constraints, let us give the expression of W_p in terms of W such that :

$$U = e^{-iW} \quad , \quad U_p U = e^{-iW_p} \qquad p=1,\ldots,n \quad ,$$

where both W and W_p fulfil Eq. (4–10) and U_p is an element of the centre of SU(n). One obtains :

$$W_p = W + \frac{2\pi}{p} V(pk_j+1)V^+ \quad , \quad p \sum_{j=1}^{n} k_j + n = 0 \qquad k_j \in Z \quad ,$$

where V is a unitary matrix which diagonalizes W. For instance, for n=2 one has :

$$U = e^{-iW} \quad , \quad -U = e^{-iW'} \qquad W' = W + (2k+1)\pi V\begin{pmatrix}1 & 0\\ 0 & -1\end{pmatrix} V^+ \qquad k \in Z \quad . \tag{4-11}$$

ii) Let us denote by $\mathcal{W}(n)$ the set of $n \times n$ complex Hermitian and traceless matrices. It is obviously a vector space over the real field, of dimension n^2-1. For the group SU(2) :

$$W = \begin{pmatrix} a & b+ic \\ b-ic & -a \end{pmatrix} \qquad a,b,c \in R \quad . \tag{4-12}$$

In Subsecs. 4.3.2 and 5.4.2 one endows $\mathcal{W}(n)$ with, respectively, a Lie algebra structure and an Euclidean vector space structure, both closely related to the group structure of SU(n). It is worthwhile to notice that studying these unitary groups amounts to studying the product of exponentials of the type of Eq. (4–3). In that connection, we shall see later (Sec. 4.4 and Appendix D) the important Baker–Campbell–Hausdorff formula, which gives W_3 in terms of W_1 and W_2 such that :

$$e^{-iW_1}\, e^{-iW_2} = e^{-iW_3} \quad .$$

4.1.3. PARAMETERS OF THE GROUP

i) It has already been shown in Subsec. 1.1.4,iv) that the matrices of SU(n) depend upon n^2-1 real parameters. This is the same number of parameters as for the matrices of $\mathcal{W}(n)$, since they are related by the Eq. (4–3). There exist several useful parametrizations. For instance, any element U in SU(n) can be parametrized by the n^2-1 real components (w^i) of W with respect to a basis (W_i) of the vector space $\mathcal{W}(n)$:

$$U = e^{-iW} = e^{-i\, w^i W_i} \quad . \tag{4-13}$$

To a given set of components (w^i) corresponds a unique U, but the converse is not true since W is not uniquely defined, cf. Eq. (4–4).

ii) Let us now focus our attention to the case of SU(2) :

1–The <u>Pauli matrices</u> σ_i, i=1,2,3 :

$$\sigma_1 = \begin{pmatrix} 0 & 1 \\ 1 & 0 \end{pmatrix} \quad , \quad \sigma_2 = \begin{pmatrix} 0 & -i \\ i & 0 \end{pmatrix} \quad , \quad \sigma_3 = \begin{pmatrix} 1 & 0 \\ 0 & -1 \end{pmatrix} \quad , \tag{4-14}$$

which are Hermitian traceless matrices and linearly independent over the real field, form a basis of the three–dimensional vector space $\mathcal{W}(2)$ over the real field. With respect to this basis, one usually writes Eq. (4–13) as :

$$U(\theta,n^i) = e^{-i\,\frac{\theta}{2}\, n^i\sigma_i} \quad , \tag{4-15}$$

where the n^i are real numbers such that :

$$\sum_{i=1}^{3} (n^i)^2 = 1 \quad , \tag{4-16}$$

and θ is a real parameter (Subsec. 4.2.1).

The Eq. (4–15) is often written in terms of an Euclidean inner product as :

$$U(\theta,n^i) = U(\theta,\hat{n};a) = e^{-i\frac{\theta}{2}\,\hat{n}.\sigma} \quad , \quad \hat{n} = n^i e_i \quad , \quad \hat{n}.\sigma = n^i\sigma_i \quad , \tag{4-17}$$

where $a = (e_i)$ denotes an orthonormal basis of a three–dimensional Euclidean vector space. One comes back in Subsec. 5.4.2,ii) to the definition of the formal inner product $\hat{n}.\sigma$. It is worthwhile to emphasize that contrary to the rotation $R(\theta, \hat{n})$, Eq. (2–35), $U(\theta, \hat{n}; a)$ is not defined in terms of an intrinsic vector $\hat{n}$, but explicitly depends upon the three numbers θn^i, taken for convenience in Eq. (4–15) as the components of a vector in an orthonormal basis. The dependence of $U(\theta, \hat{n}; a)$ upon the choice of the basis a can be made explicit as follows. Let R be any rotation, then (cf. Eq. (1–64)) :

$$\hat{n} = R^{-1} R \hat{n} = (R\hat{n})^i R^{-1} e_i \qquad ,$$

therefore one gets :

$$U(\theta, R\hat{n}; a) = U(\theta, \hat{n}; R^{-1}a) \qquad . \qquad (4\text{-}18)$$

This relation is completed by the Eq. (4–34).

Expanding the exponential in Eq. (4–15) and using the property of the σ matrices given later by Eq. (4–88), one obtains :

$$U(\theta, n^i) = \cos\frac{\theta}{2}\,\mathbf{1} - i \sin\frac{\theta}{2}\, n^i \sigma_i = \begin{pmatrix} \cos\frac{\theta}{2} - i \sin\frac{\theta}{2} n^3 & -\sin\frac{\theta}{2}(n^2 + i n^1) \\ \sin\frac{\theta}{2}(n^2 - i n^1) & \cos\frac{\theta}{2} + i \sin\frac{\theta}{2} n^3 \end{pmatrix} . \qquad (4\text{-}19)$$

Or, with the same remarks as for Eq. (4–17), one has :

$$U(\theta, \hat{n}; a) = \cos\frac{\theta}{2}\,\mathbf{1} - i \sin\frac{\theta}{2}\, \hat{n}.\sigma \qquad . \qquad (4\text{-}20)$$

These Eqs. (4–19,20) can be obtained directly using the fact that the four matrices $\mathbb{1}$, σ_i form a basis of the vector space over the complex field of 2×2 complex matrices. Therefore any U in SU(2) can be uniquely written as :

$$U = a^o\,\mathbf{1} + a^i \sigma_i \qquad a^o, a^i \in C \qquad .$$

Now expressing the unitary and unimodular character of U, one gets the above results (for this calculation it is convenient to use Eqs. (4–102, 104)).

Let us now specify the space of the parameters θ, n^i which ensures a one-to-one correspondence between $U(\theta, n^i)$ and a given set of parameters θ, n^i. From Eq. (4–19) one has :

$$U(\theta', n'^i) = U(\theta, n^i) \Leftrightarrow \cos\frac{\theta'}{2} = \cos\frac{\theta}{2} \quad \text{or} \quad \sin\frac{\theta'}{2} n'^i = \sin\frac{\theta}{2} n^i \qquad ,$$

The equality of the cosines implies :

$$\theta' = \pm\theta + 4k\pi \quad k \in Z \Rightarrow n'^i = \pm n^i \qquad .$$

Therefore if the parameters n^i, which fulfil Eq. (4–16), are allowed to be positive, negative or zero, the range of θ can be restricted to $[0,2\pi]$. The space of parameters θ, n^i, or $\Theta = \theta \times n^i e_i$ with respect to a given orthonormal basis $a = (e_i)$ is thus the three–dimensional ball D^3 of radius 2π where all the surface points ($\theta = 2\pi$) are identified and correspond to $-\mathbb{1}$. The centre of the ball corresponds to the unit element, and one obtains directly :

$$U(\theta,n^i) \Rightarrow -U = U(2\pi-\theta,-n^i) \quad , \quad U^{-1} = U(\theta,-n^i) \quad , \tag{4-21}$$

see Fig. 4.1.

2–Another parametrization of the group SU(2) is the following. Expressing the unitary and unimodular character of a complex 2×2 matrix U, one obtains :

$$U = \begin{pmatrix} u_1 & u_2 \\ u_3 & u_4 \end{pmatrix} \in SU(2) \Leftrightarrow \begin{cases} \det U = u_1u_4 - u_2u_3 = 1 \\ U^+ = \begin{pmatrix} u_1^* & u_3^* \\ u_2^* & u_4^* \end{pmatrix} = U^{-1} = \begin{pmatrix} u_4 & -u_2 \\ -u_3 & u_1 \end{pmatrix} \quad . \end{cases} \tag{4-22}$$

These relations are equivalent to requiring that the two complex numbers $u_1 = x_1 + iy_1$ and $u_2 = x_2 + iy_2$ fulfil the condition :

$$\forall U \in SU(2) \quad \begin{cases} U = \begin{pmatrix} u_1 & u_2 \\ -u_2^* & u_1^* \end{pmatrix} \\ |u_1|^2 + |u_2|^2 = x_1^2 + y_1^2 + x_2^2 + y_2^2 = 1 \quad . \end{cases} \tag{4-23}$$

Thus, there is a one–to–one correspondence between the elements of SU(2) and the surface points of the four–dimensional sphere of radius one, denoted by S^3. With this parametrization, two diametrically opposite points correspond, respectively, to U and –U. The two complex numbers u_1 and u_2 are called the Cayley–Klein parameters[27].

3–In Subsec. 4.2.4,i) a still another parametrization of the elements of SU(2) by means of the Euler angles φ, θ, and ψ is defined.

4.1.4. REMARKS ABOUT TOPOLOGICAL PROPERTIES

i) Some basic topological definitions are recalled in Appendix C.5.1. It is shown in Appendix C.5.4.3,i) that the groups SO(3) and SU(2) and more generally the linear groups GL(n,R) and GL(n,C) are Lie groups with respect to the topology induced by the one of the set of their matrix elements. Similar remarks to those made in Subsec. 2.3.2.3 about the manifold structure of SO(3) can be made for the group SU(2) :

1–In the manifold language, the parametrization of the elements of SU(2) by $(\theta n^1, \theta n^2, \theta n^3)$ defines the local chart (U_1, f_1) where :

– U_1 is the open set in SU(2) of unitary and unimodular 2×2 matrices different from $-\mathbb{1}$,
– f_1 is the one–to–one mapping such that :

$$\forall U \in U_1 \qquad f_1(U) = (\theta n^1, \theta n^2, \theta n^3) \quad ,$$

– $f_1(U_1)$ is the three–dimensional open ball D^3 of radius 2π (surface points, i.e. the sphere S^2, excluded).

Thus this local chart does not cover the whole group, i.e. is not an atlas, since it is not defined for the element $-\mathbf{1}$ of the group.

2–On the other hand, the parametrization of the elements of SU(2) by four real numbers which fulfil Eq. (4–23) allows us to define a local chart (U_3,f_3) which is an atlas :

– U_3 = SU(2),

– f_3 is the one–to–one mapping such that :

$$\forall U \in SU(2) \qquad f_3(U) = (x_1, y_1, x_2, y_2) \quad ,$$

– $f_3(U_3)$ is the open set S^3 in R^4.

Therefore, the topology of SU(2), defined in Appendix C.5.4.3,i), is equivalent to the topology of the four–dimensional sphere S^3, while it is only locally equivalent to the topology of the three–dimensional open ball D^3.

3–The parametrization of the elements of SU(2) by the Euler angles also defines a local chart which does not cover the whole group, (Subsec. 4.2.4,i)).

The previous remarks are especially useful for the study of the invariant integration on the group, cf. Subsec. 7.1.3.

ii) It has been shown in Subsec. 2.3.2.1,i) that for the group SO(3) the space of parameters θ, n^i is the three–dimensional ball D^3 of radius π, where two diametrically opposite points on the surface are identified. We just saw that for SU(2) parametrized by θ, n^i all the surface points of the three–dimensional ball of radius 2π have to be identified.

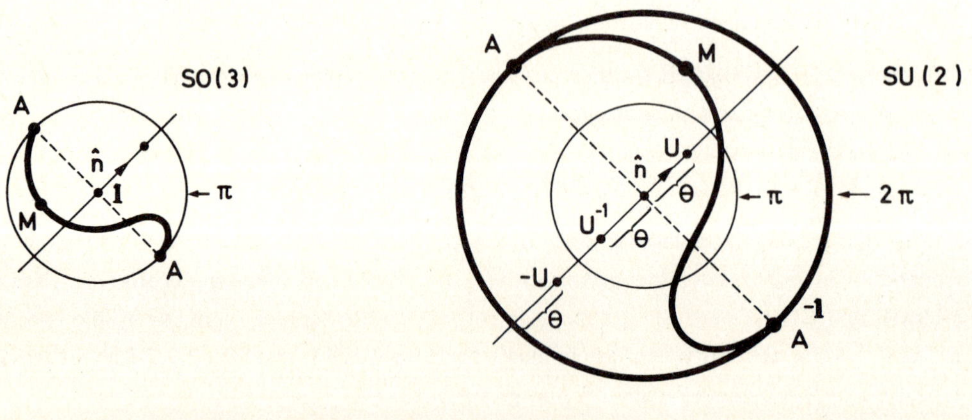

Fig. 4.1. The three–dimensional balls which are the spaces of parameters θ and $\hat{n}$ for, respectively, SO(3) and SU(2).

This difference reflects the fact that SU(2) is simply–connected while SO(3) is connected without being simply–connected. This means that, though these two groups are connected, any closed path (Def. C.44) AMA, see Fig. 4.1, in the SU(2) space of parameters can be continuously reduced to a point (one says homotope to zero, cf. remarks below), while there exist closed paths AMA in the SO(3) space of parameters for which this is not possible.

Let us now consider the factor group $SU(2)/Z_2$, i.e. the set of matrices of SU(2) defined up to a sign. The topology of this group is different from the one of SU(2). Indeed, its space of parameters can be obtained by identifying the two subsets of parameters which correspond, respectively, to U and −U :

1–With parameters θ , n^i, the space of parameters of $SU(2)/Z_2$ can be taken as the three-dimensional ball of radius π. Indeed, see Fig. 4.1, if $U(\theta, n^i)$ is such that θ belongs to $[0,\pi]$, then from Eq. (4–21) $-U = U(2\pi-\theta,-n^i)$, and $2\pi-\theta$ belongs to $[\pi,2\pi]$. If $\theta = \pi$, U and −U correspond, respectively, to diametrically opposite points on the surface of the ball of radius π, and therefore have to be identified. Thus the space of parameters of $SU(2)/Z_2$ is identical to the one of SO(3). It is shown in Subsec. 4.2.1 that these two groups are isomorphic.

2–Similar results can be obtained with the parametrization of SU(2) by means of the points of the sphere S^3. We have already noticed that in this parametrization diametrically opposite points correspond, respectively, to U and −U. Therefore it is sufficient to consider only one hemisphere to parametrize the factor group $SU(2)/Z_2$. Since the dimension is four, the boundary of a hemisphere of S^3 is the three-dimensional sphere S^2, and all the points of this hemisphere of S^3 have a unique projection inside or on the surface of the three-dimensional ball D^3 (as the boundary of a hemisphere of the three-dimensional sphere S^2 is the two-dimensional sphere S^1, i.e. the circle, and all points of the hemisphere have a unique projection inside or on the surface of the two-dimensional ball D^2). To ensure a one-to-one correspondence between $SU(2)/Z_2$ and D^3 one has to identify the diametrically opposite points, finally one finds again the same parameter space as for SO(3) parametrized by θ , n^i.

3–A similar discussion with SU(2) parametrized by the Euler angles is given in Subsec. 4.2.4,i).

Some remarks about the homotopy, cf. for instance Pontryagin[44] (Chap. 9, Sec. 49) :

A path f in a topological space E is called a null path if the function f(t) introduced in Def. C.44 is constant. The inverse path f^{-1} to a given path f is defined to be the function f(1−t) of the parameter t.

One now defines on the set of these paths in a topological space an equivalence relation and an internal law of composition as follows :

–Two paths f = AMB and g = AM'B (see Fig. 4.2) are said to be homotopic, or equivalent, if there exists a continuous deformation carrying f into g which does not change the end points. More precisely this means there exists a continuous function $\varphi(s,t)$ of real parameters s and t, $0 \leq s \leq 1$, $0 \leq t \leq 1$ such that :

$$\varphi(0,t) = f(t) \quad , \quad \varphi(1,t) = g(t) \quad ,$$

$$\varphi(s,0) = f(0) = g(0) \quad , \quad \varphi(s,1) = f(1) = g(1) \quad .$$

This defines an equivalence relation among the set of paths. The elements of the quotient set, i.e. the subsets of homotopic paths, are called the homotopy classes.

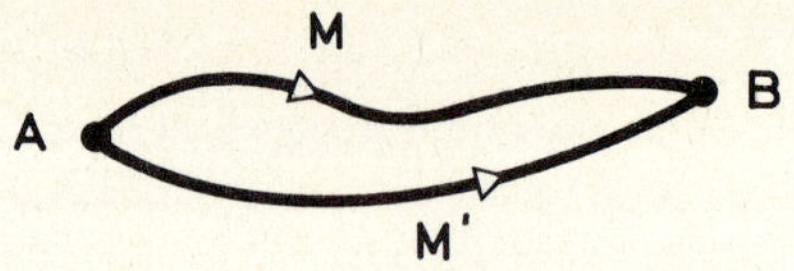

Fig. 4.2. Two homotopic paths in the plane.

–If two paths f and g are such that the terminal point of the first coincides with the initial point of the second, $f(1) = g(0)$, the product $h = f\,g$ of the paths f and g is defined as follows : $h(t) = f(2t)$ for $0 \leq t \leq \frac{1}{2}$ and $h(t) = g(2t-1)$ for $\frac{1}{2} \leq t \leq 1$.

Let us now consider the subset of closed paths about a given point O. One shows that the above multiplication law for the paths is stable under the equivalence relation which is the homotopy, and moreover that the induced multiplication law for the homotopy classes is associative. Therefore the homotopy classes, associated with the closed paths about a given point O form a group : the homotopy group for the point O, also called the fundamental group $\Pi_1(E,O)$. The unit element of this group is the subset of paths homotopic to the null path, the inverse of the homotopy class of a path $f = OM_1M_2O$ is the homotopy class of the inverse path $f^{-1} = OM_2M_1O$, $f^{-1}(t) = f(1-t)$.

If the considered topological space is simply–connected, the homotopy group consists only of the unit element, whereas this group is nontrivial for a connected but not simply–connected topological space. For connected Lie group there exist important results :

Theorem 4.3. Cf. for instance Pontryagin[44] (Chap. 9, Secs. 49F,G and Sec. 50E). Let G be a connected Lie group, then :

–The homotopy group is independent of the reference point O, and is Abelian.

–Let H be the homotopy group of G, then there exists a simply–connected Lie group G^*, called the universal covering group of G, such that G is the image of G^* in a homomorphism, the kernel of which is isomorphic to H.

This universal covering group plays an important role in the study of the representations up to a phase of connected Lie groups (Subsec. 3.2.3,ii)).

In our case G is the group SO(3) which is a connected but not simply–connected group. One shows that its homotopy group H is isomorphic to $Z_2 = \{\mathbb{1}, -\mathbb{1}\}$. For instance, if the space of parameters θ, n^i is considered, this means that any closed path in this space is homotopic either to zero ($\mathbb{1}$) or to the paths with diametrically opposite end points on the surface of the ball ($-\mathbb{1}$), see Fig. 4.1. The covering group G^* is the group SU(2) and it is shown in Subsec. 4.2.1, that SO(3) is homomorphic to SU(2) with the kernel $\{\mathbb{1}, -\mathbb{1}\}$. Let us remark that any closed path which remains in the neighbourhood of a given point of a space of parameters, (such as θ, n^i) of SO(3) is homotopic to zero. For this reason SO(3) and SU(2) (more generally G and G^*) are said to be locally isomorphic, i.e. they have the same Lie algebra, cf. Subsec. 4.3.4,ii).

iii) Another important topological property of SU(2) and SO(3), more generally of U(n), SU(n), O(n), and SO(n), is that they are compact groups, with respect to the topology induced by the one of the set of their matrix elements. The proof is given in Appendix C.5.4.3,i). It is shown in Subsec. 7.1.2 that for compact Lie groups, an integration on the group can be defined with a left and right invariant measure, which is unique up to a multiplicative factor. This integration leads for the representations of compact Lie groups to properties similar to those of finite groups (cf. Appendix C.4 and Sec. 7.2). For compact Lie groups with a real semi-simple Lie algebra (Def. B.23) see also the definite character of the Killing form (Def. B.25 and Subsec. 5.4.2).

4.1.5. ONE-PARAMETER ABELIAN SUBGROUPS

Given any element W in $\mathcal{W}(n)$, one considers the subset G_W of elements in SU(n) of the type :

$$\forall\, \theta \in R \qquad U(\theta) = e^{-i\theta W} \quad . \tag{4-24}$$

This subset G_W is clearly an Abelian subgroup of SU(n), called a one-parameter Abelian subgroup, such that :

$$U(\theta_1)U(\theta_2) = U(\theta_2)U(\theta_1) = U(\theta_1+\theta_2) \quad , \tag{4-25a}$$

$$U(\theta)^{-1} = U(-\theta) \quad . \tag{4-25b}$$

Therefore the mapping $\theta \to U(\theta)$, for a given W, is a homomorphism of the Abelian additive group of the real field into SU(n).

Conversely, let us assume there exists such a homomorphism which fulfils Eqs. (4-25a,b). Then all the unitary matrices $U(\theta)$, commuting with one another, are diagonalizable by the same unitary matrix V, Eq. (4-7) :

$$\forall\, \theta \in R \qquad U(\theta) = V\, D(\theta)\, V^+ \quad .$$

The problem is reduced to that of one dimension since the diagonal matrices $D(\theta)$ fulfil Eqs. (4-25a,b). Assuming that $U(\theta)$, and thus $D(\theta)$, is differentiable (i.e. for all i and j, $U(\theta)^i_j$ is differentiable) one obtains from Eqs. (4-25a,b) :

$$D(\theta)^i_j = e^{-i\theta d_i} \qquad d_i \in R \quad .$$

Thus $U(\theta)$ reads :

$$U(\theta) = e^{-i\theta W} \quad ,$$

where the Hermitian matrix $W = V(d_i)V^+$ can be chosen with a trace equal to zero, but is still not uniquely defined, cf. Eq. (4-4).

4.2. THE HOMOMORPHISM BETWEEN THE GROUPS SU(2) AND SO(3)

Let us present here one of the usual proofs of the homomorphism between SU(2) and SO(3), where the σ matrices play a main, but still mysterious, role. An intrinsic and thus deeper relation between SU(2) and SO(3) (more generally between SU(n) and $SO(n^2-1)$) is given later in Subsec. 5.4.2,ii), where the special role of the σ matrices is then justified.

4.2.1. CORRESPONDENCE BETWEEN SU(2) AND SO(3)

i) Let E_3 be a three-dimensional Euclidean vector space. To any U in SU(2) and any orthonormal basis $a = (e_i)$ of E_3 one associates a linear mapping R(U;a) of E_3 onto itself by :

$$(U\,;\,a) \longrightarrow R(U\,;\,a) \quad ,$$
$$v = v^i e_i \in E_3 \quad , \quad v' = R(U;a)v = v'^i e_i \in E_3 \quad , \quad v'^i \sigma_i = U\, v^j \sigma_j U^{-1} \quad , \tag{4-26}$$

or using an Euclidean inner product notation with the same remarks as for Eq. (4-17) :

$$[R(U;a)v].\sigma = v'.\sigma = U\, v.\sigma\, U^{-1} \quad . \tag{4-27}$$

Let us point out that the mapping R(U;a) depends upon the orthonormal basis a of E_3, this dependence is made explicit below by Eq. (4-32). Actually, Eq. (4-26) associates to any U in SU(2) a three-dimensional matrix $(R(U)^i_j)$ such that, using Eq. (4-89a) :

$$v'^i \delta_{ik} = \frac{1}{2}\, Tr(\sigma_k v'.\sigma) = R(U)^i_j \delta_{ik} v^j \Rightarrow R(U)^i_j \delta_{ik} = \frac{1}{2}\, Tr(\sigma_k U \sigma_j U^{-1}) \quad . \tag{4-28}$$

This matrix is the matrix representative of the linear mapping R(U;a) with respect to the basis a.

It is now shown that $(R(U)^i_j)$ is a three-dimensional rotation matrix, i.e. belongs to SO(3). This result can be obtained directly on the matrix level by computing $R(U)^i_j$ by means of Eq. (4-28). For instance (cf. below and Appendix E.1.1) with U parametrized by θ, n^i as in Eq. (4-19), one gets the rotation matrix given in Eq. (E-4). Another way to get the result is to prove that the mapping R(U;a) is a rotation. Indeed it preserves the Euclidean norm :

$$v'^2 = -\det(v'.\sigma) = -\det(v.\sigma) = v^2 \quad , \tag{4-29}$$

and thus it is an isometric application. Moreover this isometric application is a proper one since it also preserves the three-fold product, using Eq. (4-90a) :

$$(v_1, v_2, v_3) = \frac{1}{2i}\, Tr[(v_1.\sigma)(v_2.\sigma)(v_3.\sigma)] \quad . \tag{4-30}$$

This last result can also be obtained from the connected character of SU(2). Indeed, when U is changed continuously into the unit matrix, R(U;a) goes continuously into R($\mathbb{1}$;a) which is the identity mapping on E_3 and thus a proper one.

Finally, to any U in SU(2) one associates by Eq. (4-28) a rotation matrix $(R(U)^i_j)$ in SO(3), and to any U in SU(2) and any orthonormal basis a of E_3 one associates by Eq. (4-26) the rotation R(U;a) of E_3 onto itself.

ii) Conversely, is the correspondence $U \rightarrow (R(U)^i_j)$ a surjective mapping of SU(2) onto SO(3) ? To answer this question let us show that if U is parametrized by θ , n^i as in Eq. (4-19) then :

$$[U(\theta,n^i)\ ;\ a=(e_i)] \rightarrow R(U;a)=R(\theta,\hat{n}=n^i e_i) \quad , \qquad (4\text{-}31)$$

i.e. R(U;a) is the rotation of angle θ about the axis $\hat{n}$. We have previously noticed that this result can be obtained by computing $R(U)^i_j$, using Eq. (4–28), then one gets the matrix given by Eq. (E–4). Another way to get the result is the following ; $\hat{n}=n^i e_i$ defines the rotation axis of R(U;a) since it is invariant, using Eq. (4–89b) :

$$(\cos\tfrac{\theta}{2}\,\mathbf{1} - i\sin\tfrac{\theta}{2}\,\hat{n}.\sigma)\hat{n}.\sigma(\cos\tfrac{\theta}{2}\,\mathbf{1} + i\sin\tfrac{\theta}{2}\,\hat{n}.\sigma) = \hat{n}.\sigma \quad .$$

Moreover, the rotation angle of R(U;a) is given by computing the three–fold product $(v,Rv,\hat{n})$, where v is a vector orthogonal to $\hat{n}$. Using Eq. (4–90a) and the definition of R(U;a), this product takes the form :

$$(v,Rv,\hat{n}) = \frac{1}{2i}\,\mathrm{Tr}[(v.\sigma)(Uv.\sigma U^{-1})(\hat{n}.\sigma)] \quad .$$

Expressing U and U^{-1} by Eq. (4–20), one is led to compute the trace of products of σ matrices. This can be done by using several times Eqs. (4–85, 88, and 89a), one obtains :

$$(v,Rv,\hat{n}) = \|v\|^2 \sin\theta \quad .$$

Finally, to any rotation matrix $R(\theta,n^i)$ as given by Eq. (E–4), there exists at least one element U of SU(2), i.e. $U=U(\theta,n^i)$ such that Eq. (4–28) is fulfilled. Let us remark that the dependence of R(U;a) upon $a=(e_i)$ when this orthonormal basis is changed under any rotation ρ on E_3 can be made explicit as follows :

$$\hat{n}=n^i e_i \quad , \qquad [U(\theta,n^i)\rho a=(\rho e_i)] \rightarrow R(U;\rho a)=R(\theta,n^i\rho e_i=\rho\hat{n}) \quad ,$$

and using Eq. (2–36) :

$$[U(\theta,n^i);\rho a] \rightarrow R(U;\rho a) = \rho R(U;a)\rho^{-1} \quad . \qquad (4\text{-}32)$$

iii) The correspondence between SU(2) and SO(3) is a homomorphism (Def. C.21) since it is compatible with the group laws. Indeed we have :

$$\left.\begin{array}{lll}(U;a)\rightarrow R(U;a) & , & v''=R(U;a)v' \\ (V;a)\rightarrow R(V;a) & , & v'=R(V;a)v \\ (UV;a)\rightarrow R(UV;a) & , & v''=R(UV;a)v\end{array}\right\} \Rightarrow R(UV;a)=R(U;a)R(V;a) \ ,$$

since :

$$v''.\sigma = U(v'.\sigma)U^{-1} = U(Vv.\sigma V^{-1})U^{-1} = UVv.\sigma(UV)^{-1} \quad .$$

iv) There remains the question of the multiplicity of this homomorphism. Obviously to U and −U in SU(2) there corresponds by Eq. (4–28) the same rotation matrix in SO(3). Does there exist any other element in SU(2) which corresponds to the same rotation ? Using the fact that the correspondence between SU(2) and SO(3) is a homomorphism, it is sufficient to answer the question : how many elements U in SU(2) correspond to the identity in SO(3) ? These elements U have to fulfil :

$$\forall\, W = w^i\sigma_i \qquad U\, W\, U^{-1} = W \quad ,$$

hence from Prop. C.46, a direct consequence of Schur's lemma, U is a multiple of the unity and being unitary $U = \pm\mathbf{1}$.

Finally, we have the important result, i.e. Eq. (4–28) defines a two-to-one homomorphism of SU(2) onto SO(3) :

$$\pm\, U \in SU(2) \leftrightarrow (R(U)^i_{\ j}) \in SO(3) \quad . \tag{4-33}$$

The kernel of this homomorphism is the Abelian invariant subgroup $Z_2 = \{\mathbf{1}, -\mathbf{1}\}$, i.e. the centre of SU(2). Thus SO(3) is isomorphic to the factor group $SU(2)/Z_2$ of matrices of SU(2) defined up to a sign. It has been shown in Subsec. 4.1.4,ii) that the spaces of parameters of these two groups are identical. As SO(3), $SU(2)/Z_2$ is a simple group since there exists no invariant proper subgroup.

4.2.2. CLASSES OF CONJUGATE ELEMENTS UNDER SU(2)

As a direct consequence of the correspondence between SU(2) and SO(3), let us now derive for the group SU(2) parametrized by $\theta,\hat{n};a$, Eq. (4–17), a group law similar to the one given by Eq. (2–36) for the rotations. For any V in SU(2) one has :

$$V\, U(\theta,\hat{n};a)V^{-1} = U[\theta, R(V;a)\hat{n};a] \quad , \tag{4-34}$$

where R(V;a) is the rotation associated to V by Eq. (4–26) for a given orthonormal basis a. Indeed using Eqs. (4–20,26), one obtains :

$$V\, U(\theta,\hat{n};a)V^{-1} = \cos\tfrac{\theta}{2}\,\mathbf{1} - i\,\sin\tfrac{\theta}{2}\, V\,\hat{n}.\sigma\, V^{-1} = \cos\tfrac{\theta}{2}\,\mathbf{1} - i\,\sin\tfrac{\theta}{2}\,[R(V;a)\hat{n}].\sigma \ .$$

Therefore, as for the rotation group (Subsec. 2.3.2.1,ii)), <u>the classes of conjugate elements under SU(2) are the subsets of elements in SU(2) parametrized with the same value of θ</u>. In the ball of parameters θ, n^i, the elements of conjugate classes lie on concentric spheres. Using Eq. (4–21), one should note that for all U in SU(2), U and U^{-1} belong to the same class of conjugate elements under the group (cf. consequences in Subsec. 8.2.1,i)).

4.2.3. CONSEQUENCES FOR THE σ MATRICES

The matrix representative of Eq. (4–26) reads :

$$v'^i \sigma_i = R(U)^i_j \, v^j \sigma_i = U \, v^j \sigma_j \, U^{-1} \quad .$$

As these relations are true for all vectors V, one has :

$$R(U)^i_j \, \sigma_i = U \, \sigma_j \, U^{-1} \quad . \tag{4-35}$$

This important equation can be interpreted as follows ; to any U in SU(2) one associates a linear mapping A(U) of $\mathcal{W}(2)$ into $\mathcal{W}(2)$ defined by :

$$\forall \; W \in \mathcal{W}(2) \xrightarrow{A(U)} A(U)\,W = U\,W\,U^{-1} \quad , \tag{4-36}$$

where $U\,W\,U^{-1}$ belongs to $\mathcal{W}(2)$ since it is Hermitian and traceless as is W. Then, Eq. (4–35) implies that the matrix representative of A(U) with respect to the basis (σ_i) of $\mathcal{W}(2)$ is the rotation matrix $(R(U)^i_j)$. Intrinsic properties of the mapping A(U) are discussed in Subsec. 5.4.2,ii). In particular, it is shown that with respect to the inner product in $\mathcal{W}(2)$ defined by the Killing form, A(U) is a proper isometric application.

4.2.4. PARAMETRIZATION OF SU(2) BY THE EULER ANGLES

i) Using the above two–to–one homomorphism of SU(2) onto SO(3) and Eq. (2–42), any element in SU(2) can be parametrized by three Euler's angles such that :

$$[U(\varphi,\theta,\psi) = \pm U(\varphi,\delta^i_3)U(\theta,\delta^i_2)U(\psi,\delta^i_3)\,;a] \rightarrow R(\varphi,\theta,\psi;a) \quad , \tag{4-37}$$

where the same orthonormal basis a of E_3 is used in both sides of this correspondence, and φ, θ, and ψ belong to the space of parameters defined in Eq. (2–39). To get rid of the plus or minus sign, one can allow ψ (or φ) to vary from 0 to 4π . The plus sign corresponds to $0 \leqslant \psi \leqslant 2\pi$ and the minus sign to $2\pi \leqslant \psi \leqslant 4\pi$. Thus one gets the following parametrization of SU(2) by the Euler angles :

$$U(\varphi,\theta,\psi) = e^{-i\frac{\varphi}{2}\sigma_3}\, e^{-i\frac{\theta}{2}\sigma_2}\, e^{-i\frac{\psi}{2}\sigma_3} = \begin{pmatrix} e^{-i\frac{\varphi+\psi}{2}}\cos\frac{\theta}{2} & -e^{-i\frac{\varphi-\psi}{2}}\sin\frac{\theta}{2} \\ e^{i\frac{\varphi-\psi}{2}}\sin\frac{\theta}{2} & e^{i\frac{\varphi+\psi}{2}}\cos\frac{\theta}{2} \end{pmatrix} ,$$

$$0 \leqslant \varphi \leqslant 2\pi \qquad 0 \leqslant \theta \leqslant \pi \qquad 0 \leqslant \psi \leqslant 4\pi \quad . \tag{4-38}$$

One easily checks on the above expression, that the following surface points of the parameter space of the Euler angles φ, θ, and ψ for SU(2) have to be identified, see Fig. 4.3 :

(1) $\begin{cases} \varphi,\theta,0 \\ \varphi,\theta,4\pi \end{cases}$ denoted by •

(2) $\begin{cases} 2\pi,\theta,0 \\ 0,\theta,2\pi \\ 2\pi,\theta,4\pi \end{cases}$ denoted by x

(3) $\begin{cases} 0\ ,\theta\ ,0 \\ 2\pi\ ,\theta\ ,2\pi \\ 0\ ,\theta\ ,4\pi \end{cases}$ denoted by ⊗

(4) $\theta = 0, \varphi + \psi = \text{cst}$ denoted by ---

(5) $\theta = \pi\ , \varphi - \psi = \text{cst}$ denoted by —— .

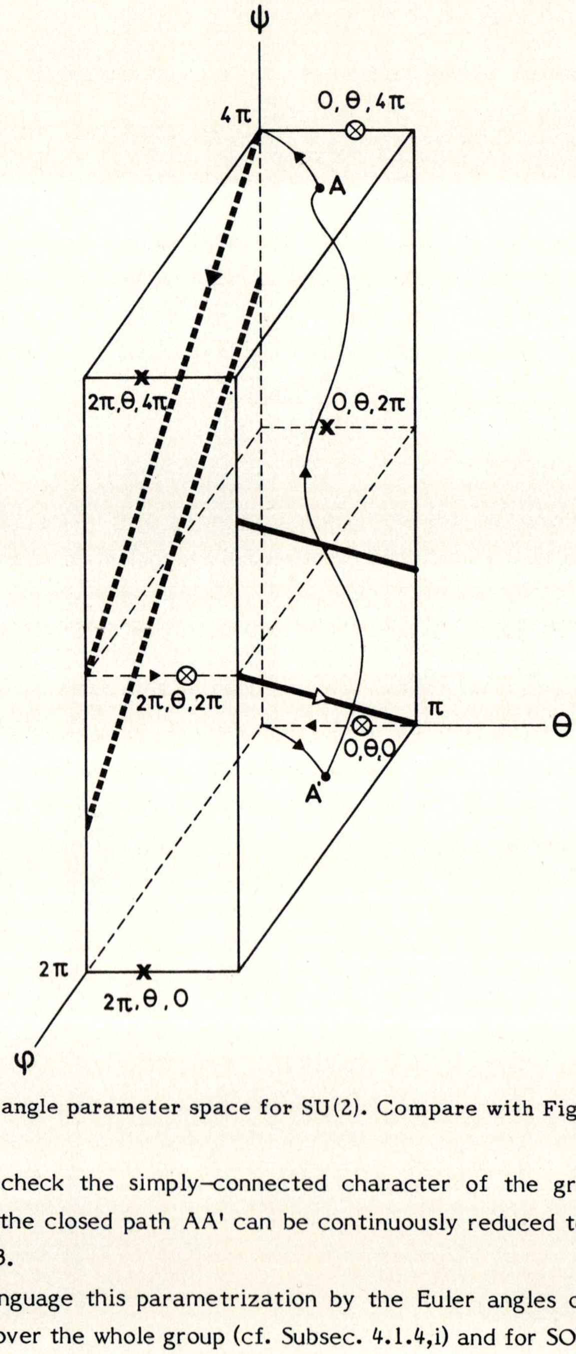

Fig. 4.3. The Euler angle parameter space for SU(2). Compare with Fig. 2.7 for SO(3).

It is also possible to check the simply-connected character of the group SU(2) (Subsec. 4.1.4,ii)). For instance the closed path AA' can be continuously reduced to a point according to the arrows in Fig. 4.3.

In the manifold language this parametrization by the Euler angles only defines a local chart which does not cover the whole group (cf. Subsec. 4.1.4,i) and for SO(3) Subsec. 2.3.2.3). This local chart is (U_2, f_2) where :

– U_2 is the open set in SU(2) of matrices $U(\varphi, \theta, \psi)$ for φ, θ, and ψ which fulfil the conditions :

$$0 < \varphi < 2\pi \qquad 0 < \theta < \pi \qquad 0 < \psi < 4\pi \quad ,$$

– f_2 is the one-to-one mapping such that :

$$\forall\, U \in U_2 \qquad f_2(U) = (\varphi,\theta,\psi) \quad ,$$

– $f_2(U_2)$ is the open set in R^3 defined by the above conditions on (φ,θ,ψ).

Thus, this local chart is not defined in particular for the unit element parametrized by $\varphi = \theta = \psi = 0$ (cf. consequences in Subsec. 7.1.3.1,iv) for the calculation of the invariant measure on the group).

ii) To the matrix $U(\varphi,\theta,\psi)$ of Eq. (4–38) corresponds by Eq. (4–26) the rotation matrix parametrized by the Euler angles, the expression of which is given in Eq. (E–14).

In the correspondence defined by Eq. (4–37) the dependence upon the orthonormal basis a, when it is changed under any rotation ρ on E_3, can be made explicit as follows, using Eq. (2–43) :

$$[U(\varphi,\theta,\psi);\rho a] \to R(\varphi,\theta,\psi;\rho a) = \rho R(\varphi,\theta,\psi;a)\rho^{-1} \quad . \tag{4-39}$$

4.3. THE INFINITESIMAL ELEMENTS OF THE GROUP AND ITS LIE ALGEBRA

The theory of Lie groups can be made in a general way, cf. for instance Chevalley[16] and Pontryagin[44]. But for the linear groups GL(n,R) or GL(n,C) and their subgroups, one only needs the more familiar topology defined over the real field (Appendix C.5.4.3,i)). Nevertheless, it is shown in Appendix C.5.4.3,ii) how the infinitesimal elements and the Lie algebra defined below in the special case of a linear group, are, respectively, related to the general notions of invariant infinitesimal transformations (Def. C.61) and to the Lie algebra of a Lie group (Def. C.62).

4.3.1. DEFINITION OF THE INFINITESIMAL ELEMENTS OF SU(n)

An <u>infinitesimal element</u> of SU(n) is by definition close to the unit matrix, i.e. :

$$U = \mathbf{1} - iW + O(W^2) \quad , \tag{4-40}$$

where the matrix elements of W are infinitesimal complex numbers of the first order. The matrix U, belonging to SU(n), has to fulfil Eq. (4–1). To the first order in W this yields :

$$W = W^{+} \quad , \qquad \mathrm{Tr}\, W = 0 \quad . \tag{4-41}$$

Conversely, these conditions are sufficient to ensure that $U = \mathbf{1} - iW$ belongs to SU(n). The factor i is introduced in Eq. (4–40) only for convenience, to get Hermitian instead of antihermitian matrices W. The set of these Hermitian and traceless matrices is the

(n^2-1)-dimensional vector space $\mathcal{W}(n)$ over the real field already considered in Subsec. 4.1.2,ii).

4.3.2. THE LIE ALGEBRA STRUCTURE OF $\mathcal{W}(n)$

Let us show that -i times the commutator of two elements in $\mathcal{W}(n)$ defines a bilinear mapping of $\mathcal{W}(n) \times \mathcal{W}(n)$ into $\mathcal{W}(n)$ which endows this vector space with a Lie algebra structure over the real field (Def. B.5). Indeed, for all W and W' in $\mathcal{W}(n)$, $-i[W,W']$ belongs to $\mathcal{W}(n)$ since it fulfils the Eq. (4–41) :

$$(-i[W,W'])^{+} = -i[W,W'] \quad , \quad \mathrm{Tr}(-i[W,W']) = 0 \quad .$$

The Lie character (anticommutativity and the Jacobi identity) is obvious from the definition of this algebra by means of a commutator (Appendix B.1.2,i)).

One shows in Appendix C.5.4.3,ii) that this Lie algebra $\mathcal{W}(n)$ of matrices is isomorphic to the Lie algebra of the group SU(n), as defined in a general frame by Def. C.62. This is the reason why abusively, $\mathcal{W}(n)$ is also often called the Lie algebra of SU(n).

Once a basis $(W_i,\ i=1,\dots,n^2-1)$ is chosen in $\mathcal{W}(n)$, the bilinear character of the mapping of $\mathcal{W}(n) \times \mathcal{W}(n)$ into $\mathcal{W}(n)$ allows us to define it by the structure constants (Def. B.4). They are the real coefficients defined by the images of the basis vectors as :

$$-i[W_i,W_j] = c_{ij}{}^{k}\, W_k \quad . \tag{4-42}$$

In the same way as when one studies the matrix representatives of a linear operator, special bases of the Lie algebra are more convenient in that they yield simple forms of the commutation relations, Eq. (4–42). Two types of such particular bases of $\mathcal{W}(n)$ are considered below.

4.3.3. CONSEQUENCES OF THE SEMI–SIMPLE CHARACTER OF $\mathcal{W}(n)$: SPECIAL BASES

i) Semi–simple Lie algebras (Def. B.23) have important properties, some of them are given in Appendix B.4. There exists a criterion due to Cartan (Prop. B.34) which states that a necessary and sufficient condition for a Lie algebra to be semi–simple is that the Killing form :

$$\forall\ W,W' \in \mathcal{W}(n) \longrightarrow g(W,W') = -\mathrm{Tr}\ (\Delta_W \Delta_{W'}) \in R \quad , \tag{4-43}$$

be nondegenerate, i.e. $\det g \neq 0$; Δ_W denotes an internal derivation or the adjoint operator of W (Def. B.13) :

$$\forall\ W' \in \mathcal{W}(n) \xrightarrow{\ \Delta_W\ } \Delta_W(W') = -i[W,W'] \in \mathcal{W}(n) \quad . \tag{4-44}$$

Using the above criterion we only check below that $\mathcal{W}(2)$ is semi–simple. Actually, one shows[13,45] that, for n greater than one, $\mathcal{W}(n)$ is moreover a simple Lie algebra. Most of the properties of semi–simple Lie algebras are based on the nondegenerate character of g. This is in particular the case for the pseudo-Euclidean structure of a semi-simple Lie algebra discussed later in Subsec. 5.4.2, and also for the existence of two special bases which are now defined.

ii) One shows (Prop. B.39) that in a basis which diagonalizes the Killing form, the structure constants are totally antisymmetrical with respect to the substitutions over the three indices. For $\mathcal{W}(2)$ one of these bases consists in the Pauli matrices (σ_i). Indeed with respect to this basis the commutation relations read :

$$-i[\sigma_i,\sigma_j] = 2\varepsilon_{ij\ell}\delta^{\ell k}\sigma_k = \sum_k \varepsilon_{ijk}\sigma_k \quad , \quad c_{ij}{}^k = 2\varepsilon_{ij\ell}\delta^{\ell k} \quad , \tag{4-45}$$

where ε_{ijk} is the totally antisymmetrical tensor of order three. Let us check that the basis (σ_i) diagonalizes the Killing form, and at the same time that g is nondegenerate, i.e. $\mathcal{W}(2)$ is semi-simple. From Prop. B.33 one has :

$$g_{ij} = g(\sigma_i,\sigma_j) = -c_{ik}{}^{\ell}\, c_{j\ell}{}^{k} = -4\sum_{k,\ell} \varepsilon_{ik\ell}\,\varepsilon_{j\ell k} = 8\,\delta_{ij} \quad . \tag{4-46}$$

Any basis (W_i) obtained from (σ_i) by a rotation matrix which belongs to SO(3) :

$$W_i = \sigma_j\, R^j_i \quad , \tag{4-47}$$

has the same properties, i.e. diagonalizes the Killing form and leads to the same totally antisymmetrical structure constants (Subsec. 4.3.4,ii)).

iii) In order to obtain another particular basis of $\mathcal{W}(n)$, which yields the standard form of the commutation relations (and more generally to a classification of semi-simple Lie algebras[13,45]) one considers the eigenvalue problem of the internal derivation Δ_W associated to a fixed arbitrary element $W = w^i W_i$ in $\mathcal{W}(n)$:

$$\Delta_W X = -i[W,X] = \lambda\, X \quad , \tag{4-48}$$

where $X = x^i W_i$ is an eigenvector corresponding to the eigenvalue λ. In the basis (W_i), the Eq. (4–48) reads :

$$(w^i\, c_{ij}{}^k - \lambda\delta^k_j)\, x^j = 0 \quad .$$

Thus, the eigenvalues λ are solutions of the secular equation :

$$\det(w^i\, c_{ij}{}^k - \lambda\delta^k_j) = 0 \quad . \tag{4-49}$$

Now two problems arise, cf. a similar discussion in Subsec. 2.2.1,ii). First, Eq. (4–49) is a polynomial of degree n^2-1 in λ with real coefficients, which may have no zero in the real field. Second, there may not exist an eigenbasis in $\mathcal{W}(n)$ over the real field. To solve these problems we are led to introduce the complex extension $\{\mathcal{W}(n)\}$ of the Lie algebra $\mathcal{W}(n)$ (Def. B.6). Then the secular Eq. (4–49) has n^2-1 solutions over the complex field.

Before solving the eigenvalue problem for $\mathcal{W}(n)$, let us make some remarks for a general semi–simple Lie algebra $\mathcal{W}$ over the real field. We have seen previously that there exist bases in which the structure constants are totally antisymmetrical. In such a basis, the matrix representative of the operator Δ_W, i.e. $(w^i c_{ij}{}^k)$, is real and antisymmetrical. Therefore the eigenvalues are zero or purely imaginary and come in complex conjugate pairs. The corresponding eigenvectors are complex and thus belong to the complex extension $\{\mathcal{W}\}$, and also come in complex conjugate pairs. For the eigenvectors X and X^*, corresponding, respectively, to the eigenvalues $-i\alpha$ and $+i\alpha$ (α real), the commutation relations read :

$$
\begin{aligned}
-i[W,X] &= -i\alpha X \quad , \quad [W,X] = \alpha X \quad , \\
-i[W,X^*] &= i\alpha X^* \quad , \quad [W,X^*] = -\alpha X^* \quad .
\end{aligned}
\tag{4-50}
$$

Furthermore, zero is clearly one of the eigenvalues of Δ_W and W belongs to its eigenspace, since W commutes with itself (Prop. B.19). The answer to the second problem, i.e. does there exist a basis of eigenvectors in $\mathcal{W}$?, is given by a theorem due to Cartan[13] :

Theorem 4.4. For a semi–simple Lie algebra over the real field, if W is chosen so that the secular Eq. (4–49) has the maximum number of different roots then only $\lambda = 0$ is degenerate, and if r is its degeneracy (also called the rank of the semi–simple Lie algebra) there exist r linearly independent eigenvectors of the eigenvalue zero.

Actually, for a linear group the problem is simpler since one handles relations between matrices. Let us show how things work for SU(n) (more generally the same study can be applied to the linear group over the complex field GL(n,C)). For this group the eigenvalue problem can be solved explicitly, i.e. W can be chosen so as to find the eigenvalues and an eigenbasis. Let X_{ij} denote the $n\times n$ matrix with its ith row and jth column element equal to 1 and all the others equal to 0. These n^2 matrices form a basis of the vector space over the complex field of $n\times n$ complex matrices. A direct calculation yields their commutation relations :

$$
[X_{ij}, X_{k\ell}] = \delta_{jk} X_{i\ell} - \delta_{i\ell} X_{kj} \quad . \tag{4-51}
$$

Let Y_i be the diagonal matrices X_{ii}, they fulfil the commutation relations :

$$
[Y_i, Y_j] = 0 \quad , \tag{4-52a}
$$

$$
[Y_i, X_{jk}] = (\delta_{ij} - \delta_{ik}) X_{jk} \quad , \tag{4-52b}
$$

$$
[X_{ij}, X_{ji}] = Y_i - Y_j \quad . \tag{4-52c}
$$

If W is chosen as a linear combination of the Y_i, $W = w^i Y_i$, then Eq. (4–52b) solves the eigenvalue problem. In the Lie algebra $\mathcal{W}(n)$ over the real field, due to the traceless condition, there exist n–1 linearly independent diagonal matrices, for instance :

$$
H_i = Y_i - Y_n \qquad i=1,\ldots,n-1 \quad . \tag{4-53}
$$

Then assuming :

$$W = \sum_{i=1}^{n-1} w^i H_i \qquad w^i \in R \quad , \tag{4-54}$$

one obtains :

– $\lambda = 0$ is an eigenvalue of order n–1 (the rank of $\mathcal{W}(n)$) and the n–1 commuting matrices H_i form a basis of the corresponding eigenspace.

–The other eigenvalues are purely imaginary and two by two complex conjugates. From Eq. (4–52b) their values are :

$$1 \leqslant j < k \leqslant n \qquad \lambda_{jk} = \mp i w^i (\delta_{ij} - \delta_{ik} - \delta_{nj} + \delta_{nk}) \quad , \tag{4-55}$$

with the corresponding eigenvectors, respectively, X_{jk} and X_{kj}.

This set of n^2–1 vectors, $\{ H_i, i=1,\dots,n-1 \; ; X_{jk}, X_{kj}, 1 \leqslant j < k \leqslant n \}$, forms a basis of $\mathcal{W}(n)$. The vectors X_{jk} and X_{kj} belong to $\{\mathcal{W}(n)\}$. From them one can construct linearly independent vectors which can be identified to vectors of the real part $\mathcal{W}(n)$ of $\{\mathcal{W}(n)\}$ (Prop. A.2 and Eq. (2–15)) :

$$Z^{(1)}_{jk} = X_{jk} + X_{kj} \quad , \quad Z^{(2)}_{jk} = i(X_{jk} - X_{kj}) \quad , \qquad 1 \leqslant j < k \leqslant n \quad .$$

Let us define $(\lambda_{jk})_i$ by :

$$(\lambda_{jk})_i = -i(\delta_{ij} - \delta_{ik} - \delta_{nj} + \delta_{nk}) \qquad i=1,\dots,n-1 \quad , \tag{4-56}$$

one then gets the standard form of the <u>commutation relations</u> :

$$-i[H_i, H_j] = 0 \qquad i,j=1,\dots,n-1 \tag{4-57a}$$

$$\left.\begin{aligned} -i[H_i, X_{jk}] &= (\lambda_{jk})_i X_{jk} && (4\text{-}57b) \\ -i[H_i, X_{kj}] &= -(\lambda_{jk})_i X_{kj} && (4\text{-}57c) \\ -i[X_{jk}, X_{kj}] &= \begin{cases} -i(H_j - H_k) & k \neq n \\ -i\, H_j & k = n \end{cases} && (4\text{-}57d) \end{aligned}\right\} \quad i=1,\dots,n-1, \quad 1 \leqslant j < k \leqslant n$$

$$-i[X_{ij}, X_{k\ell}] = -i\delta_{jk} X_{i\ell} + i\delta_{i\ell} X_{kj} \quad \text{ordered pairs } ij \text{ and } k\ell \text{ different.} \tag{4-57e}$$

One usually denotes by <u>root</u> either the form equal to the eigenvalue λ_{jk} $(j \neq k)$:

$$\lambda_{jk} = \sum_{i=1}^{n-1} w^i (\lambda_{jk})_i \quad , \tag{4-58}$$

or the n–1 (the rank of the semi–simple Lie algebra)–dimensional vector with covariant components $(\lambda_{jk})_i$.

Following Racah [45], let us quote some general features of the standard basis of a semi–simple Lie algebra. Let d be the dimension of the algebra and r its rank. Latin indices are used for the r–dimensional eigenspace of the eigenvalue zero spanned by the H_i, while Greek indices are employed for the (d–r)-dimensional subspace spanned by the eigenvectors E_α corresponding to the nonvanishing distinct eigenvalues α :

$$\Delta_W H_i = 0 \qquad i=1,\dots,r$$
$$\Delta_W E_\alpha = \alpha E_\alpha \qquad \alpha=1,\dots,d-r \qquad , \qquad (4\text{-}59)$$

W belongs to the eigenspace of the eigenvalue zero, thus :

$$W = w^i H_i \qquad . \qquad (4\text{-}60)$$

Assuming that the Lie algebra is defined by the commutator without a factor $\pm i$, from Eq. (4–59) and the Jacobi identity, one derives the standard form of the commutation relations :

$$[H_i,H_j] = 0 \qquad (4\text{-}61a)$$

$$[H_i,E_\alpha] = \alpha_i E_\alpha \qquad (4\text{-}61b)$$

$$[E_\alpha,E_{-\alpha}] = \alpha^i H_i \qquad (4\text{-}61c)$$

$$[E_\alpha,E_\beta] = \begin{cases} N_{\alpha\beta}E_{\alpha+\beta} & \text{if } \alpha+\beta \text{ is a nonvanishing root} \\ 0 & \text{otherwise} \end{cases} \quad , \qquad (4\text{-}61d)$$

where a root is defined either as the linear form :

$$\alpha = w^i \alpha_i \qquad , \qquad (4\text{-}62)$$

or as the r–dimensional vector with covariant components α_i and contravariant components :

$$\alpha^i = \alpha_j g^{ji} \qquad . \qquad (4\text{-}63)$$

In the above equation (g^{ij}) denotes the inverse matrix of (g_{ij}) which is the metric tensor of the restriction of the Killing form to the r–dimensional vector space spanned by the H_i. Indeed, the value of the structure constants in the previous basis (H_i,E_α) leads to the following form of the metric tensor of the Killing form (assuming a suitable normalization of E_α and ordering these vectors by pairs $E_\alpha, E_{-\alpha}$).

$$(g) = \begin{pmatrix} g_{ij} & & & 0 \\ & \begin{matrix} 0 & 1 \\ 1 & 0 \end{matrix} & & \\ & & \begin{matrix} 0 & 1 \\ 1 & 0 \end{matrix} & \\ 0 & & & \ddots \end{pmatrix} \qquad . \qquad (4\text{-}64)$$

Since the Lie algebra is semi–simple g is nondegenerate and thus $\det(g_{ij})$ is also nonzero.

Moreover it can be proved that (g_{ij}) is definite and therefore allows one to define in the r–dimensional vector space of the roots an Euclidean structure, the inner product of two roots α and β being :

$$(\alpha,\beta) = \alpha_i\, g^{ij} \beta_j = \alpha^i \beta_i \qquad . \qquad (4\text{-}65)$$

One then shows that the angle (Eq. (2–2)) between two roots can only take the values : 0, $\pi/6$, $\pi/4$, $\pi/3$, and $\pi/2$. This leads to a classification of semi–simple Lie algebras, which can be visualized in terms of the root diagrams where the vectors equal to the roots are drawn.

All these general properties of semi–simple Lie algebras can be checked for $\mathcal{W}(n)$ starting from the commutation relations given in Eqs. (4–57a–e), with the correspondence between notations :

$$\alpha = \lambda_{jk} \quad , \quad \alpha_i = (\lambda_{jk})_i \quad , \quad E_\alpha = X_{jk} \quad . \tag{4-66}$$

For instance, for $\mathcal{W}(2)$ of rank one, the one-dimensional vector diagram corresponds to the two opposite roots λ_+ and λ_- (cf. below) :

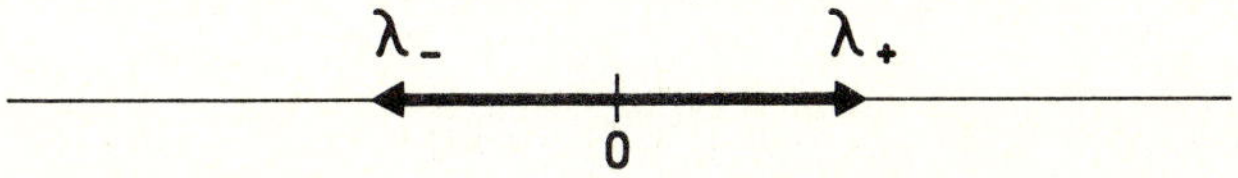

For the Lie algebra $\mathcal{W}(2)$ of dimension three and rank one, one chooses :

$$W = H_1 = \sigma_3 = \begin{pmatrix} 1 & 0 \\ 0 & -1 \end{pmatrix} \quad . \tag{4-67}$$

The nonzero eigenvalues, or roots, and the corresponding eigenvectors are :

$$\lambda_+ = \lambda_{12} = -2i \quad , \quad X_{12} = \begin{pmatrix} 0 & 1 \\ 0 & 0 \end{pmatrix} = \frac{1}{2}\sigma_+ = \frac{1}{2}(\sigma_1 + i\sigma_2) \quad ;$$

$$\lambda_- = -\lambda_{12} = 2i \quad , \quad X_{21} = \begin{pmatrix} 0 & 0 \\ 1 & 0 \end{pmatrix} = \frac{1}{2}\sigma_- = \frac{1}{2}(\sigma_1 - i\sigma_2) \quad . \tag{4-68}$$

The standard form of the commutation relations is usually written as :

$$[\sigma_3, \sigma_\pm] = \pm 2\sigma_\pm \quad , \tag{4-69a}$$

$$[\sigma_+, \sigma_-] = 4\sigma_3 \quad . \tag{4-69b}$$

With respect to the basis ($\sigma_\pm$, σ_3) of the complex extension $\{ \mathcal{W}(2) \}$ of the algebra the metric tensor of the Killing form reads :

$$(g) = \begin{pmatrix} 8 & 0 & 0 \\ 0 & 0 & 16 \\ 0 & 16 & 0 \end{pmatrix} \quad . \tag{4-70}$$

4.3.4. LIE ALGEBRAS ISOMORPHIC TO THAT OF SU(2)

i) The same study made previously for the group SU(2) can be repeated for SO(3). The infinitesimal elements A of SO(3) are of the type :

$$A = \mathbf{1} + M + 0(M^2) \quad , \tag{4-71}$$

where M is a three-dimensional matrix the elements of which are infinitesimal real numbers of first order. Since A is orthogonal ($AA^t = \mathbf{1}$), M has to be antisymmetrical. The set of these real three-dimensional antisymmetrical matrices is a three-dimensional vector space $A(3)$ over the real field. The commutator endows $A(3)$ with a Lie algebra structure since :

$$\forall\ M,M' \in A(3) \qquad [M,M']^t = -[M,M'] \in A(3) \quad .$$

More generally, the Lie algebra of the group SO(n) is isomorphic to the $n(n-1)/2$-dimensional Lie algebra $A(n)$ over the real field of $n\times n$ real antisymmetrical matrices, the bilinear form being defined by the commutator.

What is peculiar to SO(3) is that three is the only positive value of n for which the dimension of the Lie algebra of SO(n) is equal to n, cf. Subsec. 5.4.2,ii).

This Lie algebra is clearly isomorphic to the Lie algebra defined by means of the vector product in a three-dimensional Euclidean vector space E_3 (Appendix B.1.2,ii)). Indeed, any rotation of an infinitesimal angle ε about the unitary vector $\hat{n}$ takes the form given in Eq. (2-31) :

$$R(\varepsilon,\hat{n}) = \mathbf{1} + \varepsilon\hat{n}\times + O(\varepsilon^2) \quad . \tag{4-72}$$

From this relation one gets immediately bases, respectively, (M_i) in $A(3)$ and (e_i) in E_3 which yield to identical structure constants. Let (e_i) be any right-handed orthonormal basis of E_3, then the matrices M_i are defined by Eq. (E-10):

$$(M_i)^k_j = \varepsilon_{ijk} \quad ; \quad M_1 = \begin{pmatrix} 0 & 0 & 0 \\ 0 & 0 & -1 \\ 0 & 1 & 0 \end{pmatrix} , \quad M_2 = \begin{pmatrix} 0 & 0 & 1 \\ 0 & 0 & 0 \\ -1 & 0 & 0 \end{pmatrix} , \quad M_3 = \begin{pmatrix} 0 & -1 & 0 \\ 1 & 0 & 0 \\ 0 & 0 & 0 \end{pmatrix} , \tag{4-73}$$

and one has :

$$e_i \times e_j = \sum_k \varepsilon_{ijk}\, e_k \quad , \quad [M_i, M_j] = \sum_k \varepsilon_{ijk}\, M_k \quad . \tag{4-74}$$

The isomorphism between $A(3)$ and E_3 can thus be defined by :

$$\forall\, u^i \in R \qquad M_u = u^i M_i \in A(3) \longleftrightarrow u = u^i e_i \in E_3 \quad . \tag{4-75}$$

ii) Taking in the Lie algebra $\mathcal{W}(2)$ of the group SU(2) the basis $(W_i = \frac{1}{2}\sigma_i)$ yields again the same structure constants, from Eq. (4-45) :

$$-i[W_i, W_j] = \sum_k \varepsilon_{ijk}\, W_k \quad . \tag{4-76}$$

Therefore the three Lie algebras over the real field, $A(3)$, E_3, and $\mathcal{W}(2)$ are isomorphic. This isomorphism between the Lie algebras of SU(2) and SO(3) corresponds to the fact, already noticed, Subsec. 4.1.4,ii), that these two groups are locally isomorphic.

For instance, the isomorphism between E_3 and $\mathcal{W}(2)$ can be defined by :

$$\forall\, u^i \in R \qquad u = u^i e_i \in E_3 \longleftrightarrow W_u = u^i W_i = \frac{1}{2} u^i \sigma_i \in \mathcal{W}(2) \quad , \tag{4-77}$$

where the basis (e_i) fulfils Eq. (4-74). This yields a generalization of the commutation relations of the σ matrices. Let :

$$u = u^i e_i \longleftrightarrow W_u = u^i W_i \quad , \qquad v = v^i e_i \longleftrightarrow W_v = v^i W_i \quad ,$$

the isomorphism then implies that :

$$u \times v = (u \times v)^i e_i \leftrightarrow -i[W_u, W_v] = (u \times v)^i W_i \quad . \tag{4-78}$$

In terms of the σ matrices this last relation reads :

$$[u^i \sigma_i, v^j \sigma_j] = i\ 2(u \times v)^k \sigma_k \quad ,$$

or using an Euclidean inner product, cf. Eq. (4–17) $u^i \sigma_i = u.\sigma$, one has :

$$[u.\sigma, v.\sigma] = i\ 2(u \times v).\sigma \quad . \tag{4-79}$$

The isomorphism between E_3 and $\mathcal{W}(2)$ is also a way to prove that the structure constants with respect to the basis (σ_i) of $\mathcal{W}(2)$ are unchanged under an orthogonal unimodular (i.e. a rotation) change of basis $(W_i = \sigma_j R^j_i)$. Indeed we have seen, Eq. (2–12), that the vector product is stable under a proper isometric application. This property can also be obtained directly. With respect to any new basis $(W_i = \sigma_j t^j_i)$ the structure constants are given by Prop. B.1 :

$$c_{ij}{}^k = 2\varepsilon_{\ell m n'} \delta^{n'n}\, t^\ell_i\, t^m_j (t^{-1})^k_n \quad .$$

If t belongs to SO(3), $(t^{-1})^k_n = t^n_k$, and one obtains the expected result since for any matrix t :

$$\varepsilon_{ijk} \det t = \varepsilon_{\ell m n}\, t^\ell_i\, t^m_j\, t^n_k \quad . \tag{4-80}$$

4.4. CONSEQUENCES OF THE BAKER–CAMPBELL–HAUSDORFF FORMULA

In Prop. D.4 an explicit formal expression of the matrix C is derived in terms of the matrices A and B, such that :

$$e^A e^B = e^C \quad .$$

Since any element U in SU(2) takes the form given in Eq. (4–3) $U = e^{-iW}$, where W belongs to the Lie algebra $\mathcal{W}(2)$, one has :

$$U_1 U_2 = e^{-iW_1} e^{-iW_2} = U_3 = e^{-iW_3} \quad , \quad W_3 \in \mathcal{W}(2) \quad . \tag{4-81}$$

The main point of the Baker–Campbell–Hausdorff formula is that the matrix W_3 can be expressed in terms of the repeated commutators of W_1 and W_2. The first terms of this expansion read :

$$W_3 = W_1 + W_2 - i[W_1, W_2] - \frac{1}{12}([W_1,[W_1,W_2]] + [W_2,[W_2,W_1]]) + \dots \quad . \tag{4-82}$$

Therefore the group multiplication law, Eq. (4–81), is formally defined if one knows the commutation rules of any matrices W_1 and W_2 in $\mathcal{W}(2)$, i.e. if the Lie algebra of the group is known. This very important fact in connection with group theory is of course true for any group the elements of which can be written as the exponentials of matrices. In that case, quoting Racah[45] one needs not "consider the whole of the group but that part of it which lies near the identity... . Thus, instead of finite displacements of a point under a transformation, we consider the application of successive infinitesimal displacements...".

4.5. SOME PROPERTIES OF σ MATRICES

i) Let us recall the definition of the σ matrices σ_i, i=1,2,3 :

$$\sigma_1 = \begin{pmatrix} 0 & 1 \\ 1 & 0 \end{pmatrix} \quad , \quad \sigma_2 = \begin{pmatrix} 0 & -i \\ i & 0 \end{pmatrix} \quad , \quad \sigma_3 = \begin{pmatrix} 1 & 0 \\ 0 & -1 \end{pmatrix} \quad . \tag{4-83}$$

It is sometimes useful to consider the set of four matrices σ_μ , μ =0,1,2,3, where σ_0 denotes the 2 $\times$ 2 unit matrix. One also defines :

$$\sigma_+ = \sigma_1 + i\sigma_2 = \begin{pmatrix} 0 & 2 \\ 0 & 0 \end{pmatrix} \quad , \quad \sigma_- = \sigma_1 - i\sigma_2 = \begin{pmatrix} 0 & 0 \\ 2 & 0 \end{pmatrix} \quad . \tag{4-84}$$

ii) The σ matrices fulfil the properties already noticed, i.e. :

$$\sigma_i^+ = \sigma_i \quad , \quad \mathrm{Tr}\ \sigma_i = 0 \quad , \tag{4-85}$$

$$[\sigma_3,\sigma_\pm] = \pm 2\sigma_\pm \quad , \quad [\sigma_+,\sigma_-] = 4\sigma_3 \quad , \tag{4-86a}$$

$$[\sigma_i,\sigma_j] = i\ 2 \sum_k \varepsilon_{ijk}\sigma_k \Leftrightarrow \sum_{i,j} \varepsilon_{ijk}\sigma_i\sigma_j = i\,2\,\sigma_k \quad , \tag{4-86b}$$

$$[u.\sigma,v.\sigma] = i\ 2(u\times v).\sigma \Leftrightarrow [u^i\sigma_i,v^j\sigma_j] = i\,2\sum_k \varepsilon_{ijk}\ u^iv^j\sigma_k \quad , \tag{4-86c}$$

where the last equations above hold providing :

$$[u^i,v^j] = 0 \quad , \quad [u^i,\sigma_j] = 0 \quad , \quad [v^i,\sigma_j] = 0 \quad . \tag{4-86d}$$

In addition, the σ matrices also fulfil the following relations which can be checked by a direct calculation :

$$\det \sigma_i = -1 \quad , \tag{4-87}$$

$$\sigma_i\sigma_j = \delta_{ij}\ \mathbf{1} + i \sum_k \varepsilon_{ijk}\ \sigma_k \quad . \tag{4-88}$$

It is shown in Appendix C.4.4.2 that the σ matrices are, up to an equivalence, the only complex matrices which satisfy the above Eq. (4–88). It follows from this equation that :

$$\mathrm{Tr}\ (\sigma_\mu\ \sigma_\nu) = 2\ \delta_{\mu\nu} \quad , \tag{4-89a}$$

$$(u.\sigma)(v.\sigma) = u.v + i(u\times v).\sigma \Leftrightarrow u^i\sigma_iv^j\sigma_j = \sum_i u^iv^i + i\sum_k \varepsilon_{ijk}\ u^iv^j\sigma_k \ , \tag{4-89b}$$

providing one assumes :

$$[v^i, \sigma_j] = 0 \quad . \tag{4-89c}$$

By a repeated use of Eqs. (4–85,88, and 89a) one is able to compute the trace of any product of σ matrices. For instance using Eq. (4–89b) twice and then Eq. (4–85) yields a useful expression of the three-fold product, Eq. (2–8), of vectors u,v, and w :

$$(u,v,w) = \frac{1}{2i}\,\mathrm{Tr}[(u.\sigma)(v.\sigma)(w.\sigma)] \quad , \tag{4-90a}$$

providing one assumes :

$$[\sigma_i, v^j] = 0 \quad , \quad [\sigma_i, w^j] = 0 \quad . \tag{4-90b}$$

iii) It is also recalled that the σ matrices form a basis of the three-dimensional vector space over the real field of complex 2×2 Hermitian and traceless matrices X , and from Eq. (4–89a) :

$$X^+ = X \quad , \quad \mathrm{Tr}\,X = 0 \Rightarrow X = x^i\sigma_i \quad , \quad x^i = \frac{1}{2}\,\mathrm{Tr}(\sigma_i X) \in R \quad . \tag{4-91}$$

In terms of an Euclidean inner product, one has with $x_\pm = x^1 \pm ix^2$:

$$X = x.\sigma = x^i\sigma_i = \frac{1}{2}\,(x_+\sigma_- + x_-\sigma_+) + x^3\sigma_3 = \begin{pmatrix} x^3 & x_- \\ x_+ & -x^3 \end{pmatrix} \quad . \tag{4-92}$$

Furthermore, the 2×2 unit matrix σ_0 with the σ matrices form a basis of the four-dimensional vector space over the complex field of complex 2×2 matrices X, and from Eq. (4–89a) :

$$X = x^\mu\sigma_\mu \qquad x^\mu = \frac{1}{2}\,\mathrm{Tr}\,(\sigma_\mu X) \in C \quad . \tag{4-93}$$

From the above equations, the matrix element X^α_β reads :

$$X^\alpha_\beta = \delta^\alpha_\delta\,\delta^\gamma_\beta\,X^\delta_\gamma = \sum_{\mu=0}^{3} \frac{1}{2}\,(\sigma_\mu)^\gamma_\delta\,X^\delta_\gamma\,(\sigma_\mu)^\alpha_\beta \quad .$$

Therefore, since $(\sigma_0)^\gamma_\delta = \delta^\gamma_\delta$,one gets the relation :

$$\sum_{i=1}^{3} (\sigma_i)^\gamma_\delta\,(\sigma_i)^\alpha_\beta = 2\delta^\alpha_\delta\,\delta^\gamma_\beta - \delta^\gamma_\delta\,\delta^\alpha_\beta \quad . \tag{4-94}$$

Let us now consider the four-dimensional vector space Q over the real field of complex 2×2 Hermitian matrices :

$$X = x^0\,\mathbf{1} + x^j(-i\sigma_j) \quad , \quad x^0, x^j \in R \quad . \tag{4-95}$$

From Eq. (4–88), the four matrices :

$$e_0 = \mathbf{1} \quad , \quad e_j = -i\sigma_j \qquad j=1,2,3 \tag{4-96}$$

fulfil the conditions :

$$\mu = 0,1,2,3 \qquad e_0 e_\mu = e_\mu e_0 = e_\mu \quad , \quad e_i e_j = -\delta_{ij}\, e_0 + \sum_k \varepsilon_{ijk} e_k \ , \tag{4-97}$$

which characterize the algebra of real quaternions (Def. B.28). Since any element of SU(2) takes the form given by Eq. (4–20) :

$$U(\theta, n^j) = \cos\frac{\theta}{2}\, \mathbf{1} + \sin\frac{\theta}{2}\, n^j\, (-i\sigma_j) \quad , \tag{4-98}$$

SU(2) is a subset of this algebra of real quaternions Q.

Moreover, in this algebra of real quaternions, the conjugate element $\tilde{U}$ of U (Def. B.29) is equal to U^{-1} since $\tilde{U} = U^+$, or N(U) equals one in Prop. B.49.
Some remarks about the relations between the above properties of quaternions and the Eq. (4–99) below are given in Subsec. 8.2.1,ii), for details cf. Dyson[21].

iv) Let U and U^* be two conjugate complex elements in SU(2). One now proves that there exists an element C in SU(2) such that :

$$U^* = C\, U\, C^{-1} \quad , \tag{4-99}$$

where the matrix C is antisymmetrical and given by :

$$C = i\sigma_2 = \begin{pmatrix} 0 & 1 \\ -1 & 0 \end{pmatrix} \quad . \tag{4-100}$$

Actually, this is a special case of general properties of conjugate complex representations, this problem is studied in Sec. 8.2. Being unitary and unimodular the matrix C defined above belongs to SU(2), and a direct calculation shows that one has :

$$C = C^* = -C^{-1} = -C^+ = -C^t \tag{4-101}$$

$$C\, \sigma_i\, C^{-1} = -\sigma_i^{\,t} \quad , \tag{4-102}$$

and for any 2×2 matrix A :

$$\det A\, \mathbf{1} = C^{-1}\, A\, C\, A^t \quad . \tag{4-103}$$

Applying this last relation for a matrix U in SU(2) yields :

$$\mathbf{1} = C^{-1}\, U\, C\, U^t \quad , \tag{4-104}$$

and using the unitary character of U, one obtains the Eq. (4–99).

The element C in SU(2), parametrized by θ, n^i as in Eqs. (4–19,20), reads :

$$C = U(\pi,-e_2;a) = U(\pi,e_2;a)^{-1} \quad . \tag{4-105}$$

Let us remark that from Eq. (4–99) one gets the special property of the matrices of SU(2) (not true for SU(n) with n different from one and two, cf. also Subsecs. 6.1.6 and 8.2.1,i)) :

$$\forall\, U \in SU(2) \qquad \mathrm{Tr}\, U = \mathrm{Tr}\, U^* \in R \quad . \tag{4-106}$$

Actually, this result appears explicitly in the three parametrizations of the elements of SU(2) given, respectively, by Eqs. (4–19,23, and 38).

FORMULAE FOR CHAPTER 4

Definition

$SU(2) = \{2\times 2 \text{ matrices over C}, UU^+ = U^+U = \mathbf{1}, \det U = 1\}$

centre = $\{\mathbf{1},-\mathbf{1}\} \sim Z_2$.

Exponential form

$$\forall\, U \in SU(2) \qquad U = e^{-iW} \quad , \quad W = W^+ \quad , \qquad \mathrm{Tr}\, W = 0$$

$$e^{-iW} = e^{-iW'} \Leftrightarrow W' = W + 2\pi V \begin{pmatrix} k_1 & 0 \\ 0 & k_2 \end{pmatrix} V^+ \ , \quad \begin{cases} V\in SU(2) \text{ diagonalizes } U \ , & k_1,k_2 \in Z \ , \\ k_1,k_2 \in Z \ , \ k_1+k_2 = 0 \ . \end{cases}$$

Parameters of the group

1– $\mathcal{W}(2) = \{2\times 2 \text{ matrices over C}, W = W^+, \mathrm{Tr}\, W = 0\}$ = 3–dimensional vector space over R.
$(W_i,\ i=1,2,3)$ basis of $\mathcal{W}(2)$, special basis $(W_i = \sigma_i)$:

$$\forall\, W \in \mathcal{W}(2) \qquad W = w^iW_i \qquad w^i \in R \ , \quad U(w^i) = e^{-iw^iW_i}$$

$$\sum_{i=1}^{3} (n^i)^2 = 1 \quad , \quad 0 \leqslant \theta \leqslant 2\pi \quad , \quad U(\theta,n^i) = e^{-i\frac{\theta}{2}\, n^i\sigma_i}$$

$$U(\theta,n^i) = \cos\tfrac{\theta}{2}\,\mathbf{1} - i\,\sin\tfrac{\theta}{2}\, n^i\sigma_i = \begin{pmatrix} \cos\frac{\theta}{2} - i\sin\frac{\theta}{2}\, n^3 & -\sin\frac{\theta}{2}\,(n^2+in^1) \\ \sin\frac{\theta}{2}\,(n^2-in^1) & \cos\frac{\theta}{2}+i\sin\frac{\theta}{2}\, n^3 \end{pmatrix} .$$

For a given right–handed orthonormal basis :

$$a = (e_i) \quad , \quad \hat{n} = n^ie_i \ , \quad \hat{n}.\sigma = n^i\sigma_i$$

$$U(\theta,\hat{n};a) = e^{-i\frac{\theta}{2}\,\hat{n}.\sigma} = \cos\tfrac{\theta}{2}\,\mathbf{1} - i\sin\tfrac{\theta}{2}\,\hat{n}.\sigma \ , \ U(\theta,R\hat{n};a) = U(\theta,\hat{n};R^{-1}a) \ .$$

2– $0 \leqslant \varphi \leqslant 2\pi$, $0 \leqslant \theta \leqslant \pi$, $0 \leqslant \psi \leqslant 4\pi$

$$U(\varphi,\theta,\psi) = e^{-i\frac{\varphi}{2}\sigma_3}\, e^{-i\frac{\theta}{2}\sigma_2}\, e^{-i\frac{\psi}{2}\sigma_3} = \begin{pmatrix} e^{-i\frac{\varphi+\psi}{2}}\cos\frac{\theta}{2} & -e^{-i\frac{\varphi-\psi}{2}}\sin\frac{\theta}{2} \\ e^{i\frac{\varphi-\psi}{2}}\sin\frac{\theta}{2} & e^{i\frac{\varphi+\psi}{2}}\cos\frac{\theta}{2} \end{pmatrix} .$$

3– $u_1 = x_1 + iy_1$, $u_2 = x_2+iy_2 \in C$, $|u_1|^2+|u_2|^2 = (x_1)^2+(y_1)^2+(x_2)^2+(y_2)^2 = 1$.

$$U(u^1,u^2) = \begin{pmatrix} u^1 & u^2 \\ -u^{2*} & u^{1*} \end{pmatrix} .$$

Homomorphism between SU(2) and SO(3)

$a = (e_i)$ right–handed orthonormal basis of E_3,

$$[\pm U(\theta,n^i);a] \leftrightarrow R(\theta,\hat{n}=n^i e_i)=R(U;a) \qquad \text{Rotation over } E_3$$

$$\left.\begin{array}{l} v = v^i e_i \\ v' = v'^i e_i \end{array}\right\} \quad v'=R(U;a)v \ , \quad v'.\sigma=v'^i\sigma_i=Uv.\sigma U^{-1}=Uv^i\sigma_i U^{-1}$$

$$\pm U\in SU(2) \leftrightarrow R(U)\in SO(3) \ , \quad R(U)^i_j\ \delta_{ik}=\tfrac{1}{2}\,Tr(\sigma_k U\sigma_j\ U^{-1})$$

$$\forall\ V\in SU(2) \qquad VU(\theta,\hat{n};a)V^{-1} = U[\theta,R(V;a)\hat{n};a] \quad .$$

Infinitesimal elements of SU(2) and Lie algebra $\mathcal{W}(2)$

1–Infinitesimal element :

$$U = \mathbf{1}-iW + O(W^2) \quad , \quad W \in \mathcal{W}(2) \quad .$$

2–Lie algebra structure of $\mathcal{W}(2)$:

$$\forall\, W,W' \in \mathcal{W}(2) \rightarrow -i[W,W'] \in \mathcal{W}(2) \quad .$$

$$(W_i)\ \text{basis} \qquad -i[W_i,W_j] = c_{ij}{}^k\, W_k \ , \ c_{ij}{}^k \in R \quad .$$

Internal derivation or adjoint operator of W :

$$\forall\ W' \in \mathcal{W}(2) \rightarrow \Delta_W W' = -i[W,W'] \in \mathcal{W}(2) \quad .$$

3– $\mathcal{W}(2)$ is a semi–simple (even simple) Lie algebra :

–The Killing form is nondegenerate :

$$\forall\, W,W' \in \mathcal{W}(2) \qquad g(W,W')=-Tr(\Delta_W\Delta_{W'}) \ , \ \det g \neq 0 \quad .$$

–Two special bases of $\mathcal{W}(2)$:

(1) (σ_i) ⇒ completely antisymmetrical structure constants :

$$-i[\sigma_i,\sigma_j] = 2\sum_k \varepsilon_{ijk}\sigma_k \qquad \text{or} \qquad -i\sum_{i,j} \varepsilon_{ijk}\,\sigma_i\sigma_j = 2\sigma_k \quad .$$

(2) Standard basis : ($\sigma_\pm = \sigma_1 \pm i\,\sigma_2, \sigma_3$) :

$$[\sigma_3,\sigma_\pm] = \pm 2\,\sigma_\pm \qquad [\sigma_+,\sigma_-] = 4\,\sigma_3 \quad .$$

4–Euclidean structure of $\mathcal{W}(2)$:

The Killing form g is positive definite (SU(2) is compact) :

$$\forall\, W,W' \in \mathcal{W}(2) \qquad (W,W') = g(W,W') \quad .$$

(σ_i) is an orthogonal basis of $\mathcal{W}(2)$:

$$(\sigma_i,\sigma_j) = 8\ \delta_{ij} \quad .$$

Consequences of the Baker–Campbell–Hausdorff formula

$$U_1 = e^{-iW_1} \quad , \quad U_2 = e^{-iW_2} \quad , \quad U_3 = U_1U_2 = e^{-iW_3}$$

$$W_3 = W_1+W_2-i[W_1,W_2]-\frac{1}{12}([W_1,[W_1,W_2]]+[W_2,[W_2,W_1]]) + \dots \quad .$$

σ matrices

$$\sigma_1 = \begin{pmatrix} 0 & 1 \\ 1 & 0 \end{pmatrix} , \quad \sigma_2 = \begin{pmatrix} 0 & -i \\ i & 0 \end{pmatrix} , \quad \sigma_3 = \begin{pmatrix} 1 & 0 \\ 0 & -1 \end{pmatrix} \; ; \; \sigma_+ = \begin{pmatrix} 0 & 2 \\ 0 & 0 \end{pmatrix} , \quad \sigma_- = \begin{pmatrix} 0 & 0 \\ 2 & 0 \end{pmatrix}$$

$$\sigma_i = \sigma_i^+ \quad , \quad \sigma_i\sigma_i^+ = \sigma_i^2 = \mathbf{1} \quad , \quad \det \sigma_i = -1$$

$$\mathrm{Tr}\sigma_i = 0 \quad , \quad \mathrm{Tr}(\sigma_i\sigma_j) = 2\delta_{ij}$$

$$[\sigma_i,\sigma_j] = i\,2\sum_k \varepsilon_{ijk}\,\sigma_k \cdot \quad \text{or} \quad \sum_{i,j} \varepsilon_{ijk}\,\sigma_i\sigma_j = i2\,\sigma_k$$

$$\sigma_i\sigma_j = \delta_{ij}\,\mathbf{1} + i\sum_k \varepsilon_{ijk}\,\sigma_k$$

$$[u.\sigma,v.\sigma] = i\,2(u\times v).\sigma \quad \text{provided} \quad [u^i,v^j] = [u^i,\sigma_j] = [v^i,\sigma_j] = 0$$

$$(u.\sigma)(v.\sigma) = u.v + i(u\times v).\sigma \quad \text{provided } [v^i,\sigma_j] = 0$$

$$(u,v,w) = \frac{1}{2i}\,\mathrm{Tr}[(u.\sigma)(v.\sigma)(w.\sigma)] \quad \text{provided } [v^i,\sigma_j] = [w^i,\sigma_j] = 0$$

$$X = \sum_{\mu=0}^{3} x^\mu\sigma_\mu \quad , \quad \sigma_o = \mathbf{1} \quad , \quad x^\mu = \frac{1}{2}\,\mathrm{Tr}(\sigma_\mu X) \in C$$

$$X^+ = X \; , \; \mathrm{Tr}\,X = 0 \;\Rightarrow\; X = x.\sigma = x^i\sigma_i = \frac{1}{2}(x_+\sigma_- + x_-\sigma_+) + x^3\sigma_3 = \begin{pmatrix} x^3 & x_- \\ x_+ & -x^3 \end{pmatrix}$$

$$\sum_{i=1}^{3} (\sigma_i)^\gamma_\delta\,(\sigma_i)^\alpha_\beta = 2\,\delta^\alpha_\delta\,\delta^\gamma_\beta - \delta^\gamma_\delta\,\delta^\alpha_\beta$$

$$C = i\sigma_2 = \begin{pmatrix} 0 & 1 \\ -1 & 0 \end{pmatrix} = U(\pi,-e_2;a) = U(\pi,e_2;a)^{-1}$$

$$C = C^* = -C^{-1} = -C^+ = -C^t$$

$$C\sigma_i\,C^{-1} = -\sigma_i^t$$

$$\det A\,\mathbf{1} = C\,A\,C^{-1}\,A^t \quad , \quad U^* = C\,U\,C^{-1} \quad .$$

Infinitesimal elements of SO(3) and Lie algebra $A(3)$

$$R = \mathbf{1} + M + O(M^2) \quad , \quad M = -M^t \in A(3)$$

$$M = \varepsilon^i\,M_i \quad , \quad (M_i)^j_k = \varepsilon_{ikj} \quad , \quad [M_i,M_j] = \sum_k \varepsilon_{ijk}\,M_k \quad .$$

CHAPTER 5

GENERALITIES ABOUT THE REPRESENTATIONS OF SU(2) AND OF ITS LIE ALGEBRA

This chapter is closely related to the previous one. It is devoted to the study of the differentiable representations of the group SU(2) and of its Lie algebra, pointing out the relations between both. The methods which are developed give a general frame for the study of the differentiable representations of any Lie group with a semi-simple Lie algebra. In particular, as in Chap. 4, whenever it can easily be done the more general case of the group SU(n) is discussed.

We recall that some definitions and general properties of algebras, groups, and their representations are given in Appendices B and C. The Baker–Campbell–Hausdorff formula is derived in Appendix D.

5.1. INFINITESIMAL GENERATORS

The ideas developed below for the group SU(n) can be applied to any Lie group, and especially in a straightforward way to any linear group. In Appendix C.5.4.3,ii) one shows how the infinitesimal elements of the linear groups and the corresponding infinitesimal generators introduced below are related to the invariant infinitesimal transformations defined in a general frame by Defs. C.57,61.

i) In Subsec. 4.3.1, the infinitesimal elements of the group SU(n) are defined by :

$$U = \mathbf{1} - iW + O(W^2) \qquad ,$$

where W belongs to the Lie algebra $\mathcal{W}(n)$ of SU(n). Let D[SU(n)] be a differentiable representation of SU(n). Differentiable means that the matrix elements of $D[\mathbf{1} - i\varepsilon W + O(\varepsilon^2)]$ are, for all W in $\mathcal{W}(n)$, differentiable functions of ε. Then the representation of the infinitesimal elements reads :

$$D[\mathbf{1} - i\varepsilon W + O(\varepsilon^2)] = \mathbf{1} - i\varepsilon J + O(\varepsilon^2) \qquad , \qquad (5\text{-}1)$$

where the infinitesimal generator J is defined by :

$$J = i \frac{\partial}{\partial\varepsilon} D[\mathbf{1} - i\varepsilon W + O(\varepsilon^2)]\Big|_{\varepsilon=0} \qquad . \qquad (5\text{-}2)$$

Actually, for a given W, J is the infinitesimal generator of the representation of the one–

parameter Abelian subgroup associated with W (Subsec. 4.1.5).

The infinitesimal generators are linear mappings of the vector space of the representation into itself. To any basis (W_i) of the Lie algebra $\mathcal{W}(n)$, one associates by Eq.(5–2) a set of infinitesimal generators :

$$W_i \longrightarrow J_i = i\,\frac{\partial}{\partial\varepsilon}\,D[\mathbf{1}-i\varepsilon W_i + 0(\varepsilon^2)]\Big|_{\varepsilon=0} \quad . \tag{5-3}$$

If W is a linear combination of the W_i, then J defined by Eq.(5–2) is the same linear combination of the J_i :

$$W = w^i W_i \Longrightarrow J = w^i J_i \quad , \qquad w^i \in R \quad . \tag{5-4}$$

The set of infinitesimal generators is thus, as $\mathcal{W}(n)$, a (n^2-1)-dimensional vector space over the real field.

ii) One now proves that this vector space is also a Lie algebra which is a representation (Def. B.17) of the Lie algebra $\mathcal{W}(n)$. Let us consider the left (for instance) commutator (Def. C.2) of two infinitesimal elements of SU(n), $U = \mathbf{1} - i\varepsilon W + 0(\varepsilon^2)$ and $U' = \mathbf{1} - i\varepsilon' W' + 0(\varepsilon'^2)$. To the first order in ε and ε' one has :

$$UU'U^{-1}U'^{-1} = \mathbf{1}-i\varepsilon\varepsilon'(-i[W,W']) + 0(\varepsilon^2) + 0(\varepsilon'^2) \quad , \tag{5-5}$$

where $W'' = -i[W,W']$ belongs to $\mathcal{W}(n)$. The representation of the above equation reads :

$$D[\mathbf{1}-i\varepsilon W+0(\varepsilon^2)]D[\mathbf{1}-i\varepsilon' W'+0(\varepsilon'^2)]D[\mathbf{1}+i\varepsilon W+0(\varepsilon^2)]D[\mathbf{1}+i\varepsilon' W'+0(\varepsilon'^2)]=D[\mathbf{1}-i\varepsilon\varepsilon' W''+0(\varepsilon^2)+0(\varepsilon'^2)]$$

then :

$$\mathbf{1}-i\varepsilon\varepsilon'(-i[J,J']) = \mathbf{1}-i\varepsilon\varepsilon' J'' \quad ,$$

where J, J', and J'' are the infinitesimal generators which correspond, respectively, to W, W', and W''. Therefore, one obtains the result :

$$W'' = -i[W,W'] \Longrightarrow J'' = -i[J,J'] \quad , \tag{5-6}$$

which shows that the vector space of infinitesimal generators of the representation D[SU(n)] is a Lie algebra $D[\mathcal{W}(n)]$ homomorphic to (i.e. a representation of) the Lie algebra $\mathcal{W}(n)$ of SU(n). This homomorphism can be defined by Eq.(5–2).

iii) The Lie algebra $D[\mathcal{W}(n)]$ defined above has an intrinsic character. A priori the basis (J_i), related to any basis (W_i) of $\mathcal{W}(n)$ by Eq.(5–3), can be considered, and leads, respectively, in $\mathcal{W}(n)$ and $D[\mathcal{W}(n)]$ to identical structure constants :

$$-i[W_i,W_j] = c_{ij}^{\ \ k}\, W_k \Longrightarrow -i[J_i,J_j] = c_{ij}^{\ \ k}\, J_k \quad , \tag{5-7}$$

and therefore $\mathcal{W}(n)$ and $D[\mathcal{W}(n)]$ are isomorphic.

For n=2, as already noticed in Subsec. 4.2.1, the basis (σ_i) of $\mathcal{W}(2)$ plays a special role which is discussed below in Subsec. 5.4.2,ii). For this reason, and also to allow us later (Subsec. 11.1.2) to identify them with the components of the angular momentum with respect to a right-handed orthonormal basis, the commonly used basis vectors of $D[\mathcal{W}(2)]$ are those which correspond to the basis $(\frac{1}{2}\sigma_i)$ of $\mathcal{W}(2)$, i.e. :

$$J_i = i\frac{\partial}{\partial\varepsilon} D[\mathbf{1} - i\frac{\varepsilon}{2}\sigma_i + 0(\varepsilon^2)]\Big|_{\varepsilon=0} \quad . \tag{5-8}$$

They fulfill the usual commutation relations of angular momentum components, from Eqs. (4-45) and (5-7) :

$$[J_i, J_j] = i\sum_k \varepsilon_{ijk} J_k \qquad \text{or} \qquad \sum_{i,j} \varepsilon_{ijk} J_i J_j = iJ_k \quad . \tag{5-9}$$

To the standard basis $\sigma_\pm$, σ_3, Eqs. (4-67,68), of the complex extension $\{\mathcal{W}(2)\}$ of the Lie algebra of SU(2), there corresponds the standard basis of the complex extension $\{D[\mathcal{W}(2)]\}$:

$$J_\pm = J_1 \pm iJ_2 \quad , \qquad J_3 \quad , \tag{5-10}$$

where the J_i are related to the σ_i by Eq. (5-8). These infinitesimal generators fulfil the standard form of the commutation relations similar to Eqs. (4-69a,b) :

$$[J_3, J_\pm] = \pm J_\pm \quad , \tag{5-11a}$$

$$[J_+, J_-] = 2J_3 \quad . \tag{5-11b}$$

For a unitary representation D[SU(2)] and thus Hermitian generators J_i, the spectra of J_3 and $J_\pm$ are derived later in Sec. 5.5.

5.2. THE EXPONENTIAL FORM

i) Let us consider the one-parameter Abelian subgroup of SU(n) associated with any element W of the Lie algebra $\mathcal{W}(n)$ of SU(n). Using Eq. (4-25a), one has for all elements of this subgroup and all integers n :

$$U(\theta) = e^{-i\theta W} = [U(\frac{\theta}{n})]^n \quad .$$

This implies for a given representation D[SU(2)] :

$$D[U(\theta)] = \{D[U(\frac{\theta}{n})]\}^n \quad .$$

Assuming this representation is differentiable and expanding $D[U(\frac{\theta}{n})]$ to first order in $\frac{1}{n}$, one has :

$$D[U(\frac{\theta}{n})] = \mathbf{1} - i\,\frac{\theta}{n}\,J + O(\frac{1}{n^2}) \quad ,$$

where J is an infinitesimal generator defined by Eq. (5–2). Therefore, for n going to infinity one obtains the exponential form of $D[U(\theta)]$:

$$D[U(\theta)] = \lim_{n\to+\infty}\ [\mathbf{1} - i\,\frac{\theta}{n}\,J + O(\frac{1}{n^2})]^n = e^{-i\theta J} \quad . \tag{5-12}$$

For a finite–dimensional unitary representation, the exponential form can also be obtained directly using the proof of Theorem 4.1 given for U(n) in Subsec. 4.1.2,i). But the above proof is more general and applies to all differentiable representations of Lie groups for which the so–called canonical parameters exist. Using the notations of Def. C.60, canonical parameters are such that to the product of the group elements defined by, respectively, $\lambda\,a$ and $\mu\,a$, corresponds the element defined by the parameters $(\lambda+\mu)a$ for all a,λ , and μ, i.e.:

$$(\lambda+\mu)a^i = \varphi^i(\lambda a,\mu a) \quad .$$

Then the identity corresponds to the set $0 = \{0\}$.

Following the same steps as above for SU(n), one has for any differentiable representation :

$$D[g(a)] = \{D[g(\frac{a}{n})]\}^n \quad ,$$

expanding to first order in $\frac{1}{n}$:

$$D[g(a)] = [\mathbf{1} + \frac{1}{n}\,A + O(\frac{1}{n^2})]^n \quad , \quad A = \frac{\partial}{\partial\alpha}\,D[g(\alpha a)]\Big|_{\alpha=0} \quad ,$$

and finally, letting n go to infinity :

$$D[g(a)] = e^A \quad .$$

ii) Let us parametrize the elements of SU(2) by θ , n^i as in Eq. (4–15). From Eq. (5–4) one then has :

$$U(\theta,n^i) = e^{-i\frac{\theta}{2}n^i\sigma_i} \Longrightarrow D[U(\theta,n^i)]e^{-i\theta n^i J_i} \quad , \tag{5-13}$$

where the infinitesimal generators J_i are defined by Eq. (5–8). Repeating the remark about Eq. (4–17), $D[U(\theta,n^i)]$ can be written in terms of an Euclidean inner product as :

$$D[U(\theta,\hat{n};a)] = e^{-i\theta\hat{n}.J} \quad ; \quad \hat{n} = n^i e_i \quad , \quad \hat{n}.J = n^i J_i \quad . \tag{5-14}$$

An intrinsic meaning of this formal inner product $\hat{n}.J$ is discussed later in Subsec. 5.4.3,i).

If the elements of SU(2) are parametrized by the Euler angles φ, θ , and ψ , taking the representation of Eq. (4–37) one obtains :

$$D[U(\varphi,\theta,\psi)] = e^{-i\varphi J_3}\,e^{-i\theta J_2}\,e^{-i\psi J_3} \quad , \tag{5-15}$$

where J_2 and J_3 are the infinitesimal generators defined by Eq. (5–8).

5.3. CONSEQUENCES OF THE BAKER–CAMPBELL–HAUSDORFF FORMULA

It has been shown in Subsec. 5.1,ii) that to any differentiable representation D[SU(n)] of the group SU(n) corresponds a representation $D[\mathcal{W}(n)]$ of its Lie algebra $\mathcal{W}(n)$. Let us now prove, at least formally, that conversely to any representation $D[\mathcal{W}(n)]$ of the Lie algebra $\mathcal{W}(n)$ corresponds a representation of SU(n).

For any element W in $\mathcal{W}(n)$ let us denote by J = D(W) its representation. To the element $U = e^{-iW}$ in SU(n) one associates a linear mapping acting in the vector space of the representation $D[\mathcal{W}(n)]$ and formally defined by :

$$D(U) = e^{-iD(W)} = e^{-iJ} \quad . \tag{5-16}$$

Then, since the condition $D(U^{-1}) = [D(U)]^{-1}$ is clearly satisfied, the question arises ; does one have :

$$\forall U_1, U_2 \in SU(n) \qquad D(U_1)D(U_2) = D(U_1U_2) \ ?$$

The main result of the Baker–Campbell–Hausdorff formula (Prop. D.4) reads :

$$U_1U_2 = e^{-iW_1} e^{-iW_2} = e^{-iW(W_1,W_2)} \quad ,$$

$$D(U_1)D(U_2) = e^{-iJ_1} e^{-iJ_2} = e^{-iW(J_1,J_2)} \quad ,$$

where $W(W_1,W_2)$ and $W(J_1,J_2)$ belong, respectively, to the Lie algebra $\mathcal{W}(n)$ and $D[\mathcal{W}(n)]$. Indeed they are expressed in terms of, respectively W_1, W_2 and J_1, J_2 by a universal formula, Eq. (4–82), which only uses internal laws of composition of the Lie algebras, i.e. the addition, the multiplication by a real factor, and the bilinear mapping defined by minus i times the commutator. Therefore, since $D[\mathcal{W}(n)]$ is a representation of the Lie algebra $\mathcal{W}(n)$, one has formally :

$$D[W(W_1,W_2)] = W(J_1,J_2) \quad ,$$

which leads to the announced result :

$$D(U_1U_2) = e^{-iD[W(W_1,W_2)]} = e^{-iW(J_1,J_2)} = D(U_1)D(U_2) \quad .$$

If the vector space of the representation $D[\mathcal{W}(n)]$ is Hermitian and if the elements J of $D[\mathcal{W}(n)]$ are Hermitian operators acting in this vector space, then the induced representation of the group, defined above, is unitary.

This correspondence between the differentiable representations of a Lie group and the representations of its Lie algebra is of course not specific to the group SU(n). Indeed, the above derivation is only based on the existence of an exponential form (cf. general remarks in Subsec. 5.2,i)).

5.4. THE ADJOINT REPRESENTATION

The adjoint representation of a Lie group plays an important role since it yields general properties of all differentiable representations of this group.

5.4.1. DEFINITION OF THE ADJOINT REPRESENTATION

i) To each element U in SU(n) one associates a linear mapping A(U) of $\mathcal{W}(n)$ into itself, such that :

$$\forall\, W \in \mathcal{W}(n) \rightarrow A(U)W \;,\; U[\mathbf{1}-i\varepsilon W + 0(\varepsilon^2)]U^{-1} = \mathbf{1}-i\varepsilon A(U)W + 0(\varepsilon^2) \quad , \tag{5-17}$$

where ε is a real infinitesimal parameter. This is possible since for all U in SU(n), the quantity $U[\mathbf{1}-i\varepsilon W+0(\varepsilon^2)]\,U^{-1}$ is still an infinitesimal element of SU(n). Thus, at least for a linear group, A(U) is also defined by :

$$A(U)W = U\,W\,U^{-1} \quad . \tag{5-18}$$

One can check directly that $U\,W\,U^{-1}$ belongs as W to $\mathcal{W}(n)$ since it is Hermitian and traceless (this argument has already been used in Subsec. 4.2.1,i)). Clearly, the mapping $U \rightarrow A(U)$ defines a representation A[SU(n)] of SU(n) in its Lie algebra $\mathcal{W}(n)$, this is the <u>adjoint representation of the group</u>.

Let U_p denote any element of the centre of the group, Eq. (4–2), i.e. U_p commutes with all elements of the group. From Eq. (5–17) one then has :

$$U[\mathbf{1}-i\varepsilon W + 0(\varepsilon^2)]U^{-1} = (U_pU)[\mathbf{1}-i\varepsilon W + 0(\varepsilon^2)]\;(U_pU)^{-1} \quad ,$$

and since this is true for all W in the Lie algebra :

$$\forall\, U \in SU(n) \qquad A(U) = A(U_pU) \quad . \tag{5-19}$$

For SU(2) the centre is isomorphic to $Z_2 = \{\mathbf{1}, -\mathbf{1}\}$ (Subsec. 4.1.1) and thus :

$$\forall\, U \in SU(2) \qquad A(U) = A(-U) \quad . \tag{5-20}$$

ii) We have seen in Subsec. 5.1,ii) that to any differentiable representation of a Lie group corresponds a representation of its Lie algebra, let us specify here what is this representation. The infinitesimal generator J of the adjoint representation, which corresponds to W by Eq. (5–2), is such that for all W' in $\mathcal{W}$ (n) :

$$A[\mathbf{1}-i\varepsilon W+0(\varepsilon^2)]W'=[\mathbf{1}-i\varepsilon J+0(\varepsilon^2)]W'=[\mathbf{1}-i\varepsilon W+0(\varepsilon^2)]W'[\mathbf{1}+i\varepsilon W+0(\varepsilon^2)] \quad ,$$

therefore :

$$J = i(-i[W,\cdot]) = i\,\Delta_W \quad , \tag{5-21}$$

where Δ_W is the internal derivation or adjoint operator of W defined by Eq. (4–44). Thus, the adjoint representation of the Lie group SU(n) induces the adjoint representation $\Delta[\mathcal{W}(n)]$ of its Lie algebra (Def. B.21).

iii) Once a basis (W_i) is chosen in $\mathcal{W}(n)$ the matrix representative of A(U) is defined by :

$$A(U)W_i = U\,W_i\,U^{-1} = W_j\,A(U)^j_i \quad . \tag{5-22}$$

In particular for n=2, it has already been noticed in Subsec. 4.2.3 that the matrix representative of A(U) with respect to the basis (σ_i) of $\mathcal{W}(2)$ is the rotation matrix which corresponds to U by Eq. (4–26) :

$$A(U)\sigma_i = U\,\sigma_i\,U^{-1} = \sigma_j\,R(U)^j_i \quad . \tag{5-23}$$

This special role played by the Pauli σ matrices is discussed below in Subsec. 5.4.2,ii).

The matrix representative of the infinitesimal generators of the adjoint representation is closely related to the structure constants. Indeed let $(J_i = i\,\Delta_{W_i})$ be the infinitesimal generators which correspond to a basis (W_i) of the Lie algebra, then their matrix representatives with respect to this basis read :

$$J_iW_j = W_k(J_i)^k_j = i(-i[W_i,W_j]) = i\,c_{ij}{}^k\,W_k \Rightarrow (J_i)^k_j = c_{ij}{}^k \quad . \tag{5-24}$$

In particular, if one considers, as usual, the basis $(\frac{1}{2}\sigma_i)$ of $\mathcal{W}(2)$, the corresponding infinitesimal generators $(J_i = i\,\Delta_{\frac{1}{2}\sigma_i})$ fulfill the usual commutation relations of angular momentum components, Eq. (5–9), and their matrix representatives with respect to this basis $(\frac{1}{2}\sigma_i)$ read :

$$(J_i)^k_j = i\,\varepsilon_{ijk} \quad , \tag{5-25}$$

i.e. they are identical to i times the matrices M_i defined by Eq. (4–73).

5.4.2. THE EUCLIDEAN STRUCTURE OF THE SEMI-SIMPLE LIE ALGEBRA OF A COMPACT LIE GROUP

i) General results :

1–We have already noticed in Subsec. 4.3.3,i) that for a semi-simple Lie algebra the Killing form g defined by Eq. (4–43) is nondegenerate (Prop. B.34). Therefore, the inner product :

$$\forall\, W,W' \in \mathcal{W} \longrightarrow (W,W') = g(W,W') \in R \quad , \tag{5-26}$$

endows a semi-simple Lie algebra $\mathcal{W}$ over the real field with an Euclidean or pseudo-Euclidean vector space structure. Pseudo-Euclidean corresponds to the fact that g is not necessarily definite with a given sign.

With respect to any basis (W_i) of this algebra $\mathcal{W}$ the real symmetrical metric tensor of the Killing form is related to the structure constants by Prop. B.33 and can be diagonalized by a real orthogonal change of basis.

2–It is shown in Prop. B.32 that the exponential e^{Δ_W} of the adjoint operator of W is, for all W in $\mathcal{W}$, an isometric application in $\mathcal{W}$ with respect to the inner product defined by g :

$$\forall\ W,W',W''\in\mathcal{W} \qquad (e^{\Delta_W}W',\ e^{\Delta_W}W'') = (W',W'') \quad . \tag{5-27}$$

Actually, for the linear groups this mapping e^{Δ_W} is identical to the adjoint representation A(U), defined by Eq. (5–18), of the element $U = e^{-iW}$. Indeed, one shows in Prop. D.1 that formally :

$$\forall\ W,W'\in\mathcal{W} \qquad e^{-iW}\,W'\,e^{iW} = e^{\Delta_W}\,W' \quad ,$$

and therefore one has :

$$U = e^{-iW} \Longrightarrow A(U) = e^{\Delta_W} \quad . \tag{5-28}$$

More generally, let us consider any Lie group, the elements of which U have an exponential form e^{-iW} (Sec. 5.2), and such that it has a semi–simple Lie algebra over the real field. Then the adjoint representation A(U) of any U in this group is an isometric application with respect to the inner product defined by the Killing form in the Lie algebra. In a basis (W_i) of the algebra $\mathcal{W}$ the isometric character of A(U) implies that its matrix representative and the metric tensor g fulfil Eq. (2–6), i.e. in matrix notation :

$$g = A(U)^t\ g\ A(U) \quad . \tag{5-29}$$

3–Moreover, one shows that for a compact Lie group with a semi–simple Lie algebra over the real field the Killing form is positive definite (cf. for instance Pontryagin[44], Chap. 11, Sec. 61). Then, the nondegenerate and definite Killing form endows the Lie algebra with an Euclidean vector space structure.

ii) **Application to the group SU(n) and especially to the case n=2.** It is recalled that SU(n) is a compact Lie group (Appendix C.5.4.3,i)) with a semi–simple (and even simple for $n > 1$) Lie algebra $\mathcal{W}(n)$ over the real field (Subsec. 4.3.3,i)), therefore :

1–The Lie algebra $\mathcal{W}(n)$ is thus also a n^2-1-dimensional Euclidean vector space where the inner product is defined by the Killing form.

In particular $\mathcal{W}(2)$ is a three–dimensional Euclidean vector space. Let E_3 be any other three–dimensional Euclidean vector space where, as usual, the inner product is denoted by a point. It has been shown in Subsec. 4.3.4,ii) that $\mathcal{W}(2)$ and E_3 are isomorphic Lie algebras, in addition they are thus isomorphic Euclidean vector spaces. This isomorphism can be defined by :

$$\forall\ u^i\in R \qquad u = u^i e_i \in E_3 \longleftrightarrow W_u = u^i\,W_i \in \mathcal{W}(2) \quad , \tag{5-30}$$

where (e_i) and (W_i) are any bases of, respectively, E_3 and $\mathcal{W}(2)$ which yield proportional metric tensors :

$$g_{ij} = e_i . e_j = c(W_i, W_j) \quad , \quad c \in R \quad . \tag{5-31}$$

One then has :

$$\forall\, u,v \in E_3 \qquad u.v = c(W_u, W_v) \quad . \tag{5-32}$$

This isomorphism allows us now to give an intrinsic interpretation to the scalar product $\hat{n}.\sigma$ introduced in Eq. (4–17). Let (e_i) and (W_i) be two bases of, respectively, E_3 and $\mathcal{W}(2)$ and related by Eq. (5–31). Then, a formal vector W in E_3 is defined as follows :

$$W = W_i\, g^{ij}\, e_j = W^i\, e_i = W_i\, e^i \quad , \tag{5-33}$$

where (g^{ij}) is the inverse of the symmetrical metric tensor (g_{ij}), (W_i) and $(W^j = W_i\, g^{ij})$ are, respectively, the covariant and contravariant components of W with respect to the basis (e_i), Eq. (1–30), and $(e^i = g^{ij} e_j)$ is the biorthogonal basis of (e_i), Eq. (1–33). By construction, from the tensorial characters, W is thus intrinsically defined. Formal means that only the tensorial character of both sets (W_i) and (e_i) is taken into account, thus W is called a vector of E_3 although its components (W^i) do not belong to the real field. Clearly, both vector spaces E_3 and $\mathcal{W}(2)$ play a symmetrical role, and W could as well be considered as a formal vector of $\mathcal{W}(2)$. Then, the inner product in E_3 of any vector $u = u^i e_i$ with $W = W^i e_i$ reads :

$$u.W = u^i\, g_{ij}\, W^j = u^i\, W_i = u_i\, W^i \quad . \tag{5-34}$$

Since the σ matrices form an orthogonal basis (σ_i) of $\mathcal{W}(2)$, Eq. (4–46), the above result can be applied to any arbitrarily chosen orthonormal basis a $= (e_i)$ of E_3 (setting $c = \frac{1}{8}$ in Eq. (5–31)). Thus the inner product in E_3 of the unit vector $\hat{n} = n^i e_i$ with the formal vector :

$$\sigma = \sigma^i\, e_i = \sum_i \sigma_i\, e_i \quad , \tag{5-35}$$

is given by Eq. (4–17) :

$$\hat{n}.\sigma = n^i\, \sigma_i \quad . \tag{5-36}$$

2–The adjoint representation $A(U) = e^{\Delta W}$ of any $U = e^{-iW}$ in SU(n) is an isometric application in $\mathcal{W}(n)$. This mapping is a proper one. Indeed, SU(n) being a connected group (and even simply-connected : for SU(2) cf. Subsec. 4.1.4,ii) and for the general case cf. for instance Chevalley[16], Chap. II, § X, Proposition 6),and the adjoint representation being continuous, the set of isometric applications A(U), for all U in SU(n), is included in the sheet of proper isometric applications connected to the unit element $\mathbf{1} = A(\mathbf{1})$. Hence A(U) is a rotation in $\mathcal{W}(n)$. The vector W in $\mathcal{W}(n)$ is clearly invariant (rotation axis) under the rotation $A(e^{-iW})$:

$$A(e^{-iW})W = e^{-iW}\, W\, e^{iW} = W \quad .$$

Moreover, from Eq. (5–19), to every U_pU, where U_p belongs to the centre of SU(n), Eq. (4–2), there corresponds the same adjoint representation, therefore :

The correspondence $\{U_pU, p=1,...,n\} \rightarrow A(U)$ is an intrinsic homomorphism between SU(n) and the n^2-1-dimensional rotation group in $\mathcal{W}(n)$.

With respect to any basis (W_i) of $\mathcal{W}(n)$ the matrix representative of A(U) is given by Eq. (5–22). Now, choosing an orthonormal basis (i.e. such that the metric tensor be a multiple of the unit matrix) the matrix representative of A(U) belongs to $SO(n^2-1)$, cf. Eq. (5–29). The elements of this group $SO(n^2-1)$ depend upon $(n^2-1)(n^2-2)/2$ real parameters (Subsec. 2.1.1.2). What is peculiar for n=2 is that the number of real parameters for the elements of SU(n) and for those of $SO(n^2-1)$ are the same, i.e. $n^2-1 = (n^2-1)(n^2-2)/2$ only when n=2. This leads to the two–to–one homomorphism of SU(2) onto SO(3) discussed in Subsec. 4.2.1. On the other hand, the elements of SU(3) for instance depend upon 8 real parameters while those of SO(8) depend upon 28 real parameters. Therefore to any $\{U, e^{-i2\pi/3}U, e^{-i4\pi/3}U\}$ in SU(3) one associates a eight–dimensional rotation matrix but the converse is not true.

Let us emphasize the results obtained in the case n=2. The correspondence $\{U,-U\} \rightarrow A(U)$ is an intrinsic homomorphism between SU(2) and the three–dimensional rotation group in $\mathcal{W}(2)$, while the correspondence $\{U,-U\} \rightarrow R(U;a)$ defined by Eq. (4–26) depends upon the choice of an orthonormal basis a in a three–dimensional Euclidean vector space E_3. The relation between these two approaches is the following. The main property of the σ matrices which explainstheir role in the Subsec. 4.2.1 is that they form one of the orthogonal bases of $\mathcal{W}(2)$. Therefore, the matrix representative of A(U) with respect to this basis belongs to SO(3), which yields to Eq . (5–23). One has previously defined in 1– an isomorphism between $\mathcal{W}(2)$ and any three–dimensional Euclidean vector space E_3, where this isomorphism depends upon for instance the choice of an orthonormal basis $a = (e_i)$ of E_3 in correspondence by Eq. (5–31) with the orthonormal basis (σ_i) of $\mathcal{W}(2)$. Then, to the intrinsic rotation A(U) in $\mathcal{W}(2)$ corresponds the rotation R(U;a) in E_3 defined by Eq. (4–26).

We finally recall the fact, already noticed in Subsec. 4.3.4,i), that the case of the three–dimensional rotation group is peculiar since there exists only one positive value of n such that the dimension of the Lie algebra of SO(n), i.e. n(n-1)/2, is equal to n. This allows us to define the isomorphism between $\mathcal{W}(2)$, isomorphic to the Lie algebra of SO(3) (Subsec. 4.3.4,i)), and any three–dimensional Euclidean vector space on which the rotations act.

5.4.3. CONSEQUENCES FOR ANY DIFFERENTIABLE REPRESENTATION

i) It has been shown in Subsec. 5.1,iii) that $\mathcal{W}(n)$ and $D[\mathcal{W}(n)]$ are isomorphic Lie algebras such that with respect to the bases, respectively, (W_i) and (J_i) related by Eq. (5–3), the structure constants are the same, Eq. (5–7). Now from Subsec. 5.4.2,ii), $\mathcal{W}(2)$ and $D[\mathcal{W}(2)]$ are moreover isomorphic Euclidean vector spaces. Indeed, using the expression of the metric tensor of the Killing form in terms of the structure constants (Prop. B.33) one has :

$$g_{ij} = -\mathrm{Tr}(\Delta_{W_i}\Delta_{W_j}) = -c_{ik}{}^{\ell}\, c_{j\ell}{}^{k} = -\mathrm{Tr}(\Delta_{J_i}\Delta_{J_j}) \quad ,$$

where Δ_{J_i} is the adjoint operator of J_i defined by :

$$\forall J \in D[\mathcal{W}(2)] \qquad \Delta_{J_i} J = - i[J_i, J] \in D[\mathcal{W}(2)] \quad . \tag{5-37}$$

Thus the Killing form over $D[\mathcal{W}(2)]$ endows this space with an Euclidean structure where the inner product reads :

$$\forall J, J' \in D[\mathcal{W}(2)] \longrightarrow (J, J') = - \mathrm{Tr}(\Delta_J \Delta_{J'}) \in R \quad . \tag{5-38}$$

If J and J' are, respectively, the images, defined by Eq. (5–2), of W and W' in the isomorphism between $\mathcal{W}(2)$ and $D[\mathcal{W}(2)]$, one gets :

$$(W, W') = (J, J') \quad . \tag{5-39}$$

Hence the commonly used infinitesimal generators (J_i), which correspond to the basis $(\frac{1}{2}\sigma_i)$ of $\mathcal{W}(2)$, form an orthogonal basis of $D[\mathcal{W}(2)]$ such that, from Eq. (4–46) :

$$(J_i, J_j) = \left(\frac{\sigma_i}{2}, \frac{\sigma_j}{2}\right) = 2\,\delta_{ij} \quad . \tag{5-40}$$

As discussed previously in Subsec. 5.4.2,ii) for $\mathcal{W}(2)$, one defines an isomorphism between $D[\mathcal{W}(2)]$ and any three-dimensional Euclidean vector space E_3. This isomorphism is also defined by the choice of an orthonormal basis $a = (e_i)$ of E_3 such that :

$$e_i \rightarrow J_i \quad , \qquad e_i . e_j = \frac{1}{2}(J_i, J_j) = \delta_{ij} \quad . \tag{5-41}$$

Then, the inner product in E_3 of the unit vector $\hat{n} = n^i e_i$ with the formal vector, cf. Eq. (5–33) :

$$J = J_i\, g^{ij} e_j = J^i\, e_i = J_i\, e^i \quad , \tag{5-42}$$

reads :

$$\hat{n}.J = n^i\, J_i \quad . \tag{5-43}$$

Later, in Subsecs. 11.1.2,v),3,iii) the formal vector J is identified with the angular momentum.

ii) The Eq. (5–22) implies a similar relation for the infinitesimal generators of any differentiable representation $D[SU(2)]$. Indeed from Eq. (5–17) one has :

$$U[\mathbf{1} - i\varepsilon W_i + 0(\varepsilon^2)]U^{-1} = \mathbf{1} - i\varepsilon W_j A(U)^j_i + 0(\varepsilon^2) \quad ,$$

where (W_i) is any basis of $\mathcal{W}(2)$. Taking the representation of the above equation leads to :

$$D(U)D[\mathbf{1} - i\varepsilon W_i + 0(\varepsilon^2)]D(U)^{-1} = D[\mathbf{1} - i\varepsilon W_j A(U)^j_i + 0(\varepsilon^2)] \quad ,$$

or in terms of the infinitesimal generators J_i which correspond to (W_i) by Eq. (5–3) one gets :

TABLE 5.1

D[$\mathcal{W}(2)$] denotes the representation of the Lie algebra $\mathcal{W}(2)$ which corresponds to any differentiable representation D [SU(2)] of the group SU(2).

	$\mathcal{W}(2)$	$D[\mathcal{W}(2)]$
Basis	(W_i)	$(J_i = i\frac{\partial}{\partial\varepsilon} D[\mathbf{1} - i\varepsilon W_i + 0(\varepsilon^2)]\big\vert_{\varepsilon=0})$
Structure constants	$-i[W_i,W_j] = c_{ij}{}^k\, W_k$	$-i[J_i,J_j] = c_{ij}{}^k J_k$
Adjoint operator	$\Delta_W W' = -i[W,W']$	$\Delta_J J' = -i[J,J']$
Matrix representative	$\Delta_{W_i} W_j = W_k (\Delta_{W_i})^k_j$	$\Delta_{J_i} = J_k (\Delta_{J_i})^k_j$
	$(\Delta_{W_i})^k_j = c_{ij}{}^k$	$(\Delta_{J_i})^k_j = c_{ij}{}^k$
Killing's form	$U = e^{-iW}$, $U' = e^{-iW'}$	$D(U) = e^{-iJ}$, $D(U') = e^{-iJ'}$
	$(W,W') = -\mathrm{Tr}(\Delta_W \Delta_{W'})$	$=$ $(J,J') = -\mathrm{Tr}(\Delta_J \Delta_{J'})$
Metric tensor	$g_{ij} = (W_i,W_j) = -c_{ik}{}^{\ell}\, c_{j\ell}{}^k$	$=$ (J_i,J_j)
Orthogonal basis	$(W_i = \frac{\sigma_i}{2})$, $-i[\frac{\sigma_i}{2},\frac{\sigma_j}{2}] = \varepsilon_{ijk}\frac{\sigma_k}{2}$	(J_i), $-i[J_i,J_j] = \varepsilon_{ijk} J_k$
	$(\frac{\sigma_i}{2}, \frac{\sigma_j}{2}) = 2\delta_{ij}$	$=$ (J_i,J_j)
Adjoint representations are rotations.	$U = e^{-iW}$, $A(U) = e^{\Delta_W}$	$D(U) = e^{-iJ}$
	$A(U)W' = UW'U^{-1}$	$e^{\Delta_J} J' = D(U)J'D(U)^{-1}$
Matrix representative	$A(U)W_i = UW_iU^{-1} = W_j A(U)^i_j$	$e^{\Delta_J} J_i = D(U)J_iD(U)^{-1} = J_j A(U)^j_i$
	$A(U)\sigma_i = \sigma_j R(U)^j_i$	$e^{\Delta_J} J_i = J_j R(U)^j_i$, $(R(U)^j_i) \in SO(3)$

$$D(U)\; J_i\; D(U)^{-1} = J_j\; A(U)^j_i \quad . \tag{5-44}$$

In a more intrinsic way, this means that the correspondence :

$$U = e^{-iW} \in SU(n) \;\rightarrow\; D(U) = e^{-iJ} \in D[SU(n)] \;\rightarrow e^{\Delta J} \quad \text{over} \quad D[\mathcal{W}(n)]$$

defines a representation of SU(n) in $D[\mathcal{W}(n)]$ equivalent to the adjoint representation. As $e^{\Delta W}$ over $\mathcal{W}(n)$, $e^{\Delta J}$ is a proper isometric application over $D[\mathcal{W}(n)]$, i.e. a rotation. The Eq. (5–44) implies that the matrix representative of $e^{\Delta J}$ with respect to the basis (J_i) is identical to the matrix representative of the adjoint representation $e^{\Delta W}$ with respect to the basis (W_i):

$$e^{\Delta J}\; J_i = D(U) J_i D(U)^{-1} = J_j A(U)^j_i \quad .$$

In particular for n=2, with respect to the commonly used infinitesimal generators defined by Eq. (5–8), which form an orthogonal basis of $D[\mathcal{W}(2)]$ such that the metric tensor is a multiple of the unit matrix, Eq. (5–40), the matrix representative of $e^{\Delta J}$ is a rotation matrix of SO(3) :

$$D(U) J_i D(U)^{-1} = J_j\; R(U)^j_i \quad , \tag{5-45}$$

where the matrix R(U) is related to U by Eq. (4–28). By definition a law of transformation as above characterizes a vector object $J = (J_i)$ (Subsec. 8.5.4,iii)).

The previous results about the correspondence between $\mathcal{W}(2)$ and $D[\mathcal{W}(2)]$ are summed up in Table 5.1.

5.5. STANDARD BASIS OF THE REPRESENTATION SPACE

Let D[SU(2)] be a differentiable unitary representation of SU(2) in the Hermitian vector space E. The infinitesimal generators J_i defined by Eq. (5–8) are Hermitian operators on E. The standard basis of the complex extension of the Lie algebra $D[\mathcal{W}(2)]$ is given by Eq. (5–10).

In connection with the eigenvalue problem of the infinitesimal generators J_i, a standard basis of the representation space E is now derived. Before giving a particular study for the group SU(2), let us point out that the following results are only based on the existence of a standard form of commutation relations (Subsec. 4.3.3,iii)) in the semi–simple Lie algebra of the group (cf. for instance Racah[45]).

The aim of this study is to construct a set of operators which commute with every infinitesimal generator and thus also with every element of the representation of the group. Then using Prop. C.46, a direct consequence of Shur's lemma (Prop. C.45), these operators are multiple of unity on each invariant subspace of irreducible representations (Def. C.32). The set of these operators is said to be complete if it is a smallest one such that the set of their eigenvalues characterizes the irreducible representations.

For the three-dimensional rotation group the Casimir operator is defined by :

$$J^2 = \sum_i J_i^{\,2} = J.J = J_i \, \delta^{ij} J_j \quad , \tag{5-46}$$

where we have used the Euclidean inner product of the formal vector J, Eq. (5–42), with itself. From the commutation relations of the J_i, Eq. (5–9), J^2 commutes with all the infinitesimal generators :

$$[J^2, J_i] = 0 \quad , \tag{5-47}$$

and thus, using Prop. D.1, with all representations $D\,[U(\theta, n^i)] = \exp(-i\theta\, n^i J_i)$:

$$\forall\, U \in SU(2) \qquad [J^2, D(U)] = 0 \quad . \tag{5-48}$$

Any object which commutes with all the elements of a representation of a group is said to be an invariant or a scalar with respect to this group (Sec. 3.3 and Subsec. 8.3.5).

On a semi-simple Lie algebra A, in terms of the metric tensor of the Killing form the Casimir operator reads :

$$G = g^{ij} e_i e_j = - c^{ik}{}_{\ell}\, c^{j\ell}{}_{k}\, e_i e_j \quad , \tag{5-49}$$

where $c_{ij}{}^{k}$ are the structure constants with respect to a basis (e_i) of A . Actually, G does not belong to A since the objects $e_i e_j$ are not defined in the Lie algebra. G is an element of the universal enveloping algebra $\tilde{A}$ of the underlaying Lie algebra A , i.e. $\tilde{A}$ is a formal algebra over the same field as A and generated by the monomials in elements of A which fulfil among others both conditions :

$$\forall\, \tilde{u}, \tilde{v}, \tilde{w} \in \tilde{A} \qquad \tilde{u}(\tilde{v}\,\tilde{w}) = (\tilde{u}\,\tilde{v})\,\tilde{w} \quad , \tag{5-50a}$$

$$\forall\, u, v \in A \qquad [u,v] = u\,v - v\,u = a(u,v) \quad , \tag{5-50b}$$

where A is identified to a subset of $\tilde{A}$.

One shows that on $\tilde{A}$, G commutes with all elements of A :

$$\begin{aligned} [G, e_k] &= g^{ij}[e_i e_j, e_k] = g^{ij}(e_i[e_j,e_k] + [e_i,e_k]e_j) && \text{from Eq. (5-50a)} \\ &= g^{ij}\, c_{jk}{}^{\ell}\, e_i e_\ell + g^{ij}\, c_{ik}{}^{\ell}\, e_\ell e_j && \text{from Eq. (5-50b)} \\ &= (c^{i\,\ell}{}_{k} + c^{\ell\, i}{}_{k})\, e_i e_\ell = 0 \quad , \end{aligned}$$

where the antisymmetrical character of the structure constants has been used (Prop. B.38).

For SU(2), the commonly used Casimir operator J^2 defined by Eq. (5–46) is actually proportional to G given by Eq. (5–49) :

$$J^2 = 2\,G \quad . \tag{5-51}$$

A generalization of the above quadratic Casimir operator is provided by the operators :

$$G_p = c^{i_1 j_1}{}_{j_2}\, c^{i_2 j_2}{}_{j_3} \cdots c^{i_p j_p}{}_{j_1}\, e_{i_1} e_{i_2} \cdots e_{i_p} \quad , \tag{5-52}$$

which commute with every element of A (cf. Racah 45)).

The Hermitian operators J^2 and one of the J_i, for instance J_3, can be simultaneously diagonalized on E (with other operators commuting with J^2 and J_3, however these other operators can be ignored here). Let us denote for the moment by $|\lambda m>$ the normalized eigenvectors in E of J^2 and J_3 with the real eigenvalues, respectively, λ and m :

$$J^2|\lambda m> = \lambda|\lambda m> \quad , \quad J_3|\lambda m> = m|\lambda m> \quad .$$

One then establishes the important results :

(1) $\lambda = j(j+1)$, where 2j is a non-negative integer,

(2) $m = j, j-1, \ldots, -j$.

Proof. In terms of the standard basis $J_\pm$, J_3 of the Lie algebra, the Casimir operator reads :

$$J^2 = \frac{1}{2}(J_+J_- + J_-J_+) + (J_3)^2 \quad , \tag{5-53}$$

and using the commutation relations, Eqs. (5–11a,b), one gets the identities :

$$J_-J_+ = J^2 - J_3(J_3+1) \quad , \tag{5-54a}$$

$$J_+J_- = J^2 - J_3(J_3-1) \quad . \tag{5-54b}$$

Since the J_i are Hermitian operators, $(J_+)^+ = J_-$. The Hermitian norms of vectors $J_+|\lambda m>$ and $J_-|\lambda m>$, respectively, given by :

$$<\lambda m|J_-J_+|\lambda m> = \lambda - m(m+1) \quad , \quad <\lambda m|J_+J_-|\lambda m> = \lambda - m(m-1) \quad , \tag{5-55}$$

have to be non-negative. Therefore, for given non-negative λ (J^2 is a non-negative Hermitian operator), m is bounded as follows :

$$\frac{1}{2} - \frac{\sqrt{1+4\lambda}}{2} \leqslant m \leqslant -\frac{1}{2} + \frac{\sqrt{1+4\lambda}}{2} \quad . \tag{5-56}$$

From the commutation relations one gets :

$$J_3J_\pm|\lambda m> = J_\pm(J_3 \pm 1)|\lambda m> = (m\pm 1)J_\pm|\lambda m> \quad . \tag{5-57}$$

Consequently, if m_i and m_s denote, respectively, the lower and the upper values reached by m and which fulfil Eq. (5–56), one has $m = m_i, m_i+1, \ldots, m_s$, where $m_s - m_i$ is a non-negative integer, and moreover :

$$J_-|\lambda m_i> = 0 \quad , \quad J_+|\lambda m_s> = 0 \quad .$$

The norms given by Eq. (5–55) of the above two null vectors are zero, hence :

$$\lambda = m_i(m_i-1) = m_s(m_s+1) \Rightarrow (m_s+m_i)(m_s-m_i+1) = 0 \quad .$$

Then, $m_s - m_i$ being non-negative, $m_s + m_i$ has to be zero :

$$m_s = -m_i = j \Rightarrow \lambda = j(j+1) \quad , \quad 2j = m_s - m_i \quad .$$

End of the proof.

From now on the eigenvectors of J^2 and J_3 are denoted by $|jm\rangle$:

$$\begin{cases} J^2|jm\rangle = j(j+1)|jm\rangle \quad , & J_3|jm\rangle = m|jm\rangle \quad , \\ 2j \text{ non-negative integer} \quad , & m = j, j-1, \ldots, -j \quad . \end{cases} \tag{5-58}$$

Using the norms given by Eq. (5–55), one has, with a commonly used phase choice (Subsec. 6.2.3.4) :

$$J_\pm|jm\rangle = \sqrt{j(j+1)-m(m\pm 1)}\ |jm\pm 1\rangle \quad , \tag{5-59}$$

which characterizes with Eq. (5–58) a standard basis ($|jm\rangle$). One should note the identities :

$$j(j+1)-m(m\pm 1) = (j\mp m)(j\pm m+1) \quad . \tag{5-60}$$

Actually, the standard basis ($|jm\rangle$) defined above depends upon a special choice of a basis (J_i) of infinitesimal generators in D [$\mathcal{W}(2)$]. This dependence is studied later in Sec. 8.1.

For a detailed study of this problem cf. for instance Racah, loc.cit. Let us only touch on the subject with the same notations as for the remarks in Subsec. 4.3.3,iii).

Let u be simultaneously an eigenvector, in the semi–simple Lie algebra of rank r, of the r commuting infinitesimal generators of the standard basis, Eqs. (4–61a-d) :

$$i=1,\ldots,r \qquad H_i u = m_i u \quad . \tag{5-61}$$

In an r–dimensional vector space, the weight space, the set of r eigenvalues m_i are the covariant components of a vector m called the weight of u. One proves the existence and various properties of weights. For instance, the generalization of Eq. (5–58), using Eq. (4–61b) if u is a vector of weight m, E_α u is a vector of weight $m+\alpha$:

$$H_i E_\alpha u = [H_i, E_\alpha]u + E_\alpha H_i u = (\alpha_i + m_i)E_\alpha u \quad . \tag{5-62}$$

It can be shown that for any semi–simple Lie algebra A of rank r, there exist r invariant operators in $\tilde{A}$, the generalizations of the Casimir operator, which form a complete set, i.e. their eigenvalues characterize the irreducible representations. For the Lie algebra $\mathcal{W}(2)$ of rank one there exists only the quadratic Casimir operator as an invariant. In terms of the standard basis (H_i,E_α) of the algebra, the quadratic Casimir operator reads :

$$G = g^{ij}H_i H_j + \sum_\alpha E_\alpha E_{-\alpha} \quad . \tag{5-63}$$

which corresponds to Eq. (5–53).

The general study of weights leads to a classification of the irreducible representations which is out of the scope of these notes.

5.6. RELATION BETWEEN THE REPRESENTATIONS OF SU(2) AND SO(3)

The two–to–one homomorphism of SU(2) onto SO(3) (Subsec. 4.2.1) implies that the representations of these two groups are closely related :

i) From any representation $D[SO(3)]$ in a vector space E, one defines a representation $\mathcal{D}[SU(2)]$ in E as follows :

$$\forall U \in SU(2) \rightarrow R(U) \in SO(3) \ , \quad \text{Eq. (4-28)} \rightarrow \mathcal{D}(U) = D[R(U)] \quad . \tag{5-64}$$

This is clearly a representation of the group SU(2) such that :

$$\forall\ U \in SU(2) \qquad \mathcal{D}(U) = \mathcal{D}(-U) \quad . \tag{5-65}$$

ii) Conversely, let us consider any representation $\mathcal{D}[SU(2)]$ in a vector space E, one has either :

$$\forall\ U \in SU(2) \qquad \mathcal{D}(U) = \mathcal{D}(-U) \qquad \text{or} \qquad \exists\, U \in SU(2) \qquad \mathcal{D}(U) \neq \mathcal{D}(-U) \quad .$$

In the first case, a representation $D[SO(3)]$ can be defined in E according to :

$$\forall R \in SO(3) \rightarrow \exists \pm U \in SU(2) \ , \ R = R(U) \, , \ \text{Eq. (4-28)} \rightarrow D(R) = \mathcal{D}(U) = \mathcal{D}(-U) \quad . \tag{5-66}$$

Thus, with the result of i), any representation of SO(3) is of this type, i.e. :

$$D[SO(3)] \leftrightarrow \mathcal{D}[SU(2)] \quad , \quad \forall U \in SU(2) \qquad \mathcal{D}(U) = \mathcal{D}(-U) \quad . \tag{5-67}$$

For finite-dimensional irreducible representations $\mathcal{D}[SU(2)]$, one can be more specific. Since $-\mathbf{1}$ commutes with all elements of SU(2), so does $\mathcal{D}(-\mathbf{1})$ with all the elements of $\mathcal{D}[SU(2)]$. Then from Prop. C.46 $\mathcal{D}(-\mathbf{1})$ has to be a multiple of the unity. Because $(-\mathbf{1})^2 = \mathbf{1}$ in SU(2), $\mathcal{D}(-\mathbf{1})^2$ is equal to the unity of $\mathcal{D}[SU(2)]$, then either :

$$\mathcal{D}(-\mathbf{1}) = \mathbf{1} \qquad \text{or} \qquad \mathcal{D}(-\mathbf{1}) = -\mathbf{1} \quad .$$

More generally, for any finite-dimensional irreducible representation $\mathcal{D}(G)$ of a group G, the representations of the elements of the centre of the group are multiples of the unit element of $\mathcal{D}(G)$. In particular for SU(n) the elements U_p of the centre are defined by Eq. (4-2). One has :

$$p=1,\ldots,n \qquad \mathcal{D}(U_p) = \lambda_p \mathbf{1} \quad ,$$

where, since $(U_p)^n = \mathbf{1}$, the possible values of λ_p are the nth roots of the unity.

The irreducible representations $\mathcal{D}[SU(2)]$ such that $\mathcal{D}(-\mathbf{1}) = \mathbf{1}$ are called <u>even representations</u>. One then has :

$$\forall\ U \in SU(2) \qquad \mathcal{D}(-U) = \mathcal{D}(-\mathbf{1}U) = \mathcal{D}(-\mathbf{1})\,\mathcal{D}(U) = \mathcal{D}(U) \quad . \tag{5-68}$$

Therefore to any even representation there corresponds according to Eq. (5-66) a representation of SO(3).

The representations $\mathcal{D}[SU(2)]$ such that $\mathcal{D}(-\mathbf{1}) = -\mathbf{1}$ and thus for any U in SU(2) one has $\mathcal{D}(-U) = -\mathcal{D}(U)$, are called <u>odd representations</u>. From them one can only define double–valued representations of SO(3), i.e. to any element R of SO(3) there correspond two operators which differ by a sign :

$$\forall R \in SO(3) \qquad \exists U \in SU(2) \quad , \quad R = R(U) = R(-U) \quad , \quad \text{Eq. (4-28)} \quad ,$$

$$D(R) = \begin{cases} D[R(U)] = \mathcal{D}(U) \\ D[R(-U)] = \mathcal{D}(-U) = -\mathcal{D}(U) \quad . \end{cases} \tag{5-69}$$

The double–valued representations are a special case of the representations of a group up to a phase, defined in Subsec. 3.2.3,i), and for which the group law, Eq. (3–24), reads :

$$\forall R_1, R_2 \in SO(3) \qquad D(R_1)D(R_2) = \pm D(R_1R_2) \quad . \tag{5-70}$$

It is shown in Subsec. 6.1.2,iii) that the even and odd representations of SU(2) correspond, respectively, to integer and half–odd–integer values of j in Eq. (5–58).

FORMULAE FOR CHAPTER 5

Infinitesimal generators of a differentiable representation D [SU(2)]

$$\forall W \in \mathcal{W}(2) \qquad D[\mathbf{1}-i\varepsilon W + 0(\varepsilon^2)] = \mathbf{1}-i\varepsilon J+0(\varepsilon^2) \quad , \; J = i\frac{\partial}{\partial\varepsilon} D[\mathbf{1}-i\varepsilon W + 0(\varepsilon^2)]\Big|_{\varepsilon=0}$$

$$\sigma_i \rightarrow J_i = i\frac{\partial}{\partial\varepsilon} D[\mathbf{1}-i\frac{\varepsilon}{2}\sigma_i + 0(\varepsilon^2)]\Big|_{\varepsilon=0} \quad , \; [J_i,J_j] = i\sum_k \varepsilon_{ijk} J_k$$

$$(\sigma_\pm,\sigma_3) \rightarrow (J_\pm = J_1 \pm iJ_2, J_3) \quad , \quad [J_3,J_\pm] = \pm J_\pm \quad , \quad [J_+,J_-] = 2J_3 \quad .$$

Exponential form and consequences of the Baker–Campbell–Hausdorff formula

$$U = U(\theta,n^i) = e^{-i\frac{\theta}{2}n^i\sigma_i} \quad \Longrightarrow \quad D(U) = e^{-i\theta n^i J_i}$$

$$SU(2) \rightarrow D[SU(2)] \quad \Longleftrightarrow \quad \mathcal{W}(2) \rightarrow D[\mathcal{W}(2)] \quad .$$

Adjoint representation

$$\forall U \in SU(2) \rightarrow A(U) \quad , \quad \forall W \in \mathcal{W}(2) \xrightarrow{A(U)} A(U)W = U\,W\,U^{-1}$$

$$A(U) = A(-U)$$

$$A[\mathbf{1}-i\varepsilon W + 0(\varepsilon^2)] = \mathbf{1}-i\varepsilon J + 0(\varepsilon^2) \quad , \quad J = i\,\Delta_W$$

$$U = e^{-iW} \Rightarrow A = e^{\Delta_W} \quad .$$

A(U) is a rotation over the Euclidean space $\mathcal{W}(2)$:

$$A(U)\sigma_i = U\,\sigma_i\,U^{-1} = \sigma_j \cdot R(U)^j_{\;i} \quad , \quad (R(U)^j_{\;i}) \in SO(3) \quad .$$

For any differentiable representation D[SU(2)], and $\sigma_i \rightarrow J_i$:

$$D(U)J_i\,D(U)^{-1} = J_j\,R(U)^j_{\;i} \quad .$$

Standard basis of the representation space

Casimir operator J^2 :

$$J^2 = \sum_i (J_i)^2 = \frac{1}{2}(J_+J_- + J_-J_+) + (J_3)^2 = J_-J_+ + J_3(J_3+1) = J_+J_- + J_3(J_3-1)$$

$$[J^2,J_i] = [J^2,J_\pm] = 0 \quad .$$

Standard basis ($|jm>$; $j=0,\frac{1}{2},1,\ldots$, $m=j,j-1,\ldots,-j$) :

$$J_\pm|jm> = \sqrt{j(j+1)-m(m\pm1)}\,|jm\pm1> = \sqrt{(j\mp m)(j\pm m+1)}\,|jm\pm1> \quad , \quad J_3|jm> = m|jm>$$

$$J^2|jm> = j(j+1)|jm> \quad .$$

CHAPTER 6

UNITARY AND IRREDUCIBLE REPRESENTATIONS OF SU(2)

Assuming that the three-dimensional rotation group is a symmetry group, one is led to construct in Quantum Mechanics the unitary representations of the group SU(2) on the considered Hilbert space (Sec. 3.4). Actually, as for finite groups, any representation of a compact Lie group in a vector space over the complex field is equivalent to a unitary representation (Prop. C.54 and Theorem 7.8 in Subsec. 7.2.1,i)). This is mainly due to the existence of an invariant measure on the group (Sec. 7.1).

This chapter closely follows a work due to Bargmann[5)]. The method consists in constructing a unitary representation of SU(2) in a suitably chosen vector space, which yields irreducible unitary representations (Sec. 6.1). One then shows that, up to an equivalence, there are no other irreducible unitary representations of SU(2). Moreover, this special choice of the representation space allows us to give a simple proof of the "angular momentum" addition theorem in Quantum Mechanics (Sec. 6.2). A usual proof of this theorem is also recalled. One should note the different steps of the study of this special representation since they are general and used again in Chap. 10.

Notations. Throughout the remainder of this text :

(1) The up-and-down index notation is no longer used since it is unusual for the rotation matrices ($\mathcal{D}^j_{mm'}$).

(2) The Einstein convention is used for the indices m but not for the j, i.e. the sums over the indices j are always explicited. Nevertheless, any equality where there exist free indices j holds for all values of these indices.

6.1. THE UNITARY AND IRREDUCIBLE REPRESENTATIONS OF SU(2)

6.1.1. REPRESENTATION SPACE AND DETERMINATION OF THE REPRESENTATION

i) Let $\zeta = \begin{pmatrix} \xi \\ \eta \end{pmatrix}$ be a two complex component spinor which belongs to the two-dimensional complex space C^2. One considers the space F of entire functions of the complex variables ξ and η. Since any entire function ϕ can be expanded in an everywhere converging power series :

$$\phi(\zeta) = \sum_{j,m} \phi_{jm}\, \xi^{j+m}\, \eta^{j-m} \quad , \qquad \phi_{jm} \in C \quad , \tag{6-1}$$

where j and m take, respectively, the values :

$$j=0,\tfrac{1}{2},1,\ldots \qquad , \qquad m=j,j-1,\ldots,-j \quad , \tag{6-2}$$

the set of monomials ($\xi^{j+m}\ \eta^{j-m}$) forms a complete basis of the infinite-dimensional vector space F over the complex field.

ii) A Hermitian inner product of two elements ϕ and ϕ' in F is defined by :

$$(\phi,\phi') = \int d\mu(\zeta)\ \phi^*(\zeta)\ \phi'(\zeta) \quad , \tag{6-3}$$

where the integral is extended over the whole space C^2 and :

$$\zeta^+\zeta = |\xi|^2+|\eta|^2 \quad , \quad \xi = \xi'+i\xi'' \ , \qquad \eta = \eta'+i\eta'' \ , \qquad \xi',\xi'',\eta',\eta'' \in R \ ,$$

$$d\mu(\zeta) = \pi^{-2}\ e^{-\zeta^+\zeta}\ d\xi' d\xi'' d\eta' d\eta'' \quad . \tag{6-4}$$

The metric tensor with respect to the basis ($\xi^{j+m}\ \eta^{j-m}$) is defined by the inner product of any two basis vectors :

$$(\xi^{j+m}\ \eta^{j-m},\ \xi^{j'+m'}\ \eta^{j'-m'}) = \delta_{jj'}\ \delta_{mm'}\ (j+m)!\,(j-m)! \quad . \tag{6-5}$$

Therefore the set of functions e^j_m :

$$e^j_m(\zeta) = [(j+m)!\,(j-m)!]^{-\frac{1}{2}}\ \ \xi^{j+m}\ \eta^{j-m} \quad , \tag{6-6}$$

where j and m take, respectively, the values specified in Eq. (6–2), forms a complete orthonormal basis of F.

One shows that the vector space F, endowed with the finite norm $\|\phi\|=\sqrt{(\phi,\phi)} < \infty$, is a Hilbert space (indeed isomorphic to the space of complex sequences ϕ_{jm} such that $\sum_{j,m} |\phi_{jm}|^2 < \infty$).

For a given value of j, complex linear combinations of the e^j_m generate a 2j+1–dimensional subspace F_j of F, i.e. the subspace of complex coefficient homogeneous polynomials of degree 2j in ξ and η. Clearly F is the direct sum of these subspaces F_j :

$$F = \sum_{j=0}^{\infty} \oplus\ F_j \quad . \tag{6-7}$$

iii) Now let us define a representation of the group SU(2) in F. To any element U in SU(2) one associates the mapping T_U of F onto itself according to :

$$\forall\ \phi \in F \qquad \forall\ \zeta \in C^2 \qquad (T_U\phi)(\zeta) = \phi(U^t\zeta) \quad . \tag{6-8}$$

This application is clearly linear and defines a representation of SU(2) since one has :

$$\forall\ U,U' \in SU(2) \qquad T_U T_{U'} = T_{UU'} \quad , \qquad T_U^{-1} = T_{U^{-1}} \quad . \tag{6-9}$$

Moreover, this representation is unitary since T_U is an isometric application. Indeed, the measure $d\mu(\zeta)$ is invariant under unitary transformations of ζ, and thus the change of the integration variable ζ into $U^t\zeta$ yields :

$$\forall\ U \in SU(2) \qquad \forall\ \phi,\phi' \in \mathcal{F} \qquad (T_U\phi, T_U\phi') = (\phi,\phi') \quad . \tag{6-10}$$

Finally, for all U in SU(2), T_U defines a unitary representation of SU(2) in $\mathcal{F}$. This representation is clearly reducible : the mapping $\zeta \to U^t\zeta$ being homogeneous in the variables ξ and η, each of the subspaces $\mathcal{F}_j$ is as a whole invariant under the mappings T_U. Thus, from Eq. (6–7) this representation is the direct sum of finite–dimensional representations $\mathcal{D}^j[SU(2)]$ ($\mathcal{D}^j$ in short) which are, respectively, defined by the restrictions of the mappings T_U to the subspaces $\mathcal{F}_j$.

6.1.2. THE MATRICES $\mathcal{D}^j$

i) The matrix representative of the restriction of T_U to the subspace $\mathcal{F}_j$ with respect to the basis (e^j_m) defines the <u>rotation matrix elements</u> $\mathcal{D}^j_{m'm}(U)$:

$$\forall\ U \in SU(2) \qquad T_U\, e^j_m = e^j_{m'}\, \mathcal{D}^j_{m'm}(U) \quad . \tag{6-11}$$

From these matrix elements one uniquely defines a <u>rotation matrix</u> $\mathcal{D}^j(U)$ by choosing a definite order of the rows and columns, respectively labelled by m' and m, in the following way :

$$\mathcal{D}^j(U) = \begin{array}{c|c} m' \backslash m & j,j-1,\dots,-j \\ \hline j & \\ j-1 & \mathcal{D}^j_{m'm}(U) \\ \vdots & \\ -j & \end{array} \quad . \tag{6-12}$$

For a given value of j the matrices $\mathcal{D}^j(U)$ form thus a finite–dimensional unitary matrix representation of the group SU(2), and in this representation Eq. (6–9) reads :

$$\forall\ U,U' \in SU(2) \qquad \mathcal{D}^j(U)\mathcal{D}^j(U') = \mathcal{D}^j(UU') \quad , \tag{6-13a}$$

$$[\mathcal{D}^j(U)]^{-1} = \mathcal{D}^j(U^{-1}) = [\mathcal{D}^j(U)]^+ \quad . \tag{6-13b}$$

ii) The explicit form of the matrix element $\mathcal{D}^j_{m'm}(U)$ is obtained from the expression of the basis vectors given by Eq. (6–6) and from the definition, Eq. (6–8), of the representation :

$$\forall\ \zeta \in C^2 \qquad (T_U\, e^j_m)(\zeta) = e^j_m(U^t\zeta) = e^j_{m'}(\zeta)\, \mathcal{D}^j_{m'm}(U) \quad .$$

1–Setting $U = \begin{pmatrix} \alpha & \beta \\ \gamma & \delta \end{pmatrix}$ one gets :

$$e^j_m(U^t\zeta) = [\,(j+m)!\,(j-m)!\,]^{-\frac{1}{2}} \quad (\alpha\xi + \gamma\eta)^{j+m}\ (\beta\xi + \delta\eta)^{j-m} \quad ,$$

Expanding this expression in power of ξ and η, the matrix element $\mathcal{D}^j_{m'm}(U)$ is the component in this expansion on the basis vector $e^j_{m'}$. The result is :

$$\mathcal{D}^j_{m'm}(U) = [(j+m)!(j-m)!(j+m')!(j-m')!]^{\frac{1}{2}} \sum_{p+q=j-m} \frac{\alpha^{j+m'-p}\beta^p\gamma^{j-m'-q}\delta^q}{(j+m'-p)!p!(j-m'-q)!q!} \quad . \tag{6-14}$$

One sees that the matrices $\mathcal{D}^j(U)$ fulfil the relation :

$$\forall\ U \in SU(2) \qquad \mathcal{D}^j(U^*) = [\mathcal{D}^j(U)]^* \quad , \tag{6-15}$$

and then, from the unitary condition, Eq. (6–13b) :

$$\forall\ U \in SU(2) \qquad \mathcal{D}^j(U^t) = [\mathcal{D}^j(U)]^t \quad . \tag{6-16}$$

2–When U is parametrized by θ , n^i the values of α , β , γ , and δ are given by Eq. (4–19). Using the notation $U(\theta,\hat{n};a)$, Eq. (4–17), one obtains for the commonly encountered cases $\hat{n}$ = e_3 and e_2, respectively :

$$\mathcal{D}^j_{m'm}[U(\theta,e_3;a)] = e^{-im\theta}\,\delta_{m'm} \quad , \tag{6-17}$$

$$\mathcal{D}^j_{m'm}[U(\theta,e_2;a)] = d^j_{m'm}(\theta) = \tag{6-18}$$

$$[(j+m')!(j-m')!(j+m)!(j-m)!]^{\frac{1}{2}} \sum_q (-1)^{q+j-m} \frac{(\cos\frac{\theta}{2})^{2q+m'+m}(\sin\frac{\theta}{2})^{2j-2q-m'-m}}{(j-m'-q)!q!(q+m'+m)!(j-m-q)!} \quad .$$

One should remark that the matrix elements $d^j_{m'm}(\theta)$ are real homogeneous polynomials of degree 2j in $\cos(\frac{1}{2}\theta)$ and $\sin(\frac{1}{2}\theta)$. The matrix $d^j(\theta)$ is defined by the same order of the rows and columns as for $\mathcal{D}^j(U)$ in Eq. (6–12).

The definitions for the matrices $\mathcal{D}^j$ and d^j adopted by different authors are given in Appendix F.1. The matrices d^j for $j = \frac{1}{2}$, 1, and $\frac{3}{2}$ are listed in Appendix F.2. It is shown in Subsec. 8.5.2 how both three–dimensional rotation matrices $(R^j_{\ i})$ and $(\mathcal{D}^1_{m'm})$ are actually related by a simple change of basis in the complex extension of a three–dimensional Euclidean vector space. The relation between $d^j_{m'm}$ and the Jacobi polynomial is given in Appendix F.3.

In order to compute the rotation matrix $\mathcal{D}^j[U(\theta,e_1;a)]$ let us apply Eq. (4–34), $U(\theta,e_1;a)$ can then be written as :

$$U(\theta,e_1;a) = U[\theta,R(\tfrac{\pi}{2},-e_3)e_2;a] = U(\tfrac{\pi}{2},-e_3;a)U(\theta,e_2;a)U(\tfrac{\pi}{2},-e_3;a)^{-1} \quad ,$$

or, from Eq. (4–21) :

$$U(\theta,e_1;a) = U(\tfrac{\pi}{2},e_3;a)^{-1}U(\theta,e_2;a)U(\tfrac{\pi}{2},e_3;a) \quad .$$

Let us remark that if one uses the similar Eq. (2–38) for the rotation $R(\theta,e_1)$ and the two–to–one homomorphism of SU(2) onto SO(3), we only get the previous expression of $U(\theta,e_1;a)$ up to a sign.

Now, the result follows from Eqs. (6–17,18) :

$$\mathcal{D}^j_{m'm}\,[U(\theta,e_1;a)] = e^{i\frac{\pi}{2}(m'-m)}\, d^j_{m'm}(\theta) \quad . \tag{6-19}$$

One also uses the notations :

$$\mathcal{D}^j_{m'm}\,[U(\theta,\hat{n};a)] = \mathcal{D}^j_{m'm}(\theta,\hat{n};a) = \mathcal{D}^j_{m'm}(\theta,n^i) \quad , \tag{6-20}$$

never forgetting the remarks about the nonintrinsic character of the vector $\hat{n}$, cf. Eq. (4–17).

3–When U is parametrized by the Euler angles, Eq. (4–38) yields, with also a simplified notation :

$$\mathcal{D}^j_{m'm}\,[U(\varphi,\theta,\psi)] = \mathcal{D}^j_{m'm}(\varphi,\theta,\psi) = e^{-im'\varphi}\, d^j_{m'm}(\theta)\, e^{-im\psi} \quad . \tag{6-21}$$

From the previous formulae one derives several properties of the matrices d^j and $\mathcal{D}^j$ which are now summarized :

$$d^j_{m'm}(\theta) = d^{j*}_{m'm}(\theta) = d^j_{mm'}(-\theta) = (-1)^{m-m'}\, d^j_{mm'}(\theta) = (-1)^{m-m'}\, d^j_{-m'-m}(\theta)\,, \tag{6-22a}$$

$$d^\ell_{00}(\theta) = P_\ell(\cos\theta) \qquad \ell=0,1,\ldots \quad , \tag{6-22b}$$

$$d^j_{m'm}(\pi) = (-1)^{j-m}\,\delta_{m',-m} \quad , \qquad d^j(2\pi) = (-1)^{2j}\,\mathbf{1} \quad , \tag{6-22c}$$

$$\mathcal{D}^j_{m'm}(U^{-1}) = \mathcal{D}^{j*}_{mm'}(U) = (-1)^{m-m'}\,\mathcal{D}^j_{-m-m'}(U) \quad , \tag{6-23a}$$

$$\mathcal{D}^j_{m'm}(\varphi,\theta,\psi) = (-1)^{m'-m}\,\mathcal{D}^{j*}_{m'm}(-\varphi,-\theta,-\psi) \quad . \tag{6-23b}$$

iii) Remarks :

1–The matrix representations $\mathcal{D}^j(U)$ and $\mathcal{D}^j(-U)$ are related by :

$$\forall\, U \in SU(2) \qquad \mathcal{D}^j(U) = (-1)^{2j}\,\mathcal{D}^j(-U) \quad . \tag{6-24}$$

Hence, for integer values $\ell = 0,1,\ldots$ of j the even representations $\mathcal{D}^\ell$ [SU(2)] define representations of the group SO(3), Eq. (5–66), denoted by :

$$\forall\, R \in SO(3) \quad \exists \pm U \in SU(2)\,, \quad R = R(\pm U)\,, \ \text{Eq. (4-28)} \rightarrow \mathcal{D}^\ell(R) = \mathcal{D}^\ell(\pm U) \quad . \tag{6-25}$$

On the other hand, for half-odd-integer values $\frac{1}{2}$, $\frac{1}{2}+1$, ... of j, the odd representations $\mathcal{D}^j$ [SU(2)] only define double-valued representations of SO(3), Eq. (5–69).

2–A direct calculation from Eq. (6–14) gives :

$$\forall\, U \in SU(2) \qquad U = \mathcal{D}^{\frac{1}{2}}(U) \quad . \tag{6-26}$$

Whence, SU(2) is itself its smallest-dimensional representation, also called the <u>fundamental representation</u> of the group.

3–Let us now prove that the representation $\mathcal{D}^j$ and its complex conjugate $\mathcal{D}^{j*}$ are equivalent. This result also follows from the completeness of the representations $\mathcal{D}^j$ (Subsec. 6.1.5,ii)). The general problem of conjugate complex representations is studied later in Sec. 8.2. In the representation $\mathcal{D}^j$ Eq. (4–99) reads :

$$\forall U \in SU(2) \qquad \mathcal{D}^{j*}(U) = \mathcal{D}^j(C)\, \mathcal{D}^j(U)\, [\mathcal{D}^j(C)]^{-1} \quad , \tag{6-27}$$

where, from Eqs. (4–105) and (6–18) the matrix $\mathcal{D}^j(C)$ is given by :

$$\mathcal{D}^j_{m'm}(C) = (-1)^{j+m}\, \delta_{m',-m} \quad . \tag{6-28}$$

General properties of this matrix are given in Subsec. 8.2.2.1,i).

6.1.3. INFINITESIMAL GENERATORS

i) The infinitesimal generators J_i which correspond to the basis (σ_i) of the Lie algebra of SU(2) are defined by Eq. (5–8) :

$$J_i = i \frac{\partial}{\partial \varepsilon} T_{\mathbf{1} - i \frac{\varepsilon}{2} \sigma_i + O(\varepsilon^2)} \Big|_{\varepsilon=0} \quad . \tag{6-29}$$

Their expressions in terms of derivatives with respect to the variables ξ and η follow from the definition of the representation :

$$\forall \phi \in F \qquad \forall \zeta \in C^2 \qquad \left[T_{\mathbf{1}-i \frac{\varepsilon}{2} \sigma_i + O(\varepsilon^2)} \phi\right](\zeta) = \phi\{[\mathbf{1} - i \tfrac{\varepsilon}{2} \sigma_i{}^t + O(\varepsilon^2)]\zeta\} \ .$$

One gets :

$$J_i = \zeta^t \frac{\sigma_i}{2} \frac{\partial}{\partial \zeta} = (\xi, \eta) \frac{\sigma_i}{2} \begin{pmatrix} \frac{\partial}{\partial \xi} \\ \frac{\partial}{\partial \eta} \end{pmatrix} \quad , \tag{6-30a}$$

$$J_1 = \tfrac{1}{2}(\xi \frac{\partial}{\partial \eta} + \eta \frac{\partial}{\partial \xi}) \quad , \quad J_2 = \tfrac{i}{2}(\eta \frac{\partial}{\partial \xi} - \xi \frac{\partial}{\partial \eta}) \quad , \quad J_3 = \tfrac{1}{2}(\xi \frac{\partial}{\partial \xi} - \eta \frac{\partial}{\partial \eta}) \quad , \tag{6-30b}$$

$$J_+ = J_1 + iJ_2 = \xi \frac{\partial}{\partial \eta} \quad , \qquad J_- = J_1 - iJ_2 = \eta \frac{\partial}{\partial \xi} \quad . \tag{6-30c}$$

It can be checked by a direct calculation that the basis vectors e^j_m , Eq. (6–6), fulfil the characteristic equations of a standard basis i.e. :

$$J_\pm\, e^j_m = \sqrt{j(j+1) - m(m\pm 1)}\ e^j_{m\pm 1} \quad , \tag{6-31a}$$

$$J_3\, e^j_m = m\, e^j_m \quad . \tag{6-31b}$$

Then, the Casimir operator, Eq. (5–46), of this representation reads :

$$J^2 = \frac{N}{2}(\frac{N}{2} + 1) \quad , \qquad N = \xi \frac{\partial}{\partial \xi} + \eta \frac{\partial}{\partial \eta} \quad .$$

Since the e^j_m are homogeneous functions of degree 2j in ξ and η :

$$N\, e^j_m = 2j\, e^j_m \quad ,$$

and one finds again the relation :

$$J^2\, e^j_m = j(j+1)\, e^j_m \quad , \tag{6-32}$$

which follows from the characteristic Eqs. (6–31a,b) of a standard basis. Therefore, any function ϕ which is a linear combination of the e^j_m for a given value of j, i.e. which belongs to the subspace F_j, fulfils the equation :

$$J^2\phi = j(j+1)\phi \quad .$$

Another way to get the expressions given in Eqs. (6–31a,b) is to compute the matrix $\mathcal{D}^j(U)$ up to first order in ε for an infinitesimal $U = \mathbf{1} - i\frac{\varepsilon}{2}\hat{n}.\sigma + O(\varepsilon^2)$. From Eqs. (4–92) and (6–14) one immediately gets :

$$\mathcal{D}^j_{m'm}(U) = \delta_{m'm} - i\varepsilon\left[\frac{n_+}{2}\sqrt{(j+m)(j-m+1)}\,\delta_{m'm-1} + \frac{n_-}{2}\sqrt{(j-m)(j+m+1)}\,\delta_{m'm+1} + n^3 m\,\delta_{m'm}\right] + O(\varepsilon^2) \, . \tag{6-33}$$

Hence, the matrix representatives in the basis (e^j_m) of the infinitesimal generators $J_\pm$ and J_3 read :

$$J_\pm e^j_m = e^j_{m'}(J_\pm)_{m'm} \quad , \quad (J_\pm)_{m'm} = \sqrt{(j\mp m)(j\pm m+1)}\;\delta_{m'\,m\pm1} \quad , \tag{6-34a}$$

$$J_3 e^j_m = e^j_{m'}(J_3)_{m'm} \quad , \quad (J_3)_{m'm} = m\,\delta_{m'\,m} \quad . \tag{6-34b}$$

ii) Let us remark that conversely if basis vectors e^j_m fulfil the Eqs. (6–31a,b), where the J_i are always the infinitesimal generators which correspond to the σ_i by Eq. (5–8), then for all U in SU(2) the matrix representative of T_U with respect to this basis is the matrix $\mathcal{D}^j(U)$. This follows from the existence of an exponential form, Eq. (5–13) :

$$T_{U(\theta,n^i)} = e^{-i\theta n^i J_i} \quad .$$

6.1.4. COMPARISON WITH THE SCHWINGER METHOD

The Hilbert space method of Bargmann explained above is actually a realization of the Schwinger operator method[55)] more abstractly defined. It is shown that these two methods are rigorously equivalent (Bargmann[51)], Sec. 2e).

In the Schwinger method the representations $\mathcal{D}^j$ are generated by symmetrized tensorial product of the fundamental representation $\mathcal{D}^{\frac{1}{2}}[SU(2)]$, by means of the Fock construction. One considers in second quantization Bose Einstein spin particles which can exist in two different states labelled by $\mu = \pm 1$ (the elementary spin $\frac{1}{2}$ particles). The spin creation and annihilation operators, respectively a^+_μ and a_μ, fulfil the commutation relations for bosons :

$$[a_\mu, a^+_{\mu'}] = \delta_{\mu\,\mu'} \quad , \quad [a_\mu, a_{\mu'}] = [a^+_\mu, a^+_{\mu'}] = 0 \quad . \tag{6-35}$$

The particle number operators are :

$$N_+ = a^+_+ a_+ \quad , \quad N_- = a^+_- a_- \quad , \quad N = N_+ + N_- = \sum_{\mu=\pm1} a^+_\mu a_\mu \quad . \tag{6-36}$$

A direct calculation shows that the three operators defined by :

$$i=1,2,3 \qquad J_i = a^+ \frac{\sigma_i}{2} a = (a^+_+, a^+_-) \frac{\sigma_i}{2} \begin{pmatrix} a_+ \\ a_- \end{pmatrix} \quad , \tag{6-37}$$

fulfil the usual commutation relations of angular momentum, Eq. (5-9). The operators $J_\pm$ and J_3 read, respectively :

$$J_+ = J_1 + iJ_2 = a^+_+ a_- \quad , \quad J_- = J_1 - iJ_2 = a^+_- a_+ \quad , \tag{6-38a}$$

$$J_3 = \frac{1}{2}(a^+_+ a_+ - a^+_- a_-) = \frac{1}{2}(N_+ - N_-) \quad . \tag{6-38b}$$

To compute J^2, one uses the relation given by Eq. (4-94) :

$$J^2 = \sum_{i=1}^{3} (J_i)^2 = \frac{1}{4} \sum_{i=1}^{3} \sum_{\mu,\mu';\nu,\nu'=\pm1} a^+_\mu (\sigma_i)^\mu_{\mu'} a_{\mu'} a^+_\nu (\sigma_i)^\nu_{\nu'} a_{\nu'}$$

$$= \frac{1}{4} \sum_{\mu,\mu'=\pm1} (2a^+_\mu a_{\mu'} a^+_{\mu'} a_\mu - a^+_\mu a_\mu a^+_{\mu'} a_{\mu'}) \quad .$$

From the commutation relations, one has :

$$a^+_\mu a_{\mu'} a^+_{\mu'} a_\mu = a^+_\mu a_\mu a^+_{\mu'} a_{\mu'} + a^+_\mu a_\mu (1-\delta_{\mu\,\mu'}) \quad .$$

Thus, in terms of the particle number operator N, J^2 reads :

$$J^2 = \frac{N}{2}\left(\frac{N}{2} + 1\right) \quad . \tag{6-39}$$

Therefore, a state in the Fock space with a given number n=0,1,2,... of spin particles is an eigenstate of J^2 with $j = \frac{1}{2}\, n = 0, \frac{1}{2}, 1, \ldots$. Moreover, from Eqs. (6-38a,b), a state with given numbers n_+ and n_- of particles with μ equal, respectively, + 1 and − 1, is an eigenstate of J^2 and J_3 with eigenvalues, respectively :

$$j(j+1) \quad , \quad j = \frac{1}{2}(n_+ + n_-) \quad ; \quad m = \frac{1}{2}(n_+ - n_-) \quad . \tag{6-40}$$

Let us denote by $|0\rangle$ the vacuum in the Fock space :

$$a_\mu |0\rangle = 0 \quad .$$

Then, from the normalized state with prescribed occupation numbers n_+ and n_- :

$$|n_+n_-\rangle = \frac{(a_+^+)^{n_+}\,(a_-^+)^{n_-}}{\sqrt{n_+!\,n_-!}}\,|0\rangle \quad ,$$

and from Eq. (6–40), one obtains the normalized eigenstates of J^2 and J_3 :

$$|jm\rangle = \frac{(a_+^+)^{j+m}\,(a_-^+)^{j-m}}{\sqrt{(j+m)!\,(j-m)!}}\,|0\rangle \quad , \qquad (6\text{-}41)$$

which correspond to the functions e_m^j defined by Eq. (6–6).

The correspondence between Schwinger's and Bargmann's formalisms is the following, comparing Eqs. (6–30a,37) :

$$a_+ \longrightarrow \frac{\partial}{\partial\xi} \quad , \qquad a_+^+ \longrightarrow \xi \quad ,$$
$$a_- \longrightarrow \frac{\partial}{\partial\eta} \quad , \qquad a_-^+ \longrightarrow \eta \quad . \qquad (6\text{-}42)$$

Since a direct calculation shows that the action of the operators a_μ^+ and a_μ on $|jm\rangle$ is the same as the action of the corresponding derivative operators $\frac{\partial}{\partial\xi}$, $\frac{\partial}{\partial\eta}$ and the multiplicative operators ξ, η on e_m^j , this establishes the equivalence between the two methods.

6.1.5. IRREDUCIBILITY AND COMPLETENESS OF REPRESENTATIONS $\mathcal{D}^j$

i) **Irreducibility.** From Props. C.44,50, to prove the irreducibility of the unitary representation $\mathcal{D}^j[SU(2)]$ for a given j, it suffices to show that any 2j+1–dimensional matrix M which commutes with all the matrices $\mathcal{D}^j(U)$ is a multiple of the unit matrix. It is then sufficient to show that any linear mapping, also denoted M, of F_j onto itself, which commutes with a basis of infinitesimal generators, is a multiple of the unit operator. If M commutes with J_3, from Eq. (6–31b) its matrix representative with respect to the basis (e_m^j) of F_j takes the diagonal form :

$$M_{m'm} = \lambda_m\,\delta_{m'm} \quad .$$

Moreover, if M commutes with $J_\pm$, λ_m has to be independent of m, namely $M = \lambda\,\mathbf{1}$, which ends the proof.

ii) **Completeness.** By this it is meant that up to an equivalence, the representations $\mathcal{D}^j[SU(2)]$ are the only finite–dimensional unitary and irreducible representations of the group SU(2). The proof of this theorem is given later in Subsec. 7.2.2.3.

iii) **Consequences.** The basis (e_m^j) is a standard basis of the representations $\mathcal{D}^j[SU(2)]$ (Subsec. 6.1.3), and by definition the $\mathcal{D}^j_{m'm}(U)$ are the matrix elements of the isometric mapping T_U with respect to this basis. Therefore, in any unitary representation D[SU(2)], the matrix elements of the unitary operator D(U) with respect to the standard basis $(|\alpha\, jm\rangle)$, of the representation (where α denotes other possible quantum numbers) are the $\mathcal{D}^j_{m'm}(U)$:

$$D(U)|\alpha jm\rangle = |\alpha jm'\rangle\, \mathcal{D}^j_{m'm}(U) \quad . \tag{6-43}$$

Using the exponential forms for D(U) given by Eqs. (5–13,15), one has for the rotation matrix elements from Eqs. (6–18,21) :

$$d^j_{m'm}(\theta) = \langle\alpha jm'|\, e^{-i\theta J_2} |\alpha jm\rangle \quad , \tag{6-44a}$$

$$\mathcal{D}^j_{m'm}(\varphi,\theta,\psi) = \langle\alpha jm'|\, e^{-i\varphi J_3}\, e^{-i\theta J_2}\, e^{-i\psi J_3} |\alpha jm\rangle \quad , \tag{6-44b}$$

where the J_i are the infinitesimal generators defined by Eq. (5–8) of the considered representation.

6.1.6. CHARACTERS OF REPRESENTATIONS

The character of a finite-dimensional representation D[SU(2)] is defined by Def. C.31 :

$$\forall\, U \in SU(2) \qquad \chi(U) = \mathrm{Tr}\, D(U) \quad . \tag{6-45}$$

Actually the character of a representation is a class function (Prop. C.43), i.e. a function which takes the same value for all the elements of a given class of conjugate elements under the group. Since for SU(2) these classes are the subsets of unitary matrices $U(\theta, n^i)$ with the same values of θ (Subsec. 4.2.2), the characters of the representations of SU(2) are only functions of θ :

$$\forall\, \hat{n} \qquad \chi(\theta) = \chi[U(\theta,n^i)] = \mathrm{Tr}\, D[U(\theta,n^i)] \quad . \tag{6-46}$$

The calculation of the character $\chi^j(\theta)$ of the irreducible representation $\mathcal{D}^j[SU(2)]$ is the simplest on taking $\hat{n} = e_3$, then (Eq. 6–17) yields :

$$\chi^j(\theta) = \sum_{m=-j}^{+j} e^{-im\theta} = \frac{\sin(\frac{2j+1}{2}\theta)}{\sin\frac{\theta}{2}} = \begin{cases} 1+2\cos\theta+\ldots+2\cos j\theta & \text{integer } j \\ 2\cos\frac{\theta}{2}+2\cos\frac{3\theta}{2}+\ldots+2\cos j\theta & \text{half-odd-integer } j. \end{cases} \tag{6-47}$$

Either directly from Eq. (4–21), or by using the reality of the character, one gets (Subsec. 8.2.1,i)) :

$$U=U(\theta,n^i) \quad , \quad \chi^j(U^{-1}) = \chi^j(U)^* = \chi^j(U) = \chi^j(\theta) \quad . \tag{6-48}$$

Similarly, either directly from Eq. (4–21) and the expression of $\chi^j(\theta)$, or using Eq. (6–24), one obtains :

$$U=U(\theta,n^i) \quad , \quad \chi^j(-U)=\chi^j(2\pi-\theta)=(-1)^{2j}\,\chi^j(U) = (-1)^{2j}\,\chi^j(\theta) \quad . \tag{6-49}$$

6.2. DECOMPOSITION OF THE DIRECT PRODUCT OF REPRESENTATIONS

6.2.1. THE PRODUCT REPRESENTATION $\mathcal{D}^{j_1} \otimes \mathcal{D}^{j_2}$

i) Let $F^{(1)}$ and $F^{(2)}$ be two Hilbert spaces identical to the space F defined in Subsec. 6.1.1. The set $F^{(1,2)} = F^{(1)} \otimes F^{(2)}$ is the vector space over the complex field of entire functions of the four complex variables which correspond to the two complex spinors $\zeta_1 = \begin{pmatrix}\xi_1\\ \eta_1\end{pmatrix}$ and $\zeta_2 = \begin{pmatrix}\xi_2\\ \eta_2\end{pmatrix}$. Since every entire function $\phi(\zeta_1, \zeta_2)$ may be expanded in an everywhere converging power series in ξ_1, η_1, ξ_2, and η_2, the set of functions $e^{j_1}_{m_1}(\zeta_1)\, e^{j_2}_{m_2}(\zeta_2)$, where e^{j}_{m} is defined by Eq. (6–6), forms a complete basis of $F^{(1,2)}$. The Hermitian structure on F is extended to $F^{(1,2)}$ thanks to the Hermitian inner product of $\phi(\zeta_1, \zeta_2) = \phi_1(\zeta_1)\phi_2(\zeta_2)$ with $\phi'(\zeta_1, \zeta_2) = \phi'_1(\zeta_1)\, \phi'_2(\zeta_2)$ defined by :

$$(\phi,\phi') = (\phi_1,\phi'_1)(\phi_2,\phi'_2) \quad ,$$

where (ϕ_1,ϕ'_1) and (ϕ_2, ϕ'_2) are, respectively, defined on $F^{(1)}$ and $F^{(2)}$ by Eq. (6–3). Then, the set $(e^{j_1}_{m_1}(\zeta_1)\, e^{j_2}_{m_2}(\zeta_2))$ forms a complete orthonormal basis of $F^{(1,2)}$.

Finally, a unitary representation of SU(2) in $F^{(1,2)}$ is clearly defined as follows :

$$\forall \phi \in F^{(1,2)} \quad \forall \zeta_1,\zeta_2 \in C^2 \qquad (T^{(1,2)}_U\phi)(\zeta_1,\zeta_2) = \phi(U^t\zeta_1,U^t\zeta_2) \quad .$$

If $\phi(\zeta_1, \zeta_2) = \phi_1(\zeta_1)\phi_2(\zeta_2)$ one has :

$$T^{(1,2)}_U\phi = (T^{(1)}_U\phi_1)(T^{(2)}_U\phi_2) \quad ,$$

where $T^{(1)}_U$ and $T^{(2)}_U$ are the representations of U defined, respectively, in $F^{(1)}$ and $F^{(2)}$ by Eq. (6–8). Therefore, the infinitesimal generators of this representation read :

$$J^{(1,2)}_i = J^{(1)}_i \otimes \mathbf{1}^{(2)} + \mathbf{1}^{(1)} \otimes J^{(2)}_i \quad ,$$

where $J^{(1)}_i, \mathbf{1}^{(1)}$, and $J^{(2)}_i, \mathbf{1}^{(2)}$ are the infinitesimal generators defined by Eq. (6–30a) and the identity operators acting, respectively, on $F^{(1)}$ and $F^{(2)}$.

The subspace $F_{j_1 j_2} = F^{(1)}_{j_1} \otimes F^{(2)}_{j_2}$ of $F^{(1,2)}$ is spanned by the $(2j_1+1)(2j_2+1)$ orthonormal functions $e^{j_1}_{m_1}(\zeta_1)\, e^{j_2}_{m_2}(\zeta_2)$ for given values of j_1 and j_2. Then $F_{j_1 j_2}$ is, as a whole, invariant under $T^{(1,2)}_U$, and the restriction of this mapping to the subspace $F_{j_1 j_2}$ provides the product representation $\mathcal{D}^{j_1} \otimes \mathcal{D}^{j_2}$.

ii) The previous procedure can be clearly generalized to the product of more than two spaces. For example, $F_{j_1 j_2 j_3} = F^{(1)}_{j_1} \otimes F^{(2)}_{j_2} \otimes F^{(3)}_{j_3}$ is an invariant subspace of $F^{(1,2,3)} = F^{(1)} \otimes F^{(2)} \otimes F^{(3)}$ under $T^{(1,2,3)}_U$ defined as :

$$\forall \phi \in F^{(1,2,3)} \quad \forall \zeta_1,\zeta_2,\zeta_3 \in C^2 \quad (T^{(1,2,3)}_U\phi)(\zeta_1,\zeta_2,\zeta_3) = \phi(U^t\zeta_1,U^t\zeta_2,U^t\zeta_3) \quad . \qquad (6\text{–}50)$$

6.2.2. THE DECOMPOSITION OF $\mathcal{D}^{j_1} \otimes \mathcal{D}^{j_2}$

6.2.2.1. First method

Instead of studying directly the decomposition of $\mathcal{D}^{j_1} \otimes \mathcal{D}^{j_2}$, it is shown in i) and ii) that it is equivalent to solve another problem, i.e. find all the invariants which belong to $F_{j_1j_2j_3}$; this is done in iii).

i) Let us assume that the representation $\mathcal{D}^{j_3}$ is contained in $\mathcal{D}^{j_1} \otimes \mathcal{D}^{j_2}$, i.e. there are $2j_3+1$ orthonormal functions $f^{j_3}_m$ in $F_{j_1j_2}$ such that :

$$T^{(1,2)}_U \; f^{j_3}_m = f^{j_3}_{m'} \; \mathcal{D}^{j_3}_{m'm}(U) \quad . \tag{6-51}$$

Consider in $F_{j_1j_2j_3}$ the function :

$$I(\zeta_1,\zeta_2,\zeta_3) = f^{j_3}_m(\zeta_1,\zeta_2) \; \mathcal{D}^{j_3}_{mm'}(C) \; e^{j_3}_{m'}(\zeta_3) \quad ,$$

where C denotes the special matrix $i\,\sigma_2$ in SU(2) introduced in Subsec. 4.5,iv). The main role played here by this matrix C is discussed later in Subsec. 8.3.1.1,i), actually $\mathcal{D}^{j_3}(C)$ is similar to a metric tensor. From Eqs. (6–16,50, and 51), the action of $T^{(1,2,3)}_U$ on the function I reads :

$$\begin{aligned} T^{(1,2,3)}_U &= (T^{(1,2)}_U f^{j_3}_m) \; \mathcal{D}^{j_3}_{mm'}(C) \; (T^{(3)}_U \; e^{j_3}_{m'}) \\ &= f^{j_3}_\mu \; \mathcal{D}^{j_3}_{\mu m}(U) \; \mathcal{D}^{j_3}_{mm'}(C) \; e^{j_3}_{\mu'} \; \mathcal{D}^{j_3}_{\mu' m'}(U) \\ &= f^{j_3}_\mu \; \mathcal{D}^{j_3}_{\mu\mu'}(UCU^t) \; e^{j_3}_{\mu'} \quad , \end{aligned}$$

and since $UCU^t = C$, Eq. (4–104), I is invariant under $T^{(1,2,3)}_U$.

Thus, if $\mathcal{D}^{j_3}$ is contained in $\mathcal{D}^{j_1} \otimes \mathcal{D}^{j_2}$ there exists a function I in $F_{j_1j_2j_3}$ which is invariant under $T^{(1,2,3)}_U$.

ii) Conversely, let us prove that if an invariant function I in $F_{j_1j_2j_3}$ can be constructed then there exists an orthonormal set of functions $f^{j_3}_m$ in $F_{j_1j_2}$ which transform under $\mathcal{D}^{j_3}$.

As the $e^{j_3}_m$ span F_{j_3} so do the functions $\mathcal{D}^{j_3}_{mm'}(C)\, e^{j_3}_{m'}$ since $\mathcal{D}^{j_3}(C)$ is a unitary matrix. Therefore, the function I has an expansion :

$$I(\zeta_1,\zeta_2,\zeta_3) = f_m(\zeta_1,\zeta_2) \; \mathcal{D}^{j_3}_{mm'}(C) \; e^{j_3}_{m'}(\zeta_3) \quad , \tag{6-52}$$

with uniquely determined functions f_m in $F_{j_1j_2}$. One then gets :

$$\begin{aligned} T^{(1,2,3)}_U I &= (T^{(1,2)}_U f_m) \; \mathcal{D}^{j_3}_{mm'}(C) \; (T^{(3)}_U \; e^{j_3}_{m'}) \\ &= (T^{(1,2)}_U f_m) \; \mathcal{D}^{j_3}_{mm'}(CU^tC^{-1}) \; \mathcal{D}^{j_3}_{m'\mu'}(C) \; e^{j_3}_{\mu'} \quad . \end{aligned}$$

Since $CU^tC^{-1} = U^{-1}$, Eq. (4–104), and I is invariant, the comparison of the above result with Eq. (6–52) yields :

$$T_U^{(1,2)} \, f_m = f_{m'} \, \mathcal{D}^{j_3}_{m'm}(U) \quad .$$

The matrix $\mathcal{D}^{j_3}(U)$ being unitary and irreducible, one shows (theorem below) that the functions f_m are orthogonal and have the same norm in $F_{j_1j_2}$:

$$(f_m \, , \, f_{m'}) = f^2 \delta_{m\,m'} \quad .$$

Theorem 6.1. Let D(G) be a finite-dimensional unitary representation of a group G in a Hermitian vector space E, and let $u_1,\ldots,u_n$ and $v_1,\ldots,v_n$ denote two sets of vectors in E which transform under D(g), respectively, as :

$$\forall g \in G \qquad D(g)u_i = u_j A^j_i(g) \quad , \qquad D(g)v_i = v_j B^j_i(g) \quad ,$$

where the matrices A(g) and B(g) are assumed to be unitary and irreducible. The unitary character of D(G) implies for the Hermitian inner product :

$$P_{ij} = (u_i , v_j) = (D(g)u_i , D(g)v_j) = A^{k*}_i(g) P_{k\ell} B^\ell_j(g) \quad ,$$

or in matrix notation $P = A^+(g) \, P \, B(g)$, and since A(g) is unitary :

$$\forall \; g \in G \qquad A(g)P = P \, B(g) \quad .$$

Then, from Props. C.45,47 :
(1) If A(g) and B(g) are inequivalent P = 0, i.e. :

$$\forall \; i,j \qquad (u_i \, , \, v_j) = 0 \quad .$$

(2) If A(g) = B(g) (hence m=n), P is a multiple of the unit matrix. In particular, if $u_i = v_i$:

$$(u_i \, , \, u_j) = \|u_i\|^2 \; \delta_{ij} \quad .$$

If the function I is assumed to be normalized, one has :

$$(I,I) = 1 = \sum_m (f_m \, , \, f_m) = (2j_3 + 1) \, f^2 \quad .$$

Thus, the $2j_3+1$ functions in $F_{j_1j_2}$ defined up to a phase choice by :

$$f^{j_3}_m = c_{j_3m} \sqrt{2j_3+1} \;\; f_m \quad , \qquad |c_{j_3m}| = 1 \quad , \tag{6-53}$$

are orthonormal and transform under $\mathcal{D}^{j_3}$.

iii) Let us finally construct the invariant functions in $F_{j_1j_2j_3}$. From the three complex spinors ζ_1, ζ_2, and ζ_3, one defines :

$$\theta_1 = \zeta_2^t \, C \, \zeta_3 \quad , \qquad \theta_2 = \zeta_3^t \, C \, \zeta_1 \quad , \qquad \theta_3 = \zeta_1^t \, C \, \zeta_2 \quad , \tag{6-54}$$

which are, using Eqs. (6–50) and (4–104), invariant under $T_U^{(1,2,3)}$. Any monomial in θ_1, θ_2, and θ_3 is also invariant. From now on the following notations are used :

$$k=(k_1,k_2,k_3)\ k_i\in N \ , \quad \theta^{[k]}=\theta_1^{k_1}\,\theta_2^{k_2}\,\theta_3^{k_3} \quad , \ \zeta=(\zeta_1,\zeta_2,\zeta_3) \quad . \tag{6-55}$$

One then considers the normalized invariant functions :

$$I_k = n_k\ \theta^{[k]} \quad , \qquad (I_k, I_k)=1 \quad , \tag{6-56}$$

where n_k is chosen positive real. These functions belong to $F_{j_1j_2j_3}$, i.e. they are homogeneous in ζ_1, ζ_2 and ζ_3 of degree, respectively, $2j_1$, $2j_2$, and $2j_3$, provided :

$$2j_1=k_2+k_3 \quad , \quad 2j_2=k_3+k_1 \quad , \quad 2j_3=k_1+k_2 \quad . \tag{6-57}$$

Now what are the possible values of j_3 for given j_1 and j_2 ? Viz., what are the non-negative integer values of k_1,k_2, and k_3 which fulfil the above equations, or equivalently :

$$i=1,2,3 \qquad k_i=J-2j_i \quad , \quad J=k_1+k_2+k_3=j_1+j_2+j_3 \quad . \tag{6-58}$$

The answer is the following :

$$\left.\begin{aligned} k_1&=j_2+j_3-j_1\geq 0\\ k_2&=j_3+j_1-j_2\geq 0\\ k_3&=j_1+j_2-j_3\geq 0\end{aligned}\right\} \Rightarrow \ |j_1-j_2|\leq j_3\leq j_1+j_2 \ . \tag{6-59}$$

From ii), to any invariant function I_k in $F_{j_1j_2j_3}$ correspond $2j_3+1$ orthonormal functions $f_m^{j_3}$ in $F_{j_1j_2}$. Moreover, it follows from the theorem given previously in note that the $f_m^{j_3}$ which correspond to different values of j_3 are orthogonal. Hence, one gets :

$$\sum_{j_3=|j_1-j_2|}^{j_1+j_2}(2j_3+1) = (2j_1+1)(2j_2+1) \quad , \tag{6-60}$$

orthonormal functions in $F_{j_1j_2}$. Since this is precisely the dimension of $F_{j_1j_2}$, the obtained decomposition of $\mathcal{D}^{j_1}\otimes\mathcal{D}^{j_2}$ is completed :

$$\mathcal{D}^{j_1}\otimes\mathcal{D}^{j_2} = \sum_{j_3=|j_1-j_2|}^{j_1+j_2}\oplus\,\mathcal{D}^{j_3} \quad . \tag{6-61}$$

6.2.2.2. Second method

For the sake of completeness let us give another usual proof of the "angular momentum" addition theorem in Quantum Mechanics (Messiah[39],Vol.II, Chap. XIII, Subsec. 25 or Edmonds[22], Sec. 3.1). This method is based on the existence of standard bases of the representation spaces (Sec. 5.5).

Let us consider representations of the group SU(2) in two different Hermitian vector spaces $E^{(1)}$ and $E^{(2)}$. Corresponding infinitesimal generators, which fulfil the usual commutation relations, Eq. (5-9), are denoted by $(J_i^{(1)})$ and $(J_i^{(2)})$. One assumes that the eigenbases $(|j_1m_1>)$ and $(|j_2m_2>)$ of, respectively, $J^{(1)2}$, $J_3^{(1)}$ and $J^{(2)2}$, $J_3^{(2)}$ form complete orthonormal bases of $E^{(1)}$ and $E^{(2)}$. It has to be noticed that Eq. (5–58) does not uniquely define the vectors $|j_im_i>$ since there remains a phase choice to define the action of $J_\pm^{(i)}$, for instance the choice of Eq. (5–59) corresponds to a usual standard basis. These choices are not needed here, they are discussed later in Subsec. 6.2.3.4. The set $(|j_1m_1> \otimes |j_2m_2>)$ forms a complete orthonormal basis of $E = E^{(1)} \otimes E^{(2)}$. The infinitesimal generators of the representation of SU(2) in E, induced by the representation defined, respectively, in $E^{(1)}$ and $E^{(2)}$ are :

$$J_i = J_i^{(1)} \otimes \mathbf{1}^{(2)} + \mathbf{1}^{(1)} \otimes J_i^{(2)} \quad , \tag{6-62}$$

where $\mathbf{1}^{(1)}$ and $\mathbf{1}^{(2)}$ are the unit operators acting, respectively, on $E^{(1)}$ and $E^{(2)}$. Clearly the J_i fulfil the usual commutation relations given by Eq. (5–9). For given j_1 and j_2, the vectors $|j_1m_1> \otimes |j_2m_2>$ span a $(2j_1+1)(2j_2+1)$–dimensional subspace $E_{j_1j_2}$ of E. Since $J^{(1)2} \otimes \mathbf{1}^{(2)}$ and $\mathbf{1}^{(1)} \otimes J^{(2)2}$ commute with J_i, we look for the eigenvectors of J^2 and J_3 among the eigenvectors of these operators, then each subspace $E_{j_1j_2}$ may be treated separately. Let us denote by $|\alpha j_1j_2jm>$ the eigenvectors of J^2 and J_3 within this subspace and which fulfil Eq. (5–58) (with the same remark as for $|j_im_i>$ about the phase undetermination), α denotes the additional quantum numbers corresponding to a possible degeneracy of the eigenvalues j and m.

The problem is now to find the possible values of j and m as also their respective order of degeneracy N(j) and n(m), for given j_1 and j_2. From Eq. (6–62), $|j_1m_1> \otimes |j_2m_2>$ is an eigenvector of J_3 with eigenvalue m_1+m_2. The degeneracy n(m) is thus simply the number of pairs m_1, m_2 such that $m = m_1+m_2$. It is then convenient to use the diagram of Fig. 6.1 where each pair m_1, m_2 is represented by a point of abscissa m_1 and ordinate m_2.

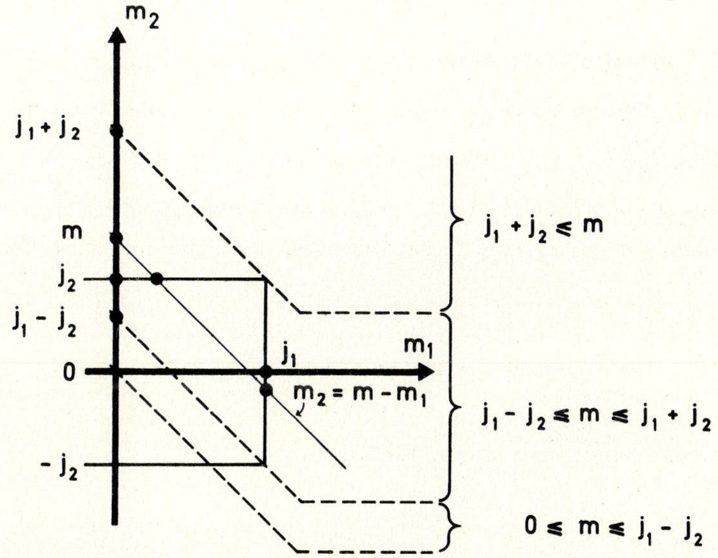

Fig. 6.1. Determination of the degeneracy n(m). Owing to symmetries the diagram is restricted to the case m $\geqslant$ 0.

Taking for instance $j_1 > j_2$, one obtains :

$$n(m) = \begin{cases} 0 & j_1+j_2 < |m| \\ j_1+j_2+1-|m| & j_1-j_2 \leqslant |m| \leqslant j_1+j_2 \\ 2j_2+1 & 0 \leqslant |m| \leqslant j_1-j_2 \end{cases} .$$

To each value of j there corresponds N(j) linearly independent series of 2j+1 eigenvectors of J^2, the vectors of a given series correspond to the 2j+1 values for m : j,j−1,...,−j. Therefore the degeneracies n(m) and N(j) are related by :

$$n(m) = \sum_{j \geqslant |m|} N(j) \Rightarrow N(j) = n(j) - n(j+1) .$$

Substituting the previous values of n(m) yields :

$$N(j) = \begin{cases} 1 & \text{if} \quad j = |j_1-j_2| , \ |j_1-j_2|+1,\ldots, j_1+j_2 \\ 0 & \text{otherwise} \end{cases} .$$

This is the "angular momentum" addition theorem, i.e. for given j_1 and j_2, in the $(2j_1+1)(2j_2+1)$–dimensional space spanned by $|j_1m_1\rangle \otimes |j_2m_2\rangle$:
(1) the possible values of j are : $|j_1-j_2|$, $|j_1-j_2|+1,\ldots, j_1+j_2$,
(2) to each of these values there corresponds only one series of 2j+1 eigenvectors characterized by j_1, j_2, j, and $m = m_1+m_2$.

The explicit construction of the eigenvectors $|j_1j_2jm\rangle$ with the usual phase choices is discussed in Subsec. 6.2.3.4.i).

6.2.3. CLEBSCH–GORDAN COEFFICIENTS AND WIGNER'S 3j SYMBOLS

6.2.3.1. Definitions and properties

i) Two orthonormal bases are now available in $F_{j_1j_2}$: the initial one $(e^{j_1}_{m_1}(\zeta_1)\, e^{j_2}_{m_2}(\zeta_2))$ and the basis $(f^{j_3}_{m_3}(\zeta_1, \zeta_2))$ defined within a phase choice by Eq. (6–53). The commonly used Clebsch–Gordon coefficients (C.G. coefficient in short) are the matrix elements of the unitary matrix which relates these two bases with the following choice of phase for the function $f^{j_3}_{m_3}$ (Subsec. 6.2.3.4) :

$$f^{j_3}_{m_3} = (-1)^{k_1} \sqrt{2j_3+1}\, f_{m_3} \quad , \quad k_1 = -j_1+j_2+j_3 . \tag{6-63}$$

The notations adopted here are the following :

$$\forall \zeta_1,\zeta_2 \in C^2 \quad f^{j_3}_{m_3}(\zeta_1,\zeta_2) = e^{j_1}_{m_1}(\zeta_1)e^{j_2}_{m_2}(\zeta_2)\langle j_1j_2m_1m_2|j_3m_3\rangle \quad , \tag{6-64a}$$

$$e^{j_1}_{m_1}(\zeta_1)e^{j_2}_{m_2}(\zeta_2) = \sum_{j_3} f^{j_3}_{m_3}(\zeta_1,\zeta_2)\langle j_3m_3|j_1j_2m_1m_2\rangle , \tag{6-64b}$$

or in terms of the Hermitian inner product in $F^{(1,2)}$:

$$\langle j_1j_2m_1m_2|j_3m_3\rangle = (e^{j_1}_{m_1}e^{j_2}_{m_2},f^{j_3}_{m_3}) = (f^{j_3}_{m_3},e^{j_1}_{m_1}e^{j_2}_{m_2})^* = \langle j_3m_3|j_1j_2m_1m_2\rangle^* . \tag{6-65}$$

Actually, it is shown that with the previous phase choice the C.G. coefficients are real (Subsecs. 6.2.3.2,4).

Some other notations and phase conventions are listed in Appendix G.1.1.

ii) The unitary character of this change of basis reads :

$$\langle j'_3m'_3|j_1j_2m_1m_2\rangle\langle j_1j_2m_1m_2|j_3m_3\rangle = \delta_{j'_3j_3}\,\delta_{m'_3m_3}\,\{j_1j_2j_3\} , \tag{6-66a}$$

$$\sum_{j_3}\langle j_1j_2m'_1m'_2|j_3m_3\rangle\langle j_3m_3|j_1j_2m_1m_2\rangle = \delta_{m'_1m_1}\,\delta_{m'_2m_2} , \tag{6-66b}$$

where $\{j_1j_2j_3\}$, called a <u>triangular delta</u>, is defined by :

$$\{j_1j_2j_3\} = \begin{cases} 1 & \text{if the } j_i \text{ are non-negative integers or} \\ & \text{half-odd-integers such that :} \\ & |j_1-j_2| \leqslant j_3 \leqslant j_1+j_2 \text{ and } j_1+j_2+j_3 \text{ integer,} \\ 0 & \text{otherwise} \end{cases} . \tag{6-67}$$

Since under the mappings $T^{(1,2)}_U$ the basis vectors $e^{j_1}_{m_1}\,e^{j_2}_{m_2}$ transform under $\mathcal{D}^{j_1}\otimes\mathcal{D}^{j_2}$ and $f^{j_3}_{m_3}$ transform under $\mathcal{D}^{j_3}$, Eqs. (6–64a,b) yield for all U in SU(2) :

$$\mathcal{D}^{j_3}_{m'_3m_3}(U) = \langle j_3m'_3|j_1j_2m'_1m'_2\rangle\,\mathcal{D}^{j_1}_{m'_1m_1}(U)\,\mathcal{D}^{j_2}_{m'_2m_2}(U)\langle j_1j_2m_1m_2|j_3m_3\rangle , \tag{6-68a}$$

$$\mathcal{D}^{j_1}_{m'_1m_1}(U)\,\mathcal{D}^{j_2}_{m'_2m_2}(U) = \sum_{j_3}\langle j_1j_2m'_1m'_2|j_3m'_3\rangle\,\mathcal{D}^{j_3}_{m'_3m_3}(U)\langle j_3m_3|j_1j_2m_1m_2\rangle . \tag{6-68b}$$

Setting $m'_1 = m_1$ and $m'_2 = m_2$ in the last above equation and summing over m_1 and m_2 one gets, using Eq. (6–66a), the multiplication law of the characters of the representations $\mathcal{D}^j$:

$$\chi^{j_1}(U)\,\chi^{j_2}(U) = \sum_{j_3}\{j_1j_2j_3\}\,\chi^{j_3}(U) . \tag{6-69}$$

From Eqs. (6–28, 52, 63, and 64a), the invariant functions I_k read :

$$I_k(\zeta) = \frac{(-1)^{j_1-j_2+m_3}}{\sqrt{2j_3+1}}\langle j_1j_2m_1m_2|j_3m_3\rangle\, e^{j_1}_{m_1}(\zeta_1)e^{j_2}_{m_2}(\zeta_2)e^{j_3}_{-m_3}(\zeta_3) , \tag{6-70}$$

where the k_i are related to the j_i by Eq. (6–58).

Either directly using the invariance of the above functions under $T^{(1,2,3)}_U$ and Eq. (6–23a), or from Eqs. (6–66,68), one obtains for all U in SU(2) :

$$\langle j_1 j_2 m_1 m_2 | j_3 m_3 \rangle = \langle j_1 j_2 m_1' m_2' | j_3 m_3' \rangle \, \mathcal{D}^{j_1 *}_{m_1' m_1}(U) \, \mathcal{D}^{j_2 *}_{m_2' m_2}(U) \, \mathcal{D}^{j_3}_{m_3' m_3}(U) \quad . \tag{6-71}$$

It is shown in Subsec. 8.3.1.1,ii) that such a relation characterizes an invariant tensorial sequence $(C^{(j_1)\,(j_2)\,[j_3]}_{m_1\;m_2\;m_3})$.

iii) The Wigner 3j symbols are the coefficients of the expansion of the invariant functions I_k in the basis $e^{j_1}_{m_1}\, e^{j_2}_{m_2}\, e^{j_3}_{m_3}$:

$$I_k(\zeta) = \begin{pmatrix} j_1 & j_2 & j_3 \\ m_1 & m_2 & m_3 \end{pmatrix} e^{j_1}_{m_1}(\zeta_1)\, e^{j_2}_{m_2}(\zeta_2)\, e^{j_3}_{m_3}(\zeta_3) \quad . \tag{6-72}$$

These symbols are thus related to the C.G. coefficients by :

$$\begin{pmatrix} j_1 & j_2 & j_3 \\ m_1 & m_2 & m_3 \end{pmatrix} = \frac{(-1)^{j_1 - j_2 - m_3}}{\sqrt{2j_3+1}} \langle j_1 j_2 m_1 m_2 | j_3 {-m_3} \rangle \quad . \tag{6-73}$$

Then, the orthogonality relations Eqs. (6–66a,b) take the forms :

$$(2j_3+1) \begin{pmatrix} j_1 & j_2 & j_3' \\ m_1 & m_2 & m_3' \end{pmatrix} \begin{pmatrix} j_1 & j_2 & j_3 \\ m_1 & m_2 & m_3 \end{pmatrix} = \delta_{j_3' j_3}\, \delta_{m_3' m_3}\, \{j_1 j_2 j_3\} \quad , \tag{6-74a}$$

$$\sum_{j_3} (2j_3+1) \begin{pmatrix} j_1 & j_2 & j_3 \\ m_1' & m_2' & m_3 \end{pmatrix} \begin{pmatrix} j_1 & j_2 & j_3 \\ m_1 & m_2 & m_3 \end{pmatrix} = \delta_{m_1' m_1}\, \delta_{m_2' m_2} \quad , \tag{6-74b}$$

and Eq. (6–71) reads :

$$\begin{pmatrix} j_1 & j_2 & j_3 \\ m_1 & m_2 & m_3 \end{pmatrix} = \begin{pmatrix} j_1 & j_2 & j_3 \\ m_1' & m_2' & m_3' \end{pmatrix} \mathcal{D}^{j_1}_{m_1' m_1}(U)\, \mathcal{D}^{j_2}_{m_2' m_2}(U)\, \mathcal{D}^{j_3}_{m_3' m_3}(U) \quad . \tag{6-75}$$

This last equation is characteristic of an invariant tensorial sequence $(W^{[j_1]\,[j_2]\,[j_3]}_{m_1\;m_2\;m_3})$ (Subsec. 8.3.1.1,ii)).

6.2.3.2. Computation of the Wigner 3j symbols

The Wigner 3j symbols can be computed by comparing the two expressions of I_k:

$$I_k(\zeta) = n_k\, \theta^{[k]} = \begin{pmatrix} j_1 & j_2 & j_3 \\ m_1 & m_2 & m_3 \end{pmatrix} e^{j_1}_{m_1}(\zeta_1)\, e^{j_2}_{m_2}(\zeta_2)\, e^{j_3}_{m_3}(\zeta_3) \quad , \tag{6-76}$$

where the k_i and j_i are always related by Eq. (6–58). The normalization factor n_k is computed in the notes below :

$$n_k = \frac{\Delta(j_1,j_2,j_3)}{k_1!k_2!k_3!} \quad , \qquad \Delta(j_1,j_2,j_3) = \left[\frac{k_1!k_2!k_3!}{(J+1)!}\right]^{\frac{1}{2}} \quad . \tag{6-77}$$

There only remains to express the monomial $\theta^{[k]}$ in the basis $(e^{j_1}_{m_1} e^{j_2}_{m_2} e^{j_3}_{m_3})$. Since :

$$\theta_1^{k_1} = (\xi_2\eta_3-\xi_3\eta_2)^{k_1} = \sum_{p_1} (-1)^{k_1-p_1} \frac{k_1!}{p_1!(k_1-p_1)!} (\xi_2\eta_3)^{p_1} (\xi_3\eta_2)^{k_1-p_1} ,$$

and the two other relations by cyclic permutations, one gets :

$$\begin{pmatrix} j_1 & j_2 & j_3 \\ m_1 & m_2 & m_3 \end{pmatrix} = (-1)^{j_1+j_2+j_3} \Delta(j_1,j_2,j_3) \prod_{i=1}^{3} \sum_{p_i} (-1)^{p_i} \frac{\sqrt{(j_i+m_i)!(j_i-m_i)!}}{p_i!(k_i-p_i)!} , \quad (6\text{-}78)$$

where the p_i, j_i, and m_i are related by :

$$m_1 = -j_2+j_3-p_2+p_3 \quad , \quad m_2 = -j_3+j_1-p_3+p_1 \quad , \quad m_3 = -j_1+j_2-p_1+p_2 \quad ,$$

which implies the selection rule :

$$m_1 + m_2 + m_3 = 0 \quad . \quad (6\text{-}79)$$

Actually, in Eq. (6–78) the sums over the p_i correspond to a simple sum since for given j_i and m_i, all the p_i can be expressed in terms of one of them, for instance $q = p_3$. This yields the formula :

$$\begin{pmatrix} j_1 & j_2 & j_3 \\ m_1 & m_2 & m_3 \end{pmatrix} = (-1)^{2j_2+j_3-m_3} \Delta(j_1,j_2,j_3) \prod_{i=1}^{3} \sqrt{(j_i+m_i)!(j_i-m_i)!} \quad (6\text{-}80)$$
$$\sum_q (-1)^q [(-j_1+j_3+m_2+q)!(-j_2+j_3-m_1+q)!q!(j_2-m_2-q)!(j_1+m_1-q)!(j_1+j_2-j_3-q)!]^{-1} ,$$

where the summation runs over all integer values of q such that the arguments of the factorials are non-negative. Therefore by inspection, q takes Q+1 values, such that :

$$Q = \min(j_1\pm m_1, j_2\pm m_2, j_3\pm m_3, -j_1+j_2+j_3, j_1-j_2+j_3, j_1+j_2-j_3) \quad . \quad (6\text{-}81)$$

Actually, there exist many other formulae for the Wigner 3j symbols which are not simply related by a change of variable in the summation over q. These different formulae can be obtained from the properties of the generalized hypergeometric (or Gauss) function[57] ${}_3F_2$, indeed Eq. (6–80) reads :

$$\begin{pmatrix} j_1 & j_2 & j_3 \\ m_1 & m_2 & m_3 \end{pmatrix} = (-1)^{2j_2+j_3-m_3} \Delta(j_1,j_2,j_3) \prod_{i=1}^{3} \sqrt{(j_i+m_i)!(j_i-m_i)!} \quad (6\text{-}82)$$
$$\times \frac{{}_3F_2(-j_2+m_2,-j_1-m_1,-j_1-j_2+j_3;-j_1+j_3+m_2+1,-j_2+j_3-m_1+1;1)}{(-j_1+j_3+m_2)!(-j_2+j_3-m_1)!(j_1+m_1)!(j_2-m_2)!(j_1+j_2-j_3)!} \quad .$$

For a detailed study of this problem cf. for instance Raynal[49].

Clearly from the above equations the Wigner 3j symbols and thus also the C.G. coefficients are real (Subsec. 6.2.3.4).

The symmetry properties of the Wigner 3j symbols are discussed below in Subsec.

6.2.3.3,iii), and the recursion formulae for the C.G. coefficients are given in Subsec. 6.2.3.4,i).

Some commonly used special values of C.G. coefficients are given in Appendix G.1.3. It should be noticed that if any one of the arguments in Eq. (6–81) vanishes, then the sum over q in Eq. (6–80) reduces to a single term, e.g. Eq. (6–104).

6.2.3.3. Generating function and symmetries of the Wigner 3j symbols

i) Definition. Following Bargmann[5] and Schwinger[55], the generating function of the Wigner 3j symbols is defined by :

$$\Phi(\tau,\xi,\eta) = \sum_{k=0}^{\infty} \theta^{[k]} \frac{\tau^{[k]}}{[k!]} = e^{\sum_{i=1}^{3} \theta_i \tau_i} \qquad , \qquad (6\text{–}83)$$

where τ, ξ, and η denote, respectively, the triple of complex variables (τ_1, τ_2, τ_3), (ξ_1, ξ_2, ξ_3), and (η_1, η_2, η_3). In addition with the shortening of writing specified in Eq. (6–55), the following notation is used :

$$[k!] = k_1\,!\,k_2\,!\,k_3\,! \qquad . \qquad (6\text{–}84)$$

Taking into account the definition of the θ_i by Eq. (6–54), one gets :

$$\Phi(\tau,\xi,\eta) = \exp \det \begin{pmatrix} \tau_1 & \xi_1 & \eta_1 \\ \tau_2 & \xi_2 & \eta_2 \\ \tau_3 & \xi_3 & \eta_3 \end{pmatrix} \qquad . \qquad (6\text{–}85)$$

From Eq. (6–76), the generating function Φ is related to the Wigner 3j symbols as follows :

$$\Phi(\tau,\xi,\eta) = \sum_k \frac{I_k(\zeta)}{n_k} \frac{\tau^{[k]}}{[k!]} = \sum_k \frac{1}{n_k} \begin{pmatrix} j_1 & j_2 & j_3 \\ m_1 & m_2 & m_3 \end{pmatrix} e^{j_1}_{m_1}(\zeta_1) e^{j_2}_{m_2}(\zeta_2) e^{j_3}_{m_3}(\zeta_3) \frac{\tau^{[k]}}{[k!]} \quad , \qquad (6\text{–}86)$$

where it is recalled that :

$$k_i = J-2j_i \in N \qquad , \qquad J = j_1+j_2+j_3 = k_1+k_2+k_3 \qquad .$$

Then, from the value of n_k, Eq. (6–77), the definition of vectors e^j_m, Eq. (6–6), and setting :

$$\kappa_i = j_i+m_i \quad , \quad \lambda_i = j_i-m_i \quad \Rightarrow \quad J = \kappa_1+\kappa_2+\kappa_3 = \lambda_1+\lambda_2+\lambda_3 \quad , \qquad (6\text{–}87)$$

one obtains :

$$\Phi(\tau,\xi,\eta) = \sum_{k,\kappa,\lambda} \left[\frac{(J+1)!}{[k!][\kappa!][\lambda!]}\right]^{\frac{1}{2}} \begin{pmatrix} j_1 & j_2 & j_3 \\ m_1 & m_2 & m_3 \end{pmatrix} \tau^{[k]}\, \xi^{[\kappa]}\, \eta^{[\lambda]} \qquad . \qquad (6\text{–}88)$$

ii) Computation of the normalization factor n_k. The value of n_k is given from Eq. (6–56) by :

$$n_k = \|\theta^{[k]}\|^{-1} \qquad .$$

The norm $||\theta^{[k]}||$ should be computed directly in $F^{(1,2,3)}$ as :

$$|| \theta^{[k]} ||^2 = \int \prod_{i=1}^{3} [d\mu(\zeta_i) \, |\theta_i|^{2k_i}] \quad .$$

Actually it is much simpler following again Bargmann, to use the previously defined generating function Φ . The outline of the calculation is the following : for given τ , $\Phi_\tau = \Phi(\tau,\xi,\eta)$ is a function in $F^{(1,2,3)}$, provided that the τ_i are small enough. One first computes in terms of τ and τ' the inner product of Φ_τ and $\Phi_{\tau'}$ in $F^{(1,2,3)}$. Then, the expansion of the result in powers of τ and τ' provides the norm $||\theta^{[k]}||$:

$$(\Phi_\tau,\Phi_{\tau'}) = (e^{\sum_i \theta_i\tau_i} , e^{\sum_i \theta_i\tau_i'}) = \sum_{k,k'} (\theta^{[k]} ,\theta^{[k']}) \frac{\tau^{[k]}}{[k!]} \frac{\tau'^{[k']}}{[k'!]} \quad . \tag{6-89}$$

Details of the calculation. From Eqs. (2–8,11) the generating function takes in vector notation the form :

$$\Phi(\tau,\xi,\eta) = e^{\tau\times\xi.\eta} \quad .$$

The measure in $F^{(1,2,3)}$ can be written as :

$$\prod_{i=1}^{3} d\mu(\zeta_i) = d\mu_3(\xi)\, d\mu_3(\eta) \quad ,$$

where for the triple $z = (z_1,z_2,z_3)$ of complex variables :

$$d\mu_3(z) = \prod_{i=1}^{3} d\mu_1(z_i) \quad , \quad d\mu_1(z_i) = \pi^{-1} e^{-|z_i|^2} dz_i \quad .$$

Separating the variables ξ and η in the integration, $(\Phi_\tau , \Phi_{\tau'})$ reads :

$$(\Phi_\tau , \Phi_{\tau'}) = \int d\mu_3(\xi) \, (e^{\tau\times\xi.\eta} , e^{\tau'\times\xi.\eta}) \quad , \tag{6-90}$$

where $(\exp\ \tau\times\xi.\eta , \exp\ \tau'\times\xi.\eta)$ denotes, for given τ , τ', and ξ , the Hermitian inner product of these two functions of η with the measure $d\mu_3(\eta)$.

Lemma : $a = (a_1,a_2,a_3)$, $b = (b_1,b_2,b_3)$, $a_i , b_i \in C$,

$$(e^{a.\eta}, e^{b.\eta}) = \int d\mu_3(\eta)\, e^{a^*.\eta^*} e^{b.\eta} = e^{a^*.b} \quad .$$

Proof. Expanding the exponentials yields :

$$(e^{a.\eta}, e^{b.\eta}) = \sum_{k,k'} \frac{a^{*[k]}}{[k!]} \frac{b^{[k']}}{[k'!]} (\eta^{[k]} , \eta^{[k']}) \quad ,$$

and with the same calculation done previously, Eq. (6–5) :

$$(\eta^{[k]} , \eta^{[k']}) = [k!]\, \delta_{kk'} \quad ,$$

which leads to the announced result.

Setting $a = \tau\times\xi$ and $b = \tau'\times\xi$ in Eq. (6–90) one has :

$$(\Phi_\tau , \Phi_{\tau'}) = \int d\mu_3(\xi)\, e^{\tau^*\times\xi^*.\tau'\times\xi} \quad .$$

Since :

$$\tau^*\times\xi^*.\tau'\times\xi = (\tau^*.\tau')(\xi^*.\xi) - (\tau^*.\xi)(\xi^*.\tau') \quad ,$$

in matrix notation the above expression reads :

$$(\Phi_\tau , \Phi_{\tau'}) = \pi^{-3} \int \prod_{i=1}^{3} d\xi_i \, e^{-\xi^+ A \xi} \quad ,$$

where the matrix elements of A are :

$$A_{ij} = (1-\tau^*.\tau')\,\delta_{ij} + \tau'_i\tau^*_j \quad . \tag{6-91}$$

Theorem 6.2. Evaluation of Gaussian (or Laplacian) integrals. Let A be an $n\times n$ complex matrix, it has a unique decomposition $A = A'+iA''$ with Hermitian A' and A''. Let z denote the n complex variables $z_j = x_j+iy_j$. If A', the Hermitian part of A, is positive definite and detA is nonzero, then :

$$I(A) = \pi^{-n} \int d_n z \, e^{-z^+ A z} = (\det A)^{-1} \quad , \tag{6-92}$$

the integral being extended over the whole space C^n and :

$$d_n z = \prod_{i=1}^{n} dx_i \, dy_i \quad , \quad z^+ A z = z^*_i A_{ij} z_j \quad .$$

Proof. One proceeds in two steps :

1–If A''= 0, the Hermitian matrix A = A' is diagonalizable by a unitary and unimodular matrix U, $A = U^+DU$, $D = (d_i)$ with real positive eigenvalues d_j. Then :

$$I(A) = \pi^{-n} \int d_n z \, e^{-z^+U^+DU\,z} \quad ,$$

and introducing new variables t = Uz, since :

$$d_n t = \det U \; d_n z = d_n z \quad ,$$

one gets the announced result :

$$I(A) = \pi^{-n} \int d_n t \, e^{-t^+Dt} = \pi^{-n} \prod_{i=1}^{n} \int dx_i dy_i \, e^{-d_i(x_i^2+y_i^2)} = \prod_{i=1}^{n} d_i^{-1} = (\det A)^{-1} \quad .$$

2–Consider now A = A'+iA'' with positive definite A' and arbitrary A''. The integral I(A) is absolutely convergent since the modulus of the integrand is $\exp(-z^+A'z)$.

Let α be a complex variable with real and imaginary parts, respectively, α' and α''. One considers the matrix :

$$A(\alpha) = A' + i\,\alpha\, A'' = A'(\alpha) + i\, A''(\alpha) \quad ,$$

where the Hermitian matrices $A'(\alpha)$ and $A''(\alpha)$ read :

$$A'(\alpha) = A' - \alpha''A'' \quad , \quad A''(\alpha) = \alpha'A'' \quad .$$

One then wants to compute the integral $I[A(\alpha)]$ for $\alpha = 1$.

It is first shown there exists a real positive β such that within the strip $|\alpha''| < \beta$ in the complex plane of α (see Fig. 6.2) $A'(\alpha)$ is positive definite. One has :

$$z^+A'(\alpha)z = k^2z^+A'z + (1-k^2)z^+A'z - \alpha''z^+A''z \quad .$$

Let us prove that for any nonzero real k with $|k| < 1$:

$$\exists\beta > 0 \; , \quad \forall|\alpha''| < \beta \;\Rightarrow\; (1-k^2)z^+A'z - \alpha''z^+A''z \geqslant 0 \quad ,$$

where the equality only holds for z = 0. The Hermitian matrices A' and A'' are diagonalizable with real eigenvalues. Let λ_i be the smallest eigenvalue of A', which is positive since A' is positive definite, and μ_i and μ_s, respectively, the smallest and the largest eigenvalues of A". For all z one then obtains :

$$0 \leqslant \alpha'' < \beta_1 = \frac{(1-k^2)\lambda_i}{|\mu_s|} \Rightarrow (1-k^2)z^+A'z-\alpha''z^+A''z \geqslant [(1-k^2)\lambda_i-\alpha''\mu_s]z^+z \geqslant 0 \quad ,$$

$$\beta_2 = \frac{(1-k^2)\lambda_i}{-|\mu_i|} < \alpha'' \leqslant 0 \Rightarrow (1-k^2)z^+A'z-\alpha''z^+A''z \geqslant [(1-k^2)\lambda_i-\alpha''\mu_i]z^+z \geqslant 0 \quad ,$$

where the equalities are fulfilled only for z = 0. Hence, one gets the announced result :

$$\forall z \qquad \forall\alpha, \qquad |\alpha''| < \beta = \min(\beta_1,-\beta_2) \qquad z^+A'(\alpha)z \geqslant k^2z^+A'z \geqslant 0 \quad ,$$

where the equalities only occur for z = 0, i.e. $A'(\alpha)$ is positive definite within this strip.

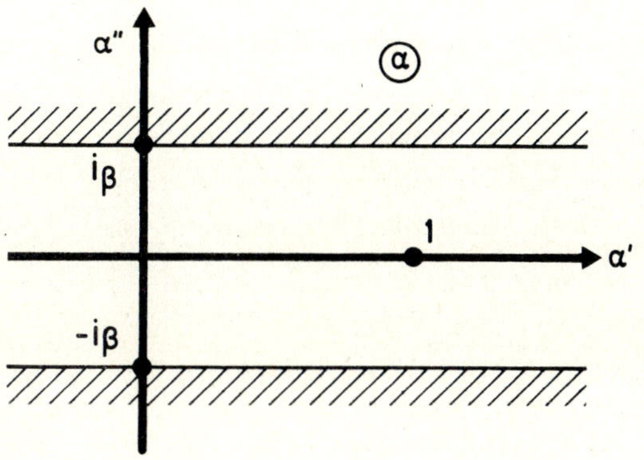

Fig. 6.2. The complex plane of the variable α .

Therefore, for purely imaginary α , $A(\alpha) = A'(\alpha)$ is Hermitian and moreover positive definite for $|\alpha''| < \beta$, thus from 1– one has :

$$\alpha = i\alpha'' \quad , \quad |\alpha''| < \beta \qquad I[A(\alpha)] = [\det A(\alpha)]^{-1} \quad .$$

It is shown now that the above result remains valid throughout the strip $|\alpha''| < \beta$ and especially for $\alpha = 1$. The integral $I[A(\alpha)]$ is analytic in α within this strip since the integrand $\exp[-z^+A(\alpha)z]$ is analytic in α and its modulus is for $|\alpha''| < \beta$ uniformly bounded by an integrable function $\exp(-k^2z^+A'z)$. Besides, $[\det A(\alpha)]^{-1}$ has only poles in α and by hypothesis $\alpha = 1$ is not one of them. This ends the proof of Theorem 6.2.

For sufficiently small τ and τ', the matrix A defined by Eq. (6–91) has a positive definite Hermitian part and the Theorem 6.2 applies :

$$(\Phi_\tau , \Phi_{\tau'}) = (\det A)^{-1} = (1-\tau^*.\tau')^{-2} \quad .$$

Expanding in a power series in τ and τ', one obtains :

$$(\Phi_\tau , \Phi_{\tau'}) = \sum_{k=0}^{\infty} (1+k)(\tau^*.\tau')^k = \sum_{k_1,k_2,k_3} (1+k_1+k_2+k_3)! \prod_{i=1}^{3} \frac{\tau_i^{*k_i} \tau_i'^{k_i}}{k_i!} \quad .$$

By comparing with Eq. (6–89), one gets :

$$(\theta^{[k]} , \theta^{[k']}) = (1+k_1+k_2+k_3)! [k!] \delta_{k k'} \quad .$$

Hence, in agreement with Eq. (6–77), the normalization factor n_k reads :

$$n_k = [(1+k_1+k_2+k_3)! [k!]]^{-\frac{1}{2}} \quad .$$

iii) Symmetries of the Wigner 3j symbols. Setting :

$$f_{k\kappa\lambda} = \left[\frac{(J+1)!}{[k!][\kappa!][\lambda!]}\right]^{\frac{1}{2}} \begin{pmatrix} j_1 & j_2 & j_3 \\ m_1 & m_2 & m_3 \end{pmatrix} \quad , \tag{6-93}$$

where k_i, κ_i, λ_i, j_i, and m_i are related by Eqs. (6–58,87), the generating function Φ reads :

$$\Phi(\tau,\xi,\eta) = \sum_{k,\kappa,\lambda} f_{k\kappa\lambda}\ \tau^{[k]}\ \xi^{[\kappa]}\ \eta^{[\lambda]} = e^{\det L} \quad , \quad L = \begin{pmatrix} \tau_1 & \xi_1 & \eta_1 \\ \tau_2 & \xi_2 & \eta_2 \\ \tau_3 & \xi_3 & \eta_3 \end{pmatrix} .$$

Then, the elementary symmetry properties of the determinant yields symmetry relations for the coefficients $f_{k\kappa\lambda}$, and thus for the Wigner 3j symbols :

The determinant is invariant, up to a sign, under permutations of the rows or of the columns, and transposition of the matrix. Let P be a permutation acting on the nine matrix elements of L, then :

$$\Phi[P(\tau,\xi,\eta)] = \sum_{k,\kappa,\lambda} f_{k\kappa\lambda} (P\tau)^{[k]} (P\xi)^{[\kappa]} (P\eta)^{[\lambda]} = \sum_{k,\kappa,\lambda} f_{PkP\kappa P\lambda} \tau^{[k]} \xi^{[\kappa]} \eta^{[\lambda]}$$

Let us consider the monomial $\prod_{i=1}^{n} x_i^{j_i}$, a permutation P , and its inverse P^{-1}. One then has :

$$P = \begin{pmatrix} 1 & \dots & n \\ P1 & & Pn \end{pmatrix} \quad , \quad \prod_{i=1}^{n} x_{Pi}^{j_i} = \prod_{i=1}^{n} x_i^{j_{P^{-1}i}} \quad .$$

Moreover, one has :

$$e^{-\det L} = \Phi(-\tau,\xi,\eta) = \sum_{k,\kappa,\lambda} (-1)^J f_{k\kappa\lambda}\ \tau^{[k]} \xi^{[\kappa]} \eta^{[\lambda]} \quad .$$

For the special permutations which change det L at most by a sign, we are led to relations among the $f_{k\kappa\lambda}$. Thus, if P is an even permutation of rows, or an even permutation of columns, or also the transposition of the matrix, then :

$$\Phi[P(\tau,\xi,\eta)] = \Phi(\tau,\xi,\eta) \Rightarrow f_{P(k,\kappa,\lambda)} = f_{k\kappa\lambda} \quad .$$

Similarly, if P is an odd permutation of rows or an odd permutation of columns, then :

$$\Phi[P(\tau,\xi,\eta)] = \Phi(-\tau,\xi,\eta) \Rightarrow f_{P(k,\kappa,\lambda)} = (-1)^J f_{k\kappa\lambda} \quad .$$

These symmetry relations for $f_{k\kappa\lambda}$ apply directly to the Wigner 3j symbols since the factor which relates them in Eq. (6–93) is clearly invariant under the considered permutations. This leads for the Wigner 3j symbols to a symmetry group discovered by Regge [51] of order 72 :

(L and L^t) × (3! permutations of rows) × (3! permutations of columns).

The commonly used symmetries of the Wigner 3j symbols are related to:
(1) permutations of the rows of L, i.e. simultaneous permutations of the j_i and the m_i,
(2) transposition of the second and third columns ($\xi \leftrightarrow \eta$), i.e. changing the sign of all the m_i.
These symmetries and the corresponding relations for the C.G. coefficients are given in Appendix G.1.2,iii).

6.2.3.4. Remarks about the phase conventions

In a first part i), following the method of decomposition of the product $\mathcal{D}^{j_1} \otimes \mathcal{D}^{j_2}$ given in Subsec. 6.2.2.2 and with the same notations, one specifies the phase choices equivalent to those of Condon and Shortley[19], (Chap. III, Sec. 10), which yield the widely used C.G. coefficients. At the same time, recursion formulae for the C.G. coefficients are given, and the explicit construction of the eigenvectors $|j_1 j_2 jm>$ is sketched. In a second part ii) it is shown that the C.G. coefficients defined and calculated above in Subsecs. 6.2.3.1, 2 correspond to these same phase choices.

i) Within the framework of Subsec. 6.2.2.2, the first two phase choices are :

C1. $(|j_1 m_1>)$ and $(|j_2 m_2>)$ are, respectively, standard bases of $E^{(1)}$ and $E^{(2)}$.
C2. $(|j_1 j_2 jm>)$ is a standard basis of $E_{j_1 j_2}$.

Consequences :
(1) From Eqs. (6–31a,b), the nonzero matrix elements of $J_\pm$ are real positive and those of J_3 are real. They are given, respectively, by Eqs. (6–34a,b). From :

$$J_1 = \frac{1}{2}(J_+ + J_-) \quad , \quad J_2 = \frac{1}{2i}(J_+ - J_-) \quad ,$$

the nonzero matrix elements of J_1 and J_2 are, respectively, real positive and purely imaginary :

$$<j'm'|J_1|jm> = \frac{1}{2}\sqrt{j(j+1)-m(m\pm1)}\ \delta_{j'j}\ \delta_{m'\,m\pm1} \quad , \tag{6-94a}$$

$$<j'm'|J_2|jm> = \frac{\pm1}{2i}\sqrt{j(j+1)-m(m\pm1)}\ \delta_{j'j}\ \delta_{m'\,m\pm1} \quad . \tag{6-94b}$$

(2) For given j_1, j_2, and j the relative phases of the $2j+1$ vectors $|j_1 j_2 jm>$, obtained by repeated action of $J_\pm$, are uniquely defined by C2.
(3) This yields recursion formulae for the C.G. coefficients by application of $J_\pm$ on both sides of their equation of definition :

$$|j_1 j_2 jm> = |j_1 m_1> \otimes |j_2 m_2> <j_1 j_2 m_1 m_2|jm> \quad .$$

From :

$$J_\pm = J_\pm^{(1)} \otimes \mathbf{1}^{(2)} + \mathbf{1}^{(1)} \otimes J_\pm^{(2)} \quad ,$$

one gets, using C1 and C2 :

$$A(j,m+1)\ \langle j_1j_2m_1m_2|jm+1\rangle = A(j_1,m_1)\ \langle j_1j_2m_1-1m_2|jm\rangle + A(j_2,m_2)\langle j_1j_2m_1m_2-1|jm\rangle \quad (6\text{-}95a)$$

$$A(j,m) \langle j_1j_2m_1m_2|jm-1\rangle = A(j_1,m_1+1)\ \langle j_1j_2m_1+1m_2|jm\rangle + A(j_2,m_2+1)\langle j_1j_2m_1m_2+1|jm\rangle\ , \quad (6\text{-}95b)$$

where :

$$A(j,m) = \sqrt{j(j+1)-m(m-1)} \quad . \quad (6\text{-}95c)$$

There still remains to specify the relative phases of vectors $|j_1j_2jm\rangle$ with different j, as also the relative phases between both sets of vectors $(|j_1j_2jm\rangle)$ and $(|j_1m_1\rangle \otimes |j_2m_2\rangle)$. These phase ambiguities are usually raised by the following third phase choice :

C3. The component of $|j_1j_2jj\rangle$ along $|j_1j_1\rangle \otimes |j_2 j{-}j_1\rangle$ is real non-negative, or in terms of C.G. coefficients, $\langle j_1j_2j_1m_2|jj\rangle$ is real non-negative.

Consequences :

(1) This phase convention is equivalent to assuming that all matrix elements of $J_3^{(1)} \otimes \mathbf{1}^{(2)}$ (or resp. $\mathbf{1}^{(1)} \otimes J_3^{(2)}$), which are nondiagonal in j, are real non-negative (resp. real nonpositive) (cf. for instance Edmonds[22], Secs. 3.3 and 3.4).

(2) Since all the C.G. coefficients with given j_1, j_2, and j are related by the recursion relations, Eqs. (6–95a,b), having real coefficients, and from C3 one of them is real, they are all real.

The explicit construction of the vectors $|j_1j_2jm\rangle$ can be sketched as follows. The vector $|j_1j_1\rangle \otimes |j_2j_2\rangle$ is an eigenstate of J_3 with the eigenvalue $m = j_1+j_2$. From the "angular momentum" addition theorem, this vector is a linear combination of eigenvectors of J^2 with different values of j and such that :

$$\sup\left(|j_1-j_2|,|m|\right) \leqslant j \leqslant j_1+j_2 \quad , \quad (6\text{-}96)$$

hence $j = j_1+j_2$. This result can also be checked using the relation :

$$J^2 = J^{(1)2} \otimes \mathbf{1}^{(2)} + \mathbf{1}^{(1)} \otimes J^{(2)2} + J_+^{(1)} \otimes J_-^{(2)} + J_-^{(1)} \otimes J_+^{(2)} + 2J_3^{(1)} \otimes J_3^{(2)} \quad . \quad (6\text{-}97)$$

From C3, one has :

$$\langle j_1\ j_2\ j_1\ j_2|j_1+j_2\ j_1+j_2\rangle = 1 \quad , \quad (6\text{-}98)$$

and consequently :

$$|j_1\ j_2\ j_1+j_2\ j_1+j_2\rangle = |j_1\ j_1\rangle \otimes |j_2\ j_2\rangle \quad . \quad (6\text{-}99)$$

From C1 and C2, one gets :

$$J_-|j_1\ j_2\ j_1+j_2\ j_1+j_2\rangle = \sqrt{2(j_1+j_2)}\,|j_1\ j_2\ j_1+j_2\ j_1+j_2-1\rangle$$
$$= \sqrt{2j_1}\,|j_1\ j_1-1\rangle \otimes |j_2\ j_2\rangle + \sqrt{2j_2}\,|j_1\ j_1\rangle \otimes |j_2\ j_2-1\rangle\ ,$$

which yield the values of the special C.G. coefficients :

$$<j_1\ j_2\ j_1-1\ j_2|j_1+j_2\ j_1+j_2-1> = \sqrt{\frac{j_1}{j_1+j_2}} \quad , \tag{6-100a}$$

$$<j_1\ j_2\ j_1\ j_2-1|j_1+j_2\ j_1+j_2-1> = \sqrt{\frac{j_2}{j_1+j_2}} \quad . \tag{6-100b}$$

The two vectors $|j_1j_1-1> \otimes |j_2j_2>$ and $|j_1j_1> \otimes |j_2j_2-1>$ span a two-dimensional subspace of eigenvectors of J_3 with eigenvalue $m = j_1+j_2-1$. Using the same argument as above for the vector $|j_1j_1> \otimes |j_2j_2>$, each vector of this subspace is a linear combination of two eigenvectors of J^2, one with $j = j_1+j_2$ and the other with $j = j_1+j_2-1$. Therefore, within this subspace, a vector orthogonal to $J_-|j_1j_2j_1+j_2j_1+j_2>$ is an eigenstate of J^2 with the eigenvalue j_1+j_2-1, i.e. :

$$|j_1\ j_2\ j_1+j_2-1\ j_1+j_2-1> \propto \sqrt{2j_1}\ |j_1\ j_1-1>\otimes|j_2\ j_2> - \sqrt{2j_2}|j_1\ j_1>\otimes|j_2\ j_2-1> \ .$$

Once again, this can be checked using Eq. (6–97). The phase ambiguity to define this normalized vector is raised by C3, which implies :

$$<j_1\ j_2\ j_1\ j_2-1|j_1+j_2-1\ j_1+j_2-1> \geq 0 \quad .$$

Therefore one has :

$$|j_1\ j_2\ j_1+j_2-1\ j_1+j_2-1> = \sqrt{\frac{j_2}{j_1+j_2}}\ |j_1\ j_1>\otimes|j_2\ j_2-1> - \sqrt{\frac{j_1}{j_1+j_2}}\ |j_1\ j_1-1>\otimes|j_2\ j_2> \ . \tag{6-101}$$

Now one may go on in this way, reducing m by one at each step, and obtaining at each step one more new value of j in the sequence j_1+j_2, j_1+j_2-1, ..., $|j_1-j_2|$. This is another way for computing the C.G. coefficients, cf. for instance Edmonds[22)] (Sec. 3.6). A third method to compute the C.G. coefficients, starting from the expression of the matrices $\mathcal{D}^j$, is given by Wigner[63)] (Chap. 17, Sec. 7; cf. below Eq. (7–53)).

ii) The phase choices which have been made to define the C.G. coefficients in Subsec. 6.2.3.1,i) by Eq. (6–64a) are the following :

C1'. The e^j_m are defined by Eq. (6–6).

C2'. The f^j_m are defined by Eq. (6–63).

Let us prove that the two sets of conventions (C1, C2, C3) and (C1', C2') yield the same C.G. coefficients.

It has been shown that (e^j_m) form a standard basis, Eqs. (6–31a,b), hence C1' implies C1. It follows from Eqs. (6–33,51) that (f^j_m) is also a standard basis, i.e. that C2' implies C2. This result can also be checked as follows. From Eqs. (6–64a,70) the function I_k reads :

$$I_k(\zeta_1,\zeta_2,\zeta_3) = \frac{(-1)^{j_1-j_2+m_3}}{\sqrt{2j_3+1}}\ f^{j_3}_{m_3}(\zeta_1,\zeta_2)\ e^{j_3}_{-m_3}(\zeta_3) \quad . \tag{6-102}$$

The invariance of I_k under the mappings $T_U^{(1,2,3)}$ implies for infinitesimal U :

$$J_{\pm,3}^{(1,2,3)} I_k = 0 \quad ,$$

and therefore :

$$(-1)^{m_3} (J_{\pm,3}^{(1,2)} f_{m_3}^{j_3})(\zeta_1,\zeta_2) e_{-m_3}^{j_3}(\zeta_3) = -(-1)^{m_3} f_{m_3}^{j_3}(\zeta_1,\zeta_2)(J_{\pm,3}^{(3)} e_{-m_3}^{j_3})(\zeta_3) \quad .$$

Hence, using the standard character of the basis (e_m^j), and since the above relations hold for all ζ_1, ζ_2, and ζ_3, one gets the announced result :

$$J_{\pm}^{(1,2)} f_m^j = \sqrt{j(j+1)-m(m\pm 1)}\, f_{m\pm 1}^j \quad , \quad J_3^{(1,2)} f_m^j = m\, f_m^j \quad . \tag{6-103}$$

Finally, since the C.G. coefficients computed in Subsec. 6.2.3.2 are all real, to fulfil C3 there remains to check that $\langle j_1 j_2 j_1 m_2 | jj\rangle$ is non–negative. Actually, from Eqs. (6–73,80) (where only $q = j_1+j_2-j_3$ contributes), one gets :

$$\langle j_1\ j_2\ j_1\ m_3-j_1 | j_3 m_3\rangle = \left[\frac{(2j_3+1)\ (-j_1+j_2+j_3)!(2j_1)!(j_1+j_2-m_3)!(j_3+m_3)!}{(j_1+j_2+j_3+1)!(j_1-j_2+j_3)!(j_1+j_2-j_3)!(-j_1+j_2+m_3)!(j_3-m_3)!}\right]^{\frac{1}{2}}, \tag{6-104}$$

which is clearly non–negative.

FORMULAE FOR CHAPTER 6

Irreducible unitary representations of SU(2)

$$\forall\, U \in SU(2) \qquad \mathcal{D}(U)|jm\rangle = |jm'\rangle\, \mathcal{D}^j_{m'm}(U) \quad \begin{cases} j=0,\tfrac{1}{2},1,\ldots \\ m=j,j-1,\ldots,-j \end{cases}$$

matrix $\mathcal{D}^j(U) = (\mathcal{D}^j_{mm'}(U)$, mth row m'th column both in the order $j,j-1,\ldots,-j)$

$$\mathcal{D}^j(U) = [\mathcal{D}^j(U^{-1})]^{-1} = \mathcal{D}^{j+}(U^{-1}) = \mathcal{D}^{j*}(U^*) = [\mathcal{D}^j(U^t)]^t = (-1)^{2j}\,\mathcal{D}^j(-U)$$

$$\mathcal{D}^{j*}(U) = \mathcal{D}^j(C)\,\mathcal{D}^j(U)\,[\mathcal{D}^j(C)]^{-1} \quad , \qquad C = i\sigma_2 = U(\pi,-e_2;a)$$

$$\mathcal{D}^{\frac{1}{2}}(U) = U$$

$$\mathcal{D}^j_{m'm}(U) = \mathcal{D}^{j*}_{mm'}(U^{-1}) = (-1)^{m'-m}\,\mathcal{D}^{j*}_{-m'-m}(U) = (-1)^{2j}\,\mathcal{D}^j_{m'm}(-U)$$

$$\mathcal{D}^j_{m'm}(\theta,\hat{n};a) = \langle\alpha jm'|e^{-i\theta J.\hat{n}}|\alpha jm\rangle$$

$$\mathcal{D}^j_{m'm}(\theta,e_1;a) = \langle\alpha jm'|e^{-i\theta J_1}|\alpha jm\rangle = e^{i\frac{\pi}{2}(m'-m)}\,d^j_{m'm}(\theta)$$

$$\mathcal{D}^j_{m'm}(\theta,e_2;a) = \langle\alpha jm'|e^{-i\theta J_2}|\alpha jm\rangle = d^j_{m'm}(\theta)$$

$$\mathcal{D}^j_{m'm}(\theta,e_3;a) = \langle\alpha jm'|e^{-i\theta J_3}|\alpha jm\rangle = e^{-im\theta}\,\delta_{m'm}$$

$$d^j_{m'm}(\theta) = d^{j*}_{m'm}(\theta) = d^j_{mm'}(-\theta) = (-1)^{m'-m}\,d^j_{mm'}(\theta) = (-1)^{m'-m}\,d^j_{-m'-m}(\theta)$$

$$d^j_{m'm}(\pi) = (-1)^{j-m}\,\delta_{m',-m} \quad , \qquad \mathcal{D}^j_{m'm}(C) = d^j_{m'm}(-\pi) = (-1)^{j+m}\,\delta_{m',-m}$$

$$d^\ell_{00}(\theta) = P_\ell(\cos\theta) \quad , \qquad \ell = 0,1,\ldots$$

$$\mathcal{D}^j_{m'm}(\varphi,\theta,\psi) = \langle\alpha jm'|e^{-i\varphi J_3}\,e^{-i\theta J_2}\,e^{-i\psi J_3}|\alpha jm\rangle = e^{-i\varphi m'}\,d^j_{m'm}(\theta)\,e^{-i\psi m}$$

$$= (-1)^{m'-m}\,\mathcal{D}^{j*}_{m'm}(-\varphi,-\theta,-\psi) \quad .$$

Characters

$$\chi(U) = \chi(U^{-1}) = \chi^*(U) \quad , \quad \chi^j(-U) = (-1)^{2j}\,\chi^j(U)$$

$$\chi^j(\theta) = \frac{\sin\frac{2j+1}{2}\theta}{\sin\frac{\theta}{2}} = \begin{cases} 1+2\cos\theta+\ldots+2\cos j\theta & \text{integer } j \\ 2\cos\frac{\theta}{2}+2\cos\frac{3}{2}\theta+\ldots+2\cos j\theta & \text{half-odd-integer } j \end{cases} .$$

Decomposition of the direct product of two irreducible representations

$$\mathcal{D}^{j_1}\otimes\mathcal{D}^{j_2} = \sum_{j_3=|j_1-j_2|}^{j_1+j_2} \oplus\,\mathcal{D}^{j_3} \quad .$$

The C.G. coefficients and the Wigner 3j symbols are real :

$$\langle j_1j_2m_1m_2|j_3'm_3'\rangle\langle j_1j_2m_1m_2|j_3m_3\rangle = \delta_{j_3'j_3}\ \delta_{m_3'm_3}\ \{j_1j_2j_3\}$$

$$\sum_{j_3} \langle j_1j_2m_1'm_2'|j_3m_3\rangle\langle j_1j_2m_1m_2|j_3m_3\rangle = \delta_{m_1'm_1}\ \delta_{m_2'm_2}$$

$$\mathcal{D}^{j_3}_{m_3'm_3}(U) = \langle j_1j_2m_1'm_2'|j_3m_3'\rangle\langle j_1j_2m_1m_2|j_3m_3\rangle\ \mathcal{D}^{j_1}_{m_1'm_1}(U)\ \mathcal{D}^{j_2}_{m_2'm_2}(U)$$

$$\mathcal{D}^{j_1}_{m_1'm_1}(U)\ \mathcal{D}^{j_2}_{m_2'm_2}(U) = \sum_{j_3} \langle j_1j_2m_1'm_2'|j_3m_3'\rangle\langle j_1j_2m_1m_2|j_3m_3\rangle\ \mathcal{D}^{j_3}_{m_3'm_3}(U)$$

$$\langle j_1j_2m_1m_2|j_3m_3\rangle = \langle j_1j_2m_1'm_2'|j_3m_3'\rangle\ \mathcal{D}^{j_1*}_{m_1'm_1}(U)\ \mathcal{D}^{j_2*}_{m_2'm_2}(U)\ \mathcal{D}^{j_3}_{m_3'm_3}(U)$$

$$\begin{pmatrix} j_1 & j_2 & j_3 \\ m_1 & m_2 & -m_3 \end{pmatrix} = \frac{(-1)^{j_1-j_2+m_3}}{\sqrt{2j_3+1}}\ \langle j_1j_2m_1m_2|j_3m_3\rangle$$

triangular delta : $$\{j_1j_2j_3\} = \begin{cases} 1 & \text{if } 2j_i \text{ non-negative integer} \\ & j_3 = |j_1-j_2|,|j_1-j_2|+1,\ldots,j_1+j_2 \\ 0 & \text{otherwise} \end{cases}$$

$$(2j_3+1)\begin{pmatrix} j_1 & j_2 & j_3' \\ m_1 & m_2 & m_3' \end{pmatrix}\begin{pmatrix} j_1 & j_2 & j_3 \\ m_1 & m_2 & m_3 \end{pmatrix} = \delta_{j_3'j_3}\ \delta_{m_3'm_3}\ \{j_1j_2j_3\}$$

$$\sum_{j_3}(2j_3+1)\begin{pmatrix} j_1 & j_2 & j_3 \\ m_1' & m_2' & m_3 \end{pmatrix}\begin{pmatrix} j_1 & j_2 & j_3 \\ m_1 & m_2 & m_3 \end{pmatrix} = \delta_{m_1'm_1}\ \delta_{m_2'm_2}$$

$$\mathcal{D}^{j_3*}_{m_3'm_3}(U) = (2j_3+1)\begin{pmatrix} j_1 & j_2 & j_3 \\ m_1' & m_2' & m_3' \end{pmatrix}\begin{pmatrix} j_1 & j_2 & j_3 \\ m_1 & m_2 & m_3 \end{pmatrix}\ \mathcal{D}^{j_1}_{m_1'm_1}(U)\ \mathcal{D}^{j_2}_{m_2'm_2}(U)$$

$$\mathcal{D}^{j_1}_{m_1'm_1}(U)\ \mathcal{D}^{j_2}_{m_2'm_2}(U) = \sum_{j_3}(2j_3+1)\begin{pmatrix} j_1 & j_2 & j_3 \\ m_1' & m_2' & m_3' \end{pmatrix}\begin{pmatrix} j_1 & j_2 & j_3 \\ m_1 & m_2 & m_3 \end{pmatrix}\ \mathcal{D}^{j_3*}_{m_3'm_3}(U)$$

$$\begin{pmatrix} j_1 & j_2 & j_3 \\ m_1 & m_2 & m_3 \end{pmatrix} = \begin{pmatrix} j_1 & j_2 & j_3 \\ m_1' & m_2' & m_3' \end{pmatrix}\ \mathcal{D}^{j_1}_{m_1'm_1}(U)\ \mathcal{D}^{j_2}_{m_2'm_2}(U)\ \mathcal{D}^{j_3}_{m_3'm_3}(U)$$

$$\chi^{j_1}(U)\ \chi^{j_2}(U) = \sum_{j_3}\{j_1j_2j_3\}\ \chi^{j_3}(U) \quad .$$

Irreducible unitary representations of SO(3)

$$\forall R\in SO(3) \longleftrightarrow \pm U\in SU(2) \quad , \qquad \mathcal{D}^{\ell}(R) = \mathcal{D}^{\ell}(\pm U) \qquad \ell=0,1,\ldots \quad .$$

CHAPTER 7

INVARIANT MEASURES, ORTHOGONALITY AND CLOSURE RELATIONS

Many properties of the representations of finite groups given in Appendix C.4 are based on the two following facts. (1) The summation over all the elements g of such a group G is well defined. (2) The rearrangement theorem (Prop. C.6) yields :

$$\forall\, g' \in G \qquad \sum_{g\in G} f(g) = \sum_{g\in G} f(g'g) = \sum_{g\in G} f(gg') \quad ,$$

where f is any function of G into the complex field. The problem is now to extend these properties to groups with an infinite number of elements. The outline of such a study is as follows. (1) To define on the group a measure $d\mu\,(g)$ which plays the role of the number of elements within an infinitesimal volume around g, and to require left and right invariance properties of this measure :

$$\forall\, g,g' \in G \qquad d\mu_\ell(g) = d\mu_\ell(g'g) \quad , \quad d\mu_r(g) = d\mu_r(gg') \quad .$$

(2) To extend the discreet sum to an integral $\int_G d\mu\,(g)$ taken over the group, such that the previous relations for finite groups take the forms :

$$\int_G d\mu_\ell(g)f(g) = \int_G d\mu_\ell(g)f(g'g) \quad , \quad \int_G d\mu_r(g)f(g) = \int_G d\mu_r(g)f(gg') \quad ,$$

where the class of functions f defined on G, for which the above integrals exist, has to be specified.

It is out of the scope of these notes to study the general problem of invariant integration on an abstract topological group, which leads to the definition of the Haar measure (cf. for instance Pontryagin[44], Chap. 5, Sec. 29). In the following, one only aims to show without claiming mathematical rigour how the previous properties for finite groups can be extended to the case of n-parameter Lie groups as defined in Appendix C.5.4, and especially compact n-parameter Lie groups. Finally, these results are applied to the special cases of the groups SU(2) and SO(3).

7.1. INVARIANT MEASURES

7.1.1. GENERALITIES FOR LIE GROUPS

The same notations as in Appendix C.5.4 are used. We now describe a direct method to construct an invariant measure on a Lie group. Another important approach to this problem, based on the existence of a Riemannian manifold structure on a Lie group, is only sketched for the group SU(2) in Subsec. 7.1.3.2.

7.1.1.1. Existence and uniqueness of a left (resp. right) invariant measure on an n-parameter Lie group

i) It is emphasized that the definition of such a measure is a local problem. It is also recalled that both the definition and the properties of the functions φ^i and ψ^i considered in Def. C.60,(2) are local and closely related to the choice of local charts (see Fig. C.4). The following derivation is mainly based on Prop. C.78, i.e., for all g in an n-parameter Lie group G, the left and right translations, L_g and R_g such that :

$$\forall g' \in G \qquad L_g g' = gg' \quad , \qquad R_g g' = g'g \quad ,$$

are analytic one-to-one mappings of G onto itself (see Fig. C.5). Then, in the same way as for topological groups the left and right translations allow us to define a local characterization of the topology (Appendix C.5.2), these translations permit, for a given choice of local charts, to construct a local density over the group in terms of its value for a given element of the group.

ii) For a given choice of an atlas such that :

$$\forall g \in G \qquad \exists a \in R^n \qquad g = g_a \quad ,$$

a necessary expression for a left invariant measure $d\mu_\ell(a)$ is now derived, assuming it exists. This measure plays the role of the number of elements within an infinitesimal volume :

$$d_n a = \prod_{i=1}^{n} da^i \quad , \tag{7-1}$$

around the point a, the density $\mu_\ell(a)$ is defined by :

$$d\mu_\ell(a) = \mu_\ell(a)\, d_n a \quad . \tag{7-2}$$

For any $g' = g_b$ the left translation L_{g_b} brings $g = g_a$ into $g'' = g_c$ (see Fig. 7.1), with according to Def. C.60, (2) :

$$g_c = L_{g_b} g_a = g_b g_a \quad , \qquad c = \ell_b(a) = \varphi(b,a) \quad ,$$

and also each element of the group parametrized by $a+\delta a$, within the volume $d_n a$, into an element of the group parametrized by $c+\delta c$, such that :

$$g_{c+\delta c} = L_{g_b}\, g_{a+\delta a} \quad , \quad c+\delta c = \ell_b(a+\delta a) = \varphi(b,a+\delta a) \quad .$$

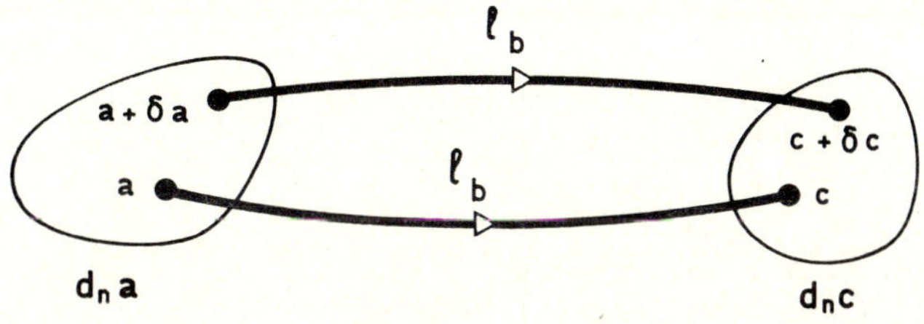

Fig. 7.1. Transformation of an infinitesimal volume element around a in R^n under a left translation ℓ_b.

Expanding $\varphi(b,a+\delta a)$ in powers of δa (for a Lie group the functions φ^i are locally analytic) yields to first order in δa :

$$c + \delta c = \varphi(b,a) + \frac{\partial\varphi(b,a)}{\partial a^j}\,\delta a^j + 0(\delta a^2) \Rightarrow \delta c^i = \frac{\partial\varphi^i(b,a)}{\partial a^j}\,\delta a^j \quad .$$

When $a+\delta a$ ranges over the entire elementary volume $d_n a$, $c+\delta c$ describes a volume $d_n c$ around the point c, and related to $d_n a$ by the absolute value of the Jacobian, denoted :

$$\frac{\partial c}{\partial a} = \left.\frac{\partial[\varphi^i(b,a')]}{\partial[a'^j]}\right|_{a'=a} \quad , \quad b = \varphi(c,\bar{a}) \quad , \tag{7-3}$$

according to :

$$d_n c = \left|\frac{\partial c}{\partial a}\right| d_n a \quad . \tag{7-4}$$

This corresponds to the usual law of transformation of the differential volume element $d_n a$ under the change of variable a into c. This locally defined Jacobian $\partial c/\partial a$ is finite and nonzero, and moreover locally analytic since the left translation is an analytic one-to-one mapping. Assuming the measure is left invariant, which means that the number of points within both volumes $d_n a$ and $d_n c$ is the same, reads :

$$d\mu_\ell(a) = \mu_\ell(a)\, d_n a = d\mu_\ell(c) = \mu_\ell(c)\, d_n c \quad .$$

Therefore, from Eq. (7–4), one gets the necessary condition :

$$\mu_\ell(c) = \mu_\ell(a) \left|\frac{\partial c}{\partial a}\right|^{-1} \quad . \tag{7-5}$$

An equivalent formula can be derived by expressing $d_n a$ in terms of $d_n c$ from :

$$g_a = g_b^{-1}\, g_c = g_{\bar{b}}\, g_c \quad , \quad a = \varphi(\bar{b},c) \quad ,$$

namely :

$$d_n a = \left|\frac{\partial a}{\partial c}\right| d_n c \quad , \quad \frac{\partial a}{\partial c} = \left.\frac{\partial[\varphi^i(\bar{b},c')]}{\partial[c'^j]}\right|_{c'=c} \quad , \quad \bar{b} = \varphi(a,\bar{c}) \quad . \tag{7-6}$$

The above relation follows from the property of Jacobians :

$$\left(\frac{\partial c}{\partial a}\right)^{-1} = \frac{\partial a}{\partial c} \quad .$$

Thus one gets another formulation of Eq. (7–5) :

$$\mu_\ell(c) = \mu_\ell(a) \left|\frac{\partial a}{\partial c}\right| \quad . \tag{7-7}$$

Since an invariant measure can be clearly determined only up to a constant multiplicative factor, the density at a point, say $\mu_\ell(a)$ at a, can be chosen arbitrarily. Then the density is defined for all g_c by Eq. (7–5) or (7–7).

The previous expressions of $\mu_\ell(c)$ are often quoted in the literature taking g_a as the unit element of the group, since the calculation of the Jacobian is then simpler. Indeed, denoting by u the parameters of the unit element, one has :

$$a = u = \bar{a} \quad , \quad c = b \quad ,$$

and Eqs. (7–5,7) read, respectively :

$$\mu_\ell(c) = \mu_\ell(u) \left|\frac{\partial[\varphi^i(c,a')]}{\partial[a'^j]}\right|^{-1}_{a'=u} = \mu_\ell(u) \left|\frac{\partial[\varphi^i(\bar{c},c')]}{\partial[c'^j]}\right|_{c'=c} \quad . \tag{7-8}$$

In practice, one must take care since the above expressions can be used only if for the chosen local parametrization the unit element is not an exceptional point. This problem occurs for instance for the group SU(2) parametrized by the Euler angles (Subsec. 7.1.3.1,iv)).

One has now to check that the previous definition of a left invariant measure is compatible with the group law (sufficient condition). Namely it has to be verified that, for all g_f such that (see Fig. 7.2) :

$$g_f = L_{g_d} g_c \quad ,$$

and the densities $\mu_\ell(c)$ and $\mu_\ell(f)$ both defined in terms of $\mu_\ell(a)$ according to Eq. (7–7) :

$$\mu_\ell(c) = \mu_\ell(a) \left|\frac{\partial a}{\partial c}\right| \quad , \quad \mu_\ell(f) = \mu_\ell(a) \left|\frac{\partial a}{\partial f}\right| \quad ,$$

one has :

$$d\mu_\ell(f) = d\mu_\ell(c) \quad ,$$

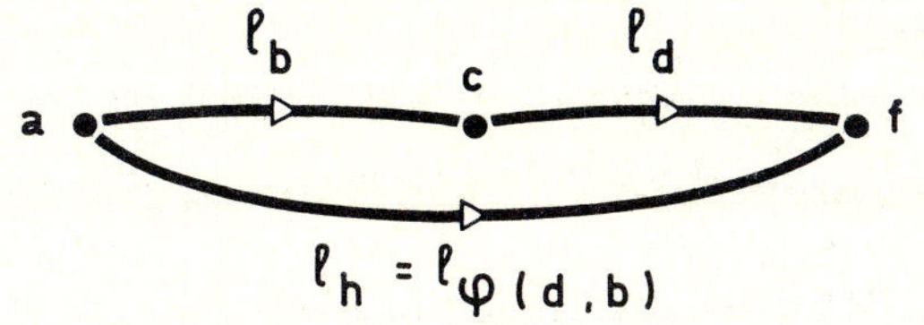

Fig. 7.2. Composition of left translations in R^n.

Using Eq. (7–6) for the pair of points f,c instead of c,a, and the chain relation of Jacobians, one gets the announced result :

$$d\mu_\ell(f) = \mu_\ell(f) d_n f = \mu_\ell(a) \left|\frac{\partial a}{\partial f}\right| d_n f = \mu_\ell(a) \left|\frac{\partial a}{\partial c}\right| \left|\frac{\partial c}{\partial f}\right| d_n f = \mu_\ell(c)\ d_n c \quad .$$

A similar calculation leads for the right invariant measure to the following expressions :

$$d\mu_r(c) = \mu_r(c)\ d_n c \quad , \quad \mu_r(c) = \mu_r(a) \left|\frac{\partial c}{\partial a}\right|^{-1} = \mu_r(a) \left|\frac{\partial a}{\partial c}\right| \quad , \tag{7-9}$$

where :

$$g_c = R_{g_b} g_a = g_a\, g_b \quad , \quad c = r_b(a) = \varphi(a,b) \quad ,$$

$$\frac{\partial c}{\partial a} = \left.\frac{\partial[\varphi^i(a',b)]}{\partial[a'^j]}\right|_{a'=a} \quad , \quad b = \varphi(\bar{a},c) \quad , \tag{7-10a}$$

$$\frac{\partial a}{\partial c} = \left.\frac{\partial[\varphi^i(c',\bar{b})]}{\partial[c'^j]}\right|_{c'=c} \quad , \quad \bar{b} = \varphi(\bar{c},a) \quad . \tag{7-10b}$$

Thus, it has been shown that on an n-parameter Lie group, with respect to a given atlas, there exist two positive invariant measures, which can be different, one is left invariant and the other right invariant. They are both uniquely defined up to a multiplicative constant.

iii) There still remains to show that these measures are independent, up to a constant multiplicative factor, from the choice of a special atlas among the set of equivalent atlases which defines the analytic n–manifold structure of the Lie group (Def. C.52). Let us denote, respectively, by the subscript i=1 or 2 all the objects related to two equivalent atlases A_1 and A_2 (see Fig. 7.3) :

$$i=1,2 \qquad d\mu_i(c_i) = \mu_i(c_i)\ d_n c_i = \mu_i(a_i) \left|\frac{\partial a_i}{\partial c_i}\right| d_n c_i \quad ,$$

where the subscript ℓ or r has been omitted since it only changes the meaning of the Jacobian, but does not affect what follows.

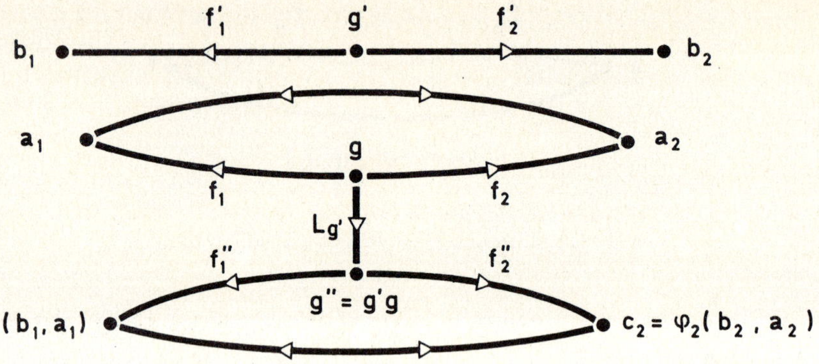

Fig. 7.3. Relation between two atlases with the same notations as in Fig. C.4.

Then the question is, assuming for $g = g_{a_1} = g_{a_2}$:

$$d\mu_2(a_2) = C\, d\mu_1(a_1) \quad , \tag{7-11}$$

where C is a constant, does one have for all $g'' = g_{c_1} = g_{c_2}$:

$$d\mu_2(c_2) = C\, d\mu_1(c_1) \quad ?$$

It follows from the properties of the Jacobian :

$$d\mu_2(c_2) = \mu_2(a_2) \left|\frac{\partial a_2}{\partial c_2}\right| d_n c_2 = \mu_2(a_2) \left|\frac{\partial a_2}{\partial a_1}\ \frac{\partial a_1}{\partial c_1}\ \frac{\partial c_1}{\partial c_2}\right| d_n c_2 \quad ,$$

where the locally defined Jacobians $\partial a_2/\partial a_1$ and $\partial c_1/\partial c_2$ are finite, nonzero and analytic since they correspond to a change of coordinates between local charts of equivalent atlases. By definition they are such that :

$$d_n a_2 = \left|\frac{\partial a_2}{\partial a_1}\right| d_n a_1 \quad , \qquad d_n c_1 = \left|\frac{\partial c_1}{\partial c_2}\right| d_n c_2 \quad .$$

Moreover from Eq. (7–11) one has :

$$\mu_2(a_2) \left|\frac{\partial a_2}{\partial a_1}\right| = C\, \mu_1(a_1) \quad .$$

Therefore, one gets the announced result :

$$d\mu_2(c_2) = C\, \mu_1(a_1) \left|\frac{\partial a_1}{\partial c_1}\right| d_n c_1 = C\, d\mu_1(c_1) \quad .$$

Finally, it has been shown that :

Theorem 7.1. On an n-parameter Lie group G, independently of any choice of atlases there exist two, respectively left and right, invariant measures, $d\mu_\ell(g)$ and $d\mu_r(g)$. Both of these measures are uniquely defined up to a multiplicative constant.

This theorem about the existence and the uniqueness of the invariant measures is very useful, since the general expressions given by Eqs. (7-5,7, and 9) are not always convenient to compute these measures, when they can sometimes be derived directly by specific methods for the considered group (Subsec. 7.1.3).

7.1.1.2. Some properties of the invariant measures

Theorem 7.2. On an n-parameter Lie group G, one has :

$$\forall g,g' \in G \qquad d\mu_\ell(gg') = C_\ell(g')d\mu_\ell(g) \quad , \quad d\mu_r(g'g) = C_r(g')d\mu_r(g) \quad , \qquad (7\text{-}12)$$

where $C_\ell(g')$ and $C_r(g')$ are some constants depending upon g'.

Proof. Clearly for given g' the measure defined on G by :

$$\forall g \in G \qquad d\rho_{g'}(g) = d\mu_\ell(gg') \quad ,$$

is left invariant since one has :

$$\forall g,g'' \in G \qquad d\rho_{g'}(g''g) = d\mu_\ell(g''gg') = d\mu_\ell(gg') = d\rho_{g'}(g) \quad .$$

Therefore, since the left invariant measure on G is uniquely defined up to a multiplicative factor, there exists $C_\ell(g')$ such that Eq. (7-12) holds. A similar proof can be given for the right invariant one. The previous results are consequences of the fact that any left translation commutes with any right one :

$$\forall g,g',g'' \in G \qquad L_{g'}R_{g''}g = g'gg'' = R_{g''}L_{g'}g \quad .$$

Consequences. If both integrals $\int_G d\mu_\ell(g)\, f(g)$ and $\int_G d\mu_r(g)\, f(g)$ exist, one gets for all g' in G :

$$\int_G d\mu_\ell(gg')f(g) = \int_G d\mu_\ell(g)f(gg'^{-1}) = C_\ell(g') \int_G d\mu_\ell(g)f(g) \qquad (7\text{-}13a)$$

$$\int_G d\mu_r(g'g)f(g) = \int_G d\mu_r(g)f(g'^{-1}g) = C_r(g') \int_G d\mu_r(g)f(g) \quad . \qquad (7\text{-}13b)$$

Theorem 7.3. On an n-parameter Lie group G, one has :

$$\forall g \in G \qquad d\mu_\ell(g^{-1}) = C\, d\mu_r(g) \quad , \qquad (7\text{-}14)$$

where C is some constant independent of g.

Proof. The measure defined on G by :

$$\forall g \in G \qquad d\rho(g) = d\mu_\ell(g^{-1}) \quad ,$$

is right invariant. Indeed, one has :

$$\forall g,g' \in G \qquad d\rho(gg') = d\mu_\ell(g'^{-1}g^{-1}) = d\mu_\ell(g^{-1}) = d\rho(g) \quad .$$

From the uniqueness up to a multiplicative constant of the right invariant measure, one gets the announced result.

Consequences. If the following integrals exist, one has :

$$\int_G d\mu_\ell(g^{-1})f(g) = \int_G d\mu_\ell(g)f(g^{-1}) = C\int_G d\mu_r(g)f(g) \quad . \qquad (7\text{-}15)$$

7.1.1.3. The special case of linear groups

Details about this subject can be found for instance in Murnaghan[42)] (Chap. VIII). The real and complex linear groups GL(n,R) and GL(n,C) are special r–parameter Lie groups with, respectively, $r = n^2$ and $r = 2n^2$ (Appendix C.5.4.3).

For GL(n,R) a natural parametrization, which also defines an atlas, is given by the n^2 real matrix elements. Both sets of functions φ^i and ψ^i of the general theory are then defined by the usual multiplication law for matrices, i.e. :

$$\forall A,B,C \in GL(n,R) \qquad C = BA \quad , \qquad C^i_j = B^i_k A^k_j \quad .$$

From the general expression given by Eq. (7–7), one gets :

$$\mu_\ell(C) = \mu_\ell(A) \left|\frac{\partial[A^i_j]}{\partial[C^{i'}_{j'}]}\right| = \mu_\ell(A) \frac{1}{|\det B|^n} = \mu_\ell(A) \frac{|\det A|^n}{|\det C|^n} \quad .$$

Therefore, up to a multiplicative constant, the left invariant measure on GL(n,R) is :

$$\forall M \in GL(n,R) \qquad d\mu(M) = \frac{1}{|\det M|^n} d_{n^2}M \quad , \quad d_{n^2}M = \prod_{i,j=1}^{n} dM^i_j \quad . \qquad (7\text{-}16)$$

A similar calculation from Eq. (7–9) shows that this measure is also right invariant. One should note that this property is true although this group is noncompact (Subsec. 7.1.2.2,i)). This measure is the generalization of the one–dimensional invariant measure under the multiplication group :

$$\forall x \neq 0 \ , \quad \forall y \neq 0 \qquad d\mu(x) = \frac{1}{|x|} dx = d\mu(yx) \quad .$$

A similar calculation yields for the left and also right invariant measure on the complex linear group to the expression :

$$\forall M = M' + iM'' \in GL(n,C) \qquad d\mu(M) = \frac{1}{|\det M|^{2n}} d_{n^2}M' \, d_{n^2}M'' \quad , \qquad (7\text{-}17)$$

where M' and M'' are real matrices.

The previous results allow us to give the expressions of the invariant measures on the subgroups O(n) or SO(n) and U(n) or SU(n) of, respectively, GL(n,R) and GL(n,C). One has :

$$\forall\ O \in O(n) \quad \text{or} \quad SO(n) \qquad d\mu(O) = d_{n^2}O\ \delta(OO^t - \mathbb{1})$$

$$\forall\ U = U' + iU'' \in U(n) \qquad d\mu(U) = d_{n^2}U'\ d_{n^2}U''\ \delta(UU^+ - \mathbb{1})$$

$$\forall\ U = U' + iU'' \in SU(n) \qquad d\mu(U) = d_{n^2}U'\ d_{n^2}U''\ \delta(UU^+ - \mathbb{1})\delta(\det U - 1)$$

where U and U' are real matrices. Indeed, it is easy to check the left and also right invariance of these measures. Actually, the above expressions are not convenient since one has to express the δ functions in order to select the independent variables upon which the elements of the considered group depend.

7.1.2. THE SPECIAL CASE OF COMPACT LIE GROUPS

7.1.2.1. Existence of finite volumes of the group

Theorem 7.4. On a compact Lie group G the integrals over the whole group :

$$V_\ell = \int_G d\mu_\ell(g) \quad , \qquad V_r = \int_G d\mu_r(g) \quad ,$$

also called the volumes of the group, exist.

Proof. It is only given for the left invariant measure, since it is similar for the right one. One denotes by F(U;g), for any subset U of the group G, the characteristic function, i.e. :

$$\forall g \in G \qquad F(U;g) = \begin{cases} 1 & \text{if } g \in U \\ 0 & \text{otherwise} \end{cases} \quad . \tag{7-18}$$

The n–manifold structure of an n–parameter Lie group implies that for all g in G, in particular for the unit element e, there exists a local chart (U,f) such that e belongs to the open set U in G and f is a one–to–one mapping of U onto the open set f(U) in R^n. One can find an n–dimensional open ball $D^n(u)$ of centre u = f(e) and included in f(U). Then, f being locally an homeomorphism, there exists an open set U_e which contains e, included in U and defined by (see Fig. 7.4) :

$$f(U_e) = D^n(u) \quad .$$

It has been shown in Subsec. 7.1.1.1 that for given local charts, the Jacobian related to the density by Eqs. (7–5,7) is locally analytic. It can then always be assumed that the ball $D^n(u)$ is chosen such that the positive density is continuous over this closed ball. Consequently, the volume of the open set U_e, defined as the integral over the group of the characteristic function $F(U_e;g)$, is bounded :

$$\int_G d\mu_\ell(g)F(U_e;g) \leq \int_{D^n(u)} d\mu_\ell(a) < \infty \quad ,$$

since the right-hand side integral is an integral of a continuous function over a compact set $D^n(U)$ (Prop. C.69) in R^n.

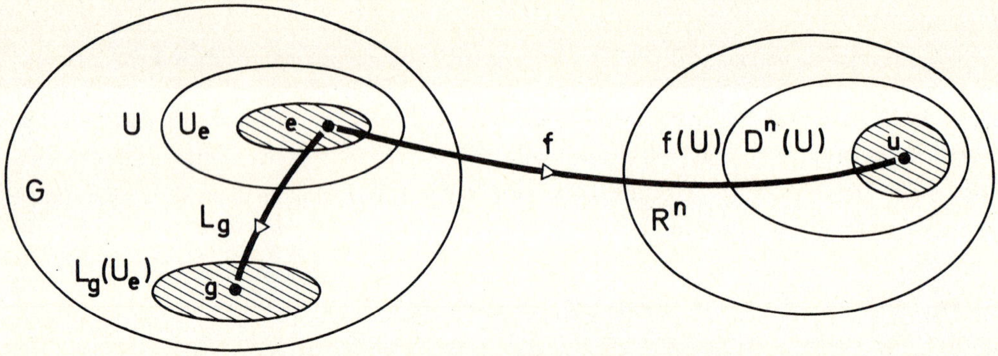

Fig. 7.4. Construction of a covering of G.

For any g in G, the left translation L_g being an homeomorphism of G onto itself, one gets an open set V_g which contains g and defined by (see Fig. 7.4) :

$$V_g = L_g(U_e) \quad .$$

The left invariance property of the measure implies that the volume of any open set V_g is bounded, indeed for all g in G :

$$g' = L_g g'' = gg'' \Rightarrow d\mu_\ell(g) = d\mu_\ell(g'') \quad , \qquad F(V_g;g') = F(U_e;g'') \quad ,$$

and thus :

$$\int_G d\mu_\ell(g')F(V_g;g') = \int_G d\mu_\ell(g'')F(U_e;g'') < \infty \quad .$$

This family of open sets $\{V_g;\ g \in G\}$ is clearly a <u>covering</u> of G. Assuming the group is compact it is possible to select in this family a finite covering of G, say : $\{V_{g_i} = L_{g_i}(U_e);\ i=1,\ldots,N\}$. One then has :

$$\forall\, g \in G \qquad \sum_{i=1}^{N} F(V_{g_i};g) \geq 1 \quad ,$$

and therefore, the measure being positive :

$$V_\ell = \int_G d\mu_\ell(g) \leq \sum_{i=1}^{N} \int_G d\mu_\ell(g)F(V_{g_i};g) \quad ,$$

the right-hand side of this inequality being itself bounded as the sum of N bounded integrals.

Since all the integrals considered above correspond to the integration of a positive measure and are bounded, they are well defined. The invariant measure $d\mu_\ell(g)$ being intrinsic, i.e. independent of the choice of an atlas, so it is with the volume V_ℓ .

7.1.2.2. Consequences

i) Setting f=1 in Eqs. (7–13a,b) yields for all g' in G :

$$\int_G d\mu_\ell(g) = C_\ell(g') \int_G d\mu_\ell(g) \quad , \quad \int_G d\mu_r(g) = C_r(g') \int_G d\mu_r(g) \quad ,$$

and therefore :

$$\forall g' \in G \qquad C_\ell(g') = C_r(g') = 1 \quad .$$

This implies, from Eq. (7–12), that :

$$\forall g,g' \in G \qquad d\mu_\ell(gg') = d\mu_\ell(g) \quad , \quad d\mu_r(g'g) = d\mu_r(g) \quad ,$$

namely $d\mu_\ell$ (resp. $d\mu_r$) is also a right (resp. left) invariant measure, hence :

Theorem 7.5. On a compact Lie group G the left invariant measure is also right invariant.

One then only says the invariant measure on G and denotes it as :

$$\forall g \in G \qquad d\mu(g) = d\mu_\ell(g) = d\mu_r(g) \quad .$$

Since the intrinsic volume of the group :

$$V = \int_G d\mu(g) \quad ,$$

is then finite, the measure can always be normalized such that this volume be equal to one.

Let us remark that the compact character of a Lie group is a sufficient but not necessary condition to get the same left and right invariant measures on a Lie group. For instance, the Abelian groups, even when noncompact, have the same left and right invariant measures. Also, we have already shown that the invariant measure on the linear group GL(n,R) is left and right invariant, although this group is noncompact.

ii) Similarly, setting f=1 in Eq. (7–15) yields C=1, and thus from Eq. (7–14) :

Theorem 7.6. On a compact Lie group G the invariant measure fulfils the relation

$$\forall g \in G \qquad d\mu(g) = d\mu(g^{-1}) \quad . \tag{7-19}$$

iii) The class of functions f of a compact Lie group G into the complex field for which the integral $\int_G d\mu(g)\, f(g)$ is well defined is $L^1(G,d\mu)$. Clearly, the bounded functions belong to this class.

7.1.2.3. Remarks about the expression of the invariant integral in terms of a parametrization

i) The problem is now to give the expression of the above intrinsic integrals in terms of a given parametrization of the elements of a compact n–parameter Lie group. It has already been noticed (Subsecs. 2.3.2.3 for SO(3), 4.1.4,i) and 4.2.4,i) for SU(2), and Appendix C.5.4.3,i) for U(1)) that in general to a special parametrization there corresponds only a local chart which does not cover the whole group. Nevertheless in practice, and in particular for the groups SU(2) and SO(3) (Subsecs. 7.1.3,4) and U(1) (ii) below), a parametrization can often be chosen such that the following theorem applies :

Theorem 7.7. Let (U,f) be a local chart on a compact n–parameter Lie group such that :
(1) the open set U does not cover the whole group :

$$U \subset G \qquad U \neq G \quad , \quad \forall g \in U \qquad f(g) = a \in R^n \quad ,$$

(2) the volume (i.e. the invariant integral of the characteristic function defined by Eq. (7–18)) of the complementary of U with respect to G, denoted by G–U, is equal to zero.

Then for any integrable function ψ of G into the complex field, one has :

$$\int_G d\mu(g)\psi(g) = \int_{\overline{f(U)}} d\mu(a)\psi(g_a) \quad , \tag{7-20}$$

where the right–hand side is a well defined integral taken over the closure (Def. C.39) $\overline{f(U)}$ of the open set f(U), i.e. a Riemann integral or its usual extensions since :
(1) the integration domain may or may not be bounded,
(2) the measure $d\mu(a)$ may or may not be well defined for a in the boundary $\overline{f(U)}-f(U)$.

Proof. The fact that the volume of G–U be zero implies :

$$\forall \varepsilon > 0 \quad \exists W_\varepsilon \text{ open set}, \quad \forall W \text{ open set}, \quad G\text{-}U \subset W \subset W_\varepsilon \Rightarrow \int_W d\mu(g) \leqslant \varepsilon \quad ,$$

or equivalently, denoting, respectively, by $V = G-W$ and $V_\varepsilon = G-W_\varepsilon$ the closed sets which are the complementary of the open sets W and W_ε with respect to G :

$$\forall \varepsilon > 0 \quad \exists V_\varepsilon \text{ closed set}, \quad \forall V \text{ closed set}, \quad V_\varepsilon \subset V \subset U \Rightarrow \int_{G-V} d\mu(g) \leqslant \varepsilon \quad ,$$

the above inclusion relations are illustrated on Fig. 7.5.

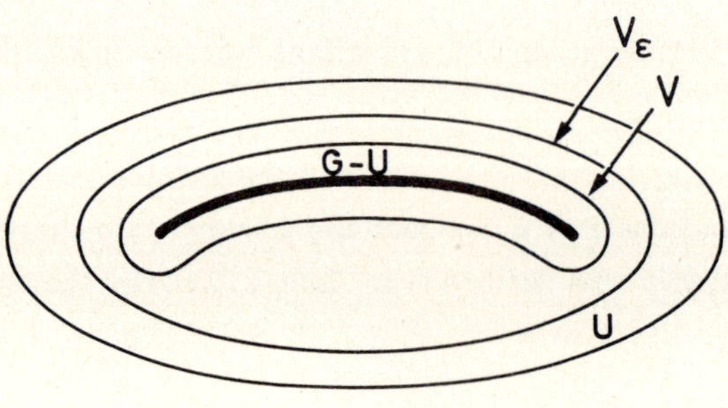

Fig. 7.5. The closed sets V and V_ε lie, respectively, between the pointed boundaries and the external boundary of the figure.

From the above equation one then gets :

$$|\int_{G-V} d\mu(g)\psi(g)| = |\int_G d\mu(g)\psi(g) - \int_V d\mu(g)\psi(g)| \leqslant \|\psi\| \int_{G-V} d\mu(g) < \|\psi\| \varepsilon \quad ,$$

where the norm of ψ is defined by :

$$\|\psi\| = \sup_{g\in G} |\psi(g)| \quad .$$

Moreover, V being a closed set included in the domain of definition of the local chart (U,f), one has :

$$\int_V d\mu(g)\psi(g) = \int_{f(V)} d\mu(a)\psi(g_a) \quad ,$$

where the right–hand side of the above equality is a well defined integral over a compact set f(V) in R^n. Therefore, one obtains finally :

$$\forall\varepsilon > 0 \qquad \exists V_\varepsilon \text{ closed set}, \quad \forall V \text{ closed set} \quad V_\varepsilon \subset V \subset U \Rightarrow$$

$$|\int_G d\mu(g)\psi(g) - \int_{f(V)} d\mu(a)\psi(g_a)| < \|\psi\| \varepsilon \quad .$$

This means that when the closed set V tends to U, the limit of the integral $\int_{f(V)} d\mu(a)\psi(g_a)$ exists and is equal to the intrinsic well defined integral over the compact Lie group $\int_G d\mu(g)\psi(g)$. Without giving details, in this limit the closed set f(V) tends to the closure $\overline{f(U)}$ of the open set f(U). If f(U) is bounded, $\overline{f(U)}$ is also bounded and being closed it is a compact set in R^n, otherwise $\overline{f(U)}$ is unbounded. Let us only illustrate these two cases for a one–parameter Lie group :

(1) $f(V_\varepsilon) = [\alpha+\varepsilon, \beta-\varepsilon] \quad , \quad f(U) =]\alpha,\beta[\quad , \quad \overline{f(U)} = [\alpha,\beta] \quad ,$

(2) $f(V_\varepsilon) = [\alpha+\varepsilon, \frac{1}{\varepsilon}] \quad , \quad f(U) =]\alpha, +\infty[\quad , \quad \overline{f(U)} = [\alpha, +\infty[\quad .$

Moreover, the measure $d\mu(a)$ may be undefined for a in $\overline{f(U)}-f(U)$. This case corresponds to the usual extension of the definition of an integral in R^n, illustrated by the following example :

$$f(U) =]0,1[\quad , \quad \overline{f(U)} = [0,1] \quad , \quad d\mu(a) = \frac{1}{\sqrt{a(1-a)}} da \quad ,$$

$$\int_0^1 da \frac{1}{\sqrt{a(1-a)}} \psi(a) = \lim_{\varepsilon\to+0} \int_\varepsilon^{1-\varepsilon} da \frac{1}{\sqrt{a(1-a)}} \psi(a) \quad .$$

ii) As a simple example let us apply the above theorem to the invariant integral over the group U(1). Indeed, it is shown in Appendix C.5.4.3,i) that U(1) is a compact one–parameter Lie group, such that to the parametrization of its elements by a real θ as : $U = e^{i\theta}$, corresponds only the local chart, not defined for the unit element :

$$(U(1)-\mathbf{1},f) \quad , \forall U \in U(1)-\mathbf{1} \qquad f(U) = \theta \in]0,2\pi[\quad .$$

In terms of this parametrization the invariant measure is clearly $d\theta$, and moreover the subset $\{\mathbf{1}\}$, on which the considered local chart is not defined, has a volume equal to zero. Therefore, for any integrable function ψ of U(1) into the complex field :

$$\int_{U(1)} d\mu(U)\psi(U) = \int_0^{2\pi} d\theta \; \psi(e^{i\theta}) \quad .$$

In particular the volume of the group (ψ =1) is :

$$V = \int_{U(1)} d\mu(U) = \int_0^{2\pi} d\theta = 2\pi \quad .$$

In this simple case the invariant integral leads to a proper Riemann integral since on one hand the integration domain is the compact set $[0,2\pi]$, and on the other hand the measure is well defined everywhere over this closed set.

7.1.3. APPLICATION TO THE GROUP SU(2)

Theorem 7.5. (Subsec. 7.1.2.2) applies to the compact Lie group SU(2) (Appendix C.5.4.3,i). Hence, there exists a uniquely defined, up to a multiplicative constant, left and also right invariant measure :

$$\forall U,V \in SU(2) \qquad d\mu(U) = d\mu(VU) = d\mu(UV) \quad .$$

The problem is now to determine this measure in terms of a given parametrization of the group. In Subsec. 7.1.3.1, applying the general formulae derived previously, $d\mu(U)$ is calculated when U is parametrized by $\{\theta n^i, i=1,2,3\}$. On the other hand, in Subsec. 7.1.3.2 another general approach to the problem of the invariant measures on a Lie group is applied to the case of SU(2).

7.1.3.1. Application of the general formulae

i) To the parametrization of the elements of SU(2) by $\{\theta^i = \theta n^i, i=1,2,3\} = \{\theta,\hat{n};a\}$ there corresponds a local chart which is only defined over the open set : SU(2), $-\mathbf{1}$ excluded (Subsec. 4.1.4,i)). There exists a one-to-one correspondence between this open set and the open three-dimensional ball D^3 of radius 2π. On the other hand, all the points of the three-dimensional sphere S^2 of radius 2π correspond to the element $-\mathbf{1}$ in SU(2), for which the local chart is not defined. These points of S^2 are called the exceptional points of the parameter space (here the closed ball D^3) for the particular parametrization adopted.

ii) Since, with respect to this parametrization, the local chart is defined for the unit element of the group (parametrized by $\theta^i = 0$, which is not an exceptional point), the more simple general formulae given by Eq. (7-8) can be applied. Let us denote this invariant measure and the corresponding density as :

$$\forall U = U(\theta,n^i) = U(\theta,\hat{n};a) \qquad d\mu(U) = d\mu(\theta,n^i) = d\mu(\theta,\hat{n};a) = \mu(\theta,\hat{n};a)d\theta d_2\hat{n}, \qquad (7\text{-}21)$$

where $d_2\hat{n}$ is the surface element of the three-dimensional sphere of radius one, such that if α and β are the polar angles of $\hat{n}$, Eq. (2–20), with respect to a given orthonormal basis $a = (e_i)$, one has :

$$\hat{n} = \sin\alpha\cos\beta\ e_1 + \sin\alpha\sin\beta\ e_2 + \cos\alpha\ e_3 \quad , \quad d_2\hat{n} = \sin\alpha\ d\alpha\ d\beta \ . \tag{7-22}$$

Actually the density $\mu(\theta,\hat{n};a)$ is independent of $\hat{n}$. Indeed, from the left and right invariant character of the measure, one has for all elements VUV^{-1} which belong to the same class as U of conjugate elements under SU(2) :

$$d\mu(VUV^{-1}) = d\mu(U) \quad .$$

Eq. (4–34) then yields :

$$d\mu(\theta,\hat{n};a) = d\mu[VU(\theta,\hat{n};a)V^{-1}] = d\mu[U(\theta,R(V;a)\hat{n};a)] \quad ,$$

where R(V;a) is the rotation associated to V by Eq. (4–26), and therefore, for any rotation R :

$$\mu(\theta,\hat{n};a)d\theta\ d_2\hat{n} = \mu(\theta,R\hat{n};a)\ d\theta\ d_2\ R\hat{n} \quad .$$

Since the surface element $d_2\hat{n}$ is invariant under all rotations R, one gets the announced result :

$$\mu(\theta,\hat{n};a) = \mu(\theta,R\hat{n};a) = \mu(\theta) \quad .$$

This property corresponds to the fact that in the parameter space, which is the three-dimensional ball of radius 2π, there does not exist any preferential direction. Therefore, let us take for simplicity $U(\theta,e_3;a)$, then in terms of the parameters θ^i the scheme of Fig. 7.1 is now given by Fig. 7.6.

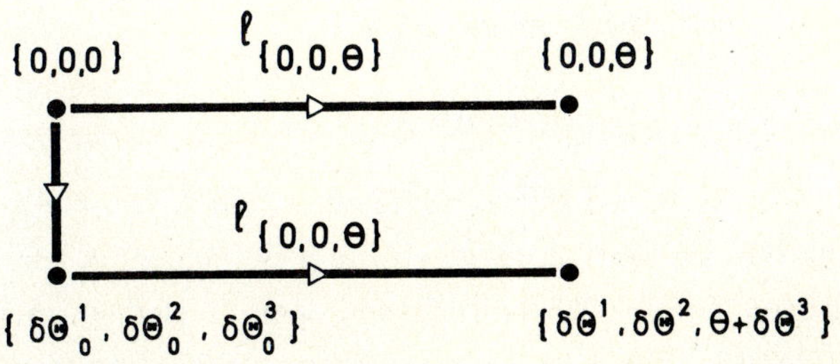

Fig. 7.6. Transformation of an infinitesimal volume element around u = {0,0,0} in the three-dimensional ball of radius 2π under a left translation parametrized by $\{0,0,\theta\}$.

Expliciting from Eq. (4–19) the relation :

$$U(\delta\theta^1,\delta\theta^2,\theta + \delta\theta^3) = U(\delta\theta^1_0,\delta\theta^2_0,\delta\theta^3_0)U(0,0,\theta) \quad ,$$

up to first order in $\delta\Theta_0^i$ and $\delta\Theta^i$, yields :

$$\delta\Theta^i = J^i_j\,\delta\Theta^j_0 \quad ,$$

$$(J^i_j) = \begin{pmatrix} \frac{\theta}{2}\cot g\frac{\theta}{2} & \frac{\theta}{2} & 0 \\ -\frac{\theta}{2} & \frac{\theta}{2}\cot g\frac{\theta}{2} & 0 \\ 0 & 0 & 1 \end{pmatrix} \quad , \quad \det J = \frac{\theta^2}{2(1-\cos\theta)} \quad .$$

Finally from Eq. (7–8), one gets the invariant measure on SU(2) up to a multiplicative factor $\mu(0)$:

$$d\mu(\theta,n^i) = \mu(0)\,\frac{2(1-\cos\theta)}{\theta^2}\prod_{i=1}^{3} d\Theta^i = \mu(0)\;2(1-\cos\theta)d\theta\;d_2\hat{n} \quad . \tag{7-23}$$

iii) Since the local chart associated to the parametrization of SU(2) by $\{\theta n^i, i=1,2,3\}$ is not defined only for the subset $\{-\mathbf{1}\}$ of SU(2), which has clearly an invariant volume equal to zero :

$$\int_{SU(2)} d\mu(U)F(\{-\mathbf{1}\};U) = 0 \quad ,$$

one can apply Theorem 7.7. (Subsec. 7.1.2.3). Thus, for any function f of SU(2) into the complex field which belongs to $L^1(SU(2),d\mu)$, the invariant integral reads, choosing the multiplicative constant $\mu(0) = 1$:

$$\int_{SU(2)} d\mu(U)f(U) = \int_0^{2\pi} d\theta\;2(1-\cos\theta)\int_{S^2} d_2\hat{n}\;f[U(\theta,\hat{n};a)] \quad . \tag{7-24}$$

Actually, as for the simple example of U(1) given in Subsec. 7.1.2.3,ii), the above integral is a proper Riemann integral since the integration domain is compact (the closed ball D^3 of radius 2π) and the density is well defined even on the boundary of this domain. With the adopted normalization, the invariant volume of the group is :

$$\int_{SU(2)} d\mu(U) = 16\,\pi^2 \quad . \tag{7-25}$$

Let us recall that the class functions defined on SU(2), i.e. the functions which take the same value for all the elements of a given class of conjugate elements under SU(2) (for instance the character, Eq. (6–47)), only depend upon θ (Subsec. 4.2.2). Therefore, for the class functions $f_c(\theta)$ in $L^1(SU(2),d\mu)$, the integration over the parameter space can first be carried out over the sphere of radius θ, one gets :

$$\int_{SU(2)} d\mu(U)f_c(U) = 8\pi\int_0^{2\pi} d\theta\;(1-\cos\theta)\;f_c(\theta) \quad . \tag{7-26}$$

iv) It should be noticed that the same method as above, using Eq. (7–8), when the elements of SU(2) are parametrized by the Euler angles, Eq. (4–38), cannot be applied to compute the invariant measure $d\mu(\varphi,\theta,\psi)$. Indeed, the local chart which corresponds to this parametrization is not defined in particular for the unit element, parametrized by $u = \{\varphi = 0, \theta = 0, \psi = 0\}$. With respect to this parametrization the point u is thus an exceptional point, as it is with other surface points of the volume of the parameter space represented on Fig. 4.3. This means that there does not exist any neighbourhood of u in which the one-to-one correspondence between the points of the parameter space and the elements of the group is realized. Indeed, up to first order in $\delta\varphi$, $\delta\theta$, and $\delta\psi$, $U(\delta\varphi, \delta\theta, \delta\psi)$ only depend upon the two quantities $\delta\theta$ and $\delta\varphi + \delta\psi$, from Eq. (4-38) :

$$U(\delta\varphi,\delta\theta,\delta\psi) = \mathbf{1} + \frac{1}{2}\begin{pmatrix} -i(\delta\varphi+\delta\psi) & -\delta\theta \\ \delta\theta & i(\delta\varphi+\delta\psi) \end{pmatrix} .$$

If nevertheless, one wants to apply the general formulae derived in Subsec. 7.1.1.1, to compute the invariant measure $d\mu(\varphi, \theta, \psi)$, one has to start from a nonexceptionnal point $a = \{\varphi, \theta, \psi\}$, but this leads to tedious calculations.

Although, once again the local chart which corresponds to the parametrization of the elements of SU(2) by the Euler angles does not cover the whole group, the invariant integration can be expressed thanks to Theorem 7.7, cf. Eq. (7-38).

7.1.3.2. Riemannian manifold structure of SU(2)

The method consists in defining a metric ds^2 on the group SU(2) so as to give to this set a Riemannian manifold structure. Then, using usual tensorial analysis, one derives from ds^2 a measure which is equal to the volume element. The invariant character of this measure is insured by considering a metric which is itself invariant.

Let us denote by dU the variation in SU(2) of an element U in the group. For a given parametrization of U by $a = \{a^i \in R, i=1,2,3\}$ one has :

$$U(a+da) = U(a) + dU(a) \quad , \quad dU(a) = \frac{\partial U(a)}{\partial a^i}\, da^i \quad . \tag{7-27}$$

The quantity $Tr(dUdU^+)$ is clearly a real positive definite quadratic form on the group. In terms of the parameters a^i it reads :

$$Tr(dU\, dU^+) = Tr\left(\frac{\partial U}{\partial a^i}\frac{\partial U^+}{\partial a^j}\right) da^i\, da^j \quad . \tag{7-28}$$

Moreover, this expression is left as well as right invariant, since from the cyclic invariance of the trace one has for any fixed V in SU(2) :

$$Tr[d(VU)d(VU)^+] = Tr(VdU\; dU^+V^+) = Tr(dUdU^+) = Tr[d(UV)d(UV)^+] \quad .$$

Hence, the expression :

$$\forall U \in SU(2) \qquad ds^2(U) = Tr(dU\, dU^+) \quad , \tag{7-29}$$

defines a left and also right invariant metric on the group SU(2) :

$$\forall\, U,V \in SU(2) \qquad ds^2(U) = ds^2(VU) = ds^2(UV) \quad ,$$

which endows this Lie group with a Riemannian manifold structure. It is then recalled that in such a manifold with the metric :

$$ds^2(a) = g_{ij}(a)\, da^i\, da^j \quad , \tag{7-30}$$

the volume element is given by :

$$d\mu(a) = \sqrt{|\det g(a)|}\, \prod_i da^i \quad . \tag{7-31}$$

In our case, the metric tensor which corresponds to a given parametrization reads :

$$a \in R^3 \qquad g_{ij}(a) = \mathrm{Tr}\left[\frac{\partial U(a)}{\partial a^i}\, \frac{\partial U(a)^+}{\partial a^j}\right] \quad . \tag{7-32}$$

From the invariance of ds^2, the measure $d\mu$ (a) is also left and right invariant. Moreover, since the invariant measure on a compact Lie group is unique up to a multiplicative constant, the expression given by Eq. (7–31) has to be proportional to the result obtained previously in Eq. (7–23).

The expression of ds^2 given by Eq. (7–29) is intrinsic, i.e. independent of the parametrization of the elements of SU(2). Since this metric is clearly invariant under the change of U into U^+, the same holds true in respect of the measure :

$$\forall\, U \in SU(2) \qquad d\mu(U) = d\mu(U^+) = d\mu(U^{-1}) \quad , \tag{7-33}$$

which corresponds to the general property for compact Lie groups given by Theorem 7.6, Eq. (7–19). Moreover, ds^2 is also invariant under the change of U into –U, thus the same property holds for $d\,\mu(U)$:

$$\forall\, U \in SU(2) \qquad d\mu(U) = d\mu(-U) \quad . \tag{7-34}$$

Let now U in SU(2) be parametrized by $\{\Theta^i = \theta\, n^i,\ i=1,2,3\}$, one has :

$$\Theta = \theta\,\hat{n} \quad , \quad \hat{n}^2 = 1 \quad , \quad \hat{n}.d\hat{n} = 0 \quad ,$$

$$d\Theta = \theta d\hat{n} + \hat{n}\, d\theta \quad , \quad (d\Theta^2) = \theta^2 (d\hat{n})^2 + (d\theta)^2, \quad \Theta.d\Theta = \theta d\theta \quad ,$$

and from Eq. (4–20) :

$$dU = -\left(\sin\tfrac{\theta}{2} + i\sigma.\hat{n}\cos\tfrac{\theta}{2}\right)\frac{d\theta}{2} - i\,\sin\tfrac{\theta}{2}\,\sigma.d\hat{n} \quad .$$

Using Eqs. (4–85,89b) and the above relations, one obtains the expression of the metric :

$$ds^2 = \frac{1}{2}(d\theta)^2 + (1-\cos\theta)(d\hat{n})^2 = \frac{1-\cos\theta}{\theta^2}(d\Theta)^2 + \left(\frac{1}{2} - \frac{1-\cos\theta}{\theta^2}\right)(\hat{n}.d\Theta)^2 \quad , \tag{7-35}$$

and thus, after a simple calculation of a determinant, the corresponding measure is :

$$d\mu(\theta, n^i) = \frac{1-\cos\theta}{4\theta^2} \prod_{i=1}^{3} d\Theta^i = \frac{1}{4}(1-\cos\theta)d\theta d_2\hat{n} \quad , \tag{7-36}$$

where $d_2\hat{n}$ is the surface element on the three-dimensional sphere of radius one. To recover the normalization of Eq. (7–24) one should have taken $ds^2 = 2Tr(dUdU^+)$. When the elements of SU(2) are parametrized by the Euler angles, Eq. (4–38), a direct calculation yields with the above normalization :

$$ds^2 = 2Tr(dUdU^+) = (d\varphi)^2 + 2\cos\theta d\varphi d\psi + (d\theta)^2 + (d\psi)^2 \quad , \tag{7-37}$$

and for the invariant measure :

$$d\mu(\varphi, \theta, \psi) = \sin\theta \; d\varphi \; d\theta \; d\psi \quad . \tag{7-38}$$

Actually, the above method, based on the existence of a Riemannian manifold structure of a Lie group is general. Let us only sketch without proof the outline of this approach for an n–parameter Lie group G with a simple Lie algebra over the real field.

One considers the tangent vector spaces (Def. C.55) $\mathcal{T}(e)$ and $\mathcal{T}(g)$ to G at, respectively, the unit element e and any other element g (see Fig. 7.7). By definition, a left (resp. right) invariant infinitesimal transformation (Def. C.61) J on G assigns to any element g a tangent vector J(g) to G at this point such that the tangent vectors J(e) and J(g) are related by the differential (Def. C.56) dL_g (resp. dR_g) of the left (resp. right) translation :

$$\forall g \in G \qquad dL_g[J(e)] = J(g) \quad .$$

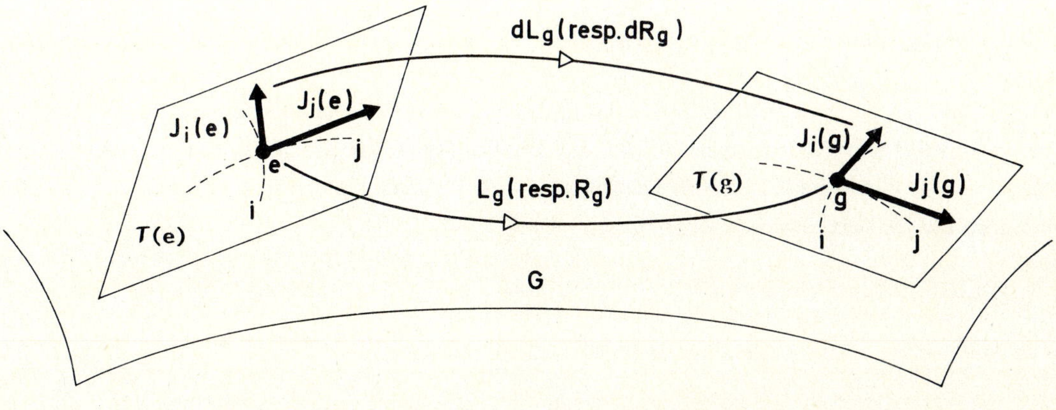

Fig. 7.7. If g in G is parametrized by $a = \{a^i \in R; i=1,...,n\}$, then only the parameter a^i is varied along the curve labelled by i.

The tangent vector space $\mathcal{T}(g)$ to G at any point g is isomorphic to the Lie algebra of the group (Prop. C.81). This Lie algebra is simple, then all the trace forms (Def. B.24) which correspond to faithful finite–dimensional representations are nondegenerate (Props. B.26,34) and moreover proportional (Prop. B.41). Thus, for instance the Killing form defines in $\mathcal{T}(e)$ an inner product which endows this vector space with a pseudo–Euclidean structure. By left (resp. right) translation it does so for every tangent vector space $\mathcal{T}(g)$. The inner product thus defined in every g induces a metric $ds^2(g)$ and by Eq. (7–31) a measure $d\mu(g)$. One shows, using the invariance of the Killing form (Props. B.31,32), that this measure is left (resp. right) invariant on G.

Let us check for the group SU(2) (more generally for SU(n) for instance) that the metric tensor $(g_{ij}(a))$ defined by Eq. (7–32) is proportional to the metric tensor of the trace form which corresponds to the faithful fundamental representation (Def. B.22, cf. Appendix B.4.2.4,ii)). Applying for instance a left translation to both sides of Eq. (7–27) so as to consider a neighbourhood of the identity, one obtains :

$$U(a)^{-1}U(a+da) = \mathbf{1}+U(a)^{-1}dU(a) = \mathbf{1}-iW(a) \quad ,$$

where :

$$W(a) = iU(a)^{-1}dU(a) = iU(a)^{-1}\frac{\partial U(a)}{\partial a^i}da^i = W_i(a)da^i \quad ,$$

belongs to the Lie algebra $\mathcal{U}(2)$. With respect to the basis $(W_i(a))$, the metric tensor of the trace form which corresponds to the fundamental representation reads :

$$(W_i(a),W_j(a))_F = Tr[W_i(a)W_j(a)] \quad .$$

Then, the elements of $\mathcal{U}(2)$ being Hermitian matrices, and from the cyclic invariance of the trace, one gets the announced result :

$$(W_i(a),W_j(a))_F = Tr[W_i(a)W_j^+(a)] = Tr\left[U(a)^{-1}\frac{\partial U(a)}{\partial a^i}\frac{\partial U(a)^+}{\partial a^j}U(a)\right] = g_{ij}(a) \quad .$$

7.1.4. APPLICATION TO THE GROUP SO(3)

i) Let us show that the invariant measure on the group SO(3) is related to the one on the group SU(2) by :

$$\forall\ R\in SO(3) \qquad d\mu(R) = d\mu(U) \quad , \quad U\in SU(2) \quad , \tag{7-39}$$

where the matrices U and R are related by Eq. (4–28). The two–to–one homomorphism between SU(2) and SO(3) (Subsec. 4.2.1), and the Eq. (7–34) imply that $d\mu(R)$ is a well defined object for all elements R in SO(3). Moreover, $d\mu(R)$ is clearly an invariant measure, since, for instance for the left invariance one has :

$$\forall U,U'\in SU(2) \qquad R=R(U) \quad , \quad R'=R(U') \quad , \quad R'R=R(U'U) \quad ,$$

$$d\mu(R'R) = d\mu(U'U) = d\mu(U) = d\mu(R) \quad .$$

The invariant measure on the compact Lie group SO(3) being unique up to a multiplicative factor, this ends the proof. In terms of the parameters θ, n^i and φ, θ, and ψ, this invariant measure on SO(3) reads, respectively, as for the group SU(2) :

$$d\mu(\theta,n^i) = 2(1-\cos\theta)d\theta d_2\hat{n} \quad , \quad d\mu(\varphi,\theta,\psi) = \sin\theta d\theta d\varphi d\psi \quad , \tag{7-40}$$

the parameter spaces of these variables being, respectively, defined in Eqs. (2–33,39).

ii) Following the method discussed for SU(2) in Subsec. 7.1.3.2, the quantity :

$$\forall R \in SO(3) \qquad ds^2(R) = \mathrm{Tr}\ (dR\ dR^t) \qquad ,$$

is a left and also right invariant metric on the group SO(3) which endows this set with the structure of a Riemannian manifold. For a given parametrization the invariant measure on SO(3) can be obtained from the metric tensor using Eq. (7–31).

7.2. ORTHOGONALITY AND CLOSURE RELATIONS

7.2.1. GENERALITIES FOR COMPACT LIE GROUPS

i) On a compact Lie group G, the general results derived in Subsec. 7.1.2 allow us to write :

$$\forall g' \in G \qquad \int_G d\mu(g) f(g) = \int_G d\mu(g) f(g'g) = \int_G d\mu(g) f(gg') \qquad , \qquad (7\text{-}41)$$

for all functions f of G into the complex field which belong to $L^1(G, d\mu)$, e.g. the bounded functions. These relations play the role of the equations for finite groups given in the introduction of this chapter. Hence, the properties proved for the finite–dimensional representations of finite groups in Appendix C.4 : Props. C.54, 55, 57, and 58 can be directly extended to the finite–dimensional representations of compact Lie groups.

Theorem 7.8. The finite–dimensional representations of a compact Lie group in a vector space over the complex field are equivalent to unitary representations. These representations are either irreducible or completely reducible (cf. Prop. C.54, this result can also be extended to infinite–dimensional representations[30]).

Theorem 7.9. Orthogonality relations for finite–dimensional unitary representations of compact Lie groups in vector spaces over the complex field (cf. Prop. C.55). The order n of the finite group is replaced by the volume V of the compact Lie group G. The classes of conjugate elements under G depend continuously upon parameters, still denoted by i as for finite groups. The number n_i of elements in the class i is replaced by the volume of the class :

$$V_i = \int_{g \in \text{class } i} d\mu(g) \qquad , \qquad (7\text{-}42)$$

and the fact that the group G is the disjoint union of classes of conjugate elements under G reads :

$$V = \int di\ V_i \qquad . \qquad (7\text{-}43)$$

Let $D^\alpha(G)$ and $D^\beta(G)$ denote two unitary irreducible representations of finite dimensions, respectively, d_α and d_β, in vector spaces over the complex field. Their characters are, respectively, χ_i^α and χ_i^β for the element g in the class i. One then has :

$$\int_G d\mu(g)\; D^\alpha(g)^i_k\; D^\beta(g)^{j*}_\ell \;=\; \frac{V}{d_\alpha}\;\delta_{k\ell}\;\delta^{ij}\;\delta^{\alpha\beta} \quad , \tag{7-44a}$$

$$\int di\; V_i\;\chi_i^\alpha\;\chi_i^{\beta*} \;=\; V\;\delta^{\alpha\beta} \quad . \tag{7-44b}$$

Theorem 7.10. Let D(G) be a finite-dimensional reducible representation of a compact Lie group G, in a vector space over the complex field. Then, the decomposition of D(G) into the finite-dimensional irreducible representations $D^\alpha(G)$:

$$D(G) = \sum_\alpha {}^+\, a_\alpha\, D^\alpha(G) \quad , \qquad \chi_i = \sum_\alpha a_\alpha\, \chi_i^\alpha \quad ,$$

where a_α are non-negative integers, is unique and completely determined by the characters χ_i of the representation. One has :

$$a_\alpha = \frac{1}{V}\int d_i V_i \chi_i^{\alpha*}\chi_i \quad , \tag{7-45a}$$

$$\sum_\alpha a_\alpha^2 = \frac{1}{V}\int d_i V_i\, |\chi_i|^2 \quad , \tag{7-45b}$$

where the sum over α runs over the values of α which occur in the expansion of D(G) (cf. Prop. C.57).

Theorem 7.11. A necessary and sufficient condition for two finite-dimensional representations of a compact Lie group, in vector spaces over the complex field, to be equivalent is that they have the same characters (cf. Prop. C.58).

ii) Completeness and closure relation :

Theorem 7.12. Let us consider the set $L^2(G,d\mu)$ of square integrable functions of a compact Lie group G into the complex field, with the invariant measure $d\mu$. Then, the functions $\sqrt{d_\alpha}\, D^\alpha(g)^i_j$, where $D^\alpha(g)^i_j$ are the matrix elements of a complete set of nonequivalent irreducible unitary representations $D^\alpha(G)$ of dimension d_α, form a complete set of functions of G into the complex field, with respect to the invariant measure on the group :

$$\sum_{\alpha,i,j} d_\alpha\, D^\alpha(g)^i_j\, D^\alpha(g')^{i*}_j \;=\; V\,\delta(g,g') \quad , \tag{7-46}$$

where $\delta(g,g')$ is the kernel of the unity operator on $L^2(G,d\mu)$, i.e. :

$$\forall f\in L^2(G,d\mu) \qquad \int_G d\mu(g')\;\delta(g,g')f(g') = f(g) \quad . \tag{7-47}$$

This is the Peter–Weyl theorem (cf. for instance Pontryagin[44], Chap. 5, Sec. 33, or Vilenkin[61], § 4.3, Theorem 1). Then, from the orthogonality relation given by Eq. (7–44a), the set of functions $\sqrt{d_\alpha/V}\, D^\alpha(g)^i_j$ forms a complete orthonormal basis of $L^2(G,d\mu)$. The proof of the above theorem is given in Subsec. 7.2.2.2 only for the group SU(2).

7.2.2. ORTHOGONALITY AND CLOSURE RELATIONS FOR THE GROUP SU(2)

7.2.2.1. Orthogonality relations

With the following normalization of the invariant measure on SU(2) :

$$\forall U \in SU(2) \qquad d\mu(U) = 2(1-\cos\theta)d\theta d_2\hat{n} = \sin\theta \; d\varphi d\theta d\psi \quad . \tag{7-48}$$

the volume of the group is given by Eq. (7–25), i.e. $16\pi^2$. The classes of conjugate elements under SU(2) are characterized by the continuous variable $\theta(0 \leqslant \theta \leqslant 2\pi)$. The volume of the class characterized by θ is defined by :

$$V(\theta) = \int_{SU(2)} d\mu(\theta';\hat{n};a)\,\delta(\theta'-\theta) = 8\pi(1-\cos\theta) \quad . \tag{7-49}$$

The irreducible representations $\mathcal{D}^j[SU(2)]$ of dimension 2j+1, and the characters $\chi^j(\theta)$ have been defined in Sec. 6.1. The representations of SU(2) are unitary and irreducible or completely reducible in agreement with these general features for representations of compact Lie groups. From Eqs. (7–44a,b) the orthogonality relations read :

$$\int_{SU(2)} d\mu(U)\; \mathcal{D}^j_{mn}(U)\; \mathcal{D}^{j'*}_{m'n'}(U) = \frac{16\pi^2}{2j+1}\,\delta_{jj'}\,\delta_{mm'}\,\delta_{nn'} \quad , \tag{7-50a}$$

$$\int_0^{2\pi} d\theta\; 8\pi(1-\cos\theta)\chi^j(\theta)\chi^{j'}(\theta) = 16\pi^2\,\delta_{jj'} \quad . \tag{7-50b}$$

Explicitly in terms of the Euler angles, the first orthogonality relation is :

$$\int_0^{2\pi} d\varphi \int_0^{\pi} d\theta\,\sin\theta \int_0^{4\pi} d\psi\; \mathcal{D}^j_{mn}(\varphi,\theta,\psi)\; \mathcal{D}^{j'*}_{m'n'}(\varphi,\theta,\psi) = \frac{16\pi^2}{2j+1}\,\delta_{jj'}\,\delta_{mm'}\,\delta_{nn'} \;. \tag{7-51}$$

This relation can be checked using the expression of the function $\mathcal{D}^j_{m'm}(\varphi,\theta,\psi)$ given by Eqs. (6–18,21). Similarly the orthogonality relation for the characters, which reads :

$$\int_0^{2\pi} d\theta(1-\cos\theta)\;\chi^j(\theta)\;\chi^{j'}(\theta) = 2\pi\,\delta_{jj'} \quad , \tag{7-52}$$

can be checked directly using the expression of characters given by Eq. (6–47).

The orthogonality relation of the functions $\mathcal{D}^j_{m'm}$, and the decomposition of the direct product of two irreducible representations expressed by Eqs. (6–68a,b), yields :

$$\frac{2j_3+1}{16\pi^2}\int_{SU(2)} d\mu(U)\mathcal{D}^{j_1}_{m_1'm_1}(U)\mathcal{D}^{j_2}_{m_2'm_2}(U)\mathcal{D}^{j_3*}_{m_3'm_3}(U) = \langle j_1j_2m_1'm_2'|j_3m_3'\rangle\langle j_1j_2m_1m_2|j_3m_3\rangle \;, \tag{7-53}$$

or in terms of the Wigner 3j symbols from Eq. (6–75) :

$$\frac{1}{16\pi^2}\int_{SU(2)} d\mu(U)\mathcal{D}^{j_1}_{m_1'm_1}(U)\mathcal{D}^{j_2}_{m_2'm_2}(U)\mathcal{D}^{j_3}_{m_3'm_3}(U) = \begin{pmatrix} j_1 & j_2 & j_3 \\ m_1' & m_2' & m_3' \end{pmatrix}\begin{pmatrix} j_1 & j_2 & j_3 \\ m_1 & m_2 & m_3 \end{pmatrix} \;. \tag{7-54}$$

Starting from the expression of the matrices $\mathcal{D}^j$ given by Eq. (6–14), the above relations provide another method to compute the C.G. coefficients or the Wigner 3j symbols (Wigner[63], Chap. 17, Sec. 7).

7.2.2.2. Closure relations

i) One considers the set $L^2(SU(2),d\mu)$ of square integrable functions of SU(2) into the complex field with the invariant measure $d\mu(U)$, i.e. the Hermitian vector space of complex functions defined on SU(2) with the Hermitian inner product :

$$\forall f,g \in L^2[SU(2),d\mu] \qquad (f,g)_{SU(2)} = \int_{SU(2)} d\mu(U) f^*(U) g(U) \quad , \tag{7-55}$$

such that :

$$\|f\|^2 = (f,f)_{SU(2)} < \infty \quad .$$

The kernel $\delta(U,U')$ of the identity operator on this function space is defined by :

$$\forall f \in L^2[SU(2),d\mu] \qquad \int_{SU(2)} d\mu(U')\delta(U,U')f(U') = f(U) \quad , \tag{7-56}$$

It is thus related to the normalization of the measure by :

$$\int_{SU(2)} d\mu(U')\delta(U,U') = 1 \quad .$$

Therefore when U is parametrized, respectively, by $\theta,\hat{n}$ and φ,θ, and ψ, the measures and the corresponding kernels read :

$$\left\{\begin{aligned} &d\mu[U(\theta,\hat{n};a)] = 2(1-\cos\theta)d\theta d_2\hat{n} = 4(\sin\tfrac{\theta}{2})^2\, d\theta d_2\hat{n} \quad , && (7\text{-}57a)\\ &\delta[U(\theta,\hat{n};a),U(\theta',\hat{n}';a)] = \frac{1}{8\sin\frac{\theta}{2}}\,\delta(\cos\tfrac{\theta}{2} - \cos\tfrac{\theta'}{2})\;\delta_2(\hat{n},\hat{n}') \quad , && (7\text{-}57b)\end{aligned}\right.$$

$$\left\{\begin{aligned} &d\mu[U(\varphi,\theta,\psi)] = \sin\theta\; d\varphi d\theta d\psi \quad , && (7\text{-}58a)\\ &\delta[U(\varphi,\theta,\psi),U(\varphi',\theta',\psi')] = \delta(\varphi-\varphi')\delta(\cos\theta-\cos\theta')\delta(\psi-\psi') \quad , && (7\text{-}58b)\end{aligned}\right.$$

where for instance $\delta(\varphi-\varphi')$ denotes the kernel of the identity mapping such that :

$$\int d\varphi'\;\delta(\varphi-\varphi')f(\varphi') = f(\varphi) \quad .$$

All these δ symbols are Dirac measures if one considers infinitely differentiable functions.

Similarly, one considers below the Hermitian vector space $L^2(S^2,d_2\hat{n})$ of square integrable complex functions defined on the two-sphere S^2 of radius one with the invariant measure under rotation. This measure and the corresponding kernel of the identity operator on $L^2(S^2,d_2\hat{n})$ are, respectively :

$$d_2\hat{n}(\alpha,\beta) = \sin\alpha\; d\alpha\; d\beta \quad , \tag{7-59a}$$

$$\delta_2(\hat{n},\hat{n}') = \delta(\cos\alpha-\cos\alpha')\ \delta(\beta-\beta') \quad . \tag{7-59b}$$

ii) Let us first derive a closure relation for the characters $\chi^j(U)$, similar to the closure relation for the characters of finite groups given by Prop. C.66. From the expression of characters given by Eq. (6–47), one gets :

$$\sum_{j=0,\frac{1}{2},\ldots} \chi^j[U(\theta,n^i)]\chi^j[U(\theta',n^i)] = \frac{1}{\sin\frac{\theta}{2}\sin\frac{\theta'}{2}} \sum_{j=0,\frac{1}{2},\ldots} \sin(\tfrac{2j+1}{2}\theta)\sin(\tfrac{2j+1}{2}\theta')$$

$$= \frac{1}{2\sin\frac{\theta}{2}\sin\frac{\theta'}{2}} \sum_{q=1}^{\infty} \left[\cos(q\tfrac{\theta-\theta'}{2}) - \cos(q\tfrac{\theta+\theta'}{2})\right] . \tag{7-60}$$

Moreover, in the sense of distributions, one has :

$$\sum_{q=1}^{\infty} \cos qx = -\frac{1}{2}+\pi \sum_{q=-\infty}^{+\infty} \delta(x-2\pi q) = -\frac{1}{2}+\frac{\pi}{2}\ \delta(\sin\tfrac{x}{2}) \quad , \tag{7-61}$$

and therefore :

$$\sum_{j=0,\frac{1}{2},\ldots} \chi^j(\theta)\chi^j(\theta') = \frac{\pi}{4\sin\frac{\theta}{2}\sin\frac{\theta'}{2}} \left[\delta(\sin\tfrac{\theta-\theta'}{4}) - \delta(\sin\tfrac{\theta+\theta'}{4})\right]$$

$$= \frac{\pi}{2\sin\frac{\theta}{2}}\ \delta(\cos\tfrac{\theta}{2} - \cos\tfrac{\theta'}{2}) \quad . \tag{7-62}$$

By comparison with Eq. (7–57b), the above relation can be written as :

$$\sum_{j=0,\frac{1}{2},\ldots} \chi^j(U)\chi^j(U') = 16\pi^2\ \delta_c(U,U') \quad , \tag{7-63}$$

which is the completeness relation of the χ^j on $L^2(SU(2),d\mu)$ restricted to the class functions, $\delta_c(U,U')$ denotes the kernel of this identity mapping on this subspace. This kernel is defined by :

$$\forall f_c[U(\theta,n^i)] = f(\theta)\in L^2(SU(2),d\mu),\ \int_{SU(2)} d\mu(U')\delta_c(U,U')f_c(U') = f_c(U) \quad . \tag{7-64}$$

iii) The closure relation for the functions $\mathcal{D}^j_{m'm}(U)$ is derived from the closure relation of the characters as follows. Setting $U' = \mathbf{1}$ in Eq. (7–63), as $\chi^j(\mathbf{1})$ is equal to the dimension of the representation, leads to :

$$\sum_{j=0,\frac{1}{2},\ldots} (2j+1)\ \chi^j(U) = \delta_c(U,\mathbf{1}) \quad . \tag{7-65}$$

Since the class of the identity only contains the element $\mathbf{1}$, the above relation can be written in terms of the kernel $\delta(U,\mathbf{1})$ of the identity mapping on $L^2(SU(2),d\mu)$, and extended to this whole space, and not only to the subset of class functions :

$$\sum_{j=0,\frac{1}{2},\ldots} (2j+1)\ \chi^j(U) = \delta(U,\mathbf{1}) \quad . \tag{7-66}$$

The representations $\mathcal{D}^j$ being unitary, one has :

$$\sum_{m,m'} \mathcal{D}^{j}_{m'm}(U)\mathcal{D}^{j*}_{m'm}(U') = \sum_{m,m'} \mathcal{D}^{j}_{m'm}(U)\mathcal{D}^{j}_{mm'}(U'^{-1}) = \sum_{m'} \mathcal{D}^{j}_{m'm'}(UU'^{-1}) = \chi^{j}(UU'^{-1}). \quad (7\text{-}67)$$

Then from Eq. (7–66), the completeness relation for the $\mathcal{D}^{j}_{m'm}$ is obtained by multiplying both sides of the above equation by 2j+1 and summing over j :

$$\sum_{j,m,m'} (2j+1)\mathcal{D}^{j}_{m'm}(U)\mathcal{D}^{j*}_{m'm}(U') = 16\pi^2\ \delta(UU'^{-1},\mathbb{1}) = 16\pi^2\ \delta(U,U') \quad . \qquad (7\text{-}68)$$

Hence, with the orthogonality relation, Eq. (7–50a) :

Theorem 7.13. The set of functions :

$$\left\{ \frac{\sqrt{2j+1}}{4\pi}\ \mathcal{D}^{j}_{m'm}(U) \quad ; \quad j=0,\tfrac{1}{2},\dots \quad , \quad m \text{ and } m' = j,j-1,\dots,-j \right\} ,$$

forms a complete orthonormal basis of $L^2(SU(2),d\mu)$. Any function f of this space can be expressed in this basis according to :

$$\forall f \in L^2(SU(2),d\mu) \ , \quad f(U) = \sum_{j,m,m'} \frac{2j+1}{16\pi^2}\mathcal{D}^{j}_{m'm}(U)\,(\mathcal{D}^{j}_{m'm},f)_{SU(2)} \quad . \qquad (7\text{-}69)$$

iv) Following Mahoux[37)], more restricted but nevertheless useful information can be extracted from the closure relation given by Eq. (7–67). The Hermitian inner product on $L^2(S^2,d_2\hat{q})$ is defined by :

$$\forall \tilde{f},\tilde{g} \in L^2(S^2,d_2\hat{q}) \qquad (\tilde{f},\tilde{q})_{S^2} = \int_{S^2} d_2\hat{q}\ \tilde{f}^*(\hat{q})\ \tilde{g}(q) \quad . \qquad (7\text{-}70)$$

To any $\tilde{f}$ in $L^2(S^2,d_2\hat{q})$ one associates a function f defined on the group SU(2), such that in terms of the Euler angles :

$$\forall\ U(\varphi,\theta,\psi) \in SU(2) \qquad f(U) = \tilde{f}(\hat{q})\ e^{-i\lambda(\varphi+\psi)} \quad , \qquad (7\text{-}71)$$

where $\hat{q}$ denotes the unit vector with polar angles θ and φ with respect to a given orthonormal basis $a = (e_i)$, and 2λ is an integer of arbitrary sign. Then f(U) belongs to $L^2(SU(2),d\mu)$ if and only if $f(\hat{q})$ belongs to $L^2(S^2,d_2\hat{q})$. To compute the inner product $(\mathcal{D}^{j}_{m'm},f)_{SU(2)}$, one notices that $\mathcal{D}^{j}_{m'm}(\varphi,\theta,\psi)$ depends upon φ and ψ through exponentials, Eq. (6–21). Therefore, since (cf. Subsec. 10.3.3.2) :

$$d\mu(\varphi,\theta,\psi) = d_2\hat{q}\ d\psi \quad , \qquad (7\text{-}72)$$

one has :

$$(\mathcal{D}^{j}_{m'm},f)_{SU(2)} = \int_{S^2} d_2\hat{q} \int_0^{4\pi} d\psi\ \mathcal{D}^{j*}_{m'm}(\varphi,\theta,\psi)\ \tilde{f}(\hat{q})\ e^{-i\lambda(\varphi+\psi)}$$

$$= 4\pi\ \delta_{m\lambda} \int_{S^2} d_2\hat{q}\ \mathcal{D}^{j*}_{m'\lambda}(\varphi,\theta,-\varphi)f(\hat{q}) = 4\pi\delta_{m\lambda}(\tilde{\mathcal{D}}^{j}_{m'\lambda},\tilde{f})_{S^2} \quad , \qquad (7\text{-}73)$$

where in this last inner product in $L^2(S^2,d_2\hat{q})$, $\tilde{\mathcal{D}}^{j}_{m'\lambda}$ is by definition the function :

$$\forall \hat{q} = \hat{q}(\theta,\varphi) \in S^2 \qquad \tilde{\mathcal{D}}^j_{m'\lambda}(\hat{q}) = \mathcal{D}^j_{m'\lambda}(\varphi,\theta,-\varphi) \quad . \tag{7-74}$$

It has already been noticed, Eqs. (2–37,42), that φ, θ, and $-\varphi$ are the Euler angles of the rotation $R_{\hat{q}}$ which brings the basis vector e_3 on the vector $\hat{q}$ with polar coordinates $(1,\theta,\varphi)$, the rotation axis being orthogonal to the plane defined by e_3 and $\hat{q}$. From Eqs. (7–69,71), one gets :

$$\forall \tilde{f} \in L^2(S^2,d_2\hat{q}) \qquad \tilde{f}(\hat{q}) = \sum_{j,m} \frac{2j+1}{4\pi} \tilde{\mathcal{D}}^j_{m\lambda}(\hat{q}) \, (\tilde{\mathcal{D}}^j_{m\lambda},\tilde{f})_{S^2} \quad . \tag{7-75}$$

This result can also be written as :

$$4\pi \, \delta_2(\hat{q},\hat{q}') = \sum_{j,m} (2j+1) \, \tilde{\mathcal{D}}^j_{m\lambda}(\hat{q}) \, \tilde{\mathcal{D}}^{j*}_{m\lambda}(\hat{q}') \quad . \tag{7-76}$$

Moreover, the functions $\tilde{\mathcal{D}}^j_{m\lambda}(\hat{q})$ are orthogonal on S^2 since Eq. (7–50a) yields :

$$(\mathcal{D}^j_{m\lambda}, \mathcal{D}^{j'}_{m'\lambda})_{SU(2)} = \frac{16\pi^2}{2j+1} \delta_{jj'} \delta_{mm'} = 4\pi (\tilde{\mathcal{D}}^j_{m\lambda},\tilde{\mathcal{D}}^{j'}_{m'\lambda})_{S^2} \quad , \tag{7-77}$$

Hence, finally :

Theorem 7.14. For any given half-odd-integer or integer λ of arbitrary sign, the set of functions :

$$\left\{ \sqrt{\frac{2j+1}{4\pi}} \, \tilde{\mathcal{D}}^j_{m\lambda}(\hat{q}) \; ; \; j=|\lambda| \, , \, |\lambda|+1,\ldots \quad , \quad m=j,j-1,\ldots,-j \right\} \quad ,$$

forms a complete orthonormal basis of $L^2(S^2,d_2\hat{n})$.

It is shown later (Subsec. 10.3.3.4) that when $\lambda=0$, and thus $j=\ell$ and m are integers, $\tilde{\mathcal{D}}^\ell_{m0}(\hat{q})$ is related to the usual spherical harmonic $Y_{\ell m}(\hat{q})$ by :

$$\tilde{\mathcal{D}}^\ell_{m0}(\hat{q}) = \sqrt{\frac{4\pi}{2\ell+1}} \, Y^*_{\ell m}(\hat{q}) \quad . \tag{7-78}$$

Then, the closure and orthogonality relations take, respectively, the more familiar forms :

$$\delta_2(\hat{q},\hat{q}') = \sum_{\ell,m} Y_{\ell m}(\hat{q}) \, Y^*_{\ell m}(\hat{q}') \quad , \tag{7-79a}$$

$$\int_{S^2} d_2\hat{q} \, Y^*_{\ell m}(\hat{q}) Y_{\ell' m'}(\hat{q}) = \int_0^{2\pi} d\varphi \int_0^{\pi} d\theta \, \sin\theta \, Y^*_{\ell m}(\theta,\varphi) Y_{\ell' m'}(\theta,\varphi) = \delta_{\ell\ell'} \delta_{mm'} \quad . \tag{7-79b}$$

7.2.2.3. Completeness of the representations $\mathcal{D}^j$

This is a direct consequence of the orthogonality and closure relations for the characters, respectively Eqs.(7–50b) and (7–63), which imply that the set of functions :

$$\left\{ \frac{1}{4\pi} \chi^j \; ; \; j = 0,\tfrac{1}{2},\ldots \right\} \quad ,$$

forms a complete orthonormal basis of the class functions. Actually this property is closely related to the Fourier expansion since :

$$\chi^0 = 1 \quad , \quad \chi^{\frac{1}{2}} = 2\cos\frac{\theta}{2} \quad , \quad \chi^1 - \chi^0 = 2\cos\theta, \ldots, \chi^{j+1} - \chi^j = 2\cos(j+1)\theta, \ldots$$

Let us assume that there exists a finite-dimensional differentiable unitary irreducible representation $D[SU(2)]$, inequivalent to the representations $\mathcal{D}^j[SU(2)]$. Then the character $\chi(\theta)$ of this representation is equal to zero since it is orthogonal to all the χ^j which form a complete set. As for the unit element of the group the character is equal to the dimension of the representation, the dimension of the representation $D[SU(2)]$ must be zero, which ends the proof of the completeness of the representations $\mathcal{D}^j$.

7.2.3. ORTHOGONALITY AND CLOSURE RELATIONS FOR THE GROUP SO(3)

i) It has been shown in Subsec. 6.1.2,iii) that only for integer values of $j=\ell$ the representations $\mathcal{D}^j[SU(2)]$ are even representations and thus allow us to define by Eq. (5–66) representations $\mathcal{D}^\ell[SO(3)]$ of the group SO(3) :

$$\forall R \in SO(3) \qquad \exists \pm U \in SU(2) \quad , \quad R = R(U) \qquad \mathcal{D}^\ell[R(U)] = \mathcal{D}^\ell(U) = \mathcal{D}^\ell(-U) \ ,$$

where R and U are related by Eq. (4–28).

The invariant measure on the group SO(3) has been defined in Subsec. 7.1.4 using Eq. (7–34) :

$$\forall R(U) \in SO(3) \quad , \quad U \in SU(2) \qquad d\mu[R(U)] = d\mu(U) = d\mu(-U) \quad .$$

Therefore in the orthogonality relation given by Eq. (7–50a,b), since the integrand takes the same value for U and –U, Eqs. (6–24,49) ,one gets :

$$\int_{SO(3)} d\mu(R)\, \mathcal{D}^\ell_{mn}(R)\, \mathcal{D}^{\ell'*}_{m'n'}(R) = \frac{8\pi^2}{2\ell+1}\, \delta_{\ell\ell'}\, \delta_{mm'}\, \delta_{nn'} \quad , \tag{7-80a}$$

$$\int_0^\pi d\theta\,(1-\cos\theta)\, \chi^\ell(\theta)\, \chi^{\ell'}(\theta) = \pi\, \delta_{\ell\ell'} \quad , \tag{7-80b}$$

ℓ and ℓ' being both integers. Explicitly, in terms of the Euler angles Eq. (7–80a) reads :

$$\int_0^{2\pi} d\varphi \int_0^\pi d\theta\, \sin\theta \int_0^{2\pi} d\psi\, \mathcal{D}^\ell_{mn}(\varphi,\theta,\psi)\mathcal{D}^{\ell'*}_{m'n'}(\varphi,\theta,\psi) = \frac{8\pi^2}{2\ell+1}\, \delta_{\ell\ell'}\, \delta_{mm'}\, \delta_{nn'} \quad . \tag{7-81}$$

The above orthogonality relation is still valid if one assumes $j=\ell$ and $j'=\ell'$ either both integers or both half-odd-integers. This relation is often quoted as the orthogonality relation of the function $\mathcal{D}^j_{m'm}$ on SU(2) (Edmonds[22], Eq. (4.6.1), Rose[54], Eq. (4–60)) forgetting to mention the previous validity conditions for j and j'.

ii) Let us denote by $\delta(R,R')$ the kernel of the identity operator over $L^2(SO(3),d\mu)$:

$$\forall f \in L^2[SO(3),d\mu] \qquad \int_{SO(3)} d\mu(R')\ \delta(R,R')\ f(R') = f(R) \quad . \tag{7-82}$$

Then, from the above orthogonality relation of the characters and the closure relation given by Eq. (7–68) for SU(2), one obtains for SO(3) :

$$\sum_{\ell,m,m'} (2\ell+1)\ \mathcal{D}^{\ell}_{m'm}(R)\ \mathcal{D}^{\ell *}_{m'm}(R') = 8\pi^2\ \delta(R,R') \quad , \tag{7-83}$$

where ℓ takes non-negative integer values. When the elements R of SO(3) are parametrized either by θ ,$\hat{n}$ or φ, θ, and ψ, $\delta(R,R')$ is equal to the expressions given for SU(2) by Eqs. (7–57b,58b). Hence :

Theorem 7.15. The set of functions :

$$\left\{ \sqrt{\frac{2\ell+1}{8\pi^2}}\ \mathcal{D}^{\ell}_{m'm}(R) \quad ; \quad \ell=0,1,\ldots \quad , \quad m',m=\ell,\ell-1,\ldots,-\ell \right\} \quad ,$$

forms a complete orthonormal basis of $L^2(SO(3),d\mu)$, and any function of this vector space can be expressed in this basis as :

$$\forall f \in L^2(SO(3),d\mu) \qquad f(R) = \sum_{\ell,m,m'} \frac{2\ell+1}{8\pi^2}\ \mathcal{D}^{\ell}_{m'm}(R)\ (\mathcal{D}^{\ell}_{m'm},f)_{SO(3)} \quad . \tag{7-84}$$

FORMULAE FOR CHAPTER 7

FORMULAE FOR THE GROUP SU(2)

Invariant measure and kernel of the identity mapping

$$\forall\, U,V \in SU(2) \qquad d\mu(U) = d\mu(UV) = d\mu(VU)$$

$$\begin{cases} d\mu(U) = d\mu(U^{+}) = d\mu(U^{-1}) = d\mu(-U) \\ \int_{SU(2)} d\mu(U')\ \delta(U,U')\ f(U') = f(U) \end{cases}$$

$$\begin{cases} d\mu(\theta,\hat{n};a) = 2(1-\cos\theta)d\theta\ d_2\hat{n} = 4(\sin\frac{\theta}{2})^2\ d\theta d_2\hat{n} \\ \delta[U(\theta,\hat{n};a),U(\theta',\hat{n}';a)] = \frac{1}{8\sin\frac{\theta}{2}}\ \delta(\cos\frac{\theta}{2} - \cos\frac{\theta'}{2})\ \delta_2(\hat{n},\hat{n}') \end{cases}$$

$$d_2\hat{n}(\alpha,\beta) = \sin\alpha\ d\alpha\ d\beta \qquad , \qquad \delta_2(\hat{n},\hat{n}') = \delta(\cos\alpha - \cos\alpha')\ \delta(\beta-\beta')$$

$$\begin{cases} d\mu[U(\varphi,\theta,\psi)] = \sin\theta\ d\varphi\ d\theta\ d\psi \\ \delta[U(\varphi,\theta,\psi),U(\varphi',\theta',\psi')] = \delta(\varphi-\varphi')\delta(\cos\theta-\cos\theta')\delta(\psi-\psi') \end{cases}$$

$$\text{Volume} = \int_{SU(2)} d\mu(U) = 16\pi^2 \quad .$$

Orthogonality relations

$$(\mathcal{D}^{j'}_{m'n'},\mathcal{D}^{j}_{mn})_{SU(2)} = \int_{SU(2)} d\mu(U)\mathcal{D}^{j}_{mn}(U)\mathcal{D}^{j'*}_{m'n'}(U) = \frac{16\pi^2}{2j+1}\ \delta_{jj'}\ \delta_{mm'}\ \delta_{nn'}$$

$$= \int_0^{2\pi} d\varphi \int_0^{\pi} \sin\theta d\theta \int_0^{4\pi} d\psi\ \mathcal{D}^{j}_{mn}(\varphi,\theta,\psi)\mathcal{D}^{j'*}_{m'n'}(\varphi,\theta,\psi)$$

$$\frac{2j_3+1}{16\pi^2} \int_{SU(2)} d\mu(U)\mathcal{D}^{j_1}_{m_1'm_1}(U)\mathcal{D}^{j_2}_{m_2'm_2}(U)\mathcal{D}^{j_3*}_{m_3'm_3}(U) = \langle j_1j_2m_1'm_2'|j_3m_3'\rangle\langle j_1j_2m_1m_2|j_3m_3\rangle$$

$$\frac{1}{16\pi^2} \int_{SU(2)} d\mu(U)\mathcal{D}^{j_1}_{m_1'm_1}(U)\mathcal{D}^{j_2}_{m_2'm_2}(U)\mathcal{D}^{j_3}_{m_3'm_3}(U) = \begin{pmatrix} j_1 & j_2 & j_3 \\ m_1' & m_2' & m_3' \end{pmatrix} \begin{pmatrix} j_1 & j_2 & j_3 \\ m_1 & m_2 & m_3 \end{pmatrix}$$

$$\int_0^{2\pi} d\theta(1-\cos\theta)\ \chi^{j}(\theta)\ \chi^{j'}(\theta) = 2\pi\ \delta_{jj'}$$

$$\int_0^{2\pi} d\theta(1-\cos\theta)\ \chi^{j_1}(\theta)\ \chi^{j_2}(\theta)\ \chi^{j_3}(\theta) = 2\pi\ \{j_1j_2j_3\} \quad .$$

Closure relations

$$\sum_{j=0,\frac{1}{2},\ldots} \chi^j(U)\,\chi^j(U') = 16\pi^2\,\delta_c\,(U,U') \quad , \quad \sum_{j=0,\frac{1}{2},\ldots} (2j+1)\,\chi^j(U) = \delta(U,\mathbf{1})$$

$$\sum_{j,m,m'} (2j+1)\,\mathcal{D}^j_{m'm}(U)\,\mathcal{D}^{j*}_{m'm}(U') = 16\pi^2\,\delta(U,U')$$

$$\forall f \in L^2\,(SU(2),d\mu) \qquad f(U) = \sum_{j,m,m'} \frac{2j+1}{16\pi^2}\,\mathcal{D}^j_{m'm}(U)\,(\mathcal{D}^j_{m'm},f)_{SU(2)} \;.$$

Special orthogonality and closure relations

$$\hat{q} \in S^2 \qquad \|\hat{q}\| = 1 \qquad \hat{q} = \hat{q}(\theta,\varphi) \quad , \quad d_2\hat{q} = \sin\theta\,d\theta\,d\varphi$$

$$\tilde{\mathcal{D}}^j_{m'm}(\hat{q}) = \mathcal{D}^j_{m'm}(\varphi,\theta,-\varphi)$$

for a given $\lambda = 0,\frac{1}{2},\ldots \qquad |\lambda| \leqslant j$

$$(\tilde{\mathcal{D}}^{j'}_{m'\lambda},\,\tilde{\mathcal{D}}^j_{m\lambda})_{S^2} = \int_{S^2} d_2\hat{q}\;\tilde{\mathcal{D}}^j_{m\lambda}(\hat{q})\;\tilde{\mathcal{D}}^{j'*}_{m'\lambda}(\hat{q}) = \frac{4\pi}{2j+1}\,\delta_{jj'}\,\delta_{mm'}$$

$$\sum_{j,m} (2j+1)\,\tilde{\mathcal{D}}^j_{m\lambda}(\hat{q})\,\tilde{\mathcal{D}}^{j*}_{m\lambda}(\hat{q}') = 4\pi\,\delta_2(\hat{q},\hat{q}') \qquad .$$

FORMULAE FOR THE GROUP SO(3)

Invariant measure

$$R = R(U) \in SO(3) \qquad U \in SU(2) \quad , \quad d\mu(R) = d\mu(U) = d\mu(R^{-1}) \quad .$$

Orthogonality relations

$\ell,\ \ell',\ m,n,\ \ m',n'$ integers

$$(\mathcal{D}^{\ell'}_{m'n'}\,,\,\mathcal{D}^{\ell}_{mn})_{SO(3)} = \int_{SO(3)} d\mu(R)\,\mathcal{D}^{\ell}_{mn}(R)\,\mathcal{D}^{\ell'*}_{m'n'}(R) = \frac{8\pi^2}{2\ell+1}\,\delta_{\ell\ell'}\,\delta_{mm'}\,\delta_{nn'}$$

$$= \int_0^{2\pi} d\varphi \int_0^{\pi} \sin\theta d\theta \int_0^{2\pi} d\psi\;\mathcal{D}^{\ell}_{mn}(\varphi,\theta,\psi)\;\mathcal{D}^{\ell'*}_{m'n'}(\varphi,\theta,\psi)$$

$$\frac{2\ell_3+1}{8\pi^2}\int_{SO(3)} d\mu(R)\;\mathcal{D}^{\ell_1}_{m_1'm_1}(R)\;\mathcal{D}^{\ell_2}_{m_2'm_2}(R)\;\mathcal{D}^{\ell_3*}_{m_3'm_3}(R) = \langle\ell_1\ell_2\,m_1'm_2'|\ell_3m_3'\rangle\langle\ell_1\ell_2m_1m_2|\ell_3m_3\rangle$$

$$\frac{1}{8\pi^2}\int_{SO(3)} d\mu(R)\;\mathcal{D}^{\ell_1}_{m_1'm_1}(R)\;\mathcal{D}^{\ell_2}_{m_2'm_2}(R)\;\mathcal{D}^{\ell_3}_{m_3'm_3}(R) = \begin{pmatrix}\ell_1 & \ell_2 & \ell_3\\ m_1' & m_2' & m_3'\end{pmatrix}\begin{pmatrix}\ell_1 & \ell_2 & \ell_3\\ m_1 & m_2 & m_3\end{pmatrix}$$

$$\int_0^\pi d\theta(1-\cos\theta)\ \chi^\ell(\theta)\ \chi^{\ell'}(\theta) = \pi\ \delta_{\ell\ell'} \quad .$$

Closure relations

$$8\pi^2\ \delta(R,R') = \sum_{\ell,m,m'} (2\ell+1)\ \mathcal{D}^\ell_{m'm}(R)\ \mathcal{D}^{\ell *}_{m'm}(R')$$

$$\forall f \in L^2(SO(3),d\mu) \qquad f(R) = \sum_{\ell,m,m'} \frac{2\ell+1}{8\pi^2}\ \mathcal{D}^\ell_{m'm}(R)\ (\mathcal{D}^\ell_{m'm},f)_{SO(3)} \quad .$$

Special orthogonality and closure relations

$$\hat{q} \in S^2 \qquad ||\hat{q}|| = 1 \qquad \hat{q} = \hat{q}(\theta,\varphi) \quad , \quad d_2\hat{q} = \sin\theta\ d\theta d\varphi$$

$$Y_{\ell m}(\hat{q}) = \sqrt{\frac{2\ell+1}{4\pi}}\ \tilde{\mathcal{D}}^{\ell *}_{mo}(\hat{q}) \qquad , . \qquad Y^*_{\ell m}(\hat{q}) = (-1)^m\ Y_{\ell -m}(\hat{q})$$

$$\int_{S^2} d_2\hat{q}\ \ Y_{\ell m}(\hat{q})\ \ Y^*_{\ell' m'}(\hat{q}) = \int_0^{2\pi} d\varphi \int_0^\pi \sin\theta d\theta\ Y_{\ell m}(\theta,\varphi)\ Y^*_{\ell' m'}(\theta,\varphi) = \delta_{\ell\ell'}\ \delta_{mm'}$$

$$\delta_2(\hat{q},\hat{q}') = \sum_{\ell,m} Y_{\ell m}(\hat{q})\ Y^*_{\ell m}(\hat{q}')$$

$$\forall \tilde{f} \in L^2(S,d_2\hat{q}) \qquad \tilde{f}(\hat{q}) = \sum_{\ell,m} Y_{\ell m}(\hat{q})\ (Y_{\ell m},\tilde{f})_S \quad .$$

CHAPTER 8

TENSOR ALGEBRA

The elementary objects we use in Quantum Mechanics are (Chap. 3) : vectors of a Hilbert space H, i.e. wave functions or kets, vectors of the dual vector space H^*, i.e. bras, and operators. Later, in Chap. 11, the special nature of these objects will be taken into account. For the time being, we will follow the tensor algebra spirit (Chap. 1), and only consider formal objects defined by a set of components and characterized by the law of transformation under the equivalent irreducible unitary representations $\mathcal{D}^j[\mathrm{SU}(2)]$ and $\mathcal{D}^{j*}[\mathrm{SU}(2)]$ of the group SU(2). This leads us to develop a new tensor algebra. It has to be noticed that, although formally equivalent to the usual tensor algebra, this new formalism is more general since it deals equally with sets of components and sets of basis vectors. According to Wigner[63] (Chap. 24) the basic ideas of this formalism were originally suggested by C. Herring. This new tensor algebra is also discussed by Fano and Racah[24].

The plan of this chapter is as follows. In the first section we study the dependence of a standard basis upon a choice of basis of infinitesimal generators, which leads to the introduction of the notion of a triad. Sec. 8.2 deals with the general problem of conjugate complex matrix representations and its application to the special case of SU(2). Sec. 8.3 is devoted to the definition and the general properties of tensors. Coupling and recoupling coefficients are studied in Sec. 8.4. Finally, Sec. 8.5 gives a brief account of the relation between spherical and Cartesian components of tensors.

The next chapter contains a discussion of graphical representations of tensor algebra. Appendix G deals with C.G. coefficients, Wigner 3j, 6j, and 9j symbols.

8.1. DEPENDENCE OF A STANDARD BASIS UPON A CHOICE OF BASIS OF INFINITESIMAL GENERATORS, TRIAD

i) For a given unitary representation D[SU(2)] of the group SU(2), the standard basis (e^j_m) of the representation vector space, defined in Sec. 5.5, is closely related to the special choice of the basis (σ_i) in the Lie algebra $\mathcal{W}(2)$. To (σ_i) there corresponds by Eq. (5–8) the uniquely defined basis (J_i) of infinitesimal generators in $D[\mathcal{W}(2)]$:

$$J_i = i \frac{\partial}{\partial\varepsilon} D[\mathbf{1} - i \frac{\varepsilon}{2} \sigma_i + O(\varepsilon^2)]\Big|_{\varepsilon=0} . \tag{8-1}$$

The J_i fulfil the relations :

$$[J_i,J_j] = i\sum_k \varepsilon_{ijk} J_k \quad , \qquad (J_i,J_j) = 2\,\delta_{ij} \quad , \tag{8-2}$$

where the inner product is defined by the Killing form. The standard basis (e^j_m) is then characterized by the relations :

$$J_\pm\, e^j_m = (J_1 \pm iJ_2)e^j_m = \sqrt{j(j+1)-m(m\pm1)}\; e^j_{m\pm1} \quad , \quad J_3 e^j_m = m\, e^j_m \quad , \tag{8-3}$$

which imply the following law of transformation under the mapping D(U) (Subsec. 6.1.3,ii)) :

$$\forall\, U \in SU(2) \qquad D(U)e^j_m = e^j_{m'}\, \mathcal{D}^j_{m'm}(U) \quad . \tag{8-4}$$

ii) Actually, the derivation of this standard basis is only based on the existence of a standard form of commutation relations of the infinitesimal generators $J_\pm$ and J_3 :

$$[J_3,J_\pm] = \pm\, J_\pm \quad , \qquad [J_+,J_-] = 2\, J_3 \quad . \tag{8-5}$$

Furthermore, it has been shown in Subsec. 5.4.3,ii) that the correspondence :

$$\forall\, U = e^{-iW} \in SU(2) \;\rightarrow\; D(U) = e^{-iJ} \in D[SU(2)] \;\rightarrow\; e^{\Delta_J} = R(U) \tag{8-6}$$

is a two-to-one homomorphism between the group SU(2) and the rotation group in the Lie algebra D $[\mathcal{W}(2)]$:

$$\forall\, \pm U \in SU(2) \;\longleftrightarrow\; R(U) = R(-U) \quad \text{rotation in} \quad D[\mathcal{W}(2)] \quad .$$

This mapping e^{Δ_J} fulfils the relation (Prop. D.1) :

$$\forall\, J' \in D[\mathcal{W}(2)] \qquad e^{\Delta_J}\, J' = D(U)J'\, D(U)^{-1} \quad , \tag{8-7}$$

therefore, it preserves the commutation relations. Indeed, one has :

$$\forall J', J'' \in D[\mathcal{W}(2)] \qquad [e^{\Delta_J} J', e^{\Delta_J} J''] = [D(U)J'D(U)^{-1}, D(U)J''D(U)^{-1}]$$

$$= D(U)[J',J'']D(U)^{-1} = e^{\Delta_J}[J',J''] \quad .$$

This result has already been obtained in Subsec. 4.3.4,ii), using the isomorphism between $\mathcal{W}(2)$ and any three-dimensional Euclidean vector space where the vector product defines a Lie algebra structure.

Let us remark that the following two properties are general for semi-simple Lie algebras :
(1) existence of a standard form of commutation relations (Subsec. 4.3.3,iii)) which leads to a standard basis of the representation vector space (Sec. 5.5),
(2) the isometric character of the mapping $\exp(\Delta_J)$ (Prop. B.32 and Subsec. 5.4.2,i)) which preserves the commutation relations.

iii) Consequently, for any rotation $R = R(\pm U)$ in $D[\mathcal{W}(2)]$ the infinitesimal generators (RJ_i) also fulfil the relations given in Eq. (8–2), and $RJ_\pm$, RJ_3 satisfy the standard form of commutation relations, Eq. (8–5). Moreover, the two bases $(D(\pm U)e^j_m)$ of the representation vector space fulfil the characteristic equations of a standard basis with respect to $RJ_\pm$ and RJ_3. Indeed, from Eqs. (8–6,7) one has :

$$\forall U \in SU(2) \quad , \quad \forall J \in D[\mathcal{W}(2)] \qquad [R(\pm U)J]D(\pm U) = D(\pm U)J \quad , \tag{8-8}$$

and thus :

$$[R(\pm U)J_3]D(\pm U)e^j_m = m\, D(\pm U)e^j_m \tag{8-9a}$$

$$[R(\pm U)J_\pm]D(\pm U)e^j_m = \sqrt{j(j+1)-m(m\pm 1)}\; D(\pm U)e^j_{m\pm 1} \quad . \tag{8-9b}$$

These two bases $(D(\pm U)e^j_m)$ are also called <u>standard bases with respect to the basis</u> $(R(\pm U)J_i)$ of infinitesimal generators. Hence, due to the two–to–one correspondence between the group SU(2) and the rotation group in $D[\mathcal{W}(2)]$, a standard basis with respect to (RJ_i) depends upon the choice of one of the two elements $\pm U$ in SU(2) which correspond to this rotation. To underline this dependence let us denote by $(e^j_{m;U})$, for any U in SU(2), a standard basis with respect to the uniquely defined basis $(R(U)J_i)$ of infinitesimal generators. Thus, $(e^j_{m;\mathbf{1}})$ denotes the standard basis (e^j_m) with respect to the infinitesimal generators defined by Eq. (8–1).

In terms of the above notation, Eqs. (8–9a,b) imply that the two unit vectors $D(U)e^j_{m;\mathbf{1}}$ and $e^j_{m;U}$ are equal up to an arbitrary phase factor. We choose it to be equal to one, then :

$$\forall U \in SU(2) \qquad D(U)e^j_{m;\mathbf{1}} = e^j_{m;U} \quad . \tag{8-10}$$

In this way the relative phase between any standard basis vectors $e^j_{m;U}$ and $e^j_{m';U'}$ is completely determined for $U = U'$ and $m \neq m'$ by Eq. (8–9), and for $U \neq U'$ and $m = m'$ by Eq. (8–10) which yields :

$$\forall U,U' \in SU(2) \qquad e^j_{m;U'} = D(U')e^j_{m;\mathbf{1}} = D(U')[D(U)^{-1}e^j_{m;U}] = D(U'U^{-1})e^j_{m;U} \; . \tag{8-11}$$

Moreover, the dependence upon the choice of U and U' can be made explicit as follows :

$$\forall U,U' \in SU(2) \qquad e^j_{m;U'} = D(U)\,[D(U^{-1}U')\; e^j_{m;\mathbf{1}}] \quad ,$$

and since $e^j_{m;\mathbf{1}} = e^j_m$ fulfils Eq. (8–4), one gets :

$$\forall U,U' \in SU(2) \qquad e^j_{m;U'} = D(U)\,[e^j_{m';\mathbf{1}}\; \mathcal{D}^j_{m'm}(U^{-1}U')] = e^j_{m';U}\; \mathcal{D}^j_{m'm}(U^{-1}U') \quad . \tag{8-12}$$

From Eq. (6–24), one has :

$$\forall U \in SU(2) \qquad e^j_{m;-U} = (-1)^{2j}\; e^j_{m;U} \quad . \tag{8-13}$$

The law of transformation under the mapping $D(U')$ of any standard basis vector $e^j_{m;U}$ is given by :

$$\forall\, U,U' \in SU(2) \qquad D(U')e^j_{m;U} = D(U)[D(U^{-1}U'U)\; e^j_{m;\mathbf{1}}] \tag{8-14}$$

$$= D(U)[e^j_{m';\mathbf{1}}\; \mathcal{D}^j_{m'm}(U^{-1}U'U)] = e^j_{m';U}\; \mathcal{D}^j_{m'm}(U^{-1}U'U) \quad .$$

Another way to get the above result is the following. Since both vectors $e^j_{m;\mathbf{1}}$ and $e^j_{m;U}$ fulfil the same Eq. (8–3), respectively, with respect to the infinitesimal generators $J_\pm$, J_3 and $R(U)J_\pm$, $R(U)J_3$, the relation :

$$U' = e^{-i\frac{\theta'}{2} n^i \sigma_i} \qquad D(U')e^j_{m;\mathbf{1}} = e^{-i\theta' n^i J_i}\; e^j_{m;\mathbf{1}} = e^j_{m';\mathbf{1}}\; \mathcal{D}^j_{m'm}(U') \quad ,$$

implies :

$$e^{-i\theta' n^i (RJ_i)}\; e^j_{m;U} = e^j_{m';U}\; \mathcal{D}^j_{m'm}(U') \quad .$$

Furthermore, one has :

$$n^i (RJ_i) = n^i R^j_i J_j = (R\hat{n})^j J_j \quad .$$

Actually, in terms of the Euclidean inner product and of the formal vector J defined in Subsec. 5.4.3,i), the above equation reads :

$$\hat{n}.R^{-1}J = R\,\hat{n}.J \quad .$$

Using Eq. (4–34) one then gets :

$$e^{-i\theta' R(U)\hat{n}.J} = D(UU'U^{-1}) \quad ,$$

and thus :

$$\forall\, U,U' \in SU(2) \qquad D(UU'U^{-1})\; e^j_{m;U} = e^j_{m';U}\; \mathcal{D}^j_{m'm}(U') \quad .$$

Setting $U^{-1}U'U$ instead of U' in this equation one gets Eq. (8–14).

iv) The index U in the standard basis $(e^j_{m;U})$ plays a similar role to the tetrad [p] introduced in the definition of the physical representations of the Lorentz group[41)] (there [p] denotes an element in SL(2,C) to which corresponds a unique Lorentz transformation). By analogy the index U in $e^j_{m;U}$ is called a triad. It denotes at the same time an element in SU(2) and the basis (RJ_i) of infinitesimal generators where R is the unique rotation in the Lie algebra which corresponds to the adjoint representation of U. Applications in Quantum Mechanics of the previous considerations are discussed later in Subsecs. 11.1.2,3.

8.2. CONJUGATE COMPLEX REPRESENTATIONS

8.2.1. GENERAL RESULTS

i) Let us consider a finite-dimensional complex matrix representation D(G) of a group G. The set of conjugate complex matrices $D^*(g)$, for all g in G, also form a matrix representation of G, denoted by $D^*(G)$. Indeed, one has :

$$\forall g,g' \in G \qquad D(g)D(g') = D(gg') \Rightarrow D^*(g)D^*(g') = D^*(gg')$$
$$D(g^{-1}) = D(g)^{-1} \Rightarrow D^*(g^{-1}) = D^*(g)^{-1} \quad .$$

Moreover, if D(G) is irreducible so is $D^*(G)$. Similarly, if D(G) is unitary so is $D^*(G)$ and vice versa. The question then arises : are the representations D(G) and $D^*(G)$ equivalent ? From, respectively, Prop. C.58 and Theorem 7.11, Subsec. 7.2.1,i) :

A necessary and sufficient condition that the two finite-dimensional matrix representations D(G) and $D^*(G)$ of a finite or a compact Lie group G, be equivalent is that they have the same character, i.e. that the character χ of the representation D(G) be real :

$$\forall g \in G \qquad \chi(g) = \chi^*(g) \in R \quad . \tag{8-15}$$

This is for instance always the case for unitary matrix representations of a group G if for all g in G, g and its inverse g^{-1} belong to the same class of conjugate elements, e.g. SU(2) cf. Subsec. 4.2.2. Since the character is a class function and from the unitarity of the representation, one then has :

$$\forall g \in G \qquad \chi(g) = \chi(g^{-1}) = \chi^*(g) \in R \quad . \tag{8-16}$$

If D(G) and $D^*(G)$ are inequivalent, one says following Wigner[63)] (Chap. 24, cf. also Frobenius and Schur[26)]) that the representation D(G) is complex. If the representations D(G) and $D^*(G)$ are equivalent, there exists a regular matrix C such that :

$$\forall g \in G \qquad D^*(g) = C\, D(g)\, C^{-1} \quad . \tag{8-17}$$

ii) **Theorem 8.1.** If D(G) and $D^*(G)$ are equivalent unitary irreducible finite-dimensional matrix representations of G, the regular matrix C is then such that :

(1) CC^+ is a multiple of the unit matrix,

(2) C is either symmetrical or antisymmetrical.

Proof. The result (1) follows from Prop. C.47 :

$$CC^+ = \lambda\, \mathbf{1} \quad .$$

Taking the complex conjugate of both sides of Eq. (8-17) yields :

$$D(g) = C^* D^*(g)\, C^{*-1} = (C^*C)\, D(g)\, (C^*C)^{-1} \quad ,$$

hence, for all g in G :

$$D(g)C^*C = C^*C\, D(g) \quad .$$

From Prop. C.46, C^*C is then a multiple of the identity :

$$C^*C = \mu\, \mathbf{1} \quad .$$

One therefore has :

$$C = \mu\, C^{*-1} = \frac{\mu}{\lambda}\, C^t = \varepsilon\, C^t \quad .$$

Taking the transpose of both sides of this equation one gets :

$$C^t = \varepsilon\, C \quad ,$$

so that ε^2 equals one and ε equals plus one or minus one, i.e. :

$$C^t = \pm\, C \quad . \tag{8-18}$$

Furthermore, the symmetry or antisymmetry of C is intrinsic to the representation, namely it remains true for all unitary matrix representations equivalent to D(G). Indeed, let D'(G) be such a representation, there exists a unitary matrix A such that :

$$\forall g \in G \qquad D'(g) = A\, D(G)\, A^{-1} \quad . \tag{8-19}$$

One then has from Eq. (8–17) :

$$D'^*(g) = A^* D^*(g) A^{*-1} = (A^* C\, A^{-1})\, D'(g)\, (A^* C\, A^{-1})^{-1} \quad ,$$

hence for all g in G :

$$D'^*(g) = C' D'(g)\, C'^{-1} \quad , \qquad C' = A^* C\, A^{-1} \quad , \tag{8-20}$$

and C' has the same symmetry as C, since :

$$C'^t = A^{-1t}\, C^t\, A^{*t} = A^*\, C^t\, A^{-1} = \pm\, C' \quad .$$

This intrinsic property of C provides a classification of the unitary irreducible finite–dimensional matrix representations with real characters[26)]. Following a terminology of Wigner, those with symmetrical C are said to be <u>potentially real</u> and those with antisymmetrical C are said to be <u>pseudoreal</u>. This terminology is based on the two following theorems :

Theorem 8.2. Let D(G) be a finite–dimensional unitary irreducible matrix representation with a real character, and such that C is symmetrical. D(G) is then equivalent to a real representation, i.e. to a representation D'(G) such that the matrices D'(g) for all g in G are real.

Proof :

1–From Prop. C.47 the matrix C can be chosen unitary, it then fulfils the necessary and sufficient condition $CC^+ = C^+C$ to ensure there exists a unitary matrix R which diagonalizes it (Theorem 4.2, Subsec. 4.1.2,i)) :

$$C = R\, \mathcal{D}\, R^{-1} \quad ,$$

where $\mathcal{D}$ is a diagonal unitary matrix and thus :

$$\mathcal{D}^{-1} = \mathcal{D}^+ = \mathcal{D}^* \quad . \tag{8-21}$$

2– $\mathcal{C}$ being unitary and symmetrical, there exists a choice of real eigenvectors and hence R can be assumed to be real orthogonal. Indeed, let v be an eigenvector of $\mathcal{C}$ with the eigenvalue λ. Multiplying both sides of the equation $\mathcal{C}v = \lambda v$ by $\mathcal{C}^{-1} = \mathcal{C}^{+} = \mathcal{C}^{*}$ leads to $v = \lambda \mathcal{C}^{*} v$. Taking the complex conjugate, and using the fact that the modulus of the eigenvalue λ of the unitary matrix $\mathcal{C}$ is equal to one, one gets :

$$\mathcal{C}\, v^{*} = \lambda\, v^{*} \quad .$$

Thus v and v^* are both eigenvectors with the eigenvalue λ. If v and v^* are linearly independent they can be replaced by the real vectors $(v+v^*)/\sqrt{2}$ and $(v-v^*)/(i\sqrt{2})$; if v and v^* are proportional they can be replaced by one or the other of the preceding vectors. In any case the matrix R, which defines the change of basis, is then real.

3–Let us now check that the matrix A defined by :

$$A = \mathcal{D}^{\frac{1}{2}}\, R^{-1} \quad ,$$

which as the product of two unitary matrices is unitary, and satisfies the equation :

$$A^{*}\, \mathcal{C}\, A^{-1} = \mathbf{1} \quad .$$

Indeed, using Eq. (8–21) and the reality of R, one gets :

$$\begin{aligned} A^{*}\, \mathcal{C}\, A^{-1} &= \mathcal{D}^{\frac{1}{2}*}\; R^{-1*}\; R\, \mathcal{D}\, R^{-1}\; R\; \mathcal{D}^{-\frac{1}{2}} \\ &= \mathcal{D}^{-\frac{1}{2}}\; R^{-1}\; R\, \mathcal{D}^{\frac{1}{2}} = \mathbf{1} \quad . \end{aligned}$$

4–Finally, from Eq. (8–20) the representation D'(G), equivalent to D(G), and defined by :

$$\forall\, g \in G \qquad D'(g) = A\, D(g)\, A^{-1} \quad ,$$

is such that :

$$\mathcal{C}' = A^{*}\, \mathcal{C}\, A^{-1} = \mathbf{1} \;\Rightarrow\; \forall\, g \in G \qquad D'^{*}(g) = D'(g) \quad .$$

Let us note that conversely, if the representation D(G) is equivalent to a real representation D'(G), $\mathcal{C}$ must then be symmetrical. Indeed, the matrix $\mathcal{C}'$ associated with the real representation D'(G) by Eq. (8–20) is a multiple of the identity, hence symmetrical, and it has been shown above that this property remains true for any equivalent unitary representation D(G). As a consequence, if $\mathcal{C}$ is antisymmetrical there does not exist any real representation equivalent to D(G).

Theorem 8.3. Let D(G) be a finite–dimensional unitary irreducible matrix representation with a real character, and such that $\mathcal{C}$ is antisymmetrical. The dimension of the representation is then even, 2n, and D(G) is equivalent to a unitary representation D'(G), such that :

$$\forall\, g \in G \qquad D'^{*}(g) = \mathcal{C}'\, D'(g)\, \mathcal{C}'^{-1} \quad ,$$

where $\mathcal{C}'$ is the real antisymmetrical orthogonal matrix :

$$\mathcal{C}' = \begin{pmatrix} 0 & \mathbf{1} \\ \mathbf{-1} & 0 \end{pmatrix} \quad ,$$

the dimension of each block being n.

Proof :

1–As for Theorem 8.2, there exists a unitary matrix U which diagonalizes $\mathcal{C}$:

$$\mathcal{C} = U \mathcal{D} U^{-1} \quad ,$$

where the diagonal matrix $\mathcal{D}$ fulfils Eq. (8–21).

2– $\mathcal{C}$ being unitary and antisymmetrical, the dimension of the representation is even and there exists a choice of basis such that $\mathcal{C}$ takes the form :

$$\mathcal{C} = R \begin{pmatrix} 0 & \begin{matrix} & & \alpha_1 \\ & \iddots & \\ \alpha_n & & \end{matrix} \\ \begin{matrix} & & -\alpha_n \\ & \iddots & \\ -\alpha_1 & & \end{matrix} & 0 \end{pmatrix} R^{-1} \quad , \tag{8-22}$$

where R is a real orthogonal matrix and the α_i are complex numbers of modulus one. Indeed, following the same arguments as for the proof of Theorem 8.2, one shows that if v is an eigenvector of $\mathcal{C}$ with the eigenvalue λ, then v^* is also an eigenvector of $\mathcal{C}$ with the eigenvalue $-\lambda$. The eigenvectors v cannot be real since the eigenvalues of the unitary matrix $\mathcal{C}$ are of modulus one and thus nonzero. The linearly independent eigenvectors v and v^* can then be replaced by the real vectors $(v+v^*)/\sqrt{2}$ and $(v-v^*)/(i\sqrt{2})$. Since one has :

$$\mathcal{C}\frac{1}{\sqrt{2}}(v+v^*) = i\lambda\frac{1}{i\sqrt{2}}(v-v^*) \quad , \quad \mathcal{C}\frac{1}{i\sqrt{2}}(v-v^*) = -i\lambda\frac{1}{\sqrt{2}}(v+v^*) \quad ,$$

with respect to the basis $\{(v_k-v_k^*)/(i\sqrt{2}), (v_k+v_k^*)/\sqrt{2} \,;\, k=1,..,n, \mathcal{C}v_k=\lambda_k v_k\}$, and a suitable ordering of the basis vectors, the matrix $\mathcal{C}$ takes the form of Eq. (8–22), where :

$$\alpha_k = i\,\lambda_k \quad ,$$

and the matrix R which defines this change of basis is real. Furthermore, R is unitary and thus orthogonal, since it is equal to the product of two unitary matrices : U and the matrix which defines the change of basis $\{v_k, v_k^*\} \rightarrow \{(v_k-v_k^*)/(i\sqrt{2}), (v_k+v_k^*)/\sqrt{2}\}$.

3–Let us denote by $\mathcal{D}'$ the diagonal unitary matrix :

$$\mathcal{D}' = \begin{pmatrix} \alpha_1 & & & & \\ & \ddots & & & \\ & & \alpha_n & & \\ & & & \alpha_n & \\ & & & & \ddots \\ & & & & & \alpha_1 \end{pmatrix} \quad , |\alpha_k|^2 = 1 \quad ,$$

which fulfils the same equation as Eq. (8–21) :

$$\mathcal{D}'^{-1} = \mathcal{D}'^{+} = \mathcal{D}'^{*} \quad . \tag{8-23}$$

Eq. (8–22) can then be written as :

$$\mathcal{C} = R\,\mathcal{D}'^{\frac{1}{2}} \begin{pmatrix} 0 & \mathbf{1} \\ \mathbf{-1} & 0 \end{pmatrix} \mathcal{D}'^{\frac{1}{2}} R^{-1} \quad .$$

4–As for Theorem 8.2, using Eq. (8–23) and the reality of R, one checks that the matrix A, defined by :

$$A = \mathcal{D}'^{\frac{1}{2}} R^{-1} \quad ,$$

is unitary, being a product of two unitary matrices, and satisfies the equation :

$$A^* \mathcal{C} A^{-1} = \begin{pmatrix} 0 & \mathbf{1} \\ \mathbf{-1} & 0 \end{pmatrix} \quad .$$

5–Finally, from Eq. (8–20) the representation D'(G), equivalent to D(G), and defined by :

$$\forall g \in G \qquad D'(g) = A\,D(g)\,A^{-1} \quad ,$$

is such that :

$$C' = A^* \; C \; A^{-1} = \begin{pmatrix} 0 & \mathbf{1} \\ \mathbf{-1} & 0 \end{pmatrix} \quad .$$

Actually, one shows[21)] that pseudoreal representations are equivalent to real quaternion matrix representations. Thus, the classification of finite-dimensional irreducible and unitary matrix representations D(G) of a group G in potentially real, complex, and pseudoreal representations is closely related to the existence of only three, up to an isomorphism, associative division algebras over the real field (Appendix B.5.2), i.e. the real numbers, the complex numbers, and the real quaternions. Details about this subject are given by Dyson[21)].

8.2.2. APPLICATION TO THE GROUPS SU(2) AND SO(3)

8.2.2.1. Application to the representations $\mathcal{D}^j$

i) The characters χ^j of the irreducible unitary representations of SU(2) are real, Eq. (6–47). Hence, from the previous general results :

(1) The irreducible unitary representations $\mathcal{D}^j$[SU(2)] and $\mathcal{D}^{j*}$ [SU(2)] are equivalent.
(2) The matrix C^j such that :

$$\mathcal{D}^{j*}[SU(2)] = C^j \; \mathcal{D}^j \; [SU(2)] \; (C^j)^{-1} \quad , \tag{8-24}$$

is either symmetrical or antisymmetrical and an arbitrary complex multiple of a unitary matrix.

Actually, these general results have already been obtained using specific methods for the group SU(2). Thus, the equivalence between the representations $\mathcal{D}^j$ and $\mathcal{D}^{j*}$ is also a consequence of the completeness of the representations $\mathcal{D}^j$ (Subsec. 7.2.2.2,iii)). Let us denote by C^j a unitary choice for C^j :

$$C^j = \lambda_j \; \mathrm{C}^j \quad , \quad \mathrm{C}^j \, \mathrm{C}^{j+} = \mathbf{1} \quad . \tag{8-25}$$

This matrix was first obtained for the smallest-dimensional representation $j = \frac{1}{2}$, i.e. for the group SU(2) itself, in Subsec. 4.5,iv) :

$$\forall U \in SU(2) \qquad U^* = \mathrm{C} \; U \; \mathrm{C}^{-1} \quad , \tag{8-26a}$$

$$\mathrm{C}^{j=\frac{1}{2}} = \mathrm{C} = i\sigma_2 = U(\pi,-e_2;a) = \begin{pmatrix} 0 & 1 \\ -1 & 0 \end{pmatrix} \quad . \tag{8-26b}$$

For all j one then gets in Subsec. 6.1.2,iii) :

$$\forall U \in SU(2) \qquad \mathcal{D}^{j*}(U) = \mathcal{D}^j(\mathrm{C}) \; \mathcal{D}^j(U) \; [\mathcal{D}^j(\mathrm{C})]^{-1} \quad , \tag{8-27a}$$

$$\mathrm{C}^j = \mathcal{D}^j(\mathrm{C}) = \mathcal{D}^j[U(\pi,-e_2;a)] \quad , \quad \mathrm{C}^j_{m'm} = d^j_{m'm}(-\pi) = (-1)^{j+m} \, \delta_{m',-m} . \tag{8-27b}$$

The properties of C, Eq. (4–101), imply for its unitary representation $\mathcal{D}^j$(C), using Eqs.

(6–15,24), that C^j is a real orthogonal matrix, antisymmetrical or symmetrical according whether j is a half–odd–integer or an integer :

$$C^j = C^{j*} \quad , \qquad (C^j)^{-1} = (C^j)^+ = (C^j)^t = (-1)^{2j}\, C^j \quad . \tag{8-28}$$

Therefore :

(1) For half–odd–integer values of j, the representations $\mathcal{D}^j$ are pseudoreal, their dimensions 2j+1 are even, and there does not exist any real representation equivalent to them. Moreover, the representation $\mathcal{D}^j$ is the one which yields for C^j the form given by Eq. (8–22). Indeed, with a suitable ordering of the basis vectors, Eq. (6–12), one has from Eq. (8–27b) :

$$j = \tfrac{1}{2}, \tfrac{3}{2}, \ldots, (C^j_{m'm}) = \begin{array}{c} m'\backslash m \\ j \\ j-1 \\ \vdots \\ \frac{1}{2} \\ -\frac{1}{2} \\ \vdots \\ -j \end{array} \begin{array}{c} j, j-1, \ldots \tfrac{1}{2}, -\tfrac{1}{2}, \ldots, -j \\ \left(\begin{array}{c|c} 0 & \mathbf{1} \\ \hline -\mathbf{1} & 0 \end{array}\right) \end{array} \quad . \tag{8-29}$$

(2) For integer values of j, the representations $\mathcal{D}^j$ are potentially real, their dimensions 2j+1 are odd, and they are equivalent to real matrix representations. For instance, for j=1 the equivalent real matrix representation is the group SO(3). The relation between the matrices (R^i_j) of SO(3) and $(\mathcal{D}^1_{m'm})$ is given later in Subsec. 8.5.2. From Eq. (8–27b), one gets :

$$j = 0,1,\ldots, \quad (C^j_{m'm}) = \begin{array}{c} m'\backslash m \\ j \\ j-1 \\ \vdots \\ 0 \\ \vdots \\ -j \end{array} \begin{array}{c} j, j-1, \ldots, 0, \ldots, -j \\ \left(\begin{array}{ccc} 0 & \Big| & \begin{array}{r} 1 \\ -1 \\ \cdot^{\cdot^{\cdot}} \end{array} \\ \hline & (-1)^j & \\ \hline \begin{array}{l} \cdot^{\cdot^{\cdot}} \\ 1 \end{array} & \Big| & 0 \end{array}\right) \end{array} \quad . \tag{8-30}$$

ii) Up to an equivalence, the only unitary and irreducible representations of SO(3) are defined by the representations $\mathcal{D}^{\ell}[SU(2)]$ for integer values of ℓ (Sec. 5.6 and Eq. (6–25)). Therefore, these representations are potentially real.

8.2.2.2. Infinitesimal generators of conjugate complex representations

Let us denote by (J_i) and $(\bar{J}_i)$ the two sets of infinitesimal generators defined by Eq. (8–1) of, respectively, a finite–dimensional complex matrix representation D [SU(2)] and the conjugate complex representation D^*[SU(2)], i.e. :

$$J_i = i\,\frac{\partial}{\partial\varepsilon}\, D[\mathbf{1} - i\,\tfrac{\varepsilon}{2}\,\sigma_i + 0(\varepsilon^2)]\Big|_{\varepsilon=0} \quad , \quad \bar{J}_i = i\,\frac{\partial}{\partial\varepsilon}\, D^*[\mathbf{1} - i\,\tfrac{\varepsilon}{2}\,\sigma_i + 0(\varepsilon^2)]\Big|_{\varepsilon=0} \quad . \tag{8-31}$$

These two representations being equivalent and related by :

$$\forall\, U \in SU(2) \qquad\qquad D^*(U) = D(C)\; D(U)\; D(C)^{-1} \qquad ,$$

one has :

$$\bar{J}_i = D(C)\, J_i\, D(C)^{-1} = R(C)\, J_i \quad , \tag{8-32}$$

where R(C) is the rotation in the Lie algebra D [$\mathcal{W}$(2)] which corresponds to the element C in SU(2) (adjoint representation of C, Eq. (8–6), Subsec. 5.4.3,ii)). From Eqs. (4–31,105), R(C) is a rotation of angle π around the vector J_2. Therefore, the basis (J_i) of D [$\mathcal{W}$(2)] being orthogonal, Eq. (8–2), one gets :

$$\bar{J}_i = R(\pi, J_2)\, J_i \quad , \tag{8-33a}$$

$$\bar{J}_1 = -J_1 \quad , \quad \bar{J}_2 = J_2 \quad , \quad \bar{J}_3 = -J_3 \quad , \tag{8-33b}$$

$$\bar{J}_\pm = \bar{J}_1 \pm i\, \bar{J}_2 = -\, J_\mp \quad . \tag{8-33c}$$

One has to take care not to confuse the matrix representatives $(\bar{J}_i)$ of $\bar{J}_i$ with the conjugate complex matrix representatives $(J_i)^*$. Indeed, for instance with respect to a standard basis of the representation space, the adopted phase conventions lead to real matrix elements for J_1 and J_3, and purely imaginary matrix elements for J_2. Hence, with respect to a standard basis one has :

$$(\bar{J}_i) = -(J_i)^* \quad .$$

8.3. DEFINITION AND GENERAL PROPERTIES OF TENSORS

8.3.1. DEFINITION, TENSORIAL CHARACTER, AND TRIAD

8.3.1.1. Tensorial sequences, tensorial character, and tensors

i) A <u>tensorial sequence</u> (T^j_m) of <u>order</u> one and <u>rank</u> j is by definition any set of 2j+1 objects T^j_m, j being a non–negative integer or half–odd–integer and m=j,j–1,...,–j, which belong to a vector space over the complex field, and are characterized by the fact that they are transformed under a representation of the group SU(2) as the standard basis vectors e^j_m, Eq. (8–4), i.e. :

$$\forall\, U \in SU(2) \qquad (T^j_m) \rightarrow ({}^U T^j_m = T^j_{m'}\, \mathcal{D}^j_{m'm}(U)) \quad . \tag{8-34}$$

Denoting (T^j_m) as a column vector T^j, the above equation reads :

$$\forall\, U \in SU(2) \qquad T^j \rightarrow {}^U T^j = [\mathcal{D}^j(U)]^t\, T^j \quad . \tag{8-35}$$

From any tensorial sequence (T^j_m), a new set of 2j+1 quantities (t^j_m) is defined as follows :

$$t^j_m = C^j_{mm'}\, T^j_{m'} \quad , \tag{8-36}$$

or in matrix notation :

$$t^j = C^j\, T^j \quad , \tag{8-37}$$

where $\mathcal{C}^j$ is the matrix considered in Eq. (8–24). This matrix being regular, the above equations can be inverted as :

$$T^j_m = (\mathcal{C}^j)^{-1}_{mm'}\ t^j_{m'} \quad , \quad T^j = (\mathcal{C}^j)^{-1}\ t^j \quad . \tag{8-38}$$

From the equivalence between the representations $\mathcal{D}^j$ and $\mathcal{D}^{j*}$, a direct calculation shows that under the linear mapping $(T^j_m) \rightarrow ({}^U T^j_m)$ defined by Eq. (8–34), the objects t^j_m are transformed by means of the matrix $\mathcal{D}^{j*}(U)$ instead of $\mathcal{D}^j(U)$. Indeed, in matrix notation one has :

$${}^U t^j = \mathcal{C}^j\ {}^U T^j = \mathcal{C}^j[\mathcal{D}^j(U)]^t\ T^j = \mathcal{C}^j[\mathcal{D}^j(U)]^t\ (\mathcal{C}^j)^{-1}\ t^j \quad ,$$

and therefore, using Eqs. (8–24,26a) one gets :

$${}^U t^j = [\mathcal{D}^{j*}(U)]^t\ t^j \quad , \quad {}^U t^j_m = t^j_{m'}\ \mathcal{D}^{j*}_{m'm}(U) \quad . \tag{8-39}$$

By analogy with the usual tensor algebra, considered in Sec. 1.1, our first definition of a tensorial sequence is extended to any set of 2j+1 objects which are transformed under a representation of the group SU(2) according to either the Eq. (8–34) or the Eq. (8–39). Which one of these two laws of transformation is satisfied defines the tensorial character of the tensorial sequence. A notation is introduced to specify the tensorial character (a similar motivation leads in the usual tensor algebra to the up–and–down index notation). Following Fano and Racah[24], $(T^{[j]}_m)$ and $(T'^{(j)}_m)$ denote two tensorial sequences which are transformed as :

$$\forall\, U \in SU(2) \qquad {}^U T^{[j]}_m = T^{[j]}_{m'}\ \mathcal{D}^j_{m'm}(U) \quad , \quad {}^U T'^{(j)}_m = T'^{(j)}_{m'}\ \mathcal{D}^{j*}_{m'm}(U) \quad . \tag{8-40}$$

It is also said that the tensorial sequence $(T^{[j]}_m)$ (resp. $(T^{(j)}_m)$) belongs to the representation $\mathcal{D}^j$ (resp. $\mathcal{D}^{j*}$), or that the index m in $T^{[j]}_m$ (resp. $T^{(j)}_m$) is contravariant (resp. covariant).

To any tensorial sequence $(T^{[j]}_m)$ one associates a tensorial sequence $(T^{(j)}_m)$ in the following way :

$$T^{(j)}_m = \mathcal{C}^j_{mm'}\ T^{[j]}_{m'} \quad , \quad T^{[j]}_m = (\mathcal{C}^j)^{-1}_{mm'}\ T^{(j)}_{m'} \quad . \tag{8-41}$$

An equivalence relation among the set of tensorial sequences is then defined by : two tensorial sequences are said to be equivalent if they correspond by the above equations to equal tensorial sequences with the same tensorial character, i.e.$(T^{[j]}_m = T'^{[j]}_m)$ or $(T^{(j)}_m = T'^{(j)}_m)$. The corresponding equivalence classes are by definition tensors of order one. The tensorial sequences $(T^{[j]}_m)$ and $(T^{(j)}_m)$ are called, respectively, the contravariant and covariant (standard) irreducible spherical components of the tensor T^j of order one and rank j. Let us emphasize that, as in the usual tensor algebra, the tensorial character (contravariant or covariant) is only a property of the set of components, but not of the tensor itself which is always the same.

Thus, a tensor T^j plays formally the role of a vector v in a vector space over the complex field with a nondegenerate inner product. It can be defined by its contravariant components ($T^{[j]}_{m}$) or by its covariant ones ($T^{(j)}_{m}$), in the same manner as the vector v by its components (v^i) or (v_i). The two sets of components are related by the regular metric tensor, the matrix C^j here plays the role of (g_{ij}) :

$$T^{(j)}_{m} = C^{j}_{mm'}\, T^{[j]}_{m'} \quad , \qquad v_i = g_{ij}\, v^j \quad .$$

The formal correspondence between the usual tensor algebra and the formalism considered in this chapter is summarized later in Table 8.1. See also Fig. 8.1 below, and compare with Fig. 1.1.

It is recalled that the matrix C^j in Eq. (8–24) is not uniquely defined, but it is a multiple of the unitary matrix C^j. Hence, from Eqs. (8–25,27b) :

$$\lambda_j \neq 0 \in C \qquad C^{j}_{mm'} = \lambda_j (-1)^{j-m}\, \delta_{m,-m'} \quad , \qquad (C^j)^{-1}_{mm'} = \frac{1}{\lambda_j}(-1)^{j+m}\, \delta_{m,-m'} \, , \tag{8-42}$$

and the general relations between the contravariant and covariant spherical components read :

$$T^{(j)}_{m} = C^{j}_{mm'}\, T^{[j]}_{m'} = \lambda_j (-1)^{j-m}\, T^{[j]}_{-m} \tag{8-43a}$$

$$T^{[j]}_{m} = (C^j)^{-1}_{mm'}\, T^{(j)}_{m'} = \frac{1}{\lambda_j}(-1)^{j+m}\, T^{(j)}_{-m} \quad . \tag{8-43b}$$

It is shown in Subsec. 8.3.2.1,iii) that to ensure the stability of the change of tensorial characters under the coupling by means of the C.G. coefficients defined in Subsec. 6.2.3, one has to choose $\lambda_j = 1$ or $(-1)^{2j}$.

ii) The previous definitions of tensorial sequences and tensors of order one are now extended in a straightforward way to any order.

A tensorial sequence ($T^{\dots[j_i]\dots(j_k)\dots}_{m_i \quad m_k}$) of order n and ranks $j_1,\dots,j_n$, each j_i being a non–negative integer or half–odd–integer, is a set of $\prod_{i=1}^{n} (2j_i+1)$ objects $T^{\dots[j_i]\dots(j_k)\dots}_{m_i \quad m_k}$ which belong to a vector space over the complex field and are labelled by the indices $m_i=j_i$, $j_i-1,\dots,-j_i$. The tensorial character of this tensorial sequence defines its law of transformation under the representation of the group SU(2). Namely, each index m_i relative to j_i is either contravariant, it is then denoted by $[j_i]$ and transformed by means of the matrix $\mathcal{D}^{j_i}(U)$, or covariant, it is then denoted by (j_i) and transformed by means of the matrix $\mathcal{D}^{j_i *}(U)$, i.e. :

$$\forall\, U \in SU(2) \qquad {}^{U}T^{\dots[j_i]\dots(j_k)\dots}_{m_i \quad m_k} = T^{\dots[j_i]\dots(j_k)\dots}_{m'_i \quad m'_k} \dots \mathcal{D}^{j_i}_{m'_i m_i}(U) \dots \mathcal{D}^{j_k *}_{m'_k m_k}(U) \dots \quad . \tag{8-44}$$

From a given tensorial sequence of order n, changing the tensorial character of each index m_i (i=1,...,n) by means of Eqs. (8–43a,b), one defines 2^n tensorial sequences related by :

$$T^{\dots(j_i)\dots}_{m_i} = C^{j}_{m_i m_i'} \, T^{\dots[j_i]\dots}_{m_i'} = \lambda_{j_i} \, (-1)^{j_i - m_i} \, T^{\dots[j_i]\dots}_{-m_i} \,, \qquad (8\text{-}45a)$$

$$T^{\dots[j_i]\dots}_{m_i} = (C^{j})^{-1}_{m_i m_i'} \, T^{\dots(j_i)\dots}_{m_i'} = \frac{1}{\lambda_{j_i}} \, (-1)^{j_i + m_i} \, T^{\dots(j_i)\dots}_{-m_i} \,. \qquad (8\text{-}45b)$$

These 2^n tensorial sequences are said to be equivalent and by definition the corresponding equivalence class is the tensor $T^{j_1 \dots j_n}$ of order n and ranks $j_1, \dots, j_n$. The tensorial sequences are also called the (standard) spherical components, which have different tensorial characters, of the same tensor. One should note that a priori the lateral order of the ranks j_i is relevant, this problem is discussed later in iii) (e.g. the important Clebsch–Gordan tensor below).

As a first example of a tensor already encountered, let $C^{j_1 j_2 j_3}$ be the tensor defined by means of the C.G. coefficients as follows :

$$C^{j_1 j_2 j_3} \,, \quad (C^{(j_1)(j_2)[j_3]}_{m_1 \; m_2 \; m_3} = \langle j_1 j_2 m_1 m_2 | j_3 m_3 \rangle) \,. \qquad (8\text{-}46)$$

We shall call it a Clebsch–Gordan tensor (C.G. tensor). Then, Eq. (6–71) shows that this tensor is invariant under the mapping $C^{j_1 j_2 j_3} \to {}^{U}C^{j_1 j_2 j_3}$ defined by Eq. (8–44). Indeed, for all U in SU(2), one has :

$$\begin{aligned} {}^{U}C^{(j_1)(j_2)[j_3]}_{m_1 \; m_2 \; m_3} &= \langle j_1 j_2 m_1' \, m_2' | j_3 m_3' \rangle \, \mathcal{D}^{j_1 *}_{m_1' m_1}(U) \, \mathcal{D}^{j_2 *}_{m_2' m_2}(U) \, \mathcal{D}^{j_3}_{m_3' m_3}(U) \\ &= \langle j_1 j_2 m_1 m_2 | j_3 m_3 \rangle = C^{(j_1)(j_2)[j_3]}_{m_1 \; m_2 \; m_3} \,. \end{aligned}$$

The general problem of invariant tensors is discussed in Subsec. 8.3.5. Changing the tensorial character of m_1 and m_2 by Eqs. (8–45a,b), and using the symmetry property of the C.G. coefficients, Eq. (G–2b) :

$$\langle j_1 j_2 m_1 m_2 | j_3 m_3 \rangle = (-1)^{j_1 + j_2 - j_3} \langle j_1 j_2 -m_1 -m_2 | j_3 -m_3 \rangle \,, \qquad (8\text{-}47)$$

the fully contravariant spherical components ($C^{[j_1] \, [j_2] \, [j_3]}_{m_1 \; m_2 \; m_3}$) of $C^{j_1 j_2 j_3}$ are given by :

$$C^{[j_1][j_2][j_3]}_{m_1 \; m_2 \; m_3} = \frac{1}{\lambda_{j_1} \lambda_{j_2}} \, (-1)^{j_3 - m_3} \, \langle j_1 j_2 m_1 m_2 | j_3 -m_3 \rangle \,.$$

In terms of the Wigner 3j symbols, Eq. (6–73), the above equations reads :

$$C^{[j_1][j_2][j_3]}_{m_1 \; m_2 \; m_3} = (-1)^{j_1 - j_2 - j_3} \, \frac{\sqrt{2j_3 + 1}}{\lambda_{j_1} \lambda_{j_2}} \begin{pmatrix} j_1 & j_2 & j_3 \\ m_1 & m_2 & m_3 \end{pmatrix} \,.$$

Whence one deduces that the C.G. coefficients and the Wigner 3j symbols are the spherical components, which have different tensorial characters, of two proportional tensors :

$$W^{j_1j_2j_3}\ ,\ \left(W^{[j_1][j_2][j_3]}_{m_1\ m_2\ m_3} = \begin{pmatrix} j_1 & j_2 & j_3 \\ m_1 & m_2 & m_3 \end{pmatrix}\right)\ ,\ C^{j_1j_2j_3} = (-1)^{j_1-j_2-j_3}\,\frac{\sqrt{2j_3+1}}{\lambda_{j_1}\lambda_{j_2}}\,W^{j_1j_2j_3}\ . \qquad (8\text{-}48)$$

We shall call $W^{j_1j_2j_3}$ a Wigner 3j tensor. In terms of the Wigner 3j symbols, the invariance of $W^{j_1j_2j_3}$ is given by Eq. (6–75).

iii) The vector space over the complex field, to which belong the spherical components of a tensor, characterizes the nature of this tensor (wave function, bra, ket, operator..., cf. Chap. 11). Clearly the subsets of tensors which have the same nature, order and ranks are vector spaces over the complex field such that :

$$\forall\lambda,\lambda'\in C \qquad T''^{j_1\cdots j_n} = \lambda T^{j_1\cdots j_n} + \lambda' T'^{j_1\cdots j_n}\ ,\ \left(T''^{[j_1]\cdots[j_n]}_{m_1\ \cdots\ m_n} = \lambda T^{[j_1]\cdots[j_n]}_{m_1\ \cdots\ m_n} + \lambda' T'^{[j_1]\cdots[j_n]}_{m_1\ \cdots\ m_n}\right) \qquad (8\text{-}49)$$

As in the usual tensor algebra, one can define the tensor product of any two of these vector spaces. The tensor product $T=T_1\otimes T_2$ of two tensors T_1 and T_2 of any order, ranks and nature is then a new tensor, a set of spherical components of which is defined by all the ordered formal products of one spherical component of T_1 by one of T_2, keeping the lateral order and the tensorial character of each index, e.g. :

$$T_1^{j_1\cdots j_n}\ ,\ \left(T^{[j_1]\cdots[j_n]}_{1\ m_1\ \cdots\ m_n}\right)\ ,\qquad T_2^{\lambda_1\cdots\lambda_p}\ ,\ \left(T^{(\lambda_1)\cdots(\lambda_p)}_{\mu_1\ \cdots\ \mu_p}\right)\ ; \qquad (8\text{-}50)$$

$$T^{j_1\cdots j_n\lambda_1\cdots\lambda_p} = T_1^{j_1\cdots j_n}\otimes T_2^{\lambda_1\cdots\lambda_p}\ ,\ \left(T^{[j_1]\cdots[j_n](\lambda_1)\cdots(\lambda_p)}_{m_1\ \cdots\ m_n\ \mu_1\ \cdots\ \mu_p} = T^{[j_1]\cdots[j_n]}_{1\ m_1\ \cdots\ m_n}\,T^{(\lambda_1)\cdots(\lambda_p)}_{2\ \mu_1\ \cdots\ \mu_p}\right)$$

Assuming the tensor product is associative, one defines step by step the tensor product of any number of tensors.

The precise nature of this formal product of components, denoted above as a usual one, depends upon the nature of the considered spherical components. If, for instance, the components of T_1 and T_2 belong to the same algebra over the complex field, the product is defined by the bilinear mapping of this algebra (usual product of complex numbers, tensor product of kets or bras, composition law of operators...). Since the formal product of components may not be commutative or even not defined in both orders (T_1T_2 and T_2T_1), so it is with the tensorial product. Consequently, for tensors $T^{j_1\cdots j_n}$ defined as the tensorial product of elementary tensors T^{j_i}, the lateral order of the ranks j_i is in general relevant.

8.3.1.2. Dependence of the spherical components upon a triad, extension of the definition of tensors

i) A tensorial sequence $(T^{[j]}_m)$ of order one has been characterized by a law of transformation under a unitary representation D[SU(2)] which is identical to the one of a standard basis (e^j_m). Consequently (Subsec. 6.1.3,i)), if (J_i) denotes the infinitesimal generators of this representation, defined from (σ_i) by Eq. (8–1), one has the relations :

$$J_{\pm}\, T^{[j]}_{m} = \sqrt{j(j+1)-m(m\pm 1)}\;\; T^{[j]}_{m\pm 1} \quad , \tag{8-51a}$$

$$J_3\, T^{[j]}_{m} = m\, T^{[j]}_{m} \quad . \tag{8-51b}$$

Using Eqs. (8–43a,b) to change the tensorial character, one immediately checks that the covariant spherical components $(T^{(j)}_{m})$ fulfil the same equations as above with respect to the infinitesimal generators $\bar{J}_i$ defined by Eq. (8–33c), that is :

$$\bar{J}_{\pm}\, T^{(j)}_{m} = \sqrt{j(j+1)-m(m\pm 1)}\;\; T^{(j)}_{m\pm 1} \quad , \tag{8-52a}$$

$$\bar{J}_3\, T^{(j)}_{m} = m\, T^{(j)}_{m} \quad . \tag{8-52b}$$

Therefore, all the results of the study made in Sec. 8.1 about the dependence of a standard basis upon a choice of basis of infinitesimal generators can be applied to any tensorial sequence of order one, and obviously more generally of any order. This leads to a new extension of the definitions of tensorial sequences and tensors. One sets for any <u>tensorial sequence</u> of order one :

$$T^{[j]}_{m;U} = {}^{U}T^{[j]}_{m;\mathbf{1}} = {}^{U}T^{[j]}_{m} \quad , \quad T^{(j)}_{m;U} = C^{j}_{mm'}\, T^{[j]}_{m';U} = {}^{U}T^{(j)}_{m;\mathbf{1}} = {}^{U}T^{(j)}_{m} \quad , \tag{8-53}$$

where the index U denotes a triad defined in Subsec. 8.1,iv), i.e. an element in SU(2) to which corresponds a unique rotation R(U) in the Lie algebra D $[\mathcal{W}(2)]$. Then, by definition $(T^{[j]}_{m;U})$ and $(T^{(j)}_{m;U})$ are the spherical components of a given <u>tensor</u> T^{j} with respect to the basis of infinitesimal generators $(R(U)J_i)$, while $(T^{[j]}_{m;\mathbf{1}})$ and $(T^{(j)}_{m;\mathbf{1}})$ denote the spherical components of the same tensor with respect to the basis (J_i) of infinitesimal generators, defined from (σ_i) by Eq. (8–1). Clearly the above definitions can be generalized to tensors of order greater than one with a similar notation, e.g. :

$$T^{j_1 \dots j_n} \quad , \quad \left(T^{[j_1]\dots(j_n)}_{m_1 \quad\; m_n;U} = {}^{U}T^{[j_1]\dots(j_n)}_{m_1 \quad\; m_n;\mathbf{1}} = {}^{U}T^{[j_1]\dots(j_n)}_{m_1 \quad\; m_n} \right) \quad . \tag{8-54}$$

From, Eqs. (8–12,14), respectively, the dependence of the spherical components upon the choice of a triad and their general law of transformation are given in the generic case of a tensor of order one by :

$$\forall\, U,U' \in SU(2) \qquad T^{[j]}_{m;U'} = T^{[j]}_{m';U}\, \mathcal{D}^{j}_{m'm}(U^{-1}U') \quad , \quad T^{(j)}_{m;U'} = T^{(j)}_{m';U}\, \mathcal{D}^{j*}_{m'm}(U^{-1}U') \quad , \tag{8-55a}$$

$${}^{U'}T^{[j]}_{m;U} = T^{[j]}_{m';U}\, \mathcal{D}^{j}_{m'm}(U^{-1}U'U) \quad , \quad {}^{U'}T^{(j)}_{m;U} = T^{(j)}_{m';U}\, \mathcal{D}^{j}_{m'm}(U^{-1}U'U) \quad . \tag{8-55b}$$

Finally with these new definitions, a tensor is characterized by its <u>nature</u>, its <u>order</u> and its <u>ranks</u>, while to characterize one set of its spherical components, i.e. a tensorial sequence one needs in addition to specify the <u>tensorial character</u> and a <u>triad</u>. An example of such a tensor of order one is the following : the standard basis $(e^{j}_{m;U})$, introduced in Subsec. 8.1,iii), are the contravariant spherical components with respect to the triad U of a tensor denoted by e^{j} :

$$e^j \;,\; \left(e^{[j]}_{m;U} = e^j_{m;U}\right) \quad . \tag{8-56}$$

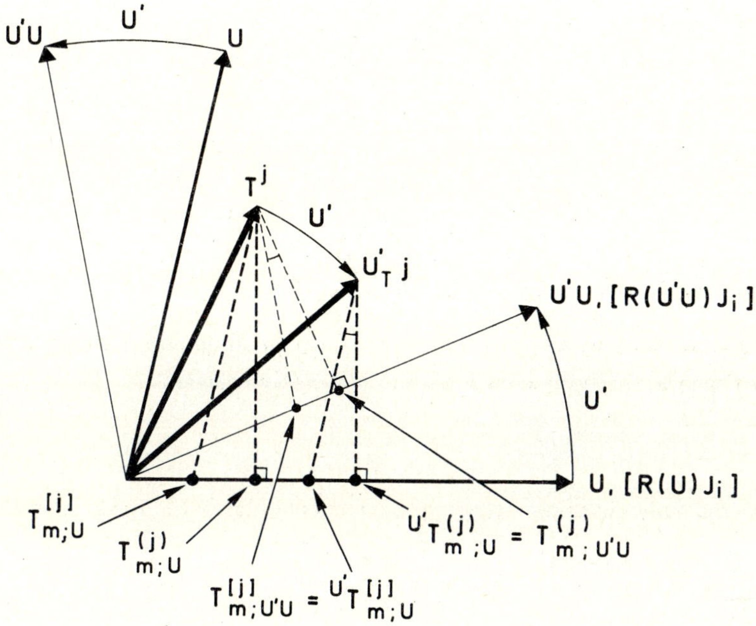

Fig. 8.1. The formal axes labelled by the triads U and U'U correspond, respectively, to the bases $(R(U)J_i)$ and $(R(U'U)J_i)$ of the Lie algebra D[$\mathcal{W}(2)$]. One should note that in this formalism, under the active point of view of the representation of U', the tensors and the axes are transformed in the opposite way (this is just a matter of definition). For the interpretation of the spherical components which have different tensorial characters compare with Fig. 1.·.

ii) The previous formalism allows us to give an active and a passive point of view of the representation of the group SU(2), similar to the one discussed in Subsec. 1.2.1. Let us consider the generic case of a tensor of order one. The active point of view of the representation corresponds to the mapping which assigns to any tensor T^j a new one such that :

$$\forall U' \in SU(2) \qquad T^j \rightarrow {}^{U'}T^j \left(\forall U \in SU(2) \;,\; {}^{U'}T^{[j]}_{m;U} = T^{[j]}_{m';U}\, \mathcal{D}^j_{m'm}(U^{-1}U'U)\right) \quad . \tag{8-57}$$

The passive point of view of the representation corresponds for a given tensor to a change of triad, to which is associated a change of basis in the Lie algebra D [$\mathcal{W}$(2)], such that :

$$T^j \;,\; \left(\forall U,U' \in SU(2) \qquad T^{[j]}_{m;U'} = T^{[j]}_{m';U}\, \mathcal{D}^j_{m'm}(U^{-1}U')\right) \;, \tag{8-58}$$

triad U(resp.U') $\rightarrow$ basis $(R(U)J_i)$ (resp.$(R(U')J_i)$) in D [$\mathcal{W}$ (2)].

As in Subsec. 1.2.1, these two points of view are said to be equivalent, since one has :

$$\forall U,U' \in SU(2) \qquad {}^{U'}T^{[j]}_{m;U} = T^{[j]}_{m;U'U} \quad . \tag{8-59}$$

This result is illustrated in Fig. 8.1. Physical interpretation and applications of the foregoing are given later in Subsecs. 11.1.2,3.

From now on, except when explicitly mentioned, the spherical components are defined with respect to the triad U = **1**, which corresponds to the basis (J_i) of infinitesimal generators, related to (σ_i) by Eq. (8–1). Then, to simplify the notation <u>the triad **1** is omitted</u>, e.g. :

$$T^{[j_1]\dots(j_n)}_{m_1\ \dots\ m_n;\mathbf{1}} = T^{[j_1]\dots(j_n)}_{m_1\ \quad m_n} \quad . \tag{8-60}$$

8.3.1.3. Symmetrical and antisymmetrical tensors

i) Let us first consider a tensor T^{jj} of order two with equal ranks, and such that its fully contravariant spherical components fulfil the symmetry relation :

$$\forall m,m' \qquad T^{[j][j]}_{m\ m'} = \varepsilon\, T^{[j][j]}_{m'\ m} \qquad \varepsilon = \pm 1 \quad . \tag{8-61}$$

Then, clearly the fully covariant spherical components also satisfy the same relation :

$$\forall m,m' \qquad T^{(j)(j)}_{m\ m'} = \varepsilon\, T^{(j)(j)}_{m'\ m} \quad . \tag{8-62}$$

Moreover, these properties are preserved under the representation of the group SU(2) ($T^{jj} \to {}^{U}T^{jj}$ in the active point of view, or a change of triad in the passive point of view) since one has :

$${}^{U}T^{[j][j]}_{m\ m'} = T^{[j][j]}_{\mu\ \mu'}\, \mathcal{D}^{j}_{\mu m}(U)\mathcal{D}^{j}_{\mu' m'}(U) = \varepsilon T^{[j][j]}_{\mu'\ \mu}\mathcal{D}^{j}_{\mu m}(U)\mathcal{D}^{j}_{\mu' m'}(U) = \varepsilon\, {}^{U}T^{[j][j]}_{m'\ m} \quad .$$

Hence, this symmetry property is intrinsic, and the tensor T^{jj} is said to be <u>symmetrical</u> or <u>antisymmetrical</u> according as ε equals +1 or –1. This is denoted by :

$$T^{jj} = \varepsilon\, T^{t,jj} \quad , \tag{8-63}$$

where $T^{t,jj}$ is a new tensor defined from any T^{jj} by :

$$T^{jj} \to T^{t,jj} \quad , \quad \left(T^{t,[j][j]}_{m\ m'} = T^{[j][j]}_{m'\ m}\right) \quad . \tag{8-64}$$

One should note that a symmetry property for the spherical components where the indices do not have the same tensorial character, such as for instance :

$$T^{[j](j)}_{m\ m'} = T^{[j](j)}_{m'\ m} \quad ,$$

is not stable under the representation of the group SU(2). Therefore, this does not correspond to an intrinsic property of the tensor itself.

ii) The previous remarks can be extended to tensors of order greater than two. Then, there may exist partial symmetries (or antisymmetries) which apply to subsets of indices provided that all these indices have the same tensorial character and equal ranks. For instance, the property :

$$\forall m_1,m_2,m_3 \qquad T^{[j][j](j')[j]}_{m_1\ m_2\ m'\ m_3} = T^{[j][j](j')[j]}_{m_3\ m_1\ m'\ m_2} \quad ,$$

is intrinsic. In particular, for a tensor of order n with n equal ranks, it may happen that its fully contravariant or covariant spherical components remain unchanged (resp. change sign) under all transpositions of any two indices. The tensor is then said to be completely symmetrical (resp. antisymmetrical). As in the usual tensor algebra the subset of completely antisymmetrical tensors of given nature, order and rank is an external algebra.

iii) Any tensor $T^{j\cdots j}$ with n equal ranks (or more generally forgetting the other possible ranks) can be decomposed into tensors which have a definite symmetry, corresponding to the irreducible representations of the symmetric group Σ_n. Let us only give the simple example of a tensor T^{jj} which can be split into a symmetrical and an antisymmetrical tensor, respectively, S^{jj} and A^{jj} defined by :

$$T^{jj} = S^{jj} + A^{jj} \quad , \quad \begin{cases} S^{jj} = \frac{1}{2}(T^{jj} + T^{t,jj}) \\ A^{jj} = \frac{1}{2}(T^{jj} - T^{t,jj}) \end{cases} \quad . \tag{8-65}$$

These symmetry properties being intrinsic, the above decompositions are a first step in the expression of any tensor $T^{j\cdots j}$ in terms of its irreducible tensors. Thus, for the previous example :

$$T^{jj} = (T^{\{jj\}\ell}\ ,\ \ell=0,1,\ldots,2j) = \sum_{\ell=0}^{2j} T_\ell \quad , \tag{8-66a}$$

$$T_\ell = (0,\ldots,0,\ T^{\{jj\}\ell}\ ,\ 0,\ldots,0) \quad , \tag{8-66b}$$

and from the symmetry property of the C.G. coefficients, Eq. (G–2a) :

$$\langle jjmn|\ell p\rangle = (-1)^{2j-\ell}\ \langle jjnm|\ell p\rangle \quad , \tag{8-67}$$

it follows that :

$$S^{jj} = \sum_{\ell=0,2j-\ell\ \text{even}}^{2j} T_\ell \quad , \quad A^{jj} = \sum_{\ell=0,2j-\ell\ \text{odd}}^{2j} T_\ell \quad . \tag{8-68}$$

All these considerations are for instance useful in Quantum Mechanics when applied to the product of creation and annihilation operators for Fermions and Bosons, Subsec. 11.2.2.2.

8.3.2. COUPLINGS AND IRREDUCIBLE SPHERICAL COMPONENTS

8.3.2.1. Couplings of two indices

i) Definition. The spherical components $(T^{[j_1][j_2]}_{m_1\ m_2})$ of a tensor $T^{j_1j_2}$ belong by definition to the product representation $\mathcal{D}^{j_1} \otimes \mathcal{D}^{j_2}$. The decomposition of this product into the irreducible representations $\mathcal{D}^{j_3}$ is realized by the C.G. coefficients, i.e. by the coupling of the indices j_1 and j_2 to j_3. By definition the C.G. coefficients introduced in Subsec. 6.2.3.1 relate indices which all belong to the representation $\mathcal{D}^j$. Indeed, in Eq. (6–64a) the vectors e^j_m as do the f^j_m transform under the mapping T_U by means of the matrix $\mathcal{D}^j(U)$, Eqs. (6–11,51). Hence, the objects defined by :

$$T^{\{[j_1][j_2]\}[j_3]}_{m_3} = T^{[j_1][j_2]}_{m_1 \; m_2} \langle j_1 j_2 m_1 m_2 | j_3 m_3 \rangle \quad , \tag{8-69}$$

really belong to the representation $\mathcal{D}^j$, as denoted by their tensorial character. As an exercise one can check this result directly. Using Eq. (6–68b) and the unitary character of the C.G. coefficients, Eq. (6–66a), one has for all U in SU(2) :

$$U_T T^{\{[j_1][j_2]\}[j_3]}_{m_3} = T^{[j_1][j_2]}_{m_1' \; m_2'} \mathcal{D}^{j_1}_{m_1' m_1}(U) \mathcal{D}^{j_2}_{m_2' m_2}(U) \langle j_1 j_2 m_1 m_2 | j_3 m_3 \rangle$$

$$= T^{[j_1][j_2]}_{m_1' \; m_2'} \langle j_1 j_2 m_1' m_2' | \lambda\mu \rangle \; \mathcal{D}^{\lambda}_{\mu\nu}(U) \; \langle \lambda\nu | j_1 j_2 m_1 m_2 \rangle \langle j_1 j_2 m_1 m_2 | j_3 m_3 \rangle$$

$$= T^{[j_1][j_2]}_{m_1' \; m_2'} \langle j_1 j_2 m_1' m_2' | \lambda\mu \rangle \; \mathcal{D}^{\lambda}_{\mu\nu}(U) \; \delta_{\lambda j_3} \delta_{\nu m_3} = T^{\{[j_1][j_2]\}[j_3]}_{\mu} \mathcal{D}^{j_3}_{\mu m_3}(U) \quad .$$

The $\sum_{j_3} \{j_1 j_2 j_3\}(2j_3+1) = (2j_1+1)(2j_2+1)$ objects $T^{\{[j_1]\,[j_2]\}[j_3]}_{m_3}$ are called the contravariant <u>irreducible</u> spherical components of the tensor $T^{j_1 j_2}$. As the C.G. coefficients have the unitary character and are chosen real, Eq. (8–69) is inverted as follows :

$$T^{[j_1][j_2]}_{m_1 \; m_2} = \sum_{j_3} T^{\{[j_1][j_2]\}[j_3]}_{m_3} \langle j_3 m_3 | j_1 j_2 m_1 m_2 \rangle \quad . \tag{8-70}$$

The irreducible spherical components are also called the <u>completely coupled spherical components</u>, and for instance ($T^{[j_1][j_2]}_{m_1 \; m_2}$) the <u>uncoupled</u> ones. Let us remark that for each given j_3 the irreducible spherical components ($T^{\{[j_1]\,[j_2]\}\,[j_3]}_{m_3}$) define a new tensor, called a <u>coupled</u> or <u>irreducible tensor</u>.

ii) Coupling scheme. The decomposition of the product representation $\mathcal{D}^{j_2} \otimes \mathcal{D}^{j_1}$, instead of $\mathcal{D}^{j_1} \otimes \mathcal{D}^{j_2}$, into the irreducible representations $\mathcal{D}^{j_3}$ leads to the coupling of the indices j_1 and j_2 in the reverse order to the one previously considered. Thus, new objects which also belong to the representation $\mathcal{D}^j$ and still called irreducible spherical components of $T^{j_1 j_2}$, can be defined by :

$$T^{j_1 j_2, \{[j_2][j_1]\}[j_3]}_{m_3} = T^{[j_1][j_2]}_{m_1 \; m_2} \langle j_2 j_1 m_2 m_1 | j_3 m_3 \rangle \quad . \tag{8-71}$$

A given order of the coupling defines a <u>coupling scheme</u>. To each coupling scheme there corresponds a set of irreducible spherical components and vice versa. From the symmetry property of the C.G. coefficients, Eq. (G–2a) :

$$\langle j_1 j_2 m_1 m_2 | j_3 m_3 \rangle = (-1)^{j_1+j_2-j_3} \langle j_2 j_1 m_2 m_1 | j_3 m_3 \rangle \quad , \tag{8-72}$$

the two sets of contravariant irreducible spherical components of $T^{j_1 j_2}$, defined, respectively, by Eqs. (8–69,71) are related by :

$$T^{\{[j_1][j_2]\}[j_3]}_{m_3} = (-1)^{j_1+j_2-j_3}\, T^{j_1 j_2, \{[j_2][j_1]\}[j_3]}_{m_3} \quad . \tag{8-73}$$

It is worthwhile to emphasize that for the irreducible spherical components of $T^{j_1 j_2}$ there exist now two independent relevant types of order of the indices j_1 and j_2 :

(1) The order of the ranks of the uncoupled spherical components, which characterizes the tensor itself (Subsec. 8.3.1.1,iii)).

(2) The order of coupling, i.e. the coupling scheme which characterizes the irreducible spherical components. The notation for the coupling is the following :

$$\{j_\alpha j_\beta\}\ j_\gamma \Leftrightarrow \left\{ \begin{array}{l} j_\alpha \text{ and } j_\beta \text{ are coupled to } j_\gamma \text{ by means of :} \\ \langle j_\alpha j_\beta m_\alpha m_\beta | j_\gamma m_\gamma \rangle \quad . \end{array} \right. \tag{8-74}$$

When both orders (1) and (2) are identical, as in Eq. (8–69), only the order of coupling is specified explicitly, while they are both specified if they are different as in Eq. (8–71).

iii) Consequences of the choice of real C.G. coefficients :

1–Since the commonly used C.G. coefficients are chosen real, they also realize the reduction of the product representation $\mathcal{D}^{j_1^*} \otimes \mathcal{D}^{j_2^*}$ into the irreducible representations $\mathcal{D}^{j_3^*}$. Indeed, the Eqs. (6–68a,b) still hold when all the matrix elements of the matrices $\mathcal{D}^j$ are replaced by their complex conjugates. Therefore, for a given tensor $T^{j_1 j_2}$, the objects defined by :

$$T^{\{(j_1)(j_2)\}(j_3)}_{m_3} = T^{(j_1)(j_2)}_{m_1\ m_2} \langle j_1 j_2 m_1 m_2 | j_3 m_3 \rangle \quad , \tag{8-75}$$

really belong to the representation $\mathcal{D}^{j^*}$ as denoted by their tensorial character. They are called covariant irreducible spherical components of the tensor $T^{j_1 j_2}$. Clearly, the remarks made previously in ii) about the order of coupling are still valid, i.e. coupling j_1 and j_2 in the reverse order, one defines :

$$T^{j_1 j_2, \{(j_2)(j_1)\}(j_3)}_{m_3} = T^{(j_1)(j_2)}_{m_1\ m_2} \langle j_2 j_1 m_2 m_1 | j_3 m_3 \rangle = (-1)^{j_1+j_2-j_3}\, T^{\{(j_1)(j_2)\}(j_3)}_{m_3} \tag{8-76}$$

Finally, the real C.G. coefficients allow us to couple two indices j_1 and j_2 to a third one j_3 only if all the three indices have the same tensorial character. The general problem of the coupling of indices with different tensorial characters is discussed below in iv).

2–By definition, both sets of spherical components ($T^{[j_1][j_2]}_{m_1\ m_2}$) and ($T^{(j_1)(j_2)}_{m_1\ m_2}$) of $T^{j_1 j_2}$ are related by the rule given in Eqs. (8–43a,b). The question then arises, are the irreducible spherical components ($T^{\{[j_1][j_2]\}[j_3]}_{m_3}$) and ($T^{\{(j_1)(j_2)\}(j_3)}_{m_3}$) still related by the same rule ? i.e. for instance :

$$T^{\{(j_1)(j_2)\}(j_3)}_{m_3} = \lambda_{j_3} (-1)^{j_3-m_3}\, T^{\{[j_1][j_2]\}[j_3]}_{-m_3} \quad ? \tag{8-77}$$

Let us show that this is true only when a suitable <u>choice of the matrix $\mathcal{C}^j$</u>, i.e. a choice of λ_j in Eq. (8–25), is made. Clearly to get a definition of the change of the tensorial character which is stable under the coupling, the matrix $\mathcal{C}^j$ itself must be one of the unitary matrices of the representation $\mathcal{D}^j[SU(2)]$. Since, from Eq. (8–25), $\mathcal{C}^j$ is a multiple of one of these matrices, i.e. C^j, Eq. (8–27b), the two possible choices of λ_j are 1 and $(-1)^{2j}$. They, respectively, correspond to :

$$\mathcal{C}^j = C^j = \mathcal{D}^j(C) \quad , \quad \mathcal{C}^j = (-1)^{2j} C^j = (C^j)^{-1} = \mathcal{D}^j(C^{-1}) \quad . \tag{8-78}$$

As an exercise, it can be checked directly that these two possibilities for $\mathcal{C}^j$ are really the only ones which lead to the desired property. Using Eqs. (8–43a,b) to change the tensorial character of indices, and the symmetry property of the C.G. coefficients, Eq. (8–47), one gets :

$$T^{\{(j_1)(j_2)\}(j_3)}_{m_3} = T^{(j_1)(j_2)}_{m_1\ m_2} \langle j_1 j_2 m_1 m_2 | j_3 m_3\rangle = \lambda_{j_1}\lambda_{j_2} (-1)^{j_1+j_2-m_1-m_2}\, T^{[j_1][j_2]}_{-m_1-m_2} \langle j_1 j_2 m_1 m_2 | j_3 m_3\rangle$$
$$= \lambda_{j_1}\lambda_{j_2} (-1)^{j_3-m_3}\, T^{[j_1][j_2]}_{-m_1-m_2} \langle j_1 j_2 -m_1 -m_2 | j_3 -m_3\rangle = \lambda_{j_1}\lambda_{j_2} (-1)^{j_3-m_3}\, T^{\{[j_1][j_2]\}[j_3]}_{-m_3} \quad .$$

Hence, to obtain Eq. (8–77), the complex coefficient λ_j has to fulfil the necessary, and clearly sufficient, functional equation :

$$\lambda_{j_3} = \lambda_{j_1}\lambda_{j_2} \quad ,$$

for all non–negative integers or half–odd–integers j_1, j_2, and j_3 such that :

$$j_3 = j_1+j_2 \quad , \quad j_1+j_2-1, \ldots, |j_1-j_2| \quad .$$

For integer j, the values $j_1=j_2=j_3=j$ fulfil the above condition, one therefore has :

$$\lambda_j = \lambda_j \lambda_j \quad \Rightarrow \quad \lambda_j = 1 \qquad \forall j \ \text{integer.}$$

Integer j_1 and half–odd–integer j_2 can only be coupled to half–odd–integer values of j_3, and from the above we therefore have :

$$\lambda_{j_3} = \lambda_{j_1}\lambda_{j_2} = \lambda_{j_2} \quad \Rightarrow \quad \lambda_j = \lambda \qquad \forall j \quad \text{half-odd-integer.}$$

Finally, two half–odd–integer values of j_1 and j_2 can only be coupled to an integer value of j_3, this determines λ :

$$\lambda_{j_3} = 1 = \lambda_{j_1}\lambda_{j_2} = \lambda^2 \quad \Rightarrow \quad \lambda = \pm 1 \qquad \forall j \quad \text{half-odd-integer.}$$

In the remainder of these notes, only these two choices of $\mathcal{C}^j$ are considered, and in what follows the change of the tensorial character of an index is defined by :

$$c_j = (-1)^{2j}\ \lambda_j = 1 \text{ or } (-1)^{2j} \quad , \tag{8-79a}$$

$$T^{(j)}_{m} = c_j (-1)^{j+m}\, T^{[j]}_{-m} \quad , \quad T^{[j]}_{m} = c_j (-1)^{j-m}\, T^{(j)}_{-m} \quad . \tag{8-79b}$$

These two choices of c_j are only different for half–odd–integer values of j for they are related by the factor $(-1)^{2j}$ which will be frequently encountered in the following. Now, without ambiguity, the tensorial character of the indices which are coupled can be omitted, e.g. :

$$T^{\{j_1j_2\}[j_3]}_{m_3} = T^{\{[j_1][j_2]\}[j_3]}_{m_3} \quad , \quad T^{\{j_1j_2\}(j_3)}_{m_3} = T^{\{(j_1)(j_2)\}(j_3)}_{m_3} \quad . \tag{8-80}$$

iv) Coupling of indices with different tensorial characters. In terms of the spherical components of a C.G. tensor, the coupling of indices which are all contravariant reads :

$$T^{\{j_1j_2\}[j_3]}_{m_3} = T^{[j_1][j_2]}_{m_1\ m_2} \langle j_1j_2m_1m_2|j_3m_3\rangle \quad . \tag{8-81}$$

Using Eqs. (8–79a,b), one is then able to define all the coupling coefficients which correspond to the different choices of the tensorial character of m_1,m_2, and m_3. These coupling coefficients are thus related to spherical components of $C^{j_1j_2j_3}$ by phase factors.

It is convenient to notice that whatever be the choice of c_j one has (Subsec. 8.3.3.1) :

$$T_1{}^{(j)}_{m}\ T_2{}^{[j]}_{m} = (-1)^{2j}\ T_1{}^{[j]}_{m}\ T_2{}^{(j)}_{m} \quad . \tag{8-82}$$

Moreover, from Eqs. (8–79a,b) and using the symmetry property of the C.G. coefficients given in Eq. (8–47), one gets useful numerical relations between the spherical components of $C^{j_1j_2j_3}$ which have different tensorial characters :

$$C^{(j_1)(j_2)[j_3]}_{m_1\ m_2\ m_3} = (-1)^{2j_3}\ C^{[j_1][j_2](j_3)}_{m_1\ m_2\ m_3} \quad , \tag{8-83a}$$

$$C^{[j_1][j_2][j_3]}_{m_1\ m_2\ m_3} = C^{(j_1)(j_2)(j_3)}_{m_1\ m_2\ m_3} \quad . \tag{8-83b}$$

We have used the equation :

$$c_{j_1}\ c_{j_2}\ c_{j_3} = \{1 \text{ or } (-1)^{2(j_1+j_2+j_3)}\} = 1 \quad , \tag{8-84}$$

since $j_1+j_2+j_3$ is an integer whenever the C.G. coefficient is nonzero. Finally, the eight possibilities of coupling are explicitly :

$$T^{\{j_1j_2\}[j_3]}_{m_3} = T^{[j_1][j_2]}_{m_1\ m_2}\, C^{(j_1)(j_2)[j_3]}_{m_1\ m_2\ m_3} = (-1)^{2j_3}\, T^{(j_1)(j_2)}_{m_1\ m_2}\, C^{[j_1][j_2][j_3]}_{m_1\ m_2\ m_3} \quad .$$

$$=(-1)^{2j_1}\, T^{(j_1)[j_2]}_{m_1\ m_2}\, C^{[j_1](j_2)[j_3]}_{m_1\ m_2\ m_3} = (-1)^{2j_2}\, T^{[j_1](j_2)}_{m_1\ m_2}\, C^{(j_1)[j_2][j_3]}_{m_1\ m_2\ m_3} , \tag{8-85}$$

and four similar equalities where the index m_3 is covariant. Actually, a compact form of these relations, as for their inversion, is discussed in Subsecs. 8.3.3.3 and 8.3.4.2.

From Eqs. (8–48,79a), the C.G. and Wigner 3j tensors are related by :

$$C^{j_1j_2j_3} = c_{j_3}\,(-1)^{j_1-j_2+j_3}\sqrt{2j_3+1}\; W^{j_1j_2j_3} . \tag{8-86}$$

Hence, in terms of the spherical components of a Wigner 3j tensor the couplings read :

$$T^{\{j_1j_2\}[j_3]}_{m_3} = c_{j_3}\,(-1)^{j_1-j_2-j_3}\sqrt{2j_3+1}\; T^{(j_1)(j_2)}_{m_1\ m_2}\, W^{[j_1][j_2][j_3]}_{m_1\ m_2\ m_3} , \tag{8-87a}$$

$$T^{\{j_1j_2\}(j_3)}_{m_3} = c_{j_3}\,(-1)^{j_1-j_2+j_3}\sqrt{2j_3+1}\; T^{[j_1][j_2]}_{m_1\ m_2}\, W^{(j_1)(j_2)(j_3)}_{m_1\ m_2\ m_3} , \tag{8-87b}$$

where :

$$W^{[j_1][j_2][j_3]}_{m_1\ m_2\ m_3} = W^{(j_1)(j_2)(j_3)}_{m_1\ m_2\ m_3} = \begin{pmatrix} j_1 & j_2 & j_3 \\ m_1 & m_2 & m_3 \end{pmatrix} . \tag{8-88}$$

Therefore, up to a multiplicative factor, the Wigner 3j symbols allow us to couple two indices which have the same tensorial character to a third one which has the other tensorial character.

The symmetry properties of the Wigner 3j symbols and the resulting one for the C.G. coefficients (Appendix G.1.2,iii)) imply symmetry relations for the corresponding tensors :

for any transposition, e.g. of j_1 and j_2
$$W^{j_1j_2j_3} = (-1)^{j_1+j_2+j_3}\, W^{j_2j_1j_3} , \tag{8-89a}$$

$$C^{j_1j_2j_3} = (-1)^{j_1+j_2-j_3}\, C^{j_2j_1j_3} = c_{j_2}\,(-1)^{j_1+j_2+j_3}\,\frac{\hat{j}_3}{\hat{j}_1}\, C^{j_3j_2j_1} . \tag{8-89b}$$

It should be noticed that the above equations must be taken with care since, exceptionally, the lateral order of the ranks is modified.

v) A criterion of tensorial character. As in the usual tensor algebra, an important problem is to decide whether a set of objects is a tensorial sequence or not. The only way to answer such a question is so far to come back to the definition of a tensor, i.e. to check explicitly that Eq. (8–44) is fulfilled. A new way is given by the following criterion of tensorial character :

A necessary and sufficient condition for a set of objects ($T_{2m_2}^{j_2}$) be a tensorial sequence, i.e. the contravariant (resp. covariant) spherical components of a tensor $T_2^{j_2}$, is that for every tensor $T_1^{j_1}$ the set of objects defined by :

$$T^{\{j_1j_2\}[j_3]}{}_{m_3} = T_1{}^{[j_1]}{}_{m_1} T_2{}^{j_2}{}_{m_2} <j_1j_2m_1m_2|j_3m_3> \quad , \tag{8-90}$$

is also a tensorial sequence, i.e. the contravariant spherical components of the tensor product $T^{j_1j_2} = T_1{}^{j_1} \otimes T_2{}^{j_2}$.

Proof. It has been shown previously in i) that this condition is necessary. Conversely, using Eq. (8–70) equivalent to Eq. (8–90), the tensorial character of the sequences $(T_1{}^{[j_1]}{}_{m_1})$ and $(T^{\{j_1j_2\}[j_3]}{}_{m_3})$ implies for all U in SU(2) :

$$^{U}T_1{}^{[j_1]}{}_{m_1} {}^{U}T_2{}^{j_2}{}_{m_2} = T^{[j_1]}{}_{m_1'} \mathcal{D}^{j_1}_{m_1'm_1}(U) {}^{U}T_2{}^{j_2}{}_{m_2} = \sum_{j_3} {}^{U}T^{\{j_1j_2\}[j_3]}{}_{m_3} <j_3m_3|j_1j_2m_1m_2> = \sum_{j_3} T^{\{j_1j_2\}[j_3]}{}_{m_3'} \mathcal{D}^{j_3}_{m_3'm_3}(U)$$
$$\times <j_3m_3|j_1j_2m_1m_2> = \sum_{j_3} T_1{}^{[j_1]}{}_{m_1'} T_2{}^{j_2}{}_{m_2'} <j_1j_2m_1'm_2'|j_3m_3'> \mathcal{D}^{j_3}_{m_3'm_3}(U) <j_3m_3|j_1j_2m_1m_2> \; .$$

Since this relation holds for every tensor $T_1{}^{j_1}$, using the properties of the matrices $\mathcal{D}^j$, Eqs. (6–23a,68a) and the symmetry property of the C.G. coefficients, Eq. (G–2d) :

$$<j_1j_2m_1m_2|j_3m_3> = (-1)^{j_2-j_3-m_1} \sqrt{\frac{2j_3+1}{2j_2+1}} <j_1j_3-m_1m_3|j_2m_2> \quad , \tag{8-91}$$

one gets the announced result :

$$^{U}T_2{}^{j_2}{}_{m_2} = \sum_{j_3} \frac{1}{2j_1+1} \mathcal{D}^{j_1}_{m_1m_1'}(U^{-1}) T_2{}^{j_2}{}_{m_2'} <j_1j_2m_1'm_2'|j_3m_3'> \mathcal{D}^{j_3}_{m_3'm_3}(U) <j_3m_3|j_1j_2m_1m_2>$$
$$= \sum_{j_3} \frac{2j_3+1}{(2j_1+1)(2j_2+1)} T_2{}^{j_2}{}_{m_2'} <j_2m_2'|j_1j_3-m_1'm_3'> \mathcal{D}^{j_1}_{-m_1'-m_1}(U) \mathcal{D}^{j_3}_{m_3'm_3}(U) <j_1j_3-m_1m_3|j_2m_2>$$
$$= T_2{}^{j_2}{}_{m_2'} \mathcal{D}^{j_2}_{m_2'm_2}(U) \quad . \tag{8-92}$$

A similar proof can be given for the covariant spherical components.

8.3.2.2. Generalization to tensors of order greater than two

i) By definition the uncoupled spherical components $(T^{[j_1]\cdots[j_n]}{}_{m_1 \cdots m_n})$ (resp. $T_{m_1 \cdots m_n}^{(j_1)\ldots(j_n)}$) of a tensor $T^{j_1\cdots j_n}$ belong to the product representation $\mathcal{D}^{j_1}\otimes\ldots\otimes\mathcal{D}^{j_n}$ (resp. $\mathcal{D}^{j_1^*}\otimes\ldots\otimes\mathcal{D}^{j_n^*}$). Using the previous procedure of coupling two indices, this product representation can be stepwise either partly reduced, which leads to coupled spherical components of $T^{j_1\cdots j_n}$, or fully reduced, and thus leads to the definition of irreducible (standard) spherical components of this tensor.

ii) Two ways of coupling generalize to several ways of reducing the product representation $\mathcal{D}^{j_1}\otimes\ldots\otimes\mathcal{D}^{j_n}$. They correspond to different coupling schemes which are characterized by the required intermediate couplings (n–2 for the irreducible spherical components), i.e. which indices are coupled and in which order. For a given tensor $T^{j_1 \dots j_n}$, all these coupling schemes are related by unitary transformations which are the product of elementary unitary transformations defined by the C.G. coefficients. Eq. (8–73) is a special case of these relations. The transformations between different coupling schemes are discussed later in Subsec. 8.4.2.

For the coupled spherical components, as in the case n=2, there exist two independent relevant types of order of the n ranks : the first one characterizes the tensor itself while the second one defines the coupling scheme of the considered components. The method specified in Eq. (8–74) is used to denote the couplings (especially both orders of the indices are made explicit only if they are different), adding the prescription that for the successive couplings of two indices the first one is defined by the most internal pair of the repeated brackets { } , and so on. As an example of coupling scheme of n indices j_i, one has :

$$T^{\{\dots\{\{j_1j_2\}k_1j_3\}k_2\dots k_{n-2}j_n\}[k_{n-1}]}{}_{n_{n-1}} = T^{[j_1]\dots[j_n]}{}_{m_1 \quad m_n}$$

$$\times \langle j_1j_2m_1m_2|k_1n_1\rangle\langle k_1j_3n_1m_3|k_2n_2\rangle\dots\langle k_{n-2}j_nn_{n-2}m_n|k_{n-1}n_{n-1}\rangle \quad . \tag{8-93}$$

This relation can be inverted as, cf. Eq. (8–70) :

$$T^{[j_1]\dots[j_n]}{}_{m_1 \quad m_n} = \sum_{k_1,\dots,k_{n-1}} T^{\{\dots\{\{j_1j_2\}k_1j_3\}k_2\dots k_{n-2}j_n\}[k_{n-1}]}{}_{n_{n-1}}$$

$$\times \langle k_{n-1}n_{n-1}|k_{n-2}j_nn_{n-2}m_n\rangle\dots\langle k_2n_2|k_1j_3n_1m_3\rangle\langle k_1n_1|j_1j_2m_1m_2\rangle \quad . \tag{8-94}$$

Later, in Subsec. 8.4.1, generalized coupling and decoupling coefficients are defined. The above equations can then be written in a compact fashion.

iii) Since a coupling scheme corresponds to successive couplings of two indices, the change of the tensorial character of one or several indices by means of the Eqs. (8–79a,b) is stable under all couplings. This is why, in Eq. (8–93) for instance, the tensorial character of the indices which are coupled is not specified. Moreover, using the method discussed in Subsec. 8.3.2.1,iv), indices which have different tensorial characters can also be coupled.

iv) Finally, the criterion of tensorial character derived in Subsec. 8.3.2.1,v) can be extended in a straightforward manner to tensors of any order and to any coupling scheme. Let us give an example : A necessary and sufficient condition for a set of objects $T_2{}^{j_2j_3}{}_{m_2m_3}$ be a tensorial sequence, is that for every tensor $T_1{}^{j_1}$ the set of objects defined by :

$$T^{\{j_1j_2\}[\lambda][j_3]}{}_{m \; m_3} = T_1{}^{[j_1]}{}_{m_1} \; T_2{}^{j_2 \; j_3}{}_{m_2 \; m_3} \langle j_1j_2m_1m_2|\lambda m\rangle \quad ,$$

be also a tensorial sequence, i.e. fully contravariant coupled spherical components of the tensor product :

$$T^{j_1 j_2 j_3} = T_1^{\;j_1} \otimes T_2^{\;j_2 j_3} \quad .$$

8.3.3. CONTRACTION OF TENSORS

8.3.3.1. Contraction of a tensor and contracted tensor product

A tensor of order greater than one is said to be contracted in the ith and kth equal ranks ($j_i = j_k = j$) if for any choice of the remaining indices m_ℓ ($\ell \neq i$ and k) the spherical components where m_i and m_k are equal and have different tensorial characters are summed up. In that way one gets two new sets of objects :

$$T'^{\;\ldots j_\ell \ldots}_{\;\;m_\ell} = \delta_{m_i m_k} \, T^{\ldots (j) \ldots [j] \ldots}_{\quad\; m_i \quad\;\; m_k} \quad , \tag{8-95a}$$

$$T''^{\;\ldots j_\ell \ldots}_{\;\;m_\ell} = \delta_{m_i m_k} \, T^{\ldots [j] \ldots (j)}_{\quad\; m_i \quad\;\; m_k} \quad . \tag{8-95b}$$

From Eq. (8–82), these two sets are related by the phase factor $(-1)^{2j}$. Let us show that, for instance, ($T^{\ldots j_\ell \ldots}_{\;m_\ell}$) is a tensorial sequence and thus the set of spherical components of a new tensor.

One way to get this result is to come back to the definition of a tensor, i.e. to check that Eq. (8–44) is fulfilled. Indeed, using the unitary character of the representation $\mathcal{D}^j$, one has :

$$\forall\, U \in SU(2) \quad {}^{U}T'^{\;\ldots j_\ell \ldots}_{\;\;m_\ell} = {}^{U}T^{\ldots (j) \ldots [j]}_{\quad\; m_i \quad\;\; m_k} \delta_{m_i m_k} = T^{\ldots (j) \ldots [j]}_{\quad\; \mu_i \quad\;\; \mu_k} \ldots \mathcal{D}^{j*}_{\mu_i m_i}(U) \ldots \mathcal{D}^{j}_{\mu_k m_k}(U) \ldots \delta_{m_i m_k}$$
$$= T^{\ldots (j) \ldots [j]}_{\quad\; \mu_i \quad\;\; \mu_k} \ldots \mathcal{D}^{j}_{\mu_i \mu_k}(\mathbf{1}) = T^{\ldots (j) \ldots [j]}_{\quad\; \mu_i \quad\;\; \mu_k} \ldots \delta_{\mu_i \mu_k} \quad .$$

Hence the law of transformation of the objects ($T'^{\;\ldots j_\ell \ldots}_{\;\;m_\ell}$) is given by the law of transformation of the n–2 remaining indices which are not contracted.

Another way to obtain the above result is to show that up to a multiplicative factor, Eqs. (8–95a,b) are nothing else but a special case of the coupling of the two contracted equal indices j_i and j_k to zero. Indeed, changing the tensorial character of one of these indices yields :

$$T'^{\;\ldots j_\ell \ldots}_{\;\;m_\ell} = c_j \sum_m (-1)^{j+m} \, T^{\ldots [j_i] \ldots [j_k] \ldots}_{\quad\;\; -m \quad\;\; m} = c_j \sum_m (-1)^{j-m} \, T^{\ldots (j_i) \ldots (j_k)}_{\quad\;\; m \quad\;\; -m} \quad .$$

Then, from the values of the C.G. coefficients, Eq. (G–4b) :

$$\langle jj\text{-}mm|00\rangle = \frac{(-1)^{j+m}}{\sqrt{2j+1}} \quad , \quad \langle jjm\text{-}m|00\rangle = \frac{(-1)^{j-m}}{\sqrt{2j+1}} \quad , \tag{8-96}$$

one gets :

$$T'^{\;\ldots j_\ell \ldots}_{\;\;m_\ell} = c_j \sqrt{2j+1}\; T^{\ldots [j] \ldots [j] \ldots}_{\quad\; -m \quad\; m} \langle jj\text{-}mm|00\rangle = c_j \sqrt{2j+1}\; T^{j_1 \ldots j_n; \ldots j_\ell \ldots \{j_i j_k\}[0] \cdot}_{\qquad\qquad\; m_\ell \qquad\quad\;\; 0} \tag{8-97a}$$

$$= c_j \sqrt{2j+1}\, T^{\cdots(j)\cdots(j)\cdots}_{\quad m \quad -m} \langle jjm\text{-}m|00\rangle = c_j \sqrt{2j+1}\, T^{j_1\cdots j_n,\cdots j_\ell\cdots\{j_ij_k\}(0)}_{\qquad m_\ell \qquad 0} \quad (8\text{-}97b)$$

In the tensorial language the above property follows from the fact that the metric tensor G^{jj}, defined below in Eq. (1) of Table 8.1 (cf. also Subsec. 8.3.1.1,i)), is actually proportional to a special C.G. tensor. Indeed from Eq. (8–96), one has :

$$G^{(j)(j)}_{m\ m'} = c_j\,(-1)^{j+m}\,\delta_{m,-m'} \quad , \quad C^{(j)(j')[0]}_{m\ m'\ 0} = \frac{(-1)^{j-m}}{\sqrt{2j+1}}\,\delta_{jj'}\,\delta_{m,-m'} \quad ,(8\text{-}98)$$

and therefore :

$$\delta_{jj'}\, G^{jj} = c_j\,(-1)^{2j}\sqrt{2j+1}\, C^{jj'0} \quad , \qquad (8\text{-}99)$$

where the first (resp. second) index of G^{jj} corresponds to the first (resp. second) index of $C^{jj'0}$. Now, a tensorial sequence of order one and rank zero is clearly invariant (Subsec. 8.3.5) since one has :

$$\forall\, U \in SU(2) \qquad \mathcal{D}^0(U) = 1 \quad . \qquad (8\text{-}100)$$

Hence $(T'^{\cdots j_\ell\cdots}_{\ m_\ell})$ is a set of spherical components of a tensor of order n–2.

This new tensor is said to be a contracted tensor in the ith and kth equal ranks. Following a notation of Wigner [63] (Chap. 24) let us denote it by $T^{j_1\cdots j_{i\bullet}\cdots j_k^{\bullet}\cdots j_n}$, where the position of the dots specifies the tensorial character of the contracted indices i.e. : the contravariant (resp. covariant) index corresponds to the upper (resp. lower) dot. Moreover, if the same rank j occurs in several pairs of contracted indices, the indices which are contracted together are then labelled by the same number of dots, e.g. :

$$T^{j_\bullet j^\bullet} = T^{(j)[j]}_{m\ m} \quad , \quad T^{j_\bullet j^{\bullet\bullet} j_{\bullet\bullet} j^\bullet} = T^{(j)[j](j)[j]}_{m\ m'\ m'\ m} \quad . \qquad (8\text{-}101)$$

Finally, from any tensor $T^{j_1\cdots j_n}$ where $j_i=j_k$, two contracted tensors in the ith and kth ranks can be defined, i.e. $T^{j_1\cdots j_{i\bullet}\cdots j_k^\bullet\cdots j_n}$ and $T^{j_1\cdots j_i^\bullet\cdots j_{k\bullet}\cdots j_n}$. They have, respectively, for spherical components $(T'^{\cdots j_\ell\cdots}_{\ m_\ell})$ and $(T''^{\cdots j_\ell\cdots}_{\ m_\ell})$, and are related by :

$$j=j_i=j_k \ , \quad T^{j_1\cdots j_{i\bullet}\cdots j_k^\bullet\cdots j_n} = (-1)^{2j}\, T^{j_1\cdots j_i^\bullet\cdots j_{k\bullet}\cdots j_n} \quad . \qquad (8\text{-}102)$$

Moreover, from Eqs. (8–97a,b), one gets the intrinsic equality between tensors :

$$j=j_i=j_k \ , \quad T^{j_1\cdots j_{i\bullet}\cdots j_k^\bullet\cdots j_n} = c_j\sqrt{2j+1}\ T^{j_1\cdots j_n,\cdots j_\ell\cdots\{j_ij_k\}0} \quad . \qquad (8\text{-}103)$$

Combining this procedure of contraction together with the tensor product one is led to the notion of contracted tensor product. Let us give an important example, which will be considered again below in Subsec. 8.3.3.3. A contracted tensor product of a C.G. tensor with $T_1^{\,j_1}$ and $T_2^{\,j_2}$ is a new tensor T^{j_3} denoted and defined by :

TABLE 8.1

This table summarizes the results about the formal correspondence between the usual tensor algebra and the one developed in Sec. 8.3. One should note the consequences of the fact that the metric tensor needs not be symmetrical. Cf. Eq. (8–64) for the definition of $G^{t,jj}$.

Nondegenerate metric tensor

$g, (g_{mm'})$	(1) $G^{jj}, \left(G^{(j)(j)}_{m\ \ m'} = C^{j}_{mm'} = c_j(-1)^{j+m}\,\delta_{m,-m'}\right)$
e.g. Euclidean case :	
$g_{mm'}$, $g^t = g$ symmetrical	(2) $G^{t,jj} = (-1)^{2j} G^{jj}$ { symmetrical for integer j; antisymmetrical for half-odd-integer j
inverse tensor : γ, $(\gamma^{mm'})$	$\Gamma^{jj}, (\Gamma^{[j][j]}_{m\ \ m'})$
$g_{mm'}\,\gamma^{m'm''} = \gamma^{m''m'} g_{m'm} = \delta^{m''}_{m}$	(3) $G^{(j)(j)}_{m\ \ m'}\Gamma^{[j][j]}_{m'\ \ m''} = \Gamma^{[j][j]}_{m''\ m'}\, G^{(j)(j)}_{m'\ \ m} = \delta_{mm''}$
	$\Gamma^{[j][j]}_{m\ \ m'} = (C^{j})^{-1}_{mm'} = c_j(-1)^{j-m}\,\delta_{m,-m'}$

Change of the tensorial character

$v_m = g_{mm'}\, v^{m'}$	(4) $T^{(j)}_{m} = G^{(j)(j)}_{m\ \ m'}\, T^{[j]}_{m'} = C^{j}_{mm'}\, T^{[j]}_{m'} = c_j(-1)^{j+m}\, T^{[j]}_{-m}$
$v^m = \gamma^{mm'}\, v_{m'}$	(5) $T^{[j]}_{m} = \Gamma^{[j][j]}_{m\ \ m'}\, T^{(j)}_{m'} = (C^{j})^{-1}_{mm'} T^{(j)}_{m'} = c_j(-1)^{j-m}\, T^{(j)}_{-m}$
$g^{m}{}_{m'} = \gamma^{mm''} g_{m''m'} = \delta^{m}_{m'}$	(6) $G^{[j](j)}_{m\ \ m'} = \Gamma^{[j][j]}_{m\ \ m''}\, G^{(j)(j)}_{m''\ m'} = \delta_{mm'}$
$g^{mm'} = \gamma^{m'm''}\, g^{m}{}_{m''} = \gamma^{m'm}$	(7) $G^{[j][j]}_{m\ \ m'} = \Gamma^{[j][j]}_{m'\ m''}\, G^{[j](j)}_{m\ \ m''} = \Gamma^{[j][j]}_{m'\ m}$
$g = \gamma^{t}$	(8) $G^{jj} = \Gamma^{t,jj}$
e.g. Euclidean case : $g = \gamma$	(9) $G^{jj} = (-1)^{2j}\,\Gamma^{jj}$
$v^m = v_{m'}\, g^{m'm}$	(10) $T^{[j]}_{m} = T^{(j)}_{m'}\, G^{[j][j]}_{m'\ m}$
$g_{m}{}^{m'} = \gamma^{m'm''}\, g_{mm''}$	(11) $G^{(j)[j]}_{m\ \ m'} = \Gamma^{[j][j]}_{m'\ m''}\, G^{(j)(j)}_{m\ \ m''}$
if $g = g^t \Rightarrow g_{m}{}^{m'} = \delta^{m'}_{m}$	$= (-1)^{2j}\,\Gamma^{[j][j]}_{m'\ m''}\, g^{(j)(j)}_{m''\ m} = (-1)^{2j}\,\delta_{mm'}$
$v^m w_m = g_{mm''}\, v_{m'} w^{m''} g^{m'm}$	(12) $S^{[j]}_{m}\, T^{(j)}_{m} = G^{(j)(j'')}_{m\ \ m''}\, S^{(j')}_{m'}\, T^{[j'']}_{m''}\, G^{[j'][j]}_{m'\ m}$
if $g = g^t \Rightarrow v^m w_m = v_m w^m$	$= (-1)^{2j}\, S^{(j)}_{m}\, T^{[j]}_{m}$

Contractions

$v^{1_\bullet m' 3^\bullet} = v_{m}{}^{m'm}$	(13) $T^{j_\bullet [j'] j^\bullet}_{m'} = T^{(j)[j'][j]}_{m\ \ m'\ \ m}$
$v^{1^\bullet m' 3_\bullet} = v^{mm'}{}_{m}$	$T^{j^\bullet [j'] j_\bullet}_{m'} = T^{[j][j'](j)}_{m\ \ m'\ \ m}$
(12) $\Rightarrow$ if $g = g^t \Rightarrow v^{1_\bullet 23^\bullet} = v^{1^\bullet 23_\bullet}$	(14) $T^{j_\bullet j' j^\bullet} = (-1)^{2j}\, T^{j^\bullet j' j_\bullet}$
$v^{1_\bullet m' 3^\bullet} = g_{mm''}\, v^{m''m'm} = g^{1_\bullet 2_{\bullet\bullet}}\, v^{1^{\bullet\bullet} m' 3^\bullet}$	(15) $T^{j_\bullet [j'] j^\bullet}_{m'} = G^{(j)(j'')}_{m\ \ m''}\, T^{[j''][j'][j]}_{m''\ m'\ m} = G^{j_\bullet j''_\bullet}\, T^{j''^\bullet [j'] j^\bullet}_{m'}$

$$T^{j_3} = T_1^{j_1^\bullet} \otimes T_2^{j_2^\bullet} \otimes C^{j_1\bullet j_2\bullet j_3} \quad , \left(T^{[j_3]}_{m_3} = T_1{}^{[j_1]}_{m_1} T_2{}^{[j_2]}_{m_2} C^{(j_1)(j_2)[j_3]}_{m_1\ m_2\ m_3}\right). \tag{8-104}$$

Since, up to a multiplicative constant, the contraction is nothing else but a special case of coupling, the criteria of tensoriality derived in Subsecs. 8.3.2.1,v) and 8.3.2.2,iv) also apply to contracted tensors.

8.3.3.2. Comparison with the contraction defined in the usual tensor algebra

There exists a complete formal analogy between the contraction defined in the usual tensor algebra and the one introduced above. More generally, this gives us the opportunity of underlining the correspondence between both formalisms. The main results are summarized in Table 8.1. One should note the consequences of the fact that the metric tensor G^{jj} is either symmetrical or anti-symmetrical according to whether j is integer or half-odd-integer.

8.3.3.3. Interpretation of changes of tensorial character and couplings in terms of contracted tensor products

The Eqs. (4,10) of Table 8.1 allow us to rewrite the change of the tensorial character of an index in terms of a contracted tensor product of the considered tensor with the metric tensor G^{jj} defined by Eq. (1) of Table 8.1. Indeed, one has :

$$\left.\begin{array}{l} T^{(j)}_{m} = G^{(j)(j)}_{m\ m'}\ T^{[j]}_{m'} \\ T^{[j]}_{m} = T^{(j)}_{m'}\ G^{[j][j]}_{m'\ m} \end{array}\right\} \Leftrightarrow T^{j} = G^{jj\bullet} \otimes T^{j^\bullet} = T^{j\,\bullet} \otimes G^{j^\bullet j} \quad . \tag{8-105}$$

As for the coupling, for each given j_3, the tensorial sequence $(T^{\{j_1j_2\}[j_3]}_{m_3})$, Eq. (8-69), is the set of spherical components of a new tensor denoted by $T^{\{j_1j_2\}j_3}$. This tensor can also be rewritten as a contracted tensor product of the tensor $T^{j_1j_2}$ with the C.G. tensor $C^{j_1j_2j_3}$. Namely one has, cf. Eq. (8-104) :

$$T^{\{j_1j_2\}[j_3]}_{m_3} = T^{[j_1][j_2]}_{m_1\ m_2}\ C^{(j_1)(j_2)[j_3]}_{m_1\ m_2\ m_3} \Leftrightarrow T^{\{j_1j_2\}j_3} = T^{j_1^\bullet j_2^\bullet} \otimes C^{j_1\bullet j_2\bullet j_3} \quad . \tag{8-106}$$

Moreover, it is recalled that the metric tensor G^{jj} is actually proportional to the special tensor C^{jj0}, Eq. (8-99). Let us check Eq. (8-103) relating the coupling to zero of two equal indices with their contraction. From Eqs. (8-106, 99, 105, and 102), one has :

$$T^{\{jj\}0} = T^{j^\bullet j^{\bullet\bullet}} \otimes C^{j\bullet j\bullet\bullet 0} = \frac{(-1)^{2j}c_j}{\sqrt{2j+1}}\ T^{j^\bullet j^{\bullet\bullet}}\ G^{j\bullet j\bullet\bullet} = \frac{c_j}{\sqrt{2j+1}}\ T^{j\bullet j^\bullet} \quad . \tag{8-107}$$

More generally, any coupling scheme of indices which have any tensorial character can be expressed in terms of contracted tensor products of the considered tensor(s) with a product of C.G. tensors (Subsec. 8.4.1). As an example, for the coupling scheme of n indices specified in Eq. (8-93), one gets :

$$T^{\{\dots\{\{j_1 j_2\}k_1 j_3\}k_2 \dots k_{n-2} j_n\}k_{n-1}} =$$

$$T^{j_1^\bullet \dots j_n^\bullet} \otimes C^{j_{1\bullet} j_{2\bullet} k_1^\bullet} \otimes C^{k_{1\bullet} j_{3\bullet} k_2^\bullet} \otimes \cdots \otimes C^{k_{n-2\bullet} j_{n\bullet} k_{n-1}} \quad . \tag{8-108}$$

The main interest of this new formulation is to provide expressions written in terms of tensors instead of spherical components which have a special tensorial character. Hence, these expressions are intrinsic and imply several equalities according to the choice of the tensorial character of the free indices. Moreover the tensorial character of the contracted indices can be changed by means of Eq. (8–82). The ranks of a contracted tensor product are clearly given by the free indices.

It has to be pointed out that, although the tensor product is a priori noncommutative, the C.G. tensors, the spherical components of which are real numbers, commute with all tensors. The same remark holds for the Wigner 3j tensors. This is the reason why, from now on the symbol ⊗ is omitted in the tensor products with these two tensors.

In Sec. 9.1 one introduces a graphical representation of couplings which is based on the above interpretation of changes of tensorial character and couplings in terms of contracted tensor products.

8.3.4. FORMAL HERMITIAN CONJUGATE OF TENSORS

8.3.4.1. Definition and general properties

i) The generic case of a tensor T^j of order one is first discussed. Let us denote by the symbol + a formal mapping acting on the set of spherical components, and which fulfils the following three conditions :

(1) this mapping is antilinear,

(2) it commutes with the representation of the group SU(2) defined by Eq. (8–44), i.e. :

$$\forall U \in SU(2) \quad \left({}^{U}T^{[j]}_{m}\right)^{+} = {}^{U}\!\left(T^{[j]+}_{m}\right) \quad , \quad \left({}^{U}T^{(j)}_{m}\right)^{+} = {}^{U}\!\left(T^{(j)+}_{m}\right) \quad , \tag{8-109}$$

(3) it is involutive, i.e. of square unity :

$$\left(T^{[j]+}_{m}\right)^{+} = T^{[j]++}_{m} = T^{[j]}_{m} \quad , \quad \left(T^{(j)+}_{m}\right)^{+} = T^{(j)++}_{m} = T^{(j)}_{m} \quad . \tag{8-110}$$

In Quantum Mechanics (Chap. 11), according to the nature of the considered tensors, this formal antilinear mapping is defined by : (1) the complex conjugate of a complex number or function, (2) the dual correspondence between the bras and the kets, (3) the Hermitian conjugate of an operator.

From the definition of the antilinear mapping one gets :

$$\forall U \in SU(2) \quad {}^{U}\!\left(T^{[j]+}_{m}\right) = \left({}^{U}T^{[j]}_{m}\right)^{+} = \left[T^{[j]}_{m'} \, \mathcal{D}^{j}_{m'm}(U)\right]^{+} = T^{[j]+}_{m'} \, \mathcal{D}^{j*}_{m'm}(U) \quad . \tag{8-111}$$

This relation implies that $(T^{[j]+}_{m})$ is a tensorial sequence, i.e. the covariant spherical components of a new tensor formally called the Hermitian conjugate tensor T^{+j} of T^j, and defined by :

$$T^{j} \;,\; \left(T^{[j]}_{m}\right) \xrightarrow{\;+\;} T^{+j} \;,\; \left(T^{+(j)}_{m} = T^{[j]+}_{m}\right) \quad . \qquad (8\text{-}112)$$

Let us emphasize that in the above notation the symbol + has its first meaning of an antilinear mapping provided it is placed on the right of the index j, while when it is on the left of this index it only recalls the definition of the new tensor. Hence the tensorial character denoted by (j) or [j] is changed if the symbol + occurs on the right of it. From the definition of T^{+j} by Eq. (8–112) and Eq. (8–79b) to change the tensorial character of the indices, one obtains the relations between the contravariant and covariant spherical components of both tensors T^j and T^{+j} :

$$T^{+[j]}_{m} = (-1)^{2j}\, T^{(j)+}_{m} = c_j\, (-1)^{j-m}\, T^{[j]+}_{-m} \quad , \qquad (8\text{-}113a)$$

$$T^{+(j)}_{m} = T^{[j]+}_{m} = c_j\, (-1)^{j-m}\, T^{(j)+}_{-m} \quad , \qquad (8\text{-}113b)$$

$$T^{[j]+}_{m} = T^{+(j)}_{m} = c_j\, (-1)^{j+m}\, T^{+[j]}_{-m} \quad , \qquad (8\text{-}113c)$$

$$T^{(j)+}_{m} = (-1)^{2j}\, T^{+[j]}_{m} = c_j\, (-1)^{j+m}\, T^{+(j)}_{-m} \quad . \qquad (8\text{-}113d)$$

Let us remark that instead of T^{+j} one could have defined another new tensor T'^{+j} by :

$$T'^{+[j]}_{m} = T^{(j)+}_{m} \;\Rightarrow\; T'^{+j} = (-1)^{2j}\, T^{+j} \quad .$$

ii) The iteration of the mapping + leads to a new tensor T^{++j} which is the Hermitian conjugate tensor of T^{+j}. According to Eq. (8–112) it is defined by :

$$T^{++j} \;,\; \left(T^{++(j)}_{m} = T^{+[j]+}_{m}\right) \quad . \qquad (8\text{-}114)$$

Therefore, from Eq. (8–113a) and using the involutive character of the mapping +, one gets :

$$T^{++(j)}_{m} = c_j\, (-1)^{j-m}\, T^{[j]++}_{-m} = c_j\, (-1)^{j-m}\, T^{[j]}_{-m} = (-1)^{2j}\, T^{(j)}_{m} \quad .$$

Hence the two tensors T^j and T^{++j} are related by :

$$T^{++j} = (-1)^{2j}\, T^{j} \quad . \qquad (8\text{-}115)$$

iii) If the nature of the tensor is not changed by the mapping +, one may ask whether T^j and T^{+j} are equal. A necessary and sufficient condition that this be so is :

$$\left(T^{[j]}_{m} = T^{+[j]}_{m}\right) \;\Leftrightarrow\; \left(T^{(j)}_{m} = T^{+(j)}_{m}\right) \quad .$$

From Eqs. (8–113c,d), one gets :

$$T^{j} = T^{+j} \;\Leftrightarrow\; T^{[j]+}_{m} = c_j\, (-1)^{j+m}\, T^{[j]}_{-m} \;\Leftrightarrow\; T^{(j)+}_{m} = c_j\, (-1)^{j+m}\, T^{(j)}_{-m} \quad . \qquad (8\text{-}116)$$

This condition can only be fulfilled for integer values of j. Indeed the mapping + being involutive, Eq. (8–113c) implies the necessary condition :

$$T^{[j]}_{m} = T^{[j]++}_{m} = c_j(-1)^{j+m}\, T^{[j]+}_{-m} = c_j(-1)^{j+m}\, c_j(-1)^{j-m}\, T^{[j]}_{m} = (-1)^{2j}\, T^{[j]}_{m} .$$

When $T^j = T^{+j}$ the tensor is said to be Hermitian.

iv) The generalization of the foregoing to tensors of any order is straightforward. The Hermitian conjugate tensor $T^{+j_1 \ldots j_n}$ of any tensor $T^{j_1 \ldots j_n}$ is defined by applying Eqs. (8–113c,d) to each index j_i :

$$T^{j_1 \ldots j_n} , \left(T^{\ldots [j_i] \ldots (j_k) \ldots}_{\quad m_i \quad\; m_k} \right) \xrightarrow{\;+\;} \tag{8-117}$$

$$T^{+j_1 \ldots j_n} , \left(T^{+\ldots (j_i) \ldots [j_k] \ldots}_{\quad m_i \quad\; m_k} \right) = \left[\prod_{(j_k)} (-1)^{2j_k} \right] T^{\ldots [j_i] \ldots (j_k) \ldots}_{\quad m_i \quad\; m_k} \Big) ,$$

where the product $\prod_{(j_k)}$ runs over all the covariant indices of $T^{\ldots [j_i] \ldots (j_k) \ldots}_{\quad m_i \quad m_k}$. Using the above definition and changing the tensorial character of any index m_i, one gets relations similar to Eqs. (8–113a,b) between the spherical components having an arbitrary tensorial character of $T^{j_1 \ldots j_n}$ and $T^{+j_1 \ldots j_n}$.

The iteration of the mapping + leads to the generalization of Eq. (8–115) :

$$T^{++j_1 \ldots j_n} = (-1)^{2(j_1 + \ldots + j_n)}\, T^{j_1 \ldots j_n} . \tag{8-118}$$

A necessary and sufficient condition for the tensor $T^{j_1 \ldots j_n}$ be Hermitian, i.e. equal to their Hermitian conjugate reads :

$$T^{j_1 \ldots j_n} = T^{+j_1 \ldots j_n} \Leftrightarrow T^{\ldots [j_i] \ldots (j_k) \ldots +}_{\quad m_i \quad\; m_k} = \left[\prod_{\ell=1}^{n} c_{j_\ell} (-1)^{j_\ell + m_\ell} \right] T^{\ldots [j_i] \ldots (j_k)}_{\quad -m_i \quad -m_k} . \tag{8-119}$$

Since the mapping + is involutive, the above condition can only be realized if $j_1 + \ldots + j_n$ is an integer.

8.3.4.2. Application to the Clebsch–Gordan and the Wigner 3j tensors

i) As an important example let us apply the foregoing considerations to the C.G. tensors. There the antilinear mapping + corresponds to the complex conjugation denoted by * . The C.G. coefficients being chosen real, one has :

$$\langle j_1 j_2 m_1 m_2 | j_3 m_3 \rangle^* = \langle j_3 m_3 | j_1 j_2 m_1 m_2 \rangle = \langle j_1 j_2 m_1 m_2 | j_3 m_3 \rangle ,$$

from the tensorial point of view Eq. (8–117) implies :

$$\langle j_3 m_3 | j_1 j_2 m_1 m_2 \rangle = C^{(j_1)(j_2)[j_3]*}_{m_1 \; m_2 \; m_3} = (-1)^{2(j_1 + j_2)}\, C^{*[j_1][j_2](j_3)}_{\; m_1 \; m_2 \; m_3} , \tag{8-120}$$

where $C^{*j_1 j_2 j_3}$ is the Hermitian conjugate tensor of $C^{j_1 j_2 j_3}$. Moreover, the symmetry property Eq. (8–47) yields :

$$C^{(j_1)(j_2)[j_3]}_{m_1\ m_2\ m_3} = (-1)^{j_1+j_2-j_3}\, C^{(j_1)(j_2)[j_3]}_{-m_1\ -m_2\ -m_3} = \Big[\prod_{i=1}^{3} c_{j_i}\,(-1)^{j_i+m_i}\Big]\, C^{(j_1)(j_2)[j_3]}_{-m_1\ -m_2\ -m_3}\ ,$$

where we have used Eq. (8–84) and the selection rule $m_1+m_2=m_3$. Therefore, a C.G. tensor is Hermitian, since it verifies Eq. (8–119) :

$$C^{j_1j_2j_3} = C^{*j_1j_2j_3}\ . \qquad (8\text{-}121)$$

This result can also be checked directly in terms of the spherical components by using Eqs. (8–83a,84) :

$$C^{[j_1][j_2](j_3)}_{m_1\ m_2\ m_3} = (-1)^{2j_3}\, \langle j_1j_2m_1m_2|j_3m_3\rangle = C^{*[j_1][j_2](j_3)}_{m_1\ m_2\ m_3}\ . \qquad (8\text{-}122)$$

Hence, from Eq. (8–120) and Eq. (11) of Table 8.1 the unitary character of the C.G. coefficients, Eqs. (6–66a,b), reads in terms of contracted tensor products :

$$C^{j_1^\bullet j_2^\bullet j_3}\ C^{j_1\bullet j_2\bullet j'_3} = \delta_{j_3j'_3}\, \{j_1j_2j_3\}\, G^{j_3j_3}\ , \qquad (8\text{-}123a)$$

$$\sum_{j_3} C^{j_1j_2j_3^\bullet}\ C^{j_1j_2j_3\bullet} = G^{j_1j_1}\ G^{j_2j_2}\ . \qquad (8\text{-}123b)$$

In these two equations the lateral order of the free indices is, as always, preserved. For instance, in the first equation the first index j_3 of the metric tensor corresponds to the index j_3 of the first C.G. tensor.

Then, for the eight different choices of the tensorial character of the three indices m_1, m_2, and m_3, the Eq. (8–85) can be inverted as follows :

$$T^{[j_1][j_2]}_{m_1\ m_2} = (-1)^{2j_3}\, C^{[j_1][j_2](j_3)}_{m_1\ m_2\ m_3}\, T^{\{j_1j_2\}[j_3]}_{m_3} = C^{[j_1][j_2][j_3]}_{m_1\ m_2\ m_3}\, T^{\{j_1j_2\}(j_3)}_{m_3}\ , \qquad (8\text{-}124)$$

and the six other expressions obtained by changing the tensorial character of the free indices m_1 and m_2. In an intrinsic way all these relations follow from :

$$T^{j_1j_2} = \sum_{j_3} C^{j_1j_2j_3^\bullet}\ T^{\{j_1j_2\}j_3\bullet}\ , \qquad (8\text{-}125)$$

which is the inversion of Eq. (8–106). This expression can also be obtained directly, using only the contracted tensor products, from Eqs. (8–106,123b, and 105) :

$$\sum_{j_3} C^{j_1j_2j_3^\bullet}\, T^{\{j_1j_2\}j_3\bullet} = \sum_{j_3} C^{j_1j_2j_3^\bullet}\, C^{j_1\bullet j_2\bullet j_3\bullet}\, T^{j_1^\bullet j_2^\bullet} = G^{j_1j_1\bullet}\, G^{j_2j_2\bullet}\, T^{j_1^\bullet j_2^\bullet} = T^{j_1j_2}.$$

More generally, as an example for a coupling scheme of n indices, the inversion of Eq. (8–108) reads (cf. Subsec. 8.4.1) :

$$T^{j_1 \cdots j_n} = \sum_{k_1, \ldots, k_{n-1}} C^{j_1 j_2 \overset{\bullet}{k}_1} C^{k_1 \bullet j_3 \overset{\bullet}{k}_2} \ldots C^{k_{n-2} \bullet j_n \overset{\bullet}{k}_{n-1}} T^{\{\ldots\{j_1 j_2\}k_1 j_3\}k_2 \ldots k_{n-2} j_n\}k_{n-1}\bullet} \qquad (8\text{-}126)$$

ii) Since the Wigner 3j tensors are real multiples of the C.G. tensors, they are also Hermitian tensors :

$$W^{j_1 j_2 j_3} = W^{*j_1 j_2 j_3} \quad . \qquad (8\text{-}127)$$

In terms of them and using Eq. (8–86), Eqs. (8–106,125) take the intrinsic forms :

$$T^{\{j_1 j_2\}j_3} = c_{j_3} (-1)^{j_1 - j_2 + j_3} \sqrt{2j_3+1}\, W^{j_1 \bullet j_2 \bullet j_3}\, T^{\overset{\bullet}{j}_1 \overset{\bullet}{j}_2} \quad , \qquad (8\text{-}128a)$$

$$T^{j_1 j_2} = \sum_{j_3} c_{j_3} (-1)^{j_1 - j_2 + j_3} \sqrt{2j_3+1}\, W^{j_1 j_2 \overset{\bullet}{j}_3}\, T^{\{j_1 j_2\}j_3 \bullet} \quad , \qquad (8\text{-}128b)$$

and Eqs. (8–123a,b) read :

$$(2j_3+1)\, W^{\overset{\bullet}{j}_1 \overset{\bullet}{j}_2 j_3}\, W^{j_1 \bullet j_2 \bullet j_3'} = \delta_{j_3 j_3'}\, \{j_1 j_2 j_3\}\, G^{j_3 j_3} \quad , \qquad (8\text{-}129a)$$

$$\sum_{j_3} (2j_3+1)\, W^{j_1 j_2 \overset{\bullet}{j}_3}\, W^{j_1 j_2 j_3 \bullet} = G^{j_1 j_1}\, G^{j_2 j_2} \quad . \qquad (8\text{-}129b)$$

iii) The two choices : real C.G. coefficients and the matrix C^j specified by Eq. (8–78), imply that the coupling and the mapping + commute. Indeed, for the generic case of the coupling of two indices, one has in an intrinsic way using Eqs. (8–113c,d, and 121) :

$$T^{\{j_1 j_2\}j_3} = C^{j_1 \bullet j_2 \bullet j_3}\, T^{\overset{\bullet}{j}_1 \overset{\bullet}{j}_2} \Leftrightarrow T^{\{j_1 j_2\}j_3 +} = C^{j_1 \bullet j_2 \bullet j_3 *}\, T^{\overset{\bullet}{j}_1 \overset{\bullet}{j}_2 +}$$

$$\Leftrightarrow T^{+\{j_1 j_2\}j_3} = (-1)^{2j_1 + 2j_2}\, C^{*\overset{\bullet}{j}_1 \overset{\bullet}{j}_2 j_3}\, T^{+j_1 \bullet j_2 \bullet} = C^{j_1 \bullet j_2 \bullet j_3}\, T^{+\overset{\bullet}{j}_1 \overset{\bullet}{j}_2} \quad . \qquad (8\text{-}130)$$

Since, up to a multiplicative factor, the contraction is a special case of coupling, it also commutes with the mapping +, e.g. :

$$T^{j_2} = T_1^{\overset{\bullet}{j}_1} \otimes T_2^{j_1 \bullet j_2} \Leftrightarrow T^{+j_2} = T_1^{+\overset{\bullet}{j}_1} \otimes T_2^{+j_1 \bullet j_2} \quad . \qquad (8\text{-}131)$$

8.3.5. INVARIANT TENSORS, THE WIGNER–ECKART THEOREM

8.3.5.1. Definition and theorem

i) A tensor $T^{j_1 \cdots j_n}$ of any order and ranks is said to be <u>invariant</u> or a <u>scalar</u> if and only if :

$$\forall\, U \in SU(2) \qquad {}^{U}T^{j_1 \cdots j_n} = T^{j_1 \cdots j_n} \quad . \qquad (8\text{-}132)$$

In the passive point of view of the representation of the group SU(2), the above condition is from Eq. (8–58) equivalent to the assertion that the spherical components of $T^{j_1 \cdots j_n}$ are independent of the choice of the triad, that is :

$$\forall\ U,U' \in SU(2) \qquad T^{\cdots[j_i]\cdots(j_k)\cdots}_{\quad m_i \quad\ m_k} ;U'U = T^{\cdots[j_i]\cdots(j_k)\cdots}_{\quad m_i \quad\ m_k} ;U \ . \tag{8-133}$$

As an example, we have already noticed that the C.G. and the Wigner 3j tensors are invariant. Actually, from the point of view of tensor algebra this is the main property of these tensors as it is emphasized below in iii).

ii) Let us now derive a theorem which allows us to characterize the invariant tensors. One should note that such a theorem, based on the existence of a finite invariant integration on the group, can be stated for any compact Lie group (Subsec. 7.1.2).

Theorem 8.4. The Wigner–Eckart theorem. A necessary and sufficient condition for a tensor $T^{j_1\cdots j_n}$ of any order and ranks to be invariant is that its irreducible spherical components belong to the representation $\mathcal{D}^j$ (or $\mathcal{D}^{j*}$) with j=0.

Proof. Since a tensor is characterized by a set of its irreducible spherical components we can restrict ourselves to the generic case of a tensor of order one. The sufficient character of the condition of the theorem is obvious from Eq. (8–100). Assuming the tensor T^j be invariant implies :

$$\forall\, U \in SU(2) \qquad {}^{U}T^{[j]}_{m} = T^{[j]}_{m'}\ \mathcal{D}^{j}_{m'm}(U) = T^{[j]}_{m} \quad .$$

This relation being true for all U in SU(2), and this group being compact, one can integrate both sides of this equation over the whole group with the invariant measure. This yields :

$$T^{[j]}_{m'} \int_{SU(2)} d\mu(U)\ \mathcal{D}^{j}_{m'm}(U) = T^{[j]}_{m}\ V \quad ,$$

where V denotes the finite volume of the group, Eq. (7–25). From the orthogonality relation Eq. (7–50a) and Eq. (8–100), one then gets the necessary condition :

$$T^{[j]}_{m'}\ \delta_{j0}\ \delta_{m'0}\ \delta_{m0} = T^{[j]}_{m} \quad ,$$

which ends the proof of the theorem.

iii) Let $T^{j_1\cdots j_n}$ be an invariant tensor. It follows from the Wigner–Eckart theorem that any of its spherical components can be expressed in terms of its only irreducible spherical component which is invariant, i.e. coupled to zero, by means of a product of C.G. coefficients. Actually, the invariant irreducible component of $T^{j_1\cdots j_n}$ is only defined up to the choice of a coupling scheme of $j_1,\ldots,j_n$ to zero. The whole dependence of the spherical components upon the indices m_i is then carried on by the C.G. coefficients. In an intrinsic way, i.e. independently of the tensorial character of the spherical components, this implies that any invariant tensor is a multiple of a contracted tensor product of C.G. tensors. These results are now made explicit for the special coupling scheme of $n \geqslant 3$ indices defined in Eq. (8–126) (cf. Eq. (8–155) for the general expression of an invariant tensor whatever be the order of couplings). A necessary and sufficient condition for a tensor $T^{j_1\cdots j_n}$ to be invariant then reads in an intrinsic way :

$$\forall U \in SU(2) \qquad U_T{}^{j_1\cdots j_n} = T^{j_1\cdots j_n} \quad \Leftrightarrow \tag{8-134}$$

$$T^{j_1\cdots j_n} = \sum_{k_1,\dots,k_{n-2}} C^{j_1 j_2 k_1^\bullet}\, C^{k_1\bullet j_3 k_2^\bullet} \cdots C^{k_{n-2\bullet} j_n 0}\, T^{\{\dots\{\{j_1 j_2\}k_1 j_3\}k_2 \cdots k_{n-2} j_n\}0} \quad .$$

Moreover, using Eqs. (8–99,102, and 105) one has :

$$C^{k_{n-3} j_{n-1} k_{n-2}^\bullet}\, C^{k_{n-2}\bullet j_n 0} = \frac{c_{j_n}}{\sqrt{2j_n+1}}\, C^{k_{n-3} j_{n-1} j_n} \quad , \tag{8-135}$$

therefore the invariant tensor $T^{j_1\cdots j_n}$ takes the form :

$$T^{j_1\cdots j_n} = \frac{c_{j_n}}{\sqrt{2j_n+1}} \sum_{k_1,\dots,k_{n-3}} C^{j_1 j_2 k_1^\bullet}\, C^{k_1\bullet j_3 k_2^\bullet} \cdots C^{k_{n-3}\bullet j_{n-1} j_n}\, T^{\{\dots\{\{j_1 j_2\}k_1 j_3\}k_2 \cdots j_n\}0} \quad . \tag{8-136}$$

For a given choice of the tensorial character of the spherical components, one gets for example :

$$T^{[j_1]\cdots[j_n]}_{m_1 \;\; m_n} = \frac{c_{j_n}}{\sqrt{2j_n+1}} \sum_{k_1,\dots,k_{n-3}} C^{[j_1][j_2][k_1]}_{m_1\, m_2\, n_1}\, C^{(k_1)[j_3][k_2]}_{n_1\, m_3\, n_2} \cdots C^{(k_{n-3})[j_{n-1}][j_n]}_{n_{n-3}\, m_{n-1}\, m_n} \times T^{\{\dots\{\{j_1 j_2\}k_1 j_3\}k_2 \cdots j_n j_n\}[0]}_{0} \tag{8-137a}$$

$$= \frac{c_{j_n}}{\sqrt{2j_n+1}} \sum_{k_1,\dots,k_{n-3}} \langle j_1 j_2 m_1 m_2 | k_1 n_1\rangle \langle k_1 j_3 n_1 m_3 | k_2 n_2\rangle \cdots \langle k_{n-3} j_{n-1} n_{n-3} m_{n-1} | j_n m_n\rangle\, T^{\{\dots\{\{j_1 j_2\}k_1 j_3\}k_2 \cdots j_n j_n\}[0]}_{0} \quad . \tag{8-137b}$$

To simplify the notations, let us from now on set :

$$T^{\{j_1\cdots j_n\}0} = T^{\{j_1\cdots j_n\}[0]}_{0} = T^{\{j_1\cdots j_n\}(0)}_{0} \quad , \tag{8-138}$$

regardless of the coupling scheme denoted symbolically by $\{j_1\cdots j_n\}0$, and moreover as usual :

$$j=0,\tfrac{1}{2},1,\dots \qquad \hat{\jmath} = \sqrt{2j+1} \quad . \tag{8-139}$$

8.3.5.2. Special cases of invariant tensors of order one, two, and three

i) Invariant tensors of order one :

$$\forall U \in SU(2) \qquad U_T{}^j = T^j \Leftrightarrow T^j = \delta_{j0}\, T^0 \Leftrightarrow T^{[j]}_m = T^{(j)}_m = \delta_{j0}\, \delta_{m0}\, T^0 \quad . \tag{8-140}$$

ii) Invariant tensors of order two :

$$\forall U \in SU(2) \qquad U_T{}^{j_1 j_2} = T^{j_1 j_2}$$

$$\Leftrightarrow T^{j_1j_2} = C^{j_1j_2 0}\, T^{\{j_1j_2\}0} = \frac{(-1)^{2j_1}}{\hat{j}_1} c_{j_1}\, T^{\{j_1j_2\}0}\, G^{j_1j_1} = \frac{\delta_{j_1j_2}}{2j_1+1} T^{j_1^\bullet j_1\bullet} G^{j_1j_1} \qquad (8\text{-}141a)$$

$$\Leftrightarrow T^{[j_1][j_2]}_{m_1\ m_2} = \langle j_1j_2m_1m_2|00\rangle\, T^{\{j_1j_2\}0} = \frac{(-1)^{j_1-m_1}}{\hat{j}_1} \delta_{j_1j_2}\, \delta_{m_1,-m_2}\, T^{\{j_1j_2\}0} \qquad (8\text{-}141b)$$

$$\Leftrightarrow T^{[j_1](j_2)}_{m_1\ m_2} = (-1)^{2j_1}\, T^{(j_1)[j_2]}_{m_1\ m_2} = \frac{(-1)^{2j_1}}{\hat{j}} c_{j_1}\, \delta_{j_1j_2}\, \delta_{m_1m_2}\, T^{\{j_1j_2\}0} \quad . \qquad (8\text{-}141c)$$

In particular, one has to notice that the spherical components $T^{[j](j)}_{m\ m}$ and $T^{(j)[j]}_{m\ m}$ are independent of m. Therefore, one has :

$$T^{[j](j)}_{m\ m} = (-1)^{2j}\, T^{(j)[j]}_{m\ m} = \frac{1}{2j+1} T^{j^\bullet j\bullet} \quad , \qquad (8\text{-}142)$$

and thus the expression of an invariant tensor $T^{j_1j_2}$ can be written into the form :

$$T^{j_1j_2} = \delta_{j_1j_2}\, T^{[j_1](j_1)}_{m\ \ m}\, G^{j_1j_1} \quad , \qquad (8\text{-}143)$$

for any allowed value of m.

iii) Invariant tensors of order three :

$$\forall\, U \in SU(2) \qquad {}^{U}T^{j_1j_2j_3} = T^{j_1j_2j_3}$$

$$\Leftrightarrow T^{j_1j_2j_3} = \frac{c_{j_3}}{\hat{j}_3} C^{j_1j_2j_3}\, T^{\{\{j_1j_2\}j_3j_3\}0} \qquad (8\text{-}144a)$$

$$\Leftrightarrow T^{[j_1][j_2][j_3]}_{m_1\ m_2\ m_3} = \frac{(-1)^{j_3+m_3}}{\hat{j}_3} \langle j_1j_2m_1m_2|j_3-m_3\rangle\, T^{\{\{j_1j_2\}j_3j_3\}0} \qquad (8\text{-}144b)$$

$$\Leftrightarrow T^{(j_1)[j_2][j_3]}_{m_1\ m_2\ m_3} = \frac{c_{j_1}}{\hat{j}_1} \langle j_2j_3m_2m_3|j_1m_1\rangle\, T^{\{\{j_1j_2\}j_3j_3\}0} \quad , \qquad (8\text{-}144c)$$

where to get the last equation above we have used the symmetry properties of the C.G. coefficients given in Eqs. (G–2a,d). Actually, to obtain this last expression it is simpler to consider another coupling scheme, i.e. $\{\, j_1\{\, j_2j_3\}\, j_1\,\}0$. Indeed, one has :

$$T^{j_1j_2j_3} = C^{j_2j_3j_1^\bullet}\, C^{j_1j_1\bullet 0}\, T^{\{j_1\{j_2j_3\}j_1\}0} = \frac{(-1)^{2j_1}}{\hat{j}_1} c_{j_1}\, C^{j_2j_3j_1^\bullet}\, G^{j_1j_1\bullet}\, T^{\{j_1\{j_2j_3\}j_1\}0\}}$$

$$= \frac{(-1)^{2j_1}\, c_{j_1}}{\hat{j}_1} C^{j_2j_3j_1}\, T^{\{j_1\{j_2j_3\}j_1\}0} \quad , \qquad (8\text{-}145)$$

which, using Eq. (8–89b), leads to the previous result once it has been shown, Eq. (G-10), that :

$$T^{\{j_1\{j_2j_3\}j_1\}0} = T^{\{\{j_1j_2\}j_3j_3\}0} \quad . \qquad (8\text{-}146)$$

In terms of a Wigner 3j tensor Eq. (8–144a) reads :

$$T^{j_1j_2j_3} = (-)^{j_1-j_2+j_3}\, W^{j_1j_2j_3}\, T^{\{\{j_1j_2\}j_3j_3\}0} \quad . \tag{8-147}$$

The most important examples of invariant tensors of order three are clearly the two proportional C.G. and Wigner 3j tensors. From, respectively, Eqs. (8–144a,147) one gets :

$$C^{\{\{j_1j_2\}j_3j_3\}0} = c_{j_3}\, \hat{j}_3 \quad , \qquad W^{\{\{j_1j_2\}j_3j_3\}0} = (-1)^{j_1-j_2+j_3} \quad . \tag{8-148}$$

Invariant tensors of order three are important in Quantum Mechanics since, as we shall see later in Subsec. 11.3.1 , they correspond to the matrix elements of irreducible tensorial operators. Especially the spherical components which have the tensorial character $T^{(j_1)[j_2][j_3]}_{m_1\ m_2\ m_3}$ correspond to the case of a matrix element $<j_1m_1|\, A^{[j_2]}_{m_2}\, |j_3m_3>$ of an irreducible tensorial operator A^{j_2}. Then, $T^{\{j_1\{j_2j_3\}j_1\}0}$ will be called a reduced matrix element.

The tensorial formulation we use may appear a bit cumbersome for those not familiar with it. Nevertheless, one must not be confused by this point of view, the ideas which for instance lead to Eq. (8–144a) are actually simple :

(1) From the Wigner–Eckart theorem any spherical component of an invariant tensor is proportional to an irreducible spherical component coupled to zero, (2) the usual C.G. coefficients can only couple two indices to a third one which all have the same tensorial character (Subsec. 8.3.2.1). Then, to project $T^{(j_1)[j_2][j_3]}_{m_1\ m_2\ m_3}$ on the irreducible spherical component $T^{\{j_1\{j_2j_3\}j_1\}0}$ by means of the C.G. coefficients, one has first to change the tensorial character of for instance m_1, i.e. :

$$\begin{aligned} T^{(j_1)[j_2][j_3]}_{m_1\ m_2\ m_3} &= c_{j_1}\, (-1)^{j_1+m_1}\, T^{[j_1][j_2][j_3]}_{-m_1\ m_2\ m_3} \\ &= c_{j_1}\, (-1)^{j_1+m_1} <j_2j_3m_2m_3|j_1m_1><j_1j_1-m_1m_1|00>\, T^{\{j_1\{j_2j_3\}j_1\}0} \\ &= \frac{c_{j_1}}{\hat{j}_1} <j_2j_3m_2m_3|j_1m_1>\, T^{\{j_1\{j_2j_3\}j_1\}0} \quad . \end{aligned} \tag{8-149}$$

8.4. COUPLING TENSORS AND RECOUPLING COEFFICIENTS

This section is concerned with only the orders of the ranks characterizing the coupling schemes, and not the tensors themselves (Subsec. 8.3.2.2,ii)). Therefore, adopting a shorter way of writing, only the ranks and their coupling schemes are specified.

8.4.1. COUPLING TENSORS

i) Corresponding to a special coupling scheme of $n \geq 2$ ranks, the Eqs. (8–108,126) can, respectively, be written as :

$$\{\ldots\{j_1j_2\}k_1\ldots k_{n-2}j_n\}k_{n-1} = C^{j_1\bullet\ \ldots j_n\bullet k_{n-1}}_{\{\ldots\{j_1j_2\}k_1\ldots k_{n-2}j_n\}}\ j_1^{\bullet}\ldots j_n^{\bullet} \quad , \tag{8-150a}$$

$$j_1\ldots j_n = \sum_{k_1,\ldots,k_{n-1}} C^{j_1\ldots j_n\ k^{\bullet}_{n-1}}_{\{\ldots\{j_1j_2\}k_1\ldots k_{n-2}j_n\}}\ \{\ldots\{j_1j_2\}k_1\ldots k_{n-2}j_n\}k_{n-1}\bullet \quad , \tag{8-150b}$$

where :

$$C^{j_1\ldots j_n\ k_{n-1}}_{\{\ldots\{j_1j_2\}k_1\ldots k_{n-2}j_n\}} = C^{j_1j_2k_1^{\bullet}}\ C^{k_{1\bullet}j_3k_2^{\bullet}}\ \ldots C^{k_{n-2\bullet}j_n\ k_{n-1}} \quad . \tag{8-151}$$

More generally, let us denote any coupling scheme of n ($n \geqslant 2$) ranks by :

$j_1,\ldots,j_n$ coupled to k_{n-1} , with
the intermediate couplings $k_1,\ldots,k_{n-2}$
in an order characterized by a Greek letter $\Big\}$ $\{j_1\ldots j_n;k_1\ldots k_{n-2},\alpha\}\ k_{n-1}$. (8-152)

One then has the general relations for $n \geqslant 2$:

$$\{j_1\ldots j_n;k_1\ldots k_{n-2},\alpha\}\ k_{n-1} = C^{j_1\bullet\ \ldots j_n\bullet k_{n-1}}_{(k_1\ldots k_{n-2},\alpha)}\ j_1^{\bullet}\ldots j_n^{\bullet} \quad , \tag{8-153a}$$

$$j_1\ldots j_n = \sum_{k_1,\ldots,k_{n-1}} C^{j_1\ldots j_n\ k^{\bullet}_{n-1}}_{(k_1\ldots k_{n-2},\alpha)}\ \{j_1\ldots j_n;k_1\ldots k_{n-2},\alpha\}\ k_{n-1}\bullet \quad , \tag{8-153b}$$

where the tensor $C^{j_1\ldots j_n\ k_{n-1}}_{(k_1\ldots k_{n-2},\alpha)}$ is defined by a contracted tensor product of n–1 C.G. tensors which depend upon the considered coupling scheme. This dependence is specified either by $(k_1\ldots k_{n-2},\alpha)$ in the general case or by the explicit coupling scheme in a special case as in Eqs. (8–150a,b). This tensor is said to be a <u>coupling tensor</u> (or also a <u>decoupling tensor</u> since the same tensor is used in both the equations above).

The coupling tensors are thus a generalization of the C.G. tensors which correspond to the special case n=2. As them, they are invariant Hermitian tensors of order n+1, the spherical components of which are all real. Since a C.G. tensor $C^{jj'j''}$ vanishes when j+j'+j" is not an integer, one must have for a coupling tensor $j_1+\ldots+j_n+k_{n-1}$ integer, otherwise it is equal to zero.

ii) The orthogonality relations for the C.G. tensors, Eqs. (8–123a,b), yield orthogonality relations for the coupling tensors, i.e. :

$$C^{j_1^{\bullet}\ldots j_n^{\bullet}\ k_{n-1}}_{(k_1\ldots k_{n-2},\alpha)}\ C^{j_{1\bullet}\ \ldots j_{n\bullet}\ k'_{n-1}}_{(k'_1\ldots k'_{n-2},\alpha)} = \prod_{i=1}^{n-1}\left[\delta_{k_ik'_i}\ \{k_i\}\right] G^{k_{n-1}k_{n-1}} \quad , \tag{8-154a}$$

$$\sum_{k_1,\ldots,k_{n-2}} C^{j_1\ldots j_n\ k^{\bullet}_{n-1}}_{(k_1\ldots k_{n-2},\alpha)}\ C^{j_1\ldots j_n\ k_{n-1\bullet}}_{(k_1\ldots k_{n-1},\alpha)} = G^{j_1j_1}\ \ldots\ G^{j_nj_n} \quad , \tag{8-154b}$$

where in the first equation, $\{k_i\}$ denotes the triangular delta which corresponds to each of the successive couplings of two indices characterizing the considered coupling scheme. As for Eqs. (8–123a,b), the lateral order of the free indices in the above equations is relevant, e.g. the index k_{n-1} of the first coupling coefficient in Eq. (8–154a) corresponds to the left index of the metric tensor.

iii) For a special order of couplings, a necessary and sufficient condition for a tensor $T^{j_1 \dots j_n}$ be invariant is given in Eq. (8–132). More generally, whatever the order of couplings, an invariant tensor of order $n \geqslant 2$ can be expressed in terms of coupling coefficients, the last rank of which is zero :

$$\forall U \in SU(2) \qquad {}^{U}T^{j_1 \dots j_n} = T^{j_1 \dots j_n} \tag{8-155}$$

$$\Leftrightarrow \quad T^{j_1 \dots j_n} = \sum_{k_1, \dots, k_{n-3}} C^{j_1 \dots j_n \; 0}_{(k_1 \dots k_{n-3}, \alpha)} \; T^{\{j_1 \dots j_n; k_1 \dots k_{n-3}, \alpha\} 0} \quad .$$

One should note that these special coupling coefficients can be expressed in terms of a contracted tensor product of n–2 (instead of n–1) C.G. tensors by using Eq. (8–135).

8.4.2. RECOUPLING COEFFICIENTS

Let us first discuss the relation between the two possible coupling schemes of two indices. This simple case is indeed sufficient to exhibit the method which can be applied to the general case. Then, some details about the most often encountered cases of three and four indices are given. The main property of the recoupling coefficients defined below is that they are multiples of completely contracted tensor products of C.G. or Wigner 3j tensors, and are thus invariant quantities or scalars.

8.4.2.1. Recoupling of two indices, 3j symbol, and general method

i) Given any tensor of order two, or more generally taking only into account two ranks of a tensor of order greater than two, the relation between the two possible coupling schemes has already been given by Eq. (8–73), i.e. :

$$\{j_1 j_2\} k_1 = (-1)^{j_1 + j_2 - k_1} \{j_2 j_1\} k_1 \quad . \tag{8-156}$$

Let us now recover this result by a method suitable for generalization to more than two indices. Using Eq. (8–106) to decouple the indices, and then Eq. (8–125) to recouple them in the desired order, one gets in an intrinsic way :

$$\begin{aligned} \{j_1 j_2\} k_1 &= \sum_{\ell_1} C^{j_1 \bullet j_2 \bullet k_1} \, C^{j_2^\bullet j_1^\bullet \ell_1^\bullet} \, \{j_2 j_1\} \, \ell_1 \bullet \\ &= \sum_{\ell_1} {}^{3}R^{k_1 \ell_1^\bullet}_{(j_1 j_2)} \, \{j_2 j_1\} \, \ell_1 \bullet \quad . \end{aligned} \tag{8-157}$$

The tensor ${}^{3}R^{k_1 \ell_1}_{(j_1 j_2)}$ thus defined is invariant since it is equal to a contracted tensor product of two invariant C.G. tensors. Hence, the Wigner–Eckart theorem, Eq. (8–141a), yields :

$$ {}^{3}R^{k_1\ell_1}_{(j_1j_2)} = \frac{\delta_{k_1\ell_1}}{2k_1+1}\; {}^{3}R^{k_1^\bullet k_1\bullet}_{(j_1j_2)}\; G^{k_1k_1} \quad . \tag{8-158} $$

Finally, from Eqs. (8–102,105) the relation between the two coupling schemes reads :

$$ \{j_1j_2\}k_1 = \frac{1}{2k_1+1}\; {}^{3}R^{k_1\bullet k_1^\bullet}_{(j_1j_2)}\; \{j_2j_1\}k_1 \quad . \tag{8-159} $$

The invariant quantity $\frac{1}{2k_1+1}\; {}^{3}R^{k_1\bullet k_1^\bullet}_{(j_1j_2)}$ is said to be a <u>recoupling coefficient</u>. We also call it a <u>square 3j symbol</u> denoted by $[j_1j_2k_1]$:

$$ \{j_1j_2\}k_1 = [j_1j_2k_1]\; \{j_2j_1\}k_1 \quad . \tag{8-160} $$

This scalar coefficient is a multiple of a completely contracted tensor product of two C.G. tensors, or from Eq. (8–86) Wigner 3j tensors, that is :

$$ [j_1j_2k_1] = \frac{1}{2k_1+1}\; {}^{3}R^{k_1\bullet k_1^\bullet}_{(j_1j_2)} = \frac{1}{2k_1+1}\, C^{j_1\bullet j_2\bullet k_1\bullet}\, C^{j_2^\bullet j_1^\bullet k_1^\bullet} = (-1)^{2k_1}\, W^{j_1\bullet j_2\bullet k_1\bullet}\, W^{j_2^\bullet j_1^\bullet k_1^\bullet} \quad . \tag{8-161} $$

In terms of the C.G. coefficients or the Wigner 3j symbols, one gets :

$$ [j_1j_2k_1] = \sum_{m_1} \langle j_1j_2m_1m_2|k_1n_1\rangle\langle j_2j_1m_2m_1|k_1n_1\rangle = (-1)^{2k_1}(2k_1+1)\sum_{m_1} \begin{pmatrix} j_1 & j_2 & k_1 \\ m_1 & m_2 & n_1 \end{pmatrix}\begin{pmatrix} j_2 & j_1 & k_1 \\ m_2 & m_1 & n_1 \end{pmatrix} , \tag{8-162} $$

for any fixed allowed value of n_1. Indeed, we have taken advantage of the consequence of the Wigner–Eckart theorem for an invariant tensor of order two given by Eq. (8–142), i.e. ${}^{3}R^{(k_1)[k_1]}_{(j_1j_2)\,n_1\;\,n_1}$ is independent of n_1. Moreover the two remaining indices of summation m_1 and m_2 have to fulfil the two selection rules for the indices m which correspond to each C.G. coefficient or Wigner 3j symbol. Actually these two constraints are identical, therefore only one independent index of summation, say m_1 remains. The numerical evaluation of a square 3j symbol is straightforward using the symmetry properties of Wigner 3j tensors, Eq. (8–89a), and their orthogonality relation, Eq. (8–129a) :

$$ [j_1j_2k_1] = (-1)^{j_1+j_2-k_1}\, W^{j_1\bullet j_2\bullet k_1\bullet}\, W^{j_1^\bullet j_2^\bullet k_1^\bullet} = (-1)^{j_1+j_2-k_1}\, \{j_1j_2k_1\} \quad . \tag{8-163} $$

Hence, a square 3j symbol is related by a phase factor to a triangular delta, Eq. (6–67). A triangular delta is also called a <u>3j symbol</u> (not to be confused with a Wigner 3j symbol). A 3j symbol is invariant under any permutation of its arguments.

ii) The above procedure for the recoupling of two indices can be carried on for a greater number of indices. Let us use the notation of Eq. (8–152) for any coupling scheme of n+1 indices $j_1,\ldots,j_{n+1}$ $(n \geqslant 1)$. Then, the relation between two different coupling schemes can be written as :

$$ \{j_1\ldots j_{n+1};k_1\ldots k_{n-1},\alpha\}k_n = \sum_{\ell_1,\ldots,\ell_n} {}^{3n}R^{k_n\;\;\ell_n^\bullet}_{(j_1\ldots j_{n+1};k_1\ldots k_{n-1},\alpha;\ell_1\ldots\ell_{n-1},\beta)}\, \{j_1\ldots j_{n+1};\ell_1\ldots\ell_{n-1},\beta\}\ell_n . \tag{8-164} $$

where, from Eqs. (8–153a,b), the tensor ${}^{3n}R^{k_n \ell_n}$ is defined by a contracted tensor product of two coupling tensors :

$$ {}^{3n}R^{k_n \ell_n}_{(j_1 \dots j_{n+1};k_1 \dots k_{n-1},\alpha;\ell_1 \dots \ell_{n-1},\beta)} = C^{j_1\bullet \dots j_{n+1}\bullet k_n}_{(k_1 \dots k_{n-1},\alpha)} \; C^{j_1^\bullet \dots j_{n+1}^\bullet \ell_n}_{(\ell_1 \dots \ell_{n-1},\beta)} \quad . \tag{8-165} $$

This tensor can also be written as a contracted tensor product of 2n C.G. tensors : n to completely decouple the n+1 indices, and n other ones to completely recouple them in the desired order. It is thus invariant, and from the Wigner–Eckart theorem, Eq. (8–141a), it reads :

$$ {}^{3n}R^{k_n \ell_n} = \frac{\delta_{k_n \ell_n}}{2k_n+1} \; {}^{3n}R^{k_n^\bullet \ell_n^\bullet} \; G^{k_n k_n} \quad . \tag{8-166} $$

Eq. (8–164) then takes the form :

$$ \{j_1 \dots j_{n+1};k_1 \dots k_{n-1},\alpha\}k_n = \frac{1}{2k_n+1} \sum_{\ell_1,\dots,\ell_{n-1}} {}^{3n}R^{k_n\bullet\, k_n^\bullet}_{(j_1 \dots j_{n+1};k_1 \dots k_{n-1},\alpha;\ell_1 \dots \ell_{n-1},\beta)} \; \{j_1 \dots j_{n+1};\ell_1 \dots \ell_{n-1},\beta\}k_n \quad . \tag{8-167} $$

The scalar $\frac{1}{2k_n+1} \; {}^{3n}R^{k_n\bullet\, k_n^\bullet}$ is a recoupling coefficient which depends upon 3n indices, i.e. : $j_1,\dots,j_{n+1},k_1,\dots,k_n,\ \ell_1,\dots,\ell_{n-1}$, and also upon the two orders of couplings denoted by α and β. We also call it a <u>square 3nj symbol</u>.

This coefficient being a multiple of a completely contracted tensor product of 2n C.G. or Wigner 3j tensors, its numerical evaluation, for instance in terms of the Wigner 3j symbols, requires n indices of summation m. Indeed, there are a priori 3n such indices, one for each index upon which the recoupling coefficient depends. To each of the 2n Wigner 3j symbols there corresponds a conservation law of the indices m, and only 2n–1 of these constraints are linearly independent. Moreover, using again the Wigner–Eckart theorem, Eq. (8–142), the invariant recoupling coefficient reads :

$$ \frac{1}{2k_n+1} \; {}^{3n}R^{k_n\bullet\, k_n^\bullet} = {}^{3n}R^{(k_n)[k_n]}_{m \quad m} \quad , \tag{8-168} $$

for any fixed allowed value of m. Hence, there remain :

$$ 3n - (2n-1) - 1 = n $$

independent indices of summation.

In a general way, these recoupling coefficients of n+1 indices ($n \geqslant 1$) or square 3nj symbols are multiple of the <u>3nj symbols</u>, which are the scalars defined by completely contracted tensor products of 2n Wigner 3j tensors, which cannot be factorized. A graphical representation of these symbols is discussed later in Subsec. 9.1.3.2.

iii) The recoupling coefficients allow us to relate coupling tensors which correspond to different coupling schemes. Indeed, either directly from the definitions, Eqs. (8–153a,b, and 167), or from Eq. (8–165) and the orthogonality relation of the coupling tensors, Eq. (8–154b), one gets :

$$C^{j_1 \dots j_{n+1}\; k_n}_{(\ell_1 \dots \ell_{n-1},\beta)} = \sum_{k_1,\dots,k_{n-1}} C^{j_1 \dots j_{n+1}\; k_n}_{(k_1 \dots k_{n-1},\alpha)} \frac{1}{2k_n+1}\, {}^{3n}R^{k_n \bullet k_n^\bullet}_{(j_1 \dots j_{n+1}; \ell_1 \dots \ell_{n-1},\alpha; k_1 \dots k_{n-1},\beta)} \; . \tag{8-169}$$

8.4.2.2. Recoupling of three and four indices, 6j and 9j symbols

The foregoing study is now applied to the special cases n=2 and 3 for, respectively, the coupling schemes :

$\alpha = \{\{j_1 j_2\}k_1 j_3\}k_2$ in terms of $\beta = \{j_1\{j_2 j_3\}\ell_1\}k_2$,

and :

$\alpha = \{\{j_1 j_2\}k_1\{j_3 j_4\}k_2\}k_3$ in terms of $\beta = \{\{j_1 j_3\}\ell_1\{j_2 j_4\}\ell_2\}k_3$.

i) Definition of the 6j and 9j symbols. One has :

$$\{\{j_1 j_2\}k_1 j_3\}k_2 = \frac{1}{2k_2+1} \sum_{\ell_1} {}^{6}R^{k_2 \bullet\, k_2^\bullet}_{(j_1 j_2 j_3; k_1,\alpha; \ell_1,\beta)} \{j_1\{j_2 j_3\}\ell_1\}k_2 \quad , \tag{8-170a}$$

$$\{\{j_1 j_2\}k_1\{j_3 j_4\}k_2\}k_3 = \frac{1}{2k_3+1} \sum_{\ell_1,\ell_2} {}^{9}R^{k_3 \bullet\, k_3^\bullet}_{(j_1 j_2 j_3 j_4; k_1 k_2,\alpha; \ell_1 \ell_2,\beta)} \{\{j_1 j_3\}\ell_1\{j_2 j_4\}\ell_2\}k_3 \; , \tag{8-170b}$$

where the recoupling coefficients are, respectively, called square 6j and 9j symbols, and denoted by :

$$\begin{bmatrix} j_1 & j_2 & k_1 \\ j_3 & k_2 & \ell_1 \end{bmatrix} = \frac{1}{2k_2+1}\, {}^{6}R^{k_2 \bullet\, k_2^\bullet}_{(j_1 j_2 j_3; k_1,\alpha; \ell_1,\beta)} \quad , \tag{8-171a}$$

$$\begin{bmatrix} j_1 & j_2 & k_1 \\ j_3 & j_4 & k_2 \\ \ell_1 & \ell_2 & k_3 \end{bmatrix} = \frac{1}{2k_3+1}\, {}^{9}R^{k_3 \bullet\, k_3^\bullet}_{(j_1 j_2 j_3 j_4; k_1 k_2,\alpha; \ell_1 \ell_2,\beta)} \quad . \tag{8-171b}$$

These coefficients are, respectively, related to the Wigner 6j and 9j symbols[63], which are identical to the 6j and 9j symbols, by :

$$\begin{Bmatrix} j_1 & j_2 & k_1 \\ j_3 & k_2 & \ell_1 \end{Bmatrix} = \frac{(-1)^{j_1+j_2+j_3+k_2}}{\hat{k}_1\, \hat{\ell}_1} \begin{bmatrix} j_1 & j_2 & k_1 \\ j_3 & k_2 & \ell_1 \end{bmatrix} \quad , \tag{8-172a}$$

$$\begin{Bmatrix} j_1 & j_2 & k_1 \\ j_3 & j_4 & k_2 \\ \ell_1 & \ell_2 & k_3 \end{Bmatrix} = \frac{1}{\hat{k}_1 \hat{k}_2 \hat{\ell}_1 \hat{\ell}_2} \begin{bmatrix} j_1 & j_2 & k_1 \\ j_3 & j_4 & k_2 \\ \ell_1 & \ell_2 & k_3 \end{bmatrix} \quad , \tag{8-172b}$$

where the choice of normalization is such as to give the symbols the maximum symmetry with respect to permutations of their arguments.

ii) Evaluation of the 6j and 9j symbols. In terms of C.G. or Wigner 3j tensors, these symbols read as multiples of completely contracted tensor products, i.e. :

$$\begin{Bmatrix} j_1 & j_2 & k_1 \\ j_3 & k_2 & \ell_1 \end{Bmatrix} = \frac{(-1)^{j_1+j_2+j_3+k_2}}{\hat{k}_1 (2k_2+1) \hat{\ell}_1} C^{k_1 \bullet j_3 \bullet k_2 \bullet} C^{j_1 \bullet j_2 \bullet k_1^\bullet} C^{j_2^\bullet j_3^\bullet \ell_1^\bullet} C^{j_1^\bullet \ell_1 \bullet k_2^\bullet}$$
$$= c_{k_1} c_{\ell_1} (-1)^{-j_1+j_2+j_3+k_2} W^{k_1 \bullet j_3 \bullet k_2 \bullet} W^{j_1 \bullet j_2 \bullet k_1^\bullet} W^{j_2^\bullet j_3^\bullet \ell_1^\bullet} W^{j_1^\bullet \ell_1 \bullet k_2^\bullet} , \qquad (8\text{-}173a)$$

$$\begin{Bmatrix} j_1 & j_2 & k_1 \\ j_3 & j_4 & k_2 \\ \ell_1 & \ell_2 & k_3 \end{Bmatrix} = \frac{1}{\hat{k}_1 \hat{k}_2 (2k_3+1) \hat{\ell}_1 \hat{\ell}_2} C^{k_1 \bullet k_2 \bullet k_3 \bullet} C^{j_1 \bullet j_2 \bullet k_1^\bullet} C^{j_3 \bullet j_4 \bullet k_2^\bullet} C^{j_1^\bullet j_3^\bullet \ell_1^\bullet} C^{j_2^\bullet j_4^\bullet \ell_2^\bullet} C^{\ell_1 \bullet \ell_2 \bullet k_3^\bullet}$$
$$= W^{k_1 \bullet k_2 \bullet k_3 \bullet} W^{j_1 \bullet j_2 \bullet k_1^\bullet} W^{j_3 \bullet j_4 \bullet k_2^\bullet} W^{j_1^\bullet j_3^\bullet \ell_1^\bullet} W^{j_2^\bullet j_4^\bullet \ell_2^\bullet} W^{\ell_1 \bullet \ell_2 \bullet k_3^\bullet} . \qquad (8\text{-}173b)$$

In terms of the C.G. coefficients, or the Wigner 3j symbols, one gets :

$$\begin{Bmatrix} j_1 & j_2 & k_1 \\ j_3 & k_2 & \ell_1 \end{Bmatrix} = \frac{(-1)^{j_1+j_2+j_3+k_2}}{\hat{k}_1 \ \hat{\ell}_1} \sum_{m_1,m_2} \langle k_1 j_3 n_1 m_3 | k_2 n_2 \rangle \langle j_1 j_2 m_1 m_2 | k_1 n_1 \rangle \langle j_2 j_3 m_2 m_3 | \ell_1 p_1 \rangle \langle j_1 \ell_1 m_1 p_1 | k_2 n_2 \rangle$$
$$= (-1)^{-j_1+j_2+j_3+k_1+k_2+\ell_1} (2k_2+1) \sum_{m_1,m_2} (-1)^{-n_1+p_1}$$
$$\times \begin{pmatrix} k_1 & j_3 & k_2 \\ n_1 & m_3 & n_2 \end{pmatrix} \begin{pmatrix} j_1 & j_2 & k_1 \\ m_1 & m_2 & -n_1 \end{pmatrix} \begin{pmatrix} j_2 & j_3 & \ell_1 \\ m_2 & m_3 & p_1 \end{pmatrix} \begin{pmatrix} j_1 & \ell_1 & k_2 \\ m_1 & -p_1 & n_2 \end{pmatrix} , \qquad (8\text{-}174a)$$

$$\begin{Bmatrix} j_1 & j_2 & k_1 \\ j_3 & j_4 & k_2 \\ \ell_1 & \ell_2 & k_3 \end{Bmatrix} = \frac{1}{\hat{k}_1 \hat{k}_2 \hat{\ell}_1 \hat{\ell}_2} \sum_{m_1,m_2,m_3} \langle k_1 k_2 n_1 n_2 | k_3 n_3 \rangle \langle j_1 j_2 m_1 m_2 | k_1 n_1 \rangle$$
$$\times \langle j_3 j_4 m_3 m_4 | k_2 n_2 \rangle \langle j_1 j_3 m_1 m_3 | \ell_1 p_1 \rangle \langle j_2 j_4 m_2 m_4 | \ell_2 p_2 \rangle \langle \ell_1 \ell_2 p_1 p_2 | k_3 n_3 \rangle$$
$$= (2k_3+1) \sum_{m_1,m_2,m_3} \begin{pmatrix} k_1 & k_2 & k_3 \\ n_1 & n_2 & n_3 \end{pmatrix} \begin{pmatrix} j_1 & j_2 & k_1 \\ m_1 & m_2 & n_1 \end{pmatrix} \begin{pmatrix} j_3 & j_4 & k_2 \\ m_3 & m_4 & n_2 \end{pmatrix}$$
$$\begin{pmatrix} j_1 & j_3 & \ell_1 \\ m_1 & m_3 & p_1 \end{pmatrix} \begin{pmatrix} j_2 & j_4 & \ell_2 \\ m_2 & m_4 & p_2 \end{pmatrix} \begin{pmatrix} \ell_1 & \ell_2 & k_3 \\ p_1 & p_2 & n_3 \end{pmatrix} , \qquad (8\text{-}174b)$$

for any fixed allowed values of, respectively, n_2 and n_3.

The numerical evaluation of the 6j and 9j symbols can be performed from the above equations, introducing the expression of the Wigner 3j symbols given by Eq. (6–80). For instance, after a tedious calculation one gets for the 6j symbols a series with only one index of summation, first obtained by Racah[46] :

$$\begin{Bmatrix} j_1 & j_2 & j_3 \\ k_1 & k_2 & k_3 \end{Bmatrix} = \Delta(j_1 j_2 j_3)\ \Delta(j_1 k_2 k_3)\ \Delta(k_1 j_2 k_3)\ \Delta(k_1 k_2 j_3)$$
$$\times \sum_{\ell} (-1)^{\ell} (\ell+1)! [(\ell-j_1-j_2-j_3)!(\ell-j_1-k_2-k_3)!(\ell-k_1-j_2-k_3)! \qquad (8\text{-}175)$$
$$\times (\ell-k_1-k_2-j_3)!(j_1+j_2+k_1+k_2-\ell)!(j_2+j_3+k_2+k_3-\ell)!(j_3+j_1+k_3+k_1-\ell)!]^{-1} ,$$

where the function Δ is defined in Eq. (6–77) and the sum over ℓ runs over all positive integer values so that the arguments of the factorials are non–negative. Therefore, by inspection, ℓ takes L+1 values where L is the smallest of the twelve numbers :

$$\begin{array}{lll} j_1+j_2-j_3 & j_1\pm(k_2-k_3) & k_1+k_2-j_3 \\ j_2+j_3-j_1 & j_2\pm(k_3-k_1) & k_2+k_3-j_1 \\ j_3+j_1-j_2 & j_3\pm(k_1-k_2) & k_3+k_1-j_2 \end{array} \quad . \tag{8-176}$$

A similar general formula has been obtained later by Schwinger [55] using another method, cf. also Bargmann [5] for the derivation of a generating function for the 6j symbols. Actually, as for the 3j symbols (Subsec. 6.2.3.2), there exist for the 6j symbols many other formulae which are not simply related by a change of variable in the previous summation over ℓ . These different formulae can be obtained from the properties of the generalized hypergeometric function [57] ${}_4F_3$ with unit argument. For a detailed study of this problem cf. for instance Raynal [50].

Values of some special 6j symbols are given in Appendix G.2.2.The values of special square 9j symbols with arguments equal to zero are listed in Appendix G.3.2,i).

iii) Properties of the 6j and 9j symbols :

1–Selection rules. The (square) 6j and 9j symbols are proportional to the following products of triangular deltas :

$$\begin{Bmatrix} j_1 & j_2 & j_3 \\ k_1 & k_2 & k_3 \end{Bmatrix} \propto \{j_1j_2j_3\}\{j_3k_1k_2\}\{j_2j_3k_3\}\{j_1k_3k_2\} \quad , \tag{8-177a}$$

$$\begin{Bmatrix} j_1 & j_2 & j_3 \\ k_1 & k_2 & k_3 \\ \ell_1 & \ell_2 & \ell_3 \end{Bmatrix} \propto \{j_1j_2j_3\}\{k_1k_2k_3\}\{j_3k_3\ell_3\}\{j_1k_1\ell_1\}\{j_2k_2\ell_2\}\{\ell_1\ell_2\ell_3\} \quad . \tag{8-177b}$$

2–Reality. The (square) 6j and 9j symbols are real.

3–Symmetry properties. The usual symmetry properties of the 6j and 9j symbols are, respectively, derived in Subsecs. 9.1.3.2,iii) and iv) as an example of application of the graphical representations. Actually, for the 6j symbols Regge [52] found wider symmetries.

4–Orthogonality relations and sum rules. The orthogonality relations of the 6j and 9j symbols follow from the unitary character and the reality of the C.G. coefficients. These relations are given later in terms of the graphical representations, Eqs. (9–30, 50a, and b). So it is with the sum rules, Eqs. (9–51a,b, and c), and the expression of the 9j symbols in terms of 6j symbols, Eqs. (9–39,52).

8.5. SPHERICAL AND CARTESIAN TENSOR ALGEBRAS

8.5.1. DEFINITIONS

Let us specify by the word spherical the tensor algebra considered so far for the group

SU(2) (i.e. spherical tensor algebra, spherical tensors, spherical tensorial sequences...). Now, a similar tensor algebra can be developed for the group SO(3), we shall call it the Cartesian tensor algebra. A Cartesian tensor of order n is then defined by a Cartesian tensorial sequence $(T_{i_1 \ldots i_n})$, i.e. by a set of 3^n objects $T_{i_1 \ldots i_n}$, $i_j = 1,2,3$, which belong to a vector space over the complex field and are characterized by their law of transformation under the following representation of SO(3) :

$$\forall R \in SO(3) \qquad T = (T_{i_1 \ldots i_n}) \xrightarrow{R} {}^{R}T = ({}^{R}T_{i_1 \ldots i_n} = T_{j_1 \ldots j_n} \prod_{k=1}^{n} R_{i_k}^{j_k}) \,, \tag{8-178}$$

compare with Eq. (8–34). The objects $T_{i_1 \ldots i_n}$ are called the Cartesian components of the Cartesian tensor T.

Once again, it has to be pointed out that the above tensor algebra is more general than the usual one, since it deals equally with sets of components or basis vectors. The usual tensor algebra defined on the tensor product of several three-dimensional Euclidean vector spaces is contained in the above formalism when the metric tensor is taken equal to (δ_{ij}).

8.5.2. RELATION BETWEEN SPHERICAL AND CARTESIAN TENSORIAL SEQUENCES

The homomorphism between SU(2) and SO(3) (Subsec. 4.2.1) implies that the spherical and the Cartesian tensor algebras are closely related. Let us consider the generic case of a tensor of order one, i.e. $(T_m^{[1]})$ and (T_i) are, respectively, a spherical and a Cartesian tensorial sequence characterized by :

$$\forall U \in SU(2) \qquad {}^{U}T_m^{[1]} = T_{m'}^{[1]} \mathcal{D}^1_{m'm}(U) \;, \quad \forall R \in SO(3) \qquad {}^{R}T_i = T_j R_i^j \;.$$

It is now shown that from any one of these two sets one can define the other one such that the two matrices U and R are related by Eq. (4–28). Four different ways to establish this result are given, each of which exhibits an interesting aspect of this problem. One should remark that according to the methods, the objects considered are either basis vectors or components.

i) The first method is based on the completeness of the representations $\mathcal{D}^j$[SU(2)] (Subsec. 7.2.2.3). The correspondence between $\pm U$ in SU(2) and R in SO(3) defines a finite-dimensional irreducible representation of SU(2). Therefore, this three-dimensional representation is equivalent to the representation $\mathcal{D}^1$[SU(2)], i.e. there exists a nonsingular matrix t such that :

$$\forall R \in SO(3) \qquad R = t\, \mathcal{D}^1(R)\, t^{-1} \quad , \quad R_j^i = t_m^i\, \mathcal{D}^1_{mm'}(R)\, (t^{-1})_j^{m'} \quad , \tag{8-179}$$

where $\mathcal{D}^1(R)$ is defined by Eq. (6–25). Let us still denote by R a rotation in an Euclidean three-dimensional vector space E_3. Then, (R_j^i) and $(\mathcal{D}^1_{mm'})$ are, respectively, the matrix representatives of R and its complex extension {R} with respect to an orthonormal basis (e_i) of E_3 and a standard basis (e_m^1) of the complex extension $\{E_3\}$ (Appendix A.2.2), such that :

$$R\, e_i = e_j\, R^j_i \quad , \quad \{R\}\, e^1_m = e^1_{m'}\, \mathcal{D}^1_{m'm}(R) \quad . \tag{8-180}$$

According to Eqs. (1–2,13), these two bases are related by :

$$e^1_m = e_i\, t^i_m \quad , \quad e_i = e^1_m\, (t^{-1})^m_i \quad . \tag{8-181}$$

From Prop. C.47 the matrix t is an arbitrary complex multiple of a unitary matrix. Then, choosing the basis (e^1_m) orthonormal with respect to the Hermitian inner product in $\{E_3\}$, the matrix elements of t are uniquely defined up to an overall arbitrary phase factor. The matrix t being independent of R, let us set $R=R(\theta, e_3)$. From Eq. (6–17), $\mathcal{D}^1[R(\theta, e_3)]$ is diagonal. Therefore, the basis vectors e^1_m are multiples of the eigenvectors of this rotation, i.e. (Subsec. 2.2.1,ii)) :

$$e^1_{\pm 1} = a_\pm e_\pm = a_\pm (e_1 \pm i e_2) \quad , \quad e^1_o = b\, e_3 \qquad a_\pm, b \in C \quad . \tag{8-182}$$

From the unitary character of t one gets :

$$\sqrt{2}\ |a_\pm| = |b| = 1 \quad .$$

Let us remove the phase factor indetermination by setting b=-i. This choice is justified later in Subsec. 8.5.4,ii). Then, setting now $R=R(\theta, e_2)$ the expressions of the two matrices (R^i_j) and $(\mathcal{D}^1_{mm'}(R))$ given, respectively, in Eq. (E–5) and Appendix F.2 yield $a_\pm = \mp 1/\sqrt{2}$ and thus :

$$t = \begin{array}{c} \\ 1 \\ 2 \\ 3 \end{array} \begin{array}{c} \begin{array}{ccc} 1 & 0 & -1 \end{array} \\ \begin{pmatrix} \frac{i}{\sqrt{2}} & 0 & \frac{-i}{\sqrt{2}} \\ \frac{-1}{\sqrt{2}} & 0 & \frac{-1}{\sqrt{2}} \\ 0 & -i & 0 \end{pmatrix} \end{array} \quad , \quad t^{-1} = \begin{array}{c} \\ 1 \\ 0 \\ -1 \end{array} \begin{array}{c} \begin{array}{ccc} 1 & 2 & 3 \end{array} \\ \begin{pmatrix} -\frac{i}{\sqrt{2}} & \frac{-1}{\sqrt{2}} & 0 \\ 0 & 0 & i \\ \frac{i}{\sqrt{2}} & \frac{-1}{\sqrt{2}} & 0 \end{pmatrix} \end{array} \quad , \tag{8-183a}$$

$$\left\{ \begin{aligned} e^1_1 &= \frac{1}{\sqrt{2}} (i e_1 - e_2) \\ e^1_0 &= -i\, e_3 \\ e^1_{-1} &= \frac{-1}{\sqrt{2}} (i e_1 + e_2) \end{aligned} \right. \quad , \quad \left\{ \begin{aligned} e_1 &= \frac{i}{\sqrt{2}} (-e^1_1 + e^1_{-1}) \\ e_2 &= \frac{-1}{\sqrt{2}} (e^1_1 + e^1_{-1}) \\ e_3 &= i\, e^1_0 \end{aligned} \right. \quad . \tag{8-183b}$$

In this method (e^1_m) and (e_i) play, respectively, the role of the tensorial sequences $(T^{[1]}_m)$ and (T_i).

ii) The second method is based on the intrinsic correspondence between the group SU(2) and the rotation group in its Lie algebra $\mathcal{W}(2)$ (Subsec. 5.4.2,ii)) : the adjoint representation A(U) of any U in SU(2) is a rotation in $\mathcal{W}(2)$, the three–dimensional Euclidean structure of this Lie algebra being defined by the Killing form. Furthermore, we have seen that, with respect to the corresponding inner product, $(e_i = \frac{1}{2}\sigma_i)$ is an orthogonal basis of $\mathcal{W}(2)$, all the e_i having the same norm, Eq. (5–40) :

$$(e_i, e_j) = 2\,\delta_{ij} \quad ,$$

such that :

$$\forall U \in SU(2) \qquad A(U)e_i = R(U)e_i = e_j\ R(U)^j_i \quad ,$$

where the rotation matrix $(R(U)^j_i)$ is related to U by Eq. (4–28). We are now looking for a new basis (e^1_m) of the complex extension $\{\mathcal{W}(2)\}$, with respect to which the matrix representative of the complex extension $\{R(U)\}$ of R(U) is $\mathcal{D}^1(U)$. From Subsec. 6.1.3,i) this basis is a standard one of the adjoint representation. It fulfils the characteristic Eqs. (6–31a,b) for j=1, i.e. :

$$J_\pm\, e^1_m = \sqrt{2-m(m\pm1)}\; e^1_{m\pm1} \quad , \quad J_3\, e^1_m = m\, e^1_m \quad , \tag{8-184}$$

where J_i denote the infinitesimal generators of the adjoint representation which correspond to σ_i by Eq. (5–8). From Eq. (5–21) these infinitesimal generators are the following adjoint operators :

$$J_\pm = [\tfrac{\sigma_\pm}{2}\,,\,.] \quad , \quad J_3 = [\tfrac{\sigma_3}{2}\,,\,.] \quad . \tag{8-185}$$

It has been pointed out in Subsec. 8.1,ii) that the derivation of the standard basis is only based on the existence of a standard form of commutation relations in the semi–simple Lie algebra $\mathcal{W}(2)$ (Subsec. 4.3.3,iii)), i.e. :

$$[\tfrac{\sigma_+}{2}\,,\,\tfrac{\sigma_-}{2}] = 2\,\tfrac{\sigma_3}{2} \quad , \quad [\tfrac{\sigma_3}{2}\,,\,\tfrac{\sigma_\pm}{2}] = \pm\,\tfrac{\sigma_\pm}{2} \quad .$$

Then, the comparison of the above relations with Eq. (8–184), where the infinitesimal generators are given by Eq. (8–185), yields :

$$e^1_{\pm1} = a\left(\mp\frac{1}{\sqrt{2}}\,\frac{\sigma_\pm}{2}\right) \quad , \quad e^1_0 = a\,\frac{\sigma_3}{2} \quad , \tag{8-186}$$

where the complex coefficient a is arbitrary. This is the result already obtained in the first method.

Let us remark that for any semi–simple Lie algebra $\mathcal{W}$:

(1) There exists a special basis (H_i,E_α) of $\mathcal{W}$ which yields the standard form of the commutation relations given by Eqs. (4–61a–d).

(2) The adjoint representation of a Lie group induces the adjoint representation $\Delta(\mathcal{W})$ of its Lie algebra $\mathcal{W}$ (Subsec. 5.4.1,ii)). Therefore, the adjoint operators of, respectively, the H_i and E_α :

$$J_i = \Delta_{H_i} = [H_i\,,\,.] \quad , \quad J_\alpha = \Delta_{E_\alpha} = [E_\alpha\,,\,.] \quad ,$$

form a basis of infinitesimal generators of the adjoint representation of the group, which also fulfil the standard form of commutation relations (Prop. B.18).

(3) Assuming a suitable normalization of E_α and ordering these vectors by pairs E_α, $E_{-\alpha}$,

the metric tensor of the Killing form takes the form given in Eq. (4–64). One should note that the vectors E_α are isotropic vectors (their Euclidean norm is zero) and thus actually belong to the complex extension of the Lie algebra. Then, diagonalizing the real symmetrical restriction (g_{ij}) of the Killing form to the r–dimensional subspace spanned by the H_i, yields an orthonormal basis $(h_i = a^j_i\, H_j)$ of this subspace.

Now, the set of vectors (h_i, E_α) clearly defines a standard basis of the adjoint representation (Sec. 5.5). For instance, one has :

$$J_i\, E_\alpha = [H_i, E_\alpha] = \alpha_i\, E_\alpha \quad ,$$

which shows that the weights of the adjoint representation are equal to the roots.

In the case of the group SU(2) and its three–dimensional Lie algebra $\mathcal{U}(2)$ of rank one, an orthogonal standard basis, all basis vectors having the same norm, is from Eq. (4–70) : $(\sigma_+/(2\sqrt{2}), \sigma_3/2)$. The relative sign between e^1_1 and e^1_{-1} in Eq. (8–186) corresponds to the special phase choice specified in Eq. (5–59).

iii) The third method is based on the fact that any irreducible representation of SU(2) is a tensorial product of its smallest–dimensional representation, i.e. SU(2) itself, Eq. (6–26). Depending upon the choice of an orthonormal basis $a=(e_i)$ of a three–dimensional Euclidean vector space E_3, one considers a mapping T of E_3 into R^3 such that :

$$T = (T_i = e^{*i}) \quad , \quad \forall v = v^i e_i \in E_3 \qquad T(v) = (T_i(v) = e^{*i}(v) = v^i) \quad , \tag{8-187}$$

where (e^{*i}) is the associated dual basis to (e_i), Eq. (1–6). With respect to the representation of the rotation group in the dual vector space E^*_3 defined by (cf. a similar definition in Eq. (8–194), and more generally Subsec. 10.1.2) :

$$\forall f \in E^*_3 \xrightarrow{\ R\ } {}^R f \quad , \quad \forall v \in E_3 \qquad {}^R f(v) = f(R^{-1}v) \quad , \tag{8-188}$$

T is a Cartesian tensor, i.e. satisfies Eq. (8–178). Indeed, one has :

$$T \xrightarrow{\ R\ } {}^R T \quad , \quad \forall v \in E_3 \qquad {}^R T(v) = ({}^R T_i(v)) = T(R^{-1}v) = (T_i(R^{-1}v)$$

$${}^R T_i(v) = T_i(R^{-1}v) = e^{*i}(R^{-1}v) = (R^{-1})^i_j\, e^{*j}(v) \;\Rightarrow\; {}^R T_i = T_j\, R^j_i \quad . \tag{8-189}$$

Let us now define from T a mapping T' of E_3 into the set of complex 2×2 matrices by :

$$C = i\sigma_2 \quad , \qquad T' = C\,\sigma.T = \sum_i C\,\sigma_i\, T_i \quad . \tag{8-190}$$

Clearly a representation of SU(2) can be defined by :

$$\forall U \in SU(2) \qquad T' \xrightarrow{\ U\ } {}^U T' = C\,\sigma.{}^{R(U;a)}T \quad , \tag{8-191}$$

where the rotation R(U;a) is related to U by Eq. (4–26). It is now shown that with respect to the above representation the matrix elements of T', cf. Eq. (4–92) :

$$T' = \begin{pmatrix} T_1 + iT_2 & -T_3 \\ -T_3 & -T_1 + iT_2 \end{pmatrix} = \begin{pmatrix} T_+ & -T_3 \\ -T_3 & -T_- \end{pmatrix} \quad ,$$

are the contravariant spherical components of a spherical tensor $T^{\frac{1}{2}\frac{1}{2}}$, i.e. fulfil Eq. (8–44). Indeed, from Eqs. (4–26,104) one gets :

$$^{U}T'(v) = C\sigma.R^{-1}(U;a)v = CU^{-1}C^{-1}C\sigma.vU = U^{t}\ T'(v)U \quad ,$$

hence with the order of rows and columns specified in Eq. (6–12) and using Eq. (6–16) :

$$\forall U \in SU(2) \qquad (^{U}T')_{mn} = (T')_{m'n'}\ \mathcal{D}^{\frac{1}{2}}_{m'm}(U)\ \mathcal{D}^{\frac{1}{2}}_{n'n}(U) \quad .$$

Furthermore, this tensor is symmetrical. Therefore, from Eq. (8–68) its irreducible tensor $T^{\{\frac{1}{2}\frac{1}{2}\}0}$ vanishes. Finally, a spherical tensor of order one was defined by :

$$T^{1} = (\ T^{[1]}_{m} = \langle\tfrac{1}{2}\ \tfrac{1}{2}\ n\ m\text{-}n|1m\rangle\ (C\sigma.T)_{n\ m-n}) \tag{8-192}$$

and using the special values of the C.G. coefficients, Eqs. (G–6a,b), a direct calculation yields:

$$T^{[1]}_{\pm 1} = \pm\ T_{\pm} = \pm\ T_{1} + iT_{2} \quad , \quad T^{[1]}_{0} = -\sqrt{2}\ T_{3} \quad . \tag{8-193}$$

This is, up to an overall multiplicative factor, the result already obtained in Eq. (8–183b). In this method, the objects T_i and $T^{[1]}_{m}$ belong to the dual vector space E_3^* or its complex extension. The image by these forms of any vector $v=v^i e_i$ in E_3 yields similar relations for the components, i.e. :

$$v^{[1]}_{\pm 1} = \pm\, v_{\pm} = \pm\ v^{1} + iv^{2} \quad , \quad v^{[1]}_{0} = -\sqrt{2}\ v^{3} \quad .$$

iv) The fourth method is based on the two following facts :

(1) One shows (Subsec. 10.33.4) that the spherical harmonics $Y_{\ell m}$ form a standard basis of the representation of SO(3) in the vector space $L^2(S^2,d_2\hat{n})$ of square integrable complex functions f on S^2, defined by :

$$\forall R \in SO(3) \qquad f \xrightarrow{\ R\ } {}^{R}f \quad , \quad \forall \hat{n} \in E_3 \quad {}^{R}f(\hat{n}) = f(R^{-1}\hat{n}) \quad , \tag{8-194}$$

where both the rotation, denoted as its matrix representative by R, and the unitary vector $\hat{n}$ are related to the choice of an orthonormal basis (e_i) of E_3. Thus the $Y_{\ell m}$ transform as :

$$\forall R \in SO(3) \qquad {}^{R}Y_{\ell m} = Y_{\ell m'}\ \mathcal{D}^{\ell}_{m'm}(R) \quad , \tag{8-195}$$

where $\mathcal{D}^{\ell}(R)$ is defined in Eq. (6–25).

(2) The spherical harmonics Y_{1m} are related to the tensor $T=(T_i=e^{*i})$ considered in the third method by :

$$N\begin{pmatrix} Y_{11} \\ Y_{10} \\ Y_{1-1} \end{pmatrix} = \sqrt{\frac{3}{4\pi}}\begin{pmatrix} -\frac{1}{\sqrt{2}}\ T_{+} \\ T_{3} \\ \frac{1}{\sqrt{2}}\ T_{-} \end{pmatrix} \quad , \tag{8-196}$$

where N is the function which assigns to any vector v its Euclidean norm.

One recovers again the linear relations between either the functions $T^{[1]}_m = N\, Y_{1m}$ and $T_i = e^{*i}$, or the corresponding images for any vector $v = v^i e_i = \|v\|\,\hat{n}$, i.e. $T^{[1]}_m = \|v\|\, Y_{1m}(\hat{n})$ and $T_i = v^i$.

8.5.3. EXTENDED MEANING OF TENSORS

i) By definition the Cartesian and spherical tensorial sequences (T_i) and $(T^{[1]}_m)$ related by :

$$T^{[1]}_m = T_i\, c t^i_m \quad , \quad T_i = T^{[1]}_m\, c^{-1} (t^{-1})^m_i \quad , \tag{8-197}$$

where the matrix t is given in Eq. (8–183a) and c is any arbitrary complex number, are called the Cartesian and spherical components of the <u>same</u> tensor $T = T^1$. The special choice of the constant c=1 is justified below in Subsec. 8.5.4,ii).The foregoing can be extended in a straightforward way to tensors of any order n and with all ranks equal to one. Thus, the two tensorial sequences $(T_{i_1 \dots i_n})$ and $(T^{[1] \dots [1]}_{m_1 \;\, m_n})$ related by :

$$T^{[1]\dots[1]}_{m_1 \;\; m_n} = T_{i_1 \dots i_n} \prod_{j=1}^{n} (c\, t^{i_j}_{m_j}) \quad , \tag{8-198}$$

are by definition, respectively, the Cartesian and spherical components of one and the same tensor $T = T^{1\dots1}$.

Finally, the two–to–one homomorphism of SU(2) onto SO(3) implies that the Cartesian tensor algebra is a special case of the spherical one, but the converse is not true.

ii) A tensor $T^{1\dots1}$ of order n can be decomposed into its irreducible tensors. For a particular coupling scheme specified by α , according to Eq. (8–150a) one has :

$$T^{\{1\dots1;k_1\dots k_{n-2},\alpha\}[\ell]}_m = T_{i_1 \dots i_n}\, A^{i_1\dots i_n; \quad [\ell]}_{(k_1\dots k_{n-2},\alpha) \;\; m} \quad , \tag{8-199a}$$

where :

$$A^{i_1\dots i_n; \quad [\ell]}_{(k_1\dots k_{n-2},\alpha) \;\; m} = C^{(1)\dots(1)[\ell]}_{(k_1\dots k_{n-2},\alpha)\, m_1 \;\, m_n \; m} \prod_{j=1}^{n} (c\, t^{i_j}_{m_j}) \quad . \tag{8-199b}$$

From Eq. (8–150b) the above relation is inverted as :

$$T_{i_1\dots i_n} = \sum_{k_1,\dots,k_{n-2},\ell} T^{\{1\dots1;k_1\dots k_{n-2},\alpha\}[\ell]}_m\, \mathcal{B}^{(k_1\dots k_{n-2},\alpha)(\ell)}_{i_1\dots i_n; \quad m} \quad , \tag{8-200a}$$

where :

$$\mathcal{B}^{(k_1\dots k_{n-2},\alpha)(\ell)}_{i_1\dots i_n; \quad m} = C^{[1]\dots[1](\ell)}_{(k_1\dots k_{n-2},\alpha) m_1 \;\, m_n \; m} \prod_{j=1}^{n} \left[c^{-1} (t^{-1})^{m_j}_{i_j} \right] \quad . \tag{8-200b}$$

The orthogonality relations for the coupling tensors, Eqs. (8–154a,b) yield :

$$\mathcal{A}^{i_1\ldots i_n;\ [\ell]}_{(k_1\ldots k_{n-2},\alpha)m}\ \mathcal{B}^{(k'_1\ldots k'_{n-2},\alpha)(\ell')}_{i_1\ldots i_n;\quad m} = \prod_{j=1}^{n-2} \delta_{k_j k'_j}\ \delta_{\ell\ell'}\ \delta_{mm'} \quad , \tag{8-201a}$$

$$\sum_{k_1,\ldots,k_{n-2},\ell} \mathcal{B}^{(k_1\ldots k_{n-2},\alpha)(\ell)}_{i_1\ldots i_n;\quad m}\ \mathcal{A}^{i'_1\ldots i'_n;\ [\ell]}_{(k_1\ldots k_{n-2},\alpha)m} = \prod_{j=1}^{n-2} \delta_{i_j i'_j} \quad . \tag{8-201b}$$

It has already been noticed in Subsec. 8.3.1.3,iii) that the allowed couplings $k_1,\ldots,k_{n-2}$, and ℓ are related to the symmetry properties. Let us make this explicit for a tensor of order two. It is first to be remarked that the symmetry properties of a tensor defined in Subsec. 8.3.1.3 in terms of the symmetry properties of its spherical components imply the same properties for its Cartesian components. Hence, for any tensor $T=T^{11}$ decomposed as in Eq. (8–65) one has :

$$S = (S_{ij} = \tfrac{1}{2}(T_{ij}+T_{ji})) = S^{11} \quad , \quad A = (A_{ij} = \tfrac{1}{2}(T_{ij}-T_{ji})) = A^{11} \quad . \tag{8-202}$$

The decomposition of this tensor into its irreducible tensors reads, cf. Eqs. (8–66a,b) :

$$T = T^{11} = (T^{\{11\}0},\ T^{\{11\}1},\ T^{\{11\}2}) = T_0 + T_1 + T_2 \quad . \tag{8-203}$$

From Eq. (8–67) one gets :

$$S = T_0 + T_2 \quad , \quad A = T_1 \quad . \tag{8-204}$$

This result follows from the symmetry property of the C.G. coefficients, Eq. (8–67), which also implies :

$$T^{\{11\}[\ell]}_m = T_{ij}\ \mathcal{A}^{ij;[\ell]}_m \quad , \quad T_{ij} = \sum_\ell T^{\{11\}[\ell]}_m\ \mathcal{B}_{ij;m}^{(\ell)} \quad , \tag{8-205a}$$

$$\mathcal{A}^{ij;[\ell]}_m = c^2 \langle 11m-n\ n|\ell m\rangle\ t^i_{m-n}\ t^j_n = (-1)^\ell\ \mathcal{A}^{ji;[\ell]}_m \quad , \tag{8-205b}$$

$$\mathcal{B}_{ij;m}^{(\ell)} = c^{-2}\langle 11m-n\ n|\ell m\rangle\ (t^{-1})^{m-n}_i\ (t^{-1})^n_j = (-1)^\ell\ \mathcal{B}_{ji;m}^{(\ell)} \quad . \tag{8-205c}$$

From the above equations and the special values of the C.G. coefficient $\langle 11{-}mm|00\rangle$, Eq. (G–4b), one gets :

$$T^{\{11\}[0]}_0 = T^{\{11\}0} = \frac{c^2}{\sqrt{3}} \sum_i T_{ii} = \frac{c^2}{\sqrt{3}}\ \mathrm{Tr}\ T \quad , \tag{8-206a}$$

$$(T^{\{11\}0},\ 0\ ,\ 0) = \frac{c^{-2}}{\sqrt{3}}\ T^{\{11\}0}\ \mathbf{1} \quad , \tag{8-206b}$$

where $\mathbf{1}$ denotes the tensor of order two with Cartesian components (δ_{ij}). Combining these two results for a given tensor T, one obtains :

$$T_0 = (T^{\{11\}0},\ 0\ ,\ 0) = \tfrac{1}{3}\ \mathrm{Tr}\ T\ \mathbf{1} \quad , \tag{8-207a}$$

and thus :

$$T_2 = (0\,,\,0\,,\,T^{\{11\}2}) = S - \frac{1}{3}\,\mathrm{Tr}\,T\,\mathbf{1} \quad . \tag{8-207b}$$

Finally, the identity of the two decompositions of any tensor of order two and ranks one is shown :

$$T = T_0 + T_1 + T_2 = \frac{1}{3}\,\mathrm{Tr}\,T\,\mathbf{1} + A + (S - \frac{1}{3}\,\mathrm{Tr}\,T\,\mathbf{1}) \quad . \tag{8-208}$$

Furthermore, a direct calculation from Eqs. (8–205a,b) and Table G.2 yields :

$$T^{\{11\}[1]}_{\pm 1} = \pm\,\frac{c^2}{2}\,[i\,A_{23} \mp A_{31}]\;,\; T^{\{11\}[1]}_{0} = -\,\frac{ic^2}{\sqrt{2}}\,A_{12} \quad , \tag{8-209a}$$

$$\left\{\begin{aligned} T^{\{11\}[2]}_{\pm 2} &= -\,\frac{c^2}{2}\,(S_{11} - S_{22} \pm i2\,S_{12}) \\ T^{\{11\}[2]}_{\pm 1} &= \pm\,c^2\,(S_{13} \pm iS_{23}) \\ T^{\{11\}[2]}_{0} &= \frac{c^2}{\sqrt{6}}\,(S_{11} + S_{22} - 2S_{33}) \end{aligned}\right. \tag{8-209b}$$

The expression of $T^{\{11\}[2]}_{m}$ in terms of $(T_2)_{ij}$ is not unique since from Eq. (8–207b) T_2 is traceless.

8.5.4. SPECIAL CASE OF TENSORS OF ORDER AND RANK ONE, VECTORS

i) By definition a <u>vector</u> v is a tensor v^1 of order and rank one. Therefore, it can be defined by any one of its three sets of components, i.e. the Cartesian ones (v_i), the contravariant spherical ones $(v^{[1]}_m)$, or the covariant spherical ones $(v^{(1)}_m)$, which are related by :

$$v^{[1]}_m = v_i\,c\,t^i_m \quad , \quad v_i = v^{[1]}_m\,c^{-1}(t^{-1})^m_i \quad , \quad v^{(1)}_m = (-1)^{1+m}\,v^{[1]}_{-m} \quad , \tag{8-210}$$

where c is any arbitrary complex number and the matrices t and t^{-1} are given in Eq. (8–183a). Hence, one explicitly has :

$$\left\{\begin{aligned} v^{[1]}_1 &= v^{(1)}_{-1} = c\,\frac{1}{\sqrt{2}}\,(iv_1 - v_2) = c\,\frac{i}{\sqrt{2}}\,v_+ \\ v^{[1]}_0 &= -v^{(1)}_0 = c\,(-i\,v_3) \\ v^{[1]}_{-1} &= v^{(1)}_1 = c\,\frac{-1}{\sqrt{2}}\,(iv_1 + v_2) = c\,\frac{-i}{\sqrt{2}}\,v_- \end{aligned}\right. \qquad \left\{\begin{aligned} v_1 &= c^{-1}\,\frac{i}{\sqrt{2}}\,(-v^{[1]}_1 + v^{[1]}_{-1}) \\ v_2 &= c^{-1}\,\frac{-1}{\sqrt{2}}\,(v^{[1]}_1 + v^{[1]}_{-1}) \\ v_3 &= c^{-1}\,i\,v^{[1]}_0 \quad . \end{aligned}\right. \tag{8-211}$$

These components are characterized by their laws of transformation under the mapping $v \to {}^R v$ for any rotation matrix R in SO(3) :

$${}^R v_i = v_j\,R^j_i \quad , \quad {}^R v^{[1]}_m = v^{[1]}_{m'}\,\mathcal{D}^1_{m'm}(R) \quad , \quad {}^R v^{(1)}_m = v^{(1)}_{m'}\,\mathcal{D}^{1*}_{m'm}(R) \quad . \tag{8-212}$$

Furthermore, it is recalled that they can be either components or basis vectors of the usual tensor algebra.

ii) Let us now show the advantages of the choice b=-i in Eq. (8–182) and c=1 in the above equations.

1–It has already been noticed that $|b|=1$ ensures the unitary character of the matrix t. Hence, if for instance the considered objects are basis vectors, (e_i) and $(e^{[1]}_m)$ then are simultaneously orthonormal bases (the first one with respect to an Euclidean inner product and the second one with respect to its complex extension).

2–The tensor product of any two vectors u and v is a tensor $T = u\otimes v$ of order two and ranks one which can be decomposed into its irreducible tensors, Eq. (8–208). The scalar $T^{\{11\}0}$ has to be a multiple of the only invariant one can construct linearly from two vectors, i.e. formally as in the usual tensor algebra, the Euclidean inner product u.v. From Eq. (8–206a) this relation reads :

$$u.v = \sum_i u_i v_i = c^{-2}\sqrt{3}\,T^{\{11\}0} = c^{-2}\sqrt{3}\,\{u^1 \otimes v^1\}^0 \quad . \tag{8-213}$$

Furthermore, it has been shown (notes in Subsec. 2.1.1.3) that in the three–dimensional case the antisymmetrical tensor product of two vectors can be identified with a vector, i.e. their vector product. Hence the vector $T^{\{11\}1}$ is a multiple of the vector product $u\times v$, from Eq. (8–209a) :

$$(u\times v)^1 = c^{-1}\sqrt{2}\,T^{\{11\}1} = c^{-1}\sqrt{2}\,\{u^1 \otimes v^1\}^1 \quad . \tag{8-214}$$

Then, the choice c=1 eliminates all superfluous factors in Eq. (8–213,214) above. Nevertheless, for the sake of generality we shall keep the factor c in what follows.

Several authors choose c=i, then :

$$v^{[1]}_1 = \frac{-1}{\sqrt{2}}(v_1 + iv_2) \quad , \quad v^{[1]}_0 = v_3 \ , \ v^{[1]}_{-1} = \frac{1}{\sqrt{2}}(v_1 - iv_2) \quad , \tag{8-215}$$

cf. Edmonds[22] Eq. (5.1.4), Messiah[39] Eq. (XIII.124), Rose[54] Eq. (4.38), de Shalit and Talmi[56] Sec. 10, and Wigner[63] Eq. (15.34). Fano and Racah[24], Eq. (5.15), take c=-1.

The spherical components of the tensor $\{u^1\otimes v^1\}^2$ are given in Eq. (8–209b) where $s_{ij} = \frac{1}{2}(u_i v_j + u_j v_i)$.

iii) Let D [SU(2)] be any differentiable representation and J_i the infinitesimal generators which are related to the σ matrices by Eq. (5–8) (or more generally a multiple of these). The J_i then form a basic example of Cartesian components of a vector denoted by $J=J^1$. Indeed, it has been shown that the σ matrices fulfil Eq. (4–35) and therefore the J_i satisfy Eq. (5–45), cf. Subsec. 5.4.3,ii) :

$$\forall U \in SU(2) \qquad J \rightarrow {}^U J \qquad {}^U J_i = D(U) J_i\, D(U)^{-1} = J_j\, R(U)^j_i \quad . \tag{8-216}$$

Hence, from Eq. (8–211) the contravariant and covariant spherical components of the vector J read :

$$J^{[1]}_{\pm1} = J^{(1)}_{\mp1} = c\,\frac{\pm1}{\sqrt{2}}\,(iJ_1 \mp J_2) = c\,\frac{\pm i}{\sqrt{2}}\,J_\pm \quad , \quad J^{[1]}_{0} = -J^{(1)}_{0} = c(-iJ_3) \quad . \qquad (8\text{-}217)$$

One should note that the infinitesimal generators $\bar{J}_i$ of the conjugate complex representation D^* [SU(2)] are also the Cartesian components of a vector $\bar{J}$ related to J by, Eqs. (8–33a–c) :

$$C = i\sigma_2 \in SU(2) \quad , \quad \bar{J} = {}^{C}J = {}^{R(\pi,e_2)}J \Rightarrow \bar{J}^{[1]}_{m} = (-1)^{1+m}\,J^{[1]}_{-m} = J^{(1)}_{m} \quad . \qquad (8\text{-}218)$$

In terms of the contravariant spherical components the commutation relations take the form :

$$[J^{[1]}_{0}\ ,\ J^{[1]}_{\pm1}] = c\ (\mp i\,J^{[1]}_{\pm1}) \quad , \qquad (8\text{-}219a)$$

$$[J^{[1]}_{1}\ ,\ J^{[1]}_{-1}] = c\ i\ J^{[1]}_{0} \quad . \qquad (8\text{-}219b)$$

iv) It is recalled that the contravariant spherical components of any tensor T^j satisfy the characteristic equations of a standard basis, Eqs. (8–51a,b). For a vector v^1 these relations read :

$$J_\pm\, v^{[1]}_{m} = \sqrt{2-m(m\pm1)}\; v^{[1]}_{m\pm1} \quad , \quad J_3\, v^{[1]}_{m} = m\, v^{[1]}_{m} \quad . \qquad (8\text{-}220)$$

Let us now establish that the Cartesian components (v_i) of v^1 fulfil the equation :

$$J_i\, v_j = i \sum_k \varepsilon_{ijk}\, v_k \quad . \qquad (8\text{-}221)$$

This relation can be obtained either from Eq. (8–220), expressing the spherical components in terms of the Cartesian ones by means of Eq. (8–211), or directly from the law of transformation of a vector. Indeed, from Eq. (8–212) and then using Eq. (E–10), one finds :

$$(\mathbf{1} - i\varepsilon J_i)v_j = v_k\, R(\varepsilon,e_i)^{k}_{j} = v_k(\delta^{k}_{j} + \varepsilon\,\varepsilon_{ijk}) \quad , \qquad (8\text{-}222)$$

thereby obtaining the announced result.

A summary of formulae for this chapter is given at the end of the next chapter, together with some graphical representations.

CHAPTER 9

GRAPHICAL REPRESENTATIONS OF TENSOR ALGEBRA AND COUPLINGS

Sec. 9.1 is devoted to the definition and to some applications of a general graphical representation of the tensor algebra developed in the foregoing chapter. A less fundamental but simpler graphical representation of couplings and recoupling transformations is discussed in Sec. 9.2.

9.1. GRAPHICAL REPRESENTATION OF THE SPHERICAL TENSOR ALGEBRA

Let us now describe the main features of a graphical representation of the spherical tensor algebra based on the interpretation of changes of tensorial character and couplings in terms of contracted tensor products (Subsec. 8.3.3.3). As for any graphical representation, at least in a first step, one has to define graphs which are in a one-to-one correspondence with the algebraic expressions. Hence, these graphs have to contain neither more nor less information than the corresponding algebraic expressions. Actually, there exist several ways to represent these informations. We follow here the notation due to Levinson[33] (cf. also Yutsis, Levinson, and Vanagas[64]). For more general graphical representations cf. for instance El-Baz[23] and Varshalovich, Moskaliev, and Khersonskii[59].

9.1.1. GRAPHICAL RULES

i) Tensors and components. Any spherical tensor is characterized by (Subsec. 8.3.1.2,i)) : (1) its nature, (2) its order n, and if n is greater than one, the order of the ranks is a priori meaningful, (3) its ranks. Furthermore, to characterize any one of its spherical components one needs in addition to specify : (4) the tensorial character of each index, and (5) the choice of a triad. Forgetting the triad, which is always assumed for simplicity to be the unity of SU(2), the other information can be represented as follows :

Rule 1. A spherical tensor of order n is graphically denoted by :
(1) an n-line vertex which is labelled by a letter (the name of the tensor),
(2) the lines are labelled from 1 to n in the order of the ranks ,
(3) each line carries in addition the index j of the rank.
Moreover, for spherical components :
(4) each line is oriented by an arrow which points toward the vertex if the index is covariant

and in the opposite sense if it is contravariant, the indices m are usually not expressed.

As an example, for any tensor $T^{jj'j''}$ one has :

$T^{jj'j''} =$ 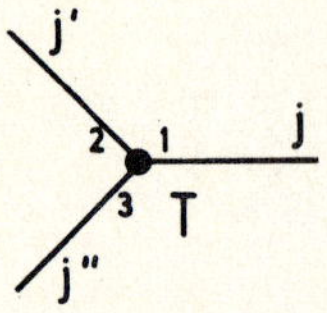

$, \quad T^{(j)(j')[j'']}_{m\ m'\ m''} =$ 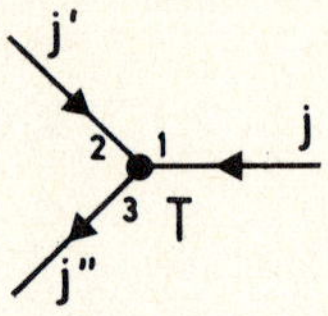

. (9-1)

ii) **Tensor products.** Since the tensor product is a priori noncommutative, the order of the different tensors in the product must be specified. This can be done by labelling each vertex in the order of the corresponding tensors. It seems to us more convenient to use the following rule :

Rule 2. A tensor product is represented by the different vertices with labelled lines as specified in Rule 1. In addition, Rule 1,(2) is applied to this new tensor defined by the tensor product, i.e. all the lines are once more labelled in the order of the corresponding ranks in the tensor product, e.g. :

$S^j \otimes T^{j'j''} =$ 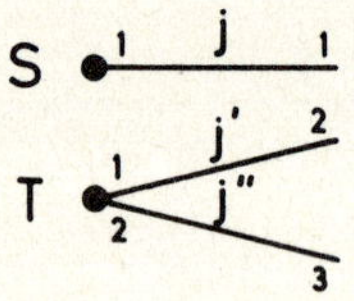

$, \quad T^{j'j''} \otimes S^j =$ 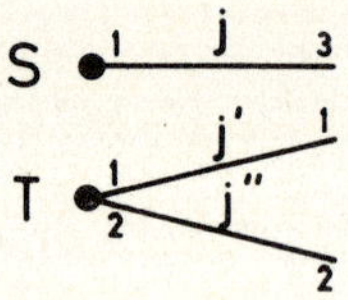

. (9-2)

Actually, this rule has to be used only for noncommutative tensor products, or to identify different expressions of the same tensor, cf. for instance Eq. (9–6).

iii) **C.G., Wigner 3j, and coupling tensors.** A C.G. coefficient and a Wigner 3j symbol are, respectively, represented by :

$\langle jj'mm'|j''m''\rangle = C^{(j)(j')[j'']}_{m\ \ m'\ \ m''} =$ 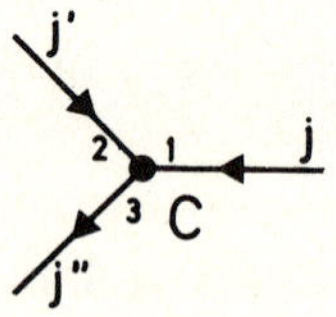

, (9-3a)

$\begin{pmatrix} j & j' & j'' \\ m & m' & m'' \end{pmatrix} = W^{[j][j'][j'']}_{m\ \ m'\ \ m''} =$ 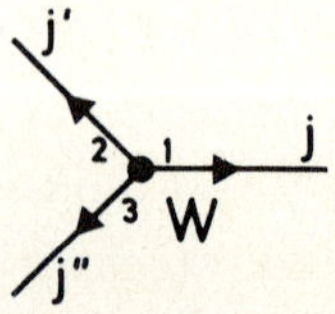

. (9-3b)

The selection rules for the above coefficients, respectively, $m+m'=m''$ and $m+m'+m''=0$ correspond graphically to an algebraic conservation of the indices m at the vertex, their signs being defined by the arrows (this is similar to the current conservation at the vertices of electric nets, or the momentum conservation at the vertices of an usual field theory with a translationally invariant Lagrangian). Indeed, this rule is fulfilled for the above equations, and the change of the tensorial character of any index yields a change of sign of m and at the same time the reversal of the corresponding arrow. More generally, it follows from the

Wigner–Eckart theorem (Subsec. 8.3.5.1) that this conservation law of the indices m holds for any invariant tensor. This is especially the case for the coupling tensors, Eqs. (8–153a,b), denoted by :

$$C^{j_1 \dots j_n\, k_{n-1}}_{(k_1 \dots k_{n-2},\alpha)} = C_{(k_1 \dots k_{n-2},\alpha)} \qquad (9\text{-}4)$$

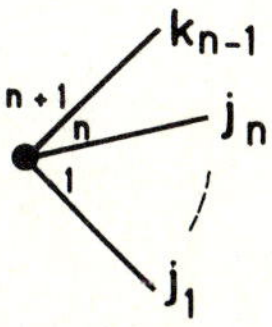

Another invariant tensor frequently encountered is the metric tensor, Eq. (8–98) :

$$G^{jj} = \qquad (9\text{-}5)$$

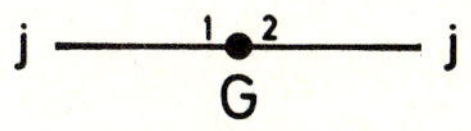

Eq. (8–99) which relates this tensor to a special C.G. tensor then reads :

$$\delta_{jj'} \qquad = c_j\,(-1)^{2j}\,\hat{j} \qquad . \qquad (9\text{-}6)$$

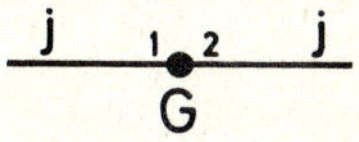

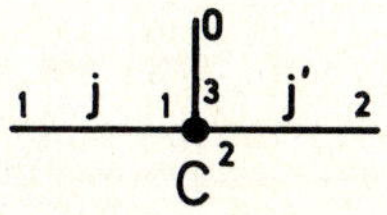

The symmetry properties of the Wigner 3j tensor, Eq. (8–89a), allow us to get rid of some information in its graphical representation. Indeed, in order to define such a tensor one only needs to specify the lateral order of the three indices up to a circular permutation. Following a notation of Yutsis, Levinson, and Vanagas[64)], this can be done graphically by :

Rule 3. For the Wigner 3j tensor an anti–clockwise (resp. clockwise) order of the three indices is denoted by means of a plus (resp. minus) sign at the vertex. Moreover, to simplify the notation, the letter W is omitted :

$$W^{jj'j''} = W^{j'j''j} = W^{j''jj'} = \qquad , \qquad (9\text{-}7a)$$

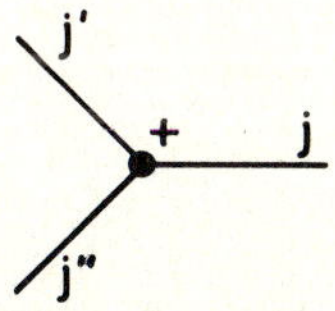

$$W^{jj''j'} = W^{j''j'j} = W^{j'jj''} = \qquad . \qquad (9\text{-}7b)$$

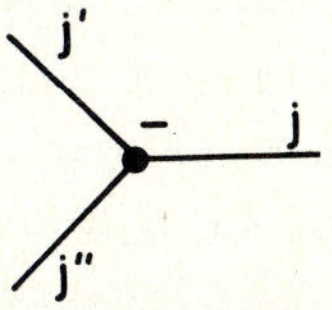

Eq. (8–89a) then reads graphically :

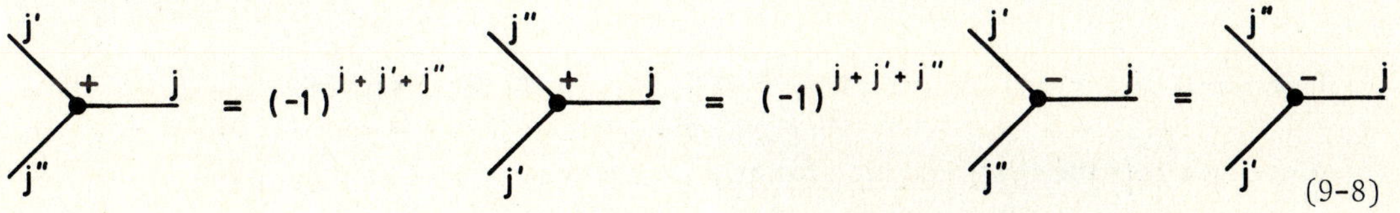

(9–8)

and the relation between the C.G. and Wigner 3j tensors, Eq. (8–86), takes the form :

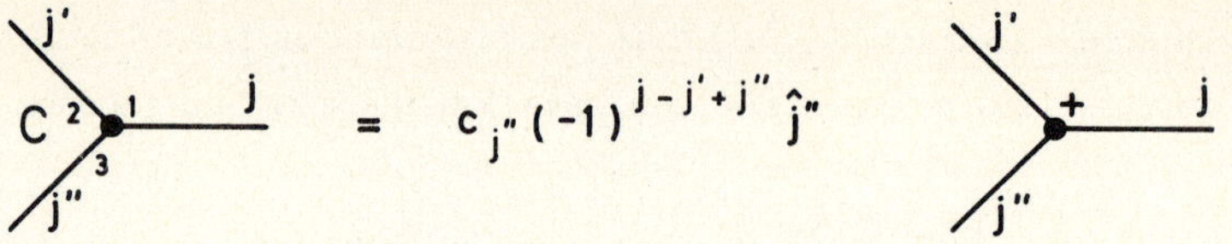

. (9-9)

iv) Contractions. It remains to specify a graphical representation of the contraction of any two equal ranks (Subsec. 8.3.3.1). Since two contracted ranks have different tensorial characters and correspond to the same rank j, the following rule is used :

Rule 4. A contraction in the ith and kth ranks of, respectively, the tensors S and T (possibly S=T) is denoted by :

$S^{\cdots j^{\bullet}\cdots} \otimes T^{\cdots j_{\bullet}\cdots} = \sum_m S^{\cdots [j] \cdots}_{\;\;\;m} \otimes T^{\cdots (j) \cdots}_{\;\;\;m} =$ 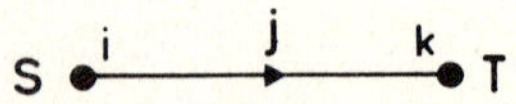

. (9-10)

Eq. (8–102) is then represented by :

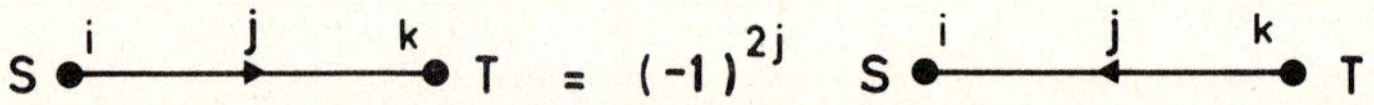

. (9-11)

Consequently, if a C.G., a Wigner 3j, or more generally any coupling tensor has all its lines contracted in a diagram, all the arrows of these lines can then be simultaneously reversed. This follows from the fact that the sum of all the ranks of each of these tensors has to be an integer, otherwise they vanish, e.g. :

. (9-12)

9.1.2. BASIC EXAMPLES AND A DECOMPOSITION THEOREM

Using the four previous rules, the graphical representations of the most important and generic relations of the spherical tensor algebra developed in Chap. 8 are collected at the end of this chapter. Let us only give here the representations of the contraction with the metric tensor in i), and of the Wigner–Eckart theorem in ii). One thus obtains useful reduction formulae which are actually special cases of a general decomposition theorem derived in iii).

i) Contraction with the metric tensor. The graphical representation of Eq. (8–105) is :

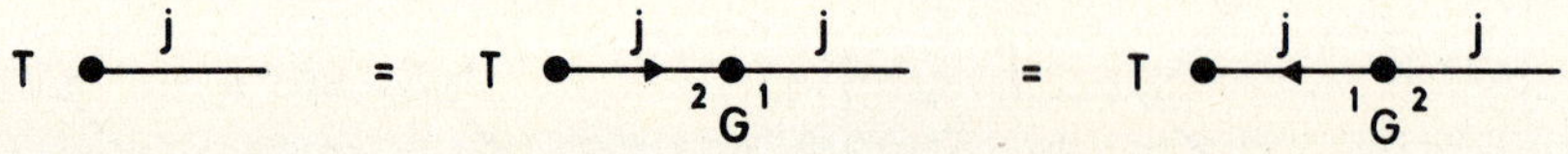

. (9-13)

Therefore, one gets the <u>first reduction formula</u> for the diagrams :

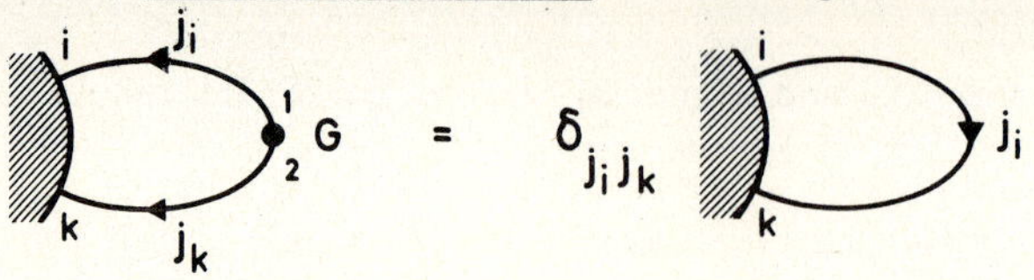

. (9-14)

Consequently, the orthogonality relations of the C.G., Wigner 3j, and more generally the coupling tensors, respectively Eqs. (8–123a,b), (8–129a,b), and (8–153a,b), which are expressed in terms of the metric tensor, provide us with a set of reduction formulae. For instance Eqs. (8–123a,129a) yield :

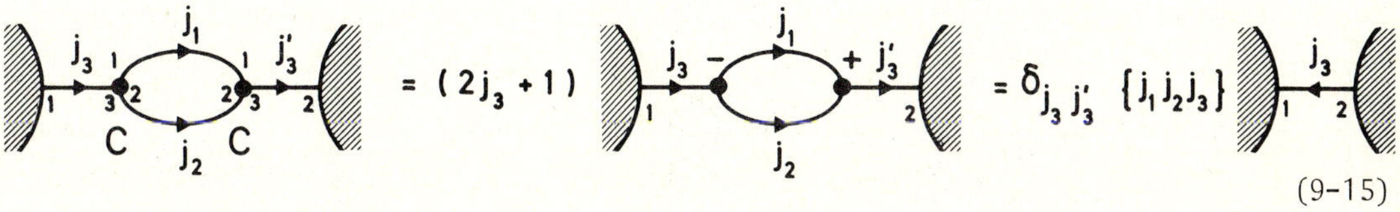

(9–15)

ii) The Wigner–Eckart theorem. For a given order of couplings specified by α, an invariant tensor of order greater than one takes the form given by Eq. (8–155), which is represented by :

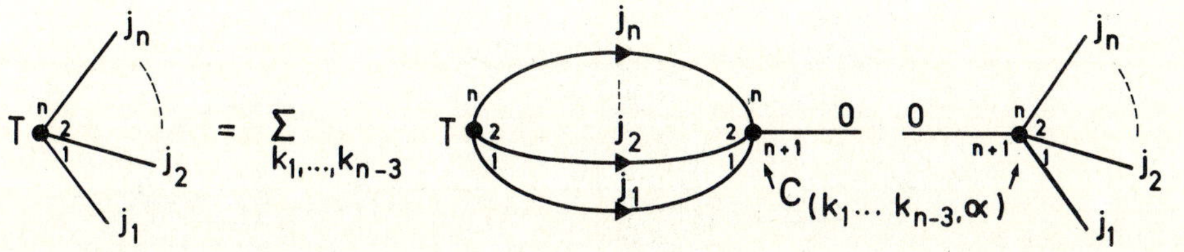

. (9–16)

For the special coupling scheme considered in Eqs. (8–137a,b), one gets in terms of the Wigner 3j tensors for $n \geqslant 3$:

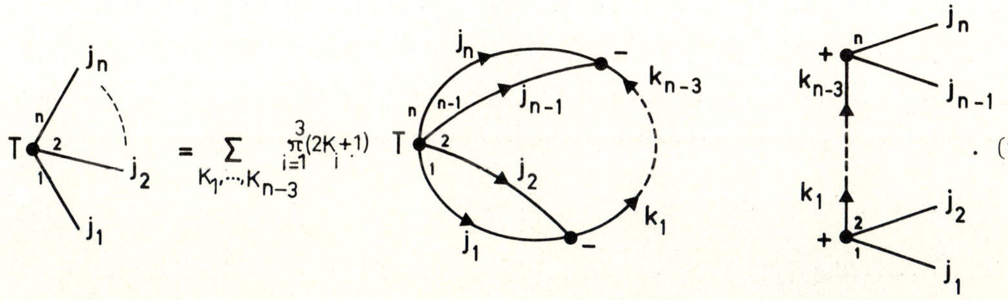

. (9–17)

For instance, an invariant tensor of order three reads, without summation over intermediate couplings, Eq. (8–147). :

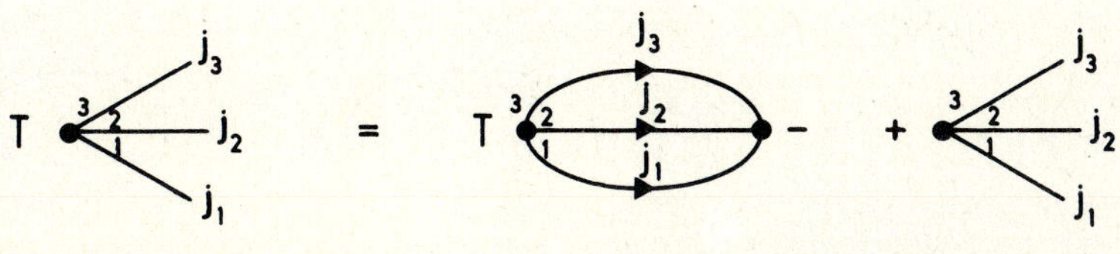

. (9–18)

For an invariant tensor of order two, one gets either from Eqs. (9–6,14, and 16) or directly from Eq. (8–141a) :

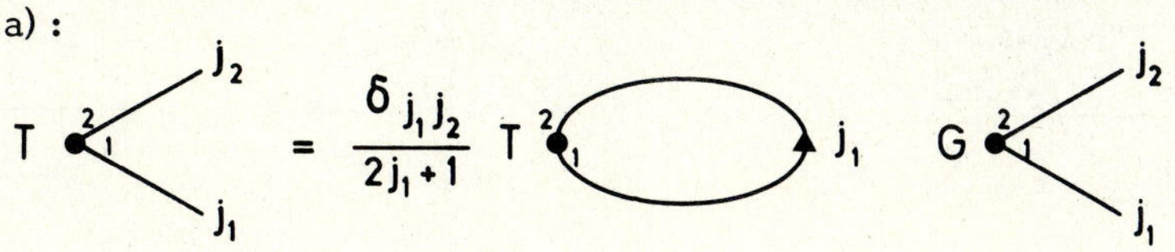

. (9–19)

iii) A decomposition theorem. Let $T^{j_1 \ldots j_n}$ be a tensor of order n, which corresponds to a diagram T with n external lines. This diagram is said to be separable on the p lines $k_1, \ldots, k_p$ when, if these lines are cut, the diagram T breaks into two diagrams T_1 and T_2, one of which, say T_2, contains all the external lines of T, i.e. :

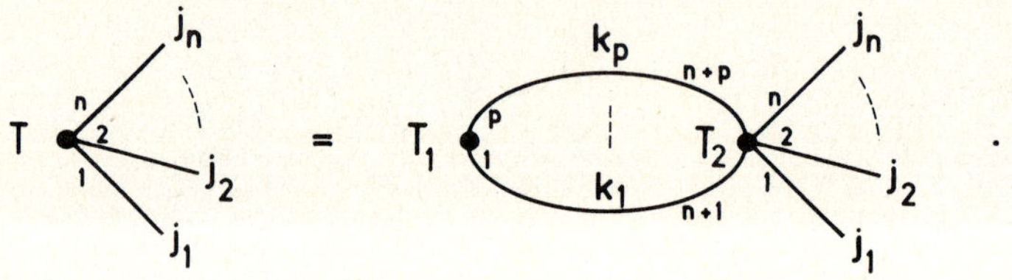

(9–20)

regardless of the orientation of the internal lines $k_1, \ldots, k_p$. Let us furthermore assume that the diagram T_1 corresponds to an invariant tensor $T_1^{k_1 \ldots k_p}$. Applying the Wigner–Eckart theorem to this tensor, Eq. (9–16), then yields a decomposition formula for diagrams :

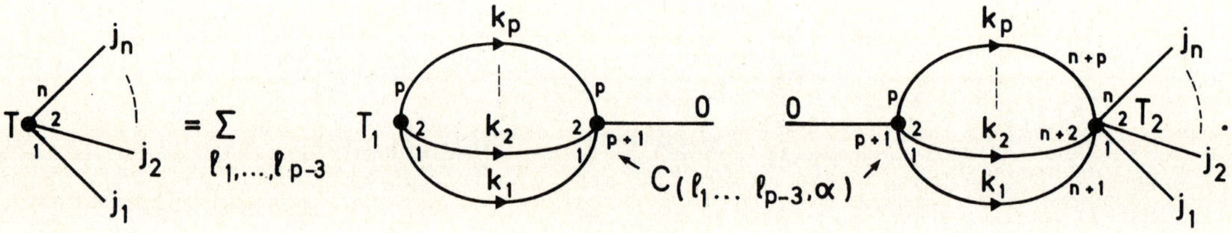

(9–21)

where the orientation of the lines $k_1, \ldots, k_p$ connected to T_2 has to be the same as in Eq. (9–20). It should be noticed that different coupling schemes for the internal lines provide in general different expressions for $T^{j_1 \ldots j_n}$.

If the diagram T is separable on less than four lines there is no index of summation ℓ , and the above decomposition is reduced to a product. One explicitly has :

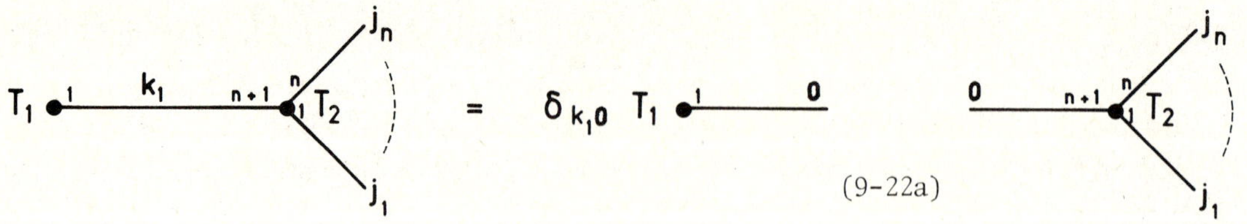

(9–22a)

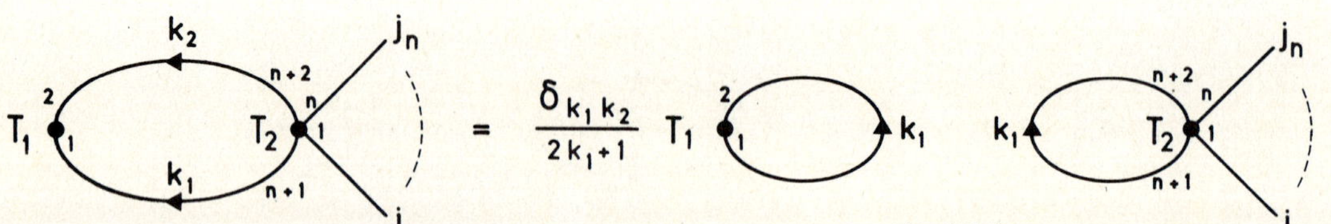

(9–22b)

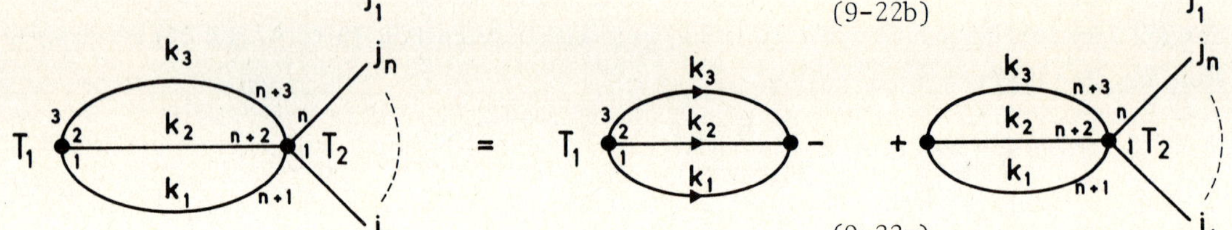

(9–22c)

As an example, let us show that the reduction formula Eq. (9–15) is a special case of the last expressions above. Indeed, one has :

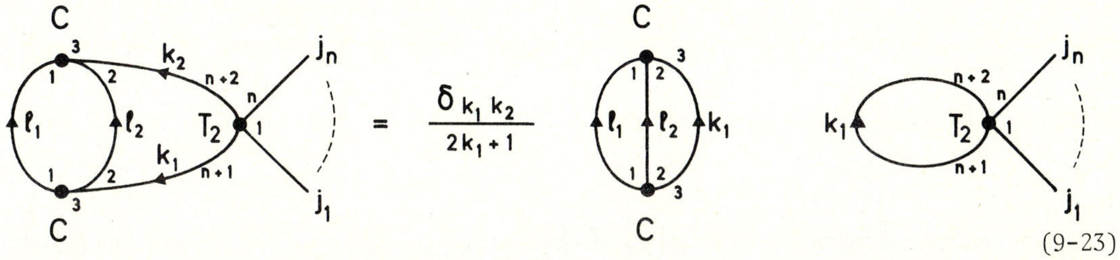

(9–23)

where from Eqs. (8–161,163), cf. also Eq. (9–24) below, the completely contracted tensor product of the two C.G. tensors is equal to $(2k_1+1)\ \{\ell_1\ \ell_2 k_1\}$.

9.1.3. APPLICATIONS

9.1.3.1. Generalities

The basic constituent of the above graphical method is the three line vertex which corresponds to the Wigner 3j tensor. In a general way, this method is well adapted to study contracted tensor products of Wigner 3j tensors and possibly other tensors. The applications are mainly of three types :

i) In order to get symmetry properties of a contracted tensor product using topological deformations and geometrical symmetries of the corresponding diagram, and also Eqs. (9–8,11) to, respectively, change the plus or minus signs at vertices and the orientation of the arrows. This method is for instance used to derive symmetry properties of the 3j, 6j, and 9j symbols in Subsec. 9.1.3.2 below.

ii) To express a contracted tensor product in terms of a sum of products of more simple contracted tensor products. Cf. for instance the expression of a 9j symbol in terms of 6j symbols given in Eq. (9–39).

iii) Inversely, to reduce a sum of products of contracted tensor products, e.g the orthogonality relations and sum rules for the 3nj symbols. Cf. Eq. (9–30) below for the 6j symbols.

These last two applications follow from the decomposition theorem given in Subsec. 9.1.2,iii). It should be noticed that for these two types of applications another graphical representation described in Sec. 9.2 is often much more simple.

9.1.3.2. Application to the 3nj symbols

i) Definition and general remarks. A 3nj symbol (n=1,2,...) is a completely contracted tensor product of 2n Wigner 3j tensors, which cannot be factorized. It is thus an invariant quantity (or a scalar), which depends upon 3n arguments. As shown in Subsec. 8.4.2.1,ii), a 3nj symbol is a multiple of a recoupling coefficient which relates two different coupling schemes of n+1 indices.

The graphical representation of a 3nj symbol is thus a connected diagram constructed from 2n vertices W and without external lines. Any line being in common with two vertices, and any vertex having three lines, the number V of vertices is related to the number L of lines by :

$$V = 2n \quad , \quad 3V = 2L \quad \Rightarrow \quad L = 3n \quad ,$$

i.e. one line for each index j. Actually, for a given n, among all the diagrams which can be drawn as specified above, only those which are not separable on less than four lines correspond to new 3nj symbols. Indeed, if it is not so, using the decomposition theorem derived in Subsec. 9.1.2,iii) the diagram can be factorized into (at least) two diagrams which both correspond to simpler coefficients, cf. Eqs. (9–22a–c). Furthermore, it is recalled that in the considered diagrams each line is labelled by an index j and oriented by an arrow, and a plus sign or minus sign is assigned to each vertex W. Nevertheless, using Eqs. (9–8,11), two diagrams which only differ by the signs at vertices or the orientations of lines, respectively, correspond to algebraic expressions related by a sign. Therefore, two such diagrams are not considered as defining different 3nj symbols. Hence, for a given n, the number of distinct 3nj symbols is determined by the number of topologically different connected diagrams which can be drawn without external lines, unlabelled internal lines, 2n unlabelled vertices with three lines, and which are not separable on less than four lines. All these diagrams can be obtained either directly, cf. Budride and Gutman[11)], or by induction from the diagrams for n–1, cf. Levinson and Chiplis[34)]. For a detailed study of this problem cf. also Yutsis, Levinson, and Vanagas[64)] loc. cit..

A symmetry property of a 3nj symbol is by definition any permutation on the 3n arguments which leaves the value of the symbol unchanged up to a sign which may depend upon the arguments. The set of these symmetry properties is clearly in a one–to–one correspondence with the subset of permutations on the 3n labels of the lines of the diagram which leave it topologically invariant without taking the arrows and the signs into account. This subset is a subgroup of the symmetry group Σ_{3n}, called the <u>subgroup of automorphisms of the diagram</u> with labelled lines and unlabelled vertices. From Prop. C.22, the order of this subgroup, and thus the number of symmetry properties of a given 3nj symbol, is a divisor of (3n)!. The change of sign which may occur in a symmetry property follows from the possible change of either the sense of the arrow carried by each line or the sign at each vertex under the symmetry of the diagram, cf. Eqs. (9–8,11). Let us remark that, since j+j'+j" has to be an integer otherwise the Wigner 3j tensor $W^{jj'j''}$ vanishes and any line being in common with two vertices, the value which corresponds to the labelled diagram of a 3nj symbol is left invariant under the following transformations (and possibly others) : (1) the simultaneous reversal of the three arrows of the lines ending on any vertex, Eq. (9–12), (2) the simultaneous change of signs at all the vertices, Eq. (9–8).

Let us now make explicit the foregoing remarks for the frequently encountered cases n=1,2, and 3, and only touch the case n=4.

ii) **The 3j symbols.**

1–Definition. For n=1, and thus two vertices, there exists only one unlabelled diagram :

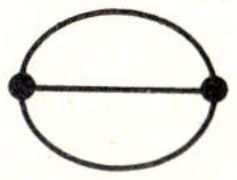

It has already been shown, Eqs. (8–161,163), that this diagram is related to the triangular delta defined by Eq. (6–67), i.e. :

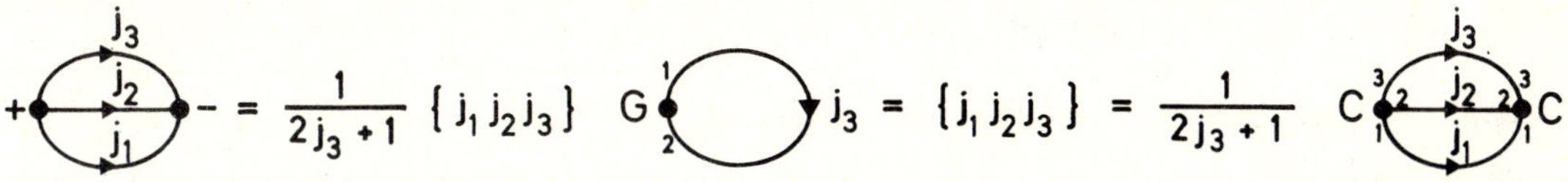

(9-24)

By definition the 3j symbol is the triangular delta $\{ j_1 j_2 j_3 \}$.

It should be noticed that exceptionally in the case n=1 the diagram is separable on three lines. Applying the reduction theorem, Eq. (9–22c), one finds that the value x of the 3j symbol is equal to x^p for all positive integer values of p. Hence, x=0 or 1 as obtained above.

2–Symmetry properties. From the general remarks, one can in the above diagram change simultaneously the two signs ; similarly for the three arrows, i.e. :

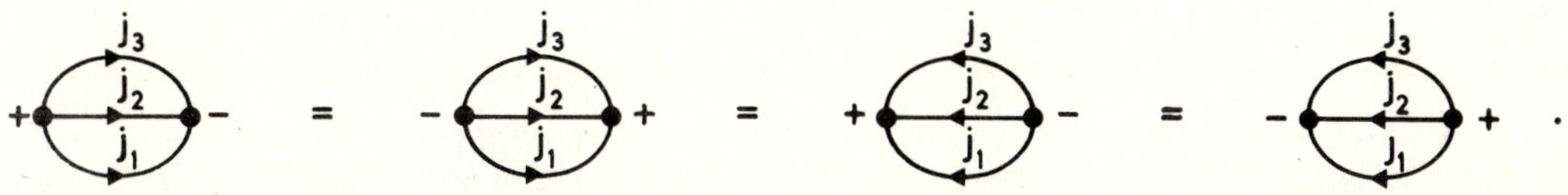

(9-25)

Therefore, a 3j symbol is invariant under any permutation of its three arguments.

iii) **The 6j symbols** of Wigner (Subsec. 8.4.2.2) :

1–Definition. For n=2, and thus four vertices, there exist two distinct unlabelled diagrams :

Only the last one, i.e. the tetrahedron, is not separable on less than four lines. This last diagram corresponds to the 6j symbols. Indeed, using more symmetrical notations in Eq. (8–173a), and then Eqs. (9–8,11) in order to give the diagram the maximum symmetry, one gets the two following convenient representations :

$$\begin{Bmatrix} j_1 & j_2 & j_3 \\ k_1 & k_2 & k_3 \end{Bmatrix} = c_{j_a} c_{k_a} (-1)^{2(j_a+k_a)}$$

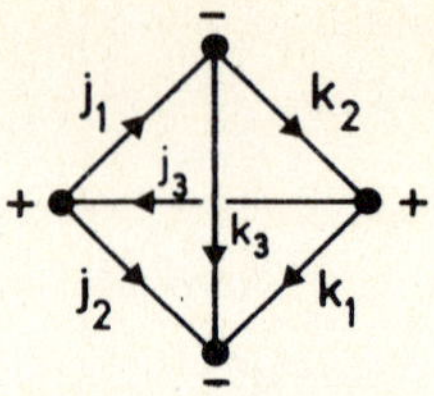

$$= c_{j_a} c_{k_a} (-1)^{2(j_a+k_a)}$$

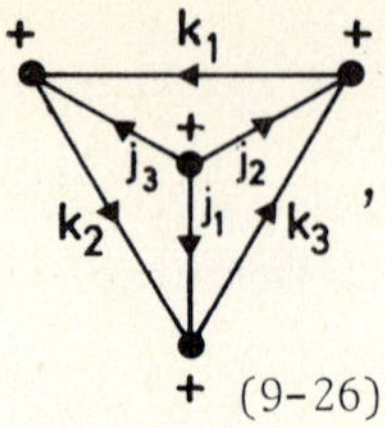

, (9–26)

where, from Eq. (8–84), the phase factor in front of these diagrams is independent of a=1,2,3.

2–Symmetry properties. From the last diagram above, we see that a 6j symbol is invariant under :

(1) any permutation of its columns since :

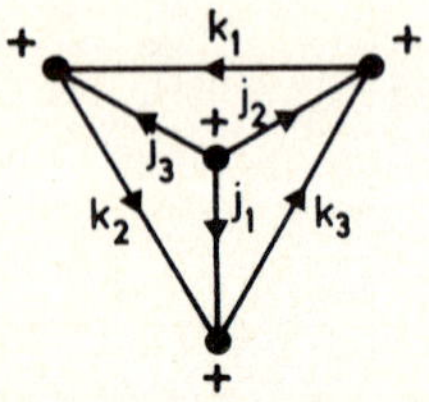

=

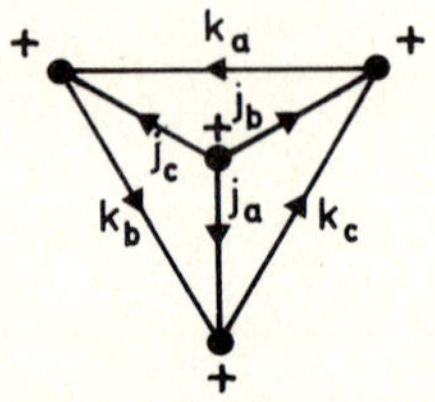

(9–27)

$$\Longrightarrow \begin{Bmatrix} j_1 & j_2 & j_3 \\ k_1 & k_2 & k_3 \end{Bmatrix} = \begin{Bmatrix} j_a & j_b & j_c \\ k_a & k_b & k_c \end{Bmatrix}$$

a, b, c = 1, 2, 3 ,

(2) the interchange of the upper and lower arguments in each of any two columns, for exchanging the central vertex with the lower one :

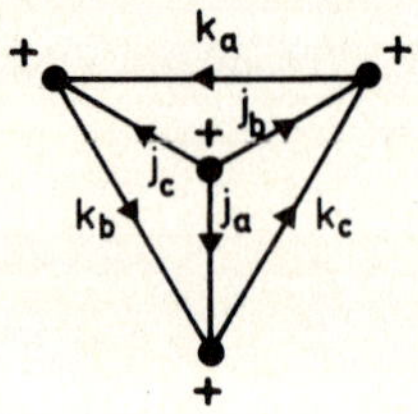

=

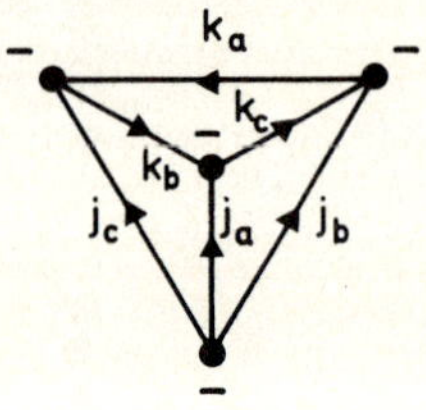

=

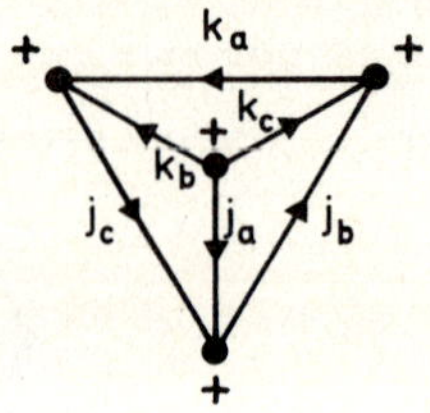

$$\Longrightarrow \begin{Bmatrix} j_1 & j_2 & j_3 \\ k_1 & k_2 & k_3 \end{Bmatrix} = \begin{Bmatrix} j_a & k_b & k_c \\ k_a & j_b & j_c \end{Bmatrix} \quad , \quad a,b,c = 1,2,3 \quad . \qquad (9\text{–}28)$$

These are the usual symmetry properties of the 6j symbols. They form a group of order 24 which is isomorphic to the symmetry group of a regular tetrahedron (Edmonds[22), Subsec. 6.2), i.e. the subgroup of the automorphisms of this diagram with labelled lines (a subgroup of Σ_6). For other symmetry properties of the 6j symbols cf. Regge[52).

A square 6j symbol, Eq. (8–172a), is a less symmetrical coefficient : the usual symmetry properties of the 6j symbols imply that it is invariant under the interchange of the upper and lower argument in any two columns and the permutation of the first two columns, e.g. :

$$\begin{bmatrix} j_1 & j_2 & j_3 \\ k_1 & k_2 & k_3 \end{bmatrix} = \begin{bmatrix} k_1 & j_2 & k_3 \\ j_1 & k_2 & j_3 \end{bmatrix} \quad . \qquad (9\text{–}29)$$

3–Orthogonality relation and sum rules. Let us derive the orthogonality relation using the decomposition theorem, Eq. (9–22c).

$$\sum_{\ell} (2\ell+1) \begin{Bmatrix} j_1 & j_2 & \ell \\ k_1 & k_2 & k_3 \end{Bmatrix} \begin{Bmatrix} j_1 & j_2 & \ell \\ k_1 & k_2 & k_3' \end{Bmatrix} = \sum_{\ell} (2\ell+1)$$

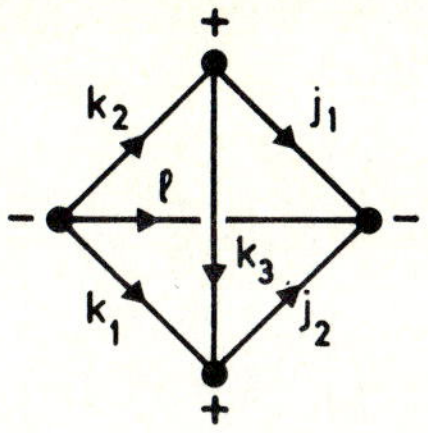

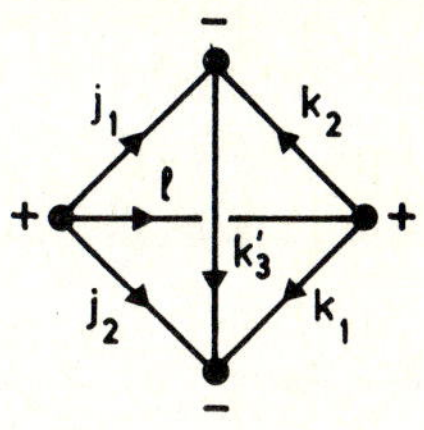

$$= \sum_{\ell} (2\ell+1)$$

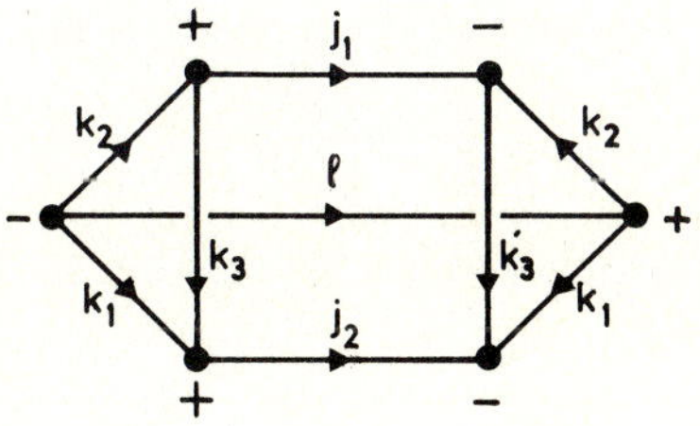

from Eq. (9-22c)

$$=$$

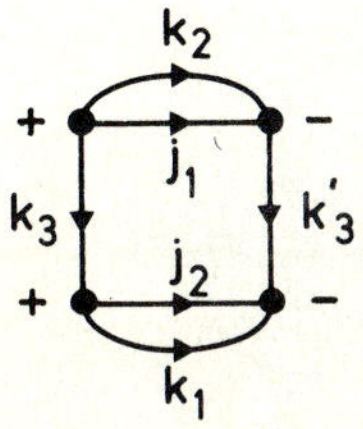

from Eq. (8-129b)

$$= \delta_{k_3 k_3'} \frac{\{j_1\, k_2\, k_3\}}{2k_3+1}$$

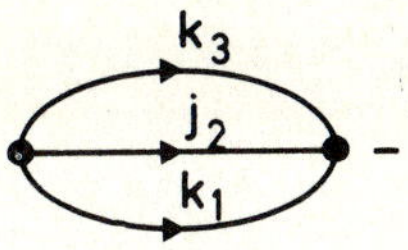

from Eq. (9-15)

$$= \delta_{k_3 k_3'} \frac{1}{2k_3+1} \{j_1\, k_2\, k_3\}\{j_2\, k_1\, k_3\} \qquad \text{from Eq. (9-24)} \quad . \qquad (9\text{-}30)$$

A more simple way to derive the above relation is given in Subsec. 9.2.2 using a second graphical representation. So it is with the sum rules, Eqs. (9–51a,b).

As an exercise, the result of which will be used in the second graphical representation given in the next section, let us show that the recoupling coefficient which relates the two coupling schemes $\{\{j_1j_2\}k_1j_3\}k_2$ and $\{\{j_1j_2\}\}\ell_1j_3\}k_2$ corresponds to the other unlabelled diagram :

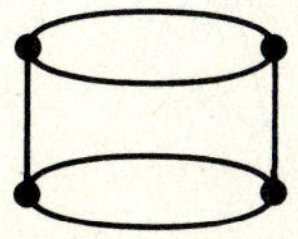

.

Indeed, following the method described in Subsec. 8.4.2.1, one obtains :

$$T^{\{\{j_1j_2\}k_1j_3\}k_2} = c_{k_1}(-1)^{2k_1}\hat{k}_1 \sum_{\ell_1} c_{\ell_1}(-1)^{2\ell_1} W^{k_1\bullet j_3\bullet k_2\bullet} W^{j_1\bullet j_2\bullet k_1^\bullet} \times W^{j_1^\bullet j_2^\bullet \ell_1^\bullet} W^{\ell_1\bullet j_3^\bullet k_2^\bullet} T^{\{\{j_1j_2\}\ell_1j_3\}k_2} \quad .$$

The graphical representation of the contracted tensor product of the four Wigner 3j tensors yields, using Eqs. (9–22c,24) :

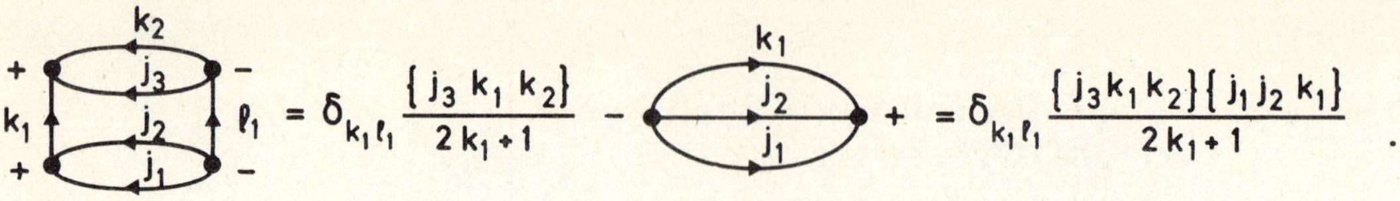

Therefore, the relation between the two considered coupling schemes reads :

$$T^{\{\{j_1j_2\}k_1j_3\}k_2} = \sum_{\ell_1} \delta_{k_1\ell_1} \{j_1j_2k_1\}\{j_3k_1k_2\}\, T^{\{\{j_1j_2\}\ell_1j_3\}k_2} \quad , \qquad (9\text{-}31)$$

which is actually a direct consequence of the orthogonality relations of the C.G. coefficients.

iv) **The 9j symbols** of Wigner (Subsec. 8.4.2.2) :

1–Definition. For n=3, and thus six vertices, there exist five unlabelled diagrams :

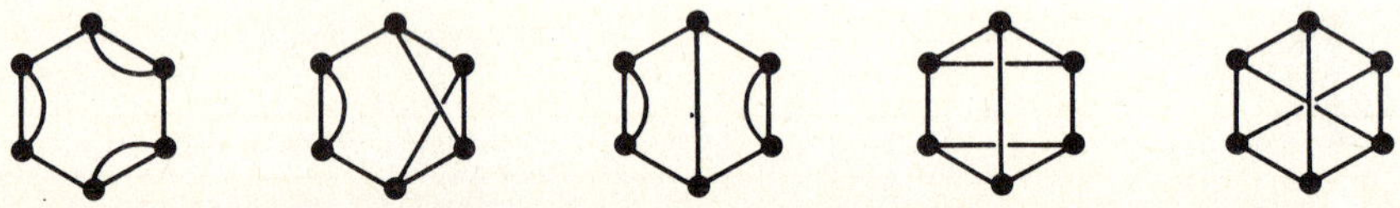

Only the last one is not separable on less than four lines. From Eq. (8–173b) this diagram is related to a 9j symbol in the following way :

$$\begin{Bmatrix} j_1 & j_2 & j_3 \\ k_1 & k_2 & k_3 \\ \ell_1 & \ell_2 & \ell_3 \end{Bmatrix} =$$

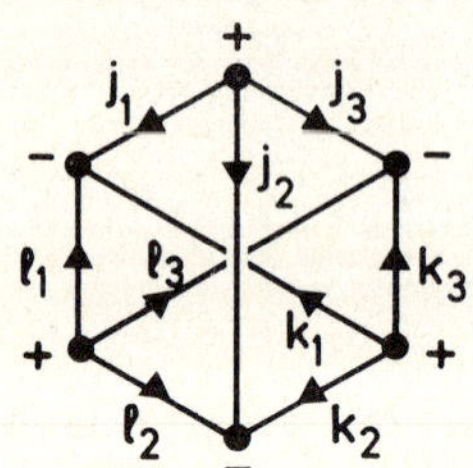

(9–32)

2–Symmetry properties. Using Eqs. (9–8,11) one obtains the useful relations : $s = \sum_{i=1}^{3} j_i+k_i+\ell_i$,

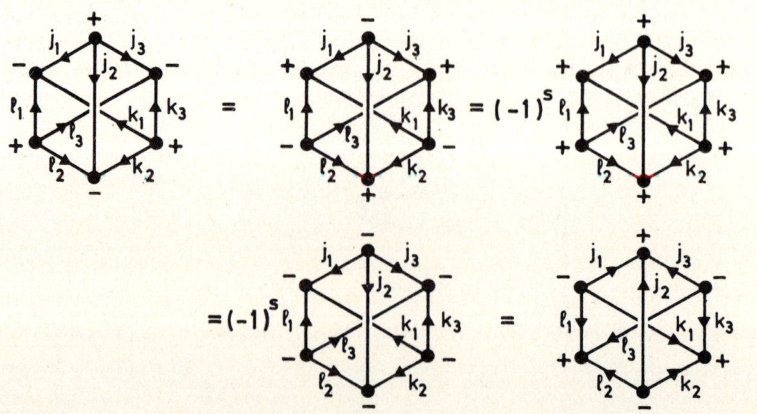

(9–33)

A 9j symbol is then :

(1) multiplied by $(-1)^s$ under the interchange of any two rows, for echanging two vertices which carry a plus sign in Eq. (9–32), e.g. :

$$\begin{Bmatrix} j_1 & j_2 & j_3 \\ k_1 & k_2 & k_3 \\ \ell_1 & \ell_2 & \ell_3 \end{Bmatrix} =$$

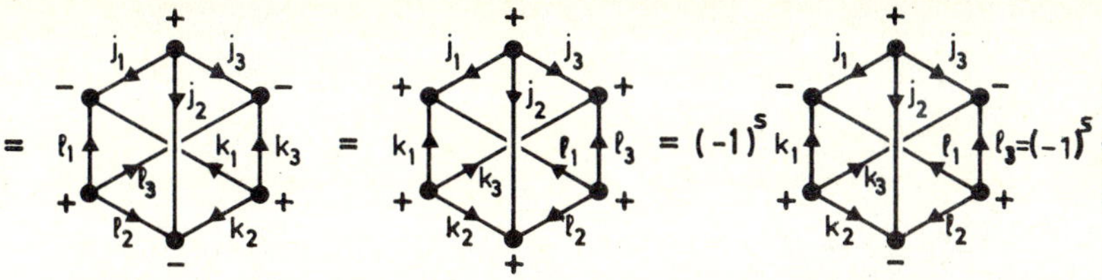

$$= (-1)^s \begin{Bmatrix} j_1 & j_2 & j_3 \\ \ell_1 & \ell_2 & \ell_3 \\ k_1 & k_2 & k_3 \end{Bmatrix} \tag{9-34}$$

(2) invariant under a symmetry with respect to its first diagonal, for by symmetry with respect to the thin dotted line below, one gets :

$$\begin{Bmatrix} j_1 & j_2 & j_3 \\ k_1 & k_2 & k_3 \\ \ell_1 & \ell_2 & \ell_3 \end{Bmatrix} =$$

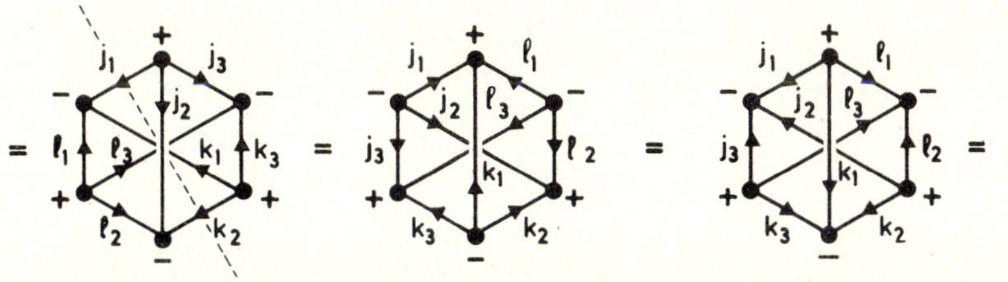

$$= \begin{Bmatrix} j_1 & k_1 & \ell_1 \\ j_2 & k_2 & \ell_2 \\ j_3 & k_3 & \ell_3 \end{Bmatrix} \tag{9-35}$$

Therefore, the 9j symbol is also multiplied by $(-1)^s$ under the interchange of any two columns, and invariant under a symmetry with respect to its second diagonal, i.e. :

$$\begin{Bmatrix} j_1 & j_2 & j_3 \\ k_1 & k_2 & k_3 \\ \ell_1 & \ell_2 & \ell_3 \end{Bmatrix} = \begin{Bmatrix} \ell_3 & k_3 & j_3 \\ \ell_2 & k_2 & j_2 \\ \ell_1 & k_1 & j_1 \end{Bmatrix} \quad . \tag{9-36}$$

The set of the above symmetry properties form a group, isomorphic to the direct product of symmetry groups $\Sigma_3 \times \Sigma_3 \times \Sigma_2$, of order 72 (a subgroup of Σ_9).

Clearly a square 9j symbol, Eq. (8–172b), is a less symmetrical coefficient. It is :

(1) multiplied by $(-1)^s$, where s is the sum of its nine arguments, under the interchange of its first two rows or columns, i.e. :

$$\begin{bmatrix} j_1 & j_2 & j_3 \\ k_1 & k_2 & k_3 \\ \ell_1 & \ell_2 & \ell_3 \end{bmatrix} = (-1)^s \begin{bmatrix} k_1 & k_2 & k_3 \\ j_1 & j_2 & j_3 \\ \ell_1 & \ell_2 & \ell_3 \end{bmatrix} = (-1)^s \begin{bmatrix} j_2 & j_1 & j_3 \\ k_2 & k_1 & k_3 \\ \ell_2 & \ell_1 & \ell_3 \end{bmatrix} \quad , \tag{9-37}$$

(2) invariant under a symmetry with respect to its first diagonal :

$$\begin{bmatrix} j_1 & j_2 & j_3 \\ k_1 & k_2 & k_3 \\ \ell_1 & \ell_2 & \ell_3 \end{bmatrix} = \begin{bmatrix} j_1 & k_1 & \ell_1 \\ j_2 & k_2 & \ell_2 \\ j_3 & k_3 & \ell_3 \end{bmatrix} \quad . \tag{9-38}$$

3–Orthogonality relation and sum rules. These formulae are easily derived in Subsec. 9.2.2 using the second graphical representation.

4–Expression in terms of the 6j symbols. Let us derive this relation as a last example of the application of the decomposition theorem :

$$\begin{Bmatrix} j_1 & j_2 & j_3 \\ k_1 & k_2 & k_3 \\ \ell_1 & \ell_2 & \ell_3 \end{Bmatrix} =$$

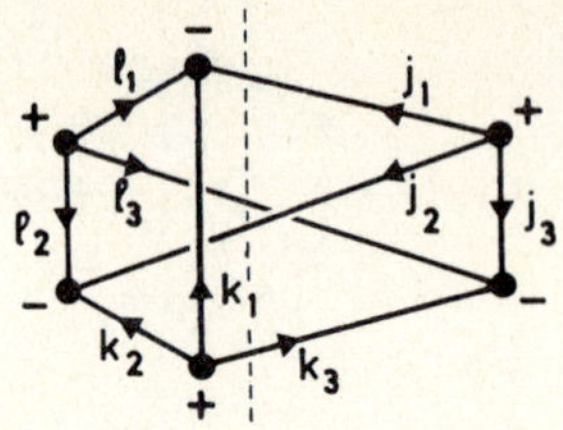

$$= \sum_{\ell} (2\ell+1)$$

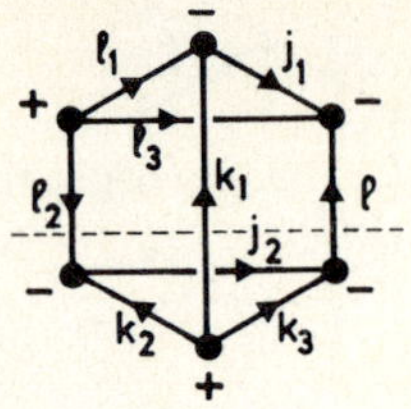

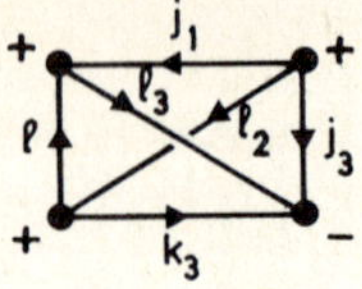

$$= \sum_{\ell} (2\ell+1)$$

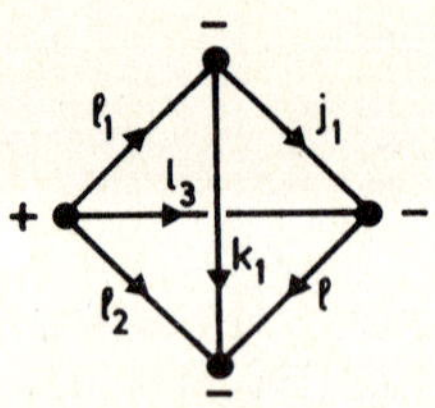

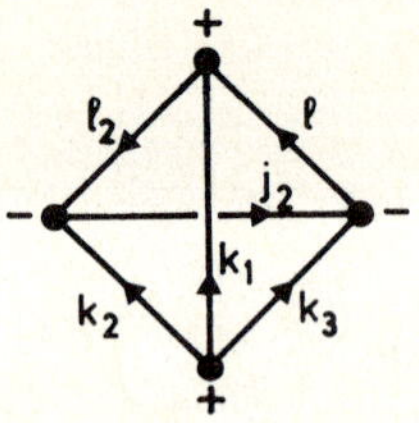

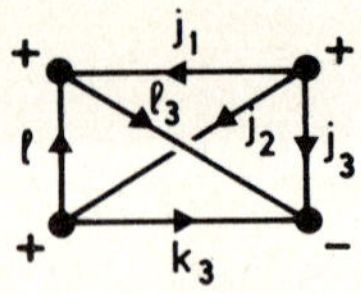

$$= \sum_{\ell} (2\ell+1) \begin{Bmatrix} j_1 & k_1 & \ell_1 \\ \ell_2 & \ell_3 & \ell \end{Bmatrix} \begin{Bmatrix} j_2 & k_2 & \ell_2 \\ k_1 & \ell & k_3 \end{Bmatrix} \begin{Bmatrix} j_1 & j_2 & j_3 \\ k_3 & \ell_3 & \ell \end{Bmatrix} \quad . \qquad (9\text{-}39)$$

where the thin dotted lines show in which way the diagrams are separable, respectively, on the four lines $k_3 j_2 \ell_3 \ell_1$ and on the three lines $\ell_2 k_1 \ell$. Actually, the above expression can also be obtained in a much more simple way thanks to the second graphical representation, Eq. (9–52).

v) **The 12j symbols.** Let us only remark that for n=4, and thus eight vertices, there exist two different unlabelled diagrams which are not separable on less than four lines, they are :

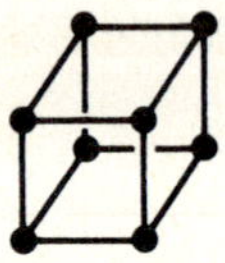

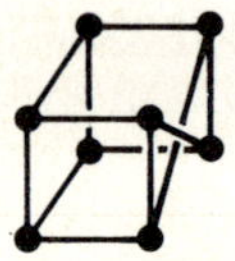

These two diagrams are, respectively, related to the 12j symbols of the first and second kinds, for details cf. for instance Ref.64.

9.2. GRAPHICAL REPRESENTATION OF RECOUPLING TRANSFORMATIONS

The elementary constituant of the foregoing graphical representation of the tensor algebra is the three line vertex W which corresponds to the Wigner 3j symbols. Actually, to study a recoupling transformation it is often much more convenient to take as elementary constituants the square 3j, 6j, and 9j symbols defined, respectively, in Eqs. (8–163, 171a, and b). It can then be helpful to visualize the recoupling transformations thanks to a simple graphical representation due to Danos[20].

It is recalled that for coupled spherical components there exist a priori two independent relevant orders of the ranks : the first one characterizes the tensor itself, and the second one defines the considered coupling scheme (Subsec. 8.3.2.2,ii)). The graphical method described below is especially well adapted to the representation of the coupling schemes. Nevertheless, it can also provide a representation of any change in the order of the ranks characterizing the tensor itself in the special case of a tensorial product of commuting or anticommuting objects. For instance, this can be applied to creation or annihilation operators for, respectively, Bosons and Fermions (Subsec. 11.2.2.2). To this end, one has to take care of the factor ε introduced by the transposition of any two indices, ε being equal to plus one or minus one according to whether the corresponding objects commute or anticommute, e.g. :

$$T^{j_1 j_2} = T_1^{\,j_1} \otimes T_2^{\,j_2} = \varepsilon\, T_2^{\,j_2} \otimes T_1^{\,j_1} = \varepsilon\, T^{j_2 j_1} \quad . \tag{9-40}$$

9.2.1. GRAPHICAL RULES

i) **Rule 1. Representation of coupling schemes.** Each index j is represented by a horizontal line which is labelled by j. The order of the lines from top to bottom corresponds to the order in which the corresponding indices appear from left to right in the coupling schemes as specified in Eq. (8–74) and Subsec. 8.3.2.2,ii). Only two adjacent lines, or two adjacent groups of already coupled lines, can be coupled. The coupling is denoted by a vertical line which crosses the coupled lines and this line is labelled by the value of the coupling, e.g. :

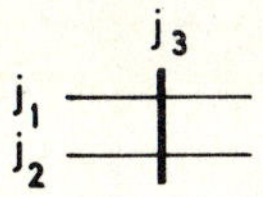

$= \{ j_1\, j_2 \}\, j_3$,

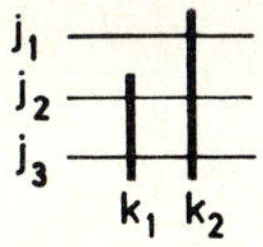

$= \{ j_1 \{ j_2\, j_3 \}\, k_1 \}\, k_2$.　(9-41)

Rule 2. Representation of the elementary recoupling transformations. The only considered recoupling coefficients and their self–explanatory graphical representation are the following :

(1) Generalized orthogonality relation :

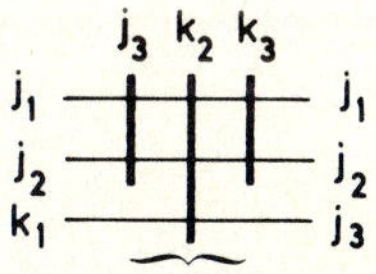

$$= \delta_{j_3 k_3}\ \{ j_1\, j_2\, j_3 \}\ \{ j_3\, k_1\, k_2 \} \quad . \tag{9-42}$$

This relation, which follows from the orthogonality of the C.G. coefficients, has been derived in Eq. (9–31).

(2) The square 3j, 6j, and 9j symbols :

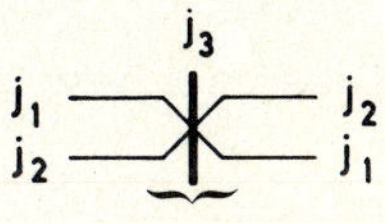

$$= [\, j_1\, j_2\, j_3 \,] = (-1)^{j_1 + j_2 - j_3}\ \{ j_1\, j_2\, j_3 \} \quad , \tag{9-43a}$$

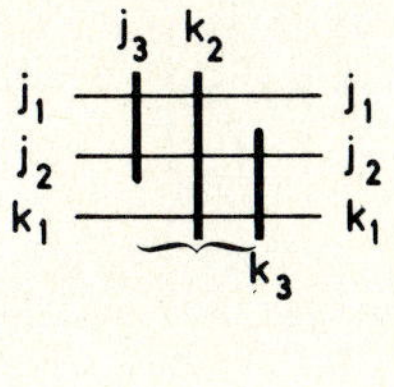

$$= \begin{bmatrix} j_1 & j_2 & j_3 \\ k_1 & k_2 & k_3 \end{bmatrix} = (-1)^{j_1 + j_2 + k_1 + k_2}\ \hat{j}_3\, \hat{k}_3\ \begin{Bmatrix} j_1 & j_2 & j_3 \\ k_1 & k_2 & k_3 \end{Bmatrix} \tag{9-43b}$$

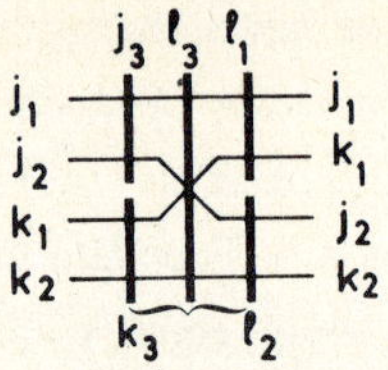

$$= \begin{bmatrix} j_1 & j_2 & j_3 \\ k_1 & k_2 & k_3 \\ \ell_1 & \ell_2 & \ell_3 \end{bmatrix} = \hat{j}_3 \hat{k}_3 \hat{\ell}_1 \hat{\ell}_2 \begin{Bmatrix} j_1 & j_2 & j_3 \\ k_1 & k_2 & k_3 \\ \ell_1 & \ell_2 & \ell_3 \end{Bmatrix} . \qquad (9\text{-}43c)$$

The lines can only cross within the above three types of recouplings. Furthermore, a set of coupled lines acts as a single one in a recoupling coefficient provided its internal couplings remain unchanged, e.g. :

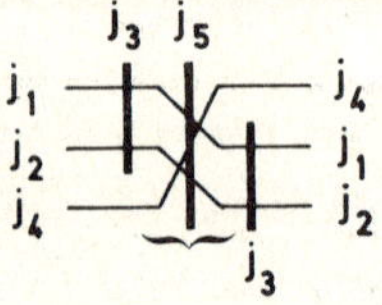

$$= [\, j_3\; j_4\; j_5 \,] \quad . \qquad (9\text{-}44)$$

Actually, the graphical method of Danos loc. cit. represents all the crossings of lines in terms of the only 9j–type recoupling. Indeed, the introduction of fictitious lines, drawn as dashed lines, with j=0 allows us to express the crossings of three and even two indices in terms of the only square 9j symbols, i.e. :

$$[\, j_1\, j_2\, j_3 \,] =$$

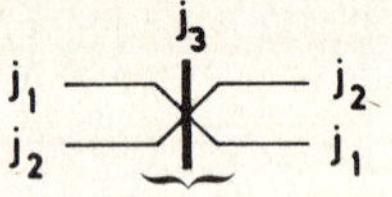

$$=$$

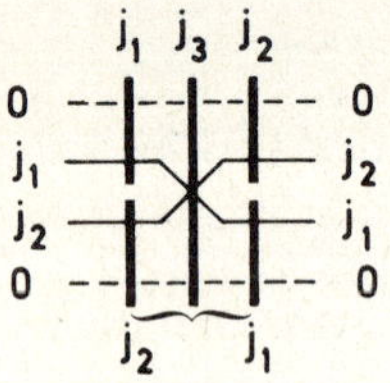

$$= \begin{bmatrix} 0 & j_1 & j_1 \\ j_2 & 0 & j_2 \\ j_2 & j_1 & j_3 \end{bmatrix} , \qquad (9\text{-}45a)$$

$$\begin{bmatrix} j_1 & j_2 & j_3 \\ k_1 & k_2 & k_3 \end{bmatrix} =$$

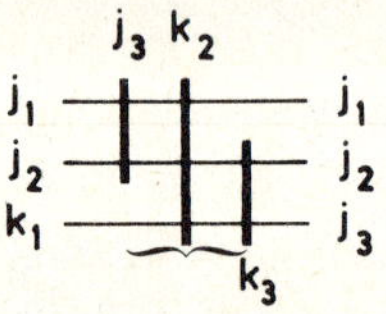

$$=$$

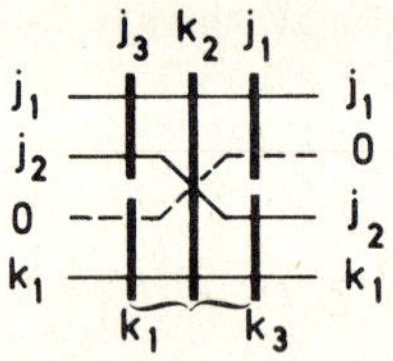

$$= \begin{bmatrix} j_1 & j_2 & j_3 \\ 0 & k_1 & k_1 \\ j_1 & k_3 & k_2 \end{bmatrix} . \qquad (9\text{-}45b)$$

This is the reason why the values of the square 9j symbols with argument(s) equal to zero are listed in Appendix G.3 (Tables G.4,5, and 6).

Rule 3. Representation of the changes of the order of the rank indices which characterizes the tensor. In the only special case where the lines correspond to commuting or anticommuting objects the crossing of any two lines introduces a factor ε equal, respectively, to plus one or minus one. The symbol ε is specified on the diagram near the crossing of the lines, e.g. :

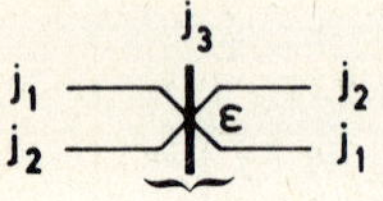

$$= \varepsilon\, [\, j_1\, j_2\, j_3 \,] \; ,$$

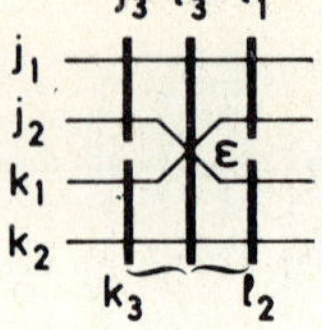

$$= \varepsilon \begin{bmatrix} j_1 & j_2 & j_3 \\ k_1 & k_2 & k_3 \\ \ell_1 & \ell_2 & \ell_3 \end{bmatrix} . \qquad (9\text{-}46)$$

It should be noticed that the above rule does not provide a representation of the order of the ranks, except when this order is the same as the one which defines the coupling scheme, e.g. Eq. (8–73) :

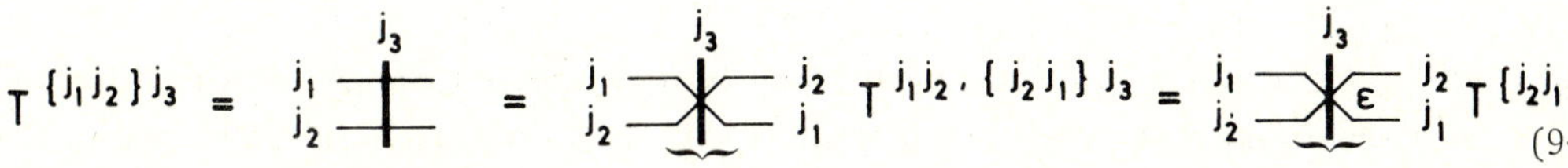

(9-47)

assuming Eq. (9–40) is satisfied.

Rule 4. Interpretations of diagrams. Any diagram can be considered either as a graphical representation of the equation which expresses the left coupling scheme of the diagram in terms of the right one, or as a graphical representation of the recoupling coefficient which relates these two coupling schemes. In both cases, the algebraic expression which corresponds to a graph is obtained by writing the product of all the recoupling coefficients. The new intermediate couplings have to be summed.

ii) **Symmetry property of the diagrams.** Let us show that the reading of any diagram from left to right as it must be by definition (Rule 4) or from right to left yields equal recoupling coefficients. Indeed, this property holds for the diagram which represents a square 9j symbol due to the symmetry of this coefficient with respect to its first diagonal, Eq. (9–38) :

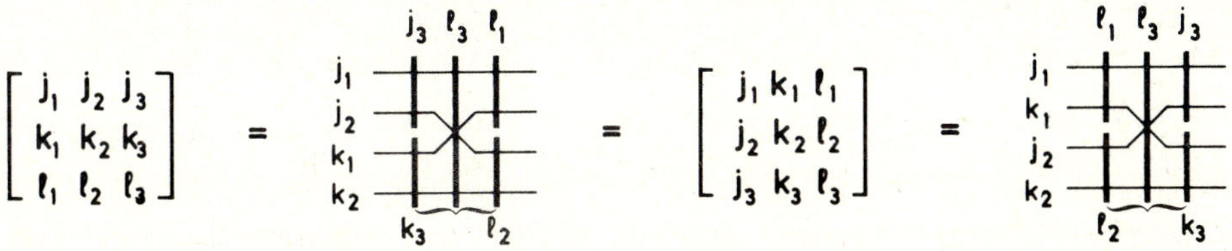

. (9-48)

From this result and the relation between the square 6j and 9j symbols given by Eq. (9–45b), the property is also true for the diagram which represents a square 6j symbol :

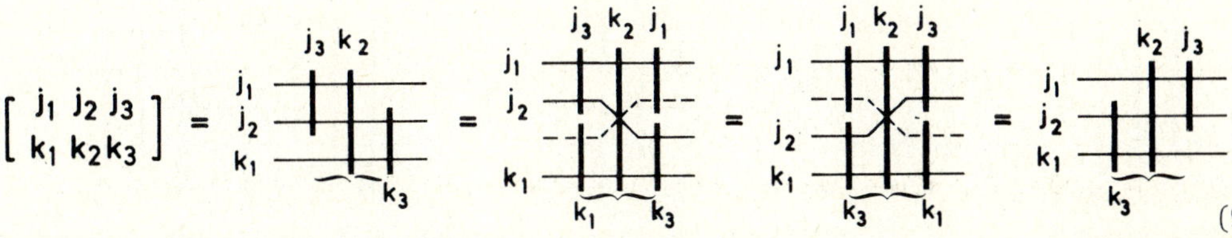

, (9-49)

where this last drawing can be taken as a fifth elementary diagram of the graphical method. Finally, since this left–right symmetry holds for the four elementary diagrams of Rule 2 (it is obviously true for the generalized orthogonality relation, Eq. (9–42), and the square 3j symbols), it applies to all diagrams.

9.2.2. APPLICATIONS

This graphical representation provides a simple method to handle most of the recoupling problems usually encountered in practice. There exist two types of applications :

i) This method can be used to only visualize in a convenient way a complicated recoupling transformation, and then obtain the expression of the corresponding recoupling coefficient in terms of the (square) 3j, 6j, and 9j symbols. Some examples are given later in Subsec. 11.3.3.3.

ii) More interesting, and which often takes place as a second step after the previous application, this graphical representation is a powerful method to derive relations between sums of products of (square) 3j, 6j, and 9j symbols, e.g. to get orthogonality relations and sum rules. Indeed, the transformation from a given coupling scheme into another one can often be performed in several ways, and two diagrams which have identical, respectively, left and right coupling schemes yield equal recoupling coefficients. Let us now give some examples of this kind.

1–Orthogonality relations. In the special case where a diagram has identical left and right coupling schemes, one gets an orthogonality relation since the corresponding recoupling coefficient is equal to one, provided the selection rules of couplings are satisfied. For, respectively, the (square) 6j (cf. Eq. (9–30)) and 9j symbols one has using Eq. (9–42) :

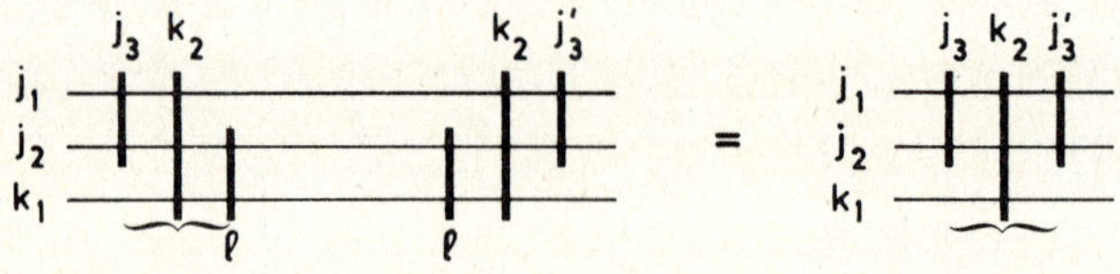

$$\sum_{\ell}\begin{bmatrix} j_1 & j_2 & j_3 \\ k_1 & k_2 & \ell \end{bmatrix}\begin{bmatrix} j_1 & j_2 & j'_3 \\ k_1 & k_2 & \ell \end{bmatrix} = (2j_3+1)\sum_{\ell}(2\ell+1)\begin{Bmatrix} j_1 & j_2 & j_3 \\ k_1 & k_2 & \ell \end{Bmatrix}\begin{Bmatrix} j_1 & j_2 & j'_3 \\ k_1 & k_2 & \ell \end{Bmatrix} = \delta_{j_3 j'_3}\{j_1 j_2 j_3\}\{j_3 k_1 k_2\}$$

(9–50a)

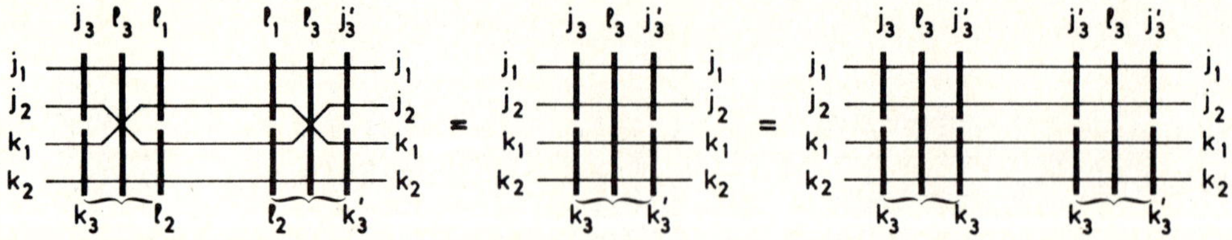

$$\sum_{\ell_1,\ell_2}\begin{bmatrix} j_1 & j_2 & j_3 \\ k_1 & k_2 & k_3 \\ \ell_1 & \ell_2 & \ell_3 \end{bmatrix}\begin{bmatrix} j_1 & k_1 & \ell_1 \\ j_2 & k_2 & \ell_2 \\ j'_3 & k'_3 & \ell_3 \end{bmatrix} = \delta_{j_3 j'_3}\{j_1 j_2 j_3\}\{j_3 k_3 \ell_3\}\delta_{k_3 k'_3}\{k_1 k_2 k_3\}$$

$$(2j_3+1)(2k_3+1)\sum_{\ell_1,\ell_2}(2\ell_1+1)(2\ell_2+1)\begin{Bmatrix} j_1 & j_2 & j_3 \\ k_1 & k_2 & k_3 \\ \ell_1 & \ell_2 & \ell_3 \end{Bmatrix}\begin{Bmatrix} j_1 & k_1 & \ell_1 \\ j_2 & k_2 & \ell_2 \\ j'_3 & k'_3 & \ell_3 \end{Bmatrix} = \delta_{j_3 j'_3}\delta_{k_3 k'_3}\{j_1 j_2 j_3\}\{k_1 k_2 k_3\}\{j_3 k_3 \ell_3\}.$$

(9–50b)

2–Sum rules. Let us give some examples for the (square) 6j and 9j symbols :

①

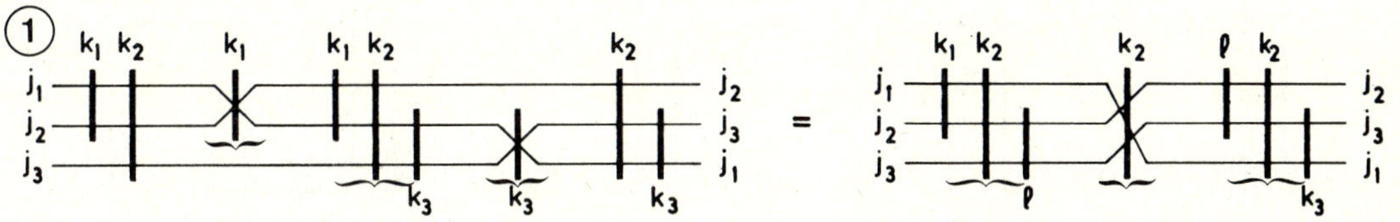

$$\begin{bmatrix} j_1 & j_2 & k_1 \end{bmatrix} \begin{bmatrix} j_2 & j_1 & k_1 \\ j_3 & k_2 & k_3 \end{bmatrix} \begin{bmatrix} j_1 & j_3 & k_3 \end{bmatrix} = \sum_{\ell} \begin{bmatrix} j_1 & j_2 & k_1 \\ j_3 & k_2 & \ell \end{bmatrix} \begin{bmatrix} j_1 & \ell & k_2 \end{bmatrix} \begin{bmatrix} j_2 & j_3 & \ell \\ j_1 & k_2 & k_3 \end{bmatrix}$$

$$\begin{Bmatrix} j_2 & j_1 & k_1 \\ j_3 & k_2 & k_3 \end{Bmatrix} = (-1)^{k_1+k_3} \sum_{\ell} (-1)^{\ell} (2\ell+1) \begin{Bmatrix} j_1 & j_2 & k_1 \\ j_3 & k_2 & \ell \end{Bmatrix} \begin{Bmatrix} j_2 & j_3 & \ell \\ j_1 & k_2 & k_3 \end{Bmatrix} , \tag{9-51a}$$

②

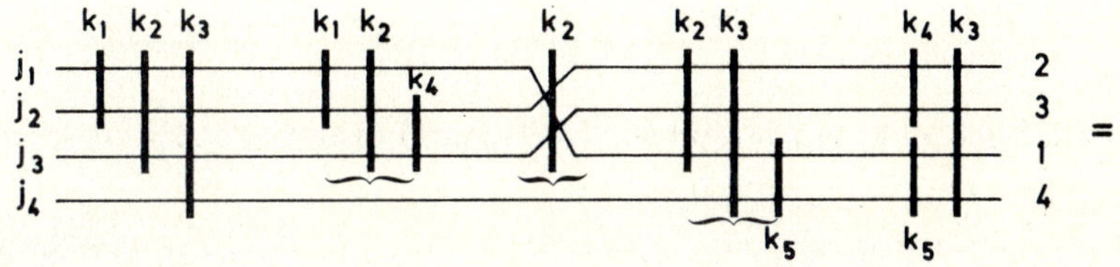

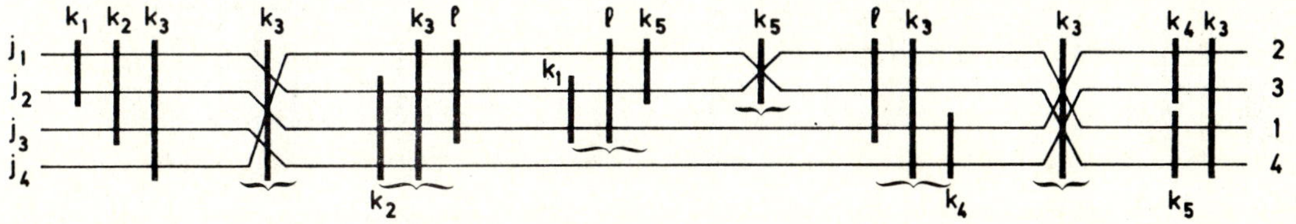

$$\begin{bmatrix} j_1 & j_2 & k_1 \\ j_3 & k_2 & k_4 \end{bmatrix} \begin{bmatrix} j_1 & k_4 & k_2 \end{bmatrix} \begin{bmatrix} k_4 & j_1 & k_2 \\ j_4 & k_3 & k_5 \end{bmatrix} = \begin{bmatrix} k_2 & j_4 & k_3 \end{bmatrix} \sum_{\ell} \begin{bmatrix} j_4 & k_1 & \ell \\ j_3 & k_3 & k_2 \end{bmatrix} \begin{bmatrix} j_4 & j_1 & k_5 \\ j_2 & \ell & k_1 \end{bmatrix} \begin{bmatrix} j_4 & j_1 & k_5 \end{bmatrix}$$

$$\begin{bmatrix} k_5 & j_2 & \ell \\ j_3 & k_3 & k_4 \end{bmatrix} \begin{bmatrix} k_5 & k_4 & k_3 \end{bmatrix} ,$$

$$\begin{Bmatrix} j_1 & j_2 & k_1 \\ j_3 & k_2 & k_4 \end{Bmatrix} \begin{Bmatrix} k_4 & j_1 & k_2 \\ j_4 & k_3 & k_5 \end{Bmatrix} = (-1)^s \sum_{\ell} (-1)^{\ell} (2\ell+1) \begin{Bmatrix} j_4 & k_1 & \ell \\ j_3 & k_3 & k_2 \end{Bmatrix} \begin{Bmatrix} j_4 & j_1 & k_5 \\ j_2 & \ell & k_1 \end{Bmatrix} \begin{Bmatrix} k_5 & j_2 & \ell \\ j_3 & k_3 & k_4 \end{Bmatrix} ,$$

$$s = j_1+j_2+j_3+j_4+k_1+k_2+k_3+k_4+k_5 \quad . \tag{9-51b}$$

Setting any one of the indices j_i or k_i equal to $\frac{1}{2}$ in the above equation, the sum over ℓ then reduces to two terms. This yields to recursion relations for the 6j symbols, cf. for instance Edmonds[22] (Subsec. 6.3).

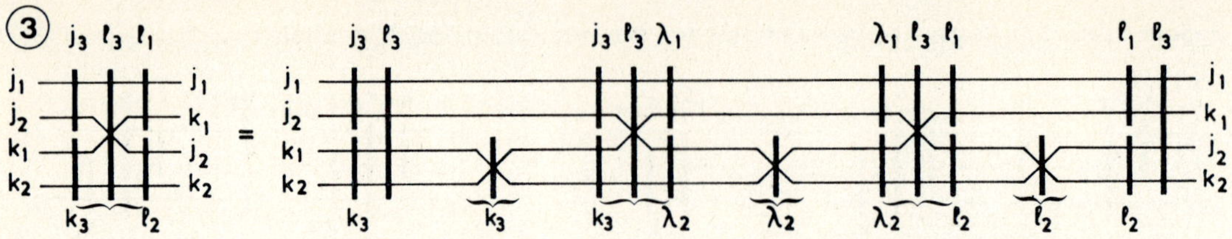

$$\begin{bmatrix} j_1 & j_2 & j_3 \\ k_1 & k_2 & k_3 \\ \ell_1 & \ell_2 & \ell_3 \end{bmatrix} = [k_1k_2k_3] \sum_{\lambda_1,\lambda_2} \begin{bmatrix} j_1 & j_2 & j_3 \\ k_2 & k_1 & k_3 \\ \lambda_1 & \lambda_2 & \ell_3 \end{bmatrix} [j_2k_1\lambda_2] \begin{bmatrix} j_1 & k_2 & \lambda_1 \\ k_1 & j_2 & \lambda_2 \\ \ell_1 & \ell_2 & \ell_3 \end{bmatrix} [k_2j_2\ell_2]$$

(9-51c)

$$\begin{Bmatrix} j_1 & j_2 & j_3 \\ k_1 & k_2 & k_3 \\ \ell_1 & \ell_2 & \ell_3 \end{Bmatrix} = (-1)^{2j_2+\ell_2-k_3} \sum_{\lambda_1,\lambda_2} (-1)^{\lambda_2}(2\lambda_1+1)(2\lambda_2+1) \begin{Bmatrix} j_1 & j_2 & j_3 \\ k_2 & k_1 & k_3 \\ \lambda_1 & \lambda_2 & \ell_3 \end{Bmatrix} \begin{Bmatrix} j_1 & k_2 & \lambda_1 \\ k_1 & j_2 & \lambda_2 \\ \ell_1 & \ell_2 & \ell_3 \end{Bmatrix} .$$

3–Expression of a (square) 9j symbol in terms of (square) 6j symbols, cf. Eq. (9–39) :

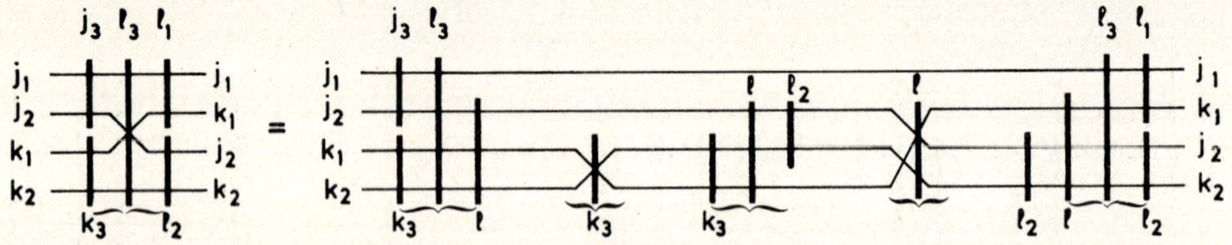

$$\begin{bmatrix} j_1 & j_2 & j_3 \\ k_1 & k_2 & k_3 \\ \ell_1 & \ell_2 & \ell_3 \end{bmatrix} = \sum_{\ell} \begin{bmatrix} j_1 & j_2 & j_3 \\ k_3 & \ell_3 & \ell \end{bmatrix} [k_1k_2k_3] \begin{bmatrix} j_2 & k_2 & \ell_2 \\ k_1 & \ell & k_3 \end{bmatrix} [\ell_2k_1\ell] \begin{bmatrix} j_1 & k_1 & \ell_1 \\ \ell_2 & \ell_3 & \ell \end{bmatrix}$$

(9-52)

$$\begin{Bmatrix} j_1 & j_2 & j_3 \\ k_1 & k_2 & k_3 \\ \ell_1 & \ell_2 & \ell_3 \end{Bmatrix} = \sum_{\ell} (-1)^{\ell} (2\ell+1) \begin{Bmatrix} j_1 & j_2 & j_3 \\ k_3 & \ell_3 & \ell \end{Bmatrix} \begin{Bmatrix} j_2 & k_2 & \ell_2 \\ k_1 & \ell & k_3 \end{Bmatrix} \begin{Bmatrix} j_1 & k_1 & \ell_1 \\ \ell_2 & \ell_3 & \ell \end{Bmatrix} .$$

FORMULAE FOR CHAPTERS 8 AND 9

Conjugate complex representations

$$\forall U \in SU(2) \qquad \mathcal{D}^{j*}(U) = C^j \mathcal{D}^j(U)(C^j)^{-1} \quad , \qquad C^j = \lambda_j C^j \qquad \lambda_j \in C$$

$$C^j = \mathcal{D}^j(C) \quad , \qquad C = i\sigma_2 = U(\pi, -e_2; a)$$

$$C^j = C^{j*} \quad , \quad (C^j)^{-1} = (C^j)^+ = (C^j)^t = (-1)^{2j}\, C^j$$

$m' \backslash m$	$j, j-1, \ldots, -j$
j, $j-1$, $\vdots$, $-j$	$C^j_{m'm} = d^j_{m'm}(-\pi) = (-1)^{j+m}\,\delta_{m',-m}$.

$$D[SU(2)],\ \sigma_i \rightarrow \begin{cases} J_i = i\,\dfrac{\partial}{\partial\varepsilon} D[\mathbf{1} - i\frac{\varepsilon}{2}\sigma_i + 0(\varepsilon^2)]\Big|_{\varepsilon=0} \\[2ex] \bar{J}_i = i\,\dfrac{\partial}{\partial\varepsilon} D^*[\mathbf{1} - i\frac{\varepsilon}{2}\sigma_i + 0(\varepsilon^2)]\Big|_{\varepsilon=0} \end{cases}$$

$$\bar{J}_i = D(C)J_i\, D(C)^{-1} = R(\pi, J_2)J_i$$

$$\bar{J}_1 = -J_1 \quad , \quad \bar{J}_2 = J_2 \quad , \quad \bar{J}_3 = -J_3$$

Spherical tensors, tensorial character

Generic case of a tensor of order one and rank j :

$$T^j = T \bullet\!\!\overset{1}{\text{———}}\, j \quad \begin{cases} T^{[j]}_m = T \bullet\!\!\overset{1}{\longrightarrow}\, j \\ T^{(j)}_m = T \bullet\!\!\overset{1}{\longleftarrow}\, j \end{cases} \qquad \forall U \in SU(2) \rightarrow U_{T^j} \begin{cases} U_T{}^{[j]}_m = T^{[j]}_{m'}\, \mathcal{D}^j_{m'm}(U) \\ U_T{}^{(j)}_m = T^{(j)}_{m'}\, \mathcal{D}^{j*}_{m'm}(U) \end{cases}$$

$$\boxed{\begin{array}{c} T^{(j)}_m = C^j_{mm'}\, T^{[j]}_{m'} \quad , \quad C^j = (-1)^{2j} c_j\, \mathcal{D}^j(C) \quad , \qquad c_j = 1 \text{ or } (-1)^{2j} \\ T^{(j)}_m = c_j(-1)^{j+m}\, T^{[j]}_{-m} \quad , \quad T^{[j]}_m = c_j(-1)^{j-m}\, T^{(j)}_{-m} \quad . \end{array}}$$

Symmetrical and antisymmetrical tensors of order two :

$$T^{jj} \rightarrow T^{t,jj} \quad , \qquad T^{t,[j][j]}_{m\ m'} = T^{[j][j]}_{m'\ m}$$

$$T^{jj} = \varepsilon\, T^{t,jj} \quad , \ \varepsilon = \pm 1 \ \Leftrightarrow \ T^{[j][j]}_{m\ m'} = \varepsilon\, T^{[j][j]}_{m'\ m} \ \Leftrightarrow \ T^{(j)(j)}_{m\ m'} = \varepsilon\, T^{(j)(j)}_{m'\ m}$$

Formal Hermitian conjugate of tensors . + is an involutive antilinear mapping which commutes with the representation of SU(2), e.g. * , + , or | > → < | :

$$T^j \xrightarrow{\ +\ } T^{+j}$$

$$T^{+[j]}_m = (-1)^{2j}\, T^{(j)+}_m = c_j(-1)^{j-m}\, T^{[j]+}_{-m} \quad , \quad T^{+(j)}_m = T^{[j]+}_m = c_j(-1)^{j-m}\, T^{(j)+}_{-m}$$

$$T^{[j]+}_{m} = T^{+(j)}_{m} = c_j(-1)^{j+m}\, T^{+[j]}_{-m} \quad , \quad T^{(j)+}_{m} = (-1)^{2j}\, T^{+[j]}_{m} = c_j(-1)^{j+m}\, T^{+(j)}_{-m}$$

$$T^{++j} = (-1)^{2j}\, T^{j} \quad .$$

Hermitian tensor, only possible for integer values of ℓ :

$$T^{\ell} = T^{+\ell} \Leftrightarrow T^{[\ell]+}_{m} = c_\ell(-1)^{\ell+m}\, T^{[\ell]}_{-m} \Leftrightarrow T^{(\ell)+}_{m} = c_\ell(-1)^{\ell+m}\, T^{(\ell)}_{-m} \quad .$$

Triad

$$T^{j} \ , \ \forall U \in SU(2) \quad \begin{cases} T^{[j]}_{m;U} = {}^{U}T^{[j]}_{m;\mathbf{1}} = {}^{U}T^{[j]}_{m} \\ T^{(j)}_{m;U} = {}^{U}T^{(j)}_{m;\mathbf{1}} = {}^{U}T^{(j)}_{m} \end{cases}$$

$$(\sigma_i) \to (J_i) \quad \text{for} \quad D[SU(2)] \quad \text{and} \quad (\bar{J}_i) \quad \text{for} \quad D^{*}[SU(2)]$$

$$J_{\pm}\ T^{[j]}_{m;\mathbf{1}} = \sqrt{j(j+1)-m(m\pm1)}\ T^{[j]}_{m\pm1;\mathbf{1}} \ , \ J_3\ T^{[j]}_{m;\mathbf{1}} = m\, T^{[j]}_{m;\mathbf{1}}$$

$$\bar{J}_{\pm}\ T^{(j)}_{m;\mathbf{1}} = \sqrt{j(j+1)-m(m\pm1)}\ T^{(j)}_{m\pm1;\mathbf{1}} \ , \ \bar{J}_3\ T^{(j)}_{m;\mathbf{1}} = m\, T^{(j)}_{m;\mathbf{1}}$$

$$\forall U \in SU(2) \quad , \quad \forall J \in D[\mathcal{W}(2)] \quad [R(\pm U)J]D(\pm U) = D(\pm U)J$$

$$\forall U \in SU(2) \qquad T^{[j]}_{m;-U} = (-1)^{2j}\, T^{[j]}_{m;U} \quad , \quad T^{(j)}_{m;-U} = (-1)^{2j}\, T^{(j)}_{m;U}$$

Active point of view of the representation : $\forall U,U' \in SU(2)$

$$T^{j} \to {}^{U}T^{j} \quad , \quad {}^{U'}T^{[j]}_{m;U} = T^{[j]}_{m';U}\, \mathcal{D}^{j}_{m'm}(U^{-1}U'U) \quad , \quad {}^{U'}T^{(j)}_{m;U} = T^{(j)}_{m';U}\, \mathcal{D}^{j}_{m'm}(U^{-1}U'U)$$

Passive point of view of the representation : $\forall U,U' \in SU(2)$

$$\text{a same } T^{j} \quad , \quad T^{[j]}_{m;U'} = T^{[j]}_{m';U}\, \mathcal{D}^{j}_{m'm}(U^{-1}U') \quad , \quad T^{(j)}_{m;U'} = T^{(j)}_{m';U}\, \mathcal{D}^{j*}_{m'm}(U^{-1}U')$$

Contractions

$$S^{j\bullet} \otimes T^{j^{\bullet}} = S^{(j)}_{m} T^{[j]}_{m} = (-1)^{2j}\, S^{j^{\bullet}} \otimes T^{j\bullet} = (-1)^{2j}\, S^{[j]}_{m}\, T^{(j)}_{m} = c_j\, \hat{\jmath}\, S^{\{j} \otimes T^{j\}0}$$

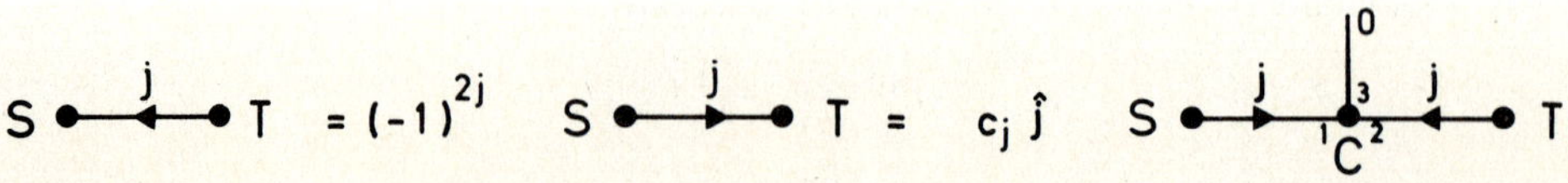

Metric tensor :

$G^{jj} =$ 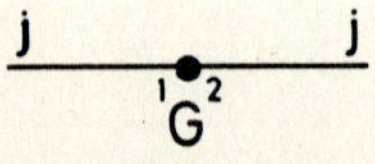

$, \ G^{(j)(j)}_{m\ m'} = C^{j}_{mm'} = c_j(-1)^{j+m}\, \delta_{m,-m'} =$ 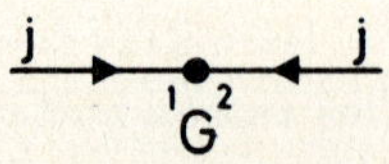

$$T^{j} = \quad T^{j^{\bullet}}\, G^{jj\bullet} \quad = T^{j\bullet}\, G^{j^{\bullet}j}$$

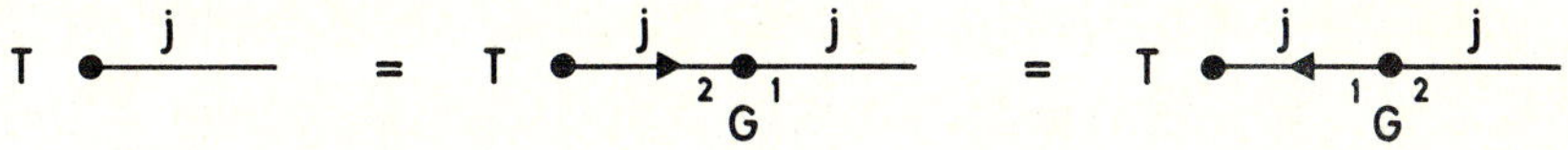

Couplings

$$T^{\{j_1j_2\}[j_3]}_{m_3} = T^{[j_1][j_2]}_{m_1\ m_2}\langle j_1j_2m_1m_2|j_3m_3\rangle\ ,\quad T^{j_1j_2,\{j_2j_1\}[j_3]}_{m_3} = T^{[j_1][j_2]}_{m_1\ m_2}\langle j_2j_1m_2m_1|j_3m_3\rangle$$

$$T^{\{j_1j_2\}j_3} = T^{j_1^\bullet j_2^\bullet}C^{j_1\bullet j_2\bullet j_3} = c_{j_3}(-1)^{j_1-j_2+j_3}\hat{j}_3\,T^{j_1^\bullet j_2^\bullet}W^{j_1\bullet j_2\bullet j_3}$$

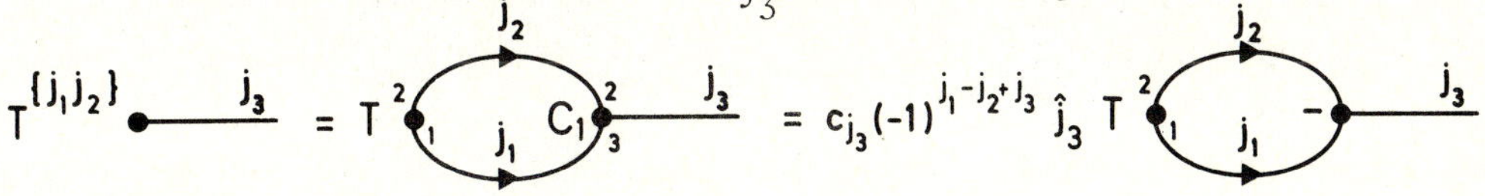

$$T^{j_1j_2} = \sum_{j_3} C^{j_1j_2j_3^\bullet}\,T^{\{j_1j_2\}j_3\bullet} = \sum_{j_3} c_{j_3}(-1)^{j_1-j_2+j_3}\hat{j}_3\,W^{j_1j_2j_3^\bullet}\,T^{\{j_1j_2\}j_3\bullet}$$

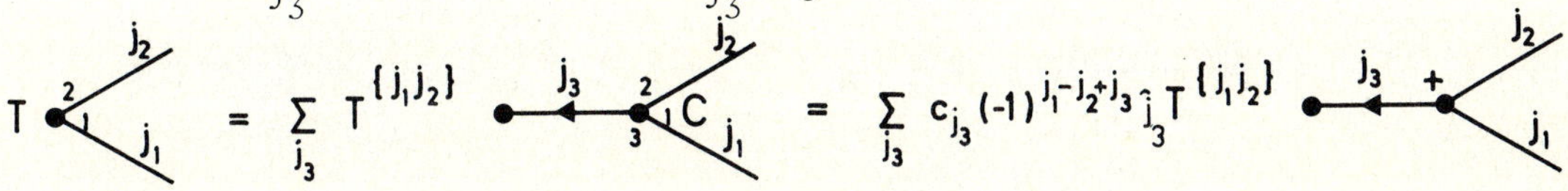

$$T^{\{j_1\dots j_n;k_1\dots k_{n-2},\alpha\}k_{n-1}} = C^{j_1\bullet\ \dots j_n\bullet\ k_{n-1}}_{(k_1\dots k_{n-2},\alpha)}\,T^{j_1^\bullet\ \dots j_n^\bullet}$$

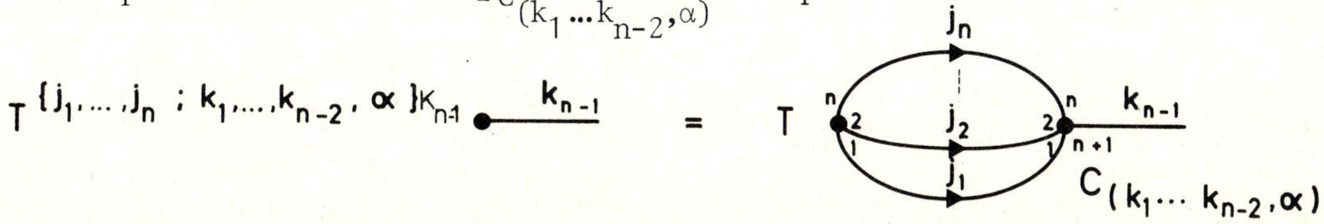

$$T^{j_1\dots j_n} = \sum_{k_1,\dots,k_{n-1}} C^{j_1\dots j_n\ k_{n-1}^\bullet}_{(k_1\dots k_{n-2},\alpha)}\,T^{\{j_1\dots j_n;k_1\dots k_{n-2},\alpha\}k_{n-1}\bullet}$$

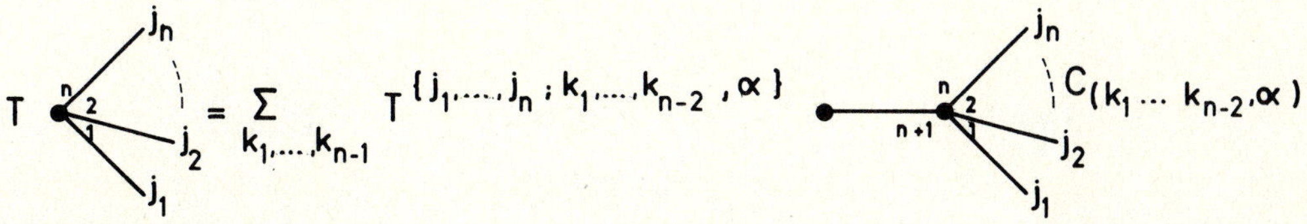

Clebsch–Gordan tensors :

$C^{j_1j_2j_3}$ = 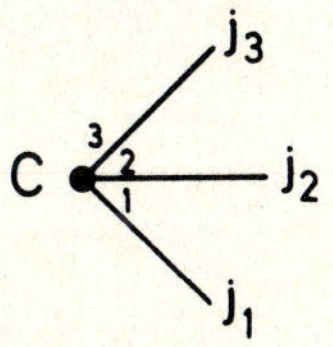

, $C^{(j_1)(j_2)[j_3]}_{m_1\ m_2\ m_3} = \langle j_1j_2m_1m_2|j_3m_3\rangle$ = 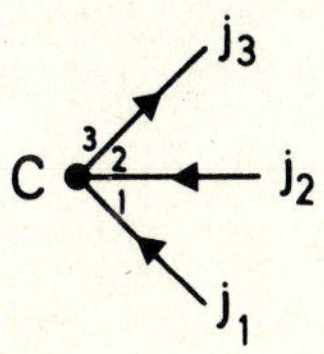

$$C^{j_1j_2j_3} = C^{*j_1j_2j_3} = (-1)^{j_1+j_2-j_3}C^{j_2j_1j_3} = c_{j_2}(-1)^{j_1+j_2+j_3}\frac{\hat{j}_3}{\hat{j}_1}C^{j_3j_2j_1}$$

$$C^{j_1^\bullet j_2^\bullet k_3}\,C^{j_1\bullet j_2\bullet k_3'} = \delta_{k_3k_3'}\{j_1j_2k_3\}\,G^{k_3k_3}$$

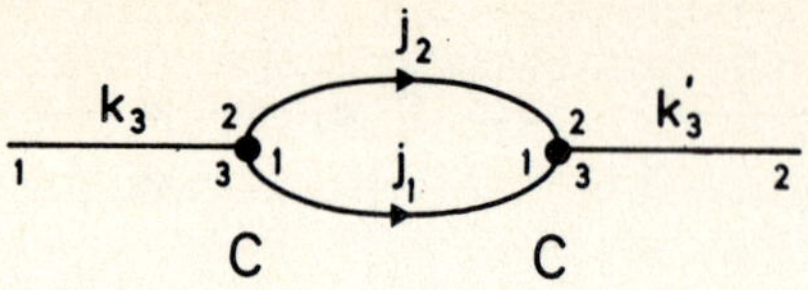 $= \delta_{k_3 k_3'} \{ j_1 j_2 k_3 \}$ 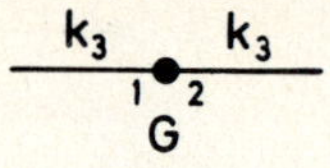

$$\sum_{k_3} C^{j_1 j_2 k_3^\bullet} C^{j_1 j_2 k_3 \bullet} = G^{j_1 j_1} G^{j_2 j_2}$$

$\sum_{k_3}$ 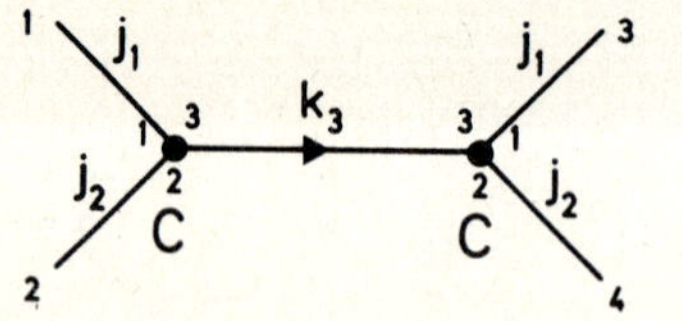 $=$ 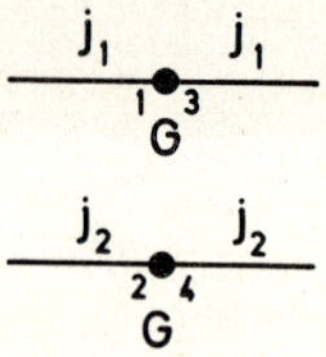

Wigner 3j tensors :

$W^{j_1 j_2 j_3} = +$ 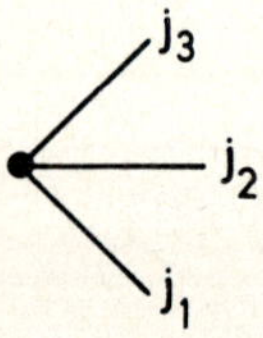 $, W^{[j_1][j_2][j_3]}_{m_1 \; m_2 \; m_3} = \begin{pmatrix} j_1 & j_2 & j_3 \\ m_1 & m_2 & m_3 \end{pmatrix} = +$ 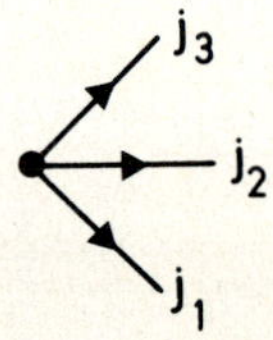

$W^{j_1 j_2 j_3} = W^{* j_1 j_2 j_3} = (-1)^{j_1+j_2+j_3} W^{j_i j_k j_\ell}$, $\begin{pmatrix} 1 & 2 & 3 \\ i & k & \ell \end{pmatrix} =$ odd number of transpositions

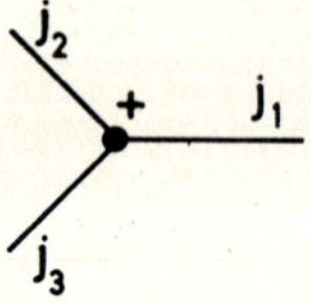 $= (-1)^{j_1+j_2+j_3}$ 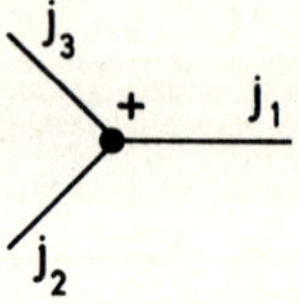 $= (-1)^{j_1+j_2+j_3}$ 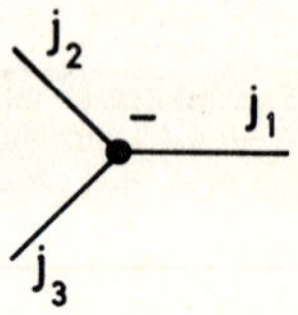 $=$ 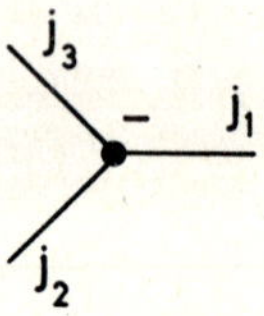

$$C^{j_1 j_2 j_3} = c_{j_3} (-1)^{j_1 - j_2 + j_3} \hat{j}_3 W^{j_1 j_2 j_3}$$

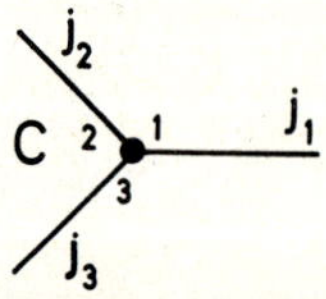 $= c_{j_3} (-1)^{j_1 - j_2 + j_3} \hat{j}_3$ 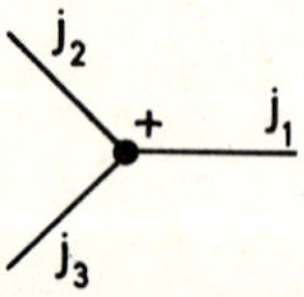

$$(2k_3+1) W^{j_1^\bullet j_2^\bullet k_3} W^{j_1 \bullet j_2 \bullet k_3'} = \delta_{k_3 k_3'} \{j_1 j_2 k_3\} G^{k_3 k_3}$$

$(2k_3+1)$ 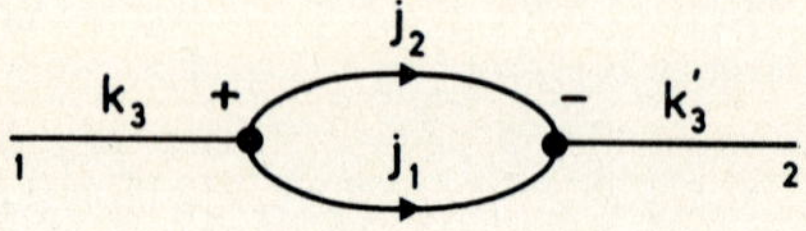 $= \delta_{k_3 k_3'} \{ j_1 j_2 k_3 \}$ 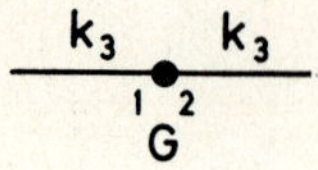

$$\sum_{k_3} (2k_3+1) W^{j_1 j_2 k_3^\bullet} W^{j_1 j_2 k_3 \bullet} = G^{j_1 j_1} G^{j_2 j_2}.$$

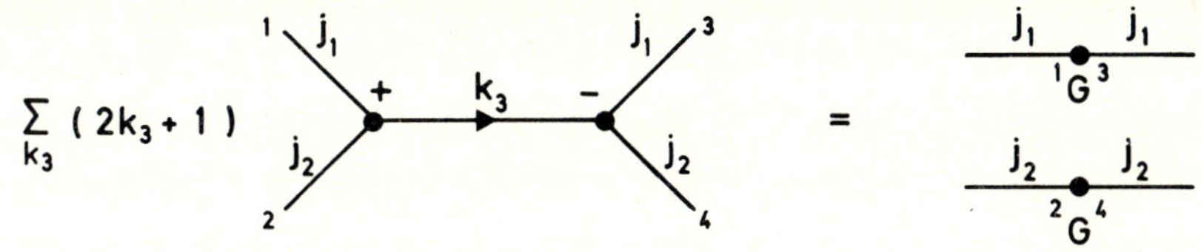

$j_1+j_2+j_3$ is an integer otherwise $C^{j_1j_2j_3}$ and $W^{j_1j_2j_3}$ vanish. Consequence, e.g. :

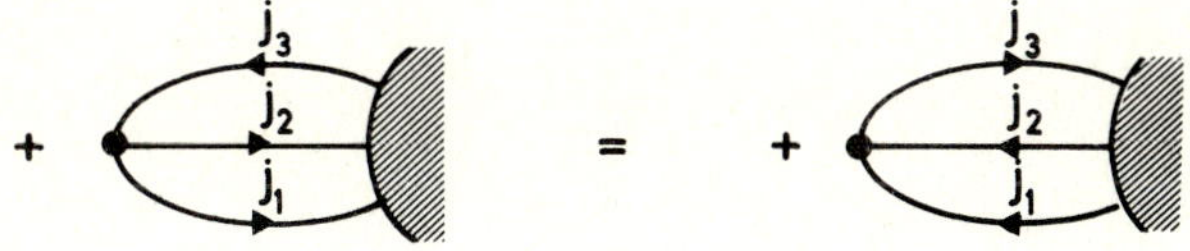

Coupling tensors :

$$C^{j_1^\bullet \ldots j_n^\bullet\, k_{n-1}}_{(k_1 \ldots k_{n-2},\alpha)} \; C^{j_1\bullet\, j_n\bullet\, k'_{n-1}}_{(k'_1 \ldots k'_{n-2},\alpha)} = \prod_{i=1}^{n-1} [\delta_{k_ik'_i} \{k_i\}] G^{k_{n-1}k_{n-1}}$$

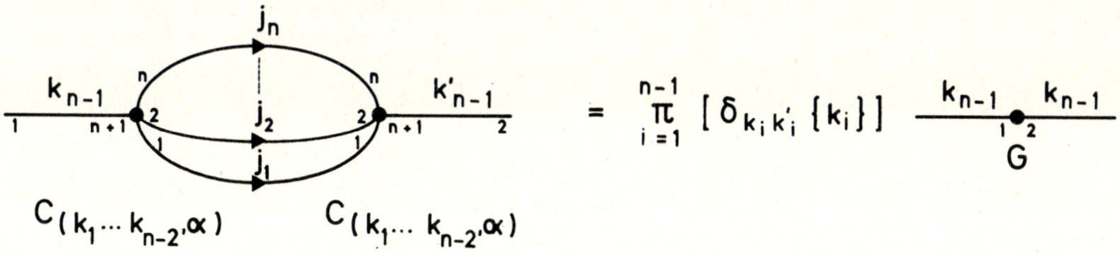

$$\sum_{k_1,\ldots,k_{n-2}} C^{j_1 \ldots j_n\, k^\bullet_{n-1}}_{(k_1 \ldots k_{n-2},\alpha)} \; C^{j_1 \ldots j_n\, k_{n-1}\bullet}_{(k_1 \ldots k_{n-2},\alpha)} = G^{j_1j_1} \ldots G^{j_nj_n}$$

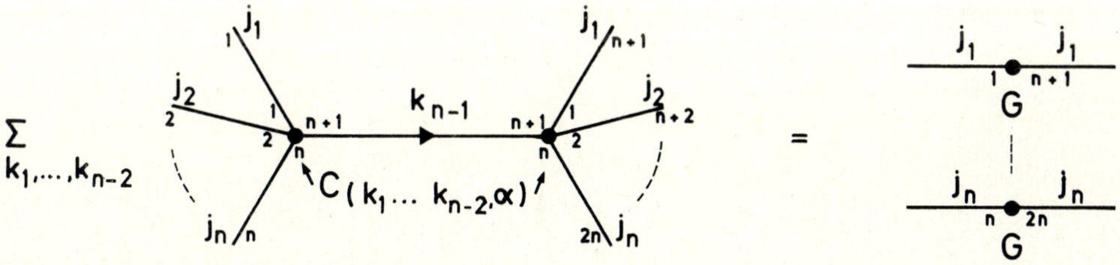

Invariant tensors, the Wigner–Eckart theorem

$\forall\, U \in SU(2) \qquad {}^{U}T^{j} = T^{j} \;\Leftrightarrow\; T^{j} = \delta_{j0}\, T^{0}$

$\forall\, U \in SU(2) \quad {}^{U}T^{j_1j_2} = T^{j_1j_2} \;\Leftrightarrow\; T^{j_1j_2} = \frac{\delta_{j_1j_2}}{2j_1+1}\, T^{j_1^\bullet j_1\bullet}\, G^{j_1j_1} = \delta_{j_1j_2}\, T^{[j_1](j_1)}_{m\quad m}\, G^{j_1j_1}$

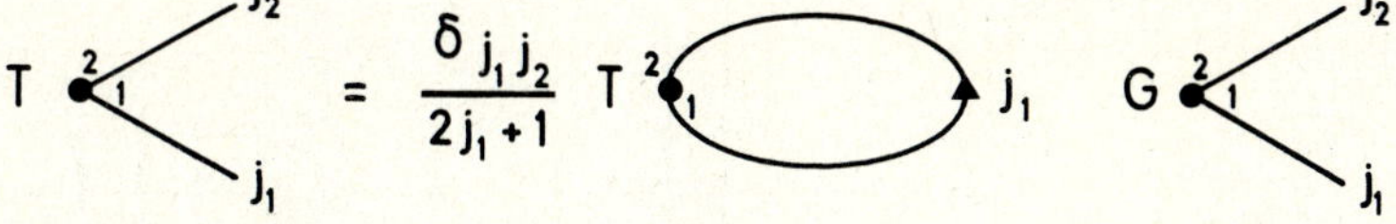

$$T^{[j](j)}_{m\quad m} = (-1)^{2j}\, T^{(j)[j]}_{m\quad m} = \frac{1}{2j+1}\, T^{j^\bullet j\bullet}$$

$$\forall U \in SU(2) \quad {}^{U}T^{j_1j_2j_3} = T^{j_1j_2j_3} \Leftrightarrow T^{j_1j_2j_3} = \frac{c_{j_3}}{\hat{j}_3}\, C^{j_1j_2j_3}\, T^{\{\{j_1j_2\}j_3j_3\}0}$$
$$= W^{j_1j_2j_3}\, T^{j_1^\bullet j_2^\bullet j_3^\bullet}\, W^{j_1\bullet j_2\bullet j_3\bullet}$$

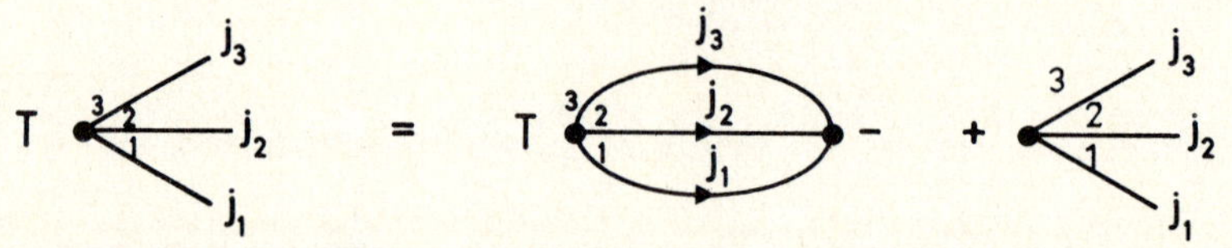

$$T^{(j_1)[j_2][j_3]}_{m_1\ m_2\ m_3} = \frac{c_{j_1}}{\hat{j}_1} \langle j_2j_3m_2m_3|j_1m_1\rangle\ T^{\{j_1\{j_2j_3\}j_1\}0}$$
$$T^{\{\{j_1j_2\}j_3j_3\}0} = T^{\{j_1\{j_2j_3\}j_1\}0}$$

Recoupling coefficients and 3 nj symbols

$$\{j_1\cdots j_{n+1};k_1\cdots k_{n-1},\alpha\}\ k_n = \frac{1}{2k_n+1} \sum_{\ell_1,\dots,\ell_{n-1}} 3n_R{}^{k_n\bullet\ k_n^\bullet}_{(j_1\cdots j_{n+1};k_1\cdots k_{n-1},\alpha;\ell_1\cdots\ell_{n-1},\beta)}\ \{j_1\cdots j_{n+1};\ell_1\cdots\ell_{n-1},\beta\}\ k_n$$

$$3n_R{}^{k_n\ell_n}_{(j_1\cdots j_{n+1};k_1\cdots k_{n-1};\alpha;\ell_1\cdots\ell_{n-1},\beta)} = C^{j_1\bullet\ \cdots j_{n+1}\bullet k_n}_{(k_1\cdots k_{n-1},\alpha)}\ C^{j_1^\bullet\cdots j_{n+1}^\bullet \ell_n}_{(\ell_1\cdots\ell_{n-1},\beta)}, \text{invariant tensor.}$$

(Square) 3j symbols :

$$\{j_1j_2\}j_3 = [j_1j_2j_3]\ \{j_2j_1\}j_3$$

square 3j symbol $[j_1j_2j_3] = (-1)^{j_1+j_2-j_3}\ \{j_1j_2\ j_3\}$

3j symbol $\{j_1j_2j_3\} = W^{j_1^\bullet j_2^\bullet j_3^\bullet}\ W^{j_1\bullet j_2\bullet j_3\bullet}$ = triangular delta

$\{j_1j_2j_3\}$ =

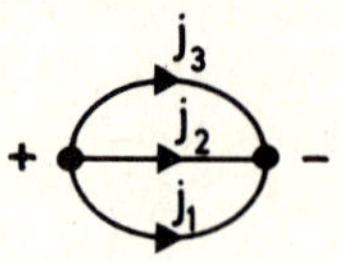

(Square) 6j symbols :

$$\{\{j_1j_2\}j_3k_1\}k_2 = \sum_{k_3} \begin{bmatrix} j_1 & j_2 & j_3 \\ k_1 & k_2 & k_3 \end{bmatrix}\ \{j_1\{j_2k_1\}k_3\}k_2$$

square 6j symbol $\begin{bmatrix} j_1 & j_2 & j_3 \\ k_1 & k_2 & k_3 \end{bmatrix} = (-1)^{j_1+j_2+k_1+k_2}\ \hat{j}_3\hat{k}_3 \begin{Bmatrix} j_1 & j_2 & j_3 \\ k_1 & k_2 & k_3 \end{Bmatrix}$

$\begin{bmatrix} j_1 & j_2 & j_3 \\ k_1 & k_2 & k_3 \end{bmatrix}$ =

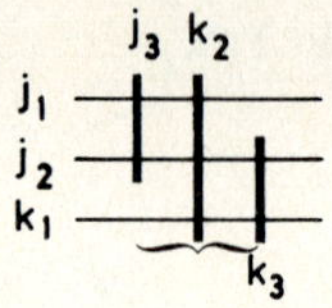

=

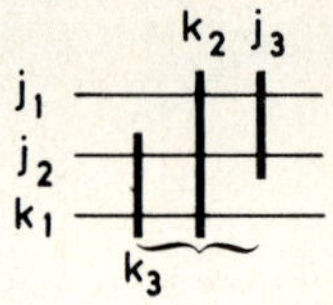

6j symbol $$\begin{Bmatrix} j_1 & j_2 & j_3 \\ k_1 & k_2 & k_3 \end{Bmatrix} = c_{j_a} c_{k_a} (-1)^{2(j_a+k_a)}\, W^{j_1^\bullet j_2^\bullet j_3 \bullet}\, W^{j_3^\bullet k_1^\bullet k_2 \bullet}\, W^{k_2^\bullet k_3^\bullet j_1 \bullet}\, W^{j_2 \bullet k_3 \bullet k_1 \bullet}$$
$a = 1,2,3$

$$\begin{Bmatrix} j_1 & j_2 & j_3 \\ k_1 & k_2 & k_3 \end{Bmatrix} = c_{j_a} c_{k_a} (-1)^{2(j_a+k_a)}$$ 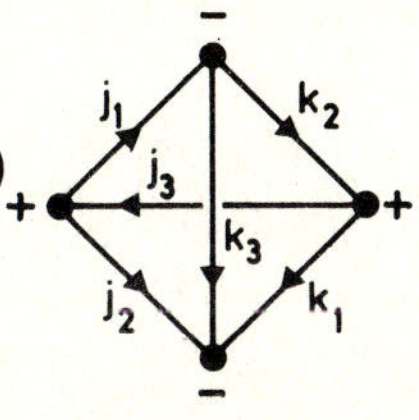

$$= c_{j_a} c_{k_a} (-1)^{2(j_a+k_a)}$$ 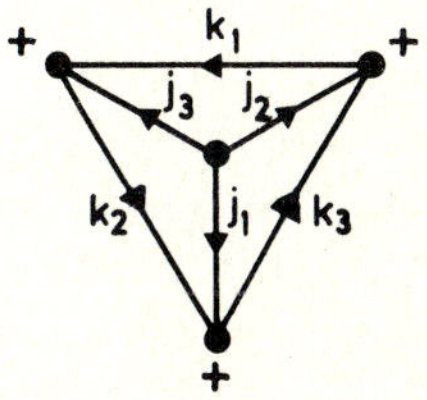

(Square) 9j symbols :

$$\{\{j_1 j_2\} j_3\ \{k_1 k_2\} k_3\} \ell_3 = \sum_{\ell_1, \ell_2} \begin{bmatrix} j_1 & j_2 & j_3 \\ k_1 & k_2 & k_3 \\ \ell_1 & \ell_2 & \ell_3 \end{bmatrix} \{\{j_1 k_1\} \ell_1 \{j_2 k_2\} \ell_2\} \ell_3$$

square 9j symbol $$\begin{bmatrix} j_1 & j_2 & j_3 \\ k_1 & k_2 & k_3 \\ \ell_1 & \ell_2 & \ell_3 \end{bmatrix} = \hat{j}_3 \hat{k}_3 \hat{\ell}_1 \hat{\ell}_2 \begin{Bmatrix} j_1 & j_2 & j_3 \\ k_1 & k_2 & k_3 \\ \ell_1 & \ell_2 & \ell_3 \end{Bmatrix}$$

$$\begin{bmatrix} j_1 & j_2 & j_3 \\ k_1 & k_2 & k_3 \\ \ell_1 & \ell_2 & \ell_3 \end{bmatrix} =$$ 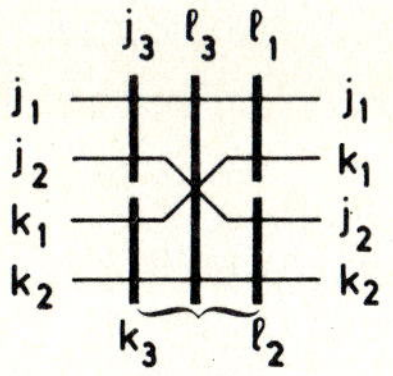

$$\begin{Bmatrix} j_1 & j_2 & j_3 \\ k_1 & k_2 & k_3 \\ \ell_1 & \ell_2 & \ell_3 \end{Bmatrix} =$$ 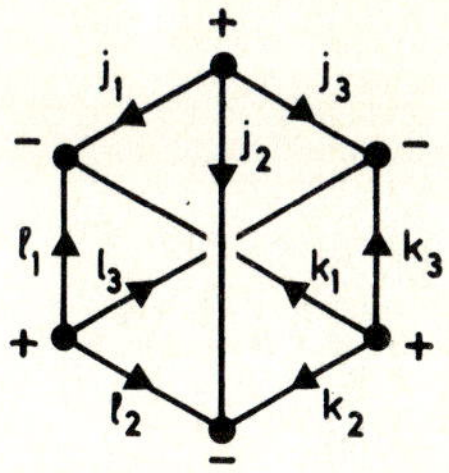

9j symbol $$\begin{Bmatrix} j_1 & j_2 & j_3 \\ k_1 & k_2 & k_3 \\ \ell_1 & \ell_2 & \ell_3 \end{Bmatrix} = W^{j_1^\bullet j_2^\bullet j_3^\bullet}\, W^{j_3 \bullet k_3 \bullet \ell_3 \bullet}\, W^{k_3^\bullet k_1^\bullet k_2^\bullet}\, W^{k_2 \bullet \ell_2 \bullet j_2 \bullet}\, W^{\ell_2^\bullet \ell_3^\bullet \ell_1^\bullet}\, W^{\ell_1 \bullet j_1 \bullet k_1 \bullet}$$

Relation between spherical and Cartesian tensor algebras

Cartesian tensors, generic case of a tensor of order one, i.e. a vector :

$$\forall R \in SO(3) \qquad T = (T_i) \xrightarrow{\ R\ } {}^R T = ({}^R T_i = T_j\, R^j_i)$$

$$T = (T_i) = (T^{[1]}_m) \quad , \quad T^{[1]}_m = c\, T_i\, t^i_m \quad , \quad T_i = c^{-1}\, T^{[1]}_m\, (t^{-1})^m_i$$

$c \in \mathbb{C}$, convenient choices $c = \pm 1$

$$T^{[1]}_1 = T_{-1} = c \frac{1}{\sqrt{2}} (iT_1 - T_2) = c \frac{i}{\sqrt{2}} T_+ \qquad T_1 = c^{-1} \frac{i}{\sqrt{2}} (-T^{[1]}_1 + T^{[1]}_{-1})$$

$$T^{[1]}_0 = -T^{(1)}_0 = c(-iT_3) \qquad T_2 = c^{-1} \frac{-1}{\sqrt{2}} (T^{[1]}_1 + T^{[1]}_{-1})$$

$$T^{[1]}_{-1} = T^{(1)}_1 = c \frac{-1}{\sqrt{2}} (iT_1 + T_2) = c \frac{-i}{\sqrt{2}} T_- \qquad T_3 = c^{-1}\, i\, T^{[1]}_0$$

$$J_i\, T_j = i \sum_k \varepsilon_{ijk}\, T_k \quad .$$

For any two vectors u and v :

$$u.v = \sum_i u_i v_i = c^{-2}\ \sqrt{3}\ \{u^1 \otimes v^1\}^0$$
$$(u\times v)^1 = c^{-1}\ \sqrt{2}\ \{u^1 \otimes v^1\}^1 \qquad .$$

The infinitesimal generators J_i, which correspond to σ_i, are the Cartesian components of a vector :

$$J^{[1]}_{\pm 1} = J^{(1)}_{\mp 1} = c\ \frac{\pm i}{\sqrt{2}}\ J_{\pm} \quad , \quad J^{[1]}_{0} = -J^{(1)}_{0} = c\ (-iJ_3)$$

$$\left[J^{[1]}_{0}\ ,\ J^{[1]}_{\pm 1}\right] = c\left(\mp i\ J^{[1]}_{\pm 1}\right)\ ,\ \left[J^{[1]}_{1}\ ,\ J^{[1]}_{-1}\right] = c\left(i\ J^{[1]}_{0}\right)$$

$$C = i\sigma_2 \quad , \quad \bar{J} = {}^{C}J = {}^{R(\pi,e_2)}J \quad , \quad \bar{J}^{[1]}_{m} = (-1)^{1+m}\ J^{[1]}_{-m} = J^{(1)}_{m} \quad .$$

CHAPTER 10

TRANSFORMATION GROUPS AND RELATED REPRESENTATIONS

This chapter is devoted to the study of a procedure to determine group representations in a functional vector space. The method is first discussed for any group G in Sec. 10.1 where, as a special case, the definitions and properties of the regular representations are given. It is then discussed in Sec. 10.2 within the framework of manifolds and Lie groups. This leads us to the consideration of the first and the second parameter groups. Finally, these general considerations are applied in Sec. 10.3 to the groups SU(2) and SO(3). In particular, this allows us to obtain representations of these groups in functional vector spaces which are frequently encountered in Quantum Mechanics as is shown in the next chapter.

10.1. A GENERAL PROCEDURE TO DETERMINE GROUP REPRESENTATIONS IN A FUNCTIONAL VECTOR SPACE

Given a group G, a set E, a vector space F, and a vector space $\mathcal{F}$ of functions of E into F, it is shown in Subsec. 10.1.2 how to get a representation $\mathcal{D}(G)$ in $\mathcal{F}$ in terms of a transformation group M(G) on E (Subsec. 10.1.1) and a representation D(G) in F. This procedure is then applied to the special case where E is the group G itself and F the real or complex field. This leads us to the definition of the left and right regular representations of a group (Subsec. 10.1.3).

10.1.1. TRANSFORMATION GROUPS

i) Let E be a set of elements p, $\Sigma(E)$ the group of one-to-one mappings of E onto itself. Any subgroup of $\Sigma(E)$ is called a <u>transformation group</u> on E. If M is a homomorphism of a group G in $\Sigma(E)$, i.e. :

$$\forall g \in G \xrightarrow{M} M(g)$$

$$\forall p \in E \longrightarrow M(g)\, p \in E \quad , \tag{10-1}$$

$$\forall g,g' \in G \qquad M(g)M(g') = M(gg') \quad , \quad M(g^{-1}) = M(g)^{-1} \quad ,$$

one then says that G has been <u>realized</u> as a transformation group M(G) on E (Def. C.26). An important special case is the one where E is a vector space and for all g in G the mappings M(g) are linear. The homomorphism M of G into the set $L(E)$ of linear applications of E into itself is then a representation of G in E.

ii) For a given p in E, the subset of elements M(g)p for all g in G is called the orbital M(G)p of p. The relation of belonging to the same orbital defines an equivalence relation on E which implies a partitioning of this set, i.e. E is a disjoint union of orbitals (also called transitivity classes). The set E is said to be homogeneous under the transformation group M(G) if there exists only one orbital, i.e. if :

$$\forall p,q \in E \qquad \exists g \in G \qquad q = M(g)\, p \quad . \tag{10-2}$$

If it is so, M(G) is said to be a transitive transformation group of E. For a given p in E the subset H_p of elements in G such that :

$$\forall g \in H_p \qquad M(g)p = p \quad , \tag{10-3}$$

is clearly a subgroup of G. It is called the isotropic group or the little group of p. If M(g)p=q, all the transformations M(g') which also map p into q are those such that g' belongs to the left coset (Def. C.10) gH_p of H_p. Indeed, one has :

$$M(g')p = q \Leftrightarrow p = M(g'^{-1})q = M(g'^{-1})M(g)p = M(g'^{-1}g)p$$

$$\Leftrightarrow g'^{-1}g \in H_p \Leftrightarrow g' \in g\,H_p \quad .$$

Consequently, for all p in E there exists a one-to-one mapping between the orbital M(G)p of p and the quotient set G/H_p, the elements of which are the left cosets of H_p (Prop. C.21). This correspondence is determined by :

$$\forall q \in M(G)p \qquad \exists g \in G \qquad q = M(g)p \longleftrightarrow g\,H_p \in G/H_p \quad . \tag{10-4}$$

Let us now show that :

$$q = M(g)p \Rightarrow \quad H_q = g\,H_p\,g^{-1} \quad , \tag{10-5}$$

namely that H_q is the transform of H_p by g (Def. C.17), also called a conjugate subgroup of H_p (Def. C.19). Indeed, if g' belongs to H_q one has :

$$p = M(g^{-1})q = M(g^{-1})M(g')q = M(g^{-1}g')M(g)p = M(g^{-1}g'g)p$$

$$\Leftrightarrow \quad g^{-1}g'g \in H_p \quad \Leftrightarrow \quad g' \in g\,H_p\,g^{-1} \quad .$$

iii) We shall later need to consider the case when E is measurable. A measure μ on E is said to be invariant under the transformation group M(G) if for all measurable subsets P in E, one has :

$$\forall g \in G \qquad \mu(P) = \mu[M(g)P] \quad , \tag{10-6}$$

where M(g)P denotes the set of elements M(g)p for all p in P.

In the special case where E is a vector space over the complex field, it can be supplied with a Hermitian structure, the inner product is then denoted by :

$$\forall u,v \in E \longrightarrow (u,v)_E \in C \quad . \tag{10-7}$$

One recalls that any finite-dimensional representation of a compact Lie-group or a finite group is equivalent to a unitary representation (Theorem 7.8, Subsec. 7.2.1 or Prop. C.54). If so, one assumes that the representation M(G) in E is unitary, i.e. :

$$\forall u,v \in E \qquad \forall g \in G \qquad (M(g)u,M(g)v)_E = (u,v)_E \quad . \tag{10-8}$$

10.1.2. REPRESENTATIONS IN A FUNCTIONAL VECTOR SPACE

i) The functional vector space. Let $\mathcal{F}$ be a vector space over the complex field of vectorial functions ϕ of a set E into a vector space F over the complex field :

$$\forall \phi \in \mathcal{F} \qquad \forall p \in E \xrightarrow{\ \phi\ } \phi(p) \in F \quad . \tag{10-9}$$

If at the same time there exists a measure μ on E and a Hermitian structure on F, a Hermitian inner product in $\mathcal{F}$ can be defined as follows :

$$\forall \phi,\phi' \in \mathcal{F} \longrightarrow (\phi,\phi')_{\mathcal{F}} = \int_E d\mu(p)\ (\phi(p),\phi'(p))_F \in C \quad , \tag{10-10}$$

provided the above integral exists.

For instance, let F be a d-dimensional Hermitian vector space and $g_{\rho\sigma} = (e_\rho, e_\sigma)_F$ the metric tensor with respect to any basis (e_ρ). One has :

$$\forall \phi \in \mathcal{F} \qquad \forall p \in E \qquad \phi(p) = \phi^\rho(p) e_\rho \in F \quad , \quad \phi^\rho(p) \in C \quad . \tag{10-11}$$

We shall then write abusively :

$$\forall \phi \in \mathcal{F} \qquad \phi = \phi^\rho\, e_\rho \quad , \tag{10-12}$$

where the ϕ^ρ are complex functions defined on E and the e_ρ are here taken as functions in $\mathcal{F}$ which, respectively, map all p in E into the basis vectors e_ρ of F. The inner product in $\mathcal{F}$ takes the form :

$$\forall \phi = \phi^\rho\, e_\rho \quad , \quad \phi' = \phi'^\rho\, e_\rho \in \mathcal{F} \qquad (\phi,\phi')_{\mathcal{F}} = \int_E d\mu(p)\ \phi^{\mu *}(p) g_{\mu\sigma} \phi'^\sigma(p) \, , \tag{10-13}$$

where the above integral exists if, for instance, $\mathcal{F}$ is taken as the space of functions ϕ the components of which ϕ^ρ , with respect to any fixed basis of F, belong to the set $L^2(E,d\mu)$ of square integrable functions on E with the measure μ .

ii) Representations of a group G in $\mathcal{F}$. Let us assume that G has been realized as a transformation group M(G) on E, and that there exists a representation D(G) of G in F. This allows us to determine three representations $\mathcal{D}_1, \mathcal{D}_2$, and $\mathcal{D}$ of G in the functional vector space $\mathcal{F}$ (see Fig. 10.1).

1–The representation D(G) of G in F induces a representation $\mathcal{D}_1(G)$ in $\mathcal{F}$:

$$\forall g \in G \xrightarrow{\mathcal{D}_1} \mathcal{D}_1(g)$$

$$\forall \phi \in \mathcal{F} \longrightarrow \mathcal{D}_1(g)\phi \in \mathcal{F} \ , \ \forall p \in E \quad [\mathcal{D}_1(g)\phi](p) = D(g)\phi(p) \ . \qquad (10\text{-}14)$$

If $\mathcal{F}$ is equiped with a Hermitian structure as specified by Eq. (10–10), and if D(G) is unitary, Eq. (10–8), so is the representation $\mathcal{D}_1(G)$.

2–The realization of G as a transformation group M(G) on E enables us to obtain a second representation $\mathcal{D}_2(G)$ in $\mathcal{F}$ by :

$$\forall g \in G \xrightarrow{\mathcal{D}_2} \mathcal{D}_2(g)$$

$$\forall \phi \in \mathcal{F} \longrightarrow \mathcal{D}_2(g)\phi = \phi \circ M(g)^{-1} \in \mathcal{F} \ . \qquad (10\text{-}15)$$

Indeed, the mappings $\mathcal{D}_2(g)$ are clearly linear and fulfil the relations, e being the unit element of G :

$$\mathcal{D}_2(e) = \mathbf{1} \ , \ \forall g,g' \in G \qquad \mathcal{D}_2(g)\mathcal{D}_2(g') = \mathcal{D}_2(gg') \ ,$$

since one has :

$$\forall g,g' \in G \qquad \mathcal{D}_2(g)[\mathcal{D}_2(g')\phi] = [\mathcal{D}_2(g')\phi]\circ M(g)^{-1} = \phi\circ M(g')^{-1}M(g)^{-1}$$

$$= \phi\circ M(gg')^{-1} = \mathcal{D}_2(gg')\phi \ . \qquad (10\text{-}16)$$

If $\mathcal{F}$ is endowed with a Hermitian structure as defined by Eq. (10–10) and if the measure on E is invariant under the transformation group M(G), Eq. (10–6), the above representation is unitary for :

$$\forall \phi,\phi' \in \mathcal{F} \ \forall g \in G \qquad (\mathcal{D}_2(g)\phi,\mathcal{D}_2(g)\phi')_{\mathcal{F}} = \int_E d\mu(p)\,([\mathcal{D}_2(g)\phi](p),[\mathcal{D}_2(g)\phi'](p))_F$$

$$= \int_E d\mu(p)\ (\phi[M(g)^{-1}p],\ \phi'[M(g)^{-1}p])_F = \int_E d\mu[M(g)p]\,(\phi(p),\phi'(p))_F$$

$$= \int_E d\mu(p)\ (\phi(p),\phi'(p))_F = (\phi,\phi')_{\mathcal{F}} \ . \qquad (10\text{-}17)$$

3–One should remark that the two representations $\mathcal{D}_1(G)$ and $\mathcal{D}_2(G)$ in $\mathcal{F}$ commute, i.e. :

$$\forall g,g' \in G \qquad \mathcal{D}_1(g)\ \mathcal{D}_2(g') = \mathcal{D}_2(g')\ \mathcal{D}_1(g) \ . \qquad (10\text{-}18)$$

The product $\mathcal{D}_1\mathcal{D}_2 = \mathcal{D}_2\mathcal{D}_1$ determines a new representation $\mathcal{D}(G)$ in $\mathcal{F}$:

$$\forall g \in G \xrightarrow{\mathcal{D}=\mathcal{D}_1\mathcal{D}_2} \mathcal{D}(g)$$

$$\forall \phi \in \mathcal{F} \longrightarrow \mathcal{D}(g)\phi = \mathcal{D}_1(g)\mathcal{D}_2(g)\phi \in \mathcal{F} \ , \qquad (10\text{-}19)$$

$$\forall p \in E \qquad [\mathcal{D}(g)\phi](p) = D(g)\phi[M(g)^{-1}p] \ .$$

If both the representations $\mathcal{D}_1(G)$ and $\mathcal{D}_2(G)$ are unitary, so is the representation $\mathcal{D}(G)$.

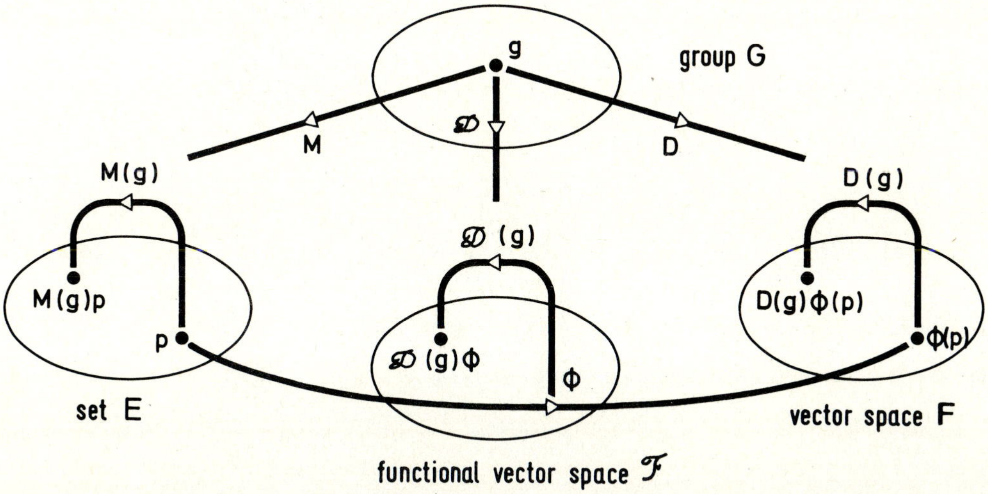

Fig. 10.1. Determination of a representation $\mathcal{D}(G)$ of a group G in a functional vector space $\mathcal{F}$ by means of the transformation group M(G) on E and the representation D(G) in F, according to Eq. (10–19).

Remarks :

1–The above method can be extended to the case where the mapping D(g) also depends upon p in E, i.e. :

$$\forall g \in G \quad \forall p \in E \longrightarrow D(g;p)$$
$$\forall u \in F \longrightarrow D(g;p)u \in F \quad . \tag{10-20}$$

It can then be checked that a necessary and sufficient condition that the mapping :

$$\forall g \in G \longrightarrow \mathcal{D}(g)$$
$$\forall \phi \in \mathcal{F} \longrightarrow \mathcal{D}(g)\phi \in \mathcal{F}\,, \quad \forall p \in E \quad [\mathcal{D}(g)\phi](p) = D(g;p)\phi[M(g)^{-1}p] \quad , \tag{10-21}$$

be a representation of G in $\mathcal{F}$ is that the applications D(g;p) satisfy the following multiplication law :

$$\forall g,g' \in G \qquad \forall p \in E \qquad D(gg';p) = D(g;p)D[g';M(g)^{-1}p] \quad . \tag{10-22}$$

Hence, for a given p, D(G;p) is not a representation of G in F.

2–Another extension of the above method amounts to considering any set E' instead of a vector space F, and assume that G has been realized as a transformation group M'(G) on E'. The set $\mathcal{F}$ of mappings ϕ of E into E' is no longer necessarily a vector space. A realization of G as a transformation group $\mathcal{M}(G)$ on $\mathcal{F}$ is then defined by :

$$\forall g \in G \longrightarrow \mathcal{M}(g)$$
$$\forall \psi \in \mathcal{F} \longrightarrow \mathcal{M}(g)\psi \in \mathcal{F} \quad , \quad \forall p \in E \quad [\mathcal{M}(g)\psi](p) = M'(g)\psi[M(g)^{-1}p] \in E' \quad . \tag{10-23}$$

The extension 1 above can also be applied to this case.

iii) **An example.** The procedure described above has already been used in Subsec. 6.1.1 to construct the unitary and irreducible representations of the group SU(2). Let us briefly establish the correspondence between the above general formalism and this special case :

(1)–E = $\{\zeta = \binom{\xi}{\eta}\ ; \xi,\eta \in C\}$ is the four–manifold C^2 of complex spinors.

–G = SU(2) is a three–parameter Lie group.

–M(G) = M [SU(2)] is defined by :

$$\forall U \in SU(2) \xrightarrow{M} M(U)$$
$$\forall \zeta \in C^2 \longrightarrow M(U)\zeta = (U^t)^{-1}\zeta \quad . \tag{10-24}$$

The realization M(G) is here a representation since E is a vector space and the mappings M(U) are linear.

–The invariant measure $d\mu(\zeta)$ on E is defined by Eq. (6–4).

(2)–F = C is the trivial one–dimensional Hermitian vector space over C :

$$\forall u,v \in C \qquad (u,v) = u^* v \in C \quad . \tag{10-25}$$

–D(G) is taken as the identity representation, i.e. :

$$\forall g \in G \qquad D(g) = \mathbf{1} \quad . \tag{10-26}$$

(3)– $\mathcal{F}$ is the vector space over the complex field of entire functions of the complex variables ξ and η.

– $\mathcal{F}$ is Hermitian with respect to the inner product defined by Eq. (6–3) which corresponds to Eq. (10–10).

– $\mathcal{D}$(G) = $\mathcal{D}$[SU(2)] is determined by Eq. (6–8) in agreement with Eq. (10–19) :

$$\forall U \in SU(2) \xrightarrow{\mathcal{D}} \mathcal{D}(U) = T_U$$
$$\forall \phi \in \mathcal{F} \longrightarrow T_U\phi \in \mathcal{F},\ \forall \zeta \in C^2\ (T_U\phi)(\zeta) = (U^t\zeta) \quad . \tag{10-27}$$

This representation is unitary. Its infinitesimal generators are given by Eqs. (6–30a,b). Cf. Subsec. 10.2.2 for the general study of the infinitesimal generators of the representation $\mathcal{D}$(G).

Another important example of the application of the general method is provided by the regular representations of a group, which are studied next.

10.1.3. THE LEFT AND RIGHT REGULAR REPRESENTATIONS OF A GROUP

10.1.3.1. Left and right translations

For all g in a group G, the left and right translations, respectively, L_g and R_g are the two one–to–one mappings of G onto itself defined by (Def. C.51) :

$$\forall g,g' \in G \qquad L_g g' = gg' \quad , \qquad R_g g' = g'g \quad . \tag{10-28}$$

It has already been remarked in Subsec. 7.1.1.2 that the left translation commutes with the right one, i.e. :

$$\forall g,g' \in G \qquad L_g R_{g'} = R_{g'} L_g \quad . \tag{10-29}$$

The same is true for the two mappings L_g and $R_{g^{-1}}$. One should note that the product $L_g R_{g^{-1}}$ is equal to the internal automorphism a(g), (Def. C.23) :

$$\forall g,g' \in G \qquad L_g R_{g^{-1}} g' = R_{g^{-1}} L_g \, g' = g g' g^{-1} = a(g) g' \quad . \tag{10-30}$$

The two applications $g \to L_g$ and $g \to R_{g^{-1}}$ of G into the set of one-to-one mappings of G onto itself are two realizations of G as transformation groups on itself. Indeed, one has :

$$\forall g,g' \in G \quad \begin{cases} L_e = \mathbf{1} \quad , & L_g L_{g'} = L_{gg'} \quad , \\ R_{e^{-1}} = \mathbf{1} \quad , & R_{g^{-1}} R_{g'^{-1}} = R_{(gg')^{-1}} \quad , \end{cases} \tag{10-31}$$

where e is the unit element of G. We denote these two realizations of G by L(G) and $R(G^{-1})$, respectively. One should remark that the set of right translations R_g for all g in G also form a group, denoted by R(G), the multiplication law of which reads :

$$\forall g,g' \in G \qquad R_g R_{g'} = R_{g'g} \quad . \tag{10-32}$$

Hence, this transformation group on G is not a realization of G. It is said to be anti-isomorphic to G, i.e. it is isomorphic to G when the elements are taken in the reverse order.

10.1.3.2. Regular representations and some of their properties

i) **Definition.** Let us apply the general method described in Subsec. 10.1.2 in order to determine representations of a group G in a vector space F over the complex field of functions ϕ taking complex values and defined on G. One thus considers the special case where :

(1) The set E is taken equal to the group G itself. Two realizations of G as transformation groups on itself are then left to our disposal, i.e. L(G) and $R(G^{-1})$. Furthermore, if for instance G is a Lie group there exists a left and right invariant measure on G (Theorem 7.1, Subsec. 7.1.1.1).

(2) As in the first example given in Subsec. 10.1.2,iii), F is here the complex field and D(G) as $\mathcal{D}_1(G)$ are the identity representations.

From Eq. (10–19) one then determines two representations of G in F :

–the left regular representation $D^\ell(G)$:

$$\begin{aligned} &\forall g \in G \quad \xrightarrow{\;D^\ell\;} \quad D^\ell(g) \\ &\qquad \forall \phi \in F \longrightarrow \quad D^\ell(g)\phi = {}^g\phi = \phi \circ L_g^{-1} \in F \\ &\forall g' \in G \qquad {}^g\phi\,(g') = \phi(L_g^{-1} g') = \phi(g^{-1} g') \quad , \end{aligned} \tag{10-33}$$

–the right regular representation $D^r(G)$:

$$\forall g \in G \xrightarrow{D^r} D^r(g)$$

$$\forall \phi \in F \longrightarrow D^r(g)\phi = \phi^g = \phi \circ R_{g^{-1}}^{-1} = \phi \circ R_g \in F \quad ,$$

$$\forall g' \in G \qquad \phi^g(g') = \phi(R_g g') = \phi(g'g) \quad . \tag{10-34}$$

A slightly different interpretation of the regular representations is given in Def. C.36.

ii) **Properties,** cf. Prop. C.63 for finite groups :

1–The regular representations are faithful. This follows from the one–to–one character of the left and right translations.

2–The commutation of L_g and $R_{g'^{-1}}$, Eq. (10–29), implies that the left regular representation commutes with the right one :

$$\forall g, g' \in G \qquad D^\ell(g)\, D^r(g') = D^r(g')\, D^\ell(g) \quad . \tag{10-35}$$

3–The left and right regular representations are equivalent. Indeed, let I be the involutive ($I^2 = \mathbf{1}$) linear mapping of F onto itself defined by :

$$\forall \phi \in F \xrightarrow{I} I\phi = \phi \circ I \in F \quad , \quad \forall g \in G \qquad I(g) = g^{-1} \quad . \tag{10-36}$$

One then gets (cf. Prop. C.18 and remark 1 after Def. C.62) :

$$\forall g \in G \qquad \forall \phi \in F \quad [I D^r(g) I]\phi = [D^r(g) I\phi] \circ I = (I\phi) \circ R_g \circ I$$

$$= \phi \circ I \circ R_g \circ I = \phi \circ L_{g^{-1}} = D^\ell(g)\, \phi \quad , \tag{10-37}$$

thereby obtaining the result :

$$\forall g \in G \qquad D^\ell(g) = I\, D^r(g) I \quad . \tag{10-38}$$

4–If there exists a left invariant measure μ_ℓ (resp. right μ_r) on G, the left (resp. right) regular representation is unitary with respect to the Hermitian inner product determined on F as in Eq. (10–10) by :

$$\forall \phi, \phi' \in F \rightarrow (\phi, \phi')_G = \int_G d\mu_\ell(g)\, \phi^*(g) \phi'(g) \in C \quad (\text{resp. with } d\mu_r(g)), \tag{10-39}$$

provided the above integral exists. This is for instance the case for a compact Lie group (Theorem 7.5, Subsec. 7.1.2.2) and taking F as the space $L^2(G, d\mu)$ of square integrable complex functions defined on G with the left and right invariant measure $d\mu$ (Subsec. 7.2.1,ii)). From the rearrangement theorem (Prop. C.6), the same result holds true for finite groups, the Hermitian inner product on F being then determined by :

$$\forall \phi, \phi' \in F \rightarrow (\phi, \phi')_G = \sum_{g \in G} \phi^*(g)\, \phi(g) \in C \quad . \tag{10-40}$$

10.1.3.3. Specific properties of the regular representations

The interest of the regular representations of a group is mainly based on the two following theorems :

Theorem 10.1. Any irreducible representation D(G) of a group G in a Hermitian vector space E is equivalent to an irreducible restriction of the left (or right, they are equivalent) regular representation of G.

Proof :

1–For a fixed nonzero vector u in E, one associates to any v in E a complex function ϕ^ℓ_v defined on G by :

$$\forall v \in E \to \phi^\ell_v , \qquad \forall g \in G \qquad \phi^\ell_v(g) = (u, D(g^{-1})v) \in C \quad . \tag{10-41}$$

The set of these functions ϕ^ℓ_v for all v in E is a vector space F^ℓ_u over the complex field since :

$$\forall \lambda, \lambda' \in C \qquad \forall v, v' \in E \qquad \lambda\, \phi^\ell_v + \lambda'\, \phi^\ell_{v'} = \phi^\ell_{\lambda v + \lambda' v'} \quad . \tag{10-42}$$

The mapping $\mathcal{D}^\ell$, such that :

$$\forall g \in G \xrightarrow{\mathcal{D}^\ell} \mathcal{D}^\ell(g)$$
$$\forall \phi^\ell_v \in F^\ell_u \longrightarrow \mathcal{D}^\ell(g)\, \phi^\ell_v = \phi^\ell_{D(g)v} \in F^\ell_u \quad , \tag{10-43}$$

determines a representation $\mathcal{D}^\ell$(G) of G in F^ℓ_u which is equal to the restriction to this subspace of the left regular representation D^ℓ(G). Indeed, one has :

$$\forall v \in E \qquad \forall g, g' \in G \qquad \phi^\ell_{D(g)v}(g') = (u, D(g'^{-1})D(g)v)$$
$$= (u, D[(g^{-1}g')^{-1}]v) = \phi^\ell_v(g^{-1}g') = {}^g\phi^\ell_v(g') \quad . \tag{10-44}$$

2–It remains to be shown that D(G) and $\mathcal{D}^\ell$(G) are equivalent representations, i.e. to show that the mapping $v \to \phi^\ell_v$ is a regular linear application of E onto F^ℓ_u. This mapping is linear, Eq. (10–42), and by construction surjective, let us show that it is also injective. Its kernel K is an invariant subspace of F. Indeed, if v belongs to K so does D(g)v for all g in G, for :

$$\forall g, g' \in G \qquad \phi^\ell_v(g') = 0 \Rightarrow \phi^\ell_{D(g)v}(g') = \phi^\ell_v(g^{-1}g') = 0 \quad .$$

The representation D(G) being assumed irreducible, K is either $\{0\}$ or E. This latter case is excluded, for setting v=u and g=e in Eq. (10–41) one gets :

$$\phi^\ell_u(e) = (u, D(e)u) = (u,u) > 0 \quad ,$$

the Hermitian inner product being positive definite. Hence K = $\{0\}$, which ends the proof.

3–It can similarly be shown that D(G) is equivalent to the restriction of the representation D^r(G) to the subspace F^r_u spanned by the functions ϕ^r_v defined as follows :

$$\forall v \in E \to \phi^r_v , \qquad \forall g \in G \quad \phi^r_v(g) = (u, D(g)v) \quad . \tag{10-45}$$

For many groups, and especially for compact groups it is now shown that all the irreducible representations can be obtained from the decomposition of a regular representation into its irreducible representations, i.e. one has (cf. Prop. C.65 for finite groups) :

Theorem 10.2. Let G be a compact group and $\{D^\alpha(G)\}$ a complete set of nonequivalent irreducible unitary representations $D^\alpha(G)$; the dimension of $D^\alpha(G)$ being d_α. To any unitary matrix representative $(D^\alpha(G)^i_j)$ of $D^\alpha(G)$ one associates the functions $(D^\alpha)^i_j$ of G into the complex field defined by :

$$\forall g \in G \qquad (D^\alpha)^i_j(g) = D^\alpha(g)^i_j \quad .$$

For a finite group as well as for a compact Lie group, $L^2(G,d\mu)$ denotes the vector space of square integrable complex functions defined on G with the invariant measure $\left(\sum_{g\in G} \text{ or } \int_G d\mu(g)\right)$.
(1) The left (resp. right) regular representation $D^\ell(G)$ (resp. $D^r(G)$) in $L^2(G,d\mu)$ is the direct sum of all the irreducible representations, $D^\alpha(G)$ being contained d_α times :

$$D^\ell(G) = \sum_\alpha \oplus\ d_\alpha\ D^\alpha(G) \qquad (\text{resp. } D^r(G) = \sum_\alpha \oplus\ d_\alpha\ D^\alpha(G)) \ . \tag{10-46}$$

(2) The restriction of $D^\ell(G)$ (resp. $D^r(G)$) to the d_α-dimensional subspace F^α_j (resp. $F^{\alpha i}$) spanned by the functions $\{(D^\alpha)^{i*}_j\ ;\ i=1,\dots,d_\alpha\}$ (resp. $\{(D^\alpha)^i_j\ ;\ j=1,\dots,d_\alpha\}$) is equivalent to the representation $D^\alpha(G)$. In addition, one has :

$$\forall \alpha,j \qquad \forall g\in G \qquad D^\ell(g)(D^\alpha)^{i*}_j = \sum_k (D^\alpha)^{k*}_j\ D^\alpha(g)^k_i \quad , \tag{10-47a}$$

$$\forall \alpha,i \qquad \forall g\in G \qquad D^r(g)(D^\alpha)^i_j = (D^\alpha)^i_k\ D^\alpha(g)^k_j \quad . \tag{10-47b}$$

Proof :

1–From the Peter–Weyl theorem (Theorem 7.12, Subsec. 7.2.1,ii)), for all α, i, and j the functions $(D^\alpha)^i_j$, as their complex conjugates, form a complete orthogonal basis of $L^2(G,d\mu)$. Therefore, this space is the direct sum of all the subspaces F^α_j, or $F^{\alpha i}$, i.e. :

$$L^2(G,d\mu) = \sum_{\alpha,j} \oplus F^\alpha_j = \sum_{\alpha,i} \oplus F^{\alpha i} \quad . \tag{10-48}$$

2–Let us now show that F^α_j is an invariant subspace of the representation $D^\ell(G)$ in $L^2(G,d\mu)$. Indeed, one has :

$$\forall g,g' \in G \qquad [D^\ell(g)(D^\alpha)^{i*}_j](g') = [(D^\alpha)^{i*}_j](g^{-1}g')$$
$$= \sum_k [(D^\alpha)^{k*}_j](g')\ D^\alpha(g)^k_i \quad . \tag{10-49}$$

These equations also show that the matrix representative of the restriction $D^{\ell\,\alpha}_{\ j}(G)$ of $D^\ell(G)$ to F^α_j with respect to the basis $((D^\alpha)^{k*}_j)$ is equal to the considered matrix representative of $D^\alpha(G)$. Hence, this restriction of $D^\ell(G)$ is equivalent to the representation $D^\alpha(G)$ and thus irreducible. Since this result holds true for all $j=1,\dots,d_\alpha$, the d_α representations $D^{\ell\,\alpha}_{\ j}$ for a given α are all equivalent.

3–One can similarly show that for any $i=1,\dots,d_\alpha$, the d_α-dimensional subspaces $F^{\alpha i}$ are invariant under the right regular representation, and that the restriction of $D^r(G)$ to each one of these subspaces is equivalent to the irreducible representation $D^\alpha(G)$. Indeed, one has :

$$\forall g,g' \in G \qquad [D^r(g)(D^\alpha)^i_j](g') = [(D^\alpha)^i_j](g'g) = [(D^\alpha)^i_k](g')D^\alpha(g)^k_j \ . \tag{10-50}$$

As a consequence of the above theorem, the regular representations contain the identity representation only once since it is one-dimensional.

10.2. APPLICATION TO MANIFOLDS AND LIE GROUPS

Before going through this section, it is recommended that Appendices C.5.3,4 be read, where several definitions and properties about manifolds and Lie groups can be found. We adopt here the notations of these appendices. As pointed out in Appendix C.5.3.1, the theory of manifolds can be thought of as a generalization of the usual geometry. This is why, when dealing with manifolds, it is worthwhile to keep always in mind some simple geometrical interpretations. Applied to the theory of Lie groups, this point of view is well developed by Cartan[13,14], cf. also Racah[45].

10.2.1. DEPENDENCE UPON THE CHOICE OF AN ATLAS ON A MANIFOLD

i) **Local parameters, reference frame.** We pointed out in Subsec. 1.1.1 (cf. also Subsec. 8.3.1.1) the intrinsic or nonintrinsic character of the considered objects. A vector is thus an intrinsic object, whereas its components are by definition nonintrinsic since they depend upon the choice of a basis. In other words, the choice of a basis $a=(e_i)$ in an n-dimensional vector space E over the real field allows us to define a one-to-one mapping f_a of E onto R^n :

$$a=(e_i)\rightarrow f_a\ ,\quad \forall v=v^i e_i \in E \xrightarrow{\ f_a\ } f_a(v)=\{f_a^i(v)=v^i\ ;\ i=1,\dots,n\}\in R^n\ . \tag{10-51}$$

The same considerations apply to an n-point space (Subsec. 1.1.5,ii)) for which, given any reference frame, a point is defined by its coordinates. An n-dimensional vector space over the real field, as an n-point space, is actually a special case of n-manifold. A similar distinction between intrinsic and nonintrinsic objects can also be made for manifolds. Thus, a point p in an n-manifold E is said to be <u>intrinsic</u>, while its local parameters are nonintrinsic for they depend upon the choice of an atlas A, i.e. a set of local charts (U,f), (Def. C.52) :

$$A=\{(U,f)\}\qquad \forall p\in U\subseteq E \xrightarrow{\ f\ } f(p)=\{f^i(p)\ ;\ i=1,\dots,n\}\in R^n\ . \tag{10-52}$$

An atlas on a manifold therefore plays the role of a basis for a vector space. By analogy with the usual geometry, one says that an atlas A on a manifold defines a <u>reference frame</u>, also denoted by A. The $f^i(p)$ in the above equations are called the local parameters of p with respect to A.

ii) **Representative of a mapping.** Let ψ be a mapping of an n-manifold E into an n'-manifold E'. This application is intrinsically defined. Now, given any atlases A on E and A' on E', there corresponds to ψ an application of a part of R^n into $R^{n'}$ which maps the local parameters of all p in E with respect to ψ into the local parameters of $\psi(p)$ with respect to A'. This mapping, denoted either by $\tilde{\psi}$ (as in Def. C.53) or $\psi_{AA'}$ (to underline its dependence upon A and A'), is said to be the <u>representative</u> of ψ with respect to A and A'. It plays the role of the matrix representative of a linear application. It is emphasized that the expression of $\psi_{AA'}$ in terms of ψ is only local, since it depends upon the local charts which determine the local parameters p and $\psi(p)$:

$$\begin{matrix} E \\ A=\{(U,f)\} \end{matrix} \xrightarrow{\psi} \begin{matrix} E' \\ A'=\{(U',f')\} \end{matrix} \Rightarrow \forall p \in U \quad \psi(p) \in U' \ \tilde{\psi} = \psi_{AA'} = f' \circ \psi \circ f^{-1} \quad . \tag{10-53}$$

Nevertheless, to avoid cumbersome notations we shall from now on always assume that the above expression of $\psi_{AA'}$ is valid for all p in E. This amounts to considering that U=E and U'=E'. One should remark that if ψ is one–to–one, so is $\psi_{AA'}$ and since :

$$(\psi_{AA'})^{-1} = (\psi^{-1})_{AA'} \quad , \tag{10-54}$$

the inverse mapping of $\psi_{AA'}$ can be denoted by $\psi^{-1}_{AA'}$ (or $\tilde{\psi}^{-1}$).

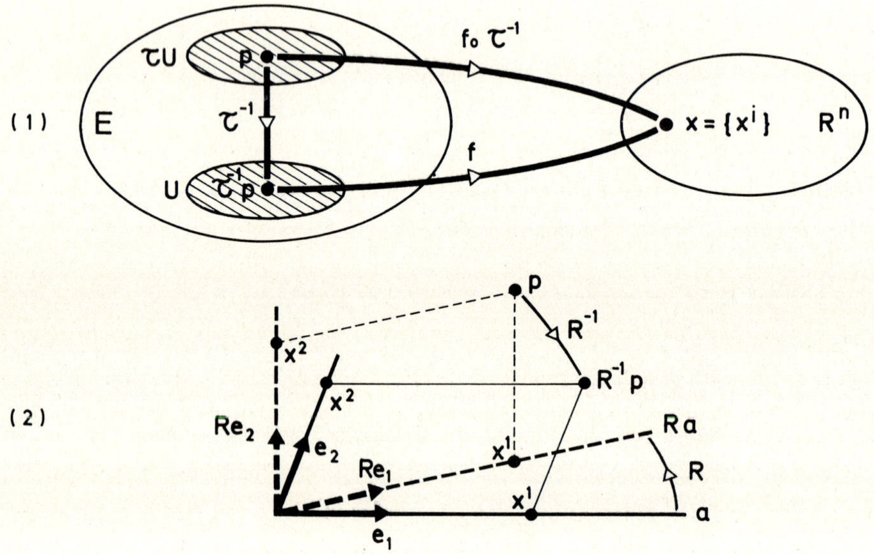

Fig. 10.2. The passive point of view of one–to–one mappings :
(1) an application τ of an n–manifold E onto itself,
(2) a linear application R of a point space $E_2^{\cdot}$ onto itself.

iii) Active and passive points of view of a mapping :

1–The active and the passive points of view of a regular linear mapping of a vector space onto itself is discussed in Subsec. 1.2.1. The so-called equivalence between these two points of view is now used to extend the above considerations to a one–to–one mapping τ of an n–manifold E onto itself. To any point p in E, τ associates a new point $\tau(p)$. This is the active point of view of τ . Its passive point of view is defined as follows. To any atlas A on E, τ associates a new atlas, denoted by τA , such that :

$$A = \{(U,f)\} \xrightarrow{\tau} \tau A = \{(\tau U, f \circ \tau^{-1})\} \quad . \tag{10-55}$$

Namely, see Fig. 10.2, the local parameters of any point p with respect to τA are equal to the local parameters of τ^{-1} p with respect to A (equivalently, the local parameters of τp with respect to τA are equal to the local parameters of p with respect to A).

2–Let us now establish the relation between the two representatives ψ_A and $\psi_{\tau A}$ of a mapping ψ of E into itself. From Eq. (10–53) one gets :

$$\psi_{\tau A} = (f\circ\tau^{-1})\circ\psi\circ(f\circ\tau^{-1})^{-1} = (f\circ\tau^{-1}\circ f^{-1})\circ(f\circ\psi\circ f^{-1})\circ(f\circ\tau\circ f^{-1})$$

$$= \tau_A^{-1} \circ \psi_A \circ \tau_A \quad . \tag{10-56}$$

The above equation is the generalization of Eq. (1–14) which relates the matrix representatives of a linear application with respect to two different bases. The representative τ_A plays the role of the matrix t.

10.2.2. MOVING REFERENCE FRAME OF A TRANSFORMATION GROUP ON A MANIFOLD

i) Definition. We assume that a group G has been realized as a transformation group M(G) on a manifold E. Let A_o be any atlas on E, we shall call it the absolute atlas or reference frame of E. Using the passive point of view of the mappings M(g) for all g in G, one generates according to Eq. (10–55) a set of atlases or reference frames in E by :

$$A_o = \{(U_o,f_o)\} \qquad \forall g\in G \quad A_g = M(g)A_o = \{(U_g=M(g)U_o, f_g=f_o\circ M(g)^{-1})\} \quad . \tag{10-57}$$

Using a geometrical language, the reference frame A_g which depends upon g in G is called the moving reference frame of the transformation group M(G) on E. Furthermore, one says that a given g defines a position of this reference frame. Since :

$$\forall\, g,g'\in G \qquad A_g = M(g)M(g')^{-1}A_{g'} = M(gg'^{-1})A_{g'} \quad , \tag{10-58}$$

all the positions of the moving reference frame can be generated from any one of them. The absolute reference frame is thus a position of the moving reference frame which is indistinguishable from the others.

One should remark that the notion of moving reference frame can more generally be defined for any set (not necessarily a group) of mappings of a manifold into itself, cf. for instance Cartan[14] (Chap. V, Sec. I).

ii) Representative of a mapping with respect to the moving reference frame. For a given atlas A on the manifold E, the representatives $M_A(g)$ of M(g) for all g in G clearly determine a realization of G as a transformation group $M_A(G)$ on the set of local parameters of E. Let ψ be any mapping of E into itself. According to Eq. (10–56), its representatives ψ_{A_o} and ψ_{A_g} with respect to, respectively, the absolute and the moving reference frames of the transformation group M(G) are related by :

$$\forall\, g\in G \qquad \psi_{A_g} = M_{A_o}(g)^{-1} \circ \psi_{A_o} \circ M_{A_o}(g) \quad . \tag{10-59}$$

In particular, setting ψ = M(g) in the above equation yields :

$$\forall\, g\in G \qquad M_{A_g}(g) = M_{A_o}(g)^{-1} \circ M_{A_o}(g) \circ M_{A_o}(g) = M_{A_o}(g) \quad . \tag{10-60}$$

Therefore, the representative of a transformation group with respect to its moving reference frame does not depend upon the position of this reference frame. In other words, the equations which determine the transformation group on the local parameters of the points in a manifold with respect to either the absolute reference frame or the moving one are the same.

10.2.3. INFINITESIMAL GENERATORS OF THE REPRESENTATIONS OF A GROUP IN A FUNCTIONAL SPACE

10.2.3.1. Hypotheses and notations

The general study made in Sec. 10.1 is now applied to the special case where both the set E and the group G are manifolds. More precisely, one makes the following hypotheses :

H1. The set E is an analytic r–manifold. With respect to an atlas $A = \{(P,h)\}$ on E, the local parameters of any p in E are labelled by the Greek indices α or β :

$$\forall p \in P \subseteq E \qquad h(p) = x \Leftrightarrow \{h^\alpha(p) = x^\alpha \;\; ; \;\; \alpha=1,\dots,r\} \in R^r \quad . \tag{10-61}$$

H2. The group G is an n–parameter Lie group. With respect to an atlas $\{(U,f)\}$ on G, the local parameters of any g in G are labelled by the Latin indices i,j, or k :

$$\forall g \in U \subseteq G \qquad f(g) = a \Leftrightarrow \{f^i(g) = a^i \;\; ; \;\; i=1,\dots,n\} \in R^n \quad . \tag{10-62}$$

The same notations as in Def. C.60 are used for the locally defined analytic functions φ and ψ which express the group laws in terms of the local parameters, i.e. :

$$g_c = g_a g_b \Leftrightarrow c = \varphi(a,b) \quad , \qquad g_a^{-1} = g_{\bar{a}} \Leftrightarrow \bar{a} = \psi(a) \quad . \tag{10-63}$$

Furthermore, since it imposes no restriction, the local parameters of the unit element e of the group are taken equal to zero :

$$g_0 = e \quad . \tag{10-64}$$

This choice implies :

$$\psi(0) = 0 \quad ; \quad \forall a \quad \varphi(0,a)=\varphi(a,0)=a \quad , \quad \varphi[a,\psi(a)]=\varphi[\psi(a),a]= 0 \quad . \tag{10-65}$$

One should remark that expanding these constraints in power series of a, one gets up to first order :

$$\frac{\partial\varphi^i(a,0)}{\partial a^j} = \frac{\partial\varphi^i(0,a)}{\partial a^j} = \delta^i_j \quad , \tag{10-66a}$$

$$\frac{\partial\varphi(a,b)}{\partial a^i} + \frac{\partial\varphi(a,b)}{\partial b^j}\frac{\partial\psi^j(a)}{\partial a^i} = 0 \quad , \qquad \frac{\partial\varphi(b,a)}{\partial b^j}\frac{\partial\psi^j(a)}{\partial a^i} + \frac{\partial\varphi(b,a)}{\partial a^i} = 0 \quad , \tag{10-66b}$$

where $b=\bar{a}=\psi(a)$.

H3. The realization of G as a transformation group M(G) on E is analytic. Namely, for all g in G the mappings M(g) of E onto itself are analytic (Def. C.53), and moreover for all p in E the mappings $g \to M(g)p$ of G into E are also analytic.

H4. The vector space F over the complex field is d-dimensional. The basis vectors and the corresponding components of any vector in F are labelled by the Greek indices ρ or σ :

$$\text{basis } (e_\rho) \qquad \forall u \in F \qquad u = u^\rho e_\rho \qquad \{u^\rho ; \rho=1,\dots,d\} \in C^d \quad . \tag{10-67}$$

H5. The representation D(G) of G in F is analytic (special case of an analytic realization, cf. H3 above).

H6. The functional vector space $\mathcal{F}$ over the complex field is such that all ϕ in $\mathcal{F}$ are analytic mappings of E into the vector space F. The representative of ϕ with respect to an atlas on E is denoted by $\tilde{\phi}$, Eq. (10-53).

The atlases on E and G, as well as the basis on F, are kept fixed all along the remaining of this Subsec. 10.2.3. All the nonintrinsic objects (local parameters, components, representatives or matrix representatives of mappings, symbols of infinitesimal transformations...) we are going to consider are always defined with respect to these given reference frames. The dependence upon the changes of atlases is discussed later in Subsec. 10.2.4.

10.2.3.2. Expressions of the infinitesimal generators

i) **Transformation group M(G) on E.** From H1, H2, and H3 there exists an analytic application m of $R^n \times R^r$ into R^r which realizes the transformation group M(G) on E in terms of the local parameters of E and G, i.e. :

$$p_{x'} = M(g_a)p_x \quad \Leftrightarrow \quad x' = m(a\; ;\; x) \quad . \tag{10-68}$$

The expression of m in terms of M is local since it depends upon the local charts (P,h) and (U,f) on E and G as follows :

$$m(a;x) = \{h \circ M[f^{-1}(a)] \circ h^{-1}\}\,(x) \quad . \tag{10-69}$$

The group laws are then expressed by :

$$\forall x \qquad m[a;m(b;x)] = m[\varphi(a,b);x] \quad , \tag{10-70a}$$

$$y = m(a;x) \Leftrightarrow x = m^{-1}(a;y) = m(\bar{a};y) = m[\psi(a);y] \quad . \tag{10-70b}$$

From the choice of the local parameters of the unit element of G, Eq. (10-64), one gets :

$$\forall x \qquad m(0\; ;\; x) = x \quad . \tag{10-71}$$

An infinitesimal element of the transformation group M(G) is a mapping $M(g_{\delta a})$ defined by infinitesimal local parameters $\delta a = \{\delta a^i\}$. One then has :

$$\forall p = p_x \in E \qquad p_{x+dx} = M(g_{\delta a})p_x \Leftrightarrow x + dx = m(\delta a;x) \quad . \tag{10-72}$$

The mapping m being analytic, the right hand side of the above equation can be expanded as power series in δa. Using Eq. (10–71), one obtains :

$$dx = \delta a^i \, \mu_i(x) + 0\,(\delta a^2) \quad , \quad \mu_i(x) = \left.\frac{\partial m(a;x)}{\partial a^i}\right|_{a=0} \quad . \tag{10-73}$$

The set of functions $\mu_i = \{\mu_i^\alpha\}$ is called a velocity field on E.

For a detailed study of the necessary and sufficient conditions that such a set of functions on a manifold E define a velocity field, i.e. be associated with a transformation group on E, cf. for instance Racah[45].

The variations dx of the local parameters x of any p in E induce on any function ϕ in F a variation (cf. remark 2 - in Def. C.56) :

$$d\phi_p = \frac{\partial\tilde{\phi}(x)}{\partial x^\alpha}\, dx^\alpha = \delta a^i \, \mu_i^\alpha(x) \frac{\partial\tilde{\phi}(x)}{\partial x^\alpha} = \delta a^i (X_i\phi)(p) \quad , \tag{10-74}$$

where the X_i are analytic infinitesimal transformations intrinsically defined on E (Defs. C.57,58). Their symbols, which depend upon the choice of local parameters on E, read (Prop. C.75) :

$$X_i = \mu_i^\alpha(x) \frac{\partial}{\partial x^\alpha} \quad . \tag{10-75}$$

From Eq. (10–74), to all infinitesimal elements of the group G there correspond infinitesimal transformations on E which are linear combinations of the X_i. These n infinitesimal transformations on E form an n–dimensional Lie algebra homomorphic to the Lie algebra of the group (Def. C.62, cf. Racah loc.cit.). One should note that the X_i are associated with a choice of local parameters on G.

ii) Representation D(G) on F. With respect to any basis (e_ρ) of F, the matrix representative of D(g) is defined by :

$$\forall g \in G \qquad D(g)e_\rho = e_\sigma \, D(g)^\sigma_\rho \quad . \tag{10-76}$$

If F is Hermitian, and if furthermore G is compact, the above matrix representative is taken unitary (Theorem 7.8, Subsec. 7.2.1).

From H2 and H5, $D(g_a)$ analytically depends upon a. Therefore, following the same method as above in i), or also in Sec. 5.1, the n infinitesimal generators J_i of the representation D(G) are determined by, using Eq. (10–64) :

$$D(g_{\delta a}) = \mathbf{1} - i\delta a^i J_i + 0\,(\delta a^2) \quad , \quad J_i = i \left.\frac{\partial D(g_a)}{\partial a^i}\right|_{a=0} . \tag{10-77}$$

The factor i is introduced here, as in Eqs. (4–40) and (5–1), in order to obtain Hermitian J_i if the representation D(G) is unitary. The matrix representative with respect to a basis (e_ρ) of F of the infinitesimal generators is given by :

$$J_i\, e_\rho = e_\sigma\, (J_i)^\sigma_\rho \quad , \quad (J_i)^\sigma_\rho = i \left.\frac{\partial D(g_a)^\sigma_\rho}{\partial a^i}\right|_{a=0} . \tag{10-78}$$

iii) **Representations $\mathcal{D}_1(G)$, $\mathcal{D}_2(G)$, and $\mathcal{D}(G)$ in F** . From the hypotheses specified in Subsec. 10.2.3.1, the mappings $\mathcal{D}_1(g_a)$, $\mathcal{D}_2(g_a)$, and $\mathcal{D}(g_a)$ analytically depend upon the local parameters a and furthermore are equal to the unit mapping for a=0. Hence, always applying the same method, one gets :

1–The infinitesimal generators $\mathcal{J}^{(1)}_i$ of the representation $\mathcal{D}_1(G)$ are determined by :

$$\mathcal{D}_1(g_{\delta a}) = \mathbf{1} - i\delta a^i\, \mathcal{J}^{(1)}_i + 0\,(\delta a^2) \quad , \quad \mathcal{J}^{(1)}_i = i \left.\frac{\partial \mathcal{D}_1(g_a)}{\partial a^i}\right|_{a=0} . \tag{10-79}$$

They are closely related to the infinitesimal generators J_i, Eq. (10–77), of the representation D(G) in F, i.e. :

$$\forall \phi \in F \qquad \forall p \in E \qquad (\mathcal{J}^{(1)}_i \phi)(p) = J_i\, \phi(p) . \tag{10-80}$$

2–The infinitesimal generators $\mathcal{J}^{(2)}_i$ of the representation $\mathcal{D}_2(G)$ are determined by :

$$\mathcal{D}_2(g_{\delta a}) = \mathbf{1} - i\delta a^i\, \mathcal{J}^{(2)}_i + 0\,(\delta a^2) \quad , \quad \mathcal{J}^{(2)}_i = i \left.\frac{\partial \mathcal{D}_2(g_a)}{\partial a^i}\right|_{a=0} . \tag{10-81}$$

They can be expressed in terms of the infinitesimal transformations X_i of the transformation group M(G) on E in the following way :

$$\forall p \in E \qquad [\mathcal{D}_2(g_{\delta a})\phi](p) = \phi[M(g_{\delta a})^{-1} p]$$

$$= \tilde{\phi}\{m[\psi(\delta a);x]\} = \phi(p) + \delta a^i \left.\frac{\partial \psi^j(a)}{\partial a^i}\right|_{a=0} (X_j\phi)(p) , \tag{10-82}$$

thereby obtaining the expression :

$$\mathcal{J}^{(2)}_i = i\, \partial_i \psi^j X_j \quad , \quad \partial_i \psi^j = \left.\frac{\partial \psi^j(a)}{\partial a^i}\right|_{a=0} . \tag{10-83}$$

From Eq. (10–75), for a given choice of local parameters on E the symbol of $\mathcal{J}^{(2)}_i$ reads :

$$\mathcal{J}^{(2)}_i = i\, \partial_i \psi^j \quad \mu^\alpha_j(x) \frac{\partial}{\partial x^\alpha} . \tag{10-84}$$

3–The infinitesimal generators $\mathcal{I}_i$ of the representation $\mathcal{D}(G)$ are determined by :

$$\mathcal{D}(g_{\delta a}) = \mathbf{1} - i\delta a^i J_i + 0(\delta a^2) \quad , \quad J_i = i \left.\frac{\partial \mathcal{D}(g_a)}{\partial a^i}\right|_{a=0} \quad . \tag{10-85}$$

Since $\mathcal{D}$ (G) is defined by the product $\mathcal{D}_1$(G) $\mathcal{D}_2$(G), one gets from Eqs. (10–79,81) :

$$\mathcal{D}(g_{\delta a}) = \mathcal{D}_1(g_{\delta a})\mathcal{D}_2(g_{\delta a}) = \mathbf{1} - i\delta a^i \,(J_i^{(1)} + J_i^{(2)}) + 0(\delta a^2) \quad , \tag{10-86}$$

and thus :

$$J_i = J_i^{(1)} + J_i^{(2)} \quad . \tag{10-87}$$

All these infinitesimal generators $J_i^{(1)}$, $J_i^{(2)}$, and J_i are Hermitian when the corresponding representations are unitary, e.g. for a compact Lie group.

10.2.3.3. Application to the regular representations of a Lie group, first and second parameter groups

i) **First and second parameter groups.** The representatives with respect to an atlas on an n–parameter Lie group G of the left and right translations, Eq. (10–28), are, respectively, the one–to–one locally defined analytic mappings ℓ_a and r_a of a part of R^n into itself, such that (Prop. C.78) :

$$g_c = L_{g_a} g_b = g_a g_b \quad \Leftrightarrow \quad c = \ell_a(b) = \varphi(a,b) \quad , \tag{10-88a}$$

$$g_c = R_{g_a} g_b = g_b g_a \quad \Leftrightarrow \quad c = r_a(b) = \varphi(b,a) \quad . \tag{10-88b}$$

Since the left and right translations commute, Eq. (10–29), so do the mappings ℓ_a and r_b :

$$[\ell_a \,,\, r_b] = 0 \quad . \tag{10-89}$$

The two sets of mappings $\{\ell_a\}$ and $\{r_a\}$ form two groups of transformations on the set of local parameters of G, for which the group laws read (see Fig. 7.2, Subsec. 7.1.1.1,ii)), respectively :

$$\ell_a \ell_b = \ell_{\varphi(a,b)} \quad , \quad \ell_a^{-1} = \ell_{\bar{a}} = \ell_{\psi(a)} \quad , \tag{10-90a}$$

$$r_a r_b = r_{\varphi(b,a)} \quad , \quad r_a^{-1} = r_{\bar{a}} = r_{\psi(a)} \quad . \tag{10-90b}$$

Let us for instance check the multiplication law of the applications ℓ_a. The result follows from the constraint on φ due to the associativity of the product on G (Def. C.60, (2)) :

$$\forall\ a,b,c \quad \ell_a[\ell_b(c)] = \varphi[a,\ell_b(c)] = \varphi[a,\varphi(b,c)] = \varphi[\varphi(a,b),c] = \ell_{\varphi(a,b)}(c) \quad . \tag{10-91}$$

These two groups are the <u>first</u> and the <u>second parameter groups</u>, P_1 and P_2, with respect to a given atlas on G. They are the expressions in terms of local parameters of the groups L(G) and R(G) defined in Subsec. 10.1.3.1. The first parameter group P_1, is, just as L(G), isomorphic to

G and thus defines a realization of this group. On the other hand, the second parameter group P_2 is, just as R(G), anti-isomorphic to G, Eq. (10–32). The function φ in Eq. (10–90a), or the function φ with its arguments taken in the reverse order in Eq. (10–90b), plays the role of the function m in Eq. (10–68). From Eq. (10–75), the symbols of the infinitesimal transformations X_i^ℓ and X_i^r of the transformation groups L(G) and R(G), respectively, read :

$$X_i^\ell = \lambda_i^j(b)\,\frac{\partial}{\partial b^j} \quad , \quad \forall\phi\in F \quad (X_i^\ell\phi)(g_b) = \lambda_i^j(b)\,\frac{\partial\tilde{\phi}(b)}{\partial b^j} \quad , \tag{10-92a}$$

$$X_i^r = \rho_i^j(b)\,\frac{\partial}{\partial b^j} \quad , \quad \forall\phi\in F \quad (X_i^r\phi)(g_b) = \rho_i^j(b)\,\frac{\partial\tilde{\phi}(b)}{\partial b^j} \quad , \tag{10-92b}$$

where F denotes a vector space of analytic mappings ϕ of G into the complex field, and λ_i and ρ_i are, respectively, the velocity fields of P_1 and P_2 which are determined from Eq. (10–73) to be :

$$b+db = \varphi(\delta a,b) \quad , \quad db = \delta a^i\lambda_i(b)+0(\delta a^2) \quad , \quad \lambda_i(b) = \left.\frac{\partial\varphi(a,b)}{\partial a^i}\right|_{a=0} \quad ; \tag{10-93a}$$

$$b+db = \varphi(b,\delta a) \quad , \quad db = \delta a^i\rho_i(b)+0(\delta a^2) \quad , \quad \rho_i(b) = \left.\frac{\partial\varphi(b,a)}{\partial a^i}\right|_{a=0} \quad . \tag{10-93b}$$

ii) Infinitesimal generators of the regular representations :

1–The left regular representation $D^\ell(G)$ is defined in Eq. (10–33) from the realization L(G) on G, the infinitesimal transformations of which are given in Eqs. (10–92a,93a). Therefore, applying Eq. (10–83), the infinitesimal generators J_i of $D^\ell(G)$ read :

$$J_i^\ell = i\,\partial_i\psi^j\,X_j^\ell \quad , \tag{10-94}$$

with the symbols :

$$J_i^\ell = i\partial_i\psi^j\lambda_j^k(b)\,\frac{\partial}{\partial b^k} \quad , \quad \forall\phi\in F \quad (J_i^\ell\phi)(g_b) = i\partial_i\psi^j\lambda_j^k(b)\,\frac{\partial\tilde{\phi}(b)}{\partial b^k} \quad . \tag{10-95}$$

The right regular representation $D^r(G)$ is defined in Eq. (10–34) from the realization $R(G^{-1})$ on G. This realization is expressed in terms of the local parameters by the mapping $r_{\bar{a}}$ such that :

$$g_c = R_{g_a}^{-1}\,g_b = g_b\,g_a^{-1} \Leftrightarrow c = r_{\bar{a}}(b) = \varphi[b,\psi(a)] \quad . \tag{10-96}$$

The corresponding velocity field ρ_i' is related to the velocity field ρ_i of R(G) as follows :

$$b+db = \varphi[b,\psi(\delta a)] \quad , \quad db = \delta a^i\partial_i\psi^j\rho_j(b) = \delta a^i\rho_i'(b)$$

$$\Rightarrow \rho_i' = \partial_i\psi^j\rho_j \quad . \tag{10-97}$$

Hence, the infinitesimal transformations Y_i of $R(G^{-1})$ are related to the infinitesimal transformations X_i^r of R(G) by :

$$Y_i = \partial_i\psi^j \, X_j^r \quad . \tag{10-98}$$

Now, applying Eq. (10–83), the infinitesimal generators J_i^r of $D^r(G)$ read :

$$J_i^r = i\,\partial_i\psi^j Y_j = i\,\partial_i\psi^j \partial_j\psi^k \, X_k^r \quad . \tag{10-99}$$

Furthermore, since ψ is involutive one has :

$$\forall a \qquad \psi[\psi(a)] = a \Rightarrow \left.\frac{\partial\psi^k(b)}{\partial b^j}\right|_{b=\psi(a)} \frac{\partial\psi^j(a)}{\partial a^i} = \delta_i^k \quad , \tag{10-100}$$

and setting a=0 in the above equation and using $\psi(0)=0$, one finds :

$$\partial_j\psi^k \, \partial_i\psi^j = \delta_i^k \quad . \tag{10-101}$$

One thereby obtains the infinitesimal generators J_i^r :

$$J_i^r = i\,X_i^r \quad , \tag{10-102}$$

the symbols of which are :

$$J_i^r = i\,\rho_i^j(b)\,\frac{\partial}{\partial b^j} \quad , \forall\phi\in F \quad (J_i^r\phi)(g_b) = i\rho_i^j(b)\,\frac{\partial\tilde{\phi}(b)}{\partial b^j} \quad . \tag{10-103}$$

2–Since the left and right regular representations commute, so do their infinitesimal generators :

$$[J_i^\ell \, , \, J_j^r] = 0 \quad . \tag{10-104}$$

This result can be checked directly using the symbols of these infinitesimal generators given in Eqs. (10–95,103). One then needs to use a constraint on φ due to the associativity, i.e. :

$$\varphi[a,\varphi(b,a)] = \varphi[\varphi(a,b),a] \quad . \tag{10-105}$$

Taking the second partial derivative with respect to a^i and a^j of both sides of the above equation, setting a=0, and using Eq. (10–65), one finds :

$$\left.\frac{\partial^2\varphi(a,b)}{\partial a^j \partial b^k}\right|_{a=0} \left.\frac{\partial\varphi^k(b,a)}{\partial a^i}\right|_{a=0} = \left.\frac{\partial^2\varphi(b,a)}{\partial a^i \partial b^k}\right|_{a=0} \left.\frac{\partial\varphi^k(a,b)}{\partial a^j}\right|_{a=0} \quad . \tag{10-106}$$

The equivalence of the left and right regular representations, Eq. (10–38), implies that their infinitesimal generators are related by :

$$J_i^\ell = I \, J_i^r \, I \quad , \tag{10-107}$$

where the definition of I is given in Eq. (10–36).

This result can also be checked using the symbols of the infinitesimal generators. One then has for all ϕ in F :

$$(IJ_i^r I\phi)(g_b) = (J_i^r I\phi)(g_{\bar{b}}) = i\rho_i^j(\bar{b}) \frac{\partial \widetilde{I\phi}(\bar{b})}{\partial \bar{b}^j}$$
$$= i\rho_i^j(\bar{b}) \frac{\partial \tilde{\phi}(b)}{\partial \bar{b}^j} = i\rho_i^j(\bar{b}) \frac{\partial \tilde{\phi}(b)}{\partial b^k} \frac{\partial b^k}{\partial \bar{b}^j} \quad .$$

The constraint on the functions φ and ψ , cf. Def. C.60, (2) :

$$\psi[\varphi(a,b)] = \varphi[\psi(b),\psi(a)] \quad , \tag{10-108}$$

implies, taking the first derivative with respect to a_k of both sides of the above equation and setting a=0 :

$$\frac{\partial \psi^j(b)}{\partial b^i} \lambda_k^i(b) = \rho_i^j(\bar{b}) \, \partial_k \psi^i \quad , \tag{10-109}$$

and therefore, from Eqs. (10–100,101) :

$$\partial_i \psi^j \lambda_k^i(b) = \rho_i^j(\bar{b}) \frac{\partial b^k}{\partial \bar{b}^j} \quad . \tag{10-110}$$

One thereby obtains the result :

$$(IJ_i^r I\phi)(g_b) = i \, \partial_i \psi^j \lambda_j^k \frac{\partial \tilde{\phi}(b)}{\partial b^k} = (J_i^\ell \phi)(g_b) \quad .$$

Finally, thanks to the factor i, the infinitesimal generators J_i^ℓ and J_i^r are Hermitian if F is supplied with a Hermitian structure and if the regular representations are unitary. This is for instance the case for compact Lie groups on taking $F = L^2(G,d\mu)$.

iii) **Intrinsic infinitesimal generators.** The left and right regular representations are related in the following way :

$$\forall \phi \in F \qquad \forall g,g' \in G \qquad [D^\ell(g')\phi](g) = \phi(g'^{-1}g) = \phi(gg^{-1}g'^{-1}g)$$
$$= [D^r(g^{-1}g'^{-1}g)\phi](g) = [D^r(g^{-1})D^r(g'^{-1})D^r(g)\phi](g) \quad , \tag{10-111a}$$

and similarly :

$$\forall \phi \in F \qquad \forall g,g' \in G \qquad [D^r(g')\phi](g) = [D^\ell(g)D^\ell(g'^{-1})D^\ell(g^{-1})\phi](g) \quad . \tag{10-111b}$$

One has to take care of the fact that in the right–hand–side of the above equations, g occurs at the same time as the argument of the function and in the mappings $D^\ell(g)$ and $D^r(g)$. Let us take g' be an infinitesimal element $g_{\delta a}$. One then obtains in terms of the infinitesimal generators J_i^ℓ and J_i^r :

$$\forall \phi \in F \qquad \forall g \in G \qquad (J_i^\ell \phi)(g) = \partial_i \psi^j [D^r(g^{-1}) J_j^r \, D^r(g)\phi](g) \quad , \tag{10-112a}$$

$$(J_i^r \phi)(g) = \partial_i \psi^j [D^\ell(g) J_j^\ell \, D^\ell(g^{-1})\phi](g) \quad , \tag{10-112b}$$

where the previous remark about the dependence upon g still applies.

In order to make precise the meaning of the above equations, let us for instance check the last one in terms of the symbols of the infinitesimal generators. One has from Eq. (10–95) :

$$[D^{\ell}(g)J^{\ell}_{j}\, D^{\ell}(g^{-1})\phi](g) = [J^{\ell}_{j}\, D^{\ell}(g^{-1})\phi](e)$$

$$= i\partial_{j}\psi^{k}\,\lambda^{m}_{k}(0)\;\left.\frac{\partial[D^{\ell}(g^{-1})\phi](g_{a})}{\partial a^{m}}\right|_{a=0} \quad .$$

Eq. (10–65) implies $\lambda^{m}_{k}(0) = \delta^{m}_{k}$; furthermore let $g = g_b$, so that using Eq. (10–101) one has :

$$\partial_{i}\psi^{j}[D^{\ell}(g)J^{\ell}_{j}\, D^{\ell}(g^{-1})\phi]I(g) = i\partial_{i}\psi^{j}\partial_{j}\psi^{k}\left.\frac{\partial\phi[g_{\varphi(b,a)}]}{\partial a^{k}}\right|_{a=0}$$

$$= i\left.\frac{\partial\phi[g_{\varphi(b,a)}]}{\partial a^{i}}\right|_{a=0} = i\left.\frac{\partial\varphi^{j}(b,a)}{\partial a^{i}}\right|_{a=0}\frac{\partial\phi(g_{b})}{\partial b^{j}} \quad ,$$

which is the expression of $(J^{r}_{i}\phi)(g_b)$ given in Eq. (10–103), q.e.d..

We define the <u>intrinsic infinitesimal generators</u> I^{ℓ}_{i} and I^{r}_{i} of the left and right regular representations, respectively, by :

$$\forall\phi\in F \qquad \forall g\in G \qquad (I^{\ell}_{i}\phi)(g) = [D^{\ell}(g)J^{\ell}_{i}D^{\ell}(g^{-1})\phi](g) \quad , \qquad (10\text{-}113a)$$

$$(I^{r}_{i}\phi)(g) = [D^{r}(g^{-1})J^{r}_{i}\, D^{r}(g)\phi](g) \quad . \qquad (10\text{-}113b)$$

From Eqs. (10–112a,b), and using Eq. (10–101), one has :

$$J^{\ell}_{i} = \partial_{i}\psi^{j}\, I^{r}_{j} \quad\Leftrightarrow\quad I^{r}_{j} = \partial_{j}\psi^{i}\, J^{\ell}_{i} \quad , \qquad (10\text{-}114a)$$

$$J^{r}_{i} = \partial_{i}\psi^{j}\, I^{\ell}_{j} \quad\Leftrightarrow\quad I^{\ell}_{j} = \partial_{j}\psi^{i}\, J^{r}_{i} \quad . \qquad (10\text{-}114b)$$

Therefore, if the commutation relations of the infinitesimal generators J^{ℓ}_{i} (or J^{r}_{i}, they are the same, cf. Eq. (5–7)) read, omitting the index ℓ (or r) :

$$i[J_{i},J_{j}] = c_{ij}{}^{k}\, J_{k} \quad , \qquad (10\text{-}115)$$

those of the intrinsic infinitesimal generators I^{ℓ}_{i} (or I^{r}_{i}) take the form :

$$i[I_{i},I_{j}] = \partial_{i}\psi^{i'}\partial_{j}\psi^{j'}c_{i'j'}{}^{k'}\partial_{k'}\psi^{k}\, I_{k} \quad . \qquad (10\text{-}116)$$

Moreover, since J^{ℓ}_{i} and J^{r}_{j} commute, one obtains :

$$[J^{\ell}_{i},I^{\ell}_{j}] = 0 \quad , \quad [J^{r}_{i},I^{r}_{j}] = 0 \quad . \qquad (10\text{-}117)$$

From Eq. (5–17), one has :

$$\forall g\in G \qquad A^{\ell}(g)J^{\ell}_{i} = D^{\ell}(g)J^{\ell}_{i}\, D^{\ell}(g^{-1}) \quad , \qquad (10\text{-}118a)$$

$$A^r(g^{-1})J_i^r = D^r(g^{-1})J_i^r\, D^r(g) \quad , \qquad (10\text{-}118b)$$

where $A^{\ell}(G)$ and $A^r(G)$ are the adjoint representations of G in the Lie algebras of the left and right regular representations, respectively, (both isomorphic to the Lie algebra of the group, cf. Table 5.1, Subsec. 5.4.3). Furthermore, it has been shown (Subsec. 5.4.2,i)) that for a connected Lie group with a semi-simple Lie algebra the adjoint representation of any element in the group is a proper isometric application on the Lie algebra, supplied with a (pseudo-) Euclidean structure by means of the Killing form. In particular, for a compact n-parameter Lie group these adjoint representations are rotations on the n-dimensional Euclidean Lie algebra of the group. The use of the word "intrinsic" follows from these considerations. This point is discussed later for the groups SU(2) and SO(3) in Subsec. 10.3.1.4.

10.2.4. COMPONENTS OF THE INFINITESIMAL DISPLACEMENT OF THE MOVING REFERENCE FRAME OF A TRANSFORMATION GROUP

For a detailed study of this problem, cf. for instance Cartan[14] (Chap. V). There, it is shown how to extend the following consideration to sets of transformations which are not necessarily possessing a group structure.

10.2.4.1. Definitions and interpretation

i) Definitions. With the same notations and hypotheses as in Subsec. 10.2.3.1, let A_a and A_{a+da} be two infinitely close positions of the moving reference frame of the transformation group M(G) on E, which correspond to g_a and g_{a+da}, i.e. from Eq. (10-57) :

$$A_a = M(g_a)A_o \quad , \qquad A_{a+da} = M(g_{a+da})A_o \quad , \qquad (10\text{-}119)$$

where A_o is any reference frame on E taken as the absolute one. These two reference frames are related by the infinitesimal element $M(g_{\delta a})$ of the transformation group, such that :

$$A_{a+da} = M(g_{\delta a})A_a \quad , \quad M(g_{\delta a}) = M(g_{a+da})M(g_a)^{-1} \quad .$$

To $M(g_{\delta a})$ there corresponds the infinitesimal transformation $\delta a^i X_i$ given in Eq. (10-74). From Eq. (10-63),one has up to first order in da :

$$\delta a^i = \varphi^i(a+da,\bar{a}) = da^j \left.\frac{\partial\varphi^i(a,a')}{\partial a^j}\right|_{a'=\bar{a}} = \Omega^i(a;da) \quad , \qquad (10\text{-}120)$$

where the Ω^i are n linear and homogeneous forms in da^j, also called Pfaff's forms. One shows (cf. for instance Cartan[14], Chap. V, § 70,71) that they are linearly independent since by definition n is the smallest number of parameters upon which any element g in G depends. The Ω^i are closely related to the velocity field $\lambda_i(a)$ of the first parameter group. Indeed, from Eq. (10-93a) one has :

$$\Omega^i(a;da) = da^j(\lambda^{-1})_j^i(a) \quad , \qquad \lambda_k^j(a)(\lambda^{-1})_j^i(a) = \delta_j^i \quad . \qquad (10\text{-}121)$$

With respect to the absolute reference frame A_o, the representative of $M(g_{\delta a})$ reads :

$$M_{A_o}(g_{\delta a}) = M_{A_o}(g_{a+da})\, M_{A_o}(g_a)^{-1} \quad , \tag{10-122}$$

and the representative of the corresponding infinitesimal element is given in terms of the symbols of the X_i, Eq. (10–75), by :

$$\Omega^i(a;da)X_{iA_o} = \Omega^i(a;da)\mu_i^\alpha(x)\,\frac{\partial}{\partial x^\alpha} \quad . \tag{10-123}$$

By definition the n forms Ω^i are said to be the absolute components of the infinitesimal displacement of the moving reference frame of M(G), or more simply of G itself since they are the same for all realizations of G.

From Eqs. (10–59,122), the representative of $M(g_{\delta a})$ with respect to the moving reference of M(G) reads :

$$\begin{aligned} M_{A_a}(g_{\delta a}) &= M_{A_o}(g_a)^{-1}\,[M_{A_o}(g_{a+da})M_{A_o}(g_a)^{-1}]M_{A_o}(g_a) \\ &= M_{A_o}(g_a)^{-1}\,M_{A_o}(g_{a+da}) = M_{A_o}(g_{\delta a'}) \quad , \end{aligned} \tag{10-124}$$

where up to first order in da, one has :

$$\delta a'^i = \varphi^i(\bar{a},a+da) = da^j\,\frac{\partial\varphi^i(a',a)}{\partial a^j}\bigg|_{a'=\bar{a}} = \omega^i(a;da) \quad . \tag{10-125}$$

Hence, the representative with respect to the moving reference frame of M(G) of the infinitesimal transformation corresponding to $M(g_{\delta a})$ is given by :

$$\Omega^i(a;da)X_{iA} = \omega^i(a;da)X_{iA_o} = \omega^i(a;da)\mu_i^\alpha(x)\,\frac{\partial}{\partial x^\alpha} \quad , \tag{10-126}$$

where the local parameters x^α are now defined with respect to A_a. By definition the ω^i are called the relative components of the infinitesimal displacement of the moving reference frame of M(G), or simply of G itself since they are the same for all realizations of G. The ω^i are also Pfaff's forms, i.e. linear and homogeneous forms in da^j. As the Ω^i, the n forms ω^i are linearly independent. Furthermore, they are closely related to the velocity field $\rho_i(a)$ of the second parameter group. Indeed, from Eq. (10–93b) one has :

$$\omega^i(a;da) = da^j(\rho^{-1})_j^i(a) \quad , \quad \rho_k^j(a)(\rho^{-1})_j^i(a) = \delta_k^i \quad . \tag{10-127}$$

ii) Geometrical interpretation of the relative components. The idea of the moving reference frame and the above considerations follow from a generalization of usual methods in kinematics. Let us for instance consider the case, on which we come back to later in Subsec. 11.4.1 , of the displacement of a rigid body allowed to rotate around a fixed point O. It is then convenient to take an intrinsic or body–fixed reference frame tied to the rigid body, and to describe the displacement in terms of the relative position of this intrinsic reference frame (which plays the role of the moving reference frame of a group) with respect to any fixed one (O,e_i), (the absolute reference frame). The group we consider is then the group of rotations $R = (O,R)$ around the fixed point O in the three-dimensional point space (Subsec. 2.3.1,ii)). Let us denote by $(O,e_i(R))$ the intrinsic reference frame, where :

$$e_i(R) = R\, e_i = e_j\, R^j_i \quad , \quad (R^j_i) \in SO(3) \quad . \tag{10-128}$$

An infinitely close position of this intrinsic reference frame is defined by the basis vectors $e_i(R+dR)$, such that :

$$e_i(R+dR) = (R+dR)e_i = e_i(R) + de_i(R) \quad , \tag{10-129}$$

where the matrix elements of dR with respect to the basis (e_i) are infinitesimal (see Fig. 10.3).

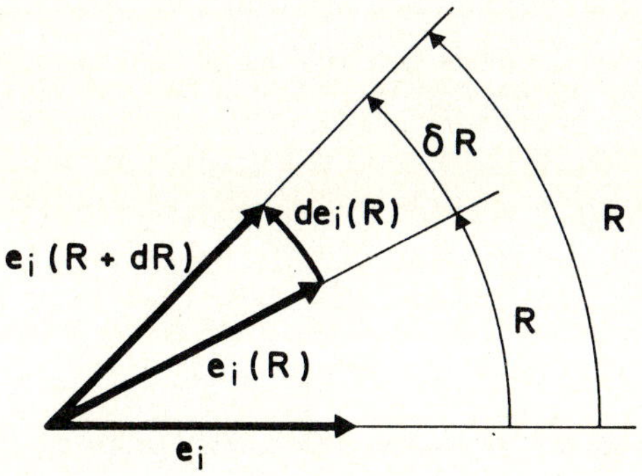

Fig. 10.3. The components of the infinitesimal variations $de_i(R)$ with respect to the intrinsic basis vector $e_i(R)$ are closely related to the relative components of the moving reference frame of the rotation group, as defined in Eq. (10–125).

It is now shown that the components of the vectors $de_i(R)$ with respect to the intrinsic basis $(e_i(R))$ are closely related to the relative components of the infinitesimal displacement of the moving reference frame (here the intrinsic one) as defined in Eq. (10–125). Indeed, from Eqs. (10–128,129) one has :

$$de_i(R) = e_j(dR)^j_i = e_k(R)(R^{-1})^k_j\,(dR)^j_i = e_j(R)(R^{-1}dR)^j_i \quad , \tag{10-130}$$

where all the matrix representatives are defined with respect to the absolute basis (e_i). The infinitesimal rotation $\delta R = R^{-1}(R+dR)$ plays here the role of $M(g_{\delta a})$ in Eq. (10–124). Choosing an infinitesimal rotation angle ε^i around each of the basis vectors e_i to parametrize δR, its matrix representative with respect to e_i takes the form, Eq. (4–72) :

$$(\delta R)^j_i = \delta^j_i + \varepsilon^k (M_k)^j_i \quad , \tag{10-131}$$

where the matrices M_k, Eq. (4–73), play the role of the representatives $X_{i\,A_0}$ of the infinitesimal transformations in Eq. (10–126). The relative components ε^k of the infinitesimal displacement of the moving reference frame are thus related to the components of $de_i(R)$ with respect to $(e_j(R))$ by :

$$(R^{-1}dR)^j_i = \varepsilon^k (M_k)^j_i = \varepsilon^k \varepsilon_{kij} \quad . \tag{10-132}$$

10.2.4.2. Some properties

i) **Relation between the absolute and relative components.** One has :

$$M(g_{a+da})M(g_a)^{-1} = M(g_{\overline{a+da}})^{-1} M(g_{\bar{a}}) \quad . \tag{10-133}$$

Furthermore, since the infinitesimal element $M(g_{a+da})M(g_a)^{-1}$ of the transformation group and the one obtained by changing a into a–da only differ by second order terms in da, one obtains up to first order :

$$M(g_{a+da})M(g_a)^{-1} = M(g_{\bar{a}})^{-1} M(g_{\overline{a-da}}) \quad . \tag{10-134}$$

Finally, setting :

$$\overline{a-da} = \bar{a} - d\bar{a} \quad , \qquad d\bar{a} = da^i \frac{\partial\psi(a)}{\partial a^i} \quad , \tag{10-135}$$

one finds the relation :

$$M(g_{a+da})M(g_a)^{-1} = M(g_{\bar{a}})^{-1} M(g_{\bar{a}-d\bar{a}}) \quad , \tag{10-136}$$

which implies the equality of the corresponding infinitesimal transformations, i.e. with respect to the absolute reference frame :

$$\Omega^i(a;da)X_{iA_o} = \omega^i(\bar{a};-d\bar{a})X_{iA_o} \Rightarrow \Omega^i(a;da) = -\omega^i(\bar{a};d\bar{a}) \quad . \tag{10-137}$$

Hence, it has been shown :

Theorem 10.3. The absolute components of the infinitesimal displacement of the moving reference frame of an n–parameter Lie group become the relative ones with the opposite sign when the parameters of any element in the group are changed into the parameters of its inverse, i.e. :

$$\Omega(a;da) = -\omega(\bar{a};d\bar{a}) \quad .$$

The above result can be checked directly using the expressions of the absolute and the relative components given, respectively, in Eqs. (10–120) and (10–125). Indeed, one has :

$$\omega^i(\bar{a};d\bar{a}) = d\bar{a}^k \left.\frac{\partial\varphi^i(a',\bar{a})}{\partial \bar{a}^k}\right|_{a'=\psi(\bar{a})=a} = d\bar{a}^k \left.\frac{\partial\varphi^i(a,b)}{\partial b^k}\right|_{b=\bar{a}}$$
$$= da^j \frac{\partial\psi^k(a)}{\partial a^j} \left.\frac{\partial\varphi^i(a,b)}{\partial b^k}\right|_{b=\bar{a}} \quad ,$$

and then, using Eq. (10–66b) :

$$\omega^i(\bar{a};d\bar{a}) = - da^j \left.\frac{\partial\varphi^i(a,b)}{\partial a^j}\right|_{b=\bar{a}} = - \Omega^i(a;da) \quad .$$

ii) Associated with the geometrical interpretation of the relative components ω^i, which characterize the relative position of two infinitely close reference frames, let us now prove the following theorem :

Theorem 10.4. Let M(G) be a realization of an n–parameter Lie group G on an r–manifold E. A necessary and sufficient condition that there exist a one–to–one mapping ψ of E onto itself such that, with respect to any local parametrization on G :

$$\forall\, g = g_a \in G \qquad \exists g' = g_{a'} \in G \qquad M(g') = \psi \circ M(g) \quad , \tag{10-138}$$

is that one has :

$$\omega(a;da) = \omega(a';da') \quad . \tag{10-139}$$

In other words, using a geometrical language, if for all g in G there exists g' in G such that the two moving reference frames $A_g = M(g)\, A_o$ and $A_{g'} = M(g')\, A_o$ have always the same relative position, then the relative components of the infinitesimal displacements of these reference frames are equal ; and vice versa.

Proof. Assuming Eq. (10–138) is satisfied, one then has :

$$\forall\, g_a, g_{a+da} \in G \qquad \exists g_{a'}, g_{a'+da'} \in G \qquad \begin{cases} M(g_{a'}) = \psi \circ M(g_a) \\ M(g_{a'+da'}) = \psi \circ M(g_{a+da}) \end{cases}$$

$$\Rightarrow\; M(g_{a'})^{-1} M(g_{a'+da'}) = M(g_a)^{-1} M(g_{a+da}) \quad , \tag{10-140}$$

which leads, from Eq. (10–125), to equal relative components :

$$\varphi(\bar{a}',a'+da') = \varphi(\bar{a},a+da) \;\Rightarrow\; \omega(a;da) = \omega(a';da') \quad .$$

Conversely, the last equality above implies Eq. (10–140), which also reads :

$$\forall\, g_a \in G \qquad \exists g_{a'} \in G \;,\quad M(g_{a'+da'})M(g_{a+da})^{-1} = M(g_{a'})M(g_a)^{-1} \quad . \tag{10-141}$$

Therefore, the mapping $M(g_{a'})M(g_a)^{-1}$ does not depend upon a, i.e. it is a fixed application ψ of E onto itself.

The interpretation of this theorem in the special case discussed previously in Subsec. 10.2.4.1,ii) is straightforward. Indeed, the components of $de_i(R)$ with respect to $e_j(R)$ are equal to the components of $\psi de_i(R)$ with respect to $\psi e_j(R)$ for any one–to–one linear mapping ψ on $E_2^{\bullet}$, i.e. a displacement.

iii) **Application to the left and right translations :**

Theorem 10.5. The relative (resp. absolute) components of the infinitesimal displacement of the moving reference frame of an n-parameter Lie group G are invariant under the left (resp. right) translations. Namely, for any local parametrization on G, one has for all g in G :

$$g_b = L_g\, g_a \quad , \quad g_{b+db} = L_g\, g_{a+da} \quad \Rightarrow \quad \omega(a;da) = \omega(b;db) \quad , \tag{10-142a}$$

$$\text{resp. } g_b = R_g\, g_a \quad , \quad g_{b+db} = R_g\, g_{a+da} \quad \Rightarrow \quad \Omega(a;da) = \Omega(b;db) \quad . \tag{10-142b}$$

Furthermore, any differential form f(a;da) which is invariant under the left (resp. right) translations is a linear combination of the n relative (resp. absolute) components ω^i (resp. Ω^i).

Proof :

1–Let us first consider the case of left translations. The first part of this theorem then follows from Theorem 10.4 in the special case where the manifold E is the group G itself, the transformation group M(G) is L(G) defined in Subsec. 10.1.3.1, and the mapping ψ is any left translation. Since in terms of local parameters the group L(G) is realized by the first parameter group P_1, Eq. (10–88a), it is also said that the relative components are invariant under P_1. This result can be checked directly using the expression of the relative components given by Eq. (10–125). Indeed, one has for $g = g_c$:

$$b = \varphi(c,a) \quad , \quad b+db = \varphi(c,a+da) \quad \Rightarrow \quad db^j = \left.\frac{\partial b^j}{\partial a^k}\right|_c da^k$$

$$\omega^i(b;db) = db^j \left.\frac{\partial\varphi^i(b',b)}{\partial b^j}\right|_{b'=\bar{b}} = da^k \left.\frac{\partial b^j}{\partial a^k}\right|_c \left.\frac{\partial\varphi^i(b',b)}{\partial a^\ell}\right|_{b'=\bar{b}} \left.\frac{\partial a^\ell}{\partial b^j}\right|_c$$

$$= da^k\delta^\ell_k \left.\frac{\partial\varphi^i[b',\varphi(c,a)]}{\partial a^\ell}\right|_{b'=\bar{b}} = da^k \left.\frac{\partial\varphi^i[\varphi(b',c),a]}{\partial a^k}\right|_{b'=\bar{b}}$$

$$= da^k \left.\frac{\partial\varphi^i(a',a)}{\partial a^k}\right|_{a'=\bar{a}} = \omega^i(a,da) \quad . \tag{10-143}$$

As for the second part of the theorem, it can be proved as follows. A priori, one has :

$$f(a;da) = \alpha_i(a)\, \omega^i(a;da) \qquad \alpha_i(a) \in R \quad .$$

The invariance of the differential forms f and ω^i under the left translations implies, with the same notations as in Eq. (10–142a) :

$$f(a;da) = f(b;db) = \alpha_i(b)\omega^i(b;db) = \alpha_i(b)\omega^i(a;da)$$

$$\Rightarrow [\alpha_i(a)-\alpha_i(b)]\; \omega^i(a;da) = 0 \quad .$$

As the n forms ω^i are linearly independent (Subsec. 10.2.4.1,i)), it follows that $\alpha_i(b) = \alpha_i(a)$ for all a and b. The functions α_i are thus constants.

2–Using the samé method as above, let us now check that the right translations, or in terms of parameters the second parameter group P_2, leave invariant the absolute components of the infinitesimal displacement of the moving reference frame of G. Indeed, using Eqs. (10–120,142b) one has for $g = g_c$:

$$b = \varphi(a,c) \quad , \quad b+db = \varphi(a+da,c) \Rightarrow \quad db^j = \left.\frac{\partial b^j}{\partial a^k}\right|_c da^k$$

$$\Omega^i(b;db) = db^j \left.\frac{\partial\varphi^i(b,b')}{\partial b^j}\right|_{b'=\bar{b}} = da^k \left.\frac{\partial\varphi^i[\varphi(a,c),b']}{\partial a^k}\right|_{b'=\bar{b}}$$

$$= da^k \left.\frac{\partial\varphi^i[a,\varphi(c,b')]}{\partial a^k}\right|_{b'=\bar{b}} = da^k \left.\frac{\partial\varphi^i(a,a')}{\partial a^k}\right|_{a'=\bar{a}} = \Omega(a;da) \quad . \tag{10-144}$$

The proof of the second part of the theorem is identical to the one given for the left translations.

10.3. APPLICATION TO THE GROUPS SU(2) AND SO(3)

Let us now apply the previous general study to the compact three-parameter Lie groups SU(2) and SO(3). We first consider the regular representations and then examples of representations which are used in Quantum Mechanics (Subsecs.11.1.2,3). This leads us to introduce the notion of spin, and also the special functions such as the spherical harmonics, the vector spherical harmonics, and more generally the spin spherical harmonics. The notations for the indices j and m are those specified in the introduction of Chap. 6. Namely, the sums over the indices j are always indicated explicitly, while the Einstein summation convention is adopted for the indices m.

10.3.1. THE REGULAR REPRESENTATIONS OF SU(2)

10.3.1.1. Determination and properties

i) The functional vector space F we consider is the set $L^2(SU(2),d\mu)$ of square integrable functions of SU(2) into the complex field with the invariant measure $d\mu$ on the group, cf. Subsecs. 7.2.2,3. It has been shown (Theorem 7.13) that the set of functions :

$$\left\{ \frac{\sqrt{2j+1}}{4\pi} \mathcal{D}^j_{m'm} \quad ; \quad j=0,\tfrac{1}{2},\ldots \quad , \quad m \text{ and } m'=j,j-1,\ldots,-j \right\} \quad ,$$

forms a complete orthonormal basis of $L^2(SU(2),d\mu)$. So does the set of their conjugate complex functions. The orthogonality and closure relations of the functions $\mathcal{D}^j_{m'm}$ are given in Eqs. (7–50a) and (7–68) with the normalization of the invariant measure on SU(2) specified in Eq. (7–48).

ii) Following the general definitions of Eqs. (10–33,34), the left and right regular representations of SU(2) in $L^2(SU(2),d\mu)$, i.e. $D^\ell[SU(2)]$ and $D^r[SU(2)]$, are determined by :

$$\forall U,U' \in SU(2) \qquad \forall f \in L^2(SU(2),d\mu) \quad \begin{cases} [D^\ell(U)f](U') = f(U^{-1}U') \quad , & (10\text{–}145a) \\ [D^r(U)f](U') = f(U'U) \quad . & (10\text{–}145b) \end{cases}$$

We recall the following properties of these representations (Subsec. 10.1.3.2,ii)) :

(1) The regular representations are faithful.

(2) The left regular representation commutes with the right one :

$$\forall U,U' \in SU(2) \qquad D^\ell(U)D^r(U') = D^r(U')D^\ell(U) \quad . \qquad (10\text{–}146)$$

(3) The left and right regular representations are equivalent :

$$\forall U \in SU(2) \qquad D^{\ell}(U) = I\, D^{r}(U)\, I \quad , \tag{10-147}$$

where the involutive linear mapping I of $L^2(SU(2),d\mu)$ onto itself is given by :

$$\forall f \in L^2(SU(2),d\mu) \qquad \forall U \in SU(2) \qquad (If)(U) = f(U^{-1}) \quad . \tag{10-148}$$

(4) The regular representations are unitary with respect to the Hermitian inner product on $L^2(SU(2),d\mu)$, Eq. (7–55).

iii) It has been shown (Subsec. 7.2.2.3) that the set of representations $\{\mathcal{D}^j,\ j=0,\frac{1}{2},...\}$ forms a complete set of nonequivalent irreducible unitary representations of SU(2). Therefore, Theorem 10.2 applies, i.e. :

(1) The left (resp. right) regular representation is the direct sum of all the irreducible representations $\mathcal{D}^j$, each one being contained 2j+1 times :

$$\mathcal{D}^{\ell}(\text{resp.}D^{r}) = \sum_{j=0,\frac{1}{2},...} \oplus\ (2j+1)\ \mathcal{D}^j \quad . \tag{10-149}$$

(2) The restriction of D^{ℓ} (resp. D^r) to the 2j+1–dimensional subspace $F^{\ell}_{jm_o}$ (resp. $F^{r}_{jm_o}$) of $L^2(SU(2),d\mu)$, spanned by the functions $\{\mathcal{D}^{j*}_{mm_o}\ ;\ m=j,j-1,...,-j\}$, (resp. $\{\mathcal{D}^{j}_{m_om}\ ;\ m=j, j-1,...,-j\}$), is equivalent to the representation $\mathcal{D}^j$. Furthermore, one has :

$$\forall U \in SU(2) \quad \begin{cases} D^{\ell}(U)\ \mathcal{D}^{j*}_{mm_o} = \mathcal{D}^{j*}_{m'm_o}\ \mathcal{D}^{j}_{m'm}(U) \quad , & (10\text{-}150a) \\ D^{r}(U)\ \mathcal{D}^{j}_{m_om} = \mathcal{D}^{j}_{m_om'}\ \mathcal{D}^{j}_{m'm}(U) \quad . & (10\text{-}150b) \end{cases}$$

Therefore, in the tensorial language introduced in Subsec. 8.3.1.1, any complex linear combination of the functions $\mathcal{D}^{j*}_{mm_o}$ for a given m, i.e. :

$$f^{\ell,[j]}_{m} = \mathcal{D}^{j*}_{mm_o}\ \alpha_{m_o} \quad , \qquad \alpha_{m_o} \in C \quad , \tag{10-151}$$

transforms under the left regular representation as a contravariant spherical component of a tensor of order one and rank j. Indeed, in agreement with Eq. (8–40) one has :

$$\forall U \in SU(2) \qquad D^{\ell}(U) f^{\ell,[j]}_{m} = f^{\ell,[j]}_{m'}\ \mathcal{D}^{j}_{m'm}(U) \quad . \tag{10-152}$$

Similarly, any function :

$$f^{r,[j]}_{m} = \alpha_{m_o}\ \mathcal{D}^{j}_{m_om} \quad , \quad \alpha_{m_o} \in C \tag{10-153}$$

transforms under the right regular representation as a contravariant spherical component of a tensor of order one and rank j, i.e. :

$$\forall U \in SU(2) \qquad D^{r}(U) f^{r,[j]}_{m} = f^{r,[j]}_{m'}\ \mathcal{D}^{j}_{m'm}(U) \quad . \tag{10-154}$$

The sum over m_o in Eqs. (10–151,153) corresponds to the fact that the representation $\mathcal{D}^j$ is contained 2j+1 times in D^{ℓ} or D^r.

For all U in SU(2) the matrices $\mathcal{D}^1(U)$ and R(U) determined by Eq. (4–28) are related by Eq. (8–179). This implies the following relation between functions in $L^2(SU(2),d\mu)$:

$$R^i_j = t^i_m \, \mathcal{D}^1_{mm'} (t^{-1})^{m'}_j \quad , \tag{10-155}$$

where the functions R_j are defined by :

$$\forall\, U \in SU(2) \xrightarrow{\ R^i_j\ } R(U)^i_j = \tfrac{1}{2}\,\mathrm{Tr}\,(\sigma_i U \sigma_j U^{-1}) \in R\ .. \tag{10-156}$$

Either from Eqs. (10–150a,155), or directly from Eqs. (10–145a,b), one derives the law of transformation of the function R^i_j, i.e. :

$$\forall\, U,U' \in SU(2) \qquad [D^\ell(U) R^i_{i_o}](U') = R(U^{-1}U')^i_{i_o} = \sum_j R(U')^j_{i_o}\, R(U)^j_i$$

$$\Rightarrow D^\ell(U)\, R^i_{i_o} = \sum_j R^j_{i_o}\, R(U)^j_i \quad , \tag{10-157a}$$

and similarly :

$$\forall\, U \in SU(2) \qquad D^r(U)\, R^{i_o}_i = R^{i_o}_j\, R(U)^j_i \quad . \tag{10-157b}$$

These relations imply that any complex linear combination of the functions $R^i_{i_o}$ (resp. $R^{i_o}_i$) for a given i, i.e. :

$$f^\ell_i = \sum_{i_o} R^i_{i_o}\, \alpha_{i_o} \quad (\text{resp. } f^r_i = R^{i_o}_i\, \alpha_{i_o}) \ , \quad \alpha_{i_o} \in C \ , \tag{10-158}$$

transforms under the left (resp. right) regular representation as a Cartesian component of a tensor of order one, i.e. a vector (Subsec. 8.5.4).

10.3.1.2. Infinitesimal generators

i) We denote by J^ℓ_i and J^r_i the infinitesimal generators of, respectively, the left and right regular representations, which correspond to the σ matrices by Eq. (5–8), i.e. :

$$U = \mathbf{1} - i\,\frac{\varepsilon^i}{2}\,\sigma_i + 0(\varepsilon^2) \Rightarrow \begin{cases} D^\ell(U) = \mathbf{1} - i\,\varepsilon^i J^\ell_i + 0(\varepsilon^2) \\ D^r(U) = \mathbf{1} - i\,\varepsilon^i J^r_i + 0(\varepsilon^2) \end{cases} \quad . \tag{10-159}$$

They fulfil the usual commutation relations, Eq. (5–9) :

$$[J^\ell_i, J^\ell_j] = i \sum_k \varepsilon_{ijk} J^\ell_k \ , \quad [J^r_i, J^r_j] = i \sum_k \varepsilon_{ijk} J^r_k \ . \tag{10-160}$$

Furthermore, the infinitesimal generators of the left regular representation commute with those of the right one, Eqs. (10–104,146) :

$$[J^\ell_i \ , J^r_j] = 0 \quad , \tag{10-161}$$

and are related by, Eqs. (10–107,147) :

$$J^{\ell}_{i} = I\, J^{r}_{i}\, I \quad . \tag{10-162}$$

Setting $U=\mathbf{1}-i\frac{1}{2}\varepsilon^{i}\sigma_{i}+O(\varepsilon^{2})$ in Eqs. (10–152,154) and using Eq. (6–33), one checks that the functions $f^{\ell,[j]}_{m}$ and $f^{r,[j]}_{m}$ behave as standard basis vectors of, respectively, the left and right regular representations. Therefore, these functions satisfy Eqs. (5–58,59). In particular, for any given m_o (=j,j–1,...,–j) one has :

$$\begin{cases} J^{\ell}_{\pm}\mathcal{D}^{j*}_{mm_o} = \sqrt{j(j+1)-m(m\pm1)}\;\mathcal{D}^{j*}_{m\pm1 m_o} \;, \quad J^{\ell}_{3}\mathcal{D}^{j*}_{mm_o} = m\,\mathcal{D}^{j*}_{mm_o} \;, \\ (J^{\ell})^{2}\mathcal{D}^{j*}_{mm_o} = j(j+1)\,\mathcal{D}^{j*}_{mm_o} \;, \end{cases} \tag{10-163a}$$

$$\begin{cases} J^{r}_{\pm}\mathcal{D}^{j}_{m_o m} = \sqrt{j(j+1)-m(m\pm1)}\;\mathcal{D}^{j}_{m_o m\pm1} \;, \quad J^{r}_{3}\mathcal{D}^{j}_{m_o m} = m\,\mathcal{D}^{j}_{m_o m} \\ (J^{r})^{2}\mathcal{D}^{j}_{m_o m} = j(j+1)\,\mathcal{D}^{j}_{m_o m} \;. \end{cases} \tag{10-163b}$$

In the same way, the functions f^{ℓ}_{i} and f^{r}_{i} given in Eq. (10–158) behaving as Cartesian components, satisfy Eq. (8–221). In particular, for any given j_o(=1,2,3) one finds :

$$J^{\ell}_{i} R^{j}_{j_o} = i\,\varepsilon_{ijk} R^{k}_{j_o} \;, \qquad J^{r}_{i} R^{j_o}_{j} = i\sum_{k}\varepsilon_{ijk} R^{j_o}_{k} \;. \tag{10-164}$$

ii) Let us now derive from Eqs. (10–95,103) the expressions of the infinitesimal generators when the elements in SU(2) are parametrized by the Euler angles φ , θ , and ψ , Eq. (4–38). The velocity fields $\lambda^{j}_{i}(\varphi,\theta,\psi)$ and $\rho^{j}_{i}(\varphi,\theta,\psi)$ of, respectively, the first and the second parameter groups are obtained in Appendix E.2.3, Eqs. (E–16 a–c) and (E–17 a–c). One should remark that in the derivation of these velocity fields, the infinitesimal element in Eqs. (E–18 a–b) is not parametrized by infinitesimal Euler angles but by infinitesimal ε^{i} as in Eq. (10–159), (it is recalled that $\varphi=0$, $\theta=0$, $\psi=0$ is an exceptional point of the parametrization by Euler angles, Subsec. 7.1.3.1,iv)). Since one has :

$$U(\varepsilon^{i}) = \mathbf{1}-i\tfrac{1}{2}\varepsilon^{i}\sigma_{i}+O(\varepsilon^{2}) \;\Rightarrow\; U(\varepsilon^{i})^{-1} = \mathbf{1}+i\tfrac{1}{2}\varepsilon^{i}\sigma_{i}+O(\varepsilon^{2}) = U(-\varepsilon^{i}) \;, \tag{10-165}$$

the quantity $\partial_{i}\psi^{j}$ defined in Eq. (10–83) is in this special case :

$$\partial_{i}\psi^{j} = -\,\delta^{j}_{i} \quad . \tag{10-166}$$

Therefore, applying Eqs. (10–95,103) one obtains :

$$J^{\ell}_{i} = i(-\delta^{j}_{i})\Big(\lambda^{1}_{j}\frac{\partial}{\partial\varphi} + \lambda^{2}_{j}\frac{\partial}{\partial\theta} + \lambda^{3}_{j}\frac{\partial}{\partial\psi}\Big) \;, \tag{10-167a}$$

$$J^{r}_{i} = i\Big(\rho^{1}_{i}\frac{\partial}{\partial\varphi} + \rho^{2}_{i}\frac{\partial}{\partial\theta} + \rho^{3}_{i}\frac{\partial}{\partial\psi}\Big) \;, \tag{10-167b}$$

and finally :

$$\begin{cases} J_1^\ell = -i\left(-\cos\varphi\,\mathrm{cotg}\theta\,\frac{\partial}{\partial\varphi} - \sin\varphi\,\frac{\partial}{\partial\theta} + \frac{\cos\varphi}{\sin\theta}\frac{\partial}{\partial\psi}\right) , & (10\text{-}168a) \\ J_2^\ell = -i\left(-\sin\varphi\,\mathrm{cotg}\theta\,\frac{\partial}{\partial\varphi} + \cos\varphi\,\frac{\partial}{\partial\theta} + \frac{\sin\varphi}{\sin\theta}\frac{\partial}{\partial\psi}\right) , & (10\text{-}168b) \\ J_3^\ell = -i\,\frac{\partial}{\partial\varphi} , & (10\text{-}168c) \\ J_\pm^\ell = J_1^\ell \pm i\,J_2^\ell = e^{\pm i\varphi}\left(i\,\mathrm{cotg}\theta\,\frac{\partial}{\partial\varphi} \pm \frac{\partial}{\partial\theta} - i\,\frac{1}{\sin\theta}\frac{\partial}{\partial\psi}\right) , & (10\text{-}168d) \end{cases}$$

$$\begin{cases} J_1^r = i\left(-\frac{\cos\psi}{\sin\theta}\frac{\partial}{\partial\varphi} + \sin\psi\,\frac{\partial}{\partial\theta} + \mathrm{cotg}\theta\,\cos\psi\,\frac{\partial}{\partial\psi}\right) , & (10\text{-}169a) \\ J_2^r = i\left(\frac{\sin\psi}{\cos\theta}\frac{\partial}{\partial\varphi} + \cos\psi\,\frac{\partial}{\partial\theta} - \mathrm{cotg}\theta\,\sin\psi\,\frac{\partial}{\partial\psi}\right) , & (10\text{-}169b) \\ J_3^r = i\,\frac{\partial}{\partial\psi} , & (10\text{-}169c) \\ J_\pm^r = J_1^r \pm i\,J_2^r = e^{\mp i\psi}\left(-i\,\frac{1}{\sin\theta}\frac{\partial}{\partial\varphi} \mp \frac{\partial}{\partial\theta} + i\,\mathrm{cotg}\theta\,\frac{\partial}{\partial\psi}\right) . & (10\text{-}169d) \end{cases}$$

On the other hand, from Eq. (4–38) one has :

$$\forall\, U(\varphi,\theta,\psi) \in SU(2) \quad \Rightarrow \quad U(\varphi,\theta,\psi)^{-1} = U(-\psi,-\theta,-\varphi) \quad . \qquad (10\text{-}170)$$

So that using Eqs. (10–148,162), the previous expressions of J_i^r and $J_\pm^r$ can also be obtained by changing (φ,θ,ψ) into $(-\psi,-\theta,-\varphi)$ in the expressions of J_i^ℓ and $J_\pm^\ell$. A direct calculation yields :

$$(J^\ell)^2 = (J^r)^2 = -\mathrm{cotg}\theta\,\frac{\partial}{\partial\theta} - \frac{1}{\sin^2\theta}\left(\frac{\partial^2}{\partial\varphi^2} - 2\cos\theta\,\frac{\partial^2}{\partial\varphi\partial\psi} + \frac{\partial^2}{\partial\psi^2}\right) - \frac{\partial^2}{\partial\theta^2} \quad . \qquad (10\text{-}171)$$

Setting these expressions of $J_\pm^\ell$, J_3^ℓ, $(J^\ell)^2$ and $J_\pm^r$, J_3^r, $(J^r)^2$ in, respectively, Eqs. (10–163a,b) one derives differential equations for the functions $\mathcal{D}^j_{mm_o}$ parametrized by the Euler angles. Since $\mathcal{D}^j_{mm_o}(\varphi,\theta,\psi)$ depends upon φ and ψ by exponentials, Eq. (6–21), we get the following differential equations for the functions $d^j_{mm_o}(\theta)$ defined in Eq. (6–18) :

$$\left(-m\,\mathrm{cotg}\theta + m_o\,\frac{1}{\sin\theta} \pm \frac{d}{d\theta}\right) d^j_{mm_o}(\theta) = \sqrt{j(j+1)-m(m\pm1)}\; d^j_{m\pm1\;m_o}(\theta) \quad , \qquad (10\text{-}172a)$$

$$\left(m\,\mathrm{cotg}\theta - m_o\,\frac{1}{\sin\theta} \mp \frac{d}{d\theta}\right) d^j_{m_o m}(\theta) = \sqrt{j(j+1)-m(m\pm1)}\; d^j_{m_o\,m\pm1}(\theta) \quad , \qquad (10\text{-}172b)$$

$$\left[j(j+1) - \frac{1}{\sin^2\theta}\,(m^2-2mm_o\cos\theta+m_o^2) + \mathrm{cotg}\theta\,\frac{d}{d\theta} + \frac{d^2}{d\theta^2}\right] d^j_{mm_o}(\theta) = 0 \quad . \qquad (10\text{-}172c)$$

Actually, the first two equations above are not distinct as can be seen by changing θ into $-\theta$ and using Eq. (6–22a) :

$$d^j_{mm_o}(\theta) = d^j_{m_o m}(-\theta) \quad . \qquad (10\text{-}173)$$

One should remark that the function $d^j_{mm_o}(\theta)$ is completely characterized by the fact that it satisfies Eq. (10–172c) with an initial condition, e.g. :

$$d^j_{mm_o}(0) = \delta_{mm_o} \quad . \tag{10-174}$$

10.3.1.3. Multiplicative operator associated with a function

i) To any function f in $L^2(SU(2),d\mu)$ one associates a multiplicative operator $\hat{f}$, i.e. a mapping of $L^2(SU(2),d\mu)$ into itself defined by :

$$\forall h \in L^2(SU(2),d\mu) \xrightarrow{\hat{f}} \hat{f}h \,, \quad \forall U \in SU(2) \quad (\hat{f}h)(U) = f(U)h(U) \quad . \tag{10-175}$$

More generally, such multiplicative operators can be defined on any space F of functions which map a manifold E into the complex field :

$$\forall f,h \in F \quad h \xrightarrow{\hat{f}} \hat{f}h \in F \,, \quad \forall p \in E \quad (\hat{f}h)(p) = f(p)h(p) \in C \quad . \tag{10-176}$$

It is now assumed that there exists an analytic representation D [SU(2)] in F, determined according to Eq. (10–15) from an analytic realization M [SU(2)] on the analytic manifold E, i.e. :

$$\forall U \in SU(2) \qquad \forall f \in F \qquad \forall p \in E \qquad [D(U)f](p) = f[M(U^{-1})p] \quad . \tag{10-177}$$

Since any infinitesimal generator J is a differentiation (Def. C.54) :

$$\forall f,h \in F \quad \forall p \in E \qquad (J\hat{f}h)(p) = (Jfh)(p) = (Jf)(p)h(p) + f(p)(Jh)(p)$$

$$= (\widehat{Jf}h)(p) + (\hat{f}Jh)(p) \quad , \tag{10-178}$$

and therefore one has :

$$\forall f \in F \qquad [J,\hat{f}] = \widehat{Jf} \quad . \tag{10-179}$$

Let us check that the infinitesimal generators J are differentiations. Up to first order in ε, one finds :

$$\forall f,h \in F \qquad \forall p \in E \qquad U = \mathbf{1} - i\varepsilon W \qquad D(U) = \mathbf{1} - i\varepsilon J$$

$$[(\mathbf{1}-i\varepsilon J)fh](p) = (fh)[M(U^{-1})p] = f[M(U^{-1})p]\; h[M(U^{-1})p]$$

$$= [(\mathbf{1}-i\varepsilon J)f](p)\; [(\mathbf{1}-i\varepsilon J)h]\,(p)$$

$$= (fh)(p) - i\varepsilon[(Jf)(p)h(p) + f(p)(Jh)(p)] \quad .$$

Consequently, for functions $f^{[j]}_m$ in F which behave like the contravariant spherical components of a tensor f^j under the representation D [SU(2)], there correspond multiplicative operators $\hat{f}^{[j]}_m$ satisfying the commutation relations :

$$\left[J_\pm,\hat{f}^{[j]}_m\right]=\sqrt{j(j+1)-m(m\pm1)}\ \hat{f}^{[j]}_{m\pm1}\quad,\quad\left[J_3,\hat{f}^{[j]}_m\right]=m\ \hat{f}^{[j]}_m\ . \tag{10-180}$$

The $\hat{f}^{[j]}_m$ are called irreducible tensor operators (Subsec. 11.2.2). Similarly, if the functions f_i (i=1,2,3) in F transform like the Cartesian components of a vector under the representation D [SU(2)], one gets from Eq. (8–221) :

$$[J_i,\ \hat{f}_j]=i\sum_k\varepsilon_{ijk}\ \hat{f}_k\ . \tag{10-181}$$

Such a relation characterizes a vector operator (Subsec. 11.2.2.1,ii).

ii) Applying the above results to the left and right regular representations of SU(2), one finds from Eqs. (10–163a,b) :

$$[J^\ell_\pm,\ \hat{\mathcal{D}}^{j*}_{mm_o}]=\sqrt{j(j+1)-m(m\pm1)}\ \hat{\mathcal{D}}^{j*}_{m\pm1m_o}\quad,\quad[J^\ell_3,\hat{\mathcal{D}}^{j*}_{mm_o}]=m\ \hat{\mathcal{D}}^{j*}_{mm_o}\ , \tag{10-182a}$$

$$[J^r_\pm,\ \hat{\mathcal{D}}^{j}_{m_om}]=\sqrt{j(j+1)-m(m\pm1)}\ \hat{\mathcal{D}}^{j}_{m_om\pm1}\quad,\quad[J^r_3,\hat{\mathcal{D}}^{j}_{m_om}]=m\ \hat{\mathcal{D}}^{j}_{m_om}\ , \tag{10-182b}$$

and from Eq. (10–164) :

$$[J^\ell_i,\ \hat{R}^j_{j_o}]=i\ \varepsilon_{ijk}\ \hat{R}^k_{j_o}\quad,\quad[J^r_i,\ \hat{R}^{j_o}_j]=i\sum_k\varepsilon_{ijk}\ \hat{R}^{j_o}_k\ . \tag{10-183}$$

Let us point out that it is worthwhile to carefully distinguish the three objects :

(1) The matrix elements $\mathcal{D}^j_{mm_o}(U)$ and $R(U)^i_{i_o}$ which are complex numbers, and thus commute with any infinitesimal generators.

(2) The functions $\mathcal{D}^j_{mm_o}$ and $R^i_{i_o}$ in $L^2(SU(2),d\mu)$ which satisfy Eqs. (10–163a,b, and 164).

(3) The multiplicative operators $\hat{\mathcal{D}}^j_{mm_o}$ and $\hat{R}^i_{i_o}$ which satisfy the commutation relations above.

10.3.1.4. Intrinsic infinitesimal generators

i) **Definitions and commutation relations.** The intrinsic infinitesimal generators of the left and right regular representations, respectively I^ℓ_i and I^r_i, are defined by, Eqs. (10–113a,b) :

$$\forall f\in L^2(SU(2),d\mu)\quad\forall U\in SU(2)\quad\begin{cases}(I^\ell_i f)(U)=[D^\ell(U)J^\ell_i D^\ell(U^{-1})f](U)\ , & (10\text{-}184a)\\ (I^r_i f)(U)=[D^r(U^{-1})J^r_i\ D^r(U)f](U)\ . & (10\text{-}184b)\end{cases}$$

From Eqs. (10–114a,b, and 166) they are related to the infinitesimal generators J^ℓ_i and J^r_i by :

$$I^\ell_i=-J^r_i\quad\Rightarrow\quad I^\ell_\pm=I^\ell_1\pm iI^\ell_2=-J^r_\pm\ , \tag{10-185a}$$

$$I^r_i=-J^\ell_i\quad\Rightarrow\quad I^r_\pm=I^r_1\pm iI^r_2=-J^\ell_\pm\ . \tag{10-185b}$$

Therefore, using Eq. (10–171), one has :

$$(I^\ell)^2=(J^r)^2=(J^\ell)^2=(I^r)^2\ . \tag{10-186}$$

From Eqs. (10–116,117, and 166) these intrinsic infinitesimal generators fulfil the following commutation rules :

$$[J_i^\ell , I_i^\ell] = 0 \quad , \qquad [J_i^r , I_i^r] = 0 \quad , \tag{10-187a}$$

$$[I_i^\ell , I_j^\ell] = -i \sum_k \varepsilon_{ijk} I_k^\ell \quad , \qquad [I_i^r , I_j^r] = -i \sum_k \varepsilon_{ijk} I_k^r \quad , \tag{10-187b}$$

where one should notice the change of sign with respect to the usual commutation relations of the J_i^ℓ or J_i^r, Eq. (10–160), cf. iv) 3– below.

ii) Action on the functions $\mathcal{D}^j_{mm_o}$ and $R^i_{i_o}$. Let us derive, using Eqs. (10–163a,b, and 185a,b), the action of the intrinsic infinitesimal generators I_i^ℓ and I_i^r on $\mathcal{D}^{j*}_{mm_o}$ and $\mathcal{D}^j_{m_om}$, respectively. From Eq. (6–23a) follows the relation between the functions :

$$\mathcal{D}^{j*}_{mm_o} = (-1)^{m-m_o} \mathcal{D}^j_{-m-m_o} \quad . \tag{10-188}$$

Therefore, one obtains :

$$I_\pm^\ell \mathcal{D}^{j*}_{mm_o} = -J_\pm^r (-1)^{m-m_o} \mathcal{D}^j_{-m-m_o} = (-1)^{m-(m_o\mp1)} \sqrt{j(j+1)-m_o(m_o\mp1)}\, \mathcal{D}^j_{-m-(m_o\mp1)}$$
$$= \sqrt{j(j+1)-m_o(m_o\mp1)}\, \mathcal{D}^{j*}_{m\,m_o\mp1} \quad , \tag{10-189a}$$

$$I_3^\ell \mathcal{D}^{j*}_{mm_o} = -J_3^r (-1)^{m-m_o} \mathcal{D}^j_{-m-m_o} = m_o (-1)^{m-m_o} \mathcal{D}^j_{-m-m_o} = m_o \mathcal{D}^{j*}_{mm_o} \quad , \tag{10-189b}$$

and similarly :

$$I_\pm^r \mathcal{D}^j_{m_om} = \sqrt{j(j+1)-m_o(m_o\mp1)}\, \mathcal{D}^j_{m_o\mp1 m} \quad , \tag{10-190a}$$

$$I_3^r \mathcal{D}^j_{m_om} = m_o \mathcal{D}^j_{m_om} \quad . \tag{10-190b}$$

In the same way, from Eqs. (10–164,185a, and b), one finds :

$$I_i^\ell R_j^{j_o} = -i \sum_k \varepsilon_{ijk} R_k^{j_o} \quad , \qquad I_i^r R_{j_o}^j = -i\, \varepsilon_{ijk} R_{j_o}^k \quad , \tag{10-191}$$

where, once again, one should note the change of sign with respect to the usual relations for infinitesimal generators, Eqs. (10–163a,b, and 164), cf. iv) 3– below.

iii) Action on the multiplicative operators $\hat{\mathcal{D}}^j_{mm_o}$ and $\hat{R}^i_{i_o}$. Applying the results derived in Subsec. 10.3.1.3, one gets from Eqs. (10–189a–191) the following commutation relations :

$$[I_\pm^\ell , \hat{\mathcal{D}}^{j*}_{mm_o}] = \sqrt{j(j+1)-m_o(m_o\mp1)}\, \hat{\mathcal{D}}^{j*}_{m\,m_o\mp1} \quad , \qquad [I_3^\ell , \hat{\mathcal{D}}^{j*}_{mm_o}] = m_o \hat{\mathcal{D}}^{j*}_{mm_o} \quad , \tag{10-192a}$$

$$[I_\pm^r , \hat{\mathcal{D}}^j_{m_om}] = \sqrt{j(j+1)-m_o(m_o\mp1)}\, \hat{\mathcal{D}}^j_{m_o\mp1\, m} \quad , \qquad [I_3^r , \hat{\mathcal{D}}^j_{m_om}] = m_o \hat{\mathcal{D}}^j_{m_om} \quad , \tag{10-192b}$$

$$[I_i^\ell , \hat{R}_j^{j_o}] = -i \sum_k \varepsilon_{ijk} \hat{R}_k^{j_o} \quad , \qquad [I_i^r , \hat{R}_{j_o}^j] = -i\, \varepsilon_{ijk} \hat{R}_{j_o}^k \quad . \tag{10-192c}$$

iv) Interpretation

1–It has been shown in Subsec. 5.4.3,ii) that the adjoint representation of any differentiable representation D[SU(2)] determines a rotation group in the Lie algebra $D[\mathcal{W}(2)]$ with respect to the Euclidean inner product defined by the Killing form. Therefore, the infinitesimal generators J^{ℓ}_i and J^{r}_i, which form orthogonal bases of, respectively, $D^{\ell}[\mathcal{W}(2)]$ and $D^{r}[\mathcal{W}(2)]$ with the metric tensor a multiple of the unit matrix, Eq. (5–40), satisfy the relations, Eqs. (5–45) and (10–118a,b) :

$$\forall\, U \in SU(2) \quad \begin{cases} A^{\ell}(U)J^{\ell}_i = D^{\ell}(U)J^{\ell}_i\, D^{\ell}(U^{-1}) = J^{\ell}_j\, R(U)^j_i \quad , & (10\text{-}193a) \\ A^{r}(U^{-1})J^{r}_i = D^{r}(U^{-1})J^{r}_i D^{r}(U) = J^{r}_j\, R(U^{-1})^j_i \quad , & (10\text{-}193b) \end{cases}$$

where the rotation matrix R(U) is related to U by Eq. (4–28). In terms of the spherical components of the infinitesimal generators (Subsec. 8.5.4,iii)), the above equations read :

$$A^{\ell}(U)J^{\ell[1]}_m = J^{\ell[1]}_{m'}\, \mathcal{D}^1_{m'm}(U) \quad , \qquad A^{r}(U)J^{r[1]}_m = J^{r[1]}_{m'}\, \mathcal{D}^1_{m'm}(U) \quad . \qquad (10\text{-}194)$$

2–The isomorphism between the Lie algebra $D[\mathcal{W}(2)]$ and any three-dimensional Euclidean vector space E_3 allowsus in Subsec. 5.4.3,i) to determine a formal vector J in E_3 by, Eq. (5–42) :

$$J = J_i\, e^i \quad , \qquad (10\text{-}195)$$

where $a=(e_i)$ is a given orthonormal basis of E_3. Let us now define an orthonormal moving basis $(e_i(U))$ in E_3, obtained by rotation of the basis (e_i), i.e. :

$$\forall\, U \in SU(2) \qquad e_i(U) = R(U;a)e_i = e_j\, R(U)^j_i \quad . \qquad (10\text{-}196)$$

One then finds that $A^{\ell}(U)J^{\ell}_i$ (resp. $A^{r}(U^{-1})J^{r}_i$) is the component of the formal vector J^{ℓ} (resp. J^{r}), along the moving basis vector $e_i(U)$ (resp. $e_i(U^{-1})$), i.e. :

$$J^{\ell}.e_i(U) = J^{\ell}_j\, e^j.e_i(U) = J^{\ell}_j\, R(U)^j_i = A^{\ell}(U)J^{\ell}_i \quad . \qquad (10\text{-}197a)$$

$$J^{r}.e_i(U^{-1}) = J_j\, e^j.e_i(U^{-1}) = J^{r}_j\, R(U^{-1})^j_i = A^{r}(U^{-1})J^{r}_i \quad . \qquad (10\text{-}197b)$$

It should be noticed that the component $e_{j_o j}(U)$ (resp. $e_{j_o j}(U^{-1})$) of the moving vector $e_{j_o}(U)$ (resp. $e_{j_o}(U^{-1})$) is $R(U)^j_{j_o}$ (resp. $R(U)^{j_o}_j$). Therefore, it follows from Eq. (10–164), that $\overset{o}{e}_{j_o j}$ (resp. $e_{j_o j}\circ I$) taken as a function in $L^2(SU(2),d\mu)$ satisfies the relations :

$$J^{\ell}_i\, e_{j_o j} = i\sum_k \varepsilon_{ijk}\, e_{j_o k} \quad , \qquad J^{r}_i(e_{j_o j}\circ I) = i\sum_k \varepsilon_{ijk}(e_{j_o k}\circ I) \quad . \qquad (10\text{-}198)$$

3–Let us now point out that I^{ℓ}_i, Eq. (10–184a) and $A^{\ell}(U)J^{\ell}_i$, Eq. (10–193a) have to be distinguished. Indeed, I^{ℓ}_i is defined assuming that the same value of U occurs in A^{ℓ} and as the argument of the function in the equation :

$$\forall\, f \in L^2(SU(2),d\mu) \qquad \forall\, U \in SU(2) \qquad (I^{\ell}_i f)(U) = [A^{\ell}(U)J^{\ell}_i f](U) \quad . \qquad (10\text{-}199)$$

The same remarks apply to I^{r}_i and $A^{r}(U)J^{r}_i$ for the right regular representation.

Nevertheless, in terms of multiplicative operators, the intrinsic infinitesimal generators can be written as :

$$I^{\ell}_{i} = J^{\ell}_{j}\,\hat{R}^{j}_{i} \quad , \quad I^{\ell[1]}_{m} = J^{\ell[1]}_{m'}\,\hat{\mathcal{D}}^{1}_{m'm} \quad , \tag{10-200a}$$

$$I^{r}_{i} = J^{r}_{j}\,\hat{R}^{j}_{i} \quad , \quad I^{r[1]}_{m} = J^{r[1]}_{m'}\,\hat{\mathcal{D}}^{1}_{m'm} \quad . \tag{10-200b}$$

It may seem that in the equations above the position of the multiplicative mappings $\hat{R}^{j}_{i}$ or $\hat{\mathcal{D}}^{1}_{m'm}$ is questionable. Actually, using Eqs. (10–182a,b, and 183) one immediately checks that :

$$J^{\ell}_{j}\,\hat{R}^{j}_{i} = \hat{R}^{j}_{i}\,J^{\ell}_{j} \quad , \quad J^{\ell[1]}_{m'}\,\hat{\mathcal{D}}^{1}_{m'm} = \hat{\mathcal{D}}^{1}_{m'm}\,J^{\ell[1]}_{m'} \quad . \tag{10-201}$$

Similar relations hold for the right regular representation. Let us also remark that although the adjoint representations, Eqs. (10–193a,b), preserve the commutation relations (Sec. 8.1,i)), the expressions given in Eqs. (10–200a,b) lead, as it may be directly checked, to the commutation relations of I^{ℓ}_{i} and I^{r}_{i} given in Eqs. (10–187b).

The intrinsic infinitesimal generators can also be expressed in terms of the multiplicative operators associated with the components of the moving basis with respect to (e_i). Thus, from Eqs. (10–197a,b) one has :

$$I^{\ell}_{i} = J^{\ell}.\hat{e}_{i} = \sum_{j} J^{\ell}_{j}\,\hat{e}_{ij} \quad , \quad I^{r}_{i} = J^{r}.(\hat{e}_{i}\circ I) = \sum_{j} J^{r}_{j}\,(\hat{e}_{ij}\circ I) \tag{10-202}$$

where, using Eq. (10–198), the expressions above are independent of the ordering of the infinitesimal generators J^{ℓ} or J^{r} and the multiplicative operators.

We come back later on the interpretation of the intrinsic infinitesimal generators when discussing the problem of a rigid rotator in Quantum Mechanics (Subsec. 11. 4.2).

10.3.1.5. Absolute and relative components of the infinitesimal displacement of the moving reference frame

i) Let us give the expressions in terms of the Euler angles of the absolute and relative components, Ω^{i} and ω^{i} respectively, of the infinitesimal displacement of the moving reference frame of SU(2). We use for that the relations given in Eqs. (10–121,127) between Ω^{i} and ω^{i} with the velocity fields λ^{j}_{i} and ρ^{j}_{i} of the first and the second parameter groups. Thus, inverting the linear Eqs. (E–16 a–c) and (E–17 a–c), one finds :

$$\begin{cases} \Omega^{1}(\varphi,\theta,\psi;d\varphi,d\theta,d\psi) = -\sin\varphi\, d\theta + \cos\varphi\,\sin\theta\, d\psi \quad , & (10\text{-}203a) \\ \Omega^{2}(\varphi,\theta,\psi;d\varphi,d\theta,d\psi) = \cos\varphi\, d\theta + \sin\varphi\,\sin\theta\, d\psi \quad , & (10\text{-}203b) \\ \Omega^{3}(\varphi,\theta,\psi;d\varphi,d\theta,d\psi) = d\varphi + \cos\theta\, d\psi \quad , & (10\text{-}203c) \end{cases}$$

$$\begin{cases} \omega^1(\varphi,\theta,\psi;d\varphi,d\theta,d\psi) = -\sin\theta\cos\psi\, d\varphi + \sin\psi\, d\theta \quad , & (10\text{-}204a) \\ \omega^2(\varphi,\theta,\psi;d\varphi,d\theta,d\psi) = \sin\theta\sin\psi\, d\varphi + \cos\psi\, d\theta \quad , & (10\text{-}204b) \\ \omega^3(\varphi,\theta,\psi;d\varphi,d\theta,d\psi) = \cos\theta\, d\varphi + d\psi \quad . & (10\text{-}204c) \end{cases}$$

Using Eq. (10–165), one checks the relation between Ω^i and ω^i given by Theorem 10.3, Eq. (10–137),i.e. :

$$\Omega^i(\varphi,\theta,\psi;d\varphi,d\theta,d\psi) = -\omega^i(-\psi,-\theta,-\varphi;-d\psi,-d\theta,-d\varphi) \quad . \qquad (10\text{-}205)$$

ii) From Theorem 10.5 (Subsec.10.2.4.2,iii)) these relative (resp. absolute) components are invariant under the left (resp. right) translations. Namely, for all U in SU(2) one has :

$$U(\varphi',\theta',\psi') = UU(\varphi,\theta,\psi) \ , \ U(\varphi'+d\varphi',\theta'+d\theta',\psi'+d\psi') = UU(\varphi+d\varphi,\theta+d\theta,\psi+d\psi)$$

$$\Rightarrow \quad \omega^i(\varphi',\theta',\psi';d\varphi',d\theta',d\psi') = \omega^i(\varphi,\theta,\psi;d\varphi,d\theta,d\psi) \quad , \qquad (10\text{-}206a)$$

and :

$$U(\varphi',\theta',\psi') = U(\varphi,\theta,\psi)U \ , \ U(\varphi'+d\varphi',\theta'+d\theta',\psi'+d\psi') = U(\varphi+d\varphi,\theta+d\theta,\psi+d\psi)\, U$$

$$\Rightarrow \quad \Omega^i(\varphi',\theta',\psi';d\varphi',d\theta',d\psi') = \Omega^i(\varphi,\theta,\psi;d\varphi,d\theta,d\psi) \quad . \qquad (10\text{-}206b)$$

10.3.2. THE REGULAR REPRESENTATIONS OF SO(3)

From the two–to–one homomorphism of SU(2) onto SO(3) (Subsec. 4.2.1 and Sec. 5.6), everything said previously about the regular representations of SU(2) apply to SO(3) provided that one restricts the irreducible representations $\mathcal{D}^j$ to those with integer values of j (Subsec. 6.1.2,iii)), and that one uses the orthogonality and closure relations in L^2 (SO(3)dμ), Eqs. (7–80a,83).

10.3.3. REPRESENTATIONS OF SO(3) AND SU(2) IN THE SPACE OF COMPLEX FUNCTIONS DEFINED ON R^3

10.3.3.1. Determination of the representations

i) Representations of SO(3). Let E_3 be a three–dimensional Euclidean vector space and E_3 an associated three–dimensional Euclidean point space (Subsec. 2.3.1). We consider the groups of rotations R on E_3 and rotations $\mathcal{R} = (O,R)$ on $E_3^\bullet$ for a given fixed point O, i.e. :

$$\forall R \begin{cases} \forall v \in E_3 \xrightarrow{\ R\ } v' = Rv \in E_3 \quad , \\ \forall M \in E_3^\bullet \xrightarrow{\ \mathcal{R}\ } M' = \mathcal{R}M \ , \quad OM' = ROM \in E_3 \quad . \end{cases} \qquad (10\text{-}207)$$

Following the general procedure described in Subsec. 10.1.2, a representation D of the rotation group in the functional vector space $\mathcal{H}$ over the complex field of complex functions defined on E_3 (resp. $\dot{E}_3$) is determined by, cf. Eq. (10–15) :

$$\forall f \in H \quad \forall R \quad \forall v \in E_3 \quad [D(R)f](v) = f(R^{-1}v) \quad ,$$
$$\forall R \quad \forall M \in \dot{E}_3 \quad [D(R)f](M) = f(R^{-1}M) \quad . \tag{10-208}$$

The set E_3 (resp. $\dot{E}_3$) is a special case of a three–manifold for which the parameters are, for instance, determined by the components (resp. coordinates) with respect to an orthonormal basis (e_i) (resp. reference frame $(0,e_i)$). In the following, to avoid cumbersome notations always we denote by f a function defined either on E_3 or $\dot{E}_3$, or its representative with respect to a given choice of parameters. As an example, in terms of the Cartesian and the spherical polar coordinates, Eqs. (2–18,20), the representatives of Eq. (10–208) read :

$$[D(R)f]\ (x^i) = f[(R^{-1})^i_j\ x^j] \quad , \quad (R^i_j) \in SO(3)$$
$$[D(R)f]\ (r,\theta,\varphi) = f\ [R^{-1}(r,\theta,\varphi)] \quad , \tag{10-209}$$

where (r,θ,φ) and $(r',\theta',\varphi') = R^{-1}(r,\theta,\varphi)$ denote the spherical polar coordinates of, respectively, M and $R^{-1}M$.

There exists on E_3 (resp. $\dot{E}_3$) an invariant measure under the rotation group, cf. Subsec. 10.3.3.2,ii). In terms of the Cartesian and the spherical polar coordinates this measure reads :

$$d_3x = \prod_{i=1}^{3} dx^i = r^2 dr\ \sin\theta\ d\theta\ d\varphi \quad . \tag{10-210}$$

Therefore, the representations determined above are unitary with respect to the Hermitian inner product on H, cf. Eq. (10–10) :

$$\forall f,f' \in H \quad (f,f')_H = \int_{R^3} d_3x\ f^*(x)\ f'(x) \quad . \tag{10-211}$$

In the following we restrict H to the vector space $L^2(R^3,d_3x)$ of square integrable functions on E_3 (resp. $\dot{E}_3$ or R^3) with the invariant measure d_3x. Actually, since $\|v\| \doteq \|OM\| = r$ is invariant under the rotation group, one only needs to consider the representation of this group in the space $L^2(S^2,d_2\hat{n})$ (Subsec. 7.2.2.2,i)) of square integrable functions with the invariant measure $d_2\hat{n}$, Eq. (7–59a), and defined on the set of unit vectors $\hat{n}$ or equivalently on the two–sphere S^2, i.e. :

$$\forall R, \quad \forall f \in L^2(S^2,d_2\hat{n}), \forall \hat{n} \qquad [D(R)f](\hat{n}) = f(R^{-1}\ \hat{n}) \quad . \tag{10-212}$$

ii) Representations of SU(2). As shown in Sec. 5.6,i), to the representation D [SO(3)] considered above there corresponds a representation $\mathcal{D}_2$[SU(2)] in the same space H and determined by :

$$\forall U \in SU(2) \rightarrow R(U) \in SO(3),\ \text{Eq. (4-28)}.\ \ \mathcal{D}_2(U) = D[R(U)] \quad . \tag{10-213}$$

Since this unitary representation of SU(2) is even, i.e. :

$$\forall\, U \in SU(2) \qquad \mathcal{D}_2(U) = \mathcal{D}_2(-U) \quad , \tag{10-214}$$

only the integer values of j occur in its standard basis (Subsec. 6.1.2,iii)), cf. Subsec. 10.3.3.4.

10.3.3.2. Relation with the regular representations

i) Let us consider the Abelian one-parameter subgroup $R_{\hat{n}} = \{ R(\alpha,\hat{n}) ; \forall\alpha \}$ of rotations around any unit vector $\hat{n}$ (Subsec. 4.1.5). Then, the relation of belonging to the same left (resp. right) coset $RR_{\hat{n}}$ (resp. $R_{\hat{n}}R$) of $R_{\hat{n}}$ determines an equivalence relation which yields a partitioning of the rotation group in cosets (Prop. C.21). A necessary and sufficient condition for two rotations R and R' to belong to the same left (resp. right) coset of $R_{\hat{n}}$ is (Prop. C.20) :

$$R \sim R' \Rightarrow \exists\, R(\alpha,\hat{n}) \in R_{\hat{n}} \qquad R = R'R(\alpha,\hat{n}) \quad , \quad \text{resp. } R = R(\alpha\hat{n})R' \quad . \tag{10-215}$$

Since one has :

$$\forall\, R' \in R\,R_{\hat{n}} \Rightarrow R'\hat{n} = R\hat{n} \quad , \quad \text{resp.} \forall R' \in R_{\hat{n}}R \Rightarrow R'^{-1}\hat{n} = R^{-1}\hat{n} \quad , \tag{10-216}$$

therefore there exists a one-to-one correspondence between the left (resp. right) cosets $RR_{\hat{n}}$ (resp. $R_{\hat{n}}R$) and the unit vectors $R\hat{n}$ (resp. $R^{-1}\hat{n}$) :

$$RR_{\hat{n}} \leftrightarrow R\hat{n} \quad , \qquad R_{\hat{n}}R \leftrightarrow R^{-1}\hat{n} \quad . \tag{10-217}$$

The relation of belonging to the same left (resp. right) coset of $R_{\hat{n}}$ is stable under the left (resp. right) translations, i.e. for all rotations ρ one has :

$$R,R' \in RR_{\hat{n}} \Leftrightarrow L_\rho R, L_\rho R' \in (L_\rho R)R_{\hat{n}} = (\rho R)R_{\hat{n}} \quad , \tag{10-218a}$$

$$R,R' \in R_{\hat{n}}R \Leftrightarrow R_\rho R, R_\rho R' \in R_{\hat{n}}(R_\rho R) = R_{\hat{n}}(R\rho) \quad . \tag{10-218b}$$

Hence, the left (resp. the inverse of the right) translations induce a realization of the rotation group in the set of left (resp. right) cosets of $R_{\hat{n}}$:

$$\forall\, \rho \qquad \tilde{L}_\rho(RR_{\hat{n}}) = (L_\rho R)R_{\hat{n}} \quad , \qquad \tilde{R}_{\rho^{-1}}(R_{\hat{n}}R) = R_{\hat{n}}(R_{\rho^{-1}}R) \quad . \tag{10-219}$$

Furthermore, this realization is isomorphic to the rotation group on the set of unit vectors. Indeed, in the correspondence determined by Eq. (10-217) one has :

$$\forall\, \rho \qquad \tilde{L}_\rho(RR_{\hat{n}}) \leftrightarrow \rho(R\hat{n}) \quad , \qquad \tilde{R}_{\rho^{-1}}(R_{\hat{n}}R) \leftrightarrow \rho(R^{-1}\hat{n}) \quad . \tag{10-220}$$

The equivalence relation on the rotation group, Eq. (10-215), induces an equivalence relation on the vector space $\mathcal{F}$ of complex functions defined on this group, such that :

$$\phi,\phi' \in \mathcal{F} \quad \phi \sim \phi' \Leftrightarrow \phi(R') = \phi(R) \qquad \forall\, R' \in RR_{\hat{n}} \text{ (resp. } R_{\hat{n}}R) \quad . \tag{10-221}$$

The corresponding equivalence classes are then complex functions $\tilde{\phi}^{\ell}$ (resp. $\tilde{\phi}^{r}$) defined on the set of left (resp. right) cosets of $R_{\hat{n}}$ by :

$$\forall R \qquad \tilde{\phi}^{\ell}(RR_{\hat{n}}) = \phi(R) \quad , \qquad \tilde{\phi}^{r}(R_{\hat{n}}R) = \phi(R) \quad . \tag{10-222}$$

It follows from Eq. (10–217) that there exists a one-to-one correspondence between the vector space $\tilde{F}^{\ell}$ (resp. $\tilde{F}^{r}$) of mappings $\tilde{\phi}^{\ell}$ (resp. $\tilde{\phi}^{r}$) and the vector space H of complex functions f defined on the set of unit vectors, i.e. :

$$\forall \tilde{\phi}^{\ell} \in \tilde{F}^{\ell} \leftrightarrow \forall f \in H \qquad \forall R\ \tilde{\phi}^{\ell}(RR_{\hat{n}}) = f(R\hat{n}) \quad , \tag{10-223a}$$

$$\forall \tilde{\phi}^{r} \in \tilde{F}^{r} \leftrightarrow \forall f \in H \qquad \forall R\ \tilde{\phi}^{r}(R_{\hat{n}}R) = f(R^{-1}\hat{n}) \quad . \tag{10-223b}$$

One should remark that if F is restricted to $L^2(SO(3),d\mu)$, $\tilde{F}$ is then isomorphic to $L^2(S^2,d_2\hat{n})$, cf. Eq. (10–228).

Finally, the left (resp. right) regular representation D^{ℓ} (resp. D^{r}) of the rotation group in F induces a representation $\tilde{D}^{\ell}$ (resp. $\tilde{D}^{r}$) of this group in $\tilde{F}$ according to :

$$\forall \rho \qquad \forall \tilde{\phi}^{\ell} \in \tilde{F}^{\ell} \qquad [\tilde{D}^{\ell}(\rho)\tilde{\phi}^{\ell}](RR_{\hat{n}}) = \tilde{\phi}^{\ell}[\tilde{L}_{\rho}^{-1}(RR_{\hat{n}})] \quad , \tag{10-224a}$$

$$\forall \tilde{\phi}^{r} \in \tilde{F}^{r} \qquad [\tilde{D}^{r}(\rho)\tilde{\phi}^{r}](R_{\hat{n}}R) = \tilde{\phi}^{r}[\tilde{R}_{\rho}(R_{\hat{n}}R)] \quad . \tag{10-224b}$$

From Eqs. (10–220,223a, and b) these two representations can be identified with the representation of the rotation group considered in Eq. (10–212), i.e. one has :

$$\tilde{\phi}^{\ell} \leftrightarrow f \Leftrightarrow \forall \rho \qquad \tilde{D}^{\ell}(\rho)\tilde{\phi}^{\ell} \leftrightarrow D(\rho)f \quad , \tag{10-225a}$$

$$\tilde{\phi}^{r} \leftrightarrow f \Leftrightarrow \forall \rho \qquad \tilde{D}^{r}(\rho)\tilde{\phi}^{r} \leftrightarrow D(\rho)f \quad . \tag{10-225b}$$

ii) Since the usual choice of a standard basis of infinitesimal generators picks out the axis e_3 it is convenient to consider the left (resp. right) cosets of the Abelian one-parameter subgroup R_{e_3} of rotations around e_3. Furthermore, the Euler angles are well adapted to this study since, arising from the multiplication law given in Eq. (2–36), one has, Eq. (2–42) :

$$R(\varphi,\theta,\psi;a) = R(\varphi,e_3)\ R(\theta,e_2)\ R(\psi,e_3) \quad . \tag{10-226}$$

Hence, a left (resp. right) coset of R_{e_3} is the subset of rotations parametrized by the Euler angles for given values of φ and θ (resp. θ and ψ). In the one-to-one correspondence determined by Eq. (10–217) this left (resp. right) coset is associated with the unit vector :

$$R(\varphi,\theta,\psi;a)R_{e_3} \leftrightarrow R(\varphi,\theta,\psi;a)e_3 = \hat{v}(\theta,\varphi) \quad , \tag{10-227a}$$

$$R_{e_3}\ R(\varphi,\theta,\psi;a) \leftrightarrow R(\varphi,\theta,\psi;a)^{-1}e_3 = R(-\psi,-\theta,-\varphi;a)e_3 = \hat{v}(-\theta,-\psi) \quad . \tag{10-227b}$$

with spherical polar components θ , φ (resp. $-\theta,-\psi$). Consequently, forgetting the dependence upon ψ (resp. φ), and restricting j to integer values ℓ , the results derived in Subsecs. 10.3.1,2 for the left (resp. right) regular representation of SO(3) in $L^2(SO(3),d\mu)$ give in a straightforward way the corresponding results for the representation D[SO(3)] in $L^2(S^2,d_2\hat{n})$, cf. Subsecs. 10.3.3.3,4 below. One should remark that the above procedure also provides us with the relation between the two invariant measures $d\mu$ and $d_2\hat{n}$ under the rotation group, i.e. from Eqs. (7–22,38) :

$$d\mu(\varphi,\theta,\psi) = \sin\theta\ d\varphi\ d\theta\ d\psi \leftrightarrow d_2\hat{n} = \sin\theta\ d\varphi\ d\theta \quad (\text{resp.} \sin\theta\ d\theta\ d\psi)\ . \tag{10-228}$$

It has already been noticed in Subsec. 2.3.2.2,i) that for $R(\varphi,\theta,\psi;a)$ the Euler angles φ and θ are the spherical polar components of the axis X_3 (see Fig. 2.6) with respect to the reference frame a; cf. the matrix representative of $R(\varphi,\theta,\psi;a)$ with respect to the basis $a=(e_i)$, Eq. (E–14), where the last column is equal to the Cartesian components of the unit vector with spherical polar components θ,φ. More generally, it has been shown that this result follows from the property that the quotient set O(n)/O(n–1) is isomorphic to the set of unit n–dimensional vectors. One should note that the procedure described previously in i) to induce realizations of a group in quotient sets is general.

10.3.3.3. Infinitesimal generators

According to custom we denote by L_i the infinitesimal generators of the representation D[SO(3)] (resp. the associated even representation $\mathcal{D}$[SU(2)]) determined in Subsec. 10.3.3.1, and which correspond to the matrices M_i, Eq. (4–73) (resp. $\frac{1}{2}\sigma_i$, Eq. (5–8)) :

$$L_i = \frac{\partial}{\partial\varepsilon} D[\mathbf{1}+\varepsilon M_i + O(\varepsilon^2)]\Big|_{\varepsilon=0} \quad , \text{ resp. } L_i = i\frac{\partial}{\partial\varepsilon}\mathcal{D}[\mathbf{1}-i\frac{\varepsilon}{2}\sigma_i + O(\varepsilon^2)]\Big|_{\varepsilon=0}\ . \tag{10-229}$$

The L_i are Hermitian operators with respect to the Hermitian inner product on $L^2(R^3,d_3x)$ or $L^2(S^2,d_2\hat{n})$, and they fulfil the usual commutation relations, Eqs. (5–9,11a, and b) :

$$[L_i\ ,L_j] = i\sum_k \varepsilon_{ijk}\ L_k \quad , \tag{10-230a}$$

$$L_\pm = L_1 \pm iL_2 \quad , \quad [L_3,L_\pm] = \pm L_\pm \quad , \quad [L_+,L_-] = 2L_3\ . \tag{10-230b}$$

Let us now derive from Eqs. (10–73,84) the expressions for the infinitesimal generators when one chooses either the Cartesian coordinates x^i, or the spherical polar ones r,θ,φ. The corresponding velocity fields are, respectively, given in Eqs. (E–12) and (E–13a,b). A calculation similar to the one developed in Subsec. 10.3.1.2,ii) leads to :

$$\begin{cases} L_i = -i\ \varepsilon_{ijk}\ x^j \dfrac{\partial}{\partial x^k} & (10\text{-}231a) \\[2ex] L_\pm = \mp\ (x^1 \pm ix^2)\dfrac{\partial}{\partial x^3} \pm 2x^3\dfrac{\partial}{\partial x^1 \mp ix^2}\ , & (10\text{-}231b) \end{cases}$$

$$\begin{cases} L_1 = -i\left(-\sin\varphi\dfrac{\partial}{\partial\theta} - \cos\varphi\ \text{cotg}\theta\dfrac{\partial}{\partial\varphi}\right)\ , & (10\text{-}232a) \\[2ex] L_2 = -i\left(\cos\varphi\dfrac{\partial}{\partial\theta} - \sin\varphi\ \text{cotg}\theta\dfrac{\partial}{\partial\varphi}\right)\ , & (10\text{-}232b) \\[2ex] L_3 = -i\dfrac{\partial}{\partial\varphi}\ , & (10\text{-}232c) \end{cases}$$

$$L_\pm = e^{\pm i\varphi}\left(\pm \frac{\partial}{\partial\theta} + i \,\mathrm{cotg}\theta \frac{\partial}{\partial\varphi}\right) \quad , \qquad (10\text{-}232d)$$

$$L^2 = \sum_i (L_i)^2 = -\mathrm{cotg}\theta \frac{\partial}{\partial\theta} - \frac{\partial^2}{\partial\theta^2} - \frac{1}{\sin^2\theta}\frac{\partial^2}{\partial\varphi^2}$$

$$= -\frac{1}{\sin\theta}\frac{\partial}{\partial\theta}\left(\sin\theta \frac{\partial}{\partial\theta}\right) - \frac{1}{\sin^2\theta}\frac{\partial^2}{\partial\varphi^2} \quad . \qquad (10\text{-}232e)$$

Actually, a simple way to get the above Eqs. (10–232a–e) is to use the relation with the regular representations of SO(3) established in Subsec. 10.3.3.2. Namely, they can be obtained from the expressions of the infinitesimal generators of the left (resp. right) regular representation, Eqs. (10–168a–d) (resp. Eqs. (10–169a–d)) forgetting the dependence upon ψ (resp. φ, and then from Eq. (10–227b) changing θ into $-\theta$ and ψ into $-\varphi$).

The infinitesimal operators L_i are identified later (Subsec. 11.1.2) with the Cartesian components of the orbital angular momentum.

10.3.3.4. Definition of the spherical harmonics

Let us for a while denote by f^ℓ_m a standard basis (Sec. 5.5) of the representation space $L^2(S^2,d_2\hat{n})$. Two ways of obtaining these complex functions of θ and φ are now given.

i) The first method consists in solving the characteristic equations of a standard basis of a representation of SO(3), i.e. with integer values of ℓ and m :

$$L_\pm f^\ell_m = \sqrt{\ell(\ell+1)-m(m\pm1)}\; f^\ell_{m\pm1} \quad , \qquad (10\text{-}233a)$$

$$L_3 f^\ell_m = m f^\ell_m \quad , \qquad (10\text{-}233b)$$

where $L_\pm$ and L_3 are given by Eqs. (10–232c,d). From the last equation one finds :

$$-i\frac{\partial}{\partial\varphi} f^\ell_m(\theta,\varphi) = m\, f^\ell_m(\theta,\varphi) \Rightarrow f^\ell_m(\theta,\varphi) = g_{\ell m}(\theta)\, e^{im\varphi} \quad . \qquad (10\text{-}234)$$

Eq. (10–233a) implies that all the functions f^ℓ_m can be obtained by a repeated action of L_+ (resp. L_-) on $f^\ell_{-\ell}$ (resp. f^ℓ_ℓ) which fulfils :

$$L_- f^\ell_{-\ell} = 0 \quad , \quad \text{resp. } L_+ f^\ell_\ell = 0 \quad . \qquad (10\text{-}235)$$

These equations imply :

$$\frac{d}{d\theta} g_{\ell\pm\ell} - \ell\, \mathrm{cotg}\theta\; g_{\ell\pm\ell} = 0 \quad , \qquad (10\text{-}236)$$

from which one obtains :

$$g_{\ell\pm\ell} = n_\ell \sin^\ell\theta \quad , \qquad (10\text{-}237)$$

where n_ℓ is a constant. Requiring that $f^\ell_{\pm\ell}$ be normalized to one in $L^2(S^2,d_2\hat{n})$ yields the normalization factor n_ℓ :

$$1 = \int_0^{\pi} \sin\theta \, d\theta \int_0^{\pi} d\varphi \; f^{\ell *}_{\pm\ell}(\theta,\varphi) f^{\ell}_{\pm\ell}(\theta,\varphi) = 2\pi \, n_\ell^2 \int_0^{2\pi} \sin^{\ell+1}\theta \, d\theta$$

$$\Rightarrow \; n_\ell = \frac{1}{2^\ell \ell!} \left[\frac{(2\ell+1)!}{4\pi}\right]^{\frac{1}{2}} \quad . \qquad (10\text{-}238)$$

All the orthonormalized f^ℓ_m, which are called the usual spherical harmonics $Y_{\ell\, m}$, then follow from Eq. (H–22a). After rearrangement one obtains :

$$Y_{\ell m}(\theta,\varphi) = (-1)^{\ell+m} \frac{1}{2^\ell \ell!} \left[\frac{(2\ell+1)(\ell-m)!}{4\pi(\ell+m)!}\right]^{\frac{1}{2}} \sin^m\theta \left(\frac{\partial}{\partial\cos\theta}\right)^{\ell+m} \sin^{2\ell}\theta \; e^{im\varphi} \, . \qquad (10\text{-}239)$$

The spherical harmonics defined as above correspond to the most commonly adopted convention, cf. iii) below.

ii) The second method is based on the relation with the regular representations of SO(3) established in Subsec. 10.3.3.2. Arising from Eq. (10–226), the dependence of $\mathcal{D}^\ell_{mm'}(\varphi,\theta,\psi)$ upon φ and ψ is given by the exponentials $e^{-im\varphi} \; e^{-im'\psi}$, Eq. (6–21). The class functions $\tilde{\phi}^{\ell,\ell}_m$ (resp. $\tilde{\phi}^{r,\ell}_m$) which according to Eq. (10–222) correspond to the standard basis functions $\phi^{\ell,\ell}_m = \mathcal{D}^{\ell *}_{mm_o}$ (resp. $\phi^{r,\ell}_m = \mathcal{D}^\ell_{0m}$), Eqs. (10–150a,b) of the left (resp. right) regular representation of SO(3) in $L^2(SO(3),d\mu)$ are therefore obtained by setting $m_o = 0$ to eliminate the dependence upon ψ (resp. φ) :

$$\tilde{\phi}^{\ell,\ell}_m(\theta,\varphi) = \mathcal{D}^{\ell *}_{m0}(\varphi,\theta,\psi) \quad , \quad \tilde{\phi}^{r,\ell}_m(\theta,\psi) = \mathcal{D}^\ell_{0m}(\varphi,\theta,\psi) \quad . \qquad (10\text{-}240)$$

These functions form a standard basis of the representation $\tilde{D}^\ell$ (resp. D^r), to which according to Eqs. (10–225a,b) there corresponds a standard basis of the representation of SO(3) in $L^2(S^2,d_2\hat{n})$. One thereby obtains using Eqs. (10–227a,b) :

$$f^\ell_m(\theta,\varphi) \propto \mathcal{D}^{\ell *}_{m0}(\varphi,\theta,\psi) \quad , \quad f^\ell_m(-\theta,-\psi) \propto \mathcal{D}^\ell_{0m}(\varphi,\theta,\psi) \quad . \qquad (10\text{-}241)$$

It can be checked that these two expressions are the same since :

$$\mathcal{D}^{\ell *}_{m\,0}(\varphi,\theta,\psi) = \mathcal{D}^\ell_{0m}(-\psi,-\theta,-\varphi) \quad . \qquad (10\text{-}242)$$

Normalizing these functions to one in $L^2(S^2,d_2\hat{n})$, one obtains from Eqs. (7–80a) and (10–228) :

$$f^\ell_m(\theta,\varphi) = \frac{\hat{\ell}}{\sqrt{4\pi}} \mathcal{D}^{\ell *}_{m0}(\varphi,\theta,\psi) \quad . \qquad (10\text{-}243)$$

These normalized functions are thus related by a phase factor to the spherical harmonics defined by Eq. (10–239). From Eqs. (F–13) and (H–15) one finds that they are equal :

$$Y_{\ell m}(\theta,\varphi) = \frac{\hat{\ell}}{\sqrt{4\pi}} \mathcal{D}^{\ell *}_{m0}(\varphi,\theta,\psi) \quad . \qquad (10\text{-}244)$$

iii) We recall the relations (Subsec. 6.1.2,ii)) :

$$\mathcal{D}^{\ell *}_{m0}(\varphi,\theta,\psi) = \mathcal{D}^{\ell}_{0m}(-\psi,-\theta,-\varphi) = (-1)^m \mathcal{D}^{\ell *}_{0m}(\psi,\theta,\varphi) = (-1)^m \mathcal{D}^{\ell}_{-m0}(\varphi,\theta,\psi) \ . \qquad (10\text{-}245)$$

Hence, the conjugate complex functions of the spherical harmonics $Y_{\ell m}$ read :

$$Y^*_{\ell m}(\theta,\varphi) = \frac{\hat{\ell}}{\sqrt{4\pi}} \mathcal{D}^{\ell}_{m0}(\varphi,\theta,\psi) = (-1)^m \frac{\hat{\ell}}{\sqrt{4\pi}} \mathcal{D}^{\ell *}_{-m0}(\varphi,\theta,\psi) = (-1)^m Y_{\ell -m}(\theta,\varphi) \ . \qquad (10\text{-}246)$$

Actually, from the tensorial point of view it is more convenient to define the spherical harmonics $Y^{[\ell]}_m$ which are related to the usual ones by :

$$Y^{[\ell]}_m = (-ic)^{\ell} Y_{\ell m} \quad , \quad |c| = 1 \ . \qquad (10\text{-}247)$$

Indeed, choosing $c = \pm 1$ these $Y^{[\ell]}_m$ are then the contravariant spherical components of a Hermitian tensor Y^{ℓ} of order one and rank ℓ (Subsec. 8.3.4.1,iii)), since one has, cf. Eq. (8–116) :

$$Y^{[\ell]*}_m = c^{-2\ell} (-1)^{\ell+m} Y^{[\ell]}_{-m} \ . \qquad (10\text{-}248)$$

For $\ell=1$, the factor c in the above equations is the same as the one which occurs in Eq. (8–210). Namely, for any vector v with spherical polar components $\|v\|$, θ , and φ, one has (Eq. (8–196) and Table H.2) :

$$v^{[1]}_m = \sqrt{\frac{4\pi}{3}} \|v\| \ Y^{[1]}_m(\theta,\varphi) \ . \qquad (10\text{-}249)$$

The definitions for the spherical harmonics adopted by different authors are given in Appendix H.2,i). In the remainder of this chapter, and also in the next chapter, we only consider the spherical harmonics $Y^{[\ell]}_m$. They are thus related to the rotation matrix elements by :

$$Y^{[\ell]}_m(\varphi,\theta) = (-ic)^{\ell} \frac{\hat{\ell}}{\sqrt{4\pi}} \mathcal{D}^{\ell *}_{m0}(\varphi,\theta,\psi) \ . \qquad (10\text{-}250)$$

and :

(1) It is always assumed that : $|c| = 1$.

(2) From the tensorial point of view the most convenient choices are : $c=\pm 1$.

(3) The usual spherical harmonics $Y_{\ell m}$ are obtained by setting : $c=i$.

10.3.3.5. Some properties of the spherical harmonics

Let us derive properties of the spherical harmonics $Y^{[\ell]}_m$ which follow from their relation with the rotation matrix elements.

i) The law of transformation under finite rotations. From the definition of the representation $D[SO(3)]$ in $L^2(S^2,d_2\hat{n})$, Eq. (10–212) and since the spherical harmonics form a standard basis (Subsec. 6.1.3,ii)) one has :

$$\forall R \qquad [D(R)\, Y^{[\ell]}_m](\theta,\varphi) = Y^{[\ell]}_m[R^{-1}(\theta,\varphi)] = Y^{[\ell]}_{m'}(\theta,\varphi)\, \mathcal{D}^{\ell}_{m'm}(R) \quad . \tag{10-251}$$

ii) **The closure and the orthogonality relations.** They have been written in Eqs. (7–79a b) for the usual spherical harmonics, hence :

$$\sum_{\ell,m} Y^{[\ell]}_m(\theta,\varphi)\, Y^{[\ell]*}_m(\theta',\varphi') = \delta(\cos\theta-\cos\theta')\,\delta(\varphi-\varphi') \quad , \tag{10-252a}$$

$$\int_0^{\pi}\sin\theta\, d\theta \int_0^{2\pi} d\varphi\, Y^{[\ell]*}_m(\theta,\varphi)\, Y^{[\ell']}_{m'}(\theta,\varphi) = \delta_{\ell\ell'}\,\delta_{mm'} \quad . \tag{10-252b}$$

Using Eqs. (10–248) and (G–4b), the closure relation takes the form :

$$\sum_{\ell} c^{-2\ell}\,\hat{\ell}\,\{Y^{\ell}(\theta,\varphi)\; Y^{\ell}(\theta',\varphi')\}^0 = \delta(\cos\theta-\cos\theta')\,\delta(\varphi-\varphi') \quad . \tag{10-253}$$

iii) **The addition theorem.** One has, cf. above :

$$\sum_m Y^{[\ell]}_m(\theta,\varphi)\, Y^{[\ell]*}_m(\theta',\varphi') = c^{-2\ell}\,\hat{\ell}\,\{Y^{\ell}(\theta,\varphi)\; Y^{\ell}(\theta',\varphi')\}^0 \quad . \tag{10-254}$$

Being coupled to zero, this expression is invariant under the rotation group, i.e. :

$$\forall R \qquad \sum_m Y^{[\ell]}_m(\theta,\varphi)\, Y^{[\ell]*}_m(\theta',\varphi') = \sum_m Y^{[\ell]}_m[R(\theta,\varphi)]\; Y^{[\ell]*}_m[R(\theta',\varphi')] \quad . \tag{10-255}$$

Let $\hat{q}$ and $\hat{q}'$ be, respectively, the two unit vectors with spherical polar components θ, φ and θ', φ' with respect to any orthonormal basis (e_i). Choosing R as a rotation which brings $\hat{q}$ onto e_3 (e.g. $R^{-1}_{\hat{q}}$, Eq. (2–37)), one obtains from Eq. (H–25a) :

$$Y^{[\ell]}_m[R(\theta,\varphi)] = (-ic)^{\ell}\,\frac{\hat{\ell}}{\sqrt{4\pi}}\,\delta_{m0} \quad . \tag{10-256}$$

Furthermore, one has :

$$e_3.R\hat{q}' = R^{-1}e_3.\hat{q}' = \hat{q}.\hat{q}' = \cos\alpha \quad , \tag{10-257}$$

where α is the angle between $\hat{q}$ and $\hat{q}'$, which implies from Eq. (H–24a) :

$$Y^{[\ell]}_0[R(\theta',\varphi')] = (-ic)^{\ell}\,\frac{\hat{\ell}}{\sqrt{4\pi}}\,P_{\ell}(\cos\alpha) \quad . \tag{10-258}$$

One thereby finds the addition theorem for the spherical harmonics :

$$\sum_m Y^{[\ell]}_m(\theta,\varphi)\, Y^{[\ell]*}_m(\theta',\varphi') = c^{-2\ell}\,\hat{\ell}\,\{Y^{\ell}(\theta,\varphi)\; Y^{\ell}(\theta',\varphi')\}^0 = \frac{2\ell+1}{4\pi}\,P_{\ell}(\cos\alpha) \quad . \tag{10-259}$$

iv) **The coupling of two spherical harmonics.** From Eqs. (6–68b) and (10–250) one obtains :

$$Y^{[\ell_1]}_{m_1}(\theta,\varphi)\; Y^{[\ell_2]}_{m_2}(\theta,\varphi) =$$

$$\sum_{\ell} (-ic)^{\ell_1+\ell_2-\ell}\,\frac{\hat{\ell}_1\hat{\ell}_2}{\sqrt{4\pi}\,\hat{\ell}}\,\langle \ell_1\ell_2 m_1 m_2|\ell m\rangle\langle \ell_1\ell_2 00|\ell 0\rangle\; Y^{[\ell]}_m(\theta,\varphi) \quad , \tag{10-260}$$

where the symmetry property Eq. (G–2a) of the C.G. coefficients implies that $\ell_1+\ell_2+\ell$ be even. Using the unitary character of the C.G. coefficients, Eq. (6–66a), one derives from the above :

$$\{Y^{\ell_1}(\theta,\varphi)\, Y^{\ell_2}(\theta,\varphi)\}^{[\ell]}_m = C(\ell_1,\ell_2,\ell)\, Y^{[\ell]}_m(\theta,\varphi) \quad , \tag{10-261a}$$

where :

$$C(\ell_1,\ell_2,\ell) = c^{\ell_1+\ell_2-\ell}\,(-1)^{\frac{1}{2}(\ell_1+\ell_2-\ell)}\,\frac{\hat{\ell}_1\hat{\ell}_2}{\sqrt{4\pi}\,\hat{\ell}}<\ell_1\ell_2 00|\ell 0> \tag{10-261b}$$

$$= c^{\ell_1+\ell_2-\ell}\,(-1)^{\frac{1}{2}(\ell_1+\ell_2+\ell)}\,\frac{\hat{\ell}_1\hat{\ell}_2}{\sqrt{4\pi}}\begin{pmatrix}\ell_1 & \ell_2 & \ell\\ 0 & 0 & 0\end{pmatrix} \quad . \tag{10-261c}$$

The value of the C.G. coefficients $<\ell_1\ell_2 00|\ell 0>$ is given in Eq. (G–5).

v) The gradient formula. This formula gives the gradient of any function f in H expressed in terms of the spherical polar coordinates by :

$$f(r,\theta,\varphi) = g(r)\, Y^{[\ell]}_m(\theta,\varphi) \quad . \tag{10-262}$$

The gradient operator ∇ is a special case of a vector operator ∇^1, the contravariant spherical components of which are related to its Cartesian ones by Eq. (8–211) :

$$\nabla^{[1]}_{\pm 1} = c\,\frac{\pm 1}{\sqrt{2}}\left(i\,\frac{\partial}{\partial x^1} \mp \frac{\partial}{\partial x^2}\right) \quad , \quad \nabla^{[1]}_0 = c\left(-i\,\frac{\partial}{\partial x^3}\right) \quad . \tag{10-263}$$

From the closure relation of the spherical harmonics, Eq. (10–252a), one has :

$$\left(\nabla^{[1]}_\mu f\right)(r,\theta,\varphi) = \sum_{\ell',m'} Y^{[\ell']}_{m'}(\theta,\varphi)\,<r(\ell')m'|\nabla^{[1]}_\mu|g[\ell]m> \quad , \tag{10-264}$$

where the matrix element :

$$<r(\ell')m'|\nabla^{[1]}_\mu|g[\ell]m> = \int_{S^2}\sin\theta\,d\theta\,d\varphi\,Y^{[\ell']*}_{m'}(\theta,\varphi)\left(\nabla^{[1]}_\mu g\,Y^{[\ell]}_m\right)(r,\theta,\varphi) \tag{10-265}$$

is the spherical component $T^{(\ell')\,[1]\,[\ell]}_{m'\;\mu\;m}$ of an invariant tensor $T^{\ell' 1 \ell}$ of order three (Subsec. 11.3.1). It then follows from the Wigner–Eckart theorem, Eq. (8–149), that the whole dependence upon m', μ, and m is carried by a C.G. coefficient, e.g. :

$$<r(\ell')m'|\nabla^{[1]}_\mu|g[\ell]m> = \frac{<1\ell\mu m|\ell' m'>}{<1\ell 00|\ell' 0>}\,<r(\ell')0|\nabla^{[1]}_0|g[\ell]0> \quad . \tag{10-266}$$

We are thus led to evaluate the simpler matrix element $<r(\ell')0|\,\nabla^{[1]}_0\,|g\,[\ell]\,0>$, where in terms of spherical polar coordinates :

$$\nabla^{[1]}_0 = -ic\left(\cos\theta\,\frac{\partial}{\partial r} - \frac{\sin\theta}{r}\,\frac{\partial}{\partial\theta}\right) = -ic\left(\cos\theta\,\frac{\partial}{\partial r} + \frac{1-\cos^2\theta}{r}\,\frac{\partial}{\partial\cos\theta}\right) \quad . \tag{10-267}$$

Then, from Eq. (10–258) and using the functional relations for the Legendre polynomials given by Eqs. (H–7,9), one obtains for $\ell \geqslant 1$:

$$\left(\nabla^{[1]}_{0}\, g\, Y^{[\ell]}_{0}\right)(r,\theta,\varphi) = -c^{2}\,\frac{\ell}{\hat{\ell}\,\widehat{\ell-1}}\left(\frac{d}{dr}+\frac{\ell+1}{r}\right) g(r)\, Y^{[\ell-1]}_{0}(\theta,\varphi)$$

$$+\frac{\ell+1}{\hat{\ell}\,\widehat{\ell+1}}\left(\frac{d}{dr}-\frac{\ell}{r}\right) g(r)\, Y^{[\ell+1]}_{0}(\theta,\varphi) \quad . \qquad (10\text{–}268)$$

From the orthogonality relation, Eq. (10–252b) the only nonzero matrix elements read with $\ell \geqslant 1$:

$$\langle r(\ell-1)0|\nabla^{[1]}_{0}|g[\ell]0\rangle = c^{2}\,\frac{\ell}{\hat{\ell}\,\widehat{\ell-1}}\left(\frac{d}{dr}+\frac{\ell+1}{r}\right) g(r) \quad , \qquad (10\text{–}269a)$$

$$\langle r(\ell+1)0|\nabla^{[1]}_{0}|g[\ell]0\rangle = \frac{\ell+1}{\hat{\ell}\,\widehat{\ell+1}}\left(\frac{d}{dr}-\frac{\ell}{r}\right) g(r) \quad . \qquad (10\text{–}269b)$$

Eq. (G–5) yields :

$$\langle 1\ell 00|\ell-1\ 0\rangle = -\frac{\sqrt{\ell}}{\hat{\ell}} \quad , \qquad \langle 1\ell 00|\ell+1\ 0\rangle = \frac{\sqrt{\ell+1}}{\hat{\ell}} \quad , \qquad (10\text{–}270)$$

and from Table G.2 one finally finds for $\ell \geqslant 1$:

$$\langle r(\ell-1)\mu+m|\nabla^{[1]}_{\mu}|g[\ell]m\rangle = c^{2}\,\langle 1\ell\mu m|\ell-1\mu+m\rangle\,\frac{\sqrt{\ell}}{\widehat{\ell-1}}\left(\frac{d}{dr}+\frac{\ell+1}{r}\right) g(r) \qquad (10\text{–}271a)$$

$$= c^{2}\,\frac{(-1)^{\ell+m}\,C^{(-)}_{\ell\mu m}}{\sqrt{2}\,\hat{\ell}\,\widehat{\ell-1}}\left(\frac{d}{dr}+\frac{\ell+1}{r}\right) g(r) \qquad (10\text{–}271b)$$

$$\langle r(\ell+1)\mu+m|\nabla^{[1]}_{\mu}|g[\ell]m\rangle = \langle 1\ell\mu m|\ell+1\mu+m\rangle\,\frac{\sqrt{\ell+1}}{\widehat{\ell+1}}\left(\frac{d}{dr}-\frac{\ell}{r}\right) g(r) \qquad (10\text{–}271c)$$

$$= \frac{(-1)^{\ell+m}\,C^{(+)}_{\ell\mu m}}{\sqrt{2}\,\hat{\ell}\,\widehat{\ell+1}}\left(\frac{d}{dr}-\frac{\ell}{r}\right) g(r) \qquad (10\text{–}271d)$$

where $C^{(\pm)}_{\ell\mu m}$ is given in Table 10.1.

TABLE 10.1

The coefficients $C^{(\pm)}_{\ell\mu m}$ occurring in the gradient formula, Eqs. (10–271b,d).

	$\mu = 1$	$\mu = 0$	$\mu = -1$
$C^{(-)}_{\ell\mu m}$	$[(\ell-m-1)(\ell-m)]^{\frac{1}{2}}$	$[2(\ell+m)(\ell-m)]^{\frac{1}{2}}$	$[(\ell+m-1)(\ell+m)]^{\frac{1}{2}}$
$C^{(+)}_{\ell\mu m}$	$[(\ell+m+1)(\ell+m+2)]^{\frac{1}{2}}$	$-[2(\ell+m+1)(\ell-m+1)]^{\frac{1}{2}}$	$[(\ell-m+1)(\ell-m+2)]^{\frac{1}{2}}$

An expression of the gradient formula in terms of vector spherical harmonics is given later in Eq. (10–315b).

Other properties of the spherical harmonics are collected in Appendix H.2, i.e. : the relation with the associated Legendre functions, recursion formulae, special cases and special values. The definitions, as well as some of the properties of the harmonic polynomials and the solid harmonics are given in Appendix H.3. Appendix H.4 is devoted to some remarks about multipole expansions ; examples are given.

10.3.3.6. Multiplicative operators associated with the spherical harmonics

To the spherical harmonics $Y^{[\ell]}_m$ there correspond the multiplicative operators $\hat{Y}^{[\ell]}_m$ of $L^2(S^2,d_2\hat{n})$ into itself defined by (Subsec. 10.3.1.3,i)) :

$$\forall f \in L^2(S^2,d_2\hat{n}) \qquad \hat{Y}^{[\ell]}_m f \quad , \quad \forall \theta,\varphi \quad (\hat{Y}^{[\ell]}_m f)(\theta,\varphi) = Y^{[\ell]}_m(\theta,\varphi) f(\theta,\varphi) \quad . \tag{10-272}$$

These operators satisfy the commutation relations with the infinitesimal generators, Eq. (10–180) :

$$[L_\pm , \hat{Y}^{[\ell]}_m] = \sqrt{\ell(\ell+1)-m(m\pm 1)}\ \hat{Y}^{[\ell]}_{m\pm 1} \quad , \quad [L_3,\hat{Y}^{[\ell]}_m] = m\ \hat{Y}^{[\ell]}_m \quad . \tag{10-273}$$

10.3.4. REPRESENTATION OF SU(2) IN A SPACE OF VECTORIAL FUNCTIONS DEFINED ON R^3

What follows is a direct application of the general procedure described in Subsecs. 10.1.2 and 10.2.3 to determine a representation of a group in a vector space of vectorial functions.

10.3.4.1. Determination, infinitesimal generators, and standard basis

i) Determination of the representation. Let F_s be a (2s+1)– dimensional Hermitian vector space for $s=0,\frac{1}{2},1,\ldots$. A unitary representation $D_1\,[SU(2)] = \mathcal{D}^s\,[SU(2)]$ is defined in F_s by assuming that this vector space is spanned by the contravariant orthonormal standard basis $(e^{[s]}_m)$, i.e. :

$$\forall v \in F_s \qquad v = v_{sm}\, e^{[s]}_m \quad , \quad \left(e^{[s]}_m , e^{[s]}_{m'}\right)_{F_s} = \delta_{mm'} \quad , \tag{10-274}$$

and :

$$\forall U \in SU(2) \qquad D_1(U)\, e^{[s]}_m = e^{[s]}_{m'}\ \mathcal{D}^s_{m'm}(U) \quad . \tag{10-275}$$

The covariant vectors $e^{(s)}_m$ also form an orthonormal basis of F_s :

$$e^{(s)}_m = c_s(-1)^{s+m}\, e^{[s]}_{-m} \quad , \quad \left(e^{(s)}_m , e^{(s)}_{m'}\right)_{F_s} = \delta_{mm'} \quad . \tag{10-276}$$

One then considers the functional vector space $\mathcal{F}_s$ over the complex field of vectorial functions f of R^3 (equivalently E_3, $E_3^\cdot$, or S^2) into F_s. Hence, one has, Eq. (10–12) :

$$\forall f \in \mathcal{F}_s \qquad f = f_{sm}\, e^{[s]}_m \quad , \quad f_{sm} \in \mathcal{H} \quad , \tag{10-277}$$

where f_{sm} are complex functions defined on R^3 and the $e^{[s]}_m$ are taken here as functions in

F_s which map any x in R^3 into the basis vector $e_m^{[s]}$ of F_s. A Hermitian inner product in F_s is given by, Eq. (10–13) :

$$\forall \quad f = f_{sm}\, e_m^{[s]} \;, \quad f' = f'_{sm}\, e_m^{[s]} \in F_s \qquad (f,f')_{F_s} = \int_{R^3} d_3x\, f^*_{sm}(x) f'_{sm}(x) \;, \qquad (10\text{-}278)$$

assuming that the functions f_{sm} and f'_{sm} belong to $L^2(R^3,d_3x)$ (resp. $L^2(S^2,d_2\hat{n})$).

The representation $D_1[SU(2)]$ in F_s induces the representation $\mathcal{D}_1[SU(2)]$ in F_s according to Eq. (10–14). Applied to each of the 2s+1 components of the vectorial functions in F_s, the representation $\mathcal{D}_2[SU(2)]$ determined in Subsec. 10.3.3.1,ii) induces a representation in F_s, still denoted by $\mathcal{D}_2[SU(2)]$, and which corresponds to Eq. (10–15). The commutative product of these two unitary representations then determines the unitary representation $\mathcal{D}[SU(2)]$ in F_s, Eq. (10–19) :

$$\forall U \in SU(2) \;\rightarrow\; \mathcal{D}(U)$$
$$\forall f \in F_s \rightarrow \mathcal{D}(U)f, \quad \forall x \in R^3 \text{ or } S^2 \quad [\mathcal{D}(U)f](x) = \mathcal{D}^s(U)\; f[R(U)^{-1}x] \;. \qquad (10\text{-}279)$$

In terms of the components of f with respect to $(e_m^{[s]})$, one has :

$$\forall U \in SU(2) \quad , \quad \forall\, f = f_{sm}\, e_m^{[s]} \in F_s \quad , \quad \forall x \in R^3 \text{ or } S^2$$
$$[\mathcal{D}(U)f](x) = e_{m'}^{[s]}\, \mathcal{D}^s_{m'm}(U)\; f_{sm}[R(U)^{-1}x] \;. \qquad (10\text{-}280)$$

ii) **Infinitesimal generators.** From Eq. (10–87), the infinitesimal generators J_i of the representation $\mathcal{D}[SU(2)]$, which correspond to the σ matrices by Eq. (5–8), read :

$$J_i = L_i + S_i \quad , \qquad (10\text{-}281)$$

where S_i and L_i are, respectively, the infinitesimal generators of the representation $\mathcal{D}_1[SU(2)]$ and $\mathcal{D}_2[SU(2)]$. The expressions of the L_i are those derived in Subsec. 10.3.3.3 ; they act on each of the 2s+1 components of the vectorial functions. The S_i are determined by their matrix elements Eqs. (6–34a,b) :

$$S_\pm\, e_m^{[s]} = (S_1 \pm iS_2)\, e_m^{[s]} = \sqrt{s(s+1)-m(m\pm 1)}\; e_{m\pm 1}^{[s]} \;, \qquad (10\text{-}282a)$$

$$S_3\, e_m^{[s]} = m\, e_m^{[s]} \;, \qquad (10\text{-}282b)$$

and on F_s one has :

$$S^2 = s(s+1)\, \mathbf{1}_{F_s} \;. \qquad (10\text{-}282c)$$

The L_i, S_i, and J_i, respectively, satisfy the usual commutation relations of angular momentum components, Eq. (5–9). Furthermore, since the representations $\mathcal{D}_1$ and $\mathcal{D}_2$ commute, so do their infinitesimal generators L_i and S_i.

The S_i are identified later (Subsec. 11.1.3) with the Cartesian components of the intrinsic angular momentum or spins, and the J_i with the Cartesian components of the total angular momentum J.

iii) **Standard basis.** A contravariant standard basis of the representation $\mathcal{D}[\,SU(2)]$ in F_s is clearly obtained by the coupling of standard bases $(Y^{[\ell]}_m)$ and $(e^{[s]}_m)$ of the representations $\mathcal{D}_2$ and $\mathcal{D}_1$, respectively. One thus defines the spin spherical harmonics :

$$\mathbf{Y}^{\ell s\,[j]}_m = Y^{[\ell]}_{m_\ell}\; e^{[s]}_{m_s}\; <\ell s m_\ell m_s|jm> \quad , \tag{10-283}$$

which are eigenvectors of L^2, S^2, J^2, and J_3 with the eigenvalues $\ell(\ell+1)$, $s(s+1)$, $j(j+1)$, and m, respectively. These vectorial functions form a complete set of orthonormal functions on F_s. Indeed, one derives from Eqs. (10–252a,b, and 274) and the unitary character of the C.G. coefficients, Eqs. (6–66a,b) :

$$\left(\mathbf{Y}^{\ell s[j]}_m , \mathbf{Y}^{\ell' s[j']}_{m'}\right)_{F_s} = \delta_{\ell\ell'}\,\delta_{jj'}\,\delta_{mm'} \quad , \tag{10-284a}$$

and

$$\forall f \in F_s \qquad f = \sum_{\ell,j} f_{\ell sjm}\,\mathbf{Y}^{\ell s[j]}_m \quad , \qquad f_{\ell sjm} = \left(\mathbf{Y}^{\ell s[j]}_m , f\right)_{F_s} \quad . \tag{10-284b}$$

10.3.4.2. The special case of a three–dimensional vector field, vector spherical harmonics

i) **Identification of F_1 with $\{E_3\}$.** For s=1 the three–dimensional Hermitian vector space F_1 can be identified with the complex extension $\{E_3\}$ of the Euclidean vector space E_3 (Appendix A.2.3). From Eq. (8–211), the relation between the Cartesian basis (e_i) and the contravariant spherical basis $(e^{[1]}_m)$ of F_1 reads :

$$\begin{cases} e^{[1]}_1 = c\,\frac{1}{\sqrt{2}}\,(ie_1 - e_2) \\ e^{[1]}_0 = c\,(-ie_3) \\ e^{[1]}_{-1} = c\,\frac{-1}{\sqrt{2}}\,(ie_1 + e_2) \end{cases} \qquad \begin{cases} e_1 = c^{-1}\,\frac{i}{\sqrt{2}}\left(-e^{[1]}_1 + e^{[1]}_{-1}\right) \\ e_2 = c^{-1}\,\frac{-1}{\sqrt{2}}\left(e^{[1]}_1 + e^{[1]}_{-1}\right) \\ e_3 = c^{-1} i\, e^{[1]}_0 \end{cases} \quad . \tag{10-285}$$

The choice of the factor c in the above relations arises from the following considerations (Subsecs. 8.5.2,i) and 8.5.4,ii)) :

(1) Choosing c real implies that the $e^{[1]}_m$ are the contravariant spherical components of a Hermitian tensor, i.e. one has, Eq. (8–116) :

$$e^{[1]*}_m = \frac{c^*}{c}\,(-1)^{1+m}\,e^{[1]}_{-m} \quad , \tag{10-286}$$

where $e^{[1]*}_m$ denotes the vector which has with respect to the Cartesian basis (e_i) the conjugate complex components of $e^{[1]}_m$.

(2) Choosing $|c|=1$ implies that the Cartesian and the spherical bases are both orthonormal (the matrix t, Eq. (8–183a), is unitary) ; e.g. with an orthonormal Cartesian basis :

$$(e_i, e_j)_{F_1} = e_i . e_j = \delta_{ij} \quad , \quad \left(e^{[1]}_m , e^{[1]}_{m'}\right)_{F_1} = e^{[1]*}_m . e^{[1]}_{m'} = |c|^2\,\delta_{mm'} \; . \tag{10-287}$$

Eqs. (10–286,287) yield :

$$e^{[1]}_{m} \cdot e^{[1]}_{m'} = c^2 \ (-1)^{1+m} \ \delta_{-mm'} \qquad . \qquad (10\text{-}288)$$

From now on it is always assumed that : $|c| = 1$.

(3) Applying the results derived in Subsec. 8.5.4,ii), let us show the advantages of the choices $c = \pm 1$. The choices of c adopted by some authors are listed in Subsec. 8.5.4,ii). Any vector v in F_1 can be expressed as :

$$v = v^i \ e_i = v_m \ e^{[1]}_m \qquad . \qquad (10\text{-}289)$$

We want to express the v_m in terms of the spherical components $v^{[1]}_m$ determined from the Cartesian ones by Eq. (8–211) :

$$v^{[1]}_{\pm 1} = c \frac{\pm 1}{\sqrt{2}} \ (iv^1 \mp v^2) \quad , \quad v^{[1]}_0 = c \ (-i) \ v^3 \quad , \qquad (10\text{-}290)$$

and which satisfy the relation similar to Eq. (10–286) with $|c| = 1$:

$$v^{[1]*}_m = c^{-2} \ (-1)^{1+m} \ v^{[1]}_{-m} \qquad . \qquad (10\text{-}291)$$

Since Eq. (10–289) formally reads as the left-hand-side of Eq. (8–213), one has :

$$v = c^{-2} \ \sqrt{3} \ \{v^1 \otimes e^1\}^0 = c^{-2} \ (-1)^{1+m} \ v^{[1]}_{-m} \ e^{[1]}_m \quad , \qquad (10\text{-}292)$$

One thereby obtains :

$$v_m = c^{-2} \ (-1)^{1+m} \ v^{[1]}_{-m} = v^{[1]*}_m \quad , \qquad (10\text{-}293)$$

and from the orthogonality relations, Eqs. (10–287,288) :

$$v^{[1]}_m = \left(e^{[1]*}_m \ , \ v\right)_{F_1} = e^{[1]}_m \cdot v \qquad . \qquad (10\text{-}294)$$

Let us again point out that Eqs. (8–213,214) are independent of the nature of the considered objects. Thus, applied to Eq. (10–292), v^1 is the vector with spherical components $v^{[1]}_m$. Since each $e^{[1]}_m$ is a vector it also determines a tensor $(e^{[1]}_m)^1$ of order and rank one. Therefore, applying Eq. (8–213) to Eq. (10–294) one gets :

$$v^{[1]}_m = e^{[1]}_m \cdot v = c^{-2} \ \sqrt{3} \ \left\{\left(e^{[1]}_m\right)^0 \otimes v^1\right\}^0 \qquad . \qquad (10\text{-}295)$$

The contravariant spherical components of $(e^{[1]}_m)^1$ can be obtained for instance from Eq. (10–294). One finds using Eq. (10–288) :

$$\left(e^{[1]}_m\right)^{[1]}_{m'} = e^{[1]}_{m'} \cdot e^{[1]}_m = c^2 \ (-1)^{1+m} \ \delta_{-m,m'} \qquad . \qquad (10\text{-}296)$$

It follows from Eq. (8–214) that :

$$\left(v \times e^{[1]}_m\right)^1 = c^{-1}\sqrt{2}\left\{v^1 \otimes \left(e^{[1]}_m\right)^1\right\}^1 \quad . \tag{10-297}$$

Then, using Eqs. (10–292,297) and also Table G.5 one gets :

$$v \times e^{[1]}_m = c^{-3}\sqrt{6}\left\{\left\{v^1 \otimes \left(e^{[1]}_m\right)^1\right\}^1 \otimes e^1\right\}^0 = c^{-3}\sqrt{6}\begin{bmatrix} 1 & 1 & 1 \\ 1 & 0 & 1 \\ 1 & 1 & 0 \end{bmatrix}\left\{\left\{v^1 \otimes e^1\right\}^1 \otimes \left(e^{[1]}_m\right)^1\right\}^0$$

$$= - c^{-1}\sqrt{2}\,\{v^1 \otimes e^1\}^{[1]}_m \quad , \tag{10-298}$$

where one should notice the simultaneous use of the two tensors e^1 and $(e^{[1]}_m)^1$. A direct calculation leads to :

$$e_i \times e_j = \sum_k \varepsilon_{ijk}\, e_k \Rightarrow \begin{cases} e^{[1]}_1 \times e^{[1]}_0 = c\, e^{[1]}_1 \\ e^{[1]}_1 \times e^{[1]}_{-1} = c\, e^{[1]}_0 \\ e^{[1]}_0 \times e^{[1]}_{-1} = c\, e^{[1]}_{-1} \end{cases} \quad . \tag{10-299}$$

These relations can also be obtained from Eqs. (10–302) below.

All the expressions derived previously apply to any $f = f^i e_i = f_m e^{[1]}_m$ in F_1 where now the e_i and $e^{[1]}_m$ are mappings of R^3 into F_1, and the components f^i and f_m, or the spherical ones $f^{[1]}_m$, belong to $L^2(R^3, d_3x)$ or $L^2(S^2, d_2\hat{n})$.

ii) **Infinitesimal generators.** It follows from Eqs. (10–77,79) and (2–31) that the infinitesimal generators S_i read :

$$S_i = i\, e_i \times \quad . \tag{10-300}$$

Therefore, one has :

$$S_\pm = ie_\pm \times = c^{-1}(\pm 1)\sqrt{2}\, e^{[1]}_{\pm 1}\times \quad , \quad S_3 = c^{-1}(-1)e^{[1]}_0 \times \quad , \tag{10-301}$$

and Eqs. (10–282a,b) imply :

$$e^{[1]}_{\pm 1} \times e^{[1]}_m = c(\pm 1)\sqrt{\frac{2-m(m\pm 1)}{2}}\; e^{[1]}_{m\pm 1} \quad , \quad e^{[1]}_0 \times e^{[1]}_m = c(-1)m\, e^{[1]}_m \tag{10-302}$$

in agreement with Eq. (10–299). On F_1 (resp. F_1) the Casimir operator is a multiple of the unity :

$$S^2 = 2\, \mathbf{1}_{F_1} \quad . \tag{10-303}$$

Let us show that the curl operator $\nabla\times$ commutes with the representation $\mathcal{D}$[SU(2)] in F_1. Indeed, for all rotations R = R(U;a), one has with respect to an orthonormal basis a = (e_i) :

$$f = f^i\, e_i \in F_1 \qquad x = x^i\, e_i \in E_3$$

$$[\nabla\times\mathcal{D}(R)f](x) = \nabla\times[e_j R^j_i\, f^i(R^{-1}x)] = \sum_{j,m} e_j \varepsilon_{jk\ell}\, R^k_m\, R^\ell_i\, \partial_m f^i(R^{-1}x) \quad ,$$

$$[\mathcal{D}(R)\nabla\times f](x) = [\mathcal{D}(R) \sum_{n,m} e_n \varepsilon_{nmi}\, \partial_m f^i](x) = \sum_{n,m} e_j\, R^j_n\, \varepsilon_{nmi}\, \partial_m\, f^i(R^{-1}x) \quad .$$

Since the matrix (R^j_n) is orthogonal, one obtains from Eq. (4–80) :

$$\varepsilon_{jk\ell}\, R^k_m\, R^\ell_i = R^j_n\, \varepsilon_{nmi} \quad ,$$

which yields the announced result :

$$\forall\, U \in SU(2) \qquad [\nabla\times, \mathcal{D}(U)] = 0 \quad . \tag{10-304}$$

Setting R equals to an infinitesimal rotation in the above equation implies that the infinitesimal generators J_i also commute with the curl operator :

$$[\nabla\times, J_i] = 0 \quad . \tag{10-305}$$

This can also be checked directly from Eqs. (10–231a,300).

iii) **Vector spherical harmonics :**

1–For s=1 the spin spherical harmonics, Eq. (10–283), are called the <u>vector spherical harmonics</u>. They are denoted by :

$$\mathbf{Y}^{\ell'[\ell]}_m = Y^{[\ell']}_{m'}\, e^{[1]}_\sigma\, \langle \ell' 1 m' \sigma | \ell m \rangle \quad . \tag{10-306}$$

The definitions of the vector spherical harmonics according to some authors are given in Appendix H.5. From Eqs. (10–248,286) with $|c|=1$ and the symmetry property of the C.G. coefficient, Eq. (G–2b), one finds :

$$\mathbf{Y}^{\ell'[\ell]*}_m = c^{-2(\ell'+1)}\, (-1)^{\ell+m}\, \mathbf{Y}^{\ell'[\ell]}_{-m} \quad . \tag{10-307}$$

Hence, choosing c real implies that $\mathbf{Y}^{\ell'[\ell]}_m$, $Y^{[\ell]}_m$ and $e^{[s]}_m$ are the contravariant spherical components of Hermitian tensors $\mathbf{Y}^{\ell'\ell}$, Y^ℓ, and e^s, respectively. This property is thus preserved under the coupling as it is shown in Subsec. 8.3.4.2,iii).

2–The vector spherical harmonics form a complete orthonormal basis of F_1, cf. Eqs. (10–284a,b) setting s=1 :

$$\left(\mathbf{Y}^{\ell'[\ell]}_m, \mathbf{Y}^{\lambda'[\lambda]}_\mu\right)_{F_1} = \int_{S^2} \sin\theta d\theta d\varphi\, \mathbf{Y}^{\ell'[\ell]*}_m(\theta,\varphi)\,.\,\mathbf{Y}^{\lambda'[\lambda]}_\mu(\theta,\varphi) = \delta_{\ell'\lambda'}\delta_{\ell\lambda}\delta_{m\mu} \,, \tag{10-308a}$$

$$\forall f \in F_1 \quad f(r,\theta,\varphi) = \sum_{\ell';\ell} f_{\ell'\ell m}(r)\, \mathbf{Y}^{\ell'[\ell]}_m(\theta,\varphi) \,, \tag{10-308b}$$

$$f_{\ell'\ell m}(r) = \left(\mathbf{Y}^{\ell'[\ell]}_m, f\right)_{F_1} = \int_{S^2} \sin\theta d\theta d\varphi\, \mathbf{Y}^{\ell'[\ell]*}_m(\theta,\varphi)\,.\,f(r,\theta,\varphi) \,. \tag{10-308c}$$

3–Let us show that vector spherical harmonics can be generated by the action of a vector operator $A=A^1$ on a function $f=gY^{[\ell]}_m$ in H, Eq. (10–262). From Eqs. (10–292) and (8–170b,171b), and also Table G.5 one has :

$$A\, g\, Y^{[\ell]}_m = c^{-2}\sqrt{3}\,\{\{A^1 \otimes e^1\}^0 \otimes g\, Y^{\ell}\}^{[\ell]}_m =$$

$$= c^{-2}\sqrt{3}\sum_{\ell'} \begin{bmatrix} 1 & 1 & 0 \\ \ell & 0 & \ell \\ \ell' & 1 & \ell \end{bmatrix} \{\{A^1 \otimes g\, Y^{\ell}\}^{\ell'} \otimes e^1\}^{[\ell]}_m$$

$$= c^{-2}\sum_{\ell'} \frac{\hat{\ell}'}{\hat{\ell}} \{\{A^1 \otimes g\, Y^{\ell}\}^{\ell'} \otimes e^1\}^{[\ell]}_m \quad . \tag{10-309}$$

Following a calculation similar to the one performed for the gradient formula (Subsec. 10.3.3.5,v)), one has :

$$A^{[1]}_{\mu}\, g\, Y^{[\ell]}_m = \sum_{\ell' m'} Y^{[\ell']}_{m'} \langle r(\ell')m' | A^{[1]}_{\mu} | g[\ell]m \rangle \quad , \tag{10-310}$$

where thanks to the Wigner–Eckart theorem, Eq. (8–149), the matrix element can be expressed in terms of a C.G. coefficient and a reduced matrix element, Eq.(11–124a), which is a scalar, i.e. :

$$\langle r(\ell')m' | A^{[1]}_{\mu} | g[\ell]m \rangle = \frac{1}{\hat{\ell}'} \langle 1\ell\mu m | \ell' m' \rangle [r\ell' \,\|A^1\|\, g\ell] \quad . \tag{10-311}$$

The unitary character of the C.G. coefficients, Eq. (6–66a), then yields :

$$\{A^1 \otimes g\, Y^{\ell}\}^{[\ell']}_{m'} = \frac{1}{\hat{\ell}'} [r\ell' \,\|A^1\|\, g\ell]\; Y^{[\ell']}_{m'} \quad , \tag{10-312}$$

and thus Eq. (10–309) reads (this is actually a special case of Eq. (11–127)) :

$$A\, g\, Y^{[\ell]}_m = c^{-2}\frac{1}{\hat{\ell}}\sum_{\ell'} [r\ell' \,\|A^1\|\, g\ell]\; \mathbf{Y}^{\ell'[\ell]}_m \quad . \tag{10-313}$$

The above result is now applied to the three special vector operators acting on H : (1) the multiplicative operator $\hat{r}$ such that from Eq. (10–249) :

$$\hat{r}^1 = \sqrt{\frac{4\pi}{3}}\,\hat{Y}^1 \quad , \tag{10-314}$$

(2) the gradient operator ∇ (Subsec. 10.3.3.5,v)), and (3) the operator L associated with the infinitesimal generators L_i (Subsec. 8.5.4,iii)). The reduced matrix elements of $\hat{r}^1$ and L^1 are computed in the next chapter, Eqs. (11–144a,135), those of ∇^1 follow from Eqs. (10–271a,c). The results are collected in Table 10.2. Hence, one obtains the equalities between tensors :

$$\hat{r}\, g\, Y^{\ell} = \frac{\sqrt{\ell}}{\hat{\ell}}\, g\, \mathbf{Y}^{\ell-1\,\ell} + c^{-2}\frac{\sqrt{\ell+1}}{\hat{\ell}}\, g\, \mathbf{Y}^{\ell+1\,\ell} \quad , \tag{10-315a}$$

$$\nabla\, g\, Y^{\ell} = \frac{\sqrt{\ell}}{\hat{\ell}}\left(\frac{d}{dr} + \frac{\ell+1}{r}\right) g\, \mathbf{Y}^{\ell-1\,\ell} + c^{-2}\frac{\sqrt{\ell+1}}{\hat{\ell}}\left(\frac{d}{dr} - \frac{\ell}{r}\right) g\, \mathbf{Y}^{\ell+1\,\ell} \quad , \tag{10-315b}$$

$$L\, g\, Y^{\ell} = c^{-1}\, i\, \sqrt{\ell(\ell+1)}\; g\, \mathbf{Y}^{\ell\ell} \quad . \tag{10-315c}$$

TABLE 10.2

Reduced matrix elements of $\hat{r}^1$, ∇^1, and L^1.

A^1	$\hat{r}^1$	∇^1	L^1
$[r\ell-1 \,\|A^1\|\, g\ell]$	$c^2\sqrt{\ell}\, g$	$c^2\sqrt{\ell}\left(\frac{d}{dr}+\frac{\ell+1}{r}\right) g$	0
$[r\ell \,\|A^1\|\, g\ell]$	0	0	$ic\sqrt{\ell(\ell+1)}\,\hat{\ell}$
$[r\ell+1 \,\|A^1\|\, g\ell]$	$\sqrt{\ell+1}\, g$	$\sqrt{\ell+1}\left(\frac{d}{dr}-\frac{\ell}{r}\right) g$	0

One should note that the operators $\hat{r}$ and ∇ change the parity of the functions while L preserves it.

4–Instead of the basis in F_1 determined by the spherical components of $\mathbf{Y}^{\ell'\ell}$, it is often more convenient, as we shall see below, to consider the basis determined by the spherical components of R^ℓ, T_1^ℓ, and T_2^ℓ defined by, setting g=1 in Eqs. (10–315a–c) :

$$R^\ell = \hat{r}\, Y^\ell \quad , \quad T_1^\ell = \frac{1}{\sqrt{\ell(\ell+1)}}\, r\nabla Y^\ell \quad , \quad T_2^\ell = \frac{-i}{\sqrt{\ell(\ell+1)}}\, L\, Y^\ell \ . \tag{10-316}$$

This corresponds to an unitary change of basis in $F_1(|c|=1)$, such that in each subspace with a given ℓ : (10–317)

$$\begin{pmatrix} R^\ell \\ T_1^\ell \\ T_2^\ell \end{pmatrix} = \left(\begin{array}{c|c|c} \frac{\sqrt{\ell}}{\hat{\ell}} & 0 & c^{-2}\frac{\sqrt{\ell+1}}{\hat{\ell}} \\ \frac{\sqrt{\ell+1}}{\hat{\ell}} & 0 & -c^{-2}\frac{\sqrt{\ell}}{\hat{\ell}} \\ 0 & c^{-1} & 0 \end{array}\right) \begin{pmatrix} \mathbf{Y}^{\ell-1\,\ell} \\ \mathbf{Y}^{\ell\ell} \\ \mathbf{Y}^{\ell+1\,\ell} \end{pmatrix} , \quad \begin{pmatrix} \mathbf{Y}^{\ell-1\,\ell} \\ \mathbf{Y}^{\ell\ell} \\ \mathbf{Y}^{\ell+1\,\ell} \end{pmatrix} = \left(\begin{array}{c|c|c} \frac{\sqrt{\ell}}{\hat{\ell}} & \frac{\sqrt{\ell+1}}{\hat{\ell}} & 0 \\ 0 & 0 & c \\ c^2\frac{\sqrt{\ell+1}}{\hat{\ell}} & -c^2\frac{\sqrt{\ell}}{\hat{\ell}} & 0 \end{array}\right) \begin{pmatrix} R^\ell \\ T_1^\ell \\ T_2^\ell \end{pmatrix} .$$

Therefore, as the $Y^{\ell'[\ell]}_{m}$, the $R^{[\ell]}_{m}$, $T_{1\,m}^{[\ell]}$, and $T_{2\,m}^{[\ell]}$ form an orthonormal basis in F_1. Using :

$$L = -i\, r \times \nabla \quad , \tag{10-318a}$$

$$\nabla = \hat{r}\frac{\partial}{\partial r} - \frac{i}{r}\,\hat{r}\times L \quad , \tag{10-318b}$$

one obtains in a straighforward manner :

$$\begin{cases} \hat{r}\times R^\ell = 0 \\ \hat{r}\times T_1^\ell = -T_2^\ell \\ \hat{r}\times T_2^\ell = T_1^\ell \end{cases} \quad , \quad \begin{cases} \hat{r}.R^\ell = Y^\ell \\ \hat{r}.T_1^\ell = 0 \\ \hat{r}.T_2^\ell = 0 \end{cases} \quad . \tag{10-319}$$

These relations show that R^ℓ corresponds to radial vectors while T_1^ℓ and T_2^ℓ to transverse ones, see Fig. 10.4.

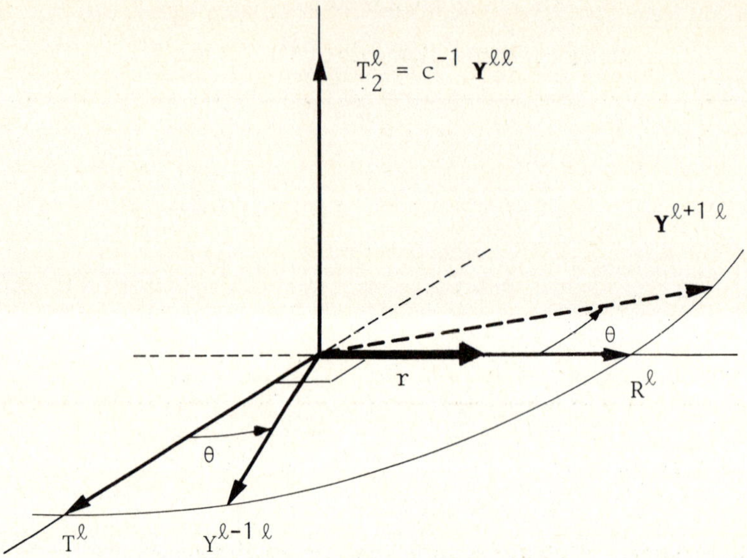

Fig. 10.4. Choosing c=1 in Eq. (10–317), we represent the two orthonormal bases in F_1 : omitting m, ($Y^{\ell-1\ell}$, $Y^{\ell\ell}$, $Y^{\ell+1\ell}$) and ($R^{\ell}, T_1^{\ell}, T_2^{\ell}$). This is only a formal drawing since F_1 is a Hermitian vector space identified with the complex extension of the Euclidean space E_3. For c=1, $Y^{\ell-1,\ell}$ and $Y^{\ell+1,\ell}$ are, respectively, obtained from T_1^{ℓ} and T_2^{ℓ} by a rotation of angle θ ($\cos\theta = \sqrt{\ell+1}/\hat{\ell}$, $\sin\theta = \sqrt{\ell}/\hat{\ell}$) around T_2^{ℓ}.

From Eqs. (10–317,319) one finds in agreement with Fig. 10.4 and the fact that $\hat{r}$ changes the parity, that :

$$\left\{\begin{aligned} \hat{r}\times\mathbf{Y}^{\ell-1\ell} &= -\,c^{-1}\,\frac{\sqrt{\ell+1}}{\hat{\ell}}\,\mathbf{Y}^{\ell\ell} \\ \hat{r}\times\mathbf{Y}^{\ell\ell} &= c\,\frac{\sqrt{\ell+1}}{\hat{\ell}}\,\mathbf{Y}^{\ell-1\ell} - c^{-1}\,\frac{\sqrt{\ell}}{\hat{\ell}}\,\mathbf{Y}^{\ell+1\ell} \\ \hat{r}\times\mathbf{Y}^{\ell+1\ell} &= c\,\frac{\sqrt{\ell}}{\hat{\ell}}\,\mathbf{Y}^{\ell\ell} \end{aligned}\right. , \qquad \left\{\begin{aligned} \hat{r}.\mathbf{Y}^{\ell-1\ell} &= \frac{\sqrt{\ell}}{\hat{\ell}}\,Y^{\ell} \\ \hat{r}.\mathbf{Y}^{\ell\ell} &= 0 \\ \hat{r}.\mathbf{Y}^{\ell+1\ell} &= c^{2}\,\frac{\sqrt{\ell+1}}{\hat{\ell}}\,Y^{\ell} \end{aligned}\right. . \tag{10-320}$$

Let us now derive the expressions of $\nabla\times g\,\mathbf{Y}^{\ell'\ell}$ and $\nabla . g\,\mathbf{Y}^{\ell'\ell}$ in the following way. One first finds the expressions of the curl and the divergence operators acting on R^{ℓ} , T_1^{ℓ}, and T_2^{ℓ}, which are easier to obtain. The change of basis, Eq. (10–317), then yields the expressions of $\nabla\times\mathbf{Y}^{\ell'\ell}$ and $\nabla .\,\mathbf{Y}^{\ell'\ell}$. As for the dependence upon the radial function g, one uses the relations for any vector v :

$$\nabla\times g\,v = (\nabla g)\times v + g\nabla\times v \quad , \qquad \nabla . g\,v = (\nabla g) . v + g\,\nabla . v \quad , \tag{10-321}$$

where :

$$\nabla g = \frac{dg}{dr}\,\hat{r} \quad , \tag{10-322}$$

$\hat{r}\times\mathbf{Y}^{\ell'\ell}$ and $\hat{r}.\mathbf{Y}^{\ell'\ell}$ are then expressed by Eq. (10–320). As a check, it is convenient to notice again that the operators $\hat{r}, \nabla, \nabla\times$, and ∇ change the parity while L preserves it, and furthermore that $\nabla\times$ commutes with the infinitesimal generators J_i, Eq. (10–305). In addition of Eqs. (10–318a,b), one needs the following relations between operators acting on F_1 :

$$L\hat{r} = 0 \qquad (a) \qquad L\times\hat{r} = -\hat{r}\times L + 2i\hat{r} \qquad (f)$$

$$\nabla\times\hat{r} = -\frac{i}{r}L \qquad (b) \qquad \nabla.\hat{r} = \frac{\partial}{\partial r} + \frac{2}{r} \qquad (g)$$

$$\nabla\times r\nabla = \hat{r}\times\nabla \qquad (c) \qquad \Delta = \frac{1}{r}\frac{\partial^2}{\partial r^2}r - \frac{L^2}{r^2} = \frac{\partial^2}{\partial r^2} + \frac{2}{r}\frac{\partial}{\partial r} - \frac{L^2}{r^2} \qquad (h)$$

$$\nabla\times L = i\frac{L^2}{r}\hat{r} + \frac{1}{r}\hat{r}\times L \qquad (d) \qquad \nabla.r\nabla = r\Delta + \hat{r}.\nabla = r\frac{\partial^2}{\partial r^2} + 3\frac{\partial}{\partial r} - \frac{L^2}{r} \qquad (i)$$

$$\hat{r}.\nabla = \frac{\partial}{\partial r} \qquad (e) \qquad \nabla.L = 0 \quad . \qquad (10\text{–}323j)$$

We thereby obtain the formulae :

$$\begin{cases} \nabla\times R^\ell = \dfrac{\sqrt{\ell(\ell+1)}}{r}T_2{}^\ell \\ \nabla\times T_1{}^\ell = -\dfrac{1}{r}T_2{}^\ell \\ \nabla\times T_2{}^\ell = \sqrt{\ell(\ell+1)}\,\dfrac{1}{r}R^\ell + \dfrac{1}{r}T_1{}^\ell \end{cases} , \qquad \begin{cases} \nabla.R^\ell = \dfrac{2}{r}Y^\ell \\ \nabla.T_1{}^\ell = -\dfrac{\sqrt{\ell(\ell+1)}}{r}Y^\ell \\ \nabla.T_2{}^\ell = 0 \end{cases} , \qquad (10\text{–}324)$$

which imply, using Eqs. (10–317,321, and 322) :

$$\begin{cases} \nabla\times g\,\mathbf{Y}^{\ell-1\ell} = -c^{-1}\dfrac{\sqrt{\ell+1}}{\hat{\ell}}\left(\dfrac{d}{dr} - \dfrac{\ell-1}{r}\right)g\,\mathbf{Y}^{\ell\ell} , \\ \nabla\times g\,\mathbf{Y}^{\ell\ell} = c\dfrac{\sqrt{\ell+1}}{\hat{\ell}}\left(\dfrac{d}{dr} + \dfrac{\ell+1}{r}\right)g\,\mathbf{Y}^{\ell-1\ell} - c^{-1}\dfrac{\sqrt{\ell}}{\hat{\ell}}\left(\dfrac{d}{dr} - \dfrac{\ell}{r}\right)g\,\mathbf{Y}^{\ell+1\ell} , \\ \nabla\times g\,\mathbf{Y}^{\ell+1\ell} = c\dfrac{\sqrt{\ell}}{\hat{\ell}}\left(\dfrac{d}{dr} + \dfrac{\ell+2}{r}\right)g\,\mathbf{Y}^{\ell\ell} , \end{cases} \qquad (10\text{–}325)$$

$$\begin{cases} \nabla.g\,\mathbf{Y}^{\ell-1\ell} = \dfrac{\sqrt{\ell}}{\hat{\ell}}\left(\dfrac{d}{dr} - \dfrac{\ell-1}{r}\right)g\,Y^\ell , \\ \nabla.g\,\mathbf{Y}^{\ell\ell} = 0 , \\ \nabla.g\,\mathbf{Y}^{\ell+1\ell} = c^2\dfrac{\sqrt{\ell+1}}{\hat{\ell}}\left(\dfrac{d}{dr} + \dfrac{\ell+2}{r}\right)g\,Y^\ell . \end{cases} \qquad (10\text{–}326)$$

Using the Eq. (10–325) twice, or the relation :

$$\nabla\times\nabla\times v = \nabla(\nabla.v) - \Delta v \quad , \tag{10-327}$$

one gets :

$$\left\{\begin{aligned}
\nabla\times\nabla\times g\,\mathbf{Y}^{\ell-1\ell} &= -\frac{\ell+1}{2\ell+1}\left[\frac{d^2}{dr^2}+\frac{2}{r}\frac{d}{dr}-\frac{\ell(\ell-1)}{r^2}\right] g\,\mathbf{Y}^{\ell-1\ell}\\
&\quad + c^{-2}\,\frac{\sqrt{\ell(\ell+1)}}{2\ell+1}\left[\frac{d^2}{dr^2}-\frac{2\ell-1}{r}\frac{d}{dr}+\frac{(\ell-1)(\ell+1)}{r^2}\right] g\,\mathbf{Y}^{\ell+1\ell}\\
\nabla\times\nabla\times g\,\mathbf{Y}^{\ell\ell} &= -\left[\frac{d^2}{dr^2}+\frac{2}{r}\frac{d}{dr}-\frac{\ell(\ell+1)}{r^2}\right] g\,\mathbf{Y}^{\ell\ell} = -\Delta g\,\mathbf{Y}^{\ell\ell}\\
\nabla\times\nabla\times g\,\mathbf{Y}^{\ell+1\ell} &= c^2\,\frac{\sqrt{\ell(\ell+1)}}{2\ell+1}\left[\frac{d^2}{dr^2}+\frac{2\ell+3}{r}\frac{d}{dr}+\frac{\ell(\ell+2)}{r^2}\right] g\,\mathbf{Y}^{\ell-1\ell}\\
&\quad - \frac{\ell}{2\ell+1}\left[\frac{d^2}{dr^2}+\frac{2}{r}\frac{d}{dr}-\frac{(\ell+1)(\ell+2)}{r^2}\right] g\,\mathbf{Y}^{\ell+1\ell}
\end{aligned}\right. \quad . \tag{10-328}$$

All these relations are especially useful for the study of the electromagnetic field.

A similar method can be used to derive the expressions of the operators L× and L. acting on R^ℓ, T_1^{ℓ}, and T_2^{ℓ}. From Eqs. (10–319,323a, and f) and the relations :

$$L.\nabla = 0 \quad , \tag{10-329a}$$

$$L\times r\nabla = -r\nabla\times L + 2\,i\,r\nabla \quad , \tag{10-329b}$$

$$L\times L = i\,L \quad , \tag{10-329c}$$

one thereby obtains :

$$\left\{\begin{aligned}
L\times R^\ell &= 2iR^\ell - i\sqrt{\ell(\ell+1)}\;T_1^{\ell}\\
L\times T_1^{\ell} &= -i\sqrt{\ell(\ell+1)}\;R^\ell + i\,T_1^{\ell}\\
L\times T_2^{\ell} &= 0
\end{aligned}\right. \quad , \quad
\left\{\begin{aligned}
L.R^\ell &= 0\\
L.T_1^{\ell} &= 0\\
L.T_2^{\ell} &= -i\sqrt{\ell(\ell+1)}\;Y^\ell
\end{aligned}\right. \quad , \tag{10-330}$$

and, using Eq. (10–317) :

$$\left\{\begin{aligned}
L\times\mathbf{Y}^{\ell-1\ell} &= -i\,(\ell-1)\,\mathbf{Y}^{\ell-1\ell}\\
L\times\mathbf{Y}^{\ell\ell} &= i\,\mathbf{Y}^{\ell\ell}\\
L\times\mathbf{Y}^{\ell+1\ell} &= i\,(\ell+2)\,\mathbf{Y}^{\ell+1\ell}
\end{aligned}\right. \quad , \quad
\left\{\begin{aligned}
L.\mathbf{Y}^{\ell-1\ell} &= 0\\
L.\mathbf{Y}^{\ell\ell} &= c\,(-i)\sqrt{\ell(\ell+1)}\;Y^\ell\\
L.\mathbf{Y}^{\ell+1\ell} &= 0
\end{aligned}\right. \quad . \tag{10-331}$$

The expressions of $L\times\mathbf{Y}^{\ell'\ell}$ can also be obtained directly using Eq. (10–300) which implies :

$$J^2 = (L+S)^2 = L^2 + 2\,\mathbf{1} + 2\,i\,L\times \quad . \tag{10-332}$$

FORMULAE FOR CHAPTER 10

The left and right regular representations of SU(2) and SO(3)

Definitions :

$$\forall U, U' \in SU(2) \qquad \forall f \in L^2(SU(2), d\mu) \quad \begin{cases} [D^{\ell}(U)f](U') = f(U^{-1}U') \\ [D^{r}(U)f](U') = f(U'U) \end{cases}$$

$$D^{\ell}(U) = \mathcal{I}\, D^{r}(U)\, \mathcal{I} \quad , \qquad (\mathcal{I}f)(U) = f(U^{-1})$$

$$D^{\ell} \sim D^{r} \sim \sum_{j=0,\frac{1}{2},1,\ldots} \oplus\ (2j+1)\ \mathcal{D}^{j} \quad .$$

Infinitesimal generators :

$$\sigma_i \to J_i^{\ell}, J_i^{r} \quad , \qquad [J_i^{\ell}, J_j^{r}] = 0 \quad , \qquad J_i^{\ell} = \mathcal{I}\, J_i^{r}\, \mathcal{I}$$

$$\begin{cases} J_1^{\ell} = -i\left(-\cos\varphi\ \mathrm{cotg}\theta\ \frac{\partial}{\partial\varphi} - \sin\varphi\ \frac{\partial}{\partial\theta} + \frac{\cos\varphi}{\sin\theta}\frac{\partial}{\partial\psi}\right) \\ J_2^{\ell} = -i\left(-\sin\varphi\ \mathrm{cotg}\theta\ \frac{\partial}{\partial\varphi} + \cos\varphi\ \frac{\partial}{\partial\theta} + \frac{\sin\varphi}{\sin\theta}\frac{\partial}{\partial\psi}\right) \\ J_3^{\ell} = -i\ \frac{\partial}{\partial\varphi} \\ J_{\pm}^{\ell} = J_1^{\ell} \pm iJ_2^{\ell} = e^{\pm i\varphi}\left(i\ \mathrm{cotg}\theta\ \frac{\partial}{\partial\varphi} \pm \frac{\partial}{\partial\theta} - i\ \frac{1}{\sin\theta}\frac{\partial}{\partial\psi}\right) \end{cases}$$

$$\begin{cases} J_1^{r} = i\left(-\frac{\cos\psi}{\sin\theta}\frac{\partial}{\partial\varphi} + \sin\psi\ \frac{\partial}{\partial\theta} + \mathrm{cotg}\theta\ \cos\psi\ \frac{\partial}{\partial\psi}\right) \\ J_2^{r} = i\left(\frac{\sin\psi}{\cos\theta}\frac{\partial}{\partial\varphi} + \cos\psi\ \frac{\partial}{\partial\theta} - \mathrm{cotg}\theta\ \sin\psi\ \frac{\partial}{\partial\psi}\right) \\ J_3^{r} = i\ \frac{\partial}{\partial\psi} \\ J_{\pm}^{r} = J_1^{r} \pm iJ_2^{r} = e^{\mp i\psi}\left(-i\ \frac{1}{\sin\theta}\frac{\partial}{\partial\varphi} \mp \frac{\partial}{\partial\theta} + i\ \mathrm{cotg}\theta\ \frac{\partial}{\partial\psi}\right) \end{cases}$$

$$(J^{\ell})^2 = (J^{r})^2 = -\mathrm{cotg}\theta\ \frac{\partial}{\partial\theta} - \frac{1}{\sin^2\theta}\left(\frac{\partial^2}{\partial\varphi^2} - 2\cos\theta\ \frac{\partial^2}{\partial\varphi\partial\psi} + \frac{\partial^2}{\partial\psi^2}\right) - \frac{\partial^2}{\partial\theta^2} \quad .$$

Standard basis :

$$J_{\pm}^{\ell}\ \mathcal{D}_{mm_o}^{j*} = \sqrt{j(j+1)-m(m\pm1)}\ \mathcal{D}_{m\pm1\, m_o}^{j*} \quad , \qquad J_3^{\ell}\ \mathcal{D}_{mm_o}^{j*} = m\ \mathcal{D}_{mm_o}^{j*}$$

$$(J^{\ell})^2\ \mathcal{D}_{mm_o}^{j*} = j(j+1)\ \mathcal{D}_{mm_o}^{j*} \quad , \qquad J_i^{\ell}\ R_{j_o}^{j} = i\ \varepsilon_{ijk}\ R_{j_o}^{k}$$

$$D^{\ell}(U)\ \mathcal{D}_{mm_o}^{j*} = \mathcal{D}_{m'm_o}^{j*}\ \mathcal{D}_{m'm}^{j}(U)$$

$$J_{\pm}^{r}\ \mathcal{D}_{m_o m}^{j} = \sqrt{j(j+1)-m(m\pm1)}\ \mathcal{D}_{m_o m\pm1}^{j} \quad , \qquad J_3^{r}\ \mathcal{D}_{m_o m}^{j} = m\ \mathcal{D}_{m_o m}^{j} \quad .$$

$$(J^r)^2\, \mathcal{D}^j_{m_o m} = j(j+1)\, \mathcal{D}^j_{m_o m} \quad , \quad J^r_i\, R^{j_o}_j = i \sum_k \varepsilon_{ijk}\, R^{j_o}_k$$

$$D^r(U)\, \mathcal{D}^j_{m_o m} = \mathcal{D}^j_{m_o m'}\, \mathcal{D}^j_{m'm}(U)$$

$$\left(-m \operatorname{cotg}\theta + m_o \frac{1}{\sin\theta} \pm \frac{d}{d\theta}\right) d^j_{mm_o}(\theta) = \sqrt{j(j+1)-m(m\pm1)}\; d^j_{m\pm1\, m_o}(\theta)$$

$$\left(m \operatorname{cotg}\theta - m_o \frac{1}{\sin\theta} \mp \frac{d}{d\theta}\right) d^j_{m_o m}(\theta) = \sqrt{j(j+1)-m(m\pm1)}\; d^j_{m_o\, m\pm1}(\theta)$$

$$\left[j(j+1) - \frac{1}{\sin^2\theta}(m^2 - 2mm_o\cos\theta + m_o^2) + \operatorname{cotg}\theta \frac{d}{d\theta} + \frac{d^2}{d\theta^2}\right] d^j_{mm_o}(\theta) = 0 \quad .$$

Multiplicative operators :

$$[J^\ell_\pm\,,\, \hat{\mathcal{D}}^{j*}_{mm_o}] = \sqrt{j(j+1)-m(m\pm1)}\; \hat{\mathcal{D}}^{j*}_{m\pm1\, m_o} \quad , \quad [J^\ell_3\,,\, \hat{\mathcal{D}}^{j*}_{mm_o}] = m\, \hat{\mathcal{D}}^{j*}_{mm_o}$$

$$[J^r_\pm\,,\, \hat{\mathcal{D}}^j_{m_o m}] = \sqrt{j(j+1)-m(m\pm1)}\; \hat{\mathcal{D}}^j_{m_o m} \quad , \quad [J^r_3\,,\, \hat{\mathcal{D}}^j_{m_o m}] = m\, \hat{\mathcal{D}}^j_{m_o m}$$

$$[J^\ell_i\,,\, \hat{R}^j_{j_o}] = i\ \varepsilon_{ijk}\, \hat{R}^k_{j_o} \quad , \quad [J^r_i\,,\, \hat{R}^{j_o}_j] = i \sum_k \varepsilon_{ijk}\, \hat{R}^{j_o}_k \quad .$$

Intrinsic infinitesimal generators :

$$I^\ell_i = J^\ell_j\, \hat{R}^j_i \quad , \quad I^{\ell[1]}_m = J^{\ell[1]}_{m'}\, \hat{\mathcal{D}}^1_{m'm}$$

$$I^r_i = I^r_j\, \hat{R}^j_i \quad , \quad I^{r[1]}_m = J^{r[1]}_{m'}\, \hat{\mathcal{D}}^1_{m'm}$$

$$I^\ell_i = -\, J^r_i \quad , \quad I^r_i = -J^\ell_i \quad , \quad (I^\ell)^2 = (J^\ell)^2 = (I^r)^2 = (J^r)^2$$

$$[J^\ell_i\,,\, I^\ell_i] = 0 \quad , \quad [J^r_i\,,\, I^r_i] = 0 \quad , \quad [I^\ell_i\,,\, I^r_j] = 0$$

$$[I^\ell_i\,,\, I^\ell_j] = -i \sum_k \varepsilon_{ijk}\, I^\ell_k \quad , \quad [I^r_i\,,\, I^r_j] = -i \sum_k \varepsilon_{ijk}\, I^r_k$$

$$I^\ell_\pm\, \mathcal{D}^{j*}_{mm_o} = \sqrt{j(j+1)-m_o(m_o\mp1)}\; \mathcal{D}^{j*}_{mm_o\mp1} \quad , \quad I^\ell_3\, \mathcal{D}^{j*}_{mm_o} = m_o\, \mathcal{D}^{j*}_{mm_o}$$

$$I^r_\pm\, \mathcal{D}^j_{m_o m} = \sqrt{j(j+1)-m_o(m_o\mp1)}\; \mathcal{D}^j_{m_o\mp1\, m} \quad , \quad I^r_3\, \mathcal{D}^j_{m_o m} = m_o\, \mathcal{D}^j_{m_o m}$$

$$I^\ell_i\, R^{j_o}_j = -i \sum_k \varepsilon_{ijk}\, R^{j_o}_k \quad , \quad I^r_i\, R^j_{j_o} = -i\ \ \varepsilon_{ijk}\, R^k_{j_o} \quad .$$

Absolute and relative components of the infinitesimal displacement of the moving reference frame :

$$\Omega^i(\varphi,\theta,\psi;d\varphi,d\theta,d\psi) \quad , \quad \omega^i(\varphi,\theta,\psi;d\varphi,d\theta,d\psi)$$

$$\begin{cases} \Omega^1 = -\sin\varphi d\theta + \cos\varphi\, \sin\theta\, d\psi \\ \Omega^2 = \cos\varphi d\theta + \sin\varphi\, \sin\theta\, d\psi \\ \Omega^3 = d\varphi + \cos\theta d\psi \end{cases} \qquad \begin{cases} \omega^1 = -\sin\theta\, \cos\psi\, d\varphi + \sin\psi\, d\theta \\ \omega^2 = \sin\theta\, \sin\psi\, d\varphi + \cos\psi\, d\theta \\ \omega^3 = \cos\theta d\varphi + d\psi \end{cases} \quad .$$

The regular representations of SO(3) : j is restricted to integer values.

Representations of SO(3) and SU(2) in the space of complex functions defined on R^3

Definitions :

$$\forall R \in SO(3) \begin{cases} \forall f \in L^2(R^3, d_3x) & [D(R)f](x) = f(R^{-1}x) \\ \forall f \in L^2(S^2, d_2\hat{n}) & [D(R)f](\hat{n}) = f(R^{-1}\hat{n}) \end{cases} .$$

Infinitesimal generators :

$$\sigma_i \to L_i = -i\,\varepsilon_{ijk}\, x^j \frac{\partial}{\partial x^k} \quad , \qquad L = -i\, r \times \nabla$$

$$L_\pm = \mp (x^1 \pm ix^2) \frac{\partial}{\partial x^3} \pm 2x^3 \frac{\partial}{\partial x^1 \mp ix^2}$$

$$L_1 = -i\left(-\sin\varphi \frac{\partial}{\partial\theta} - \cos\varphi \,\mathrm{cotg}\theta \frac{\partial}{\partial\varphi}\right)$$

$$L_2 = -i\left(\cos\varphi \frac{\partial}{\partial\theta} - \sin\varphi \,\mathrm{cotg}\theta \frac{\partial}{\partial\varphi}\right)$$

$$L_3 = -i \frac{\partial}{\partial\varphi}$$

$$L_\pm = e^{\pm i\varphi}\left(\pm \frac{\partial}{\partial\theta} + i\,\mathrm{cotg}\theta \frac{\partial}{\partial\varphi}\right)$$

$$L^2 = -\mathrm{cotg}\theta \frac{\partial}{\partial\theta} - \frac{\partial^2}{\partial\theta^2} - \frac{1}{\sin^2\theta}\frac{\partial^2}{\partial\varphi^2} = -\frac{1}{\sin\theta}\frac{\partial}{\partial\theta}\left(\sin\theta \frac{\partial}{\partial\theta}\right) - \frac{1}{\sin^2\theta}\frac{\partial^2}{\partial\varphi^2}$$

Standard basis, spherical harmonics :

$$\forall R \quad [D(R)\, Y^{[\ell]}_m](\theta,\varphi) = Y^{[\ell]}_m [R^{-1}(\theta,\varphi)] = Y^{[\ell]}_{m'}(\theta,\varphi)\, \mathcal{D}^\ell_{m'm}(R)$$

$$Y^{[\ell]}_m(\theta,\varphi) = (-ic)^\ell Y_{\ell m}(\theta,\varphi) = (-ic)^\ell \frac{\hat{\ell}}{\sqrt{4\pi}} \mathcal{D}^{\ell *}_{m0}(\varphi,\theta,\psi) \quad , \quad \boxed{|c|=1}$$

$$Y^{[\ell]*}_m = c^{-2\ell} (-1)^{\ell+m}\, Y^{[\ell]}_{-m}$$

c = i	⇒	**usual spherical harmonics**
c = ± 1	⇒	**the most convenient from the tensorial point of view**

$$v^{[1]}_m = \sqrt{\frac{4\pi}{3}}\, \|v\|\, Y^{[1]}_m(\theta,\varphi)$$

$$\sum_{\ell,m} Y^{[\ell]}_m(\theta,\varphi) Y^{[\ell]*}_m(\theta',\varphi') = \sum_\ell c^{-2\ell} \hat{\ell} \{Y^\ell(\theta,\varphi) Y^\ell(\theta',\varphi')\}^0 = \delta(\cos\theta - \cos\theta')\,\delta(\varphi-\varphi')$$

$$\int_0^\pi \sin\theta d\theta \int_0^{2\pi} d\varphi\; Y^{[\ell]*}_m(\theta,\varphi)\, Y^{[\ell']}_{m'}(\theta',\varphi') = \delta_{\ell\ell'}\,\delta_{mm'}$$

$$\sum_m Y^{[\ell]}_m(\theta,\varphi) Y^{[\ell]*}_m(\theta',\varphi') = c^{-2\ell}\, \hat{\ell} \{Y^\ell(\theta,\varphi) Y^\ell(\theta',\varphi')\}^0 = \frac{2\ell+1}{4\pi} P_\ell(\cos\alpha)$$

$$\{Y^{\ell_1}(\theta,\varphi)\, Y^{\ell_2}(\theta,\varphi)\}^{[\ell]}_m = C(\ell_1,\ell_2,\ell)\, Y^{[\ell]}_m(\theta,\varphi)$$

$$C(\ell_1,\ell_2,\ell)=(-ic)^{\ell_1+\ell_2-\ell}\ \frac{\hat{\ell}_1\hat{\ell}_2}{\sqrt{4\pi}\hat{\ell}}\ \langle \ell_1\ell_2 00|\ell 0\rangle=(-ic)^{\ell_1+\ell_2-\ell}\ (-1)^{\ell_1-\ell_2}\ \frac{\hat{\ell}_1\hat{\ell}_2}{\sqrt{4\pi}}\begin{pmatrix}\ell_1 & \ell_2 & \ell\\ 0 & 0 & 0\end{pmatrix}$$

$$OM = \alpha_1 OM_1+\alpha_2 OM_2\ ,\quad OM=(r,\theta,\varphi)\ ,\quad OM_i=(r_i,\theta_i,\varphi_i)\qquad i=1,2$$

$$r^\ell Y^{[\ell]}_m(\theta,\varphi)=\sqrt{4\pi}\sum_{\ell_1,\ell_2=0}^{\ell}\delta_{\ell_1+\ell_2,\ell}\left[\frac{(2\ell+1)!}{(2\ell_1+1)!(2\ell_2+1)!}\right]^{\frac{1}{2}}(\alpha_1 r_1)^{\ell_1}(\alpha_2 r_2)^{\ell_2}\{Y^{\ell_1}(\theta_1,\varphi_1)Y^{\ell_2}(\theta_2,\varphi_2)\}^{[\ell]}_m$$

$$\left(\nabla^{[1]}_\mu\ g\ Y^{[\ell]}_m\right)(r,\theta,\varphi)=\sum_{\ell',m'}Y^{[\ell']}_{m'}(\theta,\varphi)\langle r(\ell')m'|\nabla^{[1]}_\mu|g[\ell]m\rangle\ .$$

Multiplicative operators :

$$\forall\,\theta,\varphi\qquad \left(\hat{Y}^{[\ell]}_m\ f\right)(\theta,\varphi)=Y^{[\ell]}_m(\theta,\varphi)\ f(\theta,\varphi)$$

$$\left[L_\pm,\hat{Y}^{[\ell]}_m\right]=\sqrt{\ell(\ell+1)-m(m\pm1)}\ \hat{Y}^{[\ell]}_{m\pm1}\ ,\quad \left[L_3,\ \hat{Y}^{[\ell]}_m\right]=m\ \hat{Y}^{[\ell]}_m\ .$$

Representation of SU(2) in a space of vectorial functions defined on R^3

Definitions :

$$\forall f\in F_s\qquad f_{sm}\in L^2(R^3,d_3x)\ \text{or}\ L^2(S^2,d_2\hat{n})\quad f=f_{sm}\ e^{[s]}_m\ ,$$

$$\left(e^{[s]}_m\ ,\ e^{[s]}_{m'}\right)_{F_s}=\delta_{mm'}\ ,\qquad (f,f')_{F_s}=\int d_3x\ f^*_{sm}(x)\ f'_{sm}(x)$$

$$\forall U\in SU(2)\qquad [\mathcal{D}(U)f](x)=e^{[s]}_{m'}\ \mathcal{D}^s_{m'm}(U)\ f_{sm}[R(U)^{-1}x]$$

Infinitesimal generators and standard basis :

$$\sigma_i\ \rightarrow\ J_i=L_i+S_i$$

$$S_\pm\ e^{[s]}_m=\sqrt{s(s+1)-m(m\pm1)}\ e^{[s]}_{m\pm1}\ ,\qquad S_3\ e^{[s]}_m=m\ e^{[s]}_m$$

$$\mathbf{Y}^{\ell\,s\,[j]}_m=\{Y^\ell\otimes \varepsilon^s\}^{[j]}_m$$

The special case s = 1 : | c | = 1, cf. spherical harmonics

$$\begin{cases} e^{[1]}_1=c\,\frac{1}{\sqrt2}\,(ie_1-e_2)\\ e^{[1]}_0=c\,(-ie_3)\\ e^{[1]}_{-1}=c\,\frac{-1}{\sqrt2}\,(ie_1+e_2)\end{cases}\qquad \begin{cases} e_1=c^{-1}\,\frac{i}{\sqrt2}\left(-e^{[1]}_1+e^{[1]}_{-1}\right)\\ e_2=c^{-1}\,\frac{-1}{\sqrt2}\left(e^{[1]}_1+e^{[1]}_{-1}\right)\\ e_3=c^{-1}\ i\ e^{[1]}_0\end{cases}$$

$$e^{[1]*}_m=c^{-2}(-1)^{1+m}\ e^{[1]}_{-m}$$

$$(e_i,e_j)_{F_1}=e_i.e_j=\delta_{ij}\ ,\qquad \left(e^{[1]}_m,e^{[1]}_{m'}\right)_{F_1}=e^{[1]*}_m\ .\ e^{[1]}_{m'}=\delta_{mm'}$$

$$v=v^ie_i=v_m\ e^{[1]}_m\qquad v_m=v^{[1]*}_m=c^{-2}(-1)^{1+m}\ v^{[1]}_{-m}$$

$$v^{[1]}_m=\left(e^{[1]*}_m,v\right)_{F_1}=e^{[1]}_m.v$$

$$e^1 = \left\{e^{[1]}_m\right\} \quad , \quad e^{[1]}_m = \left\{\left(e^{[1]}_m\right)^{[1]}_{m'} = c^2(-1)^{1+m}\,\delta_{-mm'}\right\}$$

$$v = c^{-2}\sqrt{3}\,\{v^1 \otimes e^1\}^0 \quad , \quad v \times e^{[1]}_m = -c^{-1}\sqrt{2}\,\{v^1 \otimes e^1\}^{[1]}_m$$

$$v.e^{[1]}_m = c^{-2}\sqrt{3}\left\{v^1 \otimes \left(e^{[1]}_m\right)^1\right\}^0 \quad , \quad \left(v \times e^{[1]}_m\right)^1 = c^{-1}\sqrt{2}\left\{v^1 \otimes \left(e^{[1]}_m\right)^1\right\}^1$$

$$S_i = i\,e_i \times \quad , \quad S_\pm = c^{-1}(\pm 1)\sqrt{2}\,e^{[1]}_{\pm 1}\times \quad , \quad S_3 = c^{-1}(-1)\,e^{[1]}_0\times$$

$$e^{[1]}_{\pm 1} \times e^{[1]}_m = c(\pm 1)\sqrt{\frac{2-m(m\pm 1)}{2}}\,e^{[1]}_{m\pm 1} \quad , \quad e^{[1]}_0 \times e^{[1]}_m = c(-1)m\,e^{[1]}_m$$

$$\forall U \in SU(2) \qquad [\nabla\times\,,\,\mathcal{D}(U)] = 0 \quad , \quad [\nabla\times, J_i] = 0$$

$$\mathbf{Y}^{\ell'[\ell]}_m = \{Y^{\ell'} \otimes e^1\}^{[\ell]}_m \quad , \quad \mathbf{Y}^{\ell'[\ell]*}_m = c^{-2(\ell'+1)}(-)^{\ell+m}\,\mathbf{Y}^{\ell'[\ell]}_{-m}$$

$$f(r,\theta,\varphi) = \sum_{\ell',\ell} f_{\ell'\ell m}(r)\mathbf{Y}^{\ell'[\ell]}_m(\theta,\varphi)\,,\quad f_{\ell'\ell m}(r) = (\mathbf{Y}^{\ell'[\ell]}_m, f)_{F_1} = \int_{S^2}\sin\theta\,d\theta\,d\varphi\,\mathbf{Y}^{\ell'[\ell]*}_m . f(r,\theta,\varphi)$$

$$\hat{r}\,g Y^\ell = \frac{\sqrt{\ell}}{\hat{\ell}}\,g\,\mathbf{Y}^{\ell-1\ell} + c^{-2}\,\frac{\sqrt{\ell+1}}{\hat{\ell}}\,g\,\mathbf{Y}^{\ell+1\ell}$$

$$\nabla g Y^\ell = \frac{\sqrt{\ell}}{\hat{\ell}}\left(\frac{d}{dr} + \frac{\ell+1}{r}\right) g\,\mathbf{Y}^{\ell-1\ell} + c^{-2}\,\frac{\sqrt{\ell+1}}{\hat{\ell}}\left(\frac{d}{dr} - \frac{\ell}{r}\right) g\,\mathbf{Y}^{\ell+1\ell}$$

$$L\,g\,Y^\ell = c^{-1}\,i\,\sqrt{\ell(\ell+1)}\,g\,\mathbf{Y}^{\ell\ell}$$

$$\left\{\begin{aligned} \hat{r}\times\mathbf{Y}^{\ell-1\ell} &= -c^{-1}\,\frac{\sqrt{\ell+1}}{\hat{\ell}}\,\mathbf{Y}^{\ell\ell} \\ \hat{r}\times\mathbf{Y}^{\ell\ell} &= c\,\frac{\sqrt{\ell+1}}{\hat{\ell}}\,\mathbf{Y}^{\ell-1\ell} - c^{-1}\,\frac{\sqrt{\ell}}{\hat{\ell}}\,\mathbf{Y}^{\ell+1\ell} \\ \hat{r}\times\mathbf{Y}^{\ell+1\ell} &= c\,\frac{\sqrt{\ell}}{\hat{\ell}}\,\mathbf{Y}^{\ell\ell} \end{aligned}\right. \qquad \left\{\begin{aligned} \hat{r}.\mathbf{Y}^{\ell-1\ell} &= \sqrt{\ell}\,Y^\ell \\ \hat{r}.\mathbf{Y}^{\ell\ell} &= 0 \\ \hat{r}.\mathbf{Y}^{\ell+1\ell} &= c^2\,\frac{\sqrt{\ell+1}}{\hat{\ell}}\,Y^\ell \end{aligned}\right.$$

$$\left\{\begin{aligned} \nabla\times g\,\mathbf{Y}^{\ell-1\ell} &= -c^{-1}\,\frac{\sqrt{\ell+1}}{\hat{\ell}}\left(\frac{d}{dr} - \frac{\ell-1}{r}\right) g\,\mathbf{Y}^{\ell\ell} \\ \nabla\times g\,\mathbf{Y}^{\ell\ell} &= c\,\frac{\sqrt{\ell+1}}{\hat{\ell}}\left(\frac{d}{dr} + \frac{\ell+1}{r}\right) g\,\mathbf{Y}^{\ell-1\ell} - c^{-1}\,\frac{\sqrt{\ell}}{\hat{\ell}}\left(\frac{d}{dr} - \frac{\ell}{r}\right) g\,\mathbf{Y}^{\ell+1\ell} \\ \nabla\times g\,\mathbf{Y}^{\ell+1\ell} &= c\,\frac{\sqrt{\ell}}{\hat{\ell}}\left(\frac{d}{dr} + \frac{\ell+2}{r}\right) g\,\mathbf{Y}^{\ell\ell} \end{aligned}\right.$$

$$\left\{\begin{aligned} \nabla.g\,\mathbf{Y}^{\ell-1\ell} &= \frac{\sqrt{\ell}}{\hat{\ell}}\left(\frac{d}{dr} - \frac{\ell-1}{r}\right) g\,Y^\ell \\ \nabla.g\,\mathbf{Y}^{\ell\ell} &= 0 \\ \nabla.g\,\mathbf{Y}^{\ell+1} &= c^2\,\frac{\sqrt{\ell+1}}{\hat{\ell}}\left(\frac{d}{dr} + \frac{\ell+2}{r}\right) g\,Y^\ell \end{aligned}\right.$$

$$\begin{cases} L \times \mathbf{Y}^{\ell-1\ell} = -i(\ell-1)\,\mathbf{Y}^{\ell-1\ell} \\ L \times \mathbf{Y}^{\ell\ell} = i\,\mathbf{Y}^{\ell\ell} \\ L \times \mathbf{Y}^{\ell+1\ell} = i(\ell+2)\,\mathbf{Y}^{\ell+1\ell} \end{cases} \qquad \begin{cases} L.\mathbf{Y}^{\ell-1\ell} = 0 \\ L.\mathbf{Y}^{\ell\ell} = c(-i)\sqrt{\ell(\ell+1)}\,Y^{\ell} \\ L.\mathbf{Y}^{\ell+1\ell} = 0 \end{cases}$$

$$\nabla\times\nabla\times g\,\mathbf{Y}^{\ell-1\ell} = -\frac{\ell+1}{2\ell+1}\left[\frac{d^2}{dr^2}+\frac{2}{r}\frac{d}{dr}-\frac{\ell(\ell-1)}{r^2}\right] g\,\mathbf{Y}^{\ell-1\ell}$$

$$+\,c^{-2}\,\frac{\sqrt{\ell(\ell+1)}}{2\ell+1}\left[\frac{d^2}{dr^2}-\frac{2\ell-1}{r}\frac{d}{dr}+\frac{(\ell-1)(\ell+1)}{r^2}\right] g\,\mathbf{Y}^{\ell+1\ell}$$

$$\nabla\times\nabla\times g\,\mathbf{Y}^{\ell\ell} = -\Delta g\,\mathbf{Y}^{\ell\ell} = -\left[\frac{d^2}{dr^2}+\frac{2}{r}\frac{d}{dr}-\frac{\ell(\ell+1)}{r^2}\right] g\,\mathbf{Y}^{\ell\ell}$$

$$\nabla\times\nabla\times g\,\mathbf{Y}^{\ell+1\ell} = c^{2}\,\frac{\sqrt{\ell(\ell+1)}}{2\ell+1}\left[\frac{d^2}{dr^2}+\frac{2\ell+3}{r}\frac{d}{dr}+\frac{\ell(\ell+2)}{r^2}\right] g\,\mathbf{Y}^{\ell-1\ell}$$

$$-\,\frac{\ell}{2\ell+1}\left[\frac{d^2}{dr^2}+\frac{2}{r}\frac{d}{dr}-\frac{(\ell+1)(\ell+2)}{r^2}\right] g\,\mathbf{Y}^{\ell+1\ell} \quad .$$

CHAPTER 11

APPLICATIONS TO QUANTUM MECHANICS

This last chapter is devoted to the applications of the formalisms and results derived previously, mainly in Chaps. 8–10, to nonrelativistic Quantum Mechanics.

The representations of the rotation group in Quantum Mechanics are determined in Sec. 11.1 for spinless particles and also for spin–s particles. Secs. 11.2,3 discuss the tensorial properties of the tensors encountered in Quantum Mechanics, i.e. the kets, the bras, the wave functions, the operators, and the matrix elements. In particular, the expression of the Wigner–Eckart theorem in Quantum Mechanics is derived in Subsec. 11.3.1. Sec. 11.4 deals with the rigid rotator in Quantum Mechanics.

Notations. Throughout this last chapter, to avoid any ambiguity :
(1) The three–dimensional vectors are denoted by an arrow, e.g. $\vec{r}$, except the orthonormal basis vectors still denoted by e_i.
(2) The position and the momentum operators are denoted in bold–faced type, i.e. : $\vec{\mathbf{r}}, \hat{\mathbf{r}}, \mathbf{r}$; $\vec{\mathbf{p}}, \hat{\mathbf{p}}, \mathbf{p}$; as also the Euler angle operators $\boldsymbol{\varphi}$, $\boldsymbol{\theta}$, and $\boldsymbol{\psi}$.

11.1. REPRESENTATIONS OF THE ROTATION GROUP IN QUANTUM MECHANICS

11.1.1. GENERAL REMARKS

i) The general formalism of the usual Quantum Mechanics is recalled in Sec. 3.1. One should remember that the first step in the study of a physical system in Quantum Mechanics consists in enumerating all its dynamical variables and determining the algebra $\mathcal{A}$ of the associated observables. Actually, the various observables can be expressed as a function of a subset of them, which is called a <u>fundamental system of observables</u>. The algebra is then completely determined by giving the commutation relations of these <u>fundamental observables</u>. The definition of the observables is complete only when one has specified the space $\mathcal{E}$ in which they act. This can be done by determining each observable by its action on a selected basis in $\mathcal{E}$. In doing so one has to ensure that the operators representing the dynamical variables actually turn out to be observables which satisfy the commutation relations of the algebra. A basis in $\mathcal{E}$ can be determined by the eigenbasis of a <u>complete set of commuting observables</u> ; i.e. a set of observables such that none of them can be expressed as a function of the others, which all commute in pairs, and which have a uniquely defined common eigenbasis. The mere knowledge of the algebra of the observables is in general sufficient to allow an unambiguous determination of the space $\mathcal{E}$. The physical states are associated with the normalizable vectors in $\mathcal{E}$, which form a Hilbert space $\mathcal{H}$.

ii) The point space of nonrelativistic physics is the three-dimensional Euclidean point space $E_3^{\cdot}$ (Subsec. 2.3.1). From Secs. 3.2,4, assuming that the three-dimensional rotation group is a symmetry group one is led to determine the representations up to a phase of this group in the considered vector space $\mathcal{E}$. Once an orthonormal reference frame $a = (O,e_i)$ is chosen in $E_3^{\cdot}$, there exists a one-to-one correspondence between the group of rotations $\mathcal{R} = (O,R)$ around the point O and the group SO(3). The representations up to a phase of SO(3) are the true representations of its universal covering group, i.e. SU(2). The unitary and irreducible representations $\mathcal{D}^j$ of SU(2) are studied in Sec. 6.1. The relation between the representations of SU(2) and SO(3) is discussed in Sec. 5.6. Thus, to any representation $\mathcal{D}[SU(2)]$ there corresponds by Eqs. (5-66,69) a representation D [SO(3)] which is either a true representation or a double-valued representation according to whether $\mathcal{D}$ is an even ($\varepsilon = +1$) or an odd ($\varepsilon = -1$) representation :

$$\forall\, R \in SO(3) \leftrightarrow \pm U \in SU(2) \quad , \quad R = R(U) \quad \text{from Eq. (4-28)}$$

$$D(R) = \begin{cases} D[R(U)] = \mathcal{D}(U) \\ D[R(-U)] = \mathcal{D}(-U) = \varepsilon \mathcal{D}(U) \ , \quad \varepsilon = \pm 1 \end{cases} \qquad (11\text{-}1)$$

These two cases correspond, respectively, to the integer and the half-odd-integer values of j (Subsec. 6.1.2,iii)). It follows from the Wigner theorem (Subsec. 3.2.2) that the transformation under a rotation $\mathcal{R}=(O,R)$ of any pure state associated with a ray $|\underline{\varphi}\rangle$ in the Hilbert space $\mathcal{H}$, and any observable A in the algebra $\mathcal{A}$ is realized by the unitary operator $\mathcal{D}(U)$:

$$\forall\, |\varphi\rangle \in |\underline{\varphi}\rangle \subset \mathcal{H} \xrightarrow{\mathcal{R}} |{}^{\mathcal{R}}\varphi\rangle = \mathcal{D}(U)\,|\varphi\rangle \in |{}^{\mathcal{R}}\underline{\varphi}\rangle \subset \mathcal{H} \quad , \qquad (11\text{-}2a)$$

$$\forall\, A \in \mathcal{A} \xrightarrow{\mathcal{R}} {}^{\mathcal{R}}A = \mathcal{D}(U)\; A\; \mathcal{D}(U)^{-1} \in \mathcal{A} \quad . \qquad (11\text{-}2b)$$

It is recalled that $\mathcal{D}(U)$ is uniquely defined up to a multiplicative phase factor. The operator $\mathcal{D}(U)$ being unitary, the matrix elements are invariant under rotations :

$$\langle\varphi|A|\psi\rangle = \langle{}^{\mathcal{R}}\varphi|{}^{\mathcal{R}}A|{}^{\mathcal{R}}\psi\rangle \quad , \qquad (11\text{-}3)$$

and furthermore, the law of transformation of the operators preserves the commutations relations in $\mathcal{A}$.

There remains to determine from physical considerations the above representation of the rotation group for an algebra $\mathcal{A}$ and a vector space $\mathcal{E}$ associated with the description of a given physical system. The unitary operator $\mathcal{D}(U)$, for all U in SU(2), can be determined by its action on a basis of $\mathcal{E}$, e.g. the eigenbasis of a complete set of commuting observables. Hence, one only needs to specify the law of transformation under any rotation of the observables which form a complete set of commuting observables. Let us now make precise the representations of the rotation group in the spaces which correspond to the description of spinless particles in Subsec. 11.1.2, and then of spin–s particles in Subsec. 11.1.3. In Sec. 11.4 one also considers the space associated with the description of a rigid rotator.

11.1.2. SPINLESS PARTICLES

i) **Observables and states.** We intend to describe in Quantum Mechanics the classical analogue of pointlike particles. In what follows, a reference point O in the point space $E_3^{\bullet}$ is always assumed to be chosen. Furthermore, only rotations $\mathcal{R} = (O,R)$ around O are considered. Therefore, from now on we speak of vectors $\vec{r}$ in E_3 instead of points M in $E_3^{\bullet}$, and of rotations R in E_3 instead of rotations $\mathcal{R} = (O,R)$ in $E_3^{\bullet}$, such that :

$$\forall M \in E_3^{\bullet} \qquad OM = \vec{r} \in E_3 \ , \qquad M' = \mathcal{R} M \qquad OM' = R\ OM \quad . \tag{11-4}$$

A spinless particle is characterized by two dynamical variables, its position and its momentum, to which there correspond the position and the momentum operators $\vec{\mathbf{r}}$ and $\vec{\mathbf{p}}$, respectively :

$$\vec{\mathbf{r}} = \mathbf{x}^i\, e_i \quad , \qquad \vec{\mathbf{p}} = \mathbf{p}^i\, e_i \quad , \tag{11-5}$$

where a = (e_i) denotes any fixed orthonormal basis in E_3. These observables obey the following commutation relations :

$$[\mathbf{x}^i,\mathbf{x}^j] = 0 \quad , \qquad [\mathbf{p}^i,\mathbf{p}^j] = 0 \quad , \qquad [\mathbf{x}^i,\mathbf{p}^j] = i\hbar\,\delta_{ij}\,\mathbf{1} \quad . \tag{11-6}$$

Any observable is then a function $A(\vec{\mathbf{r}},\vec{\mathbf{p}})$ of the fundamental observables $\vec{\mathbf{r}}$ and $\vec{\mathbf{p}}$. It follows from the above commutation relations that (cf. for instance Messiah[39], Chap. VIII, § 6) :

(1) The position operator $\vec{\mathbf{r}}$ constitutes by itself a complete set of commuting observables.

(2) The spectrum of the vector operator $\vec{\mathbf{r}}$ is continuous, nondegenerate, and extends in the whole space E_3. The eigenvectors have an infinite norm ; we choose to define them with the normalization condition :

$$\vec{\mathbf{r}}|\vec{r}\rangle = \vec{r}|\vec{r}\rangle \quad , \qquad \langle\vec{r}\,'|\vec{r}\rangle = \delta_3(\vec{r}\,'-\vec{r}) \quad . \tag{11-7}$$

The vector space $\mathcal{E}$ is spanned by the vectors $|\vec{r}\rangle$, i.e. one has the closure relation :

$$\int_{E_3} d_3\vec{r}\ |\vec{r}\rangle\langle\vec{r}| = \mathbf{1} \quad . \tag{11-8}$$

(3) The matrix elements of $\vec{\mathbf{p}}$ in the basis ($|\vec{r}\rangle$) read symbolically in vector notation :

$$\langle\vec{r}|\vec{\mathbf{p}}|\vec{r}\,'\rangle = -i\hbar\,\delta_3'(\vec{r}-\vec{r}\,') \quad . \tag{11-9}$$

The momentum operator $\vec{\mathbf{p}}$ has also a continuous and nondegenerate spectrum extended in the whole space. Its eigenvectors have an infinite norm and are such that :

$$\vec{\mathbf{p}}|\vec{p}\rangle = \vec{p}|\vec{p}\rangle \quad , \qquad \langle\vec{r}|\vec{p}\rangle \propto e^{\frac{i}{\hbar}\vec{r}\cdot\vec{p}} \quad . \tag{11-10}$$

We choose to normalize them by :

$$<\vec{p}\,'|\vec{p}> = \delta_3(\vec{p}\,'-\vec{p}) \quad , \quad <\vec{r}|\vec{p}> = \frac{1}{(2\pi\hbar)^{\frac{3}{2}}}\, e^{\frac{i}{\hbar}\vec{r}.\vec{p}} \quad . \qquad (11\text{-}11)$$

The operator $\vec{\mathbf{p}}$ is an observable in $\mathcal{E}$, one therefore has the closure relation :

$$\int d_3\vec{p}\,|\vec{p}><\vec{p}| = \mathbf{1} \quad . \qquad (11\text{-}12)$$

The physical states are associated with normalizable vectors $|\varphi>$ in $\mathcal{E}$. For instance, with respect to the basis ($|\vec{r}>$) one has :

$$|\varphi> = \int d_3\vec{r}\;|\vec{r}>\;\varphi(\vec{r}) \quad , \quad <\varphi|\varphi> = \int d_3\vec{r}\;|\varphi(\vec{r})|^2 < \infty \quad , \qquad (11\text{-}13)$$

where the wave function $\varphi(\vec{r}) = <\vec{r}\,|\varphi>$ belongs to the Hilbert space $\mathcal{H} = L^2(E_3, d_3\vec{r})$ of square integrable functions defined on E_3. In this space the momentum operator is equal to $-i\hbar\vec{\nabla}$, indeed one has :

$$<\vec{r}|\vec{p}|\varphi> = \int d_3\vec{r}\,'\;<\vec{r}|\vec{p}|\vec{r}\,'>\;\varphi(\vec{r}\,') = -i\hbar\int d_3\vec{r}\,'\;\delta_3'(\vec{r}-\vec{r}\,')\;\varphi(\vec{r}\,')$$

$$= -i\hbar\;\vec{\nabla}\varphi(\vec{r}) \quad . \qquad (11\text{-}14)$$

ii) Determination of representations of the rotation group. It suffices to specify the law of transformation of the basis vectors $|\vec{r}>$. Although a vector $|\vec{r}>$ does not belong to the Hilbert space $\mathcal{H}$ of physical states, since it is not normalizable, it can be interpreted in the classical limit as describing a pointlike particle located at M with OM = $\vec{r}$. This leads us to the law of transformation of $|\vec{r}>$ under any rotation R :

$$|^{R}\vec{r}> = D(R)\,|\vec{r}> = |R\,\vec{r}> \quad . \qquad (11\text{-}15)$$

One thus defines a true representation of the rotation group in the space $\mathcal{E}$, since one has :

$$D(R)\;[D(R')\,|\vec{r}>] = D(R)\,|R'\vec{r}> = |RR'\vec{r}> = D(RR')\,|\vec{r}> \quad . \qquad (11\text{-}16)$$

It corresponds to an even representation of SU(2) and its irreducible representations are the $\mathcal{D}^{\ell}$ with integer values of ℓ. The law of transformation of the position operator arises from Eqs. (11–2b,15) :

$${}^{R}\vec{\mathbf{r}}\,|\vec{r}> = D(R)\vec{\mathbf{r}}\;D(R)^{-1}\,|\vec{r}> = D(R)\vec{\mathbf{r}}\;|R^{-1}\vec{r}>$$

$$= D(R)\;R^{-1}\vec{r}\;|R^{-1}\vec{r}> = R^{-1}\vec{\mathbf{r}}\,|\vec{r}> \qquad (11\text{-}17)$$

and since these equations hold for all $|\vec{r}>$, one finds :

$${}^{R}\vec{\mathbf{r}} = D(R)\;\vec{\mathbf{r}}\;D(R)^{-1} = R^{-1}\vec{\mathbf{r}} \quad , \qquad (11\text{-}18)$$

or, with respect to any orthonormal basis :

$$\vec{r} = x^i e_i \quad , \quad {}^R\vec{r} = {}^Rx^i e_i \quad ,$$

$${}^Rx^i = D(R)\, x^i\, D(R)^{-1} = x^j\, (R^{-1})^i_j = \sum_j x^j R^j_i \quad . \tag{11-19}$$

From Eqs. (11–13,15) and using the linear character of D(R) one derives the law of transformation of any vector $|\varphi>$ in E :

$$|{}^R\varphi> = D(R)\,|\varphi> = \int d_3\vec{r}\,|R\vec{r}>\,\varphi(\vec{r}) = \int d_3\vec{r}\,|\vec{r}>\,\varphi(R^{-1}\vec{r}) \quad , \tag{11-20}$$

where the last equality follows from the invariance of the measure under any rotation. In particular, for the vector $|\vec{p}>$, since $R^{-1}\vec{r}.\vec{p} = \vec{r}.R\vec{p}$, one finds from Eq. (11–11) :

$$|{}^R\vec{p}> = D(R)\,|\vec{p}> = |R\,\vec{p}> \quad . \tag{11-21}$$

Whence, repeating the derivation of the expression for ${}^R\vec{r}$, one obtains the law of transformation of the momentum operator :

$${}^R\vec{p} = D(R)\,\vec{p}\,D(R)^{-1} = R^{-1}\vec{p} \quad . \tag{11-22}$$

It should be noticed that the vector operators $\vec{r}$ and $\vec{p}$ transform under the rotation R as vectors under the rotation R^{-1}. On comparing Eqs. (8–212) and (11–19) one sees that the operators x^i, as well as p^i, behave like the Cartesian components of a tensor of order and rank one, i.e. a vector (Subsec. 8.5.4).

iii) **Transformation of wave functions.** From Eqs. (11–13,20) the law of transformation of the wave functions $\varphi(\vec{r})$ under any rotation R reads (still denoting by D(R) the unitary operator acting on these functions) :

$$\varphi \longrightarrow {}^R\varphi = D(R)\varphi \;, \;\forall\vec{r} \qquad {}^R\varphi(\vec{r}) = <\vec{r}|{}^R\varphi> = \varphi(R^{-1}\vec{r}) \quad . \tag{11-23}$$

This result also follows directly from the unitary character of D(R) :

$$<\vec{r}|{}^R\varphi> = <\vec{r}|D(R)\,|\varphi> = <R^{-1}\vec{r}\,|\varphi> \quad . \tag{11-24}$$

Thus, the representation of the rotation group in E induces a representation in the space of wave functions, which coincides with the representation of this group determined by Eq. (10–208). Therefore, <u>all of the study made in Subsec. 10.3.3 applies.</u>

iv) **Dependence upon the choice of an orthonormal basis in E_3.**

1– In practice one needs to choose a parametrization on E_3 (Subsec. 10.2.1), e.g. the Cartesian coordinates with respect to an orthonormal basis $a = (e_i)$. Let us then use the following notations :

$$\vec{r} = x^i e_i \quad , \quad x = (x^i) \quad , \quad |\vec{r}> = |a;x> \quad , \tag{11-25}$$

and furthermore denote by φ_a the representative of a wave function φ with respect to this choice of parameters :

$$\varphi_a(x) = \langle a;x|\varphi\rangle = \varphi(\vec{r}) \quad . \tag{11-26}$$

The passive and the active points of view of a rotation R in E_3 (Subsec. 1.2.1) read, respectively :

$$\vec{r} = (a;x) = (Ra;R^{-1}x) \quad , \qquad R\vec{r} = (a;Rx) = (Ra;x) \quad , \tag{11-27}$$

where :

$$Ra = (R\, e_i) \quad , \qquad Rx = (R^i_j\, x^j) \quad . \tag{11-28}$$

The active point of view of the representation of the rotation group in E is then expressed by :

$$D(R)|a;x\rangle = |a;Rx\rangle \quad , \tag{11-29}$$

which implies for the representatives of wave functions :

$${}^R\varphi_a(x) = \varphi_a(R^{-1}x) \Leftrightarrow {}^R\varphi_a = D(R)\varphi_a \quad . \tag{11-30}$$

The passive point of view of the rotation R (Subsec. 10.2.1.iii)) amounts to changing the orthonormal basis $a = (e_i)$ into the new one $Ra = (Re_i)$. One then derives from Eq. (11–27) :

$$|\vec{r}\rangle = |a;x\rangle = |Ra;R^{-1}x\rangle - D(R^{-1})|Ra;x\rangle \quad . \tag{11-31}$$

Therefore, for the representatives of the wave functions, cf. Eq. (10–56), one has :

$$\varphi_{Ra}(x) = \varphi_a(Rx) \Leftrightarrow \varphi_{Ra} = {}^{R^{-1}}\varphi_a = D(R^{-1})\varphi_a \quad . \tag{11-32}$$

The above equation expresses the so-called equivalence between the active and the passive points of view of rotations in the considered representation.

When the vectors $\vec{r}$ are parametrized by the spherical polar coordinates, Eq. (2–20) with respect to the orthonormal basis $a = (e_i)$, the following notations are used :

$$|\vec{r}\rangle = |a;r\theta\varphi\rangle \quad , \qquad \varphi_a(r,\theta,\varphi) = \langle a;r\theta\varphi|\varphi\rangle = \varphi(\vec{r}) \quad . \tag{11-33}$$

2–To the intrinsic operators A acting on the vectors $|\varphi\rangle$ there correspond operators A_a which act on the wave functions φ_a defined with respect to the orthonormal basis $a = (e_i)$. These operators are defined by :

$$(A_a\varphi_a)(x) = \langle a;x|A|\varphi\rangle = \int d_3x'\; A_a(x,x')\; \varphi_a(x') \quad , \tag{11-34}$$

where the kernel $A_a(x,x')$ reads :

$$A_a(x,x') = \langle a;x|A|a;x'\rangle \quad . \tag{11-35}$$

It then follows from Eq. (11–31) that the law of transformation of this kernel under a rotation

R of the basis a is expressed by :

$$A_{Ra}(x,x') = \langle Ra;x|A|Ra;x'\rangle = \langle a;x|D(R)^{-1}A\ D(R)|a;x'\rangle$$
$$= ({}^{R^{-1}}A)_a\ (x,x') = A_a(Rx,Rx') \tag{11-36}$$

whence one has :

$$A_{Ra} = ({}^{R^{-1}}A)_a = [D(R)^{-1}\ A\ \ D(R)]_a\quad . \tag{11-37}$$

3–As an application of the above results, let us derive the expression of the Schrödinger equation when the wave functions are relative to a basis R(t)a which rotates with a constant angular velocity $\vec{\omega}$ with respect to a Galilean basis a. In this latter basis a the Schrödinger equation reads :

$$H_a\varphi_a = i\hbar\frac{\partial}{\partial t}\varphi_a\quad ,\qquad \varphi_a(x,t) = \langle a;x|\varphi t\rangle_S\quad , \tag{11-38}$$

where the subscript S recalls that we are considering the Schrödinger representation (Subsec. 3.1,iii)). From Eqs. (11–32,49) one has :

$$\varphi_a = D[R(t)]\ \varphi_{R(t)a} = e^{-\frac{i}{\hbar}\vec{\omega}t.\vec{L}}\ \varphi_{R(t)a}\quad , \tag{11-39}$$

and therefore, since $\vec{\omega}t.\vec{L}$ commutes with $\vec{\omega}.\vec{L}$ (Prop. D.2) :

$$\frac{\partial}{\partial t}\varphi_a = D[R(t)]\left(\frac{\partial}{\partial t} - \frac{i}{\hbar}\vec{\omega}.\vec{L}\right)\varphi_{R(t)a}\quad . \tag{11-40}$$

Using Eq. (11–37) one then finds :

$$\left(H_{R(t)a} - \vec{\omega}.\vec{L}\right)\ \varphi_{R(t)a} = i\hbar\frac{\partial}{\partial t}\varphi_{R(t)a}\quad . \tag{11-41}$$

It should be remarked that $H_{R(t)a} = H_a$ if the considered Hamiltonian is rotationally invariant. The effective or <u>cranking Hamiltonian</u> :

$$H'_{R(t)a} = H_{R(t)a} - \vec{\omega}.\vec{L} \tag{11-42}$$

enables us to write the Schrödinger equation in the rotating axes in the same way as it is written in the Galilean axes. The extra term $-\vec{\omega}.\vec{L}$ is commonly called the <u>Coriolis perturbation</u> or even the <u>Coriolis force</u>, although it includes both Coriolis and centrifugal forces (cf. below).

The expression for $H'_{R(t)a}$ is also valid in Classical Mechanics. As an example, for a classical particle of mass m in a rotationally invariant potential V(r), the effective Hamiltonian H' in the axes R(t)a reads :

$$H' = \frac{\vec{p}^{\,2}}{2m} + V(r) - \vec{\omega}.\vec{L} \quad . \tag{11-43}$$

Let us then check that the extra term $-\vec{\omega}.\vec{L} = -\vec{\omega}.\vec{r}\times\vec{p}$ induces the Coriolis and the centrifugal forces. In the rotating axes the Hamilton equations of motion are :

$$\dot{\vec{r}} = \vec{\nabla}_{\vec{p}} H' = \frac{\vec{p}}{m} - \vec{\omega}\times\vec{r} \quad , \tag{11-44a}$$

$$\dot{\vec{p}} = -\vec{\nabla}_{\vec{r}} H' = -\vec{\nabla}_{\vec{r}} V + \vec{p}\times\vec{\omega} \quad , \tag{11-44b}$$

where one has to take care that the time derivatives of $\vec{r}$ and $\vec{p}$ are here defined in the axes R(t)a, e.g. :

$$\vec{r} = x^i R(t)e_i \Rightarrow \dot{\vec{r}} = \dot{x}^i R(t)e_i \quad . \tag{11-45}$$

One then recognizes in Eq. (11–44a) the expression of the relative velocity of the particle with respect to the rotating axes. The force $\vec{F}$ acting on the particle is given by :

$$\begin{aligned}\vec{F} = m\ddot{\vec{r}} &= \dot{\vec{p}} - m\vec{\omega}\times\dot{\vec{r}} && \text{from Eq. (11-44a)}\\ &= -\vec{\nabla}_{\vec{r}} V + \vec{p}\times\vec{\omega} - m\,\vec{\omega}\times\dot{\vec{r}} && \text{from Eq. (11-44b)}\\ &= -\vec{\nabla}_{\vec{r}} V - 2m\,\vec{\omega}\times\dot{\vec{r}} - m\,\vec{\omega}\times(\vec{\omega}\times\vec{r}) && \text{from Eq. (11-44a)} \quad ,\end{aligned} \tag{11-46}$$

where the second and the third terms are, respectively, the Coriolis and the centrifugal forces.

v) Infinitesimal generators, identification of the Lie algebra with E_3. The expressions of the infinitesimal generators L_i in terms of the Cartesian components or the spherical polar ones are listed in Eqs. (10–231a–232e). In an intrinsic way, using the expression of an infinitesimal rotation $R(\varepsilon,\hat{n})$ given in Eq. (E–8), one finds :

$$\begin{aligned} R\varphi(\vec{r}) &= \varphi[\vec{r}-\varepsilon\vec{n}\times\vec{r} + O(\varepsilon^2)] = \varphi(\vec{r})-\varepsilon\hat{n}.\vec{r}\times\vec{\nabla}\,\varphi(\vec{r}) + O(\varepsilon^2)\\ &= [\mathbf{1}-i\varepsilon\,\hat{n}.\vec{L} + O(\varepsilon^2)]\,\varphi(\vec{r}) \quad , \end{aligned} \tag{11-47}$$

and therefore, taking into account Eq. (11–14) :

$$\hat{n}.\vec{L} = n^i L_i = \hat{n}.\mathbf{r}\times(-i\vec{\nabla}) = \hbar^{-1}\,\hat{n}.\vec{\mathbf{r}}\times\vec{\mathbf{p}} \quad , \tag{11-48}$$

where the n^i denote the components of $\hat{n}$ with respect to an orthonormal basis (e_i). Whence, identifying the Lie algebra of the considered representation of the rotation group with the space E_3 (Subsec. 5.4.3,i)) we are led to identify the formal vector $\vec{L}$ defined by Eq. (5–42) with the angular momentum operator $\vec{\mathbf{r}}\times\vec{\mathbf{p}}$ divided by $\hbar$. In physical applications it is convenient to take into account this latter operator instead of the dimensionless infinitesimal generators considered up to now. This only amounts to changing the L_i into $\hbar^{-1}L_i$ in all equations derived until now. In other words we had chosen the units so as to have $\hbar = 1$. In the remainder of this chapter, the relations are always expressed in terms of the components of the angular momentum, hence for instance :

$$[L_i, L_j] = i\hbar \sum_k \varepsilon_{ijk} L_k \quad , \qquad D[R(\theta,\hat{n})] = e^{-\frac{i}{\hbar}\theta\hat{n}.\vec{L}} \quad . \tag{11-49}$$

vi) Standard basis, dependence upon a triad. The dependence of a standard basis upon a choice of basis of infinitesimal generators in the Lie algebra is discussed in Sec. 8.1. Having identified the Lie algebra with E_3, a choice of orthonormal basis $a = (e_i)$ in E_3 implies a choice of infinitesimal generators :

$$\vec{L} = \sum_i L_i e_i \quad . \tag{11-50}$$

The operators $\mathbf{r}$, $\vec{L}^2$, and L_3 form a complete set of commuting observables. In order to recall the dependence upon the choice of the infinitesimal generators L_i, the standard basis vectors are denoted by $|a;r\,\ell\,m>$:

$$\mathbf{r}|a;r\,\ell\,m> = r|a;r\,\ell\,m> \quad , \qquad \vec{L}^2|a;r\,\ell\,m> = \hbar^2\ell(\ell+1)|a;r\,\ell\,m> \quad , \tag{11-51a}$$

$$L_\pm|a;r\,\ell\,m> = \hbar\sqrt{\ell(\ell+1)-m(m\pm1)}\,|a;r\,\ell\,m\pm1> \ , \quad L_3|a;r\,\ell\,m> = \hbar m|a;r\,\ell\,m> \ , \tag{11-51b}$$

where a plays here the role of the triad **1** introduced in Subsecs. 8.1,iii), iv). In terms of the above notations, the active point of view of a rotation, which corresponds to a change of states, reads :

$$D(R)|a;r\,\ell\,m> = |a;r\,\ell\,m'> \mathcal{D}^{\ell}_{m'm}(R) \quad . \tag{11-52}$$

Since one deals with even representations of SU(2), it follows from Eq. (8–10) that the passive point of view of a rotation R, which corresponds to a change of quantization axes of the angular momentum components, is expressed by :

$$D(R)|a;r\,\ell\,m> = |Ra;r\,\ell\,m> \quad , \tag{11-53}$$

where Ra specifies the triad U, such that R = R(U;a). We also call Ra a triad. It expresses at the same time the rotation R which determines the vector $|Ra;r\,\ell\,m>$ by the above equation, and the infinitesimal generators :

$$RL_i = L_j R^j_i \quad , \tag{11-54}$$

with respect to which these vectors form a standard basis. More generally, from Eqs. (8–11,12) one has :

$$|R'a;r\,\ell\,m> = D(R'R^{-1})|Ra;r\,\ell\,m> = |Ra;r\,\ell\,m'> \mathcal{D}^{\ell}_{m'm}(R^{-1}R') \quad . \tag{11-55}$$

The vectors $|a;r\,\ell\,m>$ are completely determined by the following choice of phase :

$$<a;r'\theta\varphi|a;r\,\ell\,m> = \delta(r'-r)\, Y^{[\ell]}_m(\theta,\varphi) \quad . \tag{11-56}$$

where the spherical harmonics are defined by Eq. (10–247).

vii) **Many particle systems.** For a physical system which consists of several spinless particles the algebra of observables and the space on which they act are constructed in a straightforward way by tensorial products of algebras and vector spaces, respectively, as described above. Indeed, it is assumed that the observables of any particle commute with all the observables of the other particles. The representation of the rotation group is then given by :

$$D(R)\,|\vec{r}_1\rangle \otimes \ldots \otimes |\vec{r}_N\rangle = |R\vec{r}_1\rangle \otimes \ldots \otimes |R\vec{r}_N\rangle \quad . \tag{11-57}$$

11.1.3. SPIN–s PARTICLES

i) **Observables and states.** We now consider particles which in addition to their position and momentum are characterized by three new dynamical variables, without any classical analogue, and called the spin components or the components of the intrinsic angular momentum. By definition, there correspond to these variables the observables S_i which satisfy the usual commutation relations of angular momentum components, i.e. :

$$[S_i, S_j] = i\hbar \sum_k \varepsilon_{ijk}\, S_k \quad . \tag{11-58}$$

It is furthermore assumed that the S_i commute with the position and momentum operators $\vec{r}$ and $\vec{p}$. Consequently, following a very general procedure, the vector space we consider is constructed from the tensor product of the space E on which act the observables $\vec{r}$ and $\vec{p}$, cf. Subsec. 11.1.2,i), and a vector space E_s on which act the observables S_i and built up as follows. The spin space E_s is spanned by the orthonormal basis vectors $|a;sm\rangle$:

$$\sum_{m=-s}^{+s} |a;sm\rangle\langle a;sm| = \mathbf{1}_{E_s} \quad , \quad \langle a;sm|a;sm'\rangle = \delta_{mm'} \quad , \tag{11-59}$$

which form a standard basis, i.e. obey the relations (Sec. 5.5) :

$$S_\pm|a;sm\rangle = \hbar\sqrt{s(s+1)-m(m\pm1)}\,|a;sm\pm1\rangle \;, \quad S_3|a;sm\rangle = \hbar m|a;sm\rangle \quad , \tag{11-60}$$

$$S^2 = \hbar^2\, s(s+1)\, \mathbf{1}_{E_s} \quad ,$$

for a given value of s such that 2s is a non-negative integer. As pointed out in Sec. 8.1, this standard basis depends upon the choice of the observables S_i which form a basis of a Lie algebra isomorphic to the Lie algebra $\mathcal{U}(2)$ of SU(2). This dependence is underlined by a (cf. iii) below) which here replaces the triad **1** introduced in Subsec. 8.1,iii).

ii) **Determination of the representation of the rotation group.** The symmetry group of rotations leads us to consider the true unitary representations of SU(2). This allows us to understand the existence of half-odd-integer spin particles in Quantum Mechanics. One thus considers the representation $\mathcal{D}^s$ [SU(2)] in E_s determined by :

$$\forall\, U \in SU(2) \qquad \mathcal{D}(U)|a;sm\rangle = |a;sm'\rangle\, \mathcal{D}^s_{m'm}(U) \quad . \tag{11-61}$$

Any vector $|\psi>$ in $E\otimes E_s$ can be expressed as :

$$|\psi> = \int d_3\vec{r} \quad |\vec{r}> \otimes |a;sm> \quad \psi^s_{m;a}(\vec{r}) \quad . \tag{11-62}$$

It then follows from Eqs. (11–15,61) that the vector :

$$|\psi(\vec{r})> = \quad |a;sm> \psi^s_{m;a}(\vec{r}) = <\vec{r}|\psi> \quad , \tag{11-63}$$

in E_s behaves under the representation of the rotation group as a spinor field, cf. Eq. (10–279). Indeed, one has :

$$\forall U \in SU(2) \quad |^U \psi(\vec{r})> = <\vec{r}|D[R(U)]\otimes \mathcal{D}^s(U)|\psi> = <R(U)^{-1}\vec{r}|\mathcal{D}^s(U)|\psi>$$

$$= \mathcal{D}^s(U)|\psi[R(U)^{-1}\vec{r}]> \quad . \tag{11-64}$$

Therefore, all of the study of representations of SU(2) in a space of vectorial functions defined on R^3 of Subsec. 10.3.4 applies on identifying the basis vectors $|a;sm>$ with the $e^{[s]}_m$. Furthermore, the whole discussion of Subsec. 11.1.2,iv) concerning the dependence upon an orthonormal basis $a = (e_i)$ in E_3 can be repeated for $|\psi(\vec{r})>$ when $\vec{r}$ is parametrized by its Cartesian (or spherical polar) coordinates with respect to a. One then has :

$$|\psi_a(x)> = <a;x|\psi> \quad . \tag{11-65}$$

According to Eqs. (11–30,32), the active and the passive points of view of the rotation R are expressed by :

$$D(R)|\psi_a(x)> = |\psi_a(R^{-1}x)> = |\psi_{R^{-1}a}(x)> \quad , \tag{11-66}$$

where D(R) acts only on the space E.

iii) **Infinitesimal generators, standard basis, and dependence upon a triad.** Cf. Subsecs. 10.3.4.1,ii), iii).

1–On identifying the Lie algebra of the considered representation of SU(2) with the Euclidean vector space E_3, one is led to define the formal vectors in E_3, Eq. (5–42) :

$$\vec{L} = \sum_i L_i e_i \quad , \quad \vec{S} = \sum_i S_i e_i \quad , \quad \vec{J} = \vec{L}+\vec{S} = \sum_i J_i e_i \quad , \tag{11-67}$$

where $a = (e_i)$ always denotes any orthonormal basis of E_3. These vectors $\vec{L}, \vec{S}$, and $\vec{J}$ are called, respectively, the orbital, the intrinsic (or spin), and the total angular momentum of the particle. According to Eq. (10–283) the standard basis vectors are determined by the common eigenvectors of the complete set of commuting observables r, $\vec{L}^2$, $\vec{S}^2$, $\vec{J}^2$, and J_3 with the eigenvalues r, $\hbar^2 \ell(\ell+1)$, $\hbar^2 s(s+1)$, $\hbar^2 j(j+1)$, and $\hbar m$, respectively. These orthonormal vectors are determined by :

$$|a;r\ell sjm> = |a;r\ell m_\ell> \otimes |a;sm_s><\ell s m_\ell m_s|jm> \quad , \tag{11-68}$$

and one has :

$$<a;r'\theta\varphi|a;r\ell sjm> = \delta(r'-r)\ \mathbf{Y}^{\ell s[j]}_{m}(\theta,\varphi) \quad , \tag{11-69}$$

where the spin spherical harmonics are defined by Eq. (10–283) with $e^{[s]}_{m} \equiv |a;sm>$.

2–The active point of view of a rotation $R = R(\pm U;a)$ corresponds to a change of states. It reads :

$$\forall\, U \in SU(2) \qquad \mathcal{D}(U)|a;r\ell sjm> = |a;r\ell sjm'>\ \mathcal{D}^{j}_{m'm}(U) \quad . \tag{11-70}$$

On the other hand, the passive point of view of a rotation corresponds to a rotation of the quantization axes of the components of the total angular momentum. One must now take care that we are dealing with even or odd representations of SU(2) according to whether s is an integer or a half–odd–integer. It follows from Eq. (8–10) that this passive point of view is expressed by :

$$\forall\, U \in SU(2) \qquad \mathcal{D}(U)|a;r\ell sjm> = |Ua;r\ell sjm> \quad . \tag{11-71}$$

The triad Ua denotes at the same time :
(1) The element U in SU(2) which determines the vector $|Ua;r\ell sjm>$ by the above equation, especially the phase since from Eq. (8–13) one has :

$$|-Ua;r\ell sjm> = (-1)^{2j}\ |Ua;r\ell sjm> \quad . \tag{11-72}$$

(2) And the uniquely defined infinitesimal generators :

$$R(U;a)J_i = J_j\ R(U;a)^{j}_{i} \quad , \tag{11-73}$$

with respect to which the vectors $|Ua;r\ell sjm>$ form a standard basis.
More generally, arising from Eqs. (8–11,12), the dependence upon the triad can be made explicit as follows :

$$\begin{aligned} \forall\, U,U' \in SU(2) \qquad |U'a;r\ell sjm> &= \mathcal{D}(U'U^{-1})|Ua;r\ell sjm> \\ &= |Ua;r\ell sjm'>\ \mathcal{D}^{j}_{m'm}(U^{-1}U') \quad . \end{aligned} \tag{11-74}$$

3–Let us now give two examples of the application of the above relations.
(1) A gedanken <u>polarization experiment</u>, see Fig. 11.1. On going through a polarizer, particles are prepared in a state $|a;jm>$ (forgetting other quantum numbers). One now observes these particles after they pass through a similar apparatus used as an analyzer and which has been rotated with respect to the polarizer by a rotation R(U;a). What does one then observe ? The answer to this question is given by Eq. (11–74) setting $U' = \mathbf{1}$. Hence, for instance, the probability of observing a state $|Ua;jm'>$ is :

$$|<Ua;jm'|a;jm>|^2 = |\mathcal{D}^{j}_{m'm}(U^{-1})|^2 = |\mathcal{D}^{j}_{mm'}(U)|^2 \quad . \tag{11-75}$$

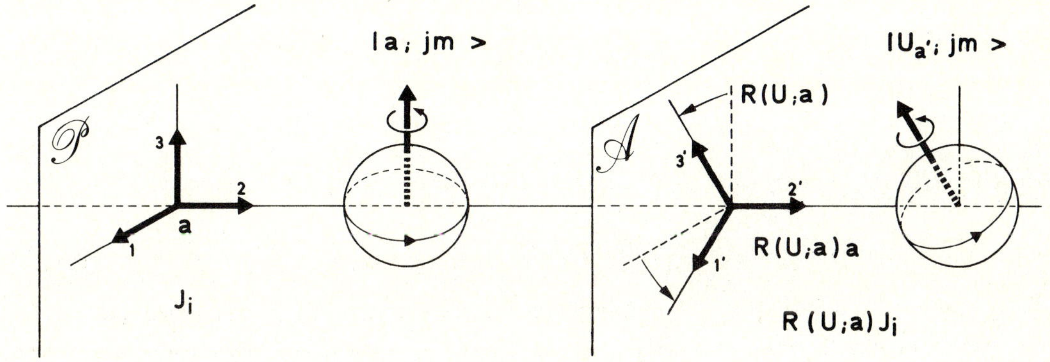

Fig.11.1. Gedanken polarization experiment. The polarizer ($\mathcal{P}$) and the analyzer ($\mathcal{A}$) determine, respectively, the orthogonal axes a and R(U;a)a. $\mathcal{P}$ and $\mathcal{A}$ prepare particles in states which are standard basis vectors of the infinitesimal generators J_i and $R(U;a)J_i$, respectively.

(2) The helicity formalism. Instead of considering the states $|\vec{p}\rangle \otimes |a;sm\rangle$ of a spin-s particle in an arbitrary orthonormal basis a, it may be convenient to choose this basis such that the third axis be along the momentum $\vec{p}$ of the particle. This is the usual helicity formalism[31] which amounts to projecting the spin $\vec{S}$ along the direction of the momentum. A similar formalism can be developed[47] on taking the position $\vec{r}$ instead of the momentum of the particle.

The definition of the helicity states is as follows. For any given orthonormal basis a = (e_i) in E_3 as reference, one chooses a definite procedure to associate with any vector $\vec{p}$ a matrix $U_{\vec{p}}$ in SU(2) such that :

$$R(U_{\vec{p}})e_3 = \hat{p} = \frac{\vec{p}}{p} \quad . \tag{11-76}$$

As an example, if p, θ, and φ are the spherical polar coordinates of $\vec{p}$ with respect to a, one may choose $U_{\vec{p}} = U(\varphi, \theta, -\varphi)$, Eq. (4–38), such that $R(U_{\vec{p}};a)$ is then the rotation $R_{\vec{p}}$ defined by Eq. (2–37). According to Eq. (11–71) one has in E_s :

$$|U_{\vec{p}}a;sm\rangle = \mathcal{D}^s(U_{\vec{p}})\,|a;sm\rangle \quad . \tag{11-77}$$

A helicity state in $E \otimes E_s$ is then defined by :

$$|U_{\vec{p}}a;\vec{p}sm\rangle = |\vec{p}\rangle \otimes |U_{\vec{p}}a;sm\rangle \quad , \tag{11-78}$$

i.e. it is an eigenstate of the operators $\vec{p}$ with the eigenvalue $\vec{p}$, and a standard basis vectors of the spin components $S_j R(U_{\vec{p}};a)^j_i$, the third one being along $\vec{p}$. A choice of $U_{\vec{p}}$ determines a phase choice for these helicity states.

The transformation law of a helicity state under any rotation R = R(U;a) follows from Eqs. (11–21,74) :

$$\forall U \in SU(2) \qquad \mathcal{D}(U)\,|U_{\vec p}a;\vec p[s]m\rangle = D[R(U;a)]\,|\vec p\rangle \otimes \mathcal{D}^s(U)\,|U_{\vec p}a;[s]m\rangle$$
$$= |R(U;a)\vec p\rangle \otimes \mathcal{D}^s(U_{R(U;a)\vec p})\mathcal{D}^s(U^{-1}_{R(U;a)\vec p}\,U\,U_{\vec p})\,|a;[s]m\rangle$$
$$= |U_{R(U;a)\vec p}a;R(U;a)\vec p[s]m'\rangle\mathcal{D}^s_{m'm}(U^{-1}_{R(U;a)\vec p}\,U\,U_{\vec p}). \qquad (11\text{-}79)$$

Actually, $R(U^{-1}_{R(U;a)\vec p}\,UU_{\vec p};a)$ is a rotation about e_3. Hence, one finally obtains :

$$\mathcal{D}(U)\,|U_{\vec p}a;\vec p[s]m\rangle = |U_{R(U;a)\vec p}a;R(U;a)\vec p[s]m\rangle\mathcal{D}^s_{mm}(U^{-1}_{R(U;a)\vec p}\,U\,U_{\vec p}) \quad , \qquad (11\text{-}80)$$

where there is no summation over m, and the matrix element $\mathcal{D}^s_{mm}$ is a phase factor. This transformation law is intuitive. Indeed, let us denote by $e_i(\vec p)$ the orthonormal basis vectors $R(U_{\vec p};a)e_i$. The state $\mathcal{D}(U)\,|U_{\vec p}a;\vec p s m\rangle$ has a momentum $R(U;a)\vec p$ and a spin m along $R(U;a)e_3(\vec p)$ which is a priori different from $e_3[R(U;a)\vec p\,]$, since for each $\vec p$ the attached orthonormal basis was arbitrarily chosen (see Fig. 11.2). According to Eq. (11–74), $U^{-1}_{R(U;a)\vec p}\,UU_{\vec p}$ is just the element in SU(2) which allows us to express a state in the triad $U\,U_{\vec p}a$ in terms of states in the triad $U_{R(U;a)\vec p}a$.

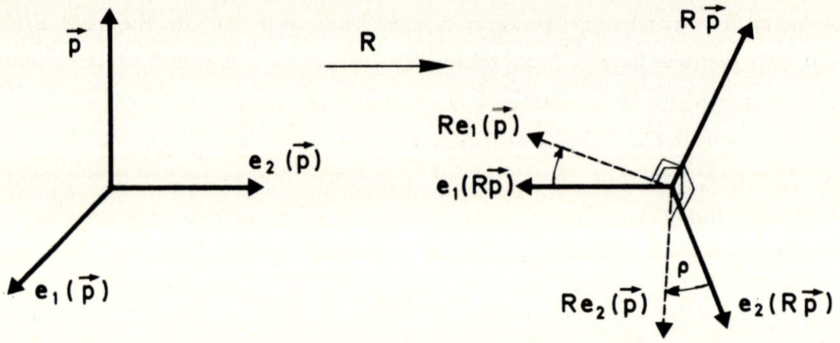

Fig. 11.2. Relative positions of the orthonormal bases $R(UU_{\vec p};a)a$ and $R(U_{R(U;a)\vec p};a)a$ associated with the triads $UU_{\vec p}a$ and $U_{R(U;a)\vec p}a$, respectively :

$$\rho = R(UU_{\vec p}\,U^{-1}_{R(U;a)\vec p};a) \quad \text{about} \quad R\,\vec p \quad .$$

A similar discussion occurs in the study of the physical representations of the Lorentz group, cf. for instance Ref. 41). There, the triad $U_{\vec p}a$ is replaced by the tetrad $[p]$, and the rotation $R(U^{-1}_{R(U;a)\vec p}\,UU_{\vec p};a)$ by the so-called Wigner rotation.

11.2. KETS, BRAS, WAVE FUNCTIONS, AND OPERATORS

Let us now make precise some tensorial features of the objects encountered in Quantum Mechanics. This section is devoted to the properties of the states and the operators ; the next one deals with the resulting properties of the matrix elements. In what follows, to avoid cumbersome notations, each time there is no ambiguity concerning the triad it will not be specified. Furthermore, it is emphasized that we only consider states and observables in the Schrödinger representation (Subsec. 3.1,iii)). For the same reason the subscript S which denotes this representation will be omitted.

11.2.1. TENSORIAL CHARACTER OF KETS, BRAS, AND WAVE FUNCTIONS

i) **Kets.** Forgetting other possible quantum numbers, the law of transformation of the kets $|jm\rangle$ in a space E under a rotation, Eqs. (11–52,70), is identical to that of the contravariant spherical components of a tensor of order one and rank j. Then, using the notations of Subsec. 8.3.1.1,i) to specify the tensorial character, these kets will from now on be denoted by $|[j]m\rangle$. From Eqs. (8–79a,b) one has :

$$|(j)m\rangle = c_j(-1)^{j+m}\,|[j]-m\rangle \quad . \tag{11-81}$$

The law of transformation under a rotation R(U;a) of $|[j]m\rangle$ and $|(j)m\rangle$ read, respectively, cf. Eq. (8–40) :

$$\forall U \in SU(2) \quad \begin{cases} |^{U}[j]m\rangle = \mathcal{D}(U)\,|[j]m\rangle = |[j]m'\rangle\, \mathcal{D}^{j}_{m'm}(U) \quad , & (11\text{-}82a) \\ |^{U}(j)m\rangle = \mathcal{D}(U)\,|(j)m\rangle = |(j)m'\rangle\, \mathcal{D}^{j*}_{m'm}(U) \quad . & (11\text{-}82b) \end{cases}$$

ii) **Bras.** To any ket $|\varphi\rangle$ in a space E of states one associates a bra $\langle\varphi|$ in the dual vector space E^*, such that :

$$\forall|\varphi\rangle \in E \xrightarrow{\ d\ } \langle\varphi| \in E^* \qquad \forall|\psi\rangle \in E \longrightarrow \langle\varphi|\psi\rangle \in C \quad . \tag{11-83}$$

The mapping d thus determined is an antilinear application of E onto E^*. It is shown in Subsec. 1.1.2 that there exists an intrinsic isomorphism between E and the dual space E^{**} of E^*. Whence, on identifying E and E^{**} the mapping d turns out to be involutive ($d^2 = 1$). The law of transformation of the bras under any rotation R(U;a) is defined by :

$$\begin{array}{ccc} |\varphi\rangle & \xrightarrow{R(U;a)} & |^{U}\varphi\rangle = \mathcal{D}(U)\,|\varphi\rangle \\ d\downarrow & & d\downarrow \\ \langle\varphi| & \xrightarrow{R(U;a)} & \langle^{U}\varphi| = \langle\varphi|\mathcal{D}(U)^{+} \end{array} \quad . \tag{11-84}$$

Hence, by construction, the mapping d commutes with the considered representation of SU(2). We are thus in a position to apply the study of Subsec. 8.3.4.1,i) where the formal mapping + is here the application d. According to Eqs. (8–112–115) and taking into account the fact that E and E^* are distinct vector spaces, we are led to the properties listed in Table 11.1, which at the same time enable us to specify our notations.

TABLE 11.1

Tensorial character of kets and bras.

$T^j = \|j\rangle$	$T^{[j]}_m = \|[j]m\rangle$ ↓ d	$T^{(j)}_m = \|(j)m\rangle$ ↓ d
$T^{dj} = \langle j\|$	$T^{d(j)}_m = \langle (j)m\|$ ↓ d	$(-1)^{2j} T^{d[j]} = (-1)^{2j} \langle [j]m\|$ ↓ d
$T^{ddj} = \|dj\rangle =$	$(-1)^{2j} T^{dd[j]}_m = (-1)^{2j} \|d[j]m\rangle$	$(-1)^{2j} T^{dd(j)}_m = (-1)^{2j} \|d(j)m\rangle$
$(-1)^{2j} T^j = (-1)^{2j} \|j\rangle$	$= \|[j]m\rangle$	$= \|(j)m\rangle$

It follows from the above that the complex conjugate of the matrix element of any operator A reads :

$$\langle (j)m|A|[j']m'\rangle^* = (-1)^{2j} \langle (j')m'|A^+|d[j]m\rangle$$

$$= \langle (j')m'|A^+|[j]m\rangle \quad . \tag{11-85}$$

iii) **Wave functions.** The tensorial character of the wave functions has already been specified in Eqs. (11–56,69), cf. also Subsecs. 10.3.3,4. From the tensorial point of view, it is convenient to deal with wave functions $f^{[\ell]}_m$ which are the contravariant spherical components of a Hermitian tensor f^ℓ. As shown in Subsec. 8.34.1, iii) this is possible only for integer values of ℓ and a necessary and sufficient condition for this to be so reads :

$$f^{[\ell]*}_m = (-1)^{\ell+m} f^{[\ell]}_{-m} \quad . \tag{11-86}$$

The spherical harmonics $Y^{[\ell]}_m$ and the vector spherical harmonics $\mathbf{Y}^{\ell'[\ell]}_m$, defined, respectively, by Eqs. (10–247) and (10–306), satisfy the above condition on choosing $c = \pm 1$, cf. Eqs. (10–248,307).

11.2.2. IRREDUCIBLE TENSOR OPERATORS

11.2.2.1. Definition and properties

i) An <u>irreducible tensor operator</u> A^j of rank j is by definition a tensor of order one and rank j, the 2j+1 components of which are operators in the Schrödinger representation acting on the considered space $\mathcal{E}$ of states. Hence, the contravariant and the covariant spherical components of A^j are operators which are, respectively, characterized by the following law of transformation under any rotation R(U;a) :

$$\forall U \in SU(2) \qquad {}^U\!A^{[j]}_m = \mathcal{D}(U) A^{[j]}_m \mathcal{D}(U)^{-1} = A^{[j]}_{m'} \mathcal{D}^j_{m'm}(U) \quad , \tag{11-87a}$$

$$\forall U \in SU(2) \qquad {}^{U}A^{(j)}_{m} = \mathcal{D}(U)\, A^{(j)}_{m} \mathcal{D}(U)^{-1} = A^{(j)}_{m'}\, \mathcal{D}^{j*}_{m'm}(U) \quad , \tag{11-87b}$$

where $\mathcal{D}$ denotes the unitary representation of SU(2) in E. As any tensorial sequence, the spherical components of an irreducible tensor operator A^{j} depend upon a triad which characterizes a choice of basis for infinitesimal generators (Subsec. 8.3.1.2), cf. below Eqs. (11–92a,b). Setting U infinitesimal in Eqs.(11–87a,b) and using Eqs. (6–33–34b) and (8–33b,c), one finds the following commutation relations :

$$[J_{\pm}, A^{[j]}_{m}] = \hbar \sqrt{j(j+1)-m(m\pm 1)}\; A^{[j]}_{m\pm 1} \ , \ [J_3, A^{[j]}_{m}] = \hbar m\, A^{[j]}_{m} \quad , \tag{11-88a}$$

$$[\bar{J}_{\pm}, A^{(j)}_{m}] = \hbar \sqrt{j(j+1)-m(m\pm 1)}\; A^{(j)}_{m\pm 1} \ , \ [\bar{J}_3, A^{(j)}_{m}] = \hbar m\, A^{(j)}_{m} \quad , \tag{11-88b}$$

where J_i and $\bar{J}_i$ are the components with respect to any orthonormal basis $a = (e_i)$ in E_3 of the angular momenta $\vec{J}$ and $\vec{\bar{J}}$, respectively :

$$\vec{J} = \sum_i J_i\, e_i \quad , \quad \vec{\bar{J}} = \sum_i \bar{J}_i\, e_i \quad . \tag{11-89}$$

Another way to derive the previous commutation relations is to note that the law of transformation of any observable A according to Eq. (11–2b) determines a representation of SU(2) in the algebra $\mathcal{A}$ of observables, which is an extension of the adjoint representation of $\mathcal{D}$ [SU(2)] (Subsec. 5.4.1). Indeed, it follows from Prop. D.1 that :

$$\mathcal{D}[U(\theta, n^{i})] = e^{-\frac{i}{\hbar}\theta n^{i} J_i} \Rightarrow \mathcal{D}(U) A\, \mathcal{D}(U)^{-1} = e^{-\frac{i}{\hbar}\theta n^{i} \mathcal{J}_i} A \quad , \tag{11-90}$$

where the $\mathcal{J}_i$ are the operators adjoint to the J_i :

$$\mathcal{J}_i = \Delta_{J_i} = [J_i \ , \ .] \quad . \tag{11-91}$$

The definition of an irreducible tensor operator A^{j} then implies (Subsec. 6.1.3,i)) that the $A^{[j]}_{m}$ behave as standard basis vectors with respect to the infinitesimal generators $\mathcal{J}_i$, thereby recovering Eq. (11–88a). It has been shown (at least formally) in Subsec. 6.1.3,ii) that conversely, any set of 2j+1 operators in $\mathcal{A}$ which obey Eq. (11–88a) transforms under any finite rotation R(U;a) according to Eq. (11–87a), and thus determines an irreducible tensor operator.

Using the same notations as for the spin-s states (Subsec. 11.1.3,iii)) the dependence of the spherical components of an irreducible tensor operator A^{j} upon the triad can be made explicit as follows, cf. Eqs. (8–57,58). For all U and U' in SU(2) one has :

$$^{U'}A^{[j]}_{m;Ua} = A^{[j]}_{m';Ua}\, \mathcal{D}^{j}_{m'm}(U^{-1}U'U) \qquad \text{active point of view} \tag{11-92a}$$

$$= A^{[j]}_{m';U'Ua} \qquad \text{passive point of view.} \tag{11-92b}$$

We now no longer specify the triad which is always assumed to be **1** a.

ii) A <u>scalar operator</u> is an irreducible tensor operator of rank zero. It corresponds to an observable which is invariant under the symmetry group of rotation, and from Eqs. (3–28) and (11–88a) its Schrödinger representative commutes with the components of the total angular momentum :

$$\forall U \in SU(2) \qquad {}^{U}A = A \quad \Leftrightarrow \quad [J_i, A] = 0 \quad . \tag{11-93}$$

A vector operator $\vec{A} = A^1$ is an irreducible tensor operator of rank one. In addition to its contravariant and covariant spherical components, respectively, $A^{[1]}_m$ and $A^{(1)}_m$, a vector operator can also be characterized by its Cartesian components A_i. These three sets of components, with respect to a same triad, are related by Eq. (8–211). In the Schrödinger representation, the Cartesian components transform under any rotation R as, Eq. (8–212) :

$${}^{R}A_i = \mathcal{D}(R) A_i \, \mathcal{D}(R)^{-1} = A_j \, R^j_i \quad . \tag{11-94}$$

It follows from this relation that the vector operator $\vec{A}$ transforms under the rotation R in the same way as the vector $\vec{A} = \sum_i A_i \vec{e}_i$ under the rotation R^{-1} :

$${}^{R}\vec{A} = R^{-1}\, \vec{A} \quad . \tag{11-95}$$

For an infinitesimal rotation one obtains, Eq. (8–221) :

$$[J_i, A_j] = i\hbar \sum_k \varepsilon_{ijk} A_k \quad , \quad [J_i, \vec{A}] = -i\hbar\, \vec{e}_i \times \vec{A} \quad . \tag{11-96}$$

iii) The general study of Subsec. 8.3.4.1 applies to the irreducible tensor operators, where the formal antilinear mapping is here the Hermitian conjugation of the operators. As always, from the tensorial point of view it is convenient to consider Hermitian irreducible tensor operators, which are by definition Hermitian tensors. It is recalled that this is only possible for integer values ℓ of j and a necessary and sufficient condition for that to be so reads, Eq. (8–116) :

$$A^{\ell} = A^{+\ell} \quad \Leftrightarrow \quad A^{[\ell]+}_m = (-1)^{\ell+m} A^{[\ell]}_{-m} \quad . \tag{11-97}$$

The observables are associated with Hermitian operators. As an example, if the Cartesian components A_i of a vector operator A^1 are Hermitian, one obtains from Eq. (8–211) :

$$A^{[1]+}_m = \left|\frac{c^*}{c}\right| (-1)^{1+m} A^{[1]}_{-m} \quad . \tag{11-98}$$

Therefore, choosing $c = \pm 1$ (more generally c real) A^1 is a Hermitian vector operator.

iv) More generally, one can define tensor operators $A^{j_1 \dots j_n}$ which are tensors of order n and ranks $j_1, \dots, j_n$, the spherical components of which are operators in the Schrödinger representation. As an example, the spherical components $A^{[j_1](j_2)}_{m_1 \; m_2}$ of a tensor operator of order two transform under any rotation R(U;a) as :

$$\forall U \in SU(2) \quad {}^{U}A^{[j_1](j_2)}_{m_1\ m_2} = \mathcal{D}(U)A^{[j_1](j_2)}_{m_1\ m_2}\mathcal{D}(U)^{-1} = A^{[j_1](j_2)}_{m_1'\ m_2'}\mathcal{D}^{j_1}_{m_1'm_1}\mathcal{D}(U)\mathcal{D}^{j_2 *}_{m_2'm_2}(U) \quad . \tag{11-99}$$

11.2.2.2. Some examples of tensor operators

i) **Vector operators.** It follows from the characterization of representations of the rotation group for spinless particles that the operators $\vec{r}$ and $\vec{p}$ are vector operators since they transform according, respectively, to Eqs. (11–18,22). The gradient operator, Eq. (11–14) :

$$\vec{\nabla} = i\hbar^{-1}\vec{p} \quad , \tag{11-100}$$

acting on the space of wave functions is also a tensor operator. This is also true for the intrinsic angular momentum $\vec{S}$ of a spin particle, as also for the orbital and the total angular momenta $\vec{L}$ and $\vec{J}$ since these operators transform under the adjoint representation of $\mathcal{D}$[SU(2)], as shown in Subsec. 5.4.3,ii) (cf. Eq. (5–45) and Subsec. 8.5.4,iii)). One should note that since the three vector operators $\vec{r}$, $\vec{p}$, and $\vec{S}$ form a fundamental system of observables in terms of which any observable can be expressed, only irreducible tensor operators with an integer rank can be constructed in the algebra of these observables.

ii) **Multiplicative operators $\hat{\mathcal{D}}^{j*}_{m_o}$, $\hat{\mathcal{D}}^{j}_{m_o}$, and $\hat{Y}^{\ell}$.** It follows from Eq. (10–180) that to standard basis functions $f^{[j]}_m$ of a representation $\mathcal{D}$[SU(2)] in a functional vector space, there correspond multiplicative operators $\hat{f}^{[j]}_m$ which determine irreducible tensor operators $\hat{f}^j$. The irreducible tensor operators :

$$\hat{\mathcal{D}}^{j*}_{m_o} = (\hat{\mathcal{D}}^{j*}_{mm_o}) \quad , \qquad \hat{\mathcal{D}}^{j}_{m_o} = (\hat{\mathcal{D}}^{j}_{m_o m}) \quad , \tag{11-101}$$

are, associated with the standard bases of, respectively, the left and the right regular representations of SU(2), Eqs. (10–182a,b). The irreducible tensor operators $\hat{Y}^{\ell}$ correspond to the standard bases of spherical harmonics $Y^{[\ell]}_m$ (Subsec. 10.3.3.6). From Eq. (10–248) one has :

$$\hat{Y}^{[\ell]+}_{m} = \left(\frac{c^*}{c}\right)^{\ell} (-1)^{\ell+m}\,\hat{Y}^{[\ell]}_{-m} \quad , \tag{11-102}$$

whence choosing c real implies that $\hat{Y}^{\ell}$ is a Hermitian irreducible tensor operator.

iii) **Creation and annihilation operators.** Let us first briefly recall the second quantization formalism. One considers the Fock space spanned by the orthonormal basis vectors $|N_1 \ldots N_m \ldots>$ characterized by the occupations numbers $\{N_m\}$ of the states $\{m\}$. These latter form a complete set of one–particle states. The occupation numbers N_m can take the values 0 or 1 for Fermions, and any non–negative integer value for Bosons. The state $|0>$ in which there is no particle at all is called the vacuum. One then defines the creation and annihilation operators a^+_m and a_m by :

$$a^+_m|N_1 \dots N_m \dots\rangle = \begin{cases} (N_m+1)^{\frac{1}{2}}\ |N_1 \dots N_m+1 \dots\rangle & \text{Bosons} , \\ (-1)^{\mathcal{N}_m}\ (1-N_m)\,|N_1 \dots N_m-1 \dots\rangle & \text{Fermions}, \end{cases} \tag{11-103a}$$

$$a_m|N_1 \dots N_m \dots\rangle = \begin{cases} N_m^{\frac{1}{2}}\ |N_1 \dots N_m-1 \dots\rangle & \text{Bosons} , \\ (-1)^{\mathcal{N}_m}\ N_m|N_1 \dots N_m-1 \dots\rangle & \text{Fermions}. \end{cases} \tag{11-103b}$$

The states are ordered arbitrarily and once for all as 1,2,3,... and $\mathcal{N}_m = \sum_{i=1}^{m} N_i$ is the sum of occupation numbers of the one-particle states preceding m. It follows from the above relations that a^+_m and a_m are conjugate Hermitian operators and furthermore obey the following (anti) commutation relations :

$$[a_m, a_n]_\varepsilon = 0 \quad , \quad [a^+_m, a^+_n]_\varepsilon = 0 \quad , \quad [a_m, a^+_n]_\varepsilon = \delta_{mn}\mathbf{1} \quad , \tag{11-104}$$

where $[\ ,\ ]_\varepsilon$ denotes the commutator ($\varepsilon = 1$) or the anticommutator ($\varepsilon = -1$) according to whether one deals with Bosons or Fermions. Iteration of Eq. (11–103a) allows us to express any basic state of the Fock space in terms of a product of creation operators acting on the vacuum :

$$|N_1 \dots N_m \dots\rangle = [N_1! \dots N_m! \dots]^{-\frac{1}{2}}\ a_1^{+N_1} \dots a_m^{+N_m} \dots |0\rangle \quad . \tag{11-105}$$

Let us now assume that the one-particle states are determined by the eigenbasis of a complete set of commuting observables including the operators $\vec{J}^2$ and J_3. These orthonormal states are labelled by α,j, and m :

$$a^+_{\alpha jm}|0\rangle = |\alpha[j]m\rangle \tag{11-106}$$

It then follows from the law of transformation of the ket $|\alpha[j]m\rangle$ under any rotation R(U;a), Eq. (11–82a), that one has :

$$\forall\, U \in SU(2) \qquad \mathcal{D}(U)a^+_{\alpha jm}\ \mathcal{D}(U)^{-1} = a^+_{\alpha jm'}\ \mathcal{D}^j_{m'm}(U) \tag{11-107}$$

Therefore, the $a^+_{\alpha jm}$ are the contravariant spherical components of an irreducible tensor of rank j. In order to make explicit the tensorial character we shall denote this operator by a^{+j}_α with :

$$a^{+[j]}_{\alpha\ m} = a^+_{\alpha jm} \quad , \tag{11-108}$$

where, according to Eq. (8–112), a^{+j}_α is the Hermitian conjugate of an irreducible tensor operator a^j_α. Since the annihilation operator $a_{\alpha jm}$ is the Hermitian conjugate of $a^+_{\alpha jm}$, one finds using Eqs. (8–114,115) :

$$a_{\alpha jm} = a^{+[j]+}_{\alpha\ m} = a^{++(j)}_{\alpha\ \ m} = (-1)^{2j}\ a^{(j)}_{\alpha\ m} \quad . \tag{11-109}$$

One can check directly taking the Hermitian conjugate of both sides of Eq. (11–107) that the

$a_{\alpha jm}$ transform as covariant spherical components of a tensor.

The above considerations are especially useful when one wants to couple creation and annihilation operators. Indeed, only spherical components which have the same tensorial character can be coupled by means of the C.G. coefficients (Subsec. 8.3.2.1). As an example a creation operator of a coupled particle–hole pair reads :

$$A^{\{j_p j_h\}[J]}_{\alpha_p \alpha_h \; M} = a^{+[j_p]}_{\alpha_p m_p} \, a^{[j_h]}_{\alpha_h m_h} \langle j_p j_h m_p m_h | JM\rangle$$
$$= c_{j_h} (-1)^{j_h + m_h} \, a^{+}_{\alpha_p j_p m_p} \, a_{\alpha_h j_h - m_h} \langle j_p j_h m_p m_h | JM\rangle \quad . \qquad (11\text{-}110)$$

iv) **Projection operators.** Let us again consider the complete orthonormal basis $\{ |\alpha[j]m\rangle \}$ of one–particle states introduced previously, Eq. (11–106) :

$$\langle \alpha(j)m|\alpha'[j']m'\rangle = \delta_{\alpha\alpha'}\delta_{jj'}\delta_{mm'} \quad , \quad \mathbf{1} = \sum_{\alpha,j,m} |\alpha[j]m\rangle\langle\alpha(j)m| \quad . \qquad (11\text{-}111)$$

The tensor characters of the kets and the bras, cf. Table 11.1, allow us to define a <u>generalized projection tensor operator</u> $P^{jj'}_{\alpha\alpha'}$, which is a tensor operator of order two and ranks j,j' with spherical components :

$$P_{\alpha\alpha'}{}^{[j](j')}_{m \;\; m'} = |\alpha[j]m\rangle\langle\alpha'(j')m'| \quad . \qquad (11\text{-}112)$$

From $P^{jj'}_{\alpha\alpha'}$, one defines the coupled generalized projection tensor operators $P^{\{jj'\}J}_{\alpha\alpha'}$. As an example using Eqs. (8–79a,b) and (G–2c) one derives :

$$P^{\{jj'\}[J]}_{\alpha\alpha' \;\; M} |\alpha''[j'']m''\rangle = \langle jj'mm'|JM\rangle c_{j'} (-1)^{j'-m'} |\alpha[j]m\rangle\langle\alpha'(j')-m'|\alpha''[j'']m''\rangle$$

$$= c_{j''}(-1)^{j+j''-J} \frac{\hat{J}}{\hat{j}} \langle Jj''Mm''|jm\rangle |\alpha[j]m\rangle \, \delta_{\alpha'\alpha''} \, \delta_{j'j''} \quad . \qquad (11\text{-}113)$$

The projection operator on the state $|\alpha[j]m\rangle$ is given by :

$$P_{\alpha}{}^{[j](j)}_{m \;\; m} = |\alpha[j]m\rangle\langle\alpha(j)m| \quad . \qquad (11\text{-}114)$$

In terms of the <u>projection tensor operator</u> P^{jj}_{α}, defined with the above spherical components, the closure relation for the states $|\alpha[j]m\rangle$ takes the form :

$$\mathbf{1} = \sum_{\alpha,j,m} P_{\alpha}{}^{[j](j)}_{m \;\; m} = \sum_{\alpha,j,m} c_j (-1)^{j+m} P_{\alpha}{}^{[j][j]}_{m \;\; -m}$$
$$= \sum_{\alpha,j} c_j (-1)^{2j} \, \hat{j} \, P^{\{jj\}0}_{\alpha} \quad , \qquad (11\text{-}115)$$

where we have used the special value of the C.G. coefficient $\langle jjm-m|00\rangle$ given by Eq. (G–4b).

Let us now establish the following expression :

$$\sum_{\alpha} P_{\alpha}{}^{[j](j)}_{m \;\; m'} = \frac{2j+1}{16\pi^2} \int_{SU(2)} d\mu(U) \, \mathcal{D}^{j*}_{mm'}(U) \, \mathcal{D}(U) \quad . \qquad (11\text{-}116)$$

Introducing the identity operator under the form of the closure relation, Eq. (11–111), one obtains :

$$\forall\, U \in SU(2) \qquad \mathcal{D}(U) = \sum_{\alpha,j,m} \mathcal{D}^{j}_{m'm}(U)\, |\alpha[j]m'\rangle\langle\alpha(j)m| \quad , \qquad (11\text{-}117)$$

The orthogonality relation, Eq. (7–50a) then implies the announced result. As special cases, the projection operators P_{jm} and P_j on the subspaces characterized, respectively, by the quantum numbers j,m, and j alone read :

$$P_{jm} = \frac{2j+1}{16\pi^2} \int_{SU(2)} d\mu(U)\, \mathcal{D}^{j*}_{mm}(U)\, \mathcal{D}(U) \quad , \qquad (11\text{-}118a)$$

$$P_{j} = \frac{2j+1}{16\pi^2} \int_{SU(2)} d\mu(U)\, \chi^{j}(U)\, \mathcal{D}(U) \quad , \qquad (11\text{-}118b)$$

where the character $\chi^j(U)$ is given by Eq. (6–47). If one considers an even representation of SU(2), and therefore only integer values of j, the integration can be carried over SO(3) changing the factor 16 into 8 in both equations above. A detailed study of the projection methods is for instance given by Raynal[48)].

v) **Examples of two–body interactions.** Let us give three examples of spin dependent two–body interactions commonly used for spin particles :

$$V_1(\mathbf{r})\, \vec{S}_1.\vec{S}_2 \quad , \qquad V_2(\mathbf{r})\, \vec{L}.\vec{S}$$

$$V_3(\mathbf{r})T_{12} = V_3(\mathbf{r})\, [3(\vec{S}_1.\hat{\mathbf{r}})(\vec{S}_2.\hat{\mathbf{r}}) - \vec{S}_1.\vec{S}_2] \quad ,$$

where $\vec{\mathbf{r}}$ is the relative position, $\hat{\mathbf{r}} = \frac{\vec{\mathbf{r}}}{\mathbf{r}}$, $\vec{\mathbf{p}}$ is the relative momentum, $\vec{L}=\vec{\mathbf{r}}\times\vec{\mathbf{p}}$ is the relative orbital angular momentum, $\vec{S}_1$ and $\vec{S}_2$ are the spins of each particle, $\vec{S} = \vec{S}_1+\vec{S}_2$, and $\vec{J} = \vec{L}+\vec{S}$ is the total angular momentum. The spin dependent operators can be written as :

$$\vec{S}_1.\vec{S}_2 = \frac{1}{2}\,(\vec{S}^2-\vec{S}_1^2-\vec{S}_2^2) = c^{-2}\sqrt{3}\,\{S_1{}^1 \otimes S_2{}^1\}^0 \quad , \qquad (11\text{-}119a)$$

$$\vec{L}.\vec{S} = \frac{1}{2}\,(\vec{J}^2-\vec{L}^2-\vec{S}^2) = c^{-2}\sqrt{3}\,\{L^1 \otimes S^1\}^0 \quad , \qquad (11\text{-}119b)$$

$$\begin{aligned} T_{12} &= c^{-4}\, 9\, \{\{S_1{}^1 \otimes \hat{r}^1\}^0 \times \{S_2{}^1 \otimes \hat{r}^1\}^0\}^0 - S_1.S_2 \\ &= c^{-4}\, 9 \sum_{\ell=0}^{2} \begin{bmatrix} 1 & 1 & 0 \\ 1 & 1 & 0 \\ \ell & \ell & 0 \end{bmatrix} \{\{S_1{}^1 \otimes S_2{}^1\}^{\ell} \otimes \{\hat{r}^1 \otimes \hat{r}^1\}^{\ell}\}^0 - \vec{S}_1.\vec{S}_2 \\ &= c^{-4}\, 3\sqrt{5}\, \{\{S_1{}^1 \otimes S_2{}^1\}^2 \otimes \{\hat{r}^1 \otimes \hat{r}^1\}^2\}^0 \quad , \end{aligned} \qquad (11\text{-}119c)$$

where we have used Eq. (8–213) and for T_{12} a 9j–type recoupling transformation, Eqs. (8–171a,171b). In the expression for T_{12} the term $\ell = 1$ vanishes since $\hat{\mathbf{r}}^1 \otimes \hat{\mathbf{r}}^1$ is a symmetrical tensor, Eq. (8–204) ; the value of the square 9j symbol is given in Table G.6. The interaction $V_3(\mathbf{r})\, T_{12}$ is called the tensor force. The three interactions considered above are rotationally invariant.

11.3. MATRIX ELEMENTS

11.3.1. THE WIGNER–ECKART THEOREM IN QUANTUM MECHANICS

i) **Theorem.** The laws of transformation under the representation $\mathcal{D}[SU(2)]$ of the bra $<(j_1)m_1|$, of the irreducible tensor operator $A^{[j_2]}_{m_2}$, and of the ket $|[j_3]m_3>$ (forgetting other possible quantum numbers) imply that the matrix element $<(j_1)m_1|A^{[j_2]}_{m_2}|[j_3]m_3>$ is a spherical component of a tensor $T^{j_1j_2j_3}$ with the following tensorial character :

$$T^{(j_1)[j_2][j_3]}_{m_1\ m_2\ m_3} = <(j_1)m_1|A^{[j_2]}_{m_2}|[j_3]m_3> \quad . \tag{11-120}$$

Furthermore, since the considered representation is unitary this tensor is clearly invariant (cf. Wigner's theorem, Subsec. 3.2.2), i.e. for all U in SU(2) one has :

$$\begin{aligned} T^{(j_1)[j_2][j_3]}_{m_1\ m_2\ m_3} &= [<(j_1)m_1|\mathcal{D}(U)^+][\mathcal{D}(U)A^{[j_2]}_{m_2}\mathcal{D}(U)^{-1}][\mathcal{D}(U)|[j_3]m_3>] \\ &= U_T\, T^{(j_1)\ [j_2][j_3]}_{m_1\ \ m_2\ m_3} \quad . \end{aligned} \tag{11-121}$$

Therefore, the general study of invariant tensors made in Subsec. 8.3.5 applies. The Wigner–Eckart (Theorem 8.4) implies that the whole dependence of the above matrix element upon m_1,m_2, and m_3 is carried by a C.G. coefficient. The results for an invariant tensor of order three are collected in Subsec. 8.3.5.2,iii). From Eq. (8–144c) one has :

$$<(j_1)m_1|A^{[j_2]}_{m_2}|[j_3]m_3> = c_{j_1}\frac{1}{\hat{j}_1}<j_2j_3m_2m_3|j_1m_1>\ T^{\{\{j_1j_2\}j_3j_3\}0} \quad . \tag{11-122}$$

The scalar $T^{\{\{j_1j_2\}j_3j_3\}0}$ is called a reduced matrix element, we denote it by :

$$[j_1\|A^{j_2}\|j_3] = T^{\{\{j_1j_2\}j_3,j_3\}0} = T^{\{j_1\{j_2j_3\}j_1\}0} \quad , \tag{11-123}$$

where the right–hand equality follows from Eq. (G–10). The Wigner–Eckart theorem for the rotation group in Quantum Mechanics then reads :

$$<(j_1)m_1|A^{[j_2]}_{m_2}|[j_3]m_3> = c_{j_1}\frac{1}{\hat{j}_1}<j_2j_3m_2m_3|j_1m_1>[j_1\|A^{j_2}\|j_3] \tag{11-124a}$$

$$= c_{j_1}(-1)^{m_1+j_2-j_3}\begin{pmatrix} j_1 & j_2 & j_3 \\ -m_1 & m_2 & m_3\end{pmatrix}[j_1\|A^{j_2}\|j_3] \quad . \tag{11-124b}$$

Here we recall that c_j is equal either to 1 or to $(-1)^{2j}$, Eq. (8–79a).

ii) **Remarks about the definition of the reduced matrix elements.** It is emphasized that the reduced matrix elements defined above are equal to scalars defined by the coupling scheme $\{\{j_1j_2\}j_3j_3\}0$ or $\{j_1\{j_2j_3\}j_1\}0$ without any additional irrelevant phase factor depending upon the j_i. This choice of definition is especially convenient in calculations involving tensors of several natures, i.e. bras, kets, operators, and matrix elements. Indeed, in calculations of this type one takes at the same time advantage of (1) the possibility to perform any coupling or recoupling transformation and (2) the nature of the considered tensors. Thus, if a recoupling transformation leads to the coupling schemes $\{\{j_1j_2\}k_1j_3\}k_2$ or $\{j_1\{j_2j_3\}k_3\}k_2$ for the three indices j_1,j_2, and j_3 associated with a matrix element, i.e. an invariant tensor, the corresponding object is a reduced matrix element $[j_1 \| A^{j_2} \| j_3]$, and one has the selection rules $k_2=0$, $k_1=j_3$, and $k_3=j_1$. This method is illustrated below in Subsecs. 11.3.2 , and 11.3.3.3,i),ii).

It should be noticed that on choosing $c_j=1$, one gets rid of all superfluous phase factors in the Wigner–Eckart theorem. This is quite convenient in practical applications to avoid any sign errors which are always very likely.

Let us now compare our definition of the reduced matrix elements with that of some other authors :

$$[j_1 \| A^{j_2} \| j_3] = c_{j_1}(-1)^{j_2+j_3-j_1} \langle j_1 \| A_{j_2} \| j_3\rangle \quad \text{- Bohr and Mottelson}^{8)}\text{, Eq. (1A-60)}$$

$$= \text{id.} \quad (j_1 \| A^{[j_2]} \| j_3) \quad \text{- Fano and Racah}^{24)}\text{, Eq. (14.4)}$$

$$= c_{j_1}(-1)^{j_1+j_2-j_3} (j_1 \| A(j_2) \| j_3) \quad \text{- Edmonds}^{22)}\text{, Eq. (5.4.1)}$$

$$= \text{id.} \quad \langle j_1 \| A^{(j_2)} \| j_3\rangle \quad \text{- Messiah}^{39)}\text{, Eq. (C.84)}$$

$$= \text{id.} \quad (j_1 \| A^{(j_2)} \| j_3) \quad \text{- de Shalit and Talmi}^{56)}\text{, Eq. (14.15)}$$

$$= c_{j_1}(-1)^{j_2+j_3-j_1} \hat{\jmath}_1 (j_1 \| A_{j_2} \| j_3) \quad \text{- Rose}^{54)}\text{, Eq. (5.14)}$$

11.3.2. EXPRESSION FOR ANY IRREDUCIBLE TENSOR OPERATOR

From the expression of the closure relation given in Eq. (11–115), the spherical component operator $A^{[j_2]}_{m_2}$ of any irreducible tensor operator reads :

$$A^{[j_2]}_{m_2} = \sum_{\alpha_1,j_1,\alpha_3,j_3} c_{j_1}c_{j_3}(-1)^{2j_1+2j_3}\hat{\jmath}_1\hat{\jmath}_3\, P^{\{j_1j_1\}0}_{\alpha_1}\, A^{[j_2]}_{m_2}\, P^{\{j_3j_3\}0}_{\alpha_3} \quad . \tag{11-125}$$

Thanks to a recoupling transformation, the diagram of which is given below (Sec. 9.2), one can put together the indices j_1,j_2, and j_3 associated with a matrix element and hence coupled to zero. One thereby finds in terms of square 6j symbols :

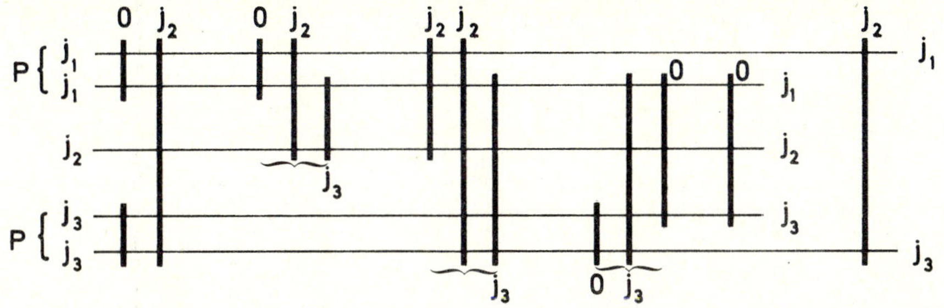

$$A^{[j_2]}_{m_2} = \sum_{\alpha_1,j_1,\alpha_3,j_3} c_{j_1} c_{j_2} (-1)^{2j_1+2j_3} \hat{j}_1\hat{j}_3 \begin{bmatrix} j_1 & j_1 & 0 \\ j_2 & j_2 & j_3\end{bmatrix}\begin{bmatrix} j_1 & j_3 & j_2 \\ 0 & j_2 & j_3\end{bmatrix}\begin{bmatrix} j_3 & j_3 & 0\\ j_3 & j_3 & 0\end{bmatrix}$$
$$[\alpha_1 j_1 \| A^{[j_2]}_{m_2} \| \alpha_3 j_3]\; P^{\{j_1j_3\}[j_2]}_{\alpha_1\alpha_3\;\;m_2} \quad , \qquad (11\text{-}126)$$

where the generalized projection tensor operator is defined in Eq. (11–112). Finally, using Eq. (8–84) and Table G.3, one finds the following expression for any irreducible tensor operator :

$$A^{j_2} = c_{j_2} \frac{1}{\hat{j}_2} \sum_{\alpha_1,j_1,\alpha_3,j_3} (-1)^{j_1-j_2+j_3} [\alpha_1 j_1 \| A^{j_2} \| \alpha_3 j_3]\; P^{\{j_1j_3\}j_2}_{\alpha_1\alpha_3} \quad . \qquad (11\text{-}127)$$

11.3.3. REDUCED MATRIX ELEMENTS

11.3.3.1. Complex conjugate of a reduced matrix element

The law of transformation of a matrix element under the complex conjugation, Eq. (11–85), and the definition of the Hermitian conjugate of an irreducible tensor operator, Eq (8–112), imply that :

$$\langle (j_1)m_1 | A^{[j_2]}_{m_2} | [j_3]m_3\rangle^* = c_{j_2} (-1)^{j_2+m_2} \langle (j_3)m_3 | A^{+[j_2]}_{-m_2} | [j_1]m_1\rangle \quad .$$

Applying the Wigner–Eckart theorem on both sides of the above equation and taking into account the reality of the C.G. coefficients yield :

$$c_{j_1} \frac{1}{\hat{j}_1} \langle j_2 j_3 m_2 m_3 | j_1 m_1\rangle\, [j_1 \| A^{j_2} \| j_3]^* = c_{j_2} c_{j_3} \frac{(-1)^{j_2+m_2}}{\hat{j}_3} \langle j_2 j_1 -m_2 m_1 | j_3 m_3\rangle [j_3 \| A^{+j_2} \| j_1] .$$

From the symmetry property, Eq. (G–2d), of the C.G. coefficients one thereby finds :

$$[j_1 \| A^{j_2} \| j_3]^* = (-1)^{j_1-j_2-j_3} [j_3 \| A^{+j_2} \| j_1] \quad , \qquad (11\text{-}128)$$

where we have used Eq. (8–84). If A^{j_2} is a Hermitian tensor, Eq. (11–97), one has :

$$A^{j_2} = A^{+j_2} \Rightarrow [j_1 \| A^{j_2} \| j_3]^* = (-1)^{j_1-j_2-j_3} [j_3 \| A^{j_2} \| j_1] \quad . \qquad (11\text{-}129)$$

11.3.3.2. Some examples of reduced matrix elements

Let us now compute some frequently encountered reduced matrix elements. The method is always the same, i.e. one evaluates the ordinary matrix element, with possibly a judicious choice of the values of m_1, m_2, and m_3, and then, from the Wigner–Eckart theorem Eq. (11–124a), one has :

$$[\alpha_1 j_1 \| A^{j_2} \| \alpha_3 j_3] = c_{j_1}\, \hat{j}_1 \,\frac{\langle \alpha_1 (j_1) m_1 | A^{[j_2]}_{m_2} | \alpha_3 [j_3] m_3 \rangle}{\langle j_2 j_3 m_2 m_3 | j_1 m_1 \rangle} \quad , \qquad (11\text{-}130)$$

assuming the C.G. coefficient is nonzero. We denote by $\{|\alpha[j]m\rangle\}$ a complete eigenbasis of $\vec{J}^2$, J_3, and other operators commuting with them and with eigenvalues specified by α .

i) **The identity operator.** It is a scalar operator, the matrix elements of which are given by the metric tensor for the considered basis :

$$\langle \alpha (j) m | \mathbf{1} | \alpha' [j'] m' \rangle = \delta_{\alpha\alpha'}\, \delta_{jj'}\, \delta_{mm'} \quad .$$

Since $\langle 0j0m | jm \rangle = 1$, one gets :

$$[\alpha j \| \mathbf{1} \| \alpha' j'] = c_j\, \hat{j}\, \delta_{\alpha\alpha'}\, \delta_{jj'} \quad . \qquad (11\text{-}131)$$

ii) **The square of the angular momentum operator.** The Casimir operator $\vec{J}^2$ is also a scalar operator since, Eqs. (8–213,217) :

$$\vec{J}^2 = c^{-2}\sqrt{3}\, \{J^1 \otimes J^1\}^0 \quad . \qquad (11\text{-}132)$$

From :

$$\langle \alpha (j) m | \vec{J}^2 | \alpha' [j'] m' \rangle = j(j+1)\, \delta_{\alpha\alpha'}\, \delta_{jj'}\, \delta_{mm'}\, \hbar^2 \quad ,$$

one finds :

$$[\alpha j \| \vec{J}^2 \| \alpha j'] = c_j\, \hat{j}\, j(j+1)\, \delta_{\alpha\alpha'}\, \delta_{jj'} \hbar^2 \quad . \qquad (11\text{-}133)$$

iii) **The angular momentum operator.** It is a special case of a vector operator J^1. The calculation of the matrix element is simplified by taking $J^{[1]}_0$. From Eq. (8–217) :

$$\langle \alpha (j) m | J^{[1]}_0 | \alpha' [j] m' \rangle = c(-i) m\, \delta_{\alpha\alpha'}\, \delta_{jj'}\, \delta_{mm'}\, \hbar \quad ,$$

and from Table G.2 :

$$\langle 1j0m | jm \rangle = -\frac{m}{\sqrt{j(j+1)}} \quad , \qquad (11\text{-}134)$$

hence one obtains :

$$[\alpha j \| J^1 \| \alpha' j'] = c\, c_j i\, \hat{j}\, \sqrt{j(j+1)}\; \delta_{\alpha\alpha'}\, \delta_{jj'}\, \hbar \quad . \tag{11-135}$$

For the spin $\frac{1}{2}$ one often considers the vector operator $\vec{\sigma} = \frac{2\vec{S}}{\hbar}$ Its reduced matrix elements then read :

$$[\alpha \tfrac{1}{2} \| \sigma^1 \| \alpha' \tfrac{1}{2}] = c\, c_{\frac{1}{2}}\, i \sqrt{6}\; \delta_{\alpha\alpha'} \quad . \tag{11-136}$$

The reduced matrix elements of J^1 allow us to express the characteristic equations of a standard basis, Eqs. (5–58,59) in terms of a particular C.G. coefficient. Indeed, from Eq. (8–217) and applying the Wigner–Eckart theorem to the matrix elements of $J_\pm$ and J_3, one obtains :

$$<\alpha(j)m|J_\pm|\alpha'[j']m'> \;=\; \pm \sqrt{2j(j+1)}\; <1j'\pm 1m'|jm>\; \delta_{\alpha\alpha'}\delta_{jj'}\,\hbar \quad , \tag{11-137a}$$

$$<\alpha(j)m|J_3|\alpha'[j']m'> \;=\; -\sqrt{j(j+1)}\; <1j'0m'|jm>\; \delta_{\alpha\alpha'}\delta_{jj'}\,\hbar \quad , \tag{11-137b}$$

and therefore :

$$J_\pm|\alpha[j]m> \;=\; \pm\sqrt{2j(j+1)}\; <1j\pm 1m|j\; m\pm 1>|\alpha[j]m\pm 1>\,\hbar \quad , \tag{11-138a}$$

$$J_3|\alpha[j]m> \;=\; -\sqrt{j(j+1)}\; <1j0m|jm>|\alpha[j]m>\,\hbar \quad . \tag{11-138b}$$

One checks using Table G.2 that these expressions are in agreement with Eqs. (5–58,59).

iv) **The gradient operator.** This is another example of a vector operator ∇^1. Its reduced matrix elements have already been derived from the gradient formula (Subsec. 10.3.3.5,v)). Namely, one has :

$$<r(\ell')m'|\nabla^{[1]}_\mu|g[\ell]m> = \int_{S^2} \sin\theta d\theta d\varphi\; Y^{[\ell']*}_{m'}(\theta,\varphi)\; \nabla^{[1]}_\mu\; g(r)\; Y^{[\ell]}_m(\theta,\varphi)$$

$$= \frac{1}{\hat{\ell}'}\; <1\ell\mu m|\ell'm'>[r\ell' \| \nabla^1 \| g\ell] \quad , \tag{11-139}$$

where the expressions of the reduced matrix element are listed in Table 10.2 (Subsec. 10.3.4.2,iii)).

v) **Reduced matrix element of "three rotation matrix elements".** One considers the matrix elements of the multiplicative operator $\hat{\mathcal{D}}^{j_2}_{m_2m_2'}$ (Subsecs. 10.3.1.3,ii) and 11.2.2.2,ii)) in the basis of functions $\mathcal{D}^{j*}_{mm'}$ in $L^2(SU(2),d\mu)$. From Eq. (7–53) one has :

$$<m_1'(j_1)m_1|\mathcal{D}^{j_2*}_{m_2m_2'}|m_3'[j_3]m_3> = \int_{SU(2)} d\mu(U)\mathcal{D}^{j_1}_{m_1m_1'}(U)\mathcal{D}^{j_2*}_{m_2m_2'}(U)\mathcal{D}^{j_3*}_{m_3m_3'}(U)$$

$$= \frac{16\pi^2}{2j_1+1}<j_2j_3m_2m_3|j_1m_1><j_2j_3m_2'm_3'|j_1m_1'> = \frac{c_{j_1}}{\hat{j}_1}<j_2j_3m_2m_3|j_1m_1>[m_1'j_1\|\mathcal{D}^{j_2*}_{m_2'}\|m_3'j_3], \tag{11-140}$$

and therefore :

$$[m_1' j_1 \| \mathcal{D}^{j_2 *}_{m_1'} \| m_3' j_3] = c_{j_1} \frac{16\pi^2}{\hat{j}_1} <j_2 j_3 m_2' m_3' | j_1 m_1'> \quad . \tag{11-141}$$

vi) Reduced matrix element of "three spherical harmonics". One now considers the matrix elements of the multiplicative operators $\hat{Y}^{[\ell_2]}_{m_2}$ in the basis of functions $Y^{[\ell]}_m$ in $L^2(S^2, d_2\hat{n})$:

$$<\alpha_1(\ell_1)m_1|\hat{Y}^{[\ell_2]}_{m_2}|\alpha_3[\ell_3]m_3> = \delta_{\alpha_1\alpha_3} \int_{S^2} \sin\theta d\theta d\varphi \, Y^{[\ell_1]*}_{m_1}(\theta,\varphi) Y^{[\ell_2]}_{m_2}(\theta,\varphi) Y^{[\ell_3]}_{m_3}(\theta,\varphi)$$

$$= \frac{1}{\hat{\ell}_1} <\ell_2\ell_3 m_2 m_3|\ell_1 m_1>[\alpha_1\ell_1 \| \hat{Y}^{\ell_2} \| \alpha_3\ell_3] \quad . \tag{11-142}$$

It follows from the relation between the functions $\mathcal{D}^{\ell *}_{m0}$ and $Y^{[\ell]}_m$, Eq. (10–250), and also between the invariant measures on SU(2) and on S^2, Eq. (10–228), that the above matrix element is related to the one of "three rotation matrix elements" by :

$$[\alpha_1\ell_1 \| \hat{Y}^{\ell_2} \| \alpha_3\ell_3] = (-ic)^{\ell_2+\ell_3-\ell_1} \frac{\hat{\ell}_1\hat{\ell}_2\hat{\ell}_3}{(4\pi)^{5/2}} [0\ell_1 \| \hat{\mathcal{D}}^{\ell_2 *}_{0} \| 0\ell_3]\delta_{\alpha_1\alpha_3} \quad ; \tag{11-143}$$

hence :

$$[\alpha_1\ell_1 \| \hat{Y}^{\ell_2} \| \alpha_3\ell_3] = (-ic)^{\ell_2+\ell_3-\ell_1} \frac{\hat{\ell}_2\hat{\ell}_3}{\sqrt{4\pi}} <\ell_2\ell_3 00|\ell_1 0> \delta_{\alpha_1}\delta_{\alpha_3} \tag{11-144a}$$

$$= (-ic)^{\ell_2+\ell_3-\ell_1} (-1)^{\ell_2+\ell_3} \frac{\hat{\ell}_1\hat{\ell}_2\hat{\ell}_3}{\sqrt{4\pi}} \begin{pmatrix} \ell_1 & \ell_2 & \ell_3 \\ 0 & 0 & 0 \end{pmatrix} \delta_{\alpha_1\alpha_3} \, . \tag{11-144b}$$

The symmetry property, Eq. (G–2b), of the C.G. coefficient implies that $\ell_1+\ell_2+\ell_3$ be even. The value of the C.G. coefficient $<\ell_2\ell_3 00|\ell_1 0>$ is given in Eq. (G–5).

vii) Reduced matrix element of the coupled generalized projection tensor operator. The result can be obtained either by computing an ordinary matrix element of a spherical component of this operator using Eq. (11–113), or more simply from Eq. (11–127). One thereby finds :

$$[\alpha j \| P^{\{j_1 j_2\}J}_{\alpha_1\alpha_2} \| \alpha' j'] = c_J (-1)^{j+j'-J} \hat{J}\{j_1 j_2 J\} \delta_{\alpha\alpha_1} \delta_{jj_1} \delta_{\alpha'\alpha_2} \delta_{j'j_2} \, . \tag{11-145}$$

11.3.3.3. Reduced matrix elements of a tensor product of operators, the Landé formula

i) Reduced matrix elements in the direct product of two spaces. The aim is to express a reduced matrix element of the type :

$$M_1 = [\{j_1(1)j_2(2)\}j \| \{A^{k_1}(1) \otimes A^{k_2}(2)\}^k \| \{j_1'(1)j_2'(2)\}j'] \quad ,$$

with (1) and (2) referring to two distinct specified spaces, in terms of a product of reduced matrix elements defined on each of these two spaces. From the tensorial point of view, M_1 is a scalar which for instance corresponds to the following coupling scheme, Eq. (11–123) :

$$\{\{\{j_1j_2\}j\{k_1k_2\}k\}j'\{j_1'j_2'\}j'\}0 \qquad .$$

This coupling scheme can be modified by means of recoupling coefficients, e.g. square 9j symbols in order to put together the objects which correspond to the same space. One thus obtains :

$$\sum_{\ell_1,\ell_2,\ell} \begin{bmatrix} j_1 & j_2 & j \\ k_1 & k_2 & k \\ \ell_1 & \ell_2 & j' \end{bmatrix} \begin{bmatrix} \ell_1 & \ell_2 & j' \\ j_1' & j_2' & j' \\ \ell & \ell & 0 \end{bmatrix} \{\{\{j_1k_1\}\ell_1j_1'\}\ell\{\{j_2k_2\}\ell_2j_2'\}\ell\}0 \quad .$$

The above considerations are independent of the nature of the considered tensor, which is here a matrix element. Let us now take it into account. Thus $\{\{j_1k_1\}\ell_1j_1'\}\ell$ is associated with a matrix element in the space (1). Therefore, this object is an invariant and one must have $\ell = 0$ and $\ell_1 = j_1'$. Similarly, for the matrix element in the second space one has $\ell_2 = j_2'$. Representing the recoupling transformations by means of the graphical representation described in Sec. 9.2, one thereby derives the reduction formula :

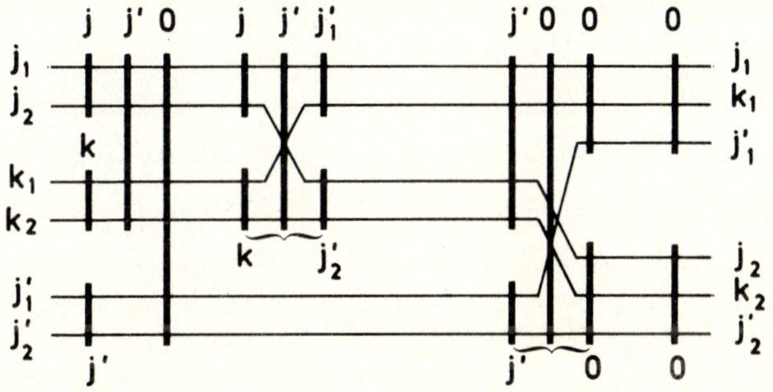

$$[\{j_1(1)j_2(2)\}j\,||\{A^{k_1}(1)\otimes A^{k_2}(2)\}^k\,||\,\{j_1'(1)j_2'(2)\}j']$$

$$= \frac{\hat{j}'}{\hat{j}_1'\hat{j}_2'} \begin{bmatrix} j_1 & j_2 & j \\ k_1 & k_2 & k \\ j_1' & j_2' & j' \end{bmatrix} [j_1||A^{k_1}||j_1'][j_2||A^{k_2}||j_2']$$

$$= \hat{j}\,\hat{k}\,\hat{j}' \begin{Bmatrix} j_1 & j_2 & j \\ k_1 & k_2 & k \\ j_1' & j_2' & j' \end{Bmatrix} [j_1||A^{k_1}||j_1']\;[j_2||A^{k_2}||j_2'] \qquad . \qquad (11\text{-}146)$$

In the special case where one of the operators A^{k_1} and A^{k_2} is the identity, one finds from above and using Table G.4 and Eq. (11–131) that :

$$[\{j_1(1)j_2(2)\}j \,\|\, A^{k_2}(2) \,\|\, \{j_1'(1)j_2'(2)\}j'] =$$

$$c_{j_1} \frac{\hat{j}'}{\hat{j}_2'} \begin{bmatrix} j_1 & j_2 & j \\ 0 & k_2 & k_2 \\ j_1 & j_2' & j' \end{bmatrix} [j_2 \| A^{k_2} \| j_2']\delta_{j_1 j_1'} = c_{j_1} (-1)^{j_1+j_2+k_2+j'} \hat{j}\hat{j}' \begin{Bmatrix} j' & j_2' & j_1 \\ j_2 & J & k_2 \end{Bmatrix} \times [j_2 \| A^{k_2} \| j_2'] \,\delta_{j_1 j_1'} \quad , \tag{11-147a}$$

$$[\{j_1(1)j_2(2)\}j \,\|\, A^{k_1}(1) \,\|\, \{j_1'(1)j_2'(2)\}j'] =$$

$$c_{j_2} \frac{\hat{j}'}{\hat{j}_1'} \begin{bmatrix} j_1 & j_2 & j \\ k_1 & 0 & k_1 \\ j_1' & j_2 & j' \end{bmatrix} [j_1 \| A^{k_1} \| j_1'] \,\delta_{j_2 j_2'} = c_{j_2} (-1)^{j_1'+k_1+j_2+j} \hat{j}\hat{j}' \begin{Bmatrix} j' & j_1' & j_2 \\ j_1 & j & k_1 \end{Bmatrix} \times [j_1 \| A^{k_1} \| j_1'] \,\delta_{j_2 j_2'} \quad . \tag{11-147b}$$

The second result can also be obtained from the first one by interchanging the subscripts 1 and 2, and then multiplying by $(-1)^{j_1+j_2-j+j_1'+j_2'-j'}$ to take into account the change of the order of coupling in the bra and the ket.

ii) Reduced matrix element of the tensor product of two tensor operators acting on the same space :

$$M_2 = [\alpha j \,\|\, \{A^{k_1} \otimes A^{k_2}\}^{k} \,\|\, \alpha' j'] \quad .$$

To express M_2 in terms of products of reduced matrix elements of the operators A^{k_1} and A^{k_2}, one introduces between these two operators the identity under the form of the closure relation given by Eq. (11–115). The reduced matrix element M_2 is then a scalar which, for instance, corresponds to the following coupling scheme :

$$\{\{j\{k_1\{j''j''\}0k_2\}k\}j'j'\}0 \quad .$$

By a recoupling transformation (see diagram below) one can put together the indices j, k_1, j'' and j'', k_2, j'. These two groups of indices are, respectively, associated with reduced matrix elements of A^{k_1} and A^{k_2}, and therefore must be coupled to zero. Using square 6j symbols, Eq. (8–171a), one thereby finds :

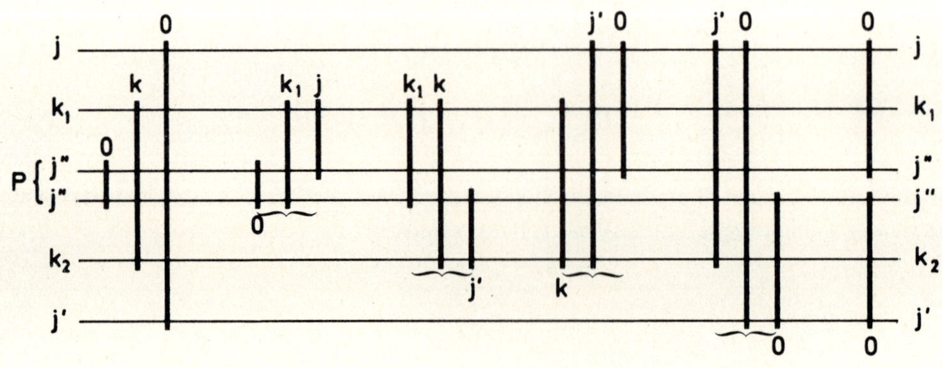

$$M_2 = \sum_{\alpha'',j''} c_{j''}(-1)^{2j''}\,\hat{j}'' \begin{bmatrix} k_1 & j'' & j \\ j'' & k_1 & 0 \end{bmatrix} \begin{bmatrix} j & j'' & k_1 \\ k_2 & k & j' \end{bmatrix} \begin{bmatrix} j & j & 0 \\ j' & j' & k \end{bmatrix} \begin{bmatrix} 0 & j' & j' \\ j' & 0 & 0 \end{bmatrix} [\alpha j \| A^{k_1} \| \alpha'' j''][\alpha'' j'' \| A^{k_2} \| \alpha' j'] \quad ,$$

and finally from Table G.3 :

$$[\alpha j \| \{A^{k_1} \otimes A^{k_2}\}^k \| \alpha j']$$

$$= \sum_{\alpha'',j''} c_{j''}(-1)^{j'+k_1-k-j''} \frac{\hat{k}}{\hat{k}_1 \hat{j}'} \begin{bmatrix} j & j'' & k_1 \\ k_2 & k & j' \end{bmatrix} [\alpha j \| A^{k_1} \| \alpha'' j''][\alpha'' j'' \| A^{k_2} \| \alpha' j'] \quad ,$$

$$= (-1)^{j+k_1+k_2+j'}\,\hat{k} \sum_{\alpha'',j''} c_{j''} \begin{Bmatrix} j & j'' & k_1 \\ k_2 & k & j' \end{Bmatrix} [\alpha j \| A^{k_1} \| \alpha'' j''][\alpha'' j'' \| A^{k_2} \| \alpha' j'] . \quad (11\text{-}148)$$

11.3.3.4. The Landé formula. In the special case $k_1 = k_2$ and $k=0$ the above reduction formula, Eq. (11–148), becomes :

$$[\alpha j \| \{A^k B^k\}^0 \| \alpha' j'] = \delta_{jj'} \sum_{\alpha'',j''} c_{j''}(-1)^{j+k-j''} \frac{1}{\hat{j}\hat{k}} [\alpha j \| A^k \| \alpha'' j''][\alpha'' j'' \| B^k \| \alpha' j'] \ . \quad (11\text{-}149)$$

Let us apply this result to the angular momentum operator J^1 and to any vector operator A^1. Using Eq. (8–213) and the value of the reduced matrix element of J^1, Eq. (11–135), one finds :

$$[\alpha j \| \vec{J}.\vec{A} \| \alpha' j'] = c^{-2}\sqrt{3}\,[\alpha j \| \{J^1 \otimes A^1\}^0 \| \alpha' j']$$

$$= c^{-1}\,(-i)\,\sqrt{j(j+1)}\,[\alpha j \| A^1 \| \alpha' j]\,\delta_{jj'}\,\hbar \ . \quad (11\text{-}150)$$

Setting $A^1 = J^1$ in the above equation one gets the reduced matrix element of $\vec{J}^2$ given in Eq. (11–133).

The Eq. (11–150) allows us to derive the Landé formula as follows. Since the C.G. coefficient which occurs in the Wigner–Eckart theorem for both matrix elements is the same, one has :

$$\frac{\langle \alpha(j)m | A^{[1]}_\mu | \alpha'[j\ ]m' \rangle}{\langle \alpha(j)m | J^{[1]}_\mu | \alpha\ [j\ ]m' \rangle} = \frac{[\alpha j \| A^1 \| \alpha' j\ ]}{[\alpha j \| J^1 \| \alpha\ j\ ]} = \frac{1}{j(j+1)} \frac{c_j}{\hat{j}} [\alpha j \| \vec{J}.\vec{A} \| \alpha' j\ ]\,\hbar^{-2} \ , \quad (11\text{-}151)$$

and therefore one obtains the Landé formula :

$$\langle \alpha(j)m | A^{[1]}_\mu | \alpha'[j\ ]m' \rangle = \frac{1}{j(j+1)} \frac{c_j}{\hat{j}} [\alpha j \| \vec{J}.\vec{A} \| \alpha' j\ ] \ \langle \alpha(j)m | J^{[1]}_\mu | \alpha\ [j\ ]m' \rangle\,\hbar^{-2}$$

$$= \frac{\langle \alpha(j)m | \vec{J}.\vec{A} | \alpha'[j\ ]m \rangle}{j(j+1)} \langle \alpha(j)m | J^{[1]}_\mu | \alpha\ [j\ ]m' \rangle\,\hbar^{-2} \ . \quad (11\text{-}152)$$

This formula states that within a subspace with given α and j any vector operator $\vec{A}$ is a multiple of the angular momentum operator with the proportionality factor

$\frac{1}{j(j+1)} <\alpha(j)m|\vec{J}.\vec{A}|\alpha\ [j]m> \hbar^{-2}$.

Applying the Landé formula to the special vector operator $J^1(\vec{J}.\vec{A})$ one finds :

$$<\alpha(j)m|J^{[1]}_{\mu}(\vec{J}.\vec{A})|\alpha'[j\ ]m'> = \frac{<\alpha(j)m|\vec{J}^2(\vec{J}.\vec{A})|\alpha'[j\ ]m>}{j(j+1)} <\alpha(j)m|J^{[1]}_{\mu}|\alpha\ [j\ ]m'>\hbar^{-2}$$

$$= <\alpha(j)m|\vec{J}.\vec{A}|\alpha'[j\ ]m><\alpha(j)m|J^{[1]}_{\mu}|\alpha\ [j\ ]m'> , \quad (11\text{-}153)$$

and using again the Landé formula :

$$<\alpha(j)m|A^{[1]}_{\mu}|\alpha'[j\ ]m'> = \frac{\hbar^{-2}}{j(j+1)} <\alpha(j)m|J^{[1]}_{\mu}(\vec{J}.\vec{A})|\alpha'[j\ ]m'> . \quad (11\text{-}154)$$

The meaning of this result is the following. The operator $\frac{\hbar^{-2}}{j(j+1)} J^{[1]}_{\mu}(\vec{J}.\vec{A})$ is the component of $\vec{A}$ along $\vec{J}$ in a subspace with a given j. Eq. (11–154) then states that only this component of $\vec{A}$ contributes to the expectation values of this vector operator. This corresponds to the precession of $\vec{A}$ around $\vec{J}$ implying that the expectation value of the component of $\vec{A}$ perpendicular to $\vec{J}$ vanishes.

As an example of application of the Landé formula let us calculate the expectation value of the magnetic moment in an eigenvector $|\ell s[j]m>$ of $\vec{L}^2,\vec{S}^2,\vec{J}^2$, and J_3. The magnetic moment reads :

$$\vec{\mu} = \mu_o\ (g_{\ell}\vec{L} + g_s\vec{S}) , \quad (11\text{-}155)$$

where μ_o is some magneton unit and g_{ℓ}, g_s are the corresponding magnetic ratios. One has :

$$<\ell s(j)m|\mu^{[1]}_{m'}|\ell s[j]m> = \frac{<\ell s(j)m|\vec{J}.\vec{\mu}|\ell s[j]m>}{j(j+1)} <\ell s(j)m|J^{[1]}_{m'}|\ell s[j]m>\hbar^{-2} . \quad (11\text{-}156)$$

From :

$$\vec{J} = \vec{L}+\vec{S} , \quad \vec{J}.\vec{\mu} = \frac{\mu_o}{2} [(g_{\ell}-g_s)\vec{L}^2+(g_s-g_{\ell})\vec{S}^2 + (g_{\ell}+g_s)\vec{J}^2] ,$$

$$<\ell s(j)m|J^{[1]}_{m'}|\ell s[j]m> = c\ (-i)\ m\ \delta_{m'0}\ \hbar , \quad (11\text{-}157)$$

one finds :

$$<\ell s(j)m|\mu^{[1]}_{m'}|\ell s[j]m> = -ic\mu_o\hbar m \left[(g_{\ell}-g_s)\ \frac{\ell(\ell+1)-s(s+1)}{2j(j+1)} + \frac{1}{2}\left(g_{\ell}+g_s\right)\right] \delta_{m'0} , \quad (11\text{-}158)$$

where the quantity in square brackets is called the Landé factor.

11.4. THE RIGID ROTATOR IN QUANTUM MECHANICS

Notation : throughout this section the prime denotes intrinsic basis vectors or components with respect to them.

A rigid body is defined in Classical Mechanics as a discrete or a continuous set of massive points subject to the constraints that the distances between all pairs of points remain constant throughout the motion. In order to take into account these constraints one is led to introduce generalized coordinates associated with each degree of freedom. A configuration of a rigid body is completely specified by locating in the three-dimensional point space an orthonormal reference frame tied to the rigid body, and called the intrinsic or body-fixed reference frame. With respect to any fixed orthonormal reference frame, one needs three coordinates to determine the origin of the intrinsic axes, and furthermore three other parameters to specify the orientation of the intrinsic axes with respect to the fixed ones. These three latter generalized coordinates may for instance be the parameters which determine the rotation bringing the fixed orthonormal basis onto the intrinsic one. We consider only the force-free rotational motion without friction of a rigid body (or rotator) about its centre of mass, also known as the motion "à la Poinsot". In order to define a rigid rotator in Quantum Mechanics one retains only the fact that a state of such a system is specified by a rotation, e.g. by the three Euler angles. Some of the main features of the motion of a rigid rotator in Classical Mechanics are recalled in Subsec. 11.4.1. The description of such a system in Quantum Mechanics is given in Subsec. 11.4.2 and its dynamics in Subsec. 11.4.3.

11.4.1. SOME REMINDERS CONCERNING THE CLASSICAL MECHANICS OF A RIGID ROTATOR

For more details concerning this problem cf. for instance Goldstein[27], Chaps. 4,5.

i) **Kinematics.** Let $\vec{\omega}$ denote the angular velocity, with respect to a Galilean orthonormal basis a, of a rigid body rotating about its fixed centre of mass O. The velocity $\vec{v}$ of any point M of this system reads :

$$\vec{OM} = \vec{r} \qquad , \qquad \vec{v} = \vec{\omega} \times \vec{r} \tag{11-159}$$

The total angular momentum about the point O is then given by :

$$\vec{L} = \int \delta m \; \vec{r} \times \vec{v} = \int \delta m \; \vec{r} \times (\vec{\omega} \times \vec{r}) = \int \delta m \; [\vec{\omega}\, r^2 - \vec{r}(\vec{r}.\vec{\omega})] \; , \tag{11-160}$$

where the Stieltjes integral is extended over the whole rigid body, and Eq. (2-26) has been used to express the double vector product. Hence, the angular momentum is related to the angular velocity by a linear mapping I such that :

$$\vec{L} = I\,\vec{\omega} \quad . \tag{11-161}$$

I is called the inertia tensor and it is characteristic of the considered rotator. With respect to any orthonormal basis (e_i) it is determined as follows :

$$\vec{r} = x^i e_i \quad , \quad \vec{\omega} = \omega^i e_i \quad , \quad \vec{L} = \sum_i L_i e_i \quad ,$$

$$L_i = I_{ij} \omega^j \quad , \quad I_{ij} = \int \delta m \, (r^2 \delta_{ij} - x^i x^j) \quad . \tag{11-162}$$

The matrix (I_{ij}) being real symmetrical there exists an intrinsic orthonormal basis, denoted by $a'=(e'_i)$ which diagonalizes it. These eigenvectors determine the principal axes of inertia and the eigenvalues are the principal moments of inertia $\mathcal{I}_i$. Let us from now on denote by a prime the components of any vector with respect to a'. One then has :

$$L'_i = \mathcal{I}_i \, \omega'^i \quad , \tag{11-163}$$

without summation over i.

The time derivative of $\vec{L}$ (the dynamical momentum $\vec{K}$ about the point O) can be written as a sum of two terms :

$$\vec{K} = \frac{d\vec{L}}{dt} = \left(\frac{d\vec{L}}{dt}\right)_{int.} + \vec{\omega} \times \vec{L} = \left(\frac{d\vec{L}}{dt}\right)_{int.} + \vec{\omega} \times I \, \vec{\omega} \quad , \tag{11-164}$$

where $\left(\frac{d\vec{L}}{dt}\right)_{int}$ denotes the vector with components with respect to a' equal to the time derivatives of the components of $\vec{L}$ in this intrinsic basis. Hence, in a' the components of $\vec{K}$ read :

$$\begin{cases} K'_1 = \mathcal{I}_1 \, \dot{\omega}'^1 - \omega'^2 \omega'^3 \, (\mathcal{I}_2 - \mathcal{I}_3) \\ K'_2 = \mathcal{I}_2 \, \dot{\omega}'^2 - \omega'^3 \omega'^1 \, (\mathcal{I}_3 - \mathcal{I}_1) \\ K'_3 = \mathcal{I}_3 \, \dot{\omega}'^3 - \omega'^1 \omega'^2 \, (\mathcal{I}_1 - \mathcal{I}_2) \end{cases} \quad . \tag{11-165}$$

From Eqs. (11–159,161) the kinetic energy of motion about the point O takes the form :

$$T = \frac{1}{2} \int \delta m \, v^2 = \frac{1}{2} \int \delta m \, \vec{v}.\vec{\omega} \times \vec{r} = \frac{1}{2} \, \vec{\omega} . \int \delta m \, \vec{r} \times \vec{v}$$

$$= \frac{1}{2} \, \vec{\omega}.\vec{L} = \frac{1}{2} \, \vec{\omega}.I \, \vec{\omega} = \frac{1}{2} \, \mathcal{I} \, \omega^2 \quad , \tag{11-166}$$

where $\mathcal{I}$ is the moment of inertia about the axis of rotation. In the intrinsic basis a' one has :

$$T = \frac{1}{2} \sum_i \mathcal{I}_i (\omega'^i)^2 = \frac{1}{2} \sum_i \frac{(L'_i)^2}{\mathcal{I}_i} \quad . \tag{11-167}$$

Let R(t) be the time dependent rotation which brings the Galilean orthonormal basis $a=(e_i)$ onto the intrinsic one $a'=(e'_i(t))$:

$$a' = R(t)a \quad , \quad e'_i(t) = R(t)e_i \quad . \tag{11-168}$$

The motion of any point M of the rotator is then determined by :

$$\vec{OM} = \vec{r} = R(t)\,\vec{r}_o \quad , \tag{11-169}$$

where $\vec{r}_o$ is some initial position of $\vec{r}$. Denoting as usual the time derivative by a point, the velocity of M reads :

$$\vec{v} = \dot{R}(t)\,\vec{r}_o = \dot{R}(t)\,R(t)^{-1}\,\vec{r} = \vec{\omega}\times\vec{r} \quad . \tag{11-170}$$

Since the above relation holds for all points in the rotator, one must have the following equality between linear operators :

$$\dot{R}(t)\,R(t)^{-1} = \vec{\omega}\times \quad . \tag{11-171}$$

Setting :

$$R(t) = R(\varphi,\theta,\psi;a) \quad , \tag{11-172}$$

let us derive the expression of the components ω'^i with respect to the intrinsic basis a' of the angular velocity in terms of the Euler angles φ,θ, and ψ. From Eqs. (1–14) and (11–168) the matrix representative of $\dot{R}(t)R(t)^{-1}$ with respect to a' is equal to the matrix representative of $R(t)^{-1}\,\dot{R}(t)$ with respect to a. One then has the matrix equation :

$$(R(-\psi,-\theta,-\varphi;a)^i_k\;R(\dot{\varphi},\dot{\theta},\dot{\psi};a)^k_j) = \begin{pmatrix} 0 & -\omega'^3 & \omega'^2 \\ \omega'^3 & 0 & -\omega'^1 \\ -\omega'^2 & \omega'^1 & 0 \end{pmatrix} \quad , \tag{11-173}$$

where $(R(\varphi,\theta,\psi;a)^i_j)$ is given by Eq. (E–14). One thereby obtains the components of $\vec{\omega}$ in the intrinsic basis :

$$\left\{\begin{aligned} \omega'^1 &= -\dot{\varphi}\sin\theta\cos\psi + \dot{\theta}\sin\psi \\ \omega'^2 &= \dot{\varphi}\sin\theta\sin\psi + \dot{\theta}\cos\psi \\ \omega'^3 &= \dot{\varphi}\cos\theta + \dot{\psi} \end{aligned}\right. \tag{11-174}$$

Actually, dt times the above components ω'^i are equal to the relative components of the infinitesimal displacement of the moving reference frame of SO(3), denoted by ω^i in Eqs. (10–204a–c). This result follows from the interpretation of these relative components, which is discussed in Subsec. 10.2.4.1,ii).

ii) **Dynamics.** The so–called Euler equations of motion for a rigid body with one point fixed arise from Eq. (11–165) where the dynamical momentum $\vec{K}$ about O is equal to the total torque about the same point. In the absence of any applied torque the equations reduce to :

$$I_1\,\dot{\omega}'^1 = \omega'^2\,\omega'^3\,(I_2-I_3) \quad ,$$

$$I_2\,\dot{\omega}'^2 = \omega'^3\,\omega'^1\,(I_3-I_1) \quad , \tag{11-175}$$

$$I_3 \; \dot{\omega}'^3 = \omega'^1 \, \omega'^2 \, (I_1 - I_2) \qquad .$$

It then follows from the conservation theorems for the total energy and for the total angular momentum that there exist two immediate integrals of the motion, i.e. the kinetic energy T and the total angular momentum $\vec{L}$. The three components of $\vec{L}$ with respect to any Galilean basis a are thus constants in time. Using the above integrals of the motion one is able to integrate Eq. (11–175) completely in terms of elliptic functions, but we are not interested here in such a treatment.

A rigid rotator with $I_1 = I_2$ is also called a symmetrical top. From Eq. (11–175), the component $I_3 \omega'^3$ of $\vec{L}$ in the intrinsic basis a' is then also a constant of the motion. Similarly, if one assumes that $I_1 = I_2 = I_3$, the three components of $\vec{L}$ in a' are constant throughout the motion.

Let us now give a brief account of the Lagrangian and Hamiltonian formalisms, since they are suitable for quantization. In terms of the generalized coordinates φ, θ , and ψ the Lagrangian of a free rigid rotator reads from Eqs. (11–167,174) :

$$\begin{aligned} L(\varphi,\theta,\psi,\dot{\varphi},\dot{\theta},\dot{\psi}) &= \frac{1}{2}\, T = \frac{1}{2}\, [I_1(-\dot{\varphi}\, \sin\theta\, \cos\psi + \dot{\theta}\, \sin\psi)^2 \\ &\quad + I_2(\dot{\varphi}\, \sin\theta\, \sin\psi + \dot{\theta}\, \cos\psi)^2 + I_3(\dot{\varphi}\, \cos\theta + \dot{\psi})^2] \; . \end{aligned} \qquad (11\text{–}176)$$

The Euler–Lagrange differential equations then provide the equations of motion for the Euler angles :

$$\frac{d}{dt}\frac{\partial L}{\partial \dot{\varphi}} = 0 \quad , \qquad \frac{d}{dt}\frac{\partial L}{\partial \dot{\theta}} - \frac{\partial L}{\partial \theta} = 0 \quad , \qquad \frac{d}{dt}\frac{\partial L}{\partial \dot{\psi}} - \frac{\partial L}{\partial \psi} = 0 \quad , \qquad (11\text{–}177)$$

where the first equation takes into account the fact that the generalized coordinate φ is cyclic, i.e. it does not appear in the Lagrangian. The generalized momenta read :

$$p_\varphi = \frac{\partial L}{\partial \dot{\varphi}} \quad , \qquad p_\theta = \frac{\partial L}{\partial \dot{\theta}} \quad , \qquad p_\psi = \frac{\partial L}{\partial \dot{\psi}} \qquad . \qquad (11\text{–}178)$$

One shows (cf. for instance Ref. 27), Sec. 2–6) that the generalized momentum conjugate to the angle of a rotation about a given axis is the component of the total angular momentum along this axis of rotation. Hence, one has :

$$p_\varphi = \vec{L} \cdot e_3 \qquad , \qquad (11\text{–}179)$$

and also :

$$p_\psi = \vec{L} \cdot e_3'(t) \qquad , \qquad (11\text{–}180)$$

Since the generalized coordinate φ is a cyclic, one it follows from the Euler–Lagrange equations that the generalized momentum p_φ is conserved. This result has already been obtained from the conservation theorem of the total angular momentum. Since the Lagrangian is quadratic in $\dot{\varphi}, \dot{\theta}$, and $\dot{\psi}$, the generalized momenta are linear homogeneous functions of these variables. Inverting these expressions one thereby derives the Hamiltonian :

$$H(\varphi,\theta,\psi,p_\varphi,p_\theta,p_\psi) = \frac{1}{2}\,T \quad , \tag{11-181}$$

which is quadratic in the generalized momenta. As an example, assuming $I_1 = I_2$ one finds :

$$H = \frac{1}{2}\left[\frac{1}{I_1\sin^2\theta}\,p_\varphi{}^2 - \frac{2\cos\theta}{I_1\sin^2\theta}\,p_\varphi p_\psi + \frac{1}{I_1}\,p_\theta{}^2 + \left(\frac{1}{I_1}\cot g^2\theta + \frac{1}{I_3}\right)p_\psi{}^2\right] . \tag{11-182}$$

It should be noticed that both φ and ψ are then cyclic variables, and therefore p_ψ, as p_φ, is constant. One thus recovers the conservation law of the component of the total angular momentum along the third intrinsic axis.

11.4.2. DESCRIPTION OF A RIGID ROTATOR IN QUANTUM MECHANICS

We now closely follow the procedure used in Subsec. 11.1.2 for the description of spinless particles in Quantum Mechanics.

i) **Observables and states.** The fundamental observables which characterize a rigid rotator in Quantum Mechanics are the Euler angle operators $\boldsymbol{\varphi}, \boldsymbol{\theta}$, and $\boldsymbol{\psi}$ defined with respect to any orthonormal basis a of E_3, and the generalized momentum operators $\mathbf{p}_\varphi$, $\mathbf{p}_\theta$, and $\mathbf{p}_\psi$. These operators obey the commutation relations :

$$[\boldsymbol{\varphi},\mathbf{p}_\varphi] = i\hbar\,\mathbf{1} \quad , \quad [\boldsymbol{\theta},\mathbf{p}_\theta] = i\hbar\,\mathbf{1} \quad , \quad [\boldsymbol{\psi},\mathbf{p}_\psi] = i\hbar\,\mathbf{1} \quad , \tag{11-183}$$

all other commutators being zero. These relations follow from the canonical procedure to quantize a classical system. Namely, it amounts to replacing the Poisson brackets of any two classical variables by $\frac{1}{i\hbar}$ times the commutator of the corresponding operators in Quantum Mechanics. Using the same method as for the observables $\vec{\mathbf{r}}$ and $\vec{\mathbf{p}}$ in Subsec. 11.1.2,i) (cf. Ref.39, Chap. VIII, § 6) one shows that :

(1) The operators $\boldsymbol{\varphi}, \boldsymbol{\theta}$, and $\boldsymbol{\psi}$ form a complete set of commuting observables.

(2) Their spectra are continuous, nondegenerate and by definition extend throughout the whole parameter space of the Euler angles for the group SU(2), Eq. (4–38), see Fig. 4.3. The eigenvectors have an infinite norm, and we choose to define them with the normalization conditions :

$$\left.\begin{matrix}\boldsymbol{\varphi}\\ \boldsymbol{\theta}\\ \boldsymbol{\psi}\end{matrix}\right\}|a;\varphi\theta\psi\rangle = \left.\begin{matrix}\varphi\\ \theta\\ \psi\end{matrix}\right\}|a;\varphi\theta\psi\rangle \quad , \tag{11-184a}$$

$$\langle a;\varphi\theta\psi|a;\varphi'\theta'\psi'\rangle = \delta(\varphi-\varphi')\delta(\cos\theta-\cos\theta')\delta(\psi-\psi') = \frac{1}{\sin\theta}\,\delta(\varphi-\varphi')\delta(\theta-\theta')\delta(\psi-\psi') . \tag{11-184b}$$

The space F on which the operators $\varphi, \theta, \psi, \mathbf{p}_\varphi, \mathbf{p}_\theta$, and $\mathbf{p}_\psi$ act is by definition spanned by the vectors $|a;\varphi\theta\psi\rangle$. Hence, one has the closure relation :

$$\int_{SU(2)} d\mu(\varphi,\theta,\psi)\,|a;\varphi\theta\psi\rangle\langle a;\varphi\theta\psi| = \mathbf{1}_F \quad , \tag{11-185}$$

where $d\mu(\varphi,\theta,\psi)$ is the invariant measure on SU(2), Eq. (7–38).

(3) The matrix elements of the operators $\mathbf{p}_\varphi$, $\mathbf{p}_\theta$, and $\mathbf{p}_\psi$ in the basis ($|a;\varphi\theta\psi\rangle$) read :

$$\begin{aligned}
\langle a;\varphi\theta\psi|\mathbf{p}_\varphi|a;\varphi'\theta'\psi'\rangle &= -i\hbar\frac{1}{\sin\theta}\,\delta'(\varphi-\varphi')\delta(\theta-\theta')\delta(\psi-\psi') \quad ,\\
\langle a;\varphi\theta\psi|\mathbf{p}_\theta|a;\varphi'\theta'\psi'\rangle &= -i\hbar\frac{1}{\sin\theta}\,\delta(\varphi-\varphi')\delta'(\theta-\theta')\delta(\psi-\psi') \quad ,\\
\langle a;\varphi\theta\psi|\mathbf{p}_\psi|a;\varphi'\theta'\psi'\rangle &= -i\hbar\frac{1}{\sin\theta}\,\delta(\varphi-\varphi')\delta(\theta-\theta')\delta'(\psi-\psi') \quad .
\end{aligned} \tag{11-186}$$

The physical states are associated with normalizable vectors $|f\rangle$ in F. With respect to the basis ($|a;\varphi\theta\psi\rangle$) one has :

$$|f\rangle = \int_{SU(2)} d\mu(\varphi,\theta,\psi)\,|a;\varphi\theta\psi\rangle\, f_a(\varphi,\theta,\psi)\,,\quad \langle f|f\rangle = \int_{SU(2)} d\mu(\varphi,\theta,\psi)\,|f_a(\varphi,\theta,\psi)|^2 < \infty \,, \tag{11-187}$$

where the wave function $f_a(\varphi,\theta,\psi) = \langle a;\varphi\theta\psi|f\rangle$ belongs to the Hilbert space $L^2(SU(2),d\mu)$ of square integrable functions defined on SU(2) with the invariant measure $d\mu$. It is recalled that the set of functions $\{(\hat{j}/4\pi)\,\mathcal{D}^j_{m'm}(\varphi,\theta,\psi)\}$ forms a complete orthonormal basis of this space (Theorem 7.13, Subsec. 7.2.2.2). The representatives of the generalized momentum operators $\mathbf{p}_\varphi$, $\mathbf{p}_\theta$, and $\mathbf{p}_\psi$ in the space of wave functions are, respectively, equal to $-i\hbar\frac{\partial}{\partial\varphi}$, $-i\hbar\frac{\partial}{\partial\theta}$, and $-i\hbar\frac{\partial}{\partial\psi}$. This result follows from Eq. (11–186), e.g. for $\mathbf{p}_\varphi$ one has :

$$\begin{aligned}
\langle a;\varphi\theta\psi|\mathbf{p}_\varphi|f\rangle &= \int_{SU(2)} d\mu(\varphi',\theta',\psi')\langle a;\varphi\theta\psi|\mathbf{p}_\varphi|a;\varphi'\theta'\psi'\rangle\, f_a(\varphi'\theta'\psi')\\
&= -i\hbar\frac{\partial}{\partial\varphi} f_a(\varphi\theta\psi) \quad .
\end{aligned} \tag{11-188}$$

ii) **Determination of the representation of the rotation group.** It suffices to specify the law of transformation of the basis vectors $|a;\varphi\theta\psi\rangle$. In order to take into account the fact that the representations up to a phase of the rotation group are true representations of SU(2), the Euler angles have been chosen as those of this latter group. It is only when φ, θ, and ψ are restricted to vary as the Euler angles of a rotation, Eq. (2–39), that the classical analogue of the considered quantum system is a rigid rotator in Classical Mechanics. In this case, although a vector $|a;\varphi\theta\psi\rangle$ does not belong to the Hilbert space of physical states, since it is not normalizable, it can be interpreted as describing a rigid rotator with its intrinsic axes determined by the Euler angles with respect to the fixed basis a. This leads us to define the law of transformation of $|a;\varphi\theta\psi\rangle$ under any rotation $R=R(U;a)$ by the unitary representation $\mathcal{D}[SU(2)]$ so that :

$$\forall\, U \in SU(2) \qquad |{}^U a;\varphi\theta\psi\rangle = \mathcal{D}(U)\,|a;\varphi\theta\psi\rangle = |a;U\{\varphi,\theta,\psi\}\rangle \quad , \tag{11-189}$$

where U $\{\varphi,\theta,\psi\}$ denotes the Euler angles of the matrix U $U(\varphi,\theta,\psi)$ in SU(2), Eq. (4–38). For any state $|f\rangle$ one then has for all U in SU(2) :

$$|{}^{U}f\rangle = \mathcal{D}(U)|f\rangle = \int_{SU(2)} d\mu(\varphi,\theta,\psi)\,|{}^{U}a;\varphi\theta\psi\rangle\, f_a(\varphi,\theta,\psi)$$

$$= \int_{SU(2)} d\mu(\varphi,\theta,\psi)\,|a;\varphi\theta\psi\rangle\, f_a(U^{-1}\{\varphi,\theta,\psi\}) \quad , \qquad (11\text{-}190)$$

where the invariance of the measure has been used. One thereby obtains the law of transformation of the wave functions f_a under any rotation R(U;a) (still denoting by $\mathcal{D}(U)$ the unitary operator acting on these functions) :

$$\forall U \in SU(2) \qquad f_a \longrightarrow {}^{U}f_a = \mathcal{D}(U) f_a \quad ,$$

$$\forall \varphi,\theta,\psi \qquad {}^{U}f_a(\varphi,\theta,\psi) = f_a(U^{-1}\{\varphi,\theta,\psi\}) \quad . \qquad (11\text{-}191)$$

This result also follows directly from the unitary character of $\mathcal{D}(U)$:

$$\langle a;\varphi\theta\psi|\mathcal{D}(U)|f\rangle = \langle {}^{U^{-1}}a;\varphi\theta\psi|f\rangle \quad . \qquad (11\text{-}192)$$

Thus, for a given orthonormal basis a the above representation of SU(2) in the space of wave functions coincides with the left regular representation of this group. Therefore all the study of this representation made in Subsec. 10.3.1 applies.

iii) **Infinitesimal generators :**

1–The expressions of the infinitesimal generators J_i, $J_\pm$ and also of the Casimir operator J^2 are given from Eqs. (10–168a–d, and 171) by :

$$J_i = \hbar J_i^{\ell} \quad , \quad J_\pm = \hbar J_\pm^{\ell} \quad , \quad J^2 = \hbar^2 (J^{\ell})^2 \quad . \qquad (11\text{-}193)$$

One should note that J_3 is identical to the generalized momentum $\mathbf{p}_\varphi$:

$$J_3 = \mathbf{p}_\varphi = -\, i\hbar \frac{\partial}{\partial\varphi} \quad , \qquad (11\text{-}194)$$

in agreement with the classical property, Eq. (11–179). Identifying the Lie algebra of the considered representation of SU(2) with the Euclidean vector space E_3 (Subsec. 5.4.3,i)) we are led to define the angular momentum operator $\vec{J}$ as the vector in E_3 with components J_i with respect to an orthonormal basis $a=(e_i)$ of this space :

$$\vec{J} = \sum_i J_i\, e_i \quad . \qquad (11\text{-}195)$$

2–The intrinsic infinitesimal generators I_i are defined from the intrinsic infinitesimal generators I_i^{ℓ} of the left regular representation of SU(2) Eqs.(10–184a,185a), by :

$$I_i = \hbar I_i^{\ell} = -\hbar J_i^{r} \quad , \qquad (11\text{-}196)$$

where the J_i^r are the infinitesimal generators of the right regular representation of SU(2). The expressions of the I_i in terms of φ, θ, and ψ then follow from Eqs. (10–169a–c). One should note that I_3 is equal to the generalized momentum p_ψ :

$$I_3 = p_\psi = -i\hbar \frac{\partial}{\partial \psi} \quad , \tag{11-197}$$

a relation which also holds in Classical Mechanics, Eq. (11–180). Furthermore, from Eq. (10–186) one has :

$$J^2 = I^2 \quad , \tag{11-198}$$

and from Eqs.(10–187a,b) the I_i obey the following commutation relations :

$$[J_i , I_j] = 0 \quad , \quad [I_i , I_j] = -i\hbar \sum_k \varepsilon_{ijk} I_k \quad . \tag{11-199}$$

Note the change of sign in these commutation relations compared to those of the J_i.

The interpretation of the intrinsic infinitesimal generators of the regular representations of SU(2) is discussed in Subsec. 10.3.1.4,iv). It follows from Eqs. (10–200a,202) that the intrinsic infinitesimal generators I_i are the operators which correspond in Quantum Mechanics to the projection of the angular momentum $\vec{J}$, Eq. (11–195), along the so-called intrinsic basis vectors defined by :

$$a' = (e_i') \quad , \quad e_i' = R(U;a)e_i \quad . \tag{11-200}$$

Namely, using Eq. (10–201), one has in terms of the multiplicative operators $\hat{e}_i'$ and $\hat{R}_i^j$:

$$I_i = \vec{J}.\hat{e}_i' = J_j \hat{R}_i^j = \hat{R}_i^j J_j \quad . \tag{11-201}$$

From Eqs. (4–34) and (10–184a) one derives :

$$e^{-\frac{i}{\hbar}\alpha I_i} = \mathcal{D}(U)\ \mathcal{D}[U(\alpha,e_i;a)]\ \mathcal{D}(U)^{-1}$$

$$= \mathcal{D}[U(\alpha,e_i' ;a)] \quad , \tag{11-202}$$

so that the I_i can be interpreted as the infinitesimal generators of the rotations about the intrinsic axes determined by the e_i'. Let us establish the following relation :

$$e^{-\frac{i}{\hbar}\alpha I_i}\ I_j\ e^{\frac{i}{\hbar}\alpha I_i} = I_k\ R(\alpha,-e_i')_j^k \quad , \tag{11-203}$$

where :

$$R(\alpha,-e_i')e_j' = e_k'\ R(\alpha,-e_i')_j^k \quad . \tag{11-204}$$

From the above interpretation of the I_i, Eq. (11–203) expresses the law of transformation of the intrinsic infinitesimal generators under a rotation of angle α about the intrinsic basis vector e'_i. From Prop. D.1 the left-hand side of Eq. (11–203) depends only upon the commutation relations of the I_i. From Eq. (11–199) the operators $-I_i$ obey the same commutation relations as the J_i. Furthermore, the J_i verify the Eq. (5–45), i.e. :

$$e^{-\frac{i}{\hbar}\alpha J_i} \; J_j \; e^{\frac{i}{\hbar}\alpha J_i} = J_k \, R(\alpha,e_i)^k_j \quad , \tag{11-205}$$

where :

$$R(\alpha,e_i)e_j = e_k \, R(\alpha,e_i)^k_j \quad . \tag{11-206}$$

Hence, one has the announced result from :

$$e^{-\frac{i}{\hbar}(-\alpha)(-I_i)} \; (-I_j) \; e^{\frac{i}{\hbar}(-\alpha)(-I_i)} = (-I_k)R(-\alpha,e_i)^k_j \quad . \tag{11-207}$$

In particular, for $\alpha = \pi$, with the notation :

$$U_i = e^{-\frac{i}{\hbar}\pi I_i} \quad , \tag{11-208}$$

one finds :

$$U_i \; I_j \; U_i^{-1} = (2\delta_{ij} - 1) \; I_j \quad . \tag{11-209}$$

Expanding the exponential series $\exp(-\frac{i}{\hbar}\pi I_j)$, it follows from the above relation that :

$$i \neq j \qquad U_i \, U_j = U_j^{-1} \, U_i \quad . \tag{11-210}$$

Other properties of the operators U_i are derived below, Eqs. (11–216–218).

Let us emphasize that in a general way the relations involving the intrinsic infinitesimal generators I_i have to be handled with care, cf. remark 3– in Subsec. 10.3.1.4,iv).

iv) **Standard basis, dependence upon a triad.** All the study of Subsec. 11.1.2,vi) concerning the dependence of a standard basis upon a triad as also that concerning the active and the passive points of view of a rotation can be repeated here. Let us only quote the properties which are specific to the considered representation.

The operators $\vec{J}^2$, J_3, and I_3 form a complete set of commuting observables. Indeed, it follows from Eqs. (10–163a, 189a, and b) that the vectors :

$$|a;[j]mk\rangle = \int_{SU(2)} d\mu(\varphi,\theta,\psi) \, |a;\varphi\theta\psi\rangle \, \frac{\hat{j}}{4\pi} \, \mathcal{D}^{j*}_{mk}(\varphi,\theta,\psi) \quad , \tag{11-211}$$

determine a common standard basis such that :

$$\left.\begin{matrix} \vec{J}^2 \\ J_3 \\ I_3 \end{matrix}\right\} |a;[j]mk\rangle = \left\{\begin{matrix} \hbar^2 \; j(j+1) \\ \hbar \;\; m \\ \hbar \;\; k \end{matrix}\right. \quad |a;[j]mk\rangle \quad , \tag{11-212a}$$

$$J_\pm \quad |a;[j]mk\rangle = \hbar \sqrt{j(j+1)-m(m\pm 1)} \; |a;[j]m\pm 1k\rangle \tag{11-212b}$$

$$I_\pm \quad |a;[j]mk\rangle \quad = \hbar \sqrt{j(j+1)-k(k\mp 1)} \quad |a;[j]mk\mp 1\rangle \quad . \tag{11-212c}$$

Furthermore, these vectors form a complete orthonormal basis of the Hilbert space H of physical states, since the set of functions $\{(\hat{j}/4\pi)\ D^{j*}_{mk}\}$ is a complete orthonormal basis of $L^2(SU(2),d\mu)$, (Theorem 7.13, Subsec. 7.2.2.2) :

$$\sum_{j,m,k} |a;[j]mk\rangle\langle a;(j)mk| = \mathbf{1}_H \quad , \tag{11-213a}$$

$$\langle a;(j)mk|a;[j']m'k'\rangle = \delta_{jj'}\ \delta_{mm'}\ \delta_{kk'} \quad . \tag{11-213b}$$

Comparing Eqs. (11–212b and c) we see that $I_\pm$ acts on the quantum number k in the same way as $J_\mp$ acts on the quantum number m. Therefore, I_1, I_2, and I_3 act on k as, respectively, J_1,$-J_2$, and J_3 act on m. One then deduces from Eqs. (6–17–19) law of transformations of the vectors $|a;jmk\rangle$ under a rotation of angle α about the intrinsic axes :

$$e^{-\frac{i}{\hbar}\alpha I_1}|a;[j]mk\rangle = |a;[j]mk'\rangle\ e^{i\frac{\pi}{2}(k'-k)}\ d^j_{k'k}(\alpha) \quad , \tag{11-214a}$$

$$e^{-\frac{i}{\hbar}\alpha I_2}|a;[j]mk\rangle = d^j_{kk'}(\alpha)|a;[j]mk'\rangle \quad , \tag{11-214b}$$

$$e^{-\frac{i}{\hbar}\alpha I_3}|a;[j]mk\rangle = e^{-i\alpha k}\ |a;[j]mk\rangle \quad . \tag{11-214c}$$

In particular for $\alpha = \pi$, using Eq. (6–22c) and the notation specified in Eq. (11–208), one gets :

$$U_1|a;[j]mk\rangle = e^{-i\pi j}\ |a;[j]m{-k}\rangle \quad , \tag{11-215a}$$

$$U_2|a;[j]mk\rangle = (-1)^{j+k}\ |a;[j]m{-k}\rangle \quad , \tag{11-215b}$$

$$U_3|a;[j]mk\rangle = e^{-i\pi k}\ |a;[j]mk\rangle \quad . \tag{11-215c}$$

From the above relations it is easily verified that for i,j, and k a cyclic permutation of 1,2, and 3 :

$$U_i\ U_j = U_k^{-1} \quad , \tag{11-216}$$

which implies, using Eq. (11–210) twice, that :

$$U_j\ U_i = U_k \quad . \tag{11-217}$$

One should also note that within a subspace for a given j, one has :

$$U_i^2 = (-1)^{2j}\ \mathbf{1} \quad . \tag{11-218}$$

Hence, for an even representation of SU(2), and thus integer values of j, the operators U_i are involutive, and being unitary they are also Hermitian.

11.4.3. DYNAMICS OF A FREE RIGID ROTATOR IN QUANTUM MECHANICS

i) As in Classical Mechanics, from the dynamical point of view a rigid rotator is by definition characterized by a real positive and symmetrical inertia tensor. From now on, it is assumed that the intrinsic vectors e'_i, Eq. (11–200), are along the principal axes of inertia. The principal moments of inertia are the I_i.

ii) Corresponding to the classical Hamiltonian, Eqs. (11–167,181), the free rotator Hamiltonian in Quantum Mechanics reads :

$$H = \frac{1}{2} \sum_i \frac{(I_i)^2}{\mathcal{I}_i} \quad . \tag{11-219}$$

It is also known as the asymmetrical top Hamiltonian. The problem is to find the eigenstates of H. The commutation relations, Eq. (11–199) imply that H commutes with $\vec{J}^2$ and J_3. This result corresponds to the invariance of the Hamiltonian under the rotation group which leads to the conservation of the angular momentum. Thus, when the three components with respect to a Galilean basis of the classical angular momentum remain constant, only the length and one component of $\vec{J}$, e.g. J_3, are conserved in Quantum Mechanics. The eigenvectors of H are then expressible in the form :

$$H|a;E[j]m\rangle = E_{jm}|a;E[j]m\rangle \; , \; |a;E[j]m\rangle = c^k_{Ejm}|a;[j]mk\rangle \quad , \tag{11-220}$$

with coefficients c^k_{Ejm} and the energies E_{jm} determined by the eigenvalue problem :

$$\langle a;(j)mk|H|a;[j]mk'\rangle \; c^{k'}_{Ejm} = E_{jm} \; c^k_{Ejm} \quad . \tag{11-221}$$

The above matrix elements of H are readily calculated from Eqs. (11–212a,c) which lead to the only nonvanishing matrix elements :

$$\langle a;(j)mk|(I_1)^2|a;[j]mk\rangle = \langle a;(j)mk|(I_2)^2|a;[j]mk\rangle$$

$$= \hbar^2 \, \frac{1}{2} \, [j(j+1)-k^2] \quad , \tag{11-222a}$$

$$\langle a;(j)mk\pm2|(I_1)^2|a;[j]mk\rangle = - \langle a;(j)mk\pm2|(I_2)^2|a;[j]mk\rangle$$

$$= \hbar^2 \, \frac{1}{4} \, [(j\mp k)(j\pm k+1)(j\mp k-1)(j\pm k+2)]^{\frac{1}{2}} \; , \tag{11-222b}$$

$$\langle a;(j)mk|(I_3)^2|a;[j]mk\rangle = \hbar^2 \, k^2 \quad . \tag{11-222c}$$

In order to proceed further in the reduction of the dimension of the space in which the eigenvalue problem has to be solved, let us take advantage of the symmetry property of the Hamiltonian under the rotations of π about the intrinsic axes. Indeed, from Eq. (11–209), H commutes with the operators U_i. Only two of these operators are independent since they are related by Eqs. (11–216,217). Furthermore, they are Hermitian, and thus can be observables, only for integer values ℓ of j, cf. Eq. (11–218). In this latter case the eigenvalue problem can be solved in each of the four subspaces spanned by the orthonormal functions :

$$\frac{1}{\sqrt{2}}\;|a;[\ell]mk\rangle + \varepsilon(-1)^{\ell+k}|a;[\ell]m-k\rangle;\quad \varepsilon = \pm 1\;,\quad \text{k even or odd,}$$

which are eigenstates of the U_i with eigenvalues $\lambda_i = \pm 1$:

$$\lambda_1 = \lambda_2\,\lambda_3\;,\quad \lambda_2 = \varepsilon\;,\quad \lambda_3 = (-1)^k\;. \tag{11-223}$$

For more details about the asymmetric rotator cf. for instance van Winter[58] and Bohr and Mottelson[8] (Secs. 4–2,3).

For a symmetrical top with $I_1 = I_2$ the Hamiltonian reads using Eq. (11–198) :

$$H = \frac{1}{2}\left[\frac{\vec{J}^2}{I_1} + \left(\frac{1}{I_3} - \frac{1}{I_1}\right)(I_3)^2\right]\;. \tag{11-224}$$

The operator I_3 then commutes with H, which implies, as in Classical Mechanics, the conservation of this intrinsic component of the angular momentum. The eigenstates of H are thus the basis vectors of the complete set of commuting observables $\vec{J}^2$, J_3, and I_3 :

$$H|a;[j]mk\rangle = E_{jk}|a;[j]mk\rangle\;, \tag{11-225a}$$

$$E_{jk} = \hbar^2\,\frac{1}{2}\left[\frac{j(j+1)}{I_1} + \left(\frac{1}{I_3} - \frac{1}{I_1}\right)k^2\right]\;. \tag{11-225b}$$

One should note that for $I_1 > I_3$ (resp. $I_1 < I_3$) the rotator is called a prolate (resp. oblate) spheroid, and the lowest eigenvalue for a given j is for k=0 (resp. k=± j). Hence, as in the classical case, the "intrinsic angular momentum tries to align" along the direction of the largest moment of inertia.

iii) The Hamiltonian which acts on the space of wave functions $f_a(\varphi,\theta,\psi)$ can be obtained either by expressing the I_i in Eq. (11–212) in terms of the Euler angles by means of Eqs. (10–169a–c) and (11–196), or by the substitutions :

$$p_\varphi \rightarrow -i\hbar\,\frac{\partial}{\partial\varphi}\;,\quad p_\theta \rightarrow -i\hbar\,\frac{\partial}{\partial\theta}\;,\quad p_\psi \rightarrow -i\hbar\,\frac{\partial}{\partial\psi}\;, \tag{11-226}$$

in the classical expression of the Hamiltonian $H(\varphi,\theta,\psi,p_\varphi,p_\theta,p_\psi)$, e.g. Eq. (11–182) for the symmetrical top.

The study of the symmetrical and even asymmetrical top in Quantum Mechanics is of great importance in the theories of rotational spectra of molecules and nuclei. For these latter, cf. for instance Bohr and Mottelson loc. cit. above. Let us once more emphasize that it is only for integer values of j or the true representations of the rotation group that the quantum rigid rotator has a classical analogue.

FORMULAE FOR CHAPTER 11

Notation : orthonormal basis $a = (e_i)$.

Spinless particles

Observables and states :

$$\vec{\mathbf{r}} = \mathbf{x}^i e_i \quad , \quad \vec{\mathbf{p}} = \mathbf{p}^i e_i$$

$$[\mathbf{x}^i,\mathbf{x}^j] = 0 \quad , \quad [\mathbf{p}^i,\mathbf{p}^j] = 0 \quad , \quad [\mathbf{x}^i,\mathbf{p}^j] = i\hbar\, \delta_{ij}\, \mathbf{1}$$

$$\vec{\mathbf{r}}|\vec{r}\rangle = \vec{r}|\vec{r}\rangle \quad \begin{cases} \langle\vec{r}'|\vec{r}\rangle = \delta_3(\vec{r}'-\vec{r}) \\ \int d_3\vec{r}\,|\vec{r}\rangle\langle\vec{r}| = \mathbf{1}_E \end{cases}$$

$$\vec{\mathbf{p}}|\vec{p}\rangle = \vec{p}|\vec{p}\rangle \quad \begin{cases} \langle\vec{p}'|\vec{p}\rangle = \delta_3(\vec{p}'-\vec{p}) \\ \int d_3\vec{p}\,|\vec{p}\rangle\langle\vec{p}| = \mathbf{1}_E \end{cases}$$

$$\langle\vec{r}|\vec{p}\rangle = \frac{1}{(2\pi\hbar)^{\frac{3}{2}}}\, e^{\frac{i}{\hbar}\vec{r}.\vec{p}}$$

$$|\varphi\rangle = \int d_3\vec{r}\,|\vec{r}\rangle\,\varphi(\vec{r}) \quad , \quad \langle\varphi|\varphi\rangle = \int d_3\vec{r}\,|\varphi(\vec{r})|^2$$

$$\langle\vec{r}|\vec{\mathbf{p}}|\vec{r}'\rangle = \frac{\hbar}{i}\,\delta_3'(\vec{r}-\vec{r}') \quad , \quad \langle\vec{r}|\vec{\mathbf{p}}|\varphi\rangle = \frac{\hbar}{i}\,\vec{\nabla}\,\varphi(\vec{r}) \quad .$$

Representation of the rotation group :

$$D(R)|\vec{r}\rangle = |R\vec{r}\rangle \Longrightarrow \begin{cases} D(R)|\vec{p}\rangle = |R\vec{p}\rangle \\ D(R)\,\vec{\mathbf{r}}\,D(R)^{-1} = R^{-1}\,\vec{\mathbf{r}} \\ D(R)\,\vec{\mathbf{p}}\,D(R)^{-1} = R^{-1}\,\vec{\mathbf{p}} \end{cases}$$

$$\varphi \xrightarrow{R} {}^R\varphi \quad , \quad {}^R\varphi(\vec{r}) = \varphi(R^{-1}\,\vec{r})$$ cf. Subsec. 10.3.3.

Dependence upon an orthonormal basis :

$$\vec{r} = x^i e_i = r(\sin\theta\cos\varphi\, e_1 + \sin\theta\sin\varphi\, e_2 + \cos\theta\, e_3) \quad , \quad x = (x^i)$$

$$|\vec{r}\rangle = |a;x\rangle = |a;r\theta\varphi\rangle \quad , \quad \varphi(\vec{r}) = \langle a;x|\varphi\rangle = \varphi_a(x) = \langle a;r\theta\varphi|\varphi\rangle = \varphi_a(r,\theta,\varphi)$$

$$D(R)|a;x\rangle = |a;Rx\rangle = |Ra;x\rangle \quad , \quad {}^R\varphi_a = D(R)\,\varphi_a = \varphi_{R^{-1}a} \quad , \quad Rx = (R^i_j\, x^j)$$

$$\langle\vec{r}|A|\varphi\rangle = (A_a\varphi_a)(x) = \int d_3x'\, A_a(x,x')\,\varphi_a(x')$$

$$A_a(x,x') = \langle a;x|A|a;x'\rangle \quad , \quad A_{Ra} = [\,D(R)^{-1}\,A_a\,D(R)]_a\,.$$

Infinitesimal generators, standard basis, and triad :

$$\vec{L} = \vec{\mathbf{r}} \times \vec{\mathbf{p}} \quad , \quad \vec{L} = \sum_i L_i \, e_i \quad , \quad [L_i, L_j] = i\hbar \sum_k \varepsilon_{ijk} L_k \quad ,$$

see the expressions of the L_i in terms of (x^i) or (θ, φ) in the formulae for Chap. 10 ,

$$<a;r'(\ell')m'|a;r'[\ell]m> = \delta(r'-r)\delta_{\ell'\ell}\,\delta_{m'm} \quad , \quad \int r^2 dr \sum_{\ell,m} |a;r[\ell]m><a;r(\ell)m| = \mathbf{1}_E$$

$$<a;r'\theta\varphi|a;r[\ell]m> = \delta(r'-r)\; Y_m^{[\ell]}(\theta,\varphi)$$

$$\mathbf{r}|a;r[\ell]m> = r|a;r[\ell]m> \quad , \quad \vec{L}^2|a;r[\ell]m> = \hbar^2\,\ell(\ell+1)|a;r[\ell]m>$$

$$L_\pm|a;r[\ell]m> = \hbar\,\sqrt{\ell(\ell+1)-m(m\pm1)}\,|a;r[\ell]m\pm1> \quad , \quad L_3|a;r[\ell]m> = \hbar m|a;r[\ell]m>$$

$$D(R')|Ra;r[\ell]m> = |Ra;r[\ell]m'>\,\mathcal{D}^{\ell}_{m'm}(R^{-1}R'R) = |R'Ra;r[\ell]m> \quad .$$

Spin–s particles

Observables and states :

$$\vec{\mathbf{r}}\;,\;\vec{\mathbf{p}}\;,\vec{S} = \sum_i S_i\,e_i \quad \Longrightarrow \quad \text{space } \;E \otimes E_s$$

$$[\vec{\mathbf{r}}\;,\;\vec{S}] = [\vec{\mathbf{p}}\;,\;\vec{S}] = 0 \quad , \quad [S_i, S_j] = i\hbar\sum_k \varepsilon_{ijk} S_k$$

$$<a;(s)m'|a;[s]m> = \delta_{m'm} \quad , \quad \sum_m |a;[s]m><a;(s)m| = \mathbf{1}_{E_s}$$

$$S_\pm|a;[s]m> = \hbar\,\sqrt{s(s+1)-m(m\pm1)}\,|a;[s]m\pm1> \;,\; S_3|a;[s]m> = \hbar m|a;[s]m> \;,$$

$$S^2 = \hbar^2\,s(s+1)\,\mathbf{1}_{E_s} \quad .$$

Representation of the rotation group :

$$\forall\, U \in SU(2) \qquad \mathcal{D}(U)|a;[s]m> = |a;[s]m'>\,\mathcal{D}^s_{m'm}(U)$$

$$|\psi> = \int d_3\vec{r}\,|\vec{r}> \otimes |\psi(\vec{r})> \rightarrow \mathcal{D}(U)|\psi> = \int d_3\vec{r}\,|\vec{r}> \otimes |{}^U\psi(\vec{r})>$$

$$|{}^U\psi(\vec{r})> = \mathcal{D}^s(U)\,|\psi[R(U;a)^{-1}\vec{r}]> \quad , \quad \text{cf. Subsec. 10.3.4.}$$

Infinitesimal generators, standard basis, and triad :

$$\vec{J} = \vec{L} + \vec{S} = \sum_i J_i\,e_i$$

$$|a;r\ell s[j]m> = |a;r[\ell]m_\ell> \otimes |a;[s]m_s><\ell s m_\ell m_s|jm>$$

$$<a;r'\ell's(j')m'|a;r\ell s[j]m> = \delta(r'-r)\;\delta_{\ell'\ell}\;\delta_{j'j}\;\delta_{m'm}$$

$$\int r^2 dr \sum_{\ell,j,m} |a;r\ell s[j]m><a;r\ell s(j)m| = \mathbf{1}_{E\otimes E_s}$$

$$\langle a;r'\theta\varphi|a;r\ell s[j]m\rangle = \delta(r'-r)\ \mathbf{Y}^{\ell s[j]}_{m}$$

$$\mathcal{D}(U')\,|Ua;r\ell s[j]m\rangle = |Ua;r\ell s[j]m'\rangle\ \mathcal{D}^{j}_{m'm}(U^{-1}U'U) = |U'Ua;r\ell s[j]m\rangle$$

$$\text{triad}\ \ Ua \Rightarrow \begin{cases} |-Ua;r\ell s[j]m\rangle = (-1)^{2j}\,|Ua;r\ell s[j]m\rangle \\ R(U;a)J_i = J_j\ R(U;a)^{j}_{i} \end{cases} .$$

Helicity formalism : definite procedure $\vec{p} \rightarrow U_{\vec{p}} \in SU(2)$, $R(U_{\vec{p}};a)e_3 = \hat{p}$

$$|U_{\vec{p}}a;\vec{p}[s]m\rangle = |\underset{\sim}{\vec{p}}\rangle \otimes |U_{\vec{p}}a;[s]m\rangle$$

$$\forall U \in SU(2)\quad \mathcal{D}(U)\,|U_{\vec{p}}a;\vec{p}[s]m\rangle = |U_{R(U;a)\vec{p}}a;R(U;a)\vec{p}[s]m\,\rangle\ \mathcal{D}^{s}_{m\,m}(U^{-1}_{R(U;a)\vec{p}}\,U\,U_{\vec{p}}) .$$

Kets, bras, wave functions, and operators

Tensorial character of kets, bras, and wave functions :

$$\mathcal{D}(U)\,|[j]m\rangle = |[j]m'\rangle\ \mathcal{D}^{j}_{m'm}(U) \quad , \quad \mathcal{D}(U)\,|(j)m\rangle = |(j)m'\rangle\ \mathcal{D}^{j*}_{m'm}(U)$$

$$|[j]m\rangle \xrightarrow{\ d\ } \langle(j)m| \xrightarrow{\ d\ } |[j]m\rangle = (-1)^{2j}\,|d[j]m\rangle$$

$$|(j)m\rangle \xrightarrow{\ d\ } (-1)^{2j}\langle[j]m| \xrightarrow{\ d\ } |(j)m\rangle = (-1)^{2j}\,|d(j)m\rangle$$

$$\langle(j)m|A|[j']m'\rangle^{*} = (-1)^{2j}\langle(j')m'|A^{+}|d[j]m\rangle = \langle(j')m'|A^{+}|[j]m\rangle$$

$$\text{Hermitian tensor}\quad f^{\ell} \Leftrightarrow f^{[\ell]*}_{m} = (-1)^{\ell+m}\ f^{[\ell]}_{-m} .$$

Irreducible tensor operators :

$$\mathcal{D}(U)A^{[j]}_{m}\ \mathcal{D}(U)^{-1} = A^{[j]}_{m'}\ \mathcal{D}^{j}_{m'm}(U) \quad , \quad \mathcal{D}(U)A^{(j)}_{m}\mathcal{D}(U)^{-1} = A^{(j)}_{m'}\ \mathcal{D}^{j*}_{m'm}(U)$$

$$[J_{\pm},A^{[j]}_{m}] = \hbar\,\sqrt{j(j+1)-m(m\pm 1)}\ A^{[j]}_{m\pm 1} \quad , \quad [J_3,A^{[j]}_{m}] = \hbar\ m\ A^{[j]}_{m}$$

$$\mathcal{D}(U')A^{[j]}_{m;Ua}\ \mathcal{D}(U')^{-1} = A^{[j]}_{m';Ua}\ \mathcal{D}^{j}_{m'm}(U^{-1}U'U) = A^{[j]}_{m;U'Ua}$$

$$A^{\ell} = A^{+\ell} \Leftrightarrow A^{[\ell]+}_{m} = (-1)^{\ell+m}\ A^{[\ell]}_{-m} .$$

Tensor operators, e.g. :

$$\mathcal{D}(U)\ A^{[j_1](j_2)}_{m_1\ \ m_2}\ \mathcal{D}(U)^{-1} = A^{[j_1](j_2)}_{m_1'\ \ m_2'}\ \mathcal{D}^{j_1}_{m_1'm_1}(U)\ \mathcal{D}^{j_2*}_{m_2'm_2}(U) .$$

Some examples of tensor operators :

Scalar operators : $\mathbf{1}$, $\vec{\mathbf{r}}^{2}$, $\vec{\mathbf{p}}^{2}$, $\vec{J}^{2} = c^{-2}\sqrt{3}\,\{J^{1}\otimes J^{1}\}^{0}$

Vector operators : $\mathbf{r}^{1}$, $\mathbf{p}^{1}$, ∇^{1}

Multiplicative operators : $\hat{\mathcal{D}}^{j*}_{m_o}$, $\hat{\mathcal{D}}^{j}_{m_o}$, $\hat{Y}^{\ell}$.

Creation and annihilation operators :

$$a^{+}_{\alpha jm}|0> = |\alpha[j]m> \quad ; \quad a^{+}_{\alpha jm} = a^{+[j]}_{\alpha\ m} \quad , \quad a_{\alpha jm} = (-1)^{2j}\, a^{(j)}_{\alpha m} \quad .$$

Generalized projection operators :

$$P_{\alpha\alpha'}{}^{[j]\ (j')}_{m\ \ m'} = |\alpha[j]m><\alpha'(j')\ m'|$$

$$\mathbf{1} = \sum_{\alpha,j,m} P_{\alpha\alpha}{}^{[j]\ (j)}_{m\ \ m} = \sum_{\alpha,j} c_j\ (-1)^{2j}\ \hat{\jmath}\ P^{\{jj\}0}_{\alpha}$$

$$\sum_{\alpha} P_{\alpha}{}^{[j](j)}_{m\ m'} = \frac{2j+1}{16\pi^2} \int_{SU(2)} d\mu(U)\ \mathcal{D}^{j*}_{mm'}(U)\ \mathcal{D}(U)$$

$$P_{jm} = \frac{2j+1}{16\pi^2} \int_{SU(2)} d\mu(U)\ \mathcal{D}^{j*}_{mm}(U)\mathcal{D}(U) \quad , \quad P_j = \frac{2j+1}{16\pi^2}\int_{SU(2)} d\mu(U)\chi^j(U)\chi(U)$$

Matrix elements

The Wigner Eckart theorem :

$$<(j_1)m_1|A^{[j_2]}_{m_2}|[j_3]m_3> = T^{(j_1)[j_2][j_3]}_{m_1\ m_2\ m_3}$$

$$[j_1\|A^{j_2}\|j_3] = T^{\{j_1\{j_2j_3\}j_1\}0} = T^{\{\{j_1j_2\}j_3j_3\}0}$$

$$<(j_1)m_1|A^{[j_2]}_{m_2}|[j_3]m_3> = \begin{cases} c_{j_1}\dfrac{1}{\hat{\jmath}_1}<j_2j_3m_2m_3|j_1m_1>[j_1\|A^{j_2}\|j_3] \\ c_{j_1}\ (-1)^{j_2-j_3+m_1}\begin{pmatrix} j_1 & j_2 & j_3 \\ -m_1 & m_2 & m_3\end{pmatrix}\ [j_1\|A^{j_2}\|j_3]\end{cases}$$

$$[j_1\|A^{j_2}\|j_3]^* = (-1)^{j_1-j_2-j_3}\ [j_3\|A^{+j_2}\|j_1]$$

$$A^{j_2} = c_{j_2}\frac{1}{\hat{\jmath}_2}\sum_{\alpha_1,j_1,\alpha_3,j_3}(-1)^{j_1-j_2+j_3}\ [\alpha_1 j_1\|A^{j_2}\|\alpha_3 j_3]\ P^{\{j_1j_3\}j_2}_{\alpha_1\alpha_3}$$

Some reduced matrix elements :

$$[\alpha j\|\mathbf{1}\|\alpha'j'] = c_j\ \hat{\jmath}\ \delta_{\alpha\alpha'}\ \delta_{jj'}$$

$$[\alpha j\|\vec{J}^2\|\alpha'j'] = c_j\ \hat{\jmath}\ j(j+1)\delta_{\alpha\alpha'}\ \delta_{jj'}\ \hbar^2$$

$$[\alpha j\|J^1\|\alpha'j'] = c\ c_j\ i\ \hat{\jmath}\ \sqrt{j(j+1)}\ \delta_{\alpha\alpha'}\ \delta_{jj'}\ \hbar$$

$$[\alpha\ \tfrac{1}{2}\|\sigma^1\|\alpha'\tfrac{1}{2}] = c\ c_{\frac{1}{2}}\ i\ \sqrt{6}\ \delta_{\alpha\alpha'}$$

$$[m_1'j_1 \| \hat{\mathcal{D}}^{j_2 *}_{m_2'} \| m_3'j_3] = c_{j_1} \frac{16\pi^2}{\hat{j}_1} \langle j_2 j_3 m_2' m_3' | j_1 m_1' \rangle$$

$$[\alpha_1 \ell_1 \| \hat{Y}^{\ell_2} \| \alpha_3 \ell_3] = \begin{cases} (-ic)^{\ell_2+\ell_3-\ell_1} \dfrac{\hat{\ell}_2 \hat{\ell}_3}{\sqrt{4\pi}} \langle \ell_2 \ell_3 00 | \ell_1 0 \rangle \, \delta_{\alpha_1 \alpha_3} \\[2ex] (-ic)^{\ell_2+\ell_3-\ell_1} (-1)^{\ell_2+\ell_3} \dfrac{\hat{\ell}_1 \hat{\ell}_2 \hat{\ell}_3}{\sqrt{4\pi}} \begin{pmatrix} \ell_1 & \ell_2 & \ell_3 \\ 0 & 0 & 0 \end{pmatrix} \delta_{\alpha_1 \alpha_3} \end{cases} .$$

Reduced matrix elements of a tensor product of operators : cf. Subsec. 11.3.3.3.

The Landé formula :

$$\langle \alpha (j) m | A^{[1]}_{\mu} | \alpha' [j \;]m' \rangle = \frac{\langle \alpha (j) m | \vec{J}.\vec{A} | \alpha' [j \;]m' \rangle}{j(j+1)} \langle \alpha (j) m | J^{[1]}_{\mu} | \alpha \; [j \;]m' \rangle \, \hbar^{2}$$

$$= \frac{1}{j(j+1)} \langle \alpha (j) m | J^{[1]}_{\mu} (\vec{J}.\vec{A}) | \alpha' [j \;]m' \rangle \, \hbar^{-2} .$$

The rigid rotator in Quantum Mechanics. Cf. Sec. 11.4 and the left regular representation of SU(2), Subsec. 10.3.1.

APPENDICES

Appendices A,B,C, and D deal with basic definitions (Def.) and properties (Prop.) of vector spaces, algebras, groups, and some useful relations such as the Baker–Campbell–Hausdorff formula. These appendices do not claim to be exhaustive. The close relation which exists between the algebras and the groups is emphasized by reference to the corresponding definitions and properties. The proofs are not always given ; the obvious ones are omitted and references are quoted for those which require too long explanations.

Appendices E to H are devoted to specific results, tables, and examples in connection with the rotation group.

APPENDIX A

VECTOR SPACES

A.1. DEFINITION OF VECTOR SPACES

Def. A.1. A vector space over a field K (assumed here to be commutative) is a set E of elements called vectors, such that there exist two laws of composition. The first one, denoted as a sum :

$$\forall\, v, v' \in E \qquad v'' = v + v' \in E \quad ,$$

gives to E the structure of an Abelian group (Defs. C.1,4).

The second one is the external composition law which associates to any pair (λ, v), λ in K and v in E, a vector v' denoted by :

$$v' = \lambda\, v \quad .$$

This product satisfies the following axioms :

$$\lambda(v+v') = \lambda v + \lambda v' \quad , \qquad (\lambda+\lambda')v = \lambda v + \lambda' v \quad ,$$

$$\lambda(\lambda' v) = (\lambda\lambda')v \quad , \qquad 1\, v = v \quad .$$

Def. A.2. A subset of E such that all the axioms of a vector space are fulfilled is a subspace of E.

Def. A.3. The vectors v_i, i=1,...,n, are said to be linearly dependent if there exists a set of elements λ^i in the field K, not all zero, such that :

$$\lambda^i\, v_i = 0 \quad .$$

In the opposite case, if the above relation implies $\lambda^i = 0$ for all i, then the vectors v_i are said to be linearly independent.

Def. A.4. A basis in a vector space E is a set of linearly independent vectors, the basis vectors, such that any vector in E is a unique linear combination of the basis vectors.

Def. A.5. A vector space E is said to be of finite dimension n if one can find at most n linearly independent vectors in E.

Prop. A.1. If the dimension of the space is finite, the number of basis vectors is the same for any basis and is equal to the dimension.

A.2. THE COMPLEX EXTENSION OF A VECTOR SPACE OVER THE REAL FIELD

A.2.1. DEFINITION AND SOME PROPERTIES

Def. A.6. Let E be a vector space over the real field R. We consider the set $\{E\} = E \times E$ of elements denoted by $W = \{u,v\}$, where u and v are vectors of E. The equality in $\{E\}$ is defined by :

$$W = \{u,v\} = W' = \{u',v'\} \Leftrightarrow u = u' \quad \text{and} \quad v = v' \quad ,$$

the sum by :

$$W + W' = \{u,v\} + \{u',v'\} = \{u+u', v+v'\} \quad ,$$

and the product by a complex number $z = x+iy$ as :

$$zW = (x+iy)\{u,v\} = \{xu-yv, yu+xv\} \quad .$$

With these composition laws all the axioms of the definition of a vector space over the complex field are fulfilled for $\{E\}$. This vector space $\{E\}$ is the complex extension of the vector space E.

Prop. A.2. The subset of $\{E\}$ of elements of the type $W = \{u,0\}$ is isomorphic to the vector space E and thus it is convenient to identify both sets. For the vectors this identification is denoted by :

$$W = \{u,0\} = u \quad .$$

Def. A.7. The space E is called the real part of the complex extension $\{E\}$. Similarly the subset of vectors of the type $W = \{0,u\} = i\{u,0\}$ is called the imaginary part of $\{E\}$, denoted by iE.

Prop. A.3. Let $E = E_n$ be a finite n-dimensional vector space and (e_i) a basis. Then the set $(e_i = \{e_i,0\})$, with the identification defined in Prop. A.2, is a basis of the complex extension $\{E_n\}$. Hence if E is of finite dimension, $\{E_n\}$ has the same dimension.

Proof. Any vector W of $\{E_n\}$ is a linear combination of the e_i :

$$\begin{aligned} W = \{u,v\} &= \{u^i e_i, v^j e_j\} = u^i\{e_i,0\} + v^j\{0,e_j\} \\ &= (u^j+iv^j)\{e_j,0\} = (u^j+iv^j)e_j \quad , \end{aligned}$$

and more precisely, the vectors e_i are linearly independent in $\{E_n\}$.

A.2.2. THE COMPLEX EXTENSION OF A LINEAR MAPPING

Def. A.8. Let A be a linear mapping of E into E. The <u>complex extension</u> $\{A\}$ of A is a mapping of $\{E\}$ into $\{E\}$ defined by :

$$\forall\ W = \{u,v\} \qquad \{A\}\ W = \{Au,Av\} \quad ,$$

$\{A\}$ is obviously a linear mapping over the complex field.

Prop. A.4. In a finite-dimensional vector space, A and $\{A\}$ have the same matrix representative, respectively, in the basis (e_i) of E and $(\{e_i,0\})$ of $\{E\}$.

A.2.3. THE COMPLEX EXTENSION OF AN EUCLIDEAN VECTOR SPACE

Def. A.9. Let E_n be an n-dimensional Euclidean vector space with the inner product defined by the symmetrical bilinear form g. The <u>complex extension</u> of g is the mapping $\{g\}$ of $\{E_n\} \times \{E_n\}$ into the complex field, defined as follows :

$$\forall\ W = \{u,v\} \quad , \quad W' = \{u',v'\} \qquad g(W,W') = g(u,u') + g(v,v')+i[g(u,v')-g(v,u')] \ .$$

Moreover, the <u>complex extension</u> $\{q\}$ of the quadratic form q is such that :

$$\{q\}(W) = \{g\}\ (W,W) \quad .$$

Prop. A.5. The properties of g imply that $\{g\}$ is Hermitian and $\{q\}$ positive definite. Hence with the inner product defined by the form $\{g\}$, and still denoted by :

$$(W,W') = \{g\}\ (W,W') \quad ,$$

the vector space $\{E_n\}$ is Hermitian.

Prop. A.6. The metric tensors of g and $\{g\}$ are the same, respectively, in the basis (e_i) of E_n and $(\{e_i,0\})$ of $\{E_n\}$. Thus if (e_i) is an orthonormal basis of E_n, $(\{e_i,0\})$ is also an orthonormal basis of $\{E_n\}$.

Proof :
$$g_{ij} = (e_i,e_j) = \{g\}_{ij} = (\{e_i,0\},\{e_j,0\}) \quad .$$

Prop. A.7. The complex extension of an isometric application A on E_n is an isometric application on $\{E_n\}$ with respect to the inner product defined by $\{g\}$.

Proof :
$$\forall\ W = \{u,v\} \quad , \quad \forall\ W' = \{u',v'\}$$

$$(\{A\}W,\ \{A\}W') = (Au,Au') + (Av,Av') + i[(Au,Av')-(Av,Au')]$$
$$= (u,u') + (v,v') + i[(u,v')-(v,u')] = (W,W') \quad .$$

APPENDIX B

ALGEBRAS

B.1. DEFINITIONS OF ALGEBRAS AND OF LIE ALGEBRAS

B.1.1. DEFINITIONS

i) **Def. B.1.** An <u>algebra</u> A over a field K is by definition a vector space over the field K, such that there exists a bilinear mapping a of $A \times A$ into A :

$$\forall\, u,v \in A \xrightarrow{\ a\ } w = a(u,v) \in A \quad .$$

Def. B.2. If in addition this law of composition is associative, i.e. if :

$$\forall\, u,v,w \in A \qquad a[u,a(v,w)] = a[a(u,v),w] \quad ,$$

then the algebra is said to be <u>associative</u>.

Def. B.3. The algebra A is said to be <u>commutative</u> or <u>Abelian</u> if :

$$\forall\, u,v \in A \qquad a(u,v) = a(v,u) \quad .$$

The bilinear character of the mapping a implies that it can be defined by specifying the images of any basis vector of A. Let us restrict ourselves to finite-dimensional vector spaces A, (e_i) denotes a basis. Then with the conventions for the indices defined in Subsec. 1.1.1 :

Def. B.4. The <u>structure constants</u> of the algebra with respect to the basis (e_i) are the coefficients $c_{ij}^{\ \ k}$, which belong to the field K, defined by :

$$a(e_i,e_j) = c_{ij}^{\ \ k}\, e_k \quad .$$

It should be clear that the algebra is intrinsically defined by the mapping a, whereas the structure constants are associated to a given basis (the same remark holds for the matrix elements of a linear application Eq. (1–10), or for the metric tensor Eq. (1–20)).

Prop. B.1. Under a change of basis defined by :

$$e'_i = e_j\, t^j_i \quad ,$$

the law of transformation of the structure constants reads, according to the tensorial character of the indices (Subsec. 1.1.1) :

$$c'_{ij}{}^{k} = c_{\ell m}{}^{n}\, t_i^\ell\, t_j^m (t^{-1})_n^k \quad .$$

ii) **Def. B.5.** A Lie algebra A is by definition an algebra such that the mapping a is anticommutative and fulfils the Jacobi identity, i.e., respectively :

$$\forall\, u,v \in A \qquad a(u,v) = -a(v,u) \quad ,$$

$$\forall\, u,v,w \in A \qquad a[u,a(v,w)] + a[v,a(w,u)] + a[w,a(u,v)] = 0 \quad .$$

Prop. B.2. A Lie algebra is not associative.

Prop. B.3. The previous conditions imply, respectively, that in all bases the structure constants satisfy the conditions :

$$c_{ij}{}^{k} = -c_{ji}{}^{k} \quad ,$$

$$c_{i\ell}{}^{m}\, c_{jk}{}^{\ell} + c_{j\ell}{}^{m}\, c_{ki}{}^{\ell} + c_{k\ell}{}^{m}\, c_{ij}{}^{\ell} = 0 \quad .$$

Prop. B.4. If the Lie algebra is moreover commutative, the anticommutativity of a implies :

$$\forall\, u,v \in A \qquad a(u,v) = 0 \quad ,$$

then the structure constants are all equal to zero.

B.1.2. EXAMPLES OF LIE ALGEBRAS

i) To any associative algebra A defined by the bilinear mapping a, one associates a Lie algebra over the same field with the following bilinear mapping :

$$\forall\, u,v \in A \longrightarrow a(u,v) - a(v,u) \quad ,$$

which is anticommutative and fulfils the Jacobi identity. If one denotes the bilinear mapping a as a product, often called the Lie product :

$$a(u,v) = u \circ v \quad ,$$

then the previous Lie algebra is defined by the commutator :

$$[u,v] = u \circ v - v \circ u \quad .$$

For instance, if a vector space A is also a ring (the ring of $n \times n$ matrices on a given field for example), then the usual commutator gives to A a Lie algebra structure.

ii) Let E_3 be an oriented three-dimensional Euclidean vector space. Then the vector product of any two vectors of E_3 defines a bilinear mapping of $E_3 \times E_3$ into E_3 :

$$\forall\, u_1,u_2 \in E_3 \longrightarrow u_3 = u_1 \times u_2 \quad .$$

Therefore with this mapping E_3 is also an algebra over the real field. In an orthonormal basis (e_i) the structure constants are given by Eq. (2–24) :

$$e_i \times e_j = \sum_k \varepsilon_{ijk}\, e_k \quad .$$

Moreover the vector product is anticommutative and fulfils the Jacobi identity :

$$u_1 \times u_2 = -u_2 \times u_1 \quad ,$$

$$u_1 \times (u_2 \times u_3) + u_2 \times (u_3 \times u_1) + u_3 \times (u_1 \times u_2) = 0 \quad .$$

This last equality can be shown by using three times the Gibbs formula, Eq. (2–26). Thus with the vector product, an oriented three-dimensional Euclidean vector space is a Lie algebra over the real field. This Lie algebra is isomorphic to the Lie algebra of the groups SO(3) and SU(2) (Subsec. 4.3.4). With respect to all bases the structure constants fulfil Prop. B.3. In particular, when choosing an orthonormal basis one obtains a property of the tensor ε_{ijk} :

$$\sum_\ell \left(\varepsilon_{i\ell m}\varepsilon_{jk\ell} + \varepsilon_{j\ell m}\varepsilon_{ki\ell} + \varepsilon_{k\ell m}\varepsilon_{ij\ell}\right) = 0 \quad .$$

One shows (Prop. B.39) that the existence of bases such that the structure constants are totally antisymmetrical with respect to the three indices is a general property of semi-simple Lie algebras.

B.1.3. THE COMPLEX EXTENSION OF AN ALGEBRA OVER THE REAL FIELD

Def. B.6. Let A be an algebra over the real field, and $\{A\}$ the complex extension of the vector space A. The complex extension $\{a\}$ of the bilinear mapping a on A is defined as follows :

$$\forall\; W = \{u,v\} \;,\; W' = \{u',v'\} \qquad W'' = \{a\}\,(W,W') = \{a(u,u')-a(v,v'),\; a(v,u')+a(u,v')\}.$$

One checks that this mapping is bilinear with respect to the multiplication law over the complex field. Therefore $\{A\}$ with the bilinear mapping $\{a\}$ is an algebra over the complex field, i.e. the complex extension of the algebra A over the real field.

Prop. B.5. The image by $\{a\}$ of two elements $W = \{u,0\}$ and $W' = \{u',0\}$ of the real part of $\{A\}$, still belongs to this real part.

Proof : $$\{a\}\,(\{u,0\},\, \{u',0\}) = \{a(u,u'),0\} = a(u,u') \quad .$$

Prop. B.6. If A is a finite-dimensional vector space, a and $\{a\}$ have the same structure constants, in the basis (e_i) of A and $(\{e_i,0\})$ of $\{A\}$, respectively.

Prop. B.7. If A is a Lie algebra, then one checks that the mapping $\{a\}$ is, as the mapping a, anticommutative and fulfils the Jacobi identity. Thus the complex extension of a Lie algebra over the real field is also a Lie algebra.

B.2. SUBALGEBRAS, COSETS, IDEALS, AND FACTOR ALGEBRAS

B.2.1. DEFINITIONS AND SOME PROPERTIES

i) **Def. B.7.** (cf. Def. C.9) A <u>subalgebra</u> $\mathcal{B}$ of an algebra A, defined by the bilinear mapping a, is a subspace of A which is closed for the restriction of a to $\mathcal{B} \times \mathcal{B}$, i.e. :

$$\forall\, u,v \in \mathcal{B} \qquad w = a(u,v) \in \mathcal{B} \quad .$$

Prop. B.8. Let (e_i) be a basis of A such that the subset of vectors with indices denoted by $\bar{i}$, i.e. $(e_{\bar{i}})$, is a basis of the subalgebra $\mathcal{B}$. With respect to such a basis the structure constants fulfil the condition :

$$\forall\, \bar{i},\bar{j} \quad , \quad \forall\, k \neq \bar{k} \qquad c_{\bar{i}\bar{j}}{}^{k} = 0 \quad .$$

ii) **Def. B.8.** (cf. Def. C.10) Let $\mathcal{B}$ be a subalgebra of an algebra A. For any element u in A, the subset of elements u+v for all v in $\mathcal{B}$, denoted by u+ $\mathcal{B}$ = $\mathcal{B}$ +u, is a <u>coset</u> of $\mathcal{B}$.

Prop. B.9. (cf. Prop. C.19) If u belongs to $\mathcal{B}$ then u+$\mathcal{B}$ is identical to $\mathcal{B}$ which is thus a coset of $\mathcal{B}$ in A. The other cosets of $\mathcal{B}$ are never subalgebra (nor even subspace) since they do not contain the element zero.

Prop. B.10. (cf. Prop. C.20) The necessary and sufficient condition for two elements u and v of an algebra A to belong to the same coset of $\mathcal{B}$ is :

$$\exists\, w \in \mathcal{B} \quad , \qquad u = v + w \quad .$$

Proof. The condition is obviously sufficient. It is also necessary since :

$$\left.\begin{array}{l} u \in t + \mathcal{B} \Rightarrow \exists\, u' \in \mathcal{B} \quad , \quad u = t+u' \\ v \in t + \mathcal{B} \Rightarrow \exists\, v' \in \mathcal{B} \quad , \quad v = t+v' \end{array}\right\} \Rightarrow \quad u = (v-v')+u' = v+(u'-v') \quad .$$

Prop. B.11. (cf. Prop. C.21) The relation of belonging to the same coset of a subalgebra $\mathcal{B}$ of an algebra A defines an equivalence relation denoted by $\sim$.

Proof. Using Prop. B.10 one checks that the relation $\sim$ is reflexive, symmetrical, and transitive. Therefore this equivalence relation allows us to consider the quotient set $A/\mathcal{B}$, the elements of which are the cosets of $\mathcal{B}$.

iii) **Def. B.9.** A <u>left (resp. right) ideal</u> J of an algebra A, defined by the bilinear mapping a, is a subspace of A such that :

$$\forall\, u \in A \quad , \quad \forall\, v \in J \qquad a(u,v) \in J \qquad (\text{resp. } a(v,u) \in J) \quad .$$

Prop. B.12. A left (resp. right) ideal is a subalgebra.

Prop. B.13. In the basis (e_i) of A defined in Prop. B.8, the structure constants satisfy the condition :

$$\forall\, i \;, \forall\, \bar{j} \quad , \quad \forall\, k \neq \bar{k} \qquad c_{i\bar{j}}{}^{k} = 0 \quad (\text{resp. } c_{\bar{j}i}{}^{k} = 0) \quad .$$

Def. B.10. (cf. Def. C.11) For a Lie algebra, due to the anticommutativity of the bilinear mapping, a left ideal is also a right ideal ; it is called an ideal of the Lie algebra.

Prop. B.14. In the previous basis (e_i), the structure constants are such that :

$$\forall i \ , \forall \bar{j} \quad , \quad \forall k \neq \bar{k} \qquad c_{i\bar{j}}{}^{k} = - c_{\bar{j}i}{}^{k} = 0 \quad .$$

iv) **Prop. B.15.** (cf. Prop. C.25) A necessary and sufficient condition that the bilinear mapping in A induce a bilinear mapping in the quotient set A/B, is that the subalgebra B be an ideal. Then, A/B is an algebra called a quotient or factor algebra.

Proof. Clearly A/B is a subspace. One now wants that if $u \sim v$ and $u' \sim v'$ then $a(u,u') \sim a(v,v')$.
Necessary condition. From Prop. B.10 :

$$u \sim v \ , \ u' \sim v' \ , \ a(u,u') \sim a(v,v') \Rightarrow \exists \ w,w',w'' \in B, \ u=v+w, u'=v'+w', a(u,u')=a(v,v')+w''$$

$$\Rightarrow \forall \ v,v' \qquad a(v,w') + a(w,v') + a(w,w') = w'' \quad .$$

Taking v' equal to zero this implies that for all v in A and w' in B, $a(v,w')$ belongs to B, hence B is an ideal of A.
Sufficient condition :

$$u \sim v \ , \quad u' \sim v' \Rightarrow \exists \ w,w' \in B \quad , \qquad a(u,u') = a(v+w, v'+w') \quad .$$

Since B is an ideal of A, $a(v,w')$ and $a(w,v')$ belong to B, hence :

$$a(u,u') = a(v,v') + w'' \ , \quad w'' = a(v,w') + a(w,v') + a(w,w') \in B \quad .$$

The bilinear mapping for the algebra A/B reads :

$$\forall \ u,v \in A \qquad a(u+B \ , \ v+B) = a(u,v) + B \quad .$$

B.2.2. DERIVATIONS, INTERNAL DERIVATIONS, AND DERIVED ALGEBRAS

i) **Def. B.11.** Let A be a Lie algebra, the set of linear mappings of A into itself is clearly an associative algebra over the same field as A. The bilinear form for this algebra is defined by the usual product of two mappings. To this associative algebra there corresponds a Lie algebra $L(A)$, where the bilinear mapping is defined by the usual commutator of two mappings (Appendix B.1.2,i)).

Def. B.12. The derivations d are special mappings which belong to $L(A)$ and fulfil the following relation :

$$\forall \ u,v \in A \qquad d[a(u,v)] = a[d(u),v] + a[u,d(v)] \quad .$$

Prop. B.16. The subset of derivations is a subalgebra $\delta(A)$ of $L(A)$.

Proof. Any linear combination of derivations is a derivation, and one checks that the commutator of two derivations is a derivation.

ii) **Def. B.13.** To any element u of an algebra A one associates a linear mapping Δ_u of A into itself, defined by :

$$\forall\, v \in A \xrightarrow{\Delta_u} \Delta_u v = a(u,v) \quad ,$$

and called an internal derivation or the adjoint operator of u.

Prop. B.17. The internal derivations are special derivations.

Proof. Consequence of the Jacobi identity.

Prop. B.18. The subset $\Delta(A)$ of internal derivations on A is a subalgebra of the Lie algebra $\delta(A)$, which is homomorphic to A (Defs. B.17,21).

Proof. The Jacobi identity implies :

$$[\Delta_u, \Delta_v] = \Delta_{a(u,v)} \quad .$$

Prop. B.19. The internal derivations on a Lie algebra A are singular mappings.

Proof : $\forall\, u \in A \quad \Delta_u(u) = a(u,u) = 0 \quad .$

Prop. B.20. With respect to a basis (e_i) of a Lie algebra A, the matrix elements of the internal derivations are closely related to the structure constants, since one has :

$$\Delta_{e_i} e_j = e_k (\Delta_{e_i})^k_j = a(e_i, e_j) = c_{ij}{}^k\, e_k \quad .$$

iii) **Def. B.14.** (cf. Def. C.14) Consider in a Lie algebra A the subset of elements a(u,v) for all u and v in A. The finite linear combinations of these elements form a subalgebra of A called the first derived algebra of A, denoted by :

$$A_1 = a(A,A) = \partial A \quad .$$

Def. B.15. (cf. Def. C.15) Proceeding in the above manner, a series of algebras called the derived series is constructed, i.e. :

$$A = A_0 \supseteq A_1 \supseteq A_2 \ldots, \quad A_{i+1} = a(A_i, A_i) = \partial^{i+1} A \quad ,$$

where A_i is the ith derived algebra of A.

Prop. B.21. (cf. Prop. C.29) A_i is an ideal of A_j for j less or equal to i, and the quotient algebras A_i / A_{i+1} are Abelian.

Proof. By construction A_1 is an ideal of A, and more generally A_{i+1} is an ideal of A_i. Let us now show the transitivity relation, A_{i+1} is also an ideal of A_{i-1}, i.e. :

$$a(A_{i-1}, A_{i+1}) \subseteq A_{i+1} \quad .$$

By the Jacobi identity, one has :

$$a(A_{i-1}, A_{i+1}) = a[A_{i-1}, a(A_i, A_i)] = a[A_i, a(A_i, A_{i-1})] + a[A_i, a(A_{i-1}, A_i)]$$

$$\subseteq a(A_i, A_i) = A_{i+1} \quad .$$

The general case : $a(A_i, A_j) \subset A_k$, $k \leqslant \max(i,j)$, can be supplied by induction.

From Prop. B.15, A_{i+1} being an ideal of A_i, A_i / A_{i+1} is an algebra. Moreover, one has :

$$\forall\, u,v \in A_i \quad a(u+A_{i+1},\ v+A_{i+1}) = a(u,v)+A_{i+1} = a(v+u-v,u+v-u) + A_{i+1}$$
$$= a(v,u)+a(v,v-u) + a(u-v,v) + A_{i+1} \quad .$$

Since A_{i+1} is the derived algebra of A_i, $a(v,v-u)$ and $a(u-v,v)$ belong to A_{i+1}, hence :

$$\forall\, u,v \in A_i \quad a(u+A_{i+1},\ v+A_{i+1}) = a(v,u) + A_{i+1} = a(v+A_{i+1},\ u+A_{i+1}) \quad .$$

Def. B.16. (cf. Def. C.16) A Lie algebra A is said to be solvable if the derived series ends with A_n :

$$A_{n-1} \neq A_n = A_{n+1} = \dots \quad ,$$

and A_n only contains the element zero.

Prop. B.22. (cf. Prop. C.30) If a Lie algebra different from $\{0\}$ is solvable, i.e. $A_n = \{0\}$, then A_{n-1} is Abelian.

Proof : $\forall\, u,v \in A_{n-1} \quad a(u,v),\ a(v,u) \in A_n \Rightarrow a(u,v)=a(v,u)=0$.

B.3. REPRESENTATIONS OF LIE ALGEBRAS

B.3.1. DEFINITIONS

Let A be a Lie algebra over the field K, and E a vector space over the same field. The set $L(E)$ of linear applications of E into itself, defined in Subsec. 1.1.3, has not only a structure of a vector space over K, but also the structure of a ring (the product $\circ$ being defined by the usual composition law of applications). Therefore, as specified in Appendix B.1.2,i), one defines on $L(E)$ a Lie algebra structure over the field K, where the bilinear mapping is the commutator of linear applications.

Def. B.17. (cf. Def. C.27) A homomorphism D of a Lie algebra A into the Lie algebra $L(E)$, i.e. such that :

$$\forall\, u,v \in A \quad D[a(u,v)] = [D(u),D(v)] = D(u) \circ D(v)-D(v) \circ D(u) \quad ,$$

is called a representation of A in the vector space E. It is also customary to improperly call representation the image of A in the homomorphism D, i.e. $D(A) = L(E)$. The dimension of E is said to be the dimension of the representation.

Def. B.18. (cf. Def. C.28) The representation is said to be faithful if D is a one-to-one mapping, i.e. an isomorphism.

Def. B.19. (cf. Def. C.30) Two representations $D(A)$ and $D'(A)$ of the same Lie algebra A in the vector spaces, respectively, E and E' (possibly E = E') over the same field, are said to be equivalent, otherwise inequivalent, if there exists a regular linear mapping A of E' into E, such that :

$$\forall\, u \in A \qquad D'(u) = A\, D(u)\, A^{-1} \quad .$$

Def. B.20. (cf. Def. C.32) If there exists a subspace of E , $\neq$ {0} or E, which is invariant for all D(u), u in A, then the representation D(A) is said to be reducible, otherwise irreducible.

B.3.2. SCHUR'S LEMMA, RELATED PROPERTIES, AND EXAMPLES

i) **Prop. B.23.** (cf. Prop. C.45) Schur's lemma : let D(A) and D'(A) be two irreducible representations of a Lie algebra A, respectively, in the vector spaces E and E' over the same field, and A a linear mapping from E' into E. If :

$$\forall\, u \in A \qquad D(u)A = A\, D'(u) \quad ,$$

then A is either zero or a bijective mapping. In the latter case D(A) and D'(A) are equivalent representations.

Proof. From the above equation, the image A(E') and the kernel $A^{-1}(0)$ are invariant subspaces, respectively, of E and E'. Let A be a nonzero mapping then, since D(A) is irreducible, A(E') = E, i.e. A is a surjective mapping ; since D'(A) is irreducible $A^{-1}(0) = \{0\}$, thus A is an injective mapping. Finally A is a one-to-one mapping and the representations D(A) and D'(A) are equivalent.

Prop. B.24. (cf. Prop. C.46) Let D(A) be a finite-dimensional irreducible representation of a Lie algebra A in a vector space over a commutative and algebraically closed field K. If there exists a nonzero mapping A such that :

$$\forall\, u \in A \qquad D(u)A = A\, D(u) \quad ,$$

then A is a multiple of the identity mapping :

$$A = \lambda \mathbf{1} \qquad \lambda \in K \quad .$$

Proof. From Schur's lemma A is a bijective mapping. Since the field of definition of the vector space is commutative the determinant of a finite-dimensional matrix is well defined. Moreover this field being algebraically closed, the characteristic equation $\det(A-\lambda\mathbf{1}) = 0$ is a polynomial in λ which has its roots in the field. Since A is bijective the eigenvalues are nonzero. Then from Prop. B.23, the singular mapping $A-\lambda\mathbf{1}$ which commutes with all D(u) is equal to zero. Cf. remarks after Prop. C.46.

One can also derive similar properties to Props. C.47–50.

ii) **Def. B.21.** (cf. Sec. 5.4) The Lie algebra $\Delta(A)$ of internal derivations on a Lie algebra A is a representation of A (from Prop. B.18), called the adjoint or regular representation of A.

Def. B.22. Let A be a Lie algebra over a field K, the elements of which are $d \times d$ matrices. The fundamental representation F(A) is the d-dimensional representation of A in the vector space K^d, where any element of A is considered as a matrix operator acting on a vector in K^d, i.e. :

$$\forall M \in A \quad , \forall x \in K^d \longrightarrow F(M)\, x = M\, x \in K^d \quad .$$

Prop. B.25. The fundamental representation is faithful.

B.4. SIMPLE AND SEMI-SIMPLE LIE ALGEBRAS

In this subsection, one only considers Lie algebras over zero characteristic fields as for instance the real or complex field.

The characteristic of a field is the smallest non-negative integer p such that :

$$p\,1 = \underbrace{1 + 1 + \ldots + 1}_{p} = 0 \quad ,$$

where 0 and 1 denote, respectively, the unit elements of the additive and multiplication laws over the field.

B.4.1. DEFINITIONS AND SOME PROPERTIES

Def. B.23. (cf. Def. C.12) A Lie algebra over a zero characteristic field is said to be simple if it is not Abelian and has no ideal other than itself and $\{0\}$. It is said to be semi-simple if it has no Abelian ideal other than $\{0\}$.

Simple and semi-simple Lie algebras are very important in algebra and group theory, especially in physical applications. Many works are devoted to them, cf. for instance Cartan[13], Chevalley[17], Bourbaki[10], and Racah[45]. Cartan[13] (cf. also Racah[45]) has given a classification of semi-simple Lie algebras which yields a canonical basis (the Weyl basis). The method used by Cartan, based on the diagonalization of an internal derivation, is only illustrated for the Lie algebra of the group SU(2) in Subsec. 4.3.3,iii).

Prop. B.26. (cf. Prop. C.24) Every simple Lie algebra is semi-simple.

Prop. B.27. (cf. Prop. C.31) A nonabelian and solvable Lie algebra is not simple. Equivalently, a simple algebra is not solvable.

Proof. From Props. B.21,22, A_{n-1} is an Abelian ideal of A different from A and $\{0\} = A_n$. Then Prop. B.26 leads to the announced result.

Prop. B.28. A necessary and sufficient condition for the adjoint representation of a Lie algebra be faithful is that this algebra be semi-simple.

Proof. Let us assume there exist u and v in the Lie algebra A such that :

$$\Delta_u = \Delta_v \Leftrightarrow \forall w \in A \qquad a(u,w) = a(v,w) \quad .$$

Therefore, from the bilinearity of a :

$$\forall\, w \in A \quad a(u-v,w) = -a(w,u-v) = 0 \quad ,$$

which implies that u-v belongs to an Abelian ideal of A.

Prop. B.29. A necessary and sufficient condition for the adjoint representation of a nonabelian Lie algebra be irreducible is that this algebra be simple.

Proof. Let us assume there exists an invariant subspace E in A, i.e. :

$$\forall\, u \in A \quad , \quad \forall\, v \in E \subseteq A \quad \Delta_u v = a(u,v) \in E \quad ,$$

then E is an ideal of A. If A is simple E is either $\{0\}$ or A itself hence the adjoint representation is irreducible. If A is not simple, there exists at least one invariant subspace other than $\{0\}$ and A itself, then the adjoint representation is reducible.

B.4.2. THE TRACE FORMS AND THE KILLING FORM

B.4.2.1. Definitions and some properties

Def. B.24. Let A be a Lie algebra over a field K. The trace form which corresponds to a finite-dimensional representation $D(A)$ is the bilinear form of $A \times A$ into K defined by :

$$\forall\, u,v \in A \longrightarrow (u,v)_D = \mathrm{Tr}[D(u)D(v)] \quad .$$

Prop. B.30. The trace form is symmetrical :

$$\forall\, u,v \in A \qquad (u,v)_D = (v,u)_D \quad .$$

Proof. Consequence of the cyclic invariance of the trace.

Prop. B.31. The trace form is said to be invariant since it fulfils the relation :

$$\forall\, u,v,w \in A \qquad (a(u,v),w)_D = (u,a(v,w))_D \quad .$$

Proof. From Def. B.17 and the cyclic invariance of the trace one has :

$$(a(u,v),w)_D = \mathrm{Tr}\{D[a(u,v)]\; D(w)\} = \mathrm{Tr}\{[D(u),D(v)]\; D(w)\}$$
$$= \mathrm{Tr}\{D(u)[D(v),D(w)]\} = \mathrm{Tr}\;\{D(u)D[a(v,w)]\} = (u,a(v,w))_D \quad .$$

Prop. B.32. With respect to the inner product defined on a Lie algebra A by a trace form, the internal derivations Δ_u are antisymmetrical mappings, and e^{Δ_u} considered as formal exponential series are isometric applications, i.e. :

$$\forall\, u,v,w \in A \qquad \begin{cases} (\Delta_u v,w)_D = -(v,\Delta_u w)_D \quad , \\ (e^{\Delta_u} v \;,\; e^{\Delta_u} w)_D = (v,w)_D \quad . \end{cases}$$

Proof. From the anticommutativity of the bilinear mapping a, the bilinear character of the trace form, and Prop. B.31, one has :

$$\forall u,v,w \in A \quad (\Delta_u v,w)_D = (a(u,v),w)_D = -(a(v,u),w)_D = -(v,a(u,w))_D = -(v,\Delta_u w)_D \quad ,$$

and :

$$(e^{\Delta_u} v, e^{\Delta_u} w)_D = \sum_{n=0}^{\infty} \frac{1}{n!} (\Delta_u^n v, e^{\Delta_u} w)_D = \sum_{n=0}^{\infty} \frac{(-1)^n}{n!} (v, \Delta_u^n e^{\Delta_u} w)_D$$

$$= (v, e^{-\Delta_u} e^{\Delta_u} w)_D = (v,w)_D \quad .$$

Therefore, with respect to a given basis of a Lie algebra, the matrix representative of the mapping exp Δ_u and the metric tensor g of the trace form fulfil the matrix Eq. (2–6) :

$$g = (e^{\Delta_u})^t \, g \, e^{\Delta_u} \quad .$$

Def. B.25. The Killing form, also called the fundamental form is up to a sign the trace form which corresponds to the adjoint representation. Therefore, it is defined for a finite–dimensional algebra A by :

$$\forall \; u,v \in A \qquad g(u,v) = -\mathrm{Tr}(\Delta_u \Delta_v) \quad ,$$

where Δ_u and Δ_v are the adjoint operators, respectively, of u and v. Some authors define the Killing form with the opposite sign.

Prop. B.33. With respect to a given basis (e_i) of a finite–dimensional Lie algebra, the metric tensor of the Killing form is expressed in terms of the structure constants by :

$$g_{ij} = g(e_i, e_j) = -c_{ik}{}^{\ell} \, c_{j\ell}{}^{k} \quad .$$

B.4.2.2. A criterion to identify semi–simple Lie algebras

Prop. B.34. Cartan's theorem : a necessary and sufficient condition for a finite–dimensional Lie algebra over a zero characteristic field to be semi–simple is that the Killing form be nondegenerate.

Proof : cf. for instance Chevalley[17)] (Chap. IV, § 2, N°2, Theorem 2). Let us only show the sufficient character of the condition. Suppose there exists an Abelian ideal different from {0}. Then with respect to the basis (e_i) of the Lie algebra defined in Prop. B.8, one has :

$$g_{i\bar{j}} = -c_{ik}{}^{\ell} \, c_{\bar{j}\ell}{}^{k} \quad ,$$

by Prop. B.14 :

$$g_{i\bar{j}} = -c_{i\bar{k}}{}^{\ell} \, c_{\bar{j}\ell}{}^{\bar{k}} = -c_{i\bar{k}}{}^{\bar{\ell}} \, c_{\bar{j}\bar{\ell}}{}^{\bar{k}} \quad ,$$

and as the ideal is Abelian $c_{\bar{j}\bar{\ell}}{}^{\bar{k}} = 0$, whence $g_{i\bar{j}} = 0$ for all i and $\bar{j}$. Thus the Killing form is degenerate (if the field is commutative, the determinant is well defined, and det g = 0).

Prop. B.35. Let $D(A)$ be a finite–dimensional representation of a semi–simple Lie algebra A. Then, the set K of elements u in A orthogonal to every v in A, with respect to the trace form which corresponds to this representation, is the kernel of the mapping D.

Proof : cf. for instance Chevalley[17)] (Chap. IV, § 3, N°1, Proposition 2).

Prop. B.36. If $D(A)$ is a faithful finite–dimensional representation of a semi–simple Lie algebra A, then the trace form which corresponds to this representation is nondegenerate.

Proof. Consequence of Prop. B.35, the kernel of D being {0}. Since from Prop. B.28 the adjoint representation of a semi–simple Lie algebra is faithful, Prop. B.36 implies that the

Killing form is then nondegenerate. This corresponds to the necessary character of the previous Cartan criterion.

B.4.2.3. Some consequences of the nondegenerate character of the Killing form

Cf. Subsec. 1.1.4,iii).

Prop. B.37. The nondegenerate inner product defined by the Killing form endows a finite-dimensional semi-simple Lie algebra with an Euclidean or pseudo-Euclidean vector space structure. Pseudo-Euclidean corresponds to the fact that the quadratic form, $q(u) = g(u,u)$, is not necessarily definite.

With respect to a given basis (e_i) of a semi-simple Lie algebra, the nondegenerate metric tensor (g_{ij}) allows us to change the tensorial character of the indices (cf. Eq. (1-27)). For example, new structure constants can be defined by :

$$c_{ijk} = g_{k\ell}\, c_{ij}^{\ \ \ell} \ .$$

The two sets of structure constants (c_{ijk}) and $(c_{ij}^{\ \ k})$ are equivalent, i.e. given any one of them the other is completely determined.

Prop. B.38. The structure constants c_{ijk} defined above are totally antisymmetrical.

Proof. From Prop. B.3 the structure constants c_{ijk} are skew in i and j, moreover using Prop. B.33 and the Jacobi identity, one has :

$$\begin{aligned} c_{ijk} &= -c_{km}^{\ \ \ n}\, c_{\ell n}^{\ \ \ m}\, c_{ij}^{\ \ \ell} = c_{km}^{\ \ \ n}\, c_{n\ell}^{\ \ \ m}\, c_{ij}^{\ \ \ell} \\ &= -c_{km}^{\ \ \ n}\, c_{i\ell}^{\ \ \ m}\, c_{jn}^{\ \ \ \ell} - c_{km}^{\ \ \ n}\, c_{j\ell}^{\ \ \ m}\, c_{ni}^{\ \ \ \ell} \\ &= -c_{km}^{\ \ \ n}\, c_{i\ell}^{\ \ \ m}\, c_{jn}^{\ \ \ \ell} - c_{mk}^{\ \ \ n}\, c_{\ell j}^{\ \ \ m}\, c_{ni}^{\ \ \ \ell} \ , \end{aligned}$$

where the right-hand side of the last equation is invariant under cyclic permutations of the indices i, j, and k.

One should remark that the objects c_{ijk} can also be defined if the Lie algebra is not semi-simple. They are still totally antisymmetrical, but some of them may vanish and there no longer exists a one-to-one correspondence between the two sets of structure constants.

Prop. B.39. In a finite-dimensional semi-simple Lie algebra there exist bases such that the structure constants $c_{ij}^{\ \ k}$ are totally antisymmetrical. They are the bases which diagonalize the Killing form.

Proof. Since the Killing form is nondegenerate (Prop. B.34) there exists a basis which diagonalizes it (Subsec. 1.1.4,iii)). Then one gets in such a basis :

$$g_{\ell k} = \delta_{\ell k} \Rightarrow c_{ij}^{\ \ k} = c_{ij\ell}\, \delta^{\ell k} \ ,$$

and thus $c_{ij}^{\ \ k}$ is totally antisymmetrical.

Prop. B.40. In a finite-dimensional semi-simple Lie algebra over the real field, with respect to an orthonormal basis (resp. pseudo-orthonormal if g is not definite) the matrix representative of $e^{\Delta u}$ is orthogonal (resp. pseudo-orthogonal).

Proof. Consequence of Props. B.32,39.

B.4.2.4. Relation between trace forms which correspond to different representations

i) **Prop. B.41.** Let $D(A)$ and $D'(A)$ be two faithful finite-dimensional representations of a simple Lie algebra A over a commutative and algebraically closed field K with zero characteristic. Then, the trace forms which correspond, respectively, to D and D' are proportional, i.e., there exists a nonzero constant λ in the field such that :

$$\forall\, u,v \in A \qquad (u,v)_D = \lambda (u,v)_{D'} \quad .$$

Proof. We follow the four steps of a proof given by Cartier[15] :

1–From Prop. B.26 A is a semi-simple Lie algebra. Prop. B.36 implies that the considered trace forms are nondegenerate.

2–There exists a nonzero linear operator A of A into A such that :

$$\forall\, u,v \in A \qquad (u,v)_D = (Au,v)_{D'} \quad .$$

Indeed, let (e_i) be a basis of A, from 1 the metric tensor $(e_i,e_j)_{D'}$ is a regular matrix. Hence, there exists a matrix $(A^i_{\ j})$ such that :

$$(e_i,e_j)_D = A^k_i \; (e_k,e_j)_{D'} \quad .$$

The intrinsic linear operator A is then defined by its matrix representative $(A^i_{\ j})$ with respect to the basis (e_i). If A equals zero, one has $(u,v)_D$ equal to zero for all u and v in A, which is inconsistent with 1.

3–The mapping A fulfils the relation :

$$\forall\, u,v \in A \qquad A\, a(u,v) = a(u,Av) \quad .$$

From 1, it is sufficient to show that :

$$\forall\, u,v,w \in A \qquad (A\, a(u,v),w)_{D'} = (a(u,Av),w)_{D'} \quad .$$

Using two times 2 and Prop. B.31, one gets :

$$\begin{aligned}(A\, a(u,v),w)_{D'} &= (a(u,v),w)_D = -(a(v,u),w)_D = -(v,a(u,w))_D \\ &= -(Av,a(u,w))_{D'} = -(a(Av,u),w)_{D'} = (a(u,Av),w)_{D'} \quad .\end{aligned}$$

4–The mapping A is a multiple of the identity. This is a consequence of the Schur lemma. Indeed, the above property 3 of the mapping A can be written in terms of the adjoint operators as :

$$\forall\, u \in A \qquad A\Delta_u = \Delta_u A \quad .$$

Since from Prop. B.29 the adjoint representation of a simple Lie algebra is irreducible, Prop. B.24 implies the result :

$$A = \lambda\, \mathbf{1} \qquad \lambda \in K \quad .$$

Prop. B.41 is then a consequence of 2.

Prop. B.42. Let A be a finite-dimensional simple Lie algebra over a commutative and algebraically closed field K of zero characteristic. Then, the trace form which corresponds to any faithful finite-dimensional representation $D(A)$ is proportional to the Killing form :

$$\forall\, u,v \in A \qquad (u,v)_D = \lambda(u,v) \qquad \lambda \in K \quad .$$

Proof. Consequence of Props. B.26, 28, and 41.

Prop. B.43. Let A be a n-dimensional simple Lie algebra over a commutative and algebraically closed field K of zero characteristic, and D(A) a faithful d-dimensional irreducible representation. Then one has :

$$\lambda\, n = \mu\, d \quad ,$$

where λ is defined above in Prop. B.42, and μ in the field is such that :

$$G = \mu\, \mathbf{1} \quad ,$$

G being the Casimir operator of the representation D(A), defined in Sec. 5.5, Eq. (5-49).

Proof. Let (g_{ij}) denote the metric tensor of the Killing form with respect to a given basis (e_i) of the Lie algebra A, then the Casimir operator of the faithful representation D(A) reads :

$$G = g^{ij}\, D(e_i)\, D(e_j) \quad , \qquad g_{ij}\, g^{kj} = \delta_i^k \quad .$$

It is shown in a remark in Sec. 5.5 that G commutes with all the elements of D(A). Therefore from Prop. B.24, G is a multiple of the identity :

$$G = \mu\, \mathbf{1} \qquad \mu \in K \quad .$$

Taking the trace of the above equation and using Prop. B.42, one gets :

$$\operatorname{Tr} G = g^{ij}\operatorname{Tr}[D(e_i)D(e_j)] = \mu d = g^{ij}(e_i,e_j)_D = \lambda\, g^{ij} g_{ij} = \lambda\, n \quad .$$

ii) Example of application to the simple Lie algebra $\mathcal{W}(2)$ of the group SU(2) : $\mathcal{W}(2)$ is the set of complex 2×2 Hermitian and traceless matrices studied in Subsec. 4.3.2. The computation of the proportionality coefficient between two trace forms, which correspond to different representations, can be performed by evaluating one special inner product in both representations (provided this inner product is nonzero). Let us for simplicity compute the inner product of the generator J_3 by itself, i.e. $\operatorname{Tr}(J_3{}^2)$ in several faithful representations :

–fundamental representation : $J_3 = \begin{pmatrix} \frac{1}{2} & 0 \\ 0 & -\frac{1}{2} \end{pmatrix}$, $(J_3,J_3)_F = \frac{1}{2}$,

–adjoint representation : $J_3 = \begin{pmatrix} 1 & 0 & 0 \\ 0 & 0 & 0 \\ 0 & 0 & -1 \end{pmatrix}$, $(J_3,J_3) = 2$,

–representation $\mathcal{D}^j$:

$$\begin{cases} J_3 = (m\delta_{mm'}) \qquad (J_3,J_3)_j = \sum\limits_{m=-j}^{j} m^2 = \frac{1}{3}\, j(j+1)(2j+1) \\ 2j \text{ non-negative integer}, \qquad m,m' = j, j-1, \dots, -j \quad . \end{cases}$$

Hence, one has for instance :

$$\forall\, W,W' \in \mathcal{W}(2) \qquad (W,W')_j = \frac{1}{6}\, j(j+1)(2j+1)\ (W,W') \quad ,$$

where the Killing form (W,W') is identical to the trace form for j=1.

Let us now check Prop. B.43 for the representation $\mathcal{D}^j$:

–dimension of the simple algebra $\mathcal{W}(2)$ over the real field : n=3,

–dimension of the faithful irreducible representation $\mathcal{D}^j$: d=2j+1,

–from the above equation : $\lambda = \frac{1}{6}\, j(j+1)(2j+1)$,

–the usual Casimir operator J^2 is related to G by Eq. (5–51) :

$$G = \frac{1}{2} J^2 \Rightarrow \mu = \frac{1}{2}\, j(j+1) \quad ,$$

thus one gets the relation :

$$\lambda n = \frac{1}{6}\, j(j+1)(2j+1).3 = \mu d = \frac{1}{2}\, j(j+1).(2j+1) \quad .$$

B.4.2.5. A criterion of solvability

Prop. B.44. Cartan's criterion : let D(A) be a finite–dimensional representation of a Lie algebra A. Then D(A) is solvable if and only if the first derived algebra A_1 is orthogonal to A with respect to the trace form which corresponds to this representation, i.e. :

$$\forall\, u,v,w \in A \qquad (a(u,v),w)_D = 0 \quad .$$

Proof : cf. for instance Bourbaki[10], (Chap. I, § 5, N°4, Theorem 2).

B.5. ASSOCIATIVE DIVISION ALGEBRAS

B.5.1. DEFINITIONS

Def. B.26. A unit element e of an algebra A is by definition such that :

$$e \in A \quad , \quad \forall\, u \in A \qquad a(e,u) = a(u,e) = u \quad .$$

Prop. B.45. If the unit element exists it is unique.

Proof. Let us assume that there exist two distinct unit elements e and e' :

$$\forall\, u \in A \qquad a(e',u) = a(u,e') = u \quad ,$$

then setting u = e' in Def. B.26, and u = e in the above equation leads to e = e'.

Def. B.27. An algebra is said to be a division algebra if there exists a unit element e, and if for any nonzero element u of the algebra there exists an inverse element, denoted u^{-1}, such that :

$$a(u,u^{-1}) = a(u^{-1},u) = e \quad .$$

Prop. B.46. In the subset {A-0} of nonzero elements of an associative division algebra A, the bilinear mapping a defines a composition law which fulfils the axioms of the definition of a group, (Def. C.1). Thus for instance, as for groups, one proves the unicity of the inverse element.

Proof. Let us assume that there exist two distinct inverse elements u_1 and u_2 of u :

$$a(u,u_1) = a(u_1,u) = e \quad , \quad a(u,u_2) = a(u_2,u) = e \quad ,$$

using the associativity of a, one has :

$$u_2 = a(u_2,e) = a[u_2,a(u,u_1)] = a[a(u_2,u),u_1] = a(e,u_1) = u_1 \quad .$$

B.5.2. ASSOCIATIVE DIVISION ALGEBRAS OVER THE REAL FIELD

i) There exists an important theorem which states that :

Prop. B.47. Frobenius' theorem[25)] : up to an isomorphism there exist only three associative division algebras over the real field, namely the real numbers, the complex numbers, and the real quaternions.

ii) The real and complex fields, R and C, are, respectively, one and two-dimensional vector spaces over the real field for which bases are for instance :

$$(e_o = 1) \quad , \quad \forall\, x \in R \qquad x = x\, e_o \quad ;$$

$$(e_o = 1 \;,\; e_1 = i \;;\; e_1^2 = -e_o) \quad , \quad \forall\, z \in C \qquad z = x^o e_o + x^1 e_1 \quad , \quad x^o, x^1 \in R \quad .$$

The product in these fields defines the bilinear mapping a which gives to these sets an algebra structure :

$$\forall\, x,x' \in R \qquad a(x,x') = xx' \quad ; \quad \forall\, z,z' \in C \qquad a(z,z') = zz' \quad .$$

These algebras are commutative, the unit element is e_o, and for any nonzero element there exists an inverse element.

The subset R' of elements of C which read $x^o e_o$ is clearly a subalgebra of C isomorphic to R.

iii) **Def. B. 28.** Let us denote by (e_μ , μ = 0,1,2,3) a basis of a four-dimensional vector space Q over the real field :

$$\forall\, q \in Q \qquad q = x^\mu\, e_\mu \qquad x^\mu \in R \quad .$$

The algebra Q of real quaternions is defined by its structure constants which fulfil the conditions, denoting a(q,q') as a product qq' :

$$\mu = 0,1,2,3 \qquad e_o e_\mu = e_\mu e_o = e_\mu \quad ,$$

$$i,j,k \in \{1,2,3\} \qquad e_i e_j = -\delta_{ij} e_o + \sum_k \varepsilon_{ijk}\, e_k \quad ,$$

where ε_{ijk} is the totally antisymmetrical tensor of order three.

Prop. B.48. The unit element of Q is e_o, and :

$$i,j,k \in \{1,2,3\} \qquad (e_i e_j)e_k = e_i(e_j e_k) \quad ,$$

hence Q is an associative but not commutative algebra.

Def. B.29. The <u>conjugate</u> $\tilde{q}$ of the quaternion q is defined by :

$$q = x^o e_o + x^i e_i \qquad \tilde{q} = x^o e_o - x^i e_i \quad .$$

Prop. B.49. The real quaternion algebra Q is a division algebra and a noncommutative field.

Proof. From Def. B.29 one gets :

$$\tilde{q}\, q = q\, \tilde{q} = \sum_{\mu=0}^{3} (x^\mu)^2 = N(q)^2\, e_o \quad ,$$

where $N(q)^2$ is a non–negative real number which is equal to zero only if q is zero. Hence, any nonzero q in Q has an inverse element :

$$q^{-1} = N(q)^{-2}\, \tilde{q} \quad .$$

Prop. B.50. The subset C' of quaternions of the type $x^o e_o + x^1 e_1$ is clearly a subalgebra of Q isomorphic to the complex algebra C. Consequently Q can be considered as a two–dimensional vector space over the complex field.

Proof. For all complex z = x + iy let us define qz, taking care of the prescribed order of q and z, by :

$$\forall\, q \in Q \qquad qz = q(xe_o + ye_1) \quad .$$

Then, the axioms of the definition of a vector space (Def. A.1) are fulfilled :

$$\forall\, q,q' \in Q \quad \forall\, z,z' \in C \qquad (q+q')z = qz + q'z \quad , \qquad q(z+z') = qz + qz' \quad ,$$

$$(qz)z' = q(zz') \quad , \qquad q\,1 = q \quad .$$

Any quaternion takes the form :

$$q = x^\mu e_\mu = e_o(x^o + x^1 e_1) + e_2(x^2 - x^3 e_1) = e_o z^o + e_2 z^2 \quad ,$$

$$z^o = x^o + x^1 i \qquad z^2 = x^2 - x^3 i \quad .$$

One should note that the conjugate element of q is given by :

$$\tilde{q} = e_o z^{o*} - e_2 z^2 \quad .$$

An example of a real quaternion algebra is given by the subset of 2 × 2 complex matrices of the type :

$$M = x^o\, \mathbf{1} + x^j(-i\sigma_j) \quad , \qquad x^\mu \in R \quad ,$$

where the σ_j are the Pauli matrices (Subsec. 4.5,iii)).

For a detailed study of the consequences of Frobenius' theorem, cf. Dyson [21], some remarks are given in Sec. 8.2.

APPENDIX C

GROUPS

C.1. DEFINITIONS AND SOME ALGEBRAIC PROPERTIES OF GROUPS

Def. C.1. A group G is a set of elements, a,b,..., for which a composition law, denoted in the following as a product, is defined and satisfies the following requirements :

(1) closure : $\forall\, a,b \in G$ ab belongs to G ,

(2) associativity : $\forall\, a,b,c \in G$ $(ab)c = a(bc)$,

(3) existence of a left unit element (or unity) e : $\forall\, a \in G$ $ea = a$,

(4) existence of a left inverse : $\forall\, a \in G \;\; \exists\, a^{-1} \in G,\; a^{-1}a = e$.

Prop. C.1. a^{-1} is a right inverse and e is a right unit element, both are unique.

Proof of the right properties :

$$\forall\, a = ea = ((a^{-1})^{-1}a^{-1})a = (a^{-1})^{-1}(a^{-1}a) = (a^{-1})^{-1}e$$

$$\Rightarrow aa^{-1} = ((a^{-1})^{-1}e)a^{-1} = (a^{-1})^{-1}(ea^{-1}) = (a^{-1})^{-1}\,a^{-1} = e \quad ,$$

$$ae = a(a^{-1}a) = (aa^{-1})a = ea = a \quad .$$

The proof of the uniqueness follows from the associativity.

Prop. C.2. Cancellation law : any element of a group G is regular on the right and on the left, i.e. if $ab = ac$ or $ba = ca$, then $b = c$.

Prop. C.3. By induction on n, one shows for the associative law that all ways of bracketing a finite ordered product $a_1a_2...a_n$, to give its value by calculating a succession of binary products, yield the same value. The brackets can thus be omitted.

Prop. C.4. Both equations $xa = b$ and $ax = b$ have unique solutions in a group.

Prop. C.5. By induction on n, one shows that the inverse of a finite ordered product $a_1a_2...a_n$ is given by $a_n^{-1}...a_2^{-1}\,a_1^{-1}$.

A group G with a countable number of elements denoted by a = 1,2,..., can be characterized by its group table. One finds in this table at the intersection of the ath row and bth column the element ab. The group table is mainly considered for finite groups (Def. C.5).

Prop. C.6. Rearrangement theorem : a given element appears in a given row (or column) once and only once.

Def. C.2. The left (resp. right) commutator of any two elements a and b of a group G is defined by :

$$c_\ell(a,b) = aba^{-1}b^{-1} \quad (\text{resp. } c_r(a,b) = a^{-1}b^{-1}ab) \quad .$$

Def. C.3. The centre C of a group G is the subset of G, the elements of which commute with all elements of the group : $c \in C \Leftrightarrow \forall g \in G \;\; cg = gc$.

Prop. C.7. If a belongs to the centre C of a group G, then a^{-1} also belongs to C, and the unit element belongs to C.

Prop. C.8. The commutator of two elements, one of them at least being in the centre C of a group G, is equal to the unit element.

Def. C.4. If the group G has the further property of commutativity, i.e. ab = ba for all a and b in G, then G is said to be commutative or Abelian.

Prop. C.9. The group table of an Abelian group is symmetrical with respect to the principal diagonal.

Prop. C.10. If G is Abelian the centre of G is the whole group.

Prop. C.11. Sufficient condition : if every element of a group G is equal to its inverse, then G is Abelian.

Def. C.5. The order O(G) of a group is the number of elements in G. If O(G) is finite the group G is said to be finite.

For finite groups the axioms of Def. C.1 can be replaced by :

Def. C.6. (1),(2) of Def. C.1, and :

(3') G has a finite number of elements,

(4') every element of G is regular on the right and on the left.

Proof. For finite groups Def. C.1 implies Def. C.6. To show the converse let $a_1,\dots,a_n$ be the n distinct elements of the finite group G. Then from the rearrangement theorem for all a_i, the elements $a_1a_i,\dots,a_na_i$ are all distinct. Therefore there exists among them the element a_i, i.e. there exists k such that $a_ka_i = a_i$. Thus a_k is a left unit element since for all elements a_j of G, one has :

$$a_ia_j = (a_ka_i)a_j = a_k(a_ia_j) \quad .$$

Similarly : $\forall\, a_i \;\exists \ell$, $a_\ell\, a_i = a_k$, so that a_ℓ is a left inverse of a_i. The axioms of Def. C.1 are thus fulfilled.

Prop. C.12. The number of elements of a finite group G, which are equal to their inverse, has the same parity as the order of the group.

Proof. Consider the partition of G into two sets : in the first one the elements are distinct from their inverse and in the second one they are equal to their inverse. The first set has an even number of elements, since :

$$(a^{-1})^{-1} = a \quad .$$

This property and also Prop. C.6 are very useful for the construction of the group table of finite groups.

Def. C.7. The order or period p of an element a of a group G is equal to the number, perhaps infinite, of distinct elements which can be generated from a in G : $a, a^2, \ldots, a^p$.

Prop. C.13. Every element of a finite group has a finite period.

Def. C.8. If a group can be generated from one element, the group is said to be cyclic.

Prop. C.14. Every cyclic group is Abelian, but the converse is not true.

C.2. SUBGROUPS, COSETS, INVARIANT SUBGROUPS, AND FACTOR GROUPS

C.2.1. DEFINITIONS AND SOME PROPERTIES

i) **Def. C.9.** (cf. Def. B.7) A subset H of a group G is called a subgroup of G if it forms a group with respect to the product defined in G.

Examples :
(1) $\{e\}$ and G are trivial or improper subgroups of G, other subgroups of G are called proper subgroups.
(2) The centre of a group is an Abelian subgroup.
(3) For an element a of period p, the subset $\{a, \ldots, a^p\}$ is an Abelian subgroup, i.e. the smallest subgroup which contains a.

Prop. C.15. The intersection of a finite number of subgroups of G is a subgroup of G.

Prop. C.16. A necessary and sufficient condition for H to be a subgroup of G is that H be closed for the product defined in G, and that for all a in H, its inverse a^{-1} be in H.

Prop. C.17. A necessary and sufficient condition for H to be a subgroup of G is that for all a and b in H, ab^{-1} be also in H.

ii) **Def. C.10.** (cf. Def. B.8) Let H be a subgroup of a group G. For any element g in G, the subset of elements hg (resp. gh) for all h in H, denoted Hg (resp. gH) is a right (resp. left) coset of H.

If g belongs to H, then Hg and gH are identical to H. Thus H is itself a right as well as a left coset of H in G.

Prop. C.18. There is a one-to-one correspondence between the right and left cosets of H.

Proof. This correspondence is given by the inverse :

$$(hg)^{-1} = g^{-1}h^{-1} \quad \text{and if} \quad h \in H \Rightarrow h^{-1} \in H \quad .$$

Prop. C.19. (cf. Prop. B.9) The cosets other than H are never subgroups since they do not contain the unit element.

Prop. C.20. (cf. Prop. B.10) The necessary and sufficient condition for two elements a and b of a group G to belong to the same right (resp. left) coset of H is :

$$\exists\ h \in H \quad , \quad a = hb \quad (\text{resp. } a = bh) \quad .$$

Proof. The condition is obviously sufficient. It is also necessary since :

$$\left.\begin{array}{lll} a\in Hg \Rightarrow & \exists h\in H & a = hg \\ b\in Hg \Rightarrow & \exists h'\in H & b = h'g \end{array}\right\} \Rightarrow \ a = h(h'^{-1}b) = (hh'^{-1})b \quad .$$

Prop. C.21. (cf. Prop. B.11) The relation of belonging to the same right coset of a subgroup H of a group G defines an equivalence relation denoted $\sim_r$ (resp. $\sim_\ell$ for left cosets).

Proof. Using Prop. C.20 one checks that the relation $\sim$ is reflexive, symmetrical, and transitive.

This equivalence relation implies a partitioning of G, i.e. G is a disjoint union of cosets of H :

$$G = H + Ha + Hb + \dots \quad ,$$

and allows us to consider the quotient set G/H, consisting of the right (or left) cosets of H. For finite groups the number of cosets is finite.

Prop. C.22. Lagrange's theorem : the order O(G) of a finite group G and the order O(H) of a subgroup H of G are such that O(G)/O(H) is an integer, called the index of H in G and denoted by [G : H].

Proof. Since every g in G has an inverse, gh = gh' (or hg = h'g) implies h = h'. Thus every left (as also right) coset of H has the same number O(H) of elements. The result then follows from the partitioning of G into cosets.
Remark : the unit element by itself is a subgroup and its cosets consist of single elements. Thus the order of a group is the index of the identity subgroup.

Prop. C.23. If the order n of a group G is prime, the only subgroups are G and {e}, and G is cyclic and therefore Abelian.

Proof. We have seen (Prop. C.13) that the period p of all elements a of a finite group G is finite and that the subset $\{a,\dots,a^p\}$ is a subgroup of G. From Lagrange's theorem p divides n. Hence p is either 1 and thus a = e, or p = n and G is cyclic and therefore Abelian (Prop. C.14).

iii) **Def. C.11.** (cf. Defs. B.9,10) A subgroup H of G is called an invariant subgroup or normal divisor of G, if for every g in G and h in H, $g^{-1}\,h\,g$ belongs to H. In other words H is an invariant subgroup of G if its right and left cosets are identical : gH = Hg.

Examples :
(1) The centre of a group is an Abelian invariant subgroup.
(2) The subset of three-dimensional rotations with a fixed rotation angle is an invariant subgroup of the rotation group (Subsec. 2.3.2.1,ii)).
(3) The improper subgroups {e} and G are invariant.

Def. C.12. (cf. Def. B.23) A group G is said to be simple if its only invariant subgroups are {e} and the whole group G. It is said to be semi-simple if its only invariant Abelian subgroup is {e}.

Prop. C.24. (cf. Prop. B.26) Every simple group G is semi-simple, except when G itself is Abelian. For instance, a cyclic group of prime order is simple but not semi-simple.

iv) **Prop. C.25.** (cf. Prop. B.15) A necessary and sufficient condition that the multiplication law in G induces a multiplication law in the quotient set G/H, is that the subgroup H be invariant. Then G/H is a group, called the quotient or factor group, with H as the unit element and Ha^{-1} (or $a^{-1}H$, they are the same) as the inverse element of Ha.

Proof. One wants to show that for all a,b,a', and b' such that $a\sim b$ and $a'\sim b'$, then $aa'\sim bb'$.
Necessary condition :

$$a\sim b\ ,\ a'\sim b'\ ,\ aa'\sim bb' \Rightarrow \exists h,h',h''\in H,\ a = hb,\ a' = h'b'\ ,\ aa' = h''bb'\ ,$$

$$aa' = hbh'b' = h''bb' \Rightarrow hbh' = h''b\ ,$$

then for all b in G and h in H, bhb^{-1} is in H.
Sufficient condition :

$$\left.\begin{array}{l} a\sim b \text{ and } a'\sim b' \Rightarrow aa' = hbh'b'\ \ h,h'\in H \\ H \text{ invariant} \Rightarrow bh'b^{-1} = h''\in H \end{array}\right\} \Rightarrow aa' = hbh'b^{-1}bb' = hh''bb'\ .$$

The multiplication law for the group G/H is :

$$(Ha)(Hb) = H\,ab\ .$$

Prop. C.26. For a finite group G, the order of G/H is equal to the index of H in G : [G : H] .

Def. C.13. A group G is called the direct product of the subgroups H_1, H_2,...,H_n, denoted by G $= H_1 \otimes ... \otimes H_n$, if :
(1) every g in G can be written in a unique way as the product of elements in H_1,...,H_n ,i.e. g $= h_1...h_n$,
(2) elements of different H_i commute.

Prop. C.27. If $G = H_1 \otimes ... \otimes H_n$, the H_i are invariant subgroups of G and the quotient group G/H_i is isomorphic (Def. C.22) to the direct product $\prod_{j\neq i} \otimes H_j$.

Prop. C.28. For a finite group, the order of G reads : $O(G) = \prod_i O(H_i)$.

C.2.2. DERIVED GROUPS

Def. C.14. (cf. Def. B.14) Consider the set of left (or right) commutators of every pair of elements of a group G. These together with the products of any finite number of them form a group, called the commutator subgroup or the first derived group of G, and denoted by G_1.

Def. C.15. (cf. Def. B.15) Proceeding in the above manner, a series of groups called the derived series is constructed, i.e. :

$$G = G_o \supseteq G_1 \supseteq G_2 \ ...\ ,$$

where G_{i+1} is the derived group of G_i. One says that G_i is the ith derived group of G.

Prop. C.29. (cf. Prop. B.21) G_i is an invariant subgroup of G_j for j less or equal to i, and the quotient groups G_i/G_{i+1} are Abelian.

Proof. $\forall g \in G$, $g\,ab\,a^{-1}b^{-1}g^{-1} = a_1 b_1 a_1^{-1} b_1^{-1}$, where $a_1 = gag^{-1}$, $b_1 = gbg^{-1}$.
Also $g\,c_1 \ldots c_n g^{-1} = gc_1 g^{-1} \ldots gc_n g^{-1}$, where $c_1,\ldots,c_n$ are commutators. Thus G_1 is invariant , and more generally G_{i+1} is an invariant subgroup of G_i. Let us now show the transitivity relation ; G_{i+1} is also an invariant subgroup of G_{i-1} :

$$\forall g \in G_{i-1} \quad \forall h \in G_{i+1} \quad ghg^{-1} = gc_1 \ldots c_n g^{-1} = gc_1 g^{-1} \ldots gc_n g^{-1} \quad ,$$

where the c_j are commutators of elements in G_i, i.e. :

$$c_j = a_j b_j a_j^{-1} b_j^{-1} \qquad a_j \; , \; b_j \in G_i \quad .$$

Since G_i is an invariant subgroup of G_{i-1}, one has :

$$\forall g \in G_{i-1} \quad ga_j g^{-1} \in G_i \quad , \quad gb_j g^{-1} \in G_i$$

$$\Rightarrow gc_j g^{-1} = (ga_j g^{-1})(gb_j g^{-1})(ga_j g^{-1})^{-1}(gb_j g^{-1})^{-1} \in G_{i+1} \quad ,$$

hence ghg^{-1} belongs to G_{i+1}.

From Prop. C.25, G_{i+1} being an invariant subgroup of G_i, G_i/G_{i+1} is a group. Moreover, since for all a and b in G_i, $a\,G_i = G_i\,a$ and $a^{-1}b^{-1}ab$ belongs to G_{i+1}, one gets :

$$aG_{i+1}bG_{i+1} = ab\,G_{i+1} = ba(a^{-1}b^{-1}ab)G_{i+1} = ba\,G_{i+1} = bG_{i+1}\,aG_{i+1} \quad .$$

Thus G_i/G_{i+1} is Abelian.

Def. C.16. (cf. Def. B.16) A group G is said to be solvable if the derived series ends with G_n :

$$G_{n-1} \neq G_n = G_{n+1} = \ldots \quad ,$$

and G_n only contains the unit element.

Prop. C.30 (cf. Prop. B.22) If a group different from $\{e\}$ is solvable, i.e. $G_n = \{e\}$, then G_{n-1} is Abelian.

Proof : $\forall g,g' \in G_{n-1}$, $gg'g^{-1}g'^{-1} \in G_n \Rightarrow gg'g^{-1}g'^{-1} = e$.

Prop. C.31. (cf. Prop. B.27) A nonabelian and solvable group is not simple. Equivalently a simple group is Abelian or not solvable.

Proof. From Props. C.29,30, G_{n-1} is an Abelian invariant subgroup of G different from G and $\{e\} = G_n$. Then Prop. C.24 leads to the announced result.

C.2.3. CONJUGATE ELEMENTS AND CLASSES OF CONJUGATE ELEMENTS

Let us distinguish the definitions and properties, respectively, for the subsets S of the group G, and the elements s of G (i.e. the special case $S = \{s\}$).

Def. C.17. Let S be a subset of a group G. The set of elements of the form gsg^{-1} for all s in S, is called the transform of S by g, and is denoted by $S_g = gS\,g^{-1}$.

Prop. C.32. S and S_g have the same number of elements. If S is a subgroup of G, then S_g is also a subgroup of G.

Def. C.17'. The transform of any element s of a group G by the element g is the element $s_g = gsg^{-1}$.

Prop. C.32'. The correspondence between s and s_g is one-to-one.

Def. C.18. For a given subset S of a group G, the subset of elements h in a subgroup H, such that $S_h = S$, is called the normalizer of S in H denoted by $N_H(S)$.

Def. C.18'. For a given subset S of a group G, the subset of elements h in a subgroup H, such that $s_h = s$ for all s in S, is called the centralizer of S in H denoted by $C_H(S)$.

Prop. C.33. $C_H(S) \subseteq N_H(S)$.

Prop. C.34. If $S = \{s\}$ then $C_H(s) = N_H(s)$.

If H = G, one speaks of the normalizer or centralizer of S.

Def. C.19. Two subsets S and S' of a group G are said to be conjugate under H, subgroup of G, if there exists an h in H such that $S' = S_h$.

Def. C.19'. Two elements s and s' of a group G are said to be conjugate under H, subgroup of G, if there exists an h in H such that $s' = s_h$.

Def. C.20. A collection of mutually conjugate subsets or elements is called a class.

Prop. C.35. Defs. C.19, 19' define an equivalence relation which yields a partitioning of the set of subsets of G or elements of G in classes.

Prop. C.36. The number of conjugates of S under H is the index in H of the normalizer of S in H : $[H : N_H(S)]$.

Proof. Two conjugates of S under H are the same if, and only if :

$$hSh^{-1} = h'Sh'^{-1} \Leftrightarrow S = (h^{-1}h')S(h^{-1}h')^{-1} \Leftrightarrow h^{-1}h' \in N_H(S) \Leftrightarrow h' \in hN_H(S) \quad ,$$

i.e. the left coset of $N_H(S)$. Therefore the number of h which generate different conjugates of S is equal to : order of H/order $N_H(S) = [H : N_H(S)]$.

Prop. C.36'. The number of elements in the class of g, conjugate under H, is equal to the index of the normalizer of $\{g\}$ in H : $[H : N_H(g)]$.

Remark : the normalizer $N_H(g)$, identical to the centralizer $C_H(g)$, of an element g is the subset of elements h in H such that $g = hgh^{-1}$, i.e. the subset of h in H which commute with g. Thus $N_H(g)$ contains at least the intersection of H with the centre of G, in particular if G is Abelian $N_H(g) = H$.

Prop. C.37. A finite group G of order n is the disjoint union of a finite number c of classes of elements conjugate under H : $G = C_1 + \ldots + C_i + \ldots + C_c$. Let n_i be the order of the class C_i, then n_i is a divisor of n :

$$n = \sum_{i=1}^{c} n_i \quad , \quad n/n_i \text{ integer} .$$

Proof. From Lagrange's theorem the index of $N_H(g)$ is a divisor of the order of H, which is itself a divisor of the order of G.

Prop. C.38. A necessary and sufficient condition for a group to be Abelian is that each class of conjugate elements under G itself contains only one element.

Prop. C.39. An invariant subgroup is a union of classes of conjugate elements under G.

C.2.4. AN EXAMPLE OF A FINITE GROUP : THE SYMMETRIC GROUP Σ_3

Let us consider the group of permutations of three distinct objects. This group is isomorphic to the group of symmetry operations which leave an equilateral triangle invariant.

i) The group table of this finite group G of order 3! = 6 is easy to establish, let us denote its elements as : e,a,a^2,b,ab,a^2b :

$G = \Sigma_3$

e	a	a^2	b	ab	a^2b
a	a^2	e	ab	a^2b	b
a^2	e	a	a^2b	b	ab
b	a^2b	ab	e	a^2	a
ab	b	a^2b	a	e	a^2
a^2b	ab	b	a^2	a	e

This group is noncommutative ($ba \neq ab$ for instance) and its centre is $C = \{e\}$.

ii) Subgroups. From Lagrange's theorem, the subgroups can be of order 1,2,3, or 6. The proper ones are :

$H_2 = \{e,b\}$, $\{e,ab\}$, $\{e,a^2b\}$ index $[G : H_2] = 6/2 = 3$,
$H_3 = \{e,a,a^2\}$ $[G : H_3] = 6/3 = 2$.

From Prop. C.23, the prime subgroups H_2 and H_3 are cyclic and thus Abelian as shown by the symmetry properties of the group table.

iii) Classes. Let us consider the classes of elements Cl_H conjugate under H first with H = G, and then $H = H_3$. The unit element e forms a class by itself. To find other members of the class of an element g one forms the products $hgh^{-1} = g'$ with h in H (we need not use for h any elèment which commutes with g since $hgh^{-1} = g$). To check Prop. C.36' one can find the elements of $N_H(g)$, i.e. the elements in H which commute with g and then one must have the order of the class of G under H equal to $[H : H_N(g)]$:

1 - H = G :
$Cl_G(e) = \{e\}$,
$Cl_G(a)$: a and a^2 commute,

$$ba\,b^{-1} = bab = a^2 \quad , \quad ab\,a(ab)^{-1} = ab\,a\,ab = a^2 \quad , \quad a^2b\,a(a^2b)^{-1} = a^2ba\,a^2b = a^2$$

$$\Rightarrow Cl_G(a) = \{a,a^2\}, \text{ order } 2 \quad .$$

One finds = $N_G(a) = \{e,a,a^2\}$, order 3 and thus :

$$\text{order of } Cl_G(a) = 2 = \frac{\text{order of } G}{\text{order of } N_G(a)} = \frac{6}{3} \quad .$$

The partitioning of G in classes of elements conjugate under G is :

$$G = \{e\} + \{a,a^2\} + \{b,ab,a^2b\} \quad .$$

2 - $H = H_3$. The partitioning of G is then :

$$G = \{e\} + \{a\} + \{a^2\} + \{b,ab,a^2b\} \quad .$$

One checks that for the Abelian subgroup H_3, each element is a class by itself (Prop. C.38).

iv) Cosets. Let us first find the right and left cosets of the subgroup $H_2 = \{e,b\}$: each coset contains two elements. Elements of a coset other than g are found by considering the products hg (or gh) with h in H_2. One obtains :

right cosets : $G = H_2 + \{a,a^2b\} + \{a^2,ab\}$,
left cosets : $G = H_2 + \{a,ab\} + \{a^2,a^2b\}$,

which are different since H_2 is not an invariant subgroup of G ; indeed, it is not a union of classes of conjugate elements under G (Prop. C.39).

On the other hand the subgroup H_3 is invariant and the right and left cosets are the same :

$$G = H_3 + \{b,ab,a^2b\} \quad .$$

The quotient group G/H_3 has thus two elements (2 = index of H in G).

C.3. GENERALITIES ABOUT REPRESENTATIONS

C.3.1. HOMOMORPHISMS, REALIZATIONS, AND REPRESENTATIONS

i) **Def. C.21.** A mapping M from a group G_1 into a group G_2 is a homomorphism if it is compatible with the group laws, i.e. :

$$\forall g,g' \in G_1 \quad M(gg') = M(g)M(g') \quad , \quad M(g^{-1}) = M(g)^{-1}, \; M(e_1) = e_2 \quad .$$

Prop. C.40. The image M (G_1) and the kernel $K = M^{-1}(e_2)$ of the homomorphism M are subgroups of, respectively, G_2 and G_1. Moreover K is an invariant subgroup of G_1.

Def. C.22. If the homomorphism M is a one-to-one mapping then it is an isomorphism. If moreover $G_1 = G_2$, M is said to be an automorphism.

Prop. C.41. If M is an homomorphism from G_1 into G_2, then the factor group G_1/K is isomorphic to G_2.

Def. C.23. The set of the automorphisms of a group G is also a group Aut(G). Among them the internal automorphisms are defined by (cf. Def. C.17') :

$$\forall\, g_1 \in G \rightarrow a(g_1) \in \mathrm{Aut}(G), \qquad \forall\, g \in G \xrightarrow{a(g_1)} g_{g_1} = g_1 g g_1^{-1} \quad .$$

ii) **Def. C.24.** If G_1 and G_2 are two groups, the product group $G = G_1 \times G_2$ is the set of pairs (g_1,g_2), g_1 in G_1 and g_2 in G_2, with the group law :

$$(g_1,g_2)(g_1',g_2') = (g_1g_1',g_2g_2') \quad .$$

Def. C.25. If there exists a homomorphism of G_1 in $\mathrm{Aut}(G_2)$, denoted by : $\forall\, g_1 \in G_1,\ \forall g_2 \in G_2 \quad g_2 \xrightarrow{g_1} {}^{g_1}g_2$, the semi-direct product of G_2 by G_1 is the set of pairs (g_1,g_2) with the following group law :

$$(g_1,g_2)(g_1',g_2') = (g_1g_1',g_2\,{}^{g_1}g_2') \quad .$$

Examples : Eqs. (1-58,62).

iii) **Def. C.26.** Let E be a set, the one-to-one mappings of E onto E form a group of generalized permutations $\Sigma(E)$, which is identical to Σ_n if E is a finite set with n elements. Let M be a homomorphism of a group G in $\Sigma(E)$, then one says that G has been realized as a transformation group on E.

Prop. C.42. Cayley's theorem : any finite group of order n is isomorphic to a subgroup of the permutation group Σ_n , i.e. the permutations which, except the identity, do not leave invariant every element. Actually, this theorem is another form of the rearrangement theorem.

Def. C.27. (cf. Def. B.17) Let $L(E)$ be the set of linear applications of a vector space E into itself. A homomorphism D of a group G into $L(E)$ is a (linear) representation of G in E. It is also customary to improperly call representation the image of G in the homomorphism D, i.e. $D(G) \subset L(E)$. The mappings, denoted D(g), g in G, are then nonsingular linear mappings of E into itself. The dimension of E is said to be the dimension of the representation.

Let us consider a finite-dimensional representation D(G), (e_i) denotes a basis of the representation space E. Then to any g in G one associates the matrix representative of the mapping D(g), defined by Eq. (1–10) :

$$D(g)e_i = e_j\, D(g)^j_i \quad .$$

The characteristic property is that to the product of elements of G corresponds the product of the matrices :

$$D(g_1g_2)^j_i = D(g_1)^j_k\, D(g_2)^k_i \quad ,$$

$$D(e)^j_i = \delta^j_i \quad , \quad D(g^{-1})^j_i = (D(g)^{-1})^j_i \quad .$$

Def. C.28. (cf. Def. B.18) The representation D(G) is said to be faithful if it is an isomorphism.

Def. C.29. The representation D(G) is unitary if for every g the mapping D(g) is an isometric application.

Def. C.30. (cf. Def. B.19) Two representations D(G) and D'(G) of the same group G in the vector spaces, respectively, E and E' (possibly E = E') over the same field are said to be equivalent (otherwise inequivalent) if there exists a regular linear mapping A such that :

$$\forall g \in G \qquad D'(g) = A\, D(g)\, A^{-1} \quad .$$

Def. C.31. For a given representation D(G) the character is the function χ of G into the field defined by :

$$\forall g \in G \qquad \chi(g) = \mathrm{Tr}[D(g)] \quad ,$$

where Tr means the trace (assuming it exists).

Prop. C.43. For finite-dimensional representations, the invariance of the trace of a product of operators under a cyclic permutation of these operators implies that the character is a class function, i.e. has the same value for all conjugate elements under the group. Also finite-dimensional equivalent representations have the same characters.

C.3.2. REDUCIBLE AND IRREDUCIBLE REPRESENTATIONS

C.3.2.1. Definitions

Def. C.32. (cf. Def. B.20) If there exists a subspace of E, not equal to {0} or E, which is invariant for all D(g), g in a group G, then the representation D(G) is said to be reducible otherwise irreducible.

For a reducible representation D(G), there exists a basis of E such that for all g in G the matrix representative of D(g) takes the form :

$$D(g) = \begin{pmatrix} A(g) & B(g) \\ 0 & C(g) \end{pmatrix}$$

Def. C.33. A representation is said to be completely reducible or decomposable if to any invariant subspace there corresponds a complementary subspace which is also invariant.

For a completely reducible representation D(G), there exists a basis of E such that for all g in G the matrix representative of D(g) reads :

$$D(g) = \begin{pmatrix} D^1(g) & & 0 \\ & D^2(g) & \\ 0 & & \ddots \end{pmatrix} ,$$

where the $D^{\alpha}(g)$ are irreducible representations. Symbolically a completely reducible representation is written as :

$$D(G) = D^1(G) \oplus D^2(G) \oplus \ldots = \sum_{\alpha} \oplus\, a_{\alpha}\, D^{\alpha}(G) \quad ,$$

where the same representation $D^{\alpha}(G)$ may appear a_{α} times (a_{α} being a non-negative integer). The character χ of D(G) is given by an ordinary addition :

$$\forall\, g \in G \qquad \chi(g) = \sum_{\alpha} a_{\alpha}\, \chi^{\alpha}(g) \quad ,$$

where χ^{α} is the character of $D^{\alpha}(G)$. Taking g equal to the unity of the group, one has : $d = \sum_{\alpha} a_{\alpha} d_{\alpha}$, where d is the dimension of the representation D(G) and d_{α} the dimension of the irreducible representation $D^{\alpha}(G)$.

Prop. C.44. A unitary representation is either irreducible or completely reducible.

Proof. If E_1 is an invariant subspace of E, its orthogonal subspace E_2 is also an invariant complementary of E_1.

C.3.2.2. Schur's lemma and related properties

Prop. C.45. (cf. Prop. B.23) Schur's lemma : let D(G) and D'(G) be two irreducible representations of a group G, respectively, in the vector spaces E and E' over the same field, and A a linear mapping from E' into E. If :

$$\forall\, g \in G \qquad D(g)A = A\, D'(g) \quad ,$$

then A is either zero or a bijective mapping. In the latter case D(G) and D'(G) are equivalent representations.

Proof. From the above equation, the image A(E') and the kernel, $\overset{-1}{A}(0)$ are invariant subspaces, respectively, of E and E'. Let A be a nonzero mapping then : since D(G) is irreducible A(E') = E, i.e. A is a surjective mapping ; since D'(G) is irreducible $\overset{-1}{A}(0) = \{0\}$, thus A is an injective mapping. Finally A is a one-to-one mapping and the representations D(G) and D'(G) are equivalent.

Prop. C.46. (cf. Prop. B.24) Let D(G) be a finite-dimensional irreducible representation of a group G in a vector space over a commutative and algebraically closed field K. If there exists a nonzero linear mapping A such that :

$$\forall\, g \in G \qquad D(g)A = A\, D(g) \quad ,$$

then A is a multiple of the identity mapping :

$$A = \lambda\, \mathbf{1} \qquad \lambda \in K \quad .$$

Proof. From Schur's lemma A is a bijective mapping. Since the field of definition of the vector space is commutative the determinant of a finite-dimensional matrix is well defined. Moreover this field being algebraically closed, the characteristic equation $\det(A - \lambda\, \mathbf{1}) = 0$, a polynomial in λ, has its roots in the field. Since A is bijective these eigenvalues are nonzero.

Then from Prop. C.45, the singular mapping $A-\lambda\,\mathbb{1}$ which commutes with all D(g) is equal to zero.

Let us emphasize that Schur's lemma is true whatever the dimension of the vector space of the representation, and the field of definition of this vector space. On the other hand, for Prop. C.46 the following sufficient conditions have been used : (1) finite-dimensional vector space, (2) commutative and algebraically closed field. This is an important remark since many of the properties of group representations (cf. Appendix C.4 and Chap. 7) are based on this theorem. In practice one considers the complex field, nevertheless it is worthwhile to note that Prop. C.46 is not true for the nonalgebraically closed real field. For instance the two-dimensional rotation matrices $\begin{pmatrix}\cos\theta & -\sin\theta\\ \sin\theta & \cos\theta\end{pmatrix}$ form an irreducible representation of the rotation group in the two-dimensional Euclidean vector space over R. And still there exist real matrices, different from multiples of the unity, which commute with all these rotation matrices :

$$\forall\, a,b \in R \qquad \begin{pmatrix}\cos\theta & -\sin\theta\\ \sin\theta & \cos\theta\end{pmatrix}\begin{pmatrix}a & -b\\ b & a\end{pmatrix} = \begin{pmatrix}a & -b\\ b & a\end{pmatrix}\begin{pmatrix}\cos\theta & -\sin\theta\\ \sin\theta & \cos\theta\end{pmatrix} \quad .$$

On the other hand, the above representation is reducible over the complex field, the equivalent representation reads : $\begin{pmatrix}\exp i\theta & 0\\ 0 & \exp -i\theta\end{pmatrix}$.

Prop. C.47. Let D(G) and D'(G) be two finite-dimensional equivalent irreducible representations of a group G in a vector space E over a commutative and algebraically closed field K. Then :

(1) The regular linear mapping A such that :

$$\forall\, g \in G \qquad D'(g) = A\, D(g)\, A^{-1} \quad ,$$

is uniquely defined up to a multiplicative constant in the field K.

(2) Furthermore, if D(G) and D'(G) are unitary representations with respect to a nondegenerate inner product in E, one has :

$$AA^{+} = A^{+}A = \lambda\,\mathbf{1} \qquad \lambda \in K = C \quad ,$$

where A^{+} denotes the adjoint mapping of A. Then A is a multiple of a linear one-to-one isometric application.

Proof. (1) Let us suppose there exist two regular linear mappings A and A' such that :

$$\forall\, g \in G \qquad D'(g) = A\, D(g)\, A^{-1} = A'\, D(g)\, A'^{-1} \quad .$$

Then one has :

$$\forall\, g \in G \qquad A'^{-1}A\, D(g) = D(g)\, A'^{-1}A \quad ,$$

and from Prop. C.46 $A'^{-1}A$ is a multiple of the identity mapping.

(2) The adjoint mapping A^{+} of the mapping A, with respect to a nondegenerate inner product in the vector space E, is defined by :

$$\forall u,v\in E \qquad (u,Av) = (A^+u,v) \quad .$$

Since the inverse of a one-to-one isometric mapping is equal to its adjoint mapping, one gets :

$$\forall g\in G \quad D'(g)^{-1} = AD(g)^{-1}A^{-1} = D'(g)^+ = A^{-1+}D(g)^+A^+ \Rightarrow \forall g\in G \quad A^+A\,D(g)^{-1} = D(g)^{-1}A^+A \ .$$

Thus, from Prop. C.46, A^+A is a multiple of the identity mapping.

Prop. C.48. Let D(G) be a finite-dimensional representation of a group G in a vector space over a commutative and algebraically closed field. If there exists a linear mapping A different from a multiple of the identity, such that :

$$\forall\, g\in G \qquad D(g)A = A\,D(g) \quad ,$$

then the representation D(G) is reducible.

Proof. Direct consequence of Prop. C.46, a representation being either irreducible or reducible.

Prop. C.49. Let D(G) be a finite-dimensional completely reducible, but not irreducible representation of a group G. Then there exists a linear mapping A, different from a multiple of the unity, and such that :

$$\forall\, g\in G \qquad D(g)A = A\,D(g) \quad .$$

Proof. There exists a basis of the representation space such that the matrix representative of D(g), for all g in G, takes the form given in Def. C.33. Then any mapping, which has with respect to the same basis, the following matrix representative :

$$A = \begin{pmatrix} \lambda_1 \mathbb{1}^1 & & 0 \\ & \lambda_2\mathbb{1}^2 & \\ 0 & & \ddots \end{pmatrix} \quad ,$$

where $\mathbb{1}^i$ are unit submatrices and λ_i elements of the field, commutes with all D(g).

Prop. C.50. Let D(G) be a finite-dimensional either irreducible or completely reducible representation of a group G. If the only linear mappings A, such that :

$$\forall g\in G \qquad D(g)A = A\,D(g) \quad ,$$

are multiple of the unity, then the representation D(G) is irreducible.

Proof. Direct consequence of Prop. C.49.

Prop. C.51. The finite-dimensional irreducible representations of an Abelian group, in a space over a commutative and algebraically closed field, are necessarily of dimension one.

Proof : $\forall\, g,g' \in G \qquad D(g)D(g') = D(g')D(g) \quad .$
Then Prop. C.46 leads to $D(g) = \lambda(g)\,\mathbb{1}$, and to be irreducible the dimension must be one.

C.4. REPRESENTATIONS OF FINITE GROUPS

Let us give some important results for the representations of finite groups since many of them can be extended to compact Lie groups (cf. Appendix C.5 and Chap. 7). Most of these properties are based on the rearrangement theorem (Prop. C.6), and on Prop. C.46 which is a direct consequence of Schur's lemma adding the, at least sufficient, conditions : finite-dimensional representation in a vector space over a commutative and algebraically closed field. In practice one considers the complex field, and it is recalled that Prop. C.46 is not true over the real field.

C.4.1. GENERAL RESULTS

Prop. C.52. The irreducible representations of a finite group of order n are at most of dimension n.

Proof. Let D(G) be a representation of a finite group G of order n in a vector space E. The relation among the vectors of E :

$$v, v' \in E \qquad v \sim v' \quad \Leftrightarrow \quad \exists\, g \in G,\ v' = D(g)v \quad ,$$

clearly defines an equivalence relation. This implies a partitioning of E in equivalence classes which are invariant subspaces of E of dimension at most n. Indeed from a nonzero vector v in E one can at most generate n distinct vectors D(g)v for all g in G.

Cf. Props. C.65,67 for other constraints for the dimensions of irreducible representations of finite groups.

Prop. C.53. The irreducible representations of a finite group of prime order, in a vector space over a commutative and algebraically closed field, are of dimension one.

Proof. Consequence of Props. C.23,51, and 52.

Prop. C.54. The representations of a finite group in a vector space over the complex field are equivalent to unitary representations. Thus, they are either irreducible or completely reducible.

Proof. From Prop. C.52, it is sufficient to prove the result for finite-dimensional representations, therefore one can consider the matrix representative. Let n be the order of the group G and consider the Hermitian positive definite matrix :

$$H = \frac{1}{n} \sum_{g \in G} D^+(g)\, D(g) \quad .$$

There exists a nonsingular matrix V such that $H = V^+V$. Using the rearrangement theorem, one has :

$$\forall\, g' \in G \qquad D^+(g')H\,D(g') = \frac{1}{n} \sum_{g \in G} D^+(gg')\, D(gg') = H \quad .$$

Therefore :

$$\forall\, g \in G \qquad D^+(g)V^+V\, D(g) = V^+V \quad ,$$

which implies that the matrix $D'(g) = VD(g)V^{-1}$ is unitary and defines a representation D'(G) of G equivalent to D(G). As a consequence of this property, one has :

$$\forall\, g \in G \qquad |\det D(g)| = 1 \quad .$$

From Prop. C.44 these unitary representations are either irreducible or completely reducible.

C.4.2. ORTHOGONALITY RELATIONS

Throughout the remainder of this Appendix C.4, only finite-dimensional unitary representations in vector spaces over the complex field are considered (cf. Prop. C.54 and remarks after Prop. C.55).

Notations. G is a finite group of order n, with elements g. This group is the disjoint union of c classes of conjugate elements under G with, respectively, n_i elements : $n = \sum_{i=1}^{c} n_i$. D^α (G) and D^β(G) are two irreducible representations of finite dimensions, respectively, d_α and d_β . Their characters are, respectively, χ_i^α and χ_i^β for the elements g in the class i.

Prop. C.55. Orthogonality relations :

$$\sum_{g\in G} D^\alpha(g)^i_k \, D^\beta(g)^{j*}_\ell = \frac{n}{d_\alpha} \delta_{k\ell} \, \delta^{ij} \delta^{\alpha\beta} \quad ,$$

$$\sum_{i=1}^{c} n_i \chi_i^\alpha \chi_i^{\beta *} = n \, \delta^{\alpha\beta} \quad .$$

Remark. More general results can be obtained in vector spaces over commutative and algebraically closed fields (to allow us to use Prop. C.46). Then the representations are no longer unitary, and one has $D^\beta(g)^{-1}$ instead of $D^\beta(g)^*$, and $\chi^\beta_{i'}$ instead of $\chi_i^{\beta *}$, where i' denotes the class of the inverse elements of class i.

Proof. Using the rearrangement theorem, one shows that for an arbitrary matrix A, the matrix B defined by :

$$B = \sum_{g\in G} D^\alpha(g) A \, D^\beta(g^{-1}) \quad ,$$

fulfils the relation :

$$\forall \; g\in G \qquad D^\alpha(g) B = B \, D^\beta(g) \quad .$$

Then from Schur's lemma and Prop. C.46 : if $\alpha \neq \beta$, D^α (G) and D^β (G) are inequivalent and B must be zero ; if $\alpha = \beta$ then B is a multiple of the identity, i.e. :

$$\sum_{g\in G} D^\alpha(g) A \, D^\beta(g^{-1}) = \lambda(A) \, \mathbb{1} \, \delta^{\alpha\beta} \quad ,$$

where the coefficient λ (A) is given by computing the trace of the above matrix :

$$\mathrm{Tr}\left[\sum_{g\in G} D^\alpha(g) A \, D^\beta(g^{-1})\right] = \lambda(A) \, d_\alpha \, \delta^{\alpha\beta} \quad .$$

The first orthogonality relation is obtained by setting :

$$A = A_{k\ell} \quad , \qquad (A_{k\ell})^m_n = \delta_{km} \delta_{\ell n} \quad .$$

Taking $i = k$ and $j = \ell$, and summing over i and j, one obtains the second one, the value of the character being the same for all g in the same class.

Prop. C.56. The number r of inequivalent irreducible representations is finite and we have the inequalities :

$$\sum_{\alpha=1}^{r} d_\alpha^2 \leqslant n \quad , \qquad r \leqslant c \quad .$$

Later (Props. C.65,67) we shall see that the equalities hold.

Proof. One considers the $D^{\alpha}(g)^{i}{}_{k}$, for given α, i, and k, as the components of an n–dimensional vector. There are $\sum_{\alpha=1}^{r} d_{\alpha}^{2}$ such vectors which are from the orthogonality relation mutually orthogonal. Their number is thus at most equal to the dimension n of the space. Similarly the r c–dimensional vectors (α = 1,...,r), with components $\sqrt{n_i}\,\chi_i^{\alpha}$ (i=1,...,c) are mutually orthogonal and thus $r \leqslant c$.

Prop. C.57. The decomposition of a finite-dimensional reducible representation D(G) into the irreducible representations $D^{\alpha}(G)$:

$$D(G) = \sum_{\alpha} \oplus a_{\alpha} D^{\alpha}(G) \quad , \quad \chi_i = \sum_{\alpha} a_{\alpha} \chi_i^{\alpha} \quad ,$$

is unique and completely determined by its character χ_i. One has :

$$a_{\alpha} = \frac{1}{n} \sum_{i=1}^{c} n_i \chi_i^{\alpha *} \chi_i \quad ,$$

$$\sum_{\alpha=1}^{r} a_{\alpha}^{2} = \frac{1}{n} \sum_{i=1}^{c} n_i |\chi_i|^2 \quad .$$

Proof and remarks. The above equations are direct consequences of the orthogonality relation of the characters. The first one uniquely determines the coefficients a_{α} from the character χ_i of the representation D(G).

A representation is irreducible if all a_{α} vanish, except one which must be one, then $n = \sum_{i=1}^{c} n_i |\chi_i|^2$. Thus, the second above equation may be used to establish whether a representation is irreducible or not. For instance if one has :

$\sum_{\alpha} a_{\alpha}^{2} = 1 \Rightarrow$ D(G) is irreducible : $a_{\alpha_1} = 1, a_{\alpha_i} = 0 \quad i \neq 1$,

$\sum_{\alpha} a_{\alpha}^{2} = 2 \Rightarrow$ D(G) contains two nonequivalent irreducible representations :

$a_{\alpha_1} = a_{\alpha_2} = 1,\ a_{\alpha_i} = 0 \quad i \neq 1$ and 2.

Prop. C.58. A necessary and sufficient condition for two finite-dimensional representations to be equivalent is that they have the same characters :

$$\forall i \qquad \chi_i = \chi_i' \quad .$$

Proof. From Prop. C.43 the condition is necessary. It is also sufficient since if $\chi_i = \chi'_i$ for all i, then from Prop. C.57 the two representations have the same decomposition in irreducible representations, and are thus equivalent.

C.4.3. THE ALGEBRA OF A FINITE GROUP AND THE REGULAR REPRESENTATIONS

i) **Def. C.34.** The set of formal linear combinations $\sum_{g \in G} z_g g$ of elements of the group G, where z_g are complex numbers, forms the algebra of the group over the complex field. The composition laws are defined by :

$$\sum_{g \in G} z_g g + \sum_{g \in G} z'_g g = \sum_{g \in G} (z_g + z'_g) g \quad , \quad \lambda \sum_{g \in G} z_g g = \sum_{g \in G} \lambda z_g g \quad ,$$

$$\sum_{g \in G} z_g g \sum_{g' \in G} z'_{g'} g' = \sum_{g, g' \in G} z_g z'_{g'} g g' \quad .$$

With respect to the basis (g) of the algebra, the structure constants $c_{gg'}{}^{g''}$ (keeping the elements of the group as indices) are defined by (Def. B.4) :

$$gg' = c_{gg'}{}^{g''} g'' ,$$

where :

$$c_{gg'}{}^{g''} = \begin{cases} 1 & \text{if } g'' = gg' \\ 0 & \text{otherwise} \end{cases} .$$

Def. C.35. The <u>class operators</u> C_i are the c elements of the algebra of the group defined by :

$$i=1,\ldots,c \qquad C_i = \sum_{g\in \text{class } i} g .$$

Prop. C.59. The class operators commute with all elements of the group algebra, and any element of the group algebra which has this property is a linear combination of the C_i.

Proof. First part of the property :

$$\forall g_o \qquad g_oC_i = g_o \sum_{g\in\text{class } i} g = \sum_{g\in\text{class } i} g_ogg_o^{-1}g_o = \sum_{g\in\text{class } i} gg_o = C_ig_o ,$$

since for all g_o, g and $g_ogg_o^{-1}$ belong to the same class.
Second part of the property : let us assume that $C = \sum_{g\in G} z_g g$ commutes with all g_o in the algebra, then from the rearrangement theorem :

$$\sum_{g\in G} z_g gg_o = \sum_{g\in G} z_g g_og = \sum_{g\in G} z_g g_og g_o^{-1} g_o = \sum_{g\in G} z_{g_o^{-1}gg_o} gg_o .$$

This implies the equality of the components z_g and $z_{g_o^{-1}gg_o}$ on the same element gg_o, for all g and g_o. Hence in the expression of C the components z_g for g belonging to the same class are the same, i.e. C is of the type :

$$C = \sum_{i=1}^{c} z_iC_i .$$

Prop. C.60. The set of linear combinations of the class operators forms a commutative algebra : the <u>class algebra</u>. The structure constants are non-negative integers $c_{ij}{}^k$, such that :

$$C_iC_j = \sum_{k=1}^{c} c_{ij}{}^k C_k .$$

Proof. Let $C_iC_j = C_jC_i = \sum_{g\in\text{class } i} \sum_{g'\in\text{class } j} gg'$. The right hand side contains complete classes. Indeed if $gg' = g''$ belongs to class k, all elements of this class are obtained by considering $hg''h^{-1}$ for all h in G. But $hg''h^{-1}$ is also in the right hand side of the above formula since :

$$hg''h^{-1} = hgh^{-1}h g'h^{-1} .$$

ii) **Prop. C.61.** There is a one-to-one correspondence between the representations of the group and the representations of the algebra of the group. The relation given in Prop. C.60 is thus true for all representations (cf. Secs. 4.3 and 5.3). For instance for an irreducible representation $D^\alpha(G)$, to the class operator C_i corresponds the matrix :

$$D_i^\alpha = \sum_{g\in\text{class } i} D^\alpha(g) ,$$

where Σ denotes the usual sum of matrices.

Prop. C.62. The matrices D_i are multiples of the unity. One deduces the relation between characters and the structure constants $c_{ij}{}^k$:

$$n_i n_j \chi_i^\alpha \chi_j^\alpha = \chi_e^\alpha \, c_{ij}{}^k \, n_k \chi_k^\alpha \quad ,$$

where e labels the class of the unit element.

Proof. From Prop. C.59 applied to the representation $D^\alpha(G)$, one has :

$$\forall g \in G \qquad D^\alpha(g) D_i^\alpha = \sum_{g' \in \text{class } i} D^\alpha(gg'g^{-1}) D^\alpha(g) = D_i^\alpha \, D^\alpha(g) \quad .$$

Then from Prop. C.46, $D_i^\alpha = \lambda_i \, \mathbf{1}$, where λ_i is determined by taking the trace :

$$\mathrm{Tr}\, D_i^\alpha = n_i \chi_i^\alpha = \lambda_i d_\alpha = \lambda_i \chi_e^\alpha \quad .$$

Furthermore, for the representation $D^\alpha(G)$ the relation given in Prop. C.60 reads :

$$D_i^\alpha \, D_j^\alpha = c_{ij}{}^k \, D_k^\alpha \Rightarrow \lambda_i \lambda_j = c_{ij}{}^k \, \lambda_k \quad ,$$

and thus one obtains the announced equation.

iii) **Def. C.36.** The <u>left and right regular representations,</u> $D^\ell(G)$ and $D^r(G)$,respectively, of a finite group G are the two representations in its Lie algebra defined by :

$$\forall\ g, g' \in G \qquad D^\ell(g) g' = gg' \quad , \qquad D^r(g) g' = g' g^{-1} \quad .$$

A slightly different definition of the regular representations of any group is given in Subsecs. 10.1.3.2,3, as well as most of their properties. These representations are actually closely related to the realizations of the group determined from the left and right translations (Def. C.51 and Subsec. 10.1.3.1).

There often exists a confusion in the literature about the words regular and adjoint. The regular representations of a finite group as defined above are also sometimes called the <u>adjoint representations.</u> Although these representations are defined in the Lie algebra of the group, they do not correspond to the adjoint representation of a Lie group in its Lie algebra as considered in Sec. 5.4.

With respect to the basis (g) of the Lie algebra of G, the matrix representatives of $D^\ell(g)$ and $D^r(g)$ read :

$$(D^\ell(g))_{g''}^{g'} = \begin{cases} 1 & \text{if } g' = gg'' \\ 0 & \text{otherwise} \end{cases} \quad , \qquad (D^r(g))_{g''}^{g'} = \begin{cases} 1 & \text{if } g' = g'' g^{-1} \\ 0 & \text{otherwise} \end{cases} \quad .$$

The matrix representative of for instance $D^\ell(g)$ can also be obtained from the group table of G established as follows : (1) order the rows by putting the elements in some convenient order, (2) replace each element in the list by its inverse and use the new list to establish the order of the columns, (3) write the products in the order row–column. The matrix representative of $D^\ell(g)$ with respect to the elements ordered as the rows is then given by setting 1 wherever g occurs in the above group table and zero everywhere else :

	e	g^{-1}	...
e	e		
g		e	
⋮			⋱

Prop. C.63. The regular representations are faithful, the left one commutes with the right one, and they are equivalent.

Proof of the equivalence : $\forall g \in G \quad D^{\ell}(g) = I D^{r}(g) I$, where $I(g) = g^{-1}$.

Prop. C.64. The character of the left, as well as of the right, regular representation reads :

$$\chi^{r}(g) = n\,\delta_{g,e} \quad .$$

Prop. C.65. Each irreducible representation appears in the regular representations a number of times a_α equal to its dimension d_α, and one has (Theorem 10.2, Subsec. 10.1.3.3) :

$$n = \sum_{\alpha=1}^{r} d_\alpha^2 \quad .$$

Proof. $\chi_i^r = \sum_{\alpha=1}^{r} a_\alpha \chi_i^\alpha$, where from Props. C.57, 64 :

$$a_\alpha = \frac{1}{n} \sum_{i=1}^{c} n_i \chi_i^{\alpha *} \chi_i^r = d_\alpha \quad .$$

Thus : $n = \sum_{\alpha=1}^{r} a_\alpha d_\alpha = \sum_{\alpha=1}^{r} d_\alpha^2$.

Prop. C.66. Closure relation for the characters :

$$\sum_{\alpha=1}^{r} \chi_i^\alpha \chi_j^{\alpha *} = \frac{n}{n_i} \delta_{ij} \quad .$$

Proof. If one denotes by the index e the class of the unit element e, one has from Props. C.64,65 :

$$\chi_i^r = n\,\delta_{ie} = \sum_{\alpha=1}^{r} a_\alpha \chi_i^\alpha = \sum_{\alpha=1}^{r} d_\alpha \chi_i^\alpha = \sum_{\alpha=1}^{r} \chi_e^\alpha \chi_i^\alpha \quad .$$

Furthermore, summing over α the equation given in Prop. C.62 one obtains :

$$n_i n_j \sum_{\alpha=1}^{r} \chi_i^\alpha \chi_j^\alpha = c_{ij}^{\ k} n_k \sum_{\alpha=1}^{r} \chi_e^\alpha \chi_k^\alpha = c_{ij}^{\ k} n_k n \delta_{ke} = n c_{ij}^{\ e} \quad .$$

But $c_{ij}^{\ e} = n_i \delta_{ij'}$, where j' denotes the class of the inverse elements of class j. Moreover $n_j = n_{j'}$ and since the representations are unitary $\chi_{j'}^\alpha = \chi_j^{\alpha *}$. Thus one obtains the closure relation for the characters.

Prop. C.67. The number of nonequivalent irreducible representations of a finite group is equal to its number of classes : $r = c$.

Proof. Using the same arguments as in Prop. C.56, one obtains that $c \leqslant r$, and since this property implies $r \leqslant c$, r equals c. The matrix with matrix elements $\sqrt{n_i/n}\,\chi_i^\alpha$, with indices α and i, is thus a unitary matrix.

As a consequence, for Abelian groups the number of nonequivalent irreducible representations equals the order of the group, since all elements of the group form a class by themselves ($r = c = n$).

Prop. C.68. The number of nonequivalent one-dimensional representations of a group G is equal to the order of G divided by the order of its commutator group G_1 : $[G : G_1]$.

Proof. It has been shown (Prop. C.29) that the factor group G/G_1 is Abelian. Therefore all irreducible representations of the factor group, which are obviously also representations of G, are one-dimensional (Prop. C.51) and in number equal to the order of G/G_1 (Props. C.38,67). Conversely, every one-dimensional representation D(G) of G defines a representation of G/G_1 as follows :

$$\forall\ gG_1 \in G/G_1 \quad \rightarrow \quad D(g) \quad ,$$

which is possible because if g is in G_1 one has :

$$D(g) = D(aba^{-1}b^{-1}) = D(a)D(b)D(a)^{-1}D(b)^{-1} = \mathbf{1} ,$$

since the representation is one-dimensional and therefore Abelian.

As a consequence of this property and Lagrange's theorem, the number of nonequivalent one-dimensional representations of a group G is a divisor of the order of the group.

Example. For the symmetric group Σ_n of permutations on n objects, the commutator subgroup is the alternating group consisting of all the even permutations. Its index is two. The symmetric group has only two nonequivalent one-dimensional representations, (+1) and ($\pm$1) (Appendix C.4.4.1).

C.4.4. EXAMPLES

Let us recall schematically the different steps to study the finite-dimensional irreducible representations of a finite group in a vector space over the complex field :

(1) The number r of nonequivalent irreducible representations equals the number c of classes : r = c (Prop. C.67).

(2) The dimensions d_α of the irreducible representations fulfil several relations :

$$\sum_{\alpha=1}^{r} d_\alpha^2 = n \qquad \text{(Prop. C.65)} \quad ,$$

–the number of nonequivalent one-dimensional representations equals $[G : G_1]$ (Prop. C.68).

–more sophisticated relations given in the literature[36)], for instance d_α is a divisor of n.

(3) The explicit construction of the irreducible representations may be performed by means of the characters, which are class functions, and fulfil the orthogonality and closure relations (respectively, Props. C.55 and C.66). It is sufficient to consider only unitary representations (Prop. C.54). One can also use the regular representationswhich contain all the irreducible representations, and the d_α-dimensional representations D^α is contained d_α times (Prop. C.65).

C.4.4.1. The symmetric group Σ_3

i) This group has been studied in Appendix C.2.4. There are three classes of conjugate elements under this group :

$$\text{class } 1 = \{e\} \ , \ \text{class } 2 = \{a,a^2\} \ , \ \text{class } 3 = \{b,ab,a^2b\} \ .$$

Thus there are three nonequivalent irreducible representations. Their dimensions have to fulfil :

$$\sum_{\alpha=1}^{3} d_\alpha^2 = 6 \quad ,$$

which has as only solution $d_1 = d_2 = 1$ and $d_3 = 2$.

ii) The one-dimensional representations. The characters of elements in the same class are equal, and since the dimension is one, we have :

$$D(a) = D(a^2) = (D(a))^2 \Rightarrow D(a) = (1) \quad , \quad D(b^2) = (D(b))^2 = D(e) = (1) \Rightarrow D(b) = (\pm 1).$$

The two one-dimensional representations are thus :

	e	a	a^2	b	ab	a^2b
D^1	(1)	(1)	(1)	(1)	(1)	(1)
D^2	(1)	(1)	(1)	(-1)	(-1)	(-1)

The characters vectors $(\sqrt{n_i}\,\chi_i^{\alpha})$ for a given α are :

$$\chi^1 = \begin{pmatrix} 1 \\ \sqrt{2} \\ \sqrt{3} \end{pmatrix} \quad , \quad \chi^2 = \begin{pmatrix} 1 \\ \sqrt{2} \\ -\sqrt{3} \end{pmatrix} \quad ,$$

which fulfil the orthogonality relation given in Prop. C.55.

iii) The two-dimensional representation. For the unit element :

$$\chi_1^3 = \mathrm{Tr}\, D^3(e) = 2 \quad .$$

From the orthogonality relations (χ^1, χ^3) and (χ^2, χ^3) equal zero, one then gets the character vector χ^3 :

$$\chi^3 = \begin{pmatrix} 2 \\ \sqrt{2}\,\chi_2^3 \\ \sqrt{3}\,\chi_3^3 \end{pmatrix} \quad , \quad \begin{aligned} \chi_2^3 &= -1 = \mathrm{Tr}\, D^3(a) = \mathrm{Tr}\, D^3(a^2) \\ \chi_3^3 &= \;\;0 = \mathrm{Tr}\, D^3(b) = \mathrm{Tr}\, D^3(ab) = \mathrm{Tr}\, D^3(a^2b) \quad . \end{aligned}$$

For a unitary representation $D^3(b)\, D^{3+}(b) = \mathbb{1}$, and since $b^2 = e$, $D^3(b)$ is Hermitian. Its eigenvalues are real and have the absolute value one. In an eigenbasis of $D^3(b)$ one has :

$$D^3(b) = \begin{pmatrix} 1 & 0 \\ 0 & -1 \end{pmatrix} \quad , \quad D^3(a) = \begin{pmatrix} D_{11} & D_{12} \\ D_{21} & D_{22} \end{pmatrix} \quad \Rightarrow \quad D^3(ab) = \begin{pmatrix} D_{11} & -D_{12} \\ D_{21} & -D_{22} \end{pmatrix} \quad ,$$

$$\left.\begin{aligned} \mathrm{Tr}\, D^3(a) &= -1 = D_{11} + D_{22} \\ \mathrm{Tr}\, D^3(ab) &= \;\;0 = D_{11} - D_{22} \end{aligned}\right\} \Rightarrow D_{11} = D_{22} = -\frac{1}{2} \quad .$$

From $(ab)^2 = e$ we get $D_{12}D_{21} = -\frac{3}{4}$. Since $D^3(ab)$ is taken to be unitary, $(ab)^2 = e$ implies that $D^3(ab)$ is Hermitian. Thus $D_{12} = -D_{21}^*$ and then $|D_{12}|^2 = \frac{3}{4}$. The general solution is $D_{12} = \frac{\sqrt{3}}{2} e^{i\varphi}$, where φ is an arbitrary real number :

$$D^3(a) = \begin{pmatrix} -1/2 & -e^{-i\varphi}\sqrt{3}/2 \\ e^{i\varphi}\sqrt{3}/2 & -1/2 \end{pmatrix} .$$

Let us choose $\varphi = 0$, since representations with different φ are equivalent and connected by the matrix :

$$A = \begin{pmatrix} e^{i\varphi/2} & 0 \\ 0 & e^{-i\varphi/2} \end{pmatrix} .$$

The two-dimensional representation D^3 is thus :

e	a	a^2	b	ab	a^2b
$\begin{pmatrix} 1 & 0 \\ 0 & 1 \end{pmatrix}$	$\begin{pmatrix} -1/2 & -\sqrt{3}/2 \\ \sqrt{3}/2 & -1/2 \end{pmatrix}$	$\begin{pmatrix} -1/2 & \sqrt{3}/2 \\ -\sqrt{3}/2 & -1/2 \end{pmatrix}$	$\begin{pmatrix} 1 & 0 \\ 0 & -1 \end{pmatrix}$	$\begin{pmatrix} -1/2 & \sqrt{3}/2 \\ \sqrt{3}/2 & 1/2 \end{pmatrix}$	$\begin{pmatrix} -1/2 & -\sqrt{3}/2 \\ -\sqrt{3}/2 & 1/2 \end{pmatrix}$.

C.4.4.2. The σ matrices

Let us follow the lecture notes of Veltman[60)] .

i) The problem is to find three complex matrices σ_i, i=1,2,3, which fulfil Eq. (4–88) :

$$\sigma_i\sigma_j = \delta_{ij}\,\mathbb{1} + i\sum_k \varepsilon_{ijk}\,\sigma_k \quad ,$$

where ε_{ijk} is the completely antisymmetrical tensor of order three. Instead of the σ_j, let us consider $\tau_j = i\,\sigma_j$ which fulfil (cf. the algebra Q of real quaternions, Def. B.28) :

$$\tau_i\tau_j = -\delta_{ij}\,\mathbb{1} - \sum_k \varepsilon_{ijk}\,\tau_k \quad , \qquad \text{(C-1)}$$

and also the five other matrices :

$$\tau'_i = -\tau_i \qquad i=1,2,3; \quad e = \mathbb{1} \ ; \quad e' = -\mathbb{1} \ .$$

These eight matrices form a group, the group table of which is easy to construct :

e	e'	τ_1	τ'_1	τ_2	τ'_2	τ_3	τ'_3
e'	e	τ'_1	τ_1	τ'_2	τ_2	τ'_3	τ_3
τ_1	τ'_1	e'	e	τ'_3	τ_3	τ_2	τ'_2
τ'_1	τ_1	e	e'	τ_3	τ'_3	τ'_2	τ_2
τ_2	τ'_2	τ_3	τ'_3	e'	e	τ'_1	τ_1
τ'_2	τ_2	τ'_3	τ_3	e	e'	τ_1	τ'_1
τ_3	τ'_3	τ'_2	τ_2	τ_1	τ'_1	e'	e
τ'_3	τ_3	τ_2	τ'_2	τ'_1	τ_1	e	e'

In order to derive the matrices σ_i, one constructs the irreducible representations of this group, and then finds which representation(s) verifies the Eq. (C–1). For that, the representation D must satisfy :

$$D(\tau_i') = -D(\tau_i) \qquad i=1,2,3 \quad , \quad D(e') = -D(e) = -\mathbb{1} \quad . \qquad \text{(C-2)}$$

ii) There are five classes of conjugate elements under the group :

class 1 = $\{e\}$,class 2 = $\{e'\}$,

class 3 = $\{\tau_1, \tau_1'\}$,class 4 = $\{\tau_2, \tau_2'\}$, class 5 = $\{\tau_3, \tau_3'\}$,

and thus there exist five nonequivalent irreducible representations. Their dimensions have to fulfil :

$$\sum_{\alpha=1}^{5} d_\alpha^2 = 8 \quad ,$$

with the only solution $d_1 = d_2 = d_3 = d_4 = 1$ and $d_5 = 2$.

The one–dimensional representations. Since these representations are Abelian they do not fulfil the Eq. (C–1). Nevertheless to study the two–dimensional representation it is useful to get their character vectors. For one–dimensional representations and τ_i and τ_i' being in the same class, one has :

$$D(\tau_i) = D(\tau_i') \qquad i=1,2,3 \quad .$$

Since $\tau_i \tau_i' = e$ and $\tau_i \tau_i = e'$, one obtains using the previous relation :

$$D(\tau_i)D(\tau_i')=D(e)= \mathbb{1} = D(\tau_i)D(\tau_i) = D(e') \Rightarrow D(e)=D(e') = \mathbb{1} \;,\; D(\tau_i)=D(\tau_i') = \pm \mathbb{1} \quad .$$

Moreover $\tau_1 \tau_2' = \tau_3$ and thus one has :

$$D(\tau_1)D(\tau_2') = D(\tau_1)D(\tau_2) = D(\tau_3) \quad .$$

The four one–dimensional representations are then :

	e	e'	τ_1	τ_1'	τ_2	τ_2'	τ_3	τ_3'
D^1	(1)	(1)	(1)	(1)	(1)	(1)	(1)	(1)
D^2	(1)	(1)	(-1)	(-1)	(-1)	(-1)	(1)	(1)
D^3	(1)	(1)	(1)	(1)	(-1)	(-1)	(-1)	(-1)
D^4	(1)	(1)	(-1)	(-1)	(1)	(1)	(-1)	(-1)

,

and the corresponding character vectors $(\sqrt{n_i}\,\chi_i^\alpha)$:

$$\chi^1 = \begin{pmatrix} 1 \\ 1 \\ \sqrt{2} \\ \sqrt{2} \\ \sqrt{2} \end{pmatrix} , \quad \chi^2 = \begin{pmatrix} 1 \\ 1 \\ -\sqrt{2} \\ -\sqrt{2} \\ \sqrt{2} \end{pmatrix} , \quad \chi^3 = \begin{pmatrix} 1 \\ 1 \\ \sqrt{2} \\ -\sqrt{2} \\ -\sqrt{2} \end{pmatrix} , \quad \chi^4 = \begin{pmatrix} 1 \\ 1 \\ -\sqrt{2} \\ \sqrt{2} \\ -\sqrt{2} \end{pmatrix} \quad .$$

The two-dimensional representation. We restrict ourselves to unitary representations. From $D^5(e) = \mathbb{1}$, $\chi_1^5 = \mathrm{Tr}\, D^5(e) = 2$, and the orthogonality relations $(\chi^\alpha \;, \chi^5) = 0$, $\alpha = 1,2,3,4$, one obtains :

$$\chi^5 = \begin{pmatrix} 2 \\ \chi_2^5 \\ \sqrt{2}\,\chi_3^5 \\ \sqrt{2}\,\chi_4^5 \\ \sqrt{2}\,\chi_5^5 \end{pmatrix} \qquad \begin{array}{l} \chi_2^5 = -2 = \mathrm{Tr}\, D^5(e') \\ \chi_i^5 = 0 \qquad i=3,4,5 \\ \mathrm{Tr}\, D^5(\tau_i) = \mathrm{Tr}\, D^5(\tau_i') = 0 \quad . \end{array}$$

Since the square modulus of any element of a unitary matrix is less than or equal to one, one deduces from $\mathrm{Tr}\, D^5(e') = -2$:

$$D^5(e') = \begin{pmatrix} -1 & 0 \\ 0 & -1 \end{pmatrix} = -\mathbb{1} \quad .$$

From the group law $\tau_i \tau_i = e'$, the above expression of $D^5(e')$ and the unitary character of the representations, one has :

$$D(\tau_i)D(\tau_i) = D(e') = -\mathbb{1} \quad , \quad D(\tau_i)D(\tau_i)^+ = \mathbb{1} \quad .$$

Therefore $D(\tau_i)$ is antihermitian, and being traceless it reads :

$$D^5(\tau_i) = \begin{pmatrix} i\theta_i & \varphi_i \\ -\varphi_i^* & -i\theta_i \end{pmatrix} \qquad i=1,2,3 \qquad \theta_i \text{ real} \quad .$$

The eigenvalues of these antihermitian and unitary matrices must be purely imaginary and with modulus one, i.e. $\pm$ i. For instance in an eigenbasis of $D(\tau_3)$ and using $\tau_2 \tau_1 = \tau_3$, one has :

$$D^5(\tau_3) = \begin{pmatrix} i & 0 \\ 0 & -i \end{pmatrix} \quad , \quad D^5(\tau_2) = \begin{pmatrix} i\theta & \varphi \\ -\varphi^* & -i\theta \end{pmatrix} \quad , \quad D^5(\tau_1) = \begin{pmatrix} -\theta & i\varphi \\ i\varphi^* & -\theta \end{pmatrix} \quad .$$

Requiring $\mathrm{Tr}\, D^5(\tau_3) = 0$ gives $\theta = 0$. As in the previous example (Appendix C.4.4.1) φ is arbitrary provided its modulus is equal to one. Let us take $\varphi = 1$. As $D^5(e') = -\mathbb{1}$ we have $D^5(\tau_i') = D^5(e'\tau_i) = -D^5(\tau_i)$, the representation D^5 fulfils therefore Eq. (C–2). With the first notation σ_i, one obtains the <u>σ matrices</u> or <u>Pauli matrices</u> :

$$\sigma_1 = \begin{pmatrix} 0 & 1 \\ 1 & 0 \end{pmatrix} \quad , \quad \sigma_2 = \begin{pmatrix} 0 & -i \\ i & 0 \end{pmatrix} \quad , \quad \sigma_3 = \begin{pmatrix} 1 & 0 \\ 0 & -1 \end{pmatrix} \quad ,$$

which are thus, up to an equivalence, the only complex matrices which fulfil the Eq. (4–88).

Remark. In a similar way, one shows that, up to an equivalence, the only four matrices which fulfil the condition :

$$\{\gamma_\mu,\gamma_\nu\} = \gamma_\mu\gamma_\nu + \gamma_\nu\gamma_\mu = 2\delta_{\mu\nu}\,\mathbb{I} \qquad \mu,\nu = 0,1,2,3,$$

are the usual four-dimensional γ matrices, i.e. for instance in a basis which diagonalizes γ_o :

$$\gamma_o = \begin{pmatrix} 1 & 0 \\ 0 & -1 \end{pmatrix} \quad , \quad \gamma_i = \begin{pmatrix} 0 & \sigma_i \\ -\sigma_i & 0 \end{pmatrix} \quad .$$

C.5. TOPOLOGICAL GROUPS AND LIE GROUPS

It is out of the scope of these lectures to give details about this vast subject (cf. for instance Refs. 44 ,16). In a general way one first considers groups which are at the same time topological spaces (Appendix C.5.1), this leads to the notion of topological group (Appendix C.5.2). One then considers groups which are at the same time manifolds (Appendix C.5.3) this leads to the notion of Lie group (Appendix C.5.4). Our purpose is only to give some definitions and remarks which are illustrated for the linear groups and mainly for the groups SO(3) and SU(2) in Chap. 4.

C.5.1. SOME BASIC TOPOLOGICAL DEFINITIONS

C.5.1.1. Definitions

Def. C.37. A topological space is the pair of a space E and a set of parts of E, called open sets, which fulfil the three conditions :
(1) any union (finite or infinite) of open sets is an open set,
(2) any finite intersection of open sets is an open set,
(3) E and $\emptyset$ are open sets.

Def. C.38. A subset of E is said to be a closed set if its complementary with respect to E is an open set.

Def. C.39. The interior U^o of a subset U of a topological space E is the largest open set contained in U. The closure $\bar{U}$ of U is the smallest closed set containing U. The boundary $U^\cdot$ of U is the complementary of U with respect to $\bar{U}$, also denoted $\bar{U}-U$.

Def. C.40. A neighbourhood of an element a in E is any subset of E which contains an open set containing a.

Def. C.41. A set $\mathcal{N}(a)$ of neighbourhoods of a point a in E is called a complete system of neighbourhoods of a, if for every open set $\mathcal{O}$ containing a there exists an open set belonging to $\mathcal{N}(a)$ and included in $\mathcal{O}$.

Def. C.42. A topological space E is said to be Hausdorff or separable if, given any two distinct points a and b in E, there exist a neighbourhood V_a of a and a neighbourhood V_b of b which do not intersect.

Only separable topological spaces are useful in practice.

Def. C.43. An application f from a topological space E into a topological space F is said to be continuous in a point a of E if for any neighbourhood V of f(a) there exists a neighbourhood U of a such that f(U) is included in V. If moreover f is one-to-one, it is called an homeomorphism.

Def. C.44. A path f in a topological space E is a continuous function f of the real parameter t, $0 \leqslant t \leqslant 1$ (with the usual topology of R, cf. ii) below) associating with each number t a point f(t) in E. The points f(0) and f(1) are the end points of the path, respectively, the initial point and the terminal point. The path f is said to join f(0) to f(1) in E. A path, the end points of which coincide, is said to be a closed path about the common end point.

Def. C.45. A topological space E is said to be (arcwise) connected if any two of its elements may be joined by a path in E.

Def. C.46. A topological space E is said to be simply-connected if it is connected and if any closed path in E can be continuously reduced to a point (one says homotopic to zero, cf. remarks in Subsec. 4.1.4).

Def. C.47. A topological space E is said to be compact if it is separable and if from any covering of E by open sets it is possible to select a finite covering.

A collection $\mathcal{C}$ of sets in a space E is called a covering of a set $M \subseteq E$ if the union of the sets of $\mathcal{C}$ contains M.

C.5.1.2. Remarks and the example of R''

i) Let us consider a metric space, i.e. such that a distance is defined. Then there is a natural definition of open sets in such a space which turns it into a topological space. Indeed, clearly the set of all points in a metric space at a distance less than a given positive number from a fixed point, fulfils the axioms of Def. C.37 of an open set. Conversely :

Def. C.48. A topological space such that there exists a metric which generates its topology as specified above is said to be metrizable.

Any metrizable topological space is separable.

ii) An important example is the set R^n, for n a positive integer and R being the real set. R^n is a metric space with for instance the Euclidean distance :

$$\forall\, x_i, y_j \in R \;;\; x = (x_1, \ldots, x_n),\; y = (y_1, \ldots, y_n) \;;\; d(x,y) = \left[\sum_{i=1}^{n} (x_i - y_i)^2\right]^{1/2}.$$

The topology defined as above is the usual simply-connected and separable topology of R^n.

Def. C.49. B^n as D^n denote the open or closed n-ball (or n-disk) of radius r, defined by the set of points x in R^n such that, respectively, $\|x\| < r$ or $\|x\| \leqslant r$. $S^{n-1} = \partial D^n$ is the (n-1)-sphere of radius r, defined by the set of points x in R^n such that $\|x\| = r$. In both cases $\|x\|$ denotes the Euclidean norm :

$$\|x\| = d(x,o) = \left[\sum_{i=1}^{n} (x_i)^2\right]^{1/2} .$$

Prop. C.69. With respect to this topology a subset of R^n is compact if and only if it is bounded and closed (closed means that the limit of every convergent sequence in the subset also belongs to this subset).

For instance any closed ball in R^n is compact while R^n itself is not compact.

C.5.2. TOPOLOGICAL GROUPS

Def. C.50. A continuous or topological group G is a group which is also a topological space, and such that for all g and g' in G the application $(g,g') \to gg'^{-1}$ of $G \times G$ onto G (or equivalently both applications $(g,g') \to gg'$ of $G \times G$ onto G and $g \to g^{-1}$ of G onto itself) is (are) continuous with respect to the topology of G.

Then, any notion which is either group theoretical or topological (as for instance those previously recalled) has a meaning when applied to a topological group.

Def. C.51. For all g in a group G, the two mappings L_g and R_g of G onto itself, respectively called the left and right translations, are defined by :

$$\forall\, g' \in G \qquad L_g g' = gg' \quad , \qquad R_g g' = g'g \quad .$$

L_g is a realization of the group G into itself closely related to the left regular representation defined for a finite group (Def. C.36). On the other hand, for a nonabelian group R_g is not a realization of G, since one has for instance :

$$R_g R_{g'} = R_{g'g} \neq R_{gg'} \quad .$$

See also the left and right regular representations of any group in Subsecs. 10.1.3.2,3.

Prop. C.70. On a topological group G, for all g in G the left and right translations L_g and R_g are one-to-one continuous mapping, i.e. homeomorphism, of G onto itself.

Consequently, if we know a complete system of neighbourhoods of an element of G, say the unit element, a complete system of neighbourhoods of any element g can be obtained by applying L_g or R_g. Thus the group structure of the topological space G implies the existence of a local characterization of the topology (nevertheless, conversely the complete system of neighbourhoods of an element cannot be given arbitrarily and has to fulfil conditions related to the group structure of G).

C.5.3. n-MANIFOLDS

C.5.3.1. Intuitive notion of manifold and an example

i) The theory of manifolds is based on simple geometrical ideas. Indeed, an intuitive notion of a manifold is well given by the set of points which form a regular two-dimensional surface in the usual three-dimensional space. The terminology of the manifold theory is even taken from geography (e.g. chart, atlas) or more generally from geometry (e.g. tangent vector, tangent space, derivative of a function in the direction of a tangent vector). Therefore, to illustrate these notions, which are next defined in an abstract way, let us first consider the example of the surface of a three-dimensional ball, i.e. S^2. This set is a topological space with the topology induced by the usual one of R^3.

ii) A local bijection from this two-sphere S^2 onto R^2 can be obtained by projecting stereographically from say the south pole onto the plane tangent to the north pole, and then defining the points in this plane by their two coordinates x_1 and y_1 with respect to a given reference frame, say $(N;e_{x_1}, e_{y_1})$, see Fig. C.1.

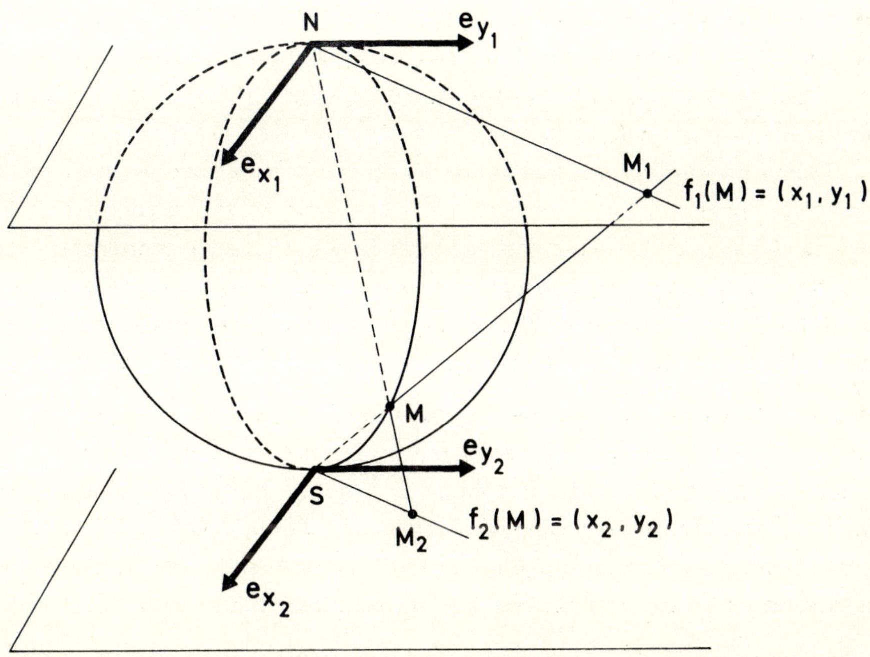

Fig. C.1. Example of two local charts.

This projection defines coordinates for the points on the surface except for the south pole. We shall say that this is a local chart (Def. C.52) which consists in : (1) the open set U_1 equal to S^2 south pole excluded, and (2) the one-to-one mapping f_1 of U_1 onto R^2 such that :

$$\forall\ M \in U_1 \subset S^2 \longrightarrow f_1(M) = (x_1, y_1) \in R^2 \quad .$$

Similarly, interchanging the roles of the poles, one gets another local bijection f_2 of the open set U_2, equal to S^2 north pole excluded, onto R^2 such that :

$$\forall\ M \in U_2 \subset S^2 \longrightarrow f_2(M) = (x_2, y_2) \in R^2 \quad ,$$

and thus this defines another local chart. For any point of the intersection $U_1 \cap U_2$, i.e. any point different from the poles, two parametrizations are thus available. Taking for simplicity the corresponding reference axes parallel and with orthonormal basis vectors, one gets the expression of the change of coordinates :

$$\left.\begin{array}{l} d = SN \\ \dfrac{x_1}{y_1} = \dfrac{x_2}{y_2} \\ NM_1 \cdot SM_2 = d^2 \end{array}\right\} \Rightarrow (x_1, y_1) \xrightarrow{f_2 f_1^{-1}} \left(x_2 = \frac{x_1 d^2}{x_1^2 + y_1^2} \;,\; y_2 = \frac{y_1 d^2}{x_1^2 + y_1^2}\right) \quad ,$$

which is an analytic one-to-one mapping from $f_1^{-1}(U_1 \cap U_2) = R^2$ zero excluded onto itself (= $f_2^{-1}(U_1 \cap U_2)$). Moreover, the union $U_1 \cup U_2$ is clearly a covering of the two-sphere. Therefore, we shall say that the two local charts defined above form an atlas and endow the two-sphere with an analytic two-manifold structure (Def. C.52).

Thus a manifold is a set of points which can be described by local coordinates and the conditions imposed on the functions $f_j f_i^{-1}$ correspond to the requirement that in the region covered by two coordinate systems one must be able to change coordinates "smoothly".

One should note that, although the points of S^2 are parametrized by two real numbers, the topology of S^2 induced by the usual one of R^3, and the usual topology of R^2 are locally identical but globally different (the first one is compact while the second one is noncompact).

iii) Let us now consider the tangent plane T to the two-sphere S^2 at the point M, see Fig. C.2. This two-dimensional vector space over the real field is spanned by the two vectors t_x and t_y, respectively, tangent at the point M to the curves y constant and x constant on S^2, which correspond to a given local parametrization. Let us denote by $F(M)$ the set of functions ϕ of S^2 into the real field which are analytic at M. Then to any tangent vector t to S^2 at M one associates a mapping T of $F(M)$ into the real field defined by :

$$\forall \quad \alpha_x, \alpha_y \in R \qquad t = \alpha_x t_x + \alpha_y t_y \longrightarrow T = \alpha_x \frac{\partial}{\partial x} + \alpha_y \frac{\partial}{\partial y} \quad .$$

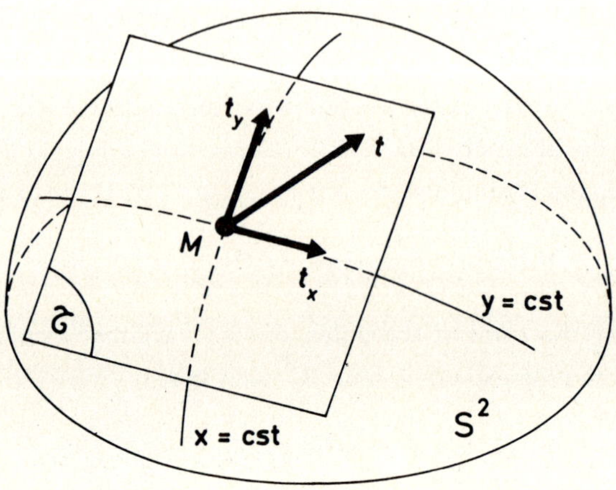

Fig. C.2. Tangent vectors and tangent space.

We shall say that T is a tangent vector to the manifold S^2 at the point M (Def. C.54), and for any function ϕ in $F(M)$, $T(\phi)$ will be called the derivative of ϕ in the direction T. This real number $T(\phi)$ is a first order variation of ϕ when M is varied on S^2. The set of the mappings T is clearly a two-dimensional vector space over the real field, called the tangent plane T .

C.5.3.2. Definitions

Def. C.52. Let E be a topological space and consider R^n with its usual topology. A local chart (U,f) on E is a bijection f from an open set U of E to an open set of R^n. An atlas on E is a family A of local charts $\{(U_i f_i), i=1,2,...\}$ such that :
(1) the U_i form a covering of E : $\cup U_i = E$,
(2) for any two local charts (U_i, f_i) and (U_j, f_j) such that $U_i \cap U_j \neq \emptyset$, the restriction of f_i^{-1} to the subset $f_i(U_i \cap U_j)$ allows us to define a one-to-one mapping $f_j f_i^{-1}$ from $f_i(U_i \cap U_j)$ to $f_j(U_i \cap U_j)$ (see Fig. C.3). One requires that this mapping be continuous (of class C^o), r-continuously differentiable (C^r), infinitely differentiable (C^∞) or analytic.

Two atlases A_1 and A_2 are equivalent if and only if $A_1 \cup A_2$ is an atlas. A differentiable structure E on E is an equivalence class of atlases on E.

Then, a respectively : continuous (C^o), r-continuously differentiable (C^r), infinitely differentiable (C^∞) or analytic n-manifold is a set E endowed with a differentiable structure E, n is said to be the dimension of the manifold.

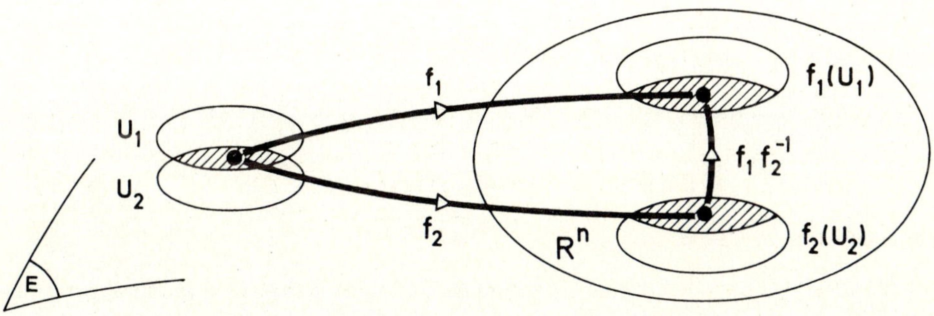

Fig. C.3. Local charts of an n-manifold E .

One should note that the topologies of E and R^n are only locally equivalent but may be globally different, as we have seen in the example of Appendix C.5.3.1,ii).

Def. C.53. A mapping ψ of an analytic n-manifold E into an analytic n'-manifold E' is said to be analytic at the point p of E if for a local chart (U,f) of E defined at p ($p \in U$), and a local chart (U',f') of E' defined at $\psi(p)$ ($\psi(p) \in U'$), the locally defined function $\tilde{\psi} = f' \circ \psi \circ f^{-1}$ of R^n into $R^{n'}$ is analytic at f(p).

Prop. C.71. The function $\tilde{\psi}$ defined above is analytic at f(p), for all local charts (U,f) and (U',f') which belong to the classes of atlases defining, respectively, the differentiable structures of E and E', and such that p belongs to U and $\psi(p)$ to U'.

Proof. Consequence of Def. C.52, (2), the mapping $f_j f_i^{-1}$ which relates two local charts being analytic.

C.5.3.3. Tangent vectors, differentials, and infinitesimal transformations

Cf. for instance Chevalley[16] (Chap. III, § IV and V).

i) **Def. C.54.** Let p be a point in an analytic n-manifold E, and $F(p)$ the set of functions of E into the real field which are analytic (Def. C.53) at p. A tangent vector T at the point p is a mapping of $F(p)$ into the real field which fulfils both conditions :

(1) T is a linear mapping :

$$\forall\ x,x' \in R \qquad \forall\ \phi,\phi' \in F(p) \qquad T(x\phi+x'\phi') = xT(\phi)+x'T(\phi') \quad .$$

(2) T is a differentiation, i.e. :

$$\forall\ \phi,\phi' \in F(p) \qquad T(\phi\phi') = T(\phi)\phi'(p)+\phi(p)T(\phi') \quad ,$$

where $\phi\phi'$ denotes the function of E into the real field, which also belongs to $F(p)$, and such that :

$$\forall\ q \in E \qquad (\phi\phi')(q) = \phi(q)\phi'(q) \quad .$$

The real number $T(\phi)$ is called the derivative of ϕ in the direction T at the point p.

Def. C.55. The set of tangent vectors at a point p of E is a vector space over the real field, called the tangent vector space to E at p.

Clearly, for all tangent vectors T and T' to E at p, and all real numbers x and x', the mapping :

$$\forall\ \phi \in F(p) \longrightarrow xT(\phi)+x'T'(\phi) \in R$$

is a tangent vector to E at p.

Prop. C.72. Let (U,f) be a local chart defined at the point p ($p \in U$) in an analytic n-manifold E. From Def. C.53, if a mapping ϕ of E into R belongs to $F(p)$, it has a locally defined expression in terms of the local coordinates, i.e. :

$$\forall\ q \in U \qquad f(q) = a = \{f^i(q) = a^i \in R;\ i=1,\dots,n\} \in R^n \Rightarrow \phi(q) = \phi[f^{-1}(a)] = \tilde{\phi}(a) .$$

To simplify the notations we shall write $\frac{\partial\phi}{\partial f^i}$ instead of $\left.\frac{\partial\tilde{\phi}}{\partial a^i}\right|_{a^i=f^i(p)}$. Then every tangent vector T at p is of the form :

$$\forall\ \phi \in F(p) \longrightarrow T(\phi) = x^i \frac{\partial\phi}{\partial f^i} \in R \quad , \quad x^i = T(f^i) \in R \quad .$$

Proof. For all real numbers x^i the above mapping of $F(p)$ into R is a tangent vector at p. Conversely, let us show that every tangent vector is of this form :

1–Every tangent vector maps a constant function into zero. Indeed, let ϕ' be such a constant function, then for all ϕ in $F(p)$ one has, since a tangent vector is a differentiation and linear :

$$T(\phi\phi') = T(\phi)\phi'(p)+\phi(p)T(\phi') = T(\phi)\phi'(p) \Rightarrow T(\phi') = 0 .$$

2–If ϕ belongs to $F(p)$, it can be expressed as :

$$\forall q\in U \quad \phi(q) = \tilde{\phi}[f(q)]=\phi(p) + \frac{\partial\phi}{\partial f^i}[f^i(q)-f^i(p)]+g_{ij}(q)[f^i(q)-f^i(p)][f^j(q)-f^j(p)] ,$$

where g_{ij} is a mapping which also belongs to $F(p)$. Then applying any tangent vector, and making use of its linear character and of the result 1–, one gets :

$$T(\phi) = \frac{\partial\phi}{\partial f^i} T(f^i) + T\{(g_{ij}(q)[f^i(q)-f^i(p)][f^j(q)-f^j(p)]\} .$$

From the differentiation character of T the last term of the above expression vanishes, which ends the proof.

Prop. C.73. The tangent vector space to an analytic n-manifold at a point is an n-dimensional vector space over the real field.

Proof. For a given local chart (U,f) defined at the point p, let us denote by T_i the tangent vector at p defined by, cf. Prop. C.72 :

$$T_i(f^j) = \delta_i^j \quad ; \quad \forall\phi\in F(p) \xrightarrow{T_i} \frac{\partial\phi}{\partial f^i} \in R .$$

These n tangent vectors at p are linearly independent since one has :

$$\forall\, x^i\in R \qquad x^iT_i = 0 \Rightarrow \forall\, j=1,\ldots,n \quad (x^iT_i)(f^j) = x^j = 0 .$$

Moreover, every tangent vector T at p is a linear combination of the T_i, for :

$$\forall\, j=1,\ldots,n \quad T(f^j) = T(f^i)\delta_i^j = [T(f^i)T_i]\,(f^j) \Rightarrow T = T(f^i)T_i ,$$

and from Prop. C.72 a tangent vector is uniquely determined by the values it assigns to the functions f^i.

ii) **Def. C.56.** Let ψ be a mapping of an analytic manifold E into an analytic manifold E', analytic at the point p in E. The <u>differential</u> of the mapping ψ at p, denoted by $d\psi$ or $d\psi_p$, is the mapping which assigns to every tangent vector T to E at p the tangent vector T' to E' at p' = $\psi(p)$ defined by :

$$T \rightarrow T' = d\psi_p T \quad , \quad \forall\,\phi'\in F'(p') \quad T'(\phi') = T(\phi'\circ\psi) .$$

Remarks :

1–One easily checks that T' is a tangent vector to E' at p'. $d\psi$ is thus a mapping of the tangent vector space to E at p into the tangent vector space E' at $\psi(p)$, which is clearly linear.

2–Special case E' = R considered as a one-dimensional manifold, i.e. ψ is a mapping of E into R. Then $d\psi$ is a mapping of the tangent vector space to E at p into the one–dimensional tangent vector space to R at x = $\psi(p)$. It is convenient to identify this last vector space with R itself as follows. Let us define the special tangent vector T_1' to E' = R at the point x by the condition :

$$T_1'(X) = 1 ,$$

where X denotes the identity mapping of R into itself :

$$\forall\, x \in R \qquad X(x) = x \quad .$$

The above condition completely determines T'_1 since for all mappings ϕ' of R into itself, analytic at x, $T'_1(\phi')$ is well defined using the analyticity of ϕ' and the properties of a tangent vector (linear and differentiation character). Now T'_1 can be taken as a basis of the tangent vector space to R at the point x. Then every tangent vector T' to R at x reads :

$$T' = \tau\, T'_1 \quad , \qquad \tau = T'(X) \in R \quad .$$

From the definition of the differential of ψ at p, one has for all mappings ϕ' of R into itself, analytic at x :

$$T'(\phi') = T'(X)T'_1(\phi') = [d\psi_p(T)]\ (\phi') = T(\phi' \circ \psi) \quad .$$

Setting $\phi' = X$ in the above equation yields :

$$T'(X)T'_1(X) = T'(X) = T(X \circ \psi) = T(\psi) \quad ,$$

one therefore has :

$$[d\psi_p(T)]\ (\phi') = T(\psi)T'_1(\phi') \quad .$$

Identifying the tangent vector T'_1 with the vector 1 in R :

$$T'_1 \equiv 1 \quad ,$$

one gets the usual definition of the differential of a mapping ψ of E into the real field, i.e. :

$$d\psi(T) = T(\psi) \quad .$$

Prop. C.74. Let E, E', and E'' be three analytic manifolds, and ψ, ψ' two mappings of, respectively, E into E' and E' into E'' analytic at the points, respectively, p in E and $p' = \psi(p)$ in E'. Then one has :

$$d(\psi' \circ \psi)_p = d\psi'_{p'} \circ d\psi_p \quad .$$

Proof. Consequence of the associativity of the composition law of applications :

$$T'' = d\psi'_{p'} \circ d\psi_p T \Rightarrow T''(\phi'') = (d\psi_p T)(\phi'' \circ \psi') = T[(\phi'' \circ \psi') \circ \psi] = T[\phi'' \circ (\psi' \circ \psi)]$$

$$\Rightarrow T'' = d(\psi' \circ \psi)_p\ T \quad .$$

iii) **Def. C.57.** Let E be an analytic manifold. A <u>vector field</u> or an <u>infinitesimal transformation</u> J on E is a mapping which assigns to every point p in E a tangent vector J(p) to E at this point.

Def. C.58. To every infinitesimal transformation J on E, one associates a mapping which assigns to every function ϕ of E into the real field a new function, denoted by $J\phi$, of E into the real field defined by :

$$\phi \rightarrow J\phi \quad , \qquad \forall\, p \in E \qquad (J\phi)(p) = [J(p)](\phi) \in R \quad .$$

If for all analytic ϕ, the function $J\phi$ is analytic, J is called an <u>analytic infinitesimal</u>

transformation.

Prop. C.75. Let (U,f) be a local chart in an analytic n-manifold E. Then, with the same notations as in Prop. C.72, every analytic infinitesimal transformation defined on U is of the form :

$$\forall p \in U \quad , \quad \forall \phi \in F(p) \qquad (J\phi)(p) = A^i(p) \frac{\partial \phi}{\partial f^i} \cdot \quad ,$$

where $A^1,\dots,A^n$ are n analytic mappings of U into the real field. $A^i \frac{\partial}{\partial f^i}$ is called the symbol of the infinitesimal transformation.

Proof. At every point p of U, J(p) is by definition a tangent vector at p, thus from Prop. C.73 it is a linear combination of the linearly independent tangent vectors T_i at p :

$$J(p) = A^i(p) J_i \quad , \quad (J_i\phi)(p) = T_i\phi = \frac{\partial \phi}{\partial f^i} \quad ,$$

where the components A^i depend upon p. Since for any analytic function ϕ , $J\phi$ is analytic, these functions A^i have also to be analytic on U.

Prop. C.76. The set of analytic infinitesimal transformations on an analytic manifold E is a Lie algebra over the real field, where the bilinear mapping is defined by the commutator.

Proof. Let J and J' be two analytic infinitesimal transformations on E. One denotes by JJ' the compositions of these transformations considered as mappings of the set of real functions analytic on some open set U of E into itself, as defined in Def. C.58. Then one easily checks that for all p in E, ([J,J'])(p) is linear and a differentiation i.e. a tangent vector to E at p. Moreover if ϕ is analytic, $[J,J']\phi$ is also analytic. With respect to a given local chart (U,f) one has with the same notations as in Prop. C.72 :

$$\forall p \in U \quad , \quad \forall \phi \in F(p) \qquad \begin{cases} (J\phi)(p) = A^i(p) \dfrac{\partial \phi}{\partial f^i} \\ (J'\phi)(p) = A'^j(p) \dfrac{\partial \phi}{\partial f^j} \end{cases}$$

$$\Rightarrow ([J,J']\phi)(p) = \left[A^i(p) \frac{\partial A'^j}{\partial f^i} - A'^i(p) \frac{\partial A^j}{\partial f^i}\right] \frac{\partial \phi}{\partial f^j} \quad .$$

The set of analytic infinitesimal transformations on E is clearly a vector space over the real field, the commutator endows this space with a Lie algebra structure (cf. Appendix B.1.2,i)).

Def. C.59. Let ψ be a mapping of an analytic manifold E into an analytic manifold E'. The infinitesimal transformations J and J' on, respectively, E and E' are said to be ψ-related if :

$$\forall p \in E \qquad d\psi_p[J(p)] = J'[\psi(p)] \quad .$$

Prop. C.77. Let ψ be an analytic mapping of an analytic manifold E into an analytic manifold E'. Let J_1, J_2 be analytic infinitesimal transformations on E, and J'_1 , J'_2 analytic infinitesimal transformations on E'. If J_i is ψ-related to J'_i (i = 1,2), then $[J_1,J_2]$ is ψ-related to $[J'_1, J'_2]$.

Proof. Let p be a point in E and ϕ' a function of E into R, analytic at the point $\psi(p)$. Then from Defs. C.56,59, J_i and J'_i are ψ-related if and only if :

$$[d\psi_p J_i(p)](\phi') = J_i(p)(\phi' \circ \psi) = \{J_i'[\psi(p)]\}(\phi') \quad .$$

In terms of the mapping associated to every infinitesimal transformation and defined in Def. C.58, the above equation reads :

$$J_i(\phi' \circ \psi) = (J_i' \phi') \circ \psi \; .$$

Hence, one gets :

$$(J_1 J_2)(\phi' \circ \psi) = J_1[J_2(\phi' \circ \psi)] = J_1[(J_2' \phi') \circ \psi] = [J_1'(J_2' \phi')] \circ \psi = [(J_1' J_2') \phi'] \circ \psi \quad ,$$

and a similar relation by interchanging the indices 1 and 2. Substracting these equations one obtains the result :

$$[J_1, J_2](\phi' \circ \psi) = ([J_1', J_2'] \phi') \circ \psi \Rightarrow d\psi_p\{[J_1, J_2](p)\} = [J_1', J_2'][\psi(p)] \quad .$$

C.5.4. LIE GROUPS

C.5.4.1. Definition

i) **Def. C.60.** An <u>n-parameter Lie group</u> is a topological group G which has also the structure of an analytic n-manifold, and such that for all g and g' in G the application $(g,g') \to gg'^{-1}$ of $G \times G$ onto G is analytic (or equivalently the applications $(g,g') \to gg'$ of $G \times G$ onto G and $g \to g^{-1}$ of G onto itself are both analytic). Namely one has :

(1) The n-manifold structure of G implies that for all g in G there exists at least one local chart (U,f) such that g belongs to the open set U in G and f(g) = a belongs to an open set in R^n. One also says that (U,f) is a local parametrization which defines the local parameters a of g :

$$g = g_a \qquad , \qquad a = f(g) = \{f^i(g) = a^i \in R \; ; \; i=1,\ldots,n\} \in R^n \quad .$$

(2) The group structure of G implies that for given local charts of an atlas, which define the local parameters of any g, g^{-1}, g', and gg' in G, there exist two sets of n locally defined functions $\varphi = \{ \varphi^i ; i = 1,\ldots,n \}$ and $\psi = \{ \psi^i ; i = 1,\ldots,n \}$ such that :

$$g = g_a \qquad g^{-1} = g_{\bar{a}} \qquad , \qquad \bar{a} = \{\bar{a}^i = \psi^i(a) \, ; \; i=1,\ldots,n\} = \psi(a) \quad ;$$

$$g' = g_b \qquad gg' = g_c \qquad , \qquad c = \{c^i = \varphi^i(a,b) \, ; \; i=1,\ldots,n\} = \varphi(a,b) \quad .$$

This is shown in Fig. C.4 where the arrows denote the mappings which are defined, a one-to-one mapping is denoted by $\longleftrightarrow$. The functions ψ^i and φ^i are, respectively, defined by the following chains of mappings :

$$a \xrightarrow{\;f^{-1}\;} g_a \xrightarrow{\;\text{inverse in G}\;} g_a^{-1} \xrightarrow{\;f_{-1}\;} \bar{a}$$

$$\bar{a} = \psi(a) = f_{-1}\{ [f^{-1}(a)]^{-1}\} \quad ,$$

$$\left.\begin{matrix} a \xrightarrow{\;f^{-1}\;} g_a \\ b \xrightarrow{\;f'^{-1}\;} g_b \end{matrix}\right\} \xrightarrow{\;\text{multiplication in G}\;} g_a g_b \xrightarrow{\;f''\;} c$$

$$c = \varphi(a,b) = f''[f^{-1}(a)f'^{-1}(b)] \quad .$$

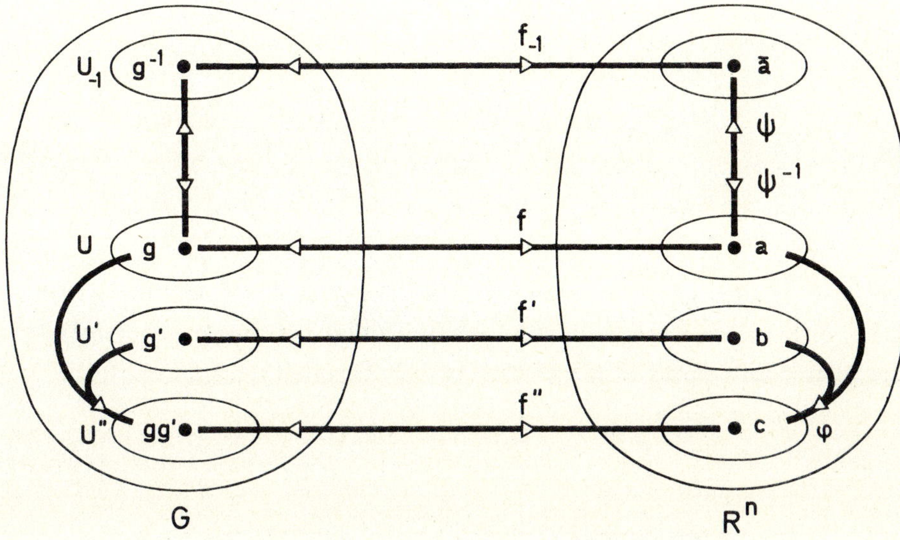

Fig. C.4. Definitions of the functions φ and ψ .

The group structure of G implies constraints on the functions φ and ψ, i.e. let u denote local parameters of the unit element e, then :

$$g_a e = e g_a = g_a \quad \Rightarrow \quad \varphi(a,u) = \varphi(u,a) = a \quad ,$$

$$e^{-1} = e \quad \Rightarrow \quad \psi(u) = u \quad ,$$

$$g_a(g_b g_c) = (g_a g_b) g_c \Rightarrow \varphi[a,\varphi(b,c)] = \varphi[\varphi(a,b),c] \quad ,$$

$$(g_a g_b)^{-1} = g_b^{-1} g_a^{-1} \quad \Rightarrow \quad \psi[\varphi(a,b)] = \varphi[\psi(b),\psi(a)] \quad .$$

Actually, for convenience we have introduced two sets of n functions φ and ψ but only one set ϕ is necessary, such that :

$$g_a g_b^{-1} = g_c \quad , \quad c = \phi(a,b) = \varphi[a,\psi(b)] \quad .$$

(3) The group G being a topological group, both mappings $g \rightarrow g^{-1}$ and $(g,g') \rightarrow gg'$ are continuous. Therefore, since G is also a continuous n-manifold the functions ψ^i and φ^i are locally continuous.

(4) The fact that both applications $g \rightarrow g^{-1}$ and $(g,g') \rightarrow gg'$ are analytic means that for given local charts of an atlas the functions ψ^i and φ^i are locally analytic (Def. C.53). Therefore one has :

Prop. C.78. For all g in G the left and right translations, respectively L_g and R_g, are one–to–one analytic mappings of G onto itself. Therefore, the two one–to–one locally defined applications ℓ_b and r_b :

$$a \xrightarrow{\ell_b} c = \ell_b(a) \quad , \quad a \xrightarrow{r_b} c' = r_b(a) \quad ,$$

such that :

$$g = g_b \quad \begin{cases} g_a \xrightarrow{L_g} L_g(g_a) = g_b g_a & , \quad c = \varphi(b,a) = \ell_b(a) \\ g_a \xrightarrow{R_g} R_g(g_a) = g_a g_b & , \quad c' = \varphi(a,b) = r_b(a) \end{cases} \quad ,$$

are locally analytic. Consequently, from any atlas and g' in G, the left translation L_g (resp. right) allows us to define a new equivalent atlas such that the new local parameters c of any g' in G are related to the old one a, according to (see Fig. C.5) :

$$c = \ell_b(a) \qquad (\text{resp.}\, c = r_b(a)) \quad .$$

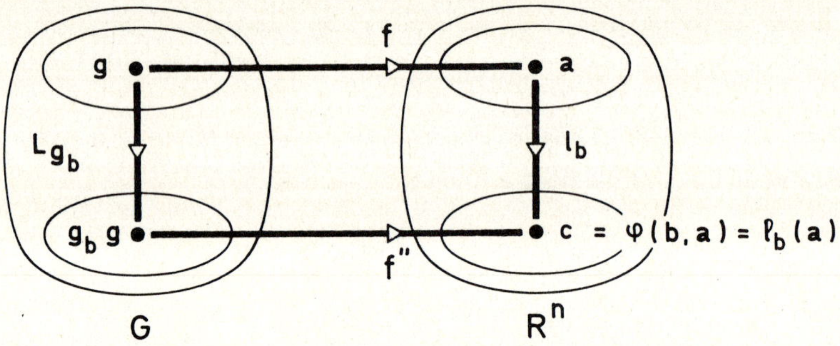

Fig. C.5. Left translation .

This property plays an important role in the construction of an invariant measure on a Lie group (Subsec. 7.1.1.1).

ii) Remarks. The usual topology of R^n induces a topology on such a Lie group G, i.e. g_a belongs to an open set of G if and only if a belongs to an open set of R^n. Nevertheless, since the correspondence between G and R^n is not in general globally one–to–one (even in the simple cases of interest as for instance the subgroups of the linear group, cf. Appendix C.5.4.3,i) below) the topology so defined on G is not in general separable, and therefore not very useful in practice. This shows that one needs to define directly on G a separable topology. The topologies of G and R^n are only locally equivalent. We emphasize that the definition and the properties of all the previously defined functions φ^i and ψ^i are local and closely related the choice of local charts (cf. Def. C.60,(2)).

C.5.4.2. Lie algebra of a Lie group

Prop. C.79. The left (resp. right) translation L_g (resp. R_g) for all g in a Lie group G has a differential dL_g (resp. dR_g) which maps isomorphically the tangent vector space to G at every point g' onto the tangent vector space to G at gg' (resp. g'g), see Fig. 7.7.

Proof. Consequence of Prop. C.78. For details, cf. for instance Chevalley[16] (Chap. III, § IV, Propositions 1,2, and 3).

Def. C.61. An infinitesimal transformation J on a Lie group G is said to be <u>left</u> (resp. <u>right</u>) <u>invariant</u> if it is L_g (resp. R_g) – related with itself, i.e. :

$$\forall g,g' \in G \qquad dL_{g'g^{-1}}[J(g)] = J(g') \qquad (\text{resp.}\, dR_{g^{-1}g'}[J(g)] = J(g')) \quad .$$

Prop. C.80. A necessary and sufficient condition for an infinitesimal transformation J on a Lie group G to be left (resp. right) invariant is that :

$$\forall\, g \in G \qquad dL_g J(e) = J(g) \qquad (\text{resp.}\, dR_g J(e) = J(g)) \quad ,$$

where e is the unit element of G.

Proof. The condition is obviously necessary, let us show that it is also sufficient. Since $L_{g^{-1}}$ is the inverse mapping of L_g, $dL_{g^{-1}}$ is the inverse mapping of dL_g, thus :

$$J(e) = dL_{g^{-1}}\, J(g) \quad .$$

Therefore, using Prop. C.74, one gets the result :

$$\forall g,g' \in G \qquad J(g') = dL_{g'}J(e) = dL_{g'} \circ dL_{g^{-1}}J(g) = d(L_{g'} \circ L_{g^{-1}})J(g) = dL_{g'g^{-1}}J(g) \quad .$$

A similar proof holds for the right invariance.

Prop. C.81. The set of left (resp. right) invariant infinitesimal transformations on an n–parameter Lie group G is an n–dimensional vector space over the real field, isomorphic to the tangent vector space to G at any point g.

Proof. Direct consequence of Def. C.61, since an invariant infinitesimal transformation is completely determined by the value it assigns to any point g. Then from Prop. C.73, for a given g, the set of tangent vectors J(g) to G at g is an n–dimensional vector space over the real field.

Prop. C.82. Every left (resp. right) invariant infinitesimal transformation on an n–parameter Lie group G is analytic.

Prop. C.83. Let J and J' be two left (resp. right) invariant infinitesimal transformations on a Lie group, then the commutator [J,J'] is also left (resp. right) invariant.

Proof. Consequence of Prop. C.77, J being L_g–related to itself, and J' L_g–related to itself, so it is for [J,J'].

Prop. C.84. The set of left (resp. right) invariant infinitesimal transformations on an n–parameter Lie group is an n–dimensional Lie algebra over the real field, the bilinear mapping being defined by the commutator.

Proof. Consequence of Props. C.76,81, and 83.

Def. C.62. This Lie algebra of left (resp. right) invariant infinitesimal transformations on an n-parameter Lie group G is called the Lie algebra of G.

Remarks :

1-One can speak of the Lie algebra of G since both Lie algebras of, respectively, left and right invariant infinitesimal transformations on G are isomorphic. Indeed, let I be the one-to-one analytic mapping of G onto itself which assigns to all g in G its inverse g^{-1}. Then if J is a left invariant infinitesimal transformation, J' defined by :

$$\forall g \in G \qquad J'(g) = dI[J(g^{-1})]$$

is right invariant, using Prop. C.74 :

$$\forall g \in G \qquad dR_g[J'(e)] = (dR_g \circ dI)[J(e)] = d(R_g \circ I)[J(e)] \quad ,$$

since :

$$\forall g \in G \qquad R_g \circ I = I \circ L_{g^{-1}} \quad ,$$

one gets the result from Prop. C.80 :

$$\forall g \in G \qquad dR_g[J'(e)] = d(I \circ L_{g^{-1}})[J(e)] = dI[J(g^{-1})] = dI[J'(g)] \quad .$$

Moreover, from Prop. C.77, one has :

$$dI\{[J_1,J_2](g^{-1})\} = [dI(J_1),dI(J_2)](g^{-1}) = [J_1',J_2'](g) \quad ,$$

hence the commutator is stable in the mapping $J \to J'$.

2-From Prop. C.81, the Lie algebra of G induces an isomorphic Lie algebra structure on the tangent vector space to G at every point. It is often convenient to consider the tangent vector space to G at the unit element e (cf. Appendix C.5.4.3,ii)).

C.5.4.3. The important example of linear groups

i) Linear groups as Lie groups. Let us consider the linear group GL(n,R) of real regular $n\times n$ matrices. In this case there exists a one-to-one correspondence between the elements of the group, the matrices A, and n^2 real parameters, their n^2 matrix elements A^i_j. Namely, the n^2 functions f^i_j of the group into the real field, which assign to any A in GL(n,R) its matrix elements A^i_j, play the role of the functions f^i associated to a local chart in the general case, Def. C.60,(1) :

$$\forall A \in GL(n,R) \longrightarrow f(A) = \{f^i_j(A) = A^i_j \in R \ ; \ i,j=1,\dots,n\} \in R^{n^2} \quad .$$

Hence, this natural local parametrization is also a global one and defines an atlas. Consequently the usual topology of R^{n^2} induces a separable topology on GL(n,R). For instance, with respect to this topology a sequence of matrices $A^{(p)} = (A^{(p)i}_j)$ converges to the matrix $A = (A^i_j)$ when p goes to infinity if and only if :

$$\forall i,j \qquad \lim_{p\to+\infty} A^{(p)i}_j = A^i_j \quad .$$

The functions φ^i and ψ^i of Def. C.60,(2), which express the group laws in terms of its parameters, are clearly defined by the usual matrix multiplication laws, and hence are analytic functions over R^{n^2}. With this topology GL(n,R) is thus an n^2-parameter Lie group.

Let us now consider the subgroup O(n) of orthogonal real $n \times n$ matrices (Subsec. 2.2.1). The topology previously defined on GL(n,R) induces a separable topology on O(n) (i.e. the open sets of O(n) are defined by the intersection of the open sets of GL(n,R) with O(n)). Thus this topology on O(n) is still induced by the usual topology of R^{n^2}, and not by the one of $R^{n(n-1)/2}$, though O(n) depends upon n(n-1)/2 real parameters. Indeed the correspondence between the matrices of O(n) and n(n-1)/2 real parameters is not one-to-one since among others there remains the indeterminacy of the value ± 1 of the determinant. This means that for O(n) considered as a manifold of dimension n(n-1)/2, we no longer have only one local chart which covers the whole group, i.e. which forms by itself an atlas. The topology of O(n) induced by the usual topology of R^{n^2} is only locally identical to the one of $R^{n(n-1)/2}$.

Actually this problem is not related to the existence of a group structure, but only to the definition of the induced topology on a submanifold. Let us illustrate this point as follows. A circle in a plane, i.e. a 1-sphere is a subset of points which depend upon only one real parameter (say a polar angle θ). Nevertheless the separable topology of S^1 induced by the usual topology of R^2, and the usual topology of R are globally different (the topology of R is noncompact and simply-connected while the topology of S^1 is compact and connected without being simply-connected). These two topologies are only locally identical.

With the previous separable topology O(n) is a compact Lie group. Indeed from Prop. C.69 it is sufficient to show that the set of matrix elements A^i_j, for all A in O(n) is bounded and closed. The orthogonality condition implies for the matrix elements A^i_j :

$$\forall\, i \qquad \sum_j (A^i_j)^2 = 1 \quad ,$$

and thus all the A^i_j are in absolute value less than or equal to one. Moreover, let $(A^{(p)})$ be a sequence of matrices in O(n) which converges (in the sense specified above) to a matrix A when p goes to infinity. Then, taking the limit for p going to infinity of :

$$A^{(p)}\, A^{(p)t} = \mathbb{1} \quad ,$$

implies the necessary condition for A :

$$A\, A^t = \mathbf{1} \quad .$$

Therefore A belongs to O(n) which is thus closed.

Since the determinant is a continuous function which does not vanish on O(n), this group is not connected. It has two connected subsets, one of which is the special orthogonal group SO(n) of matrices with determinant +1, and the other the set of orthogonal matrices of determinant −1.

Clearly SO(n) is also a compact Lie group with respect to the separable topology induced by the one of GL(n,R). Once again this topology is only locally identical to the usual one of $R^{n(n-1)/2}$. Indeed the correspondence between n(n−1)/2 real parameters and a matrix in SO(n) is only locally one–to–one. Therefore for SO(n) considered as a manifold of dimension n(n–1)/2, we no longer have only one local chart which covers the whole group. To get a one–to–one correspondence between the group and a subset Ω in $R^{n(n-1)/2}$, the parameter space, one needs to identify some points of Ω (cf. for SO(3) Subsec. 2.3.2.3, for SU(2) Subsec. 4.1.4, and for U(1) remarks below).

The above discussion can be repeated for the complex linear group GL(n,C) which is a $2n^2$–parameter Lie group with respect to the separable topology induced by the usual one of R^{2n^2}. The unitary and special unitary groups, U(n) and SU(n), are compact Lie groups with respect to the separable topologies induced by the one of GL(n,C) (the set of matrix elements is closed and bounded). These topologies of U(n) and SU(n) are again only locally identical to the usual one of, respectively, R^{n^2} and R^{n^2-1}.

The previous remarks about the topology of S^1 induced by the usual one of R^2 apply directly to the topology of U(1) induced by the topology of GL(1,C), i.e. the one of R^2. Indeed, the elements of U(1) read $e^{i\theta}$, where θ is a real parameter (the polar angle for the circle).

One should note that there exists a one–to–one correspondence between U(1) and the set $\theta \in [0,2\pi[$. But with respect to the topology on U(1) induced by the usual one of $[0,2\pi[$, this group is not a topological one since the application $(g,g') \to gg'$ is not continuous in for instance $g = g' = -\mathbf{1}$. For the topology of U(1) induced by the one of GL(1,C) the correspondence between $U \in U(1)$ and $\theta \in R$ is only locally one–to–one. Actually the parametrization of U(1) by θ only defines a local chart $(\mathcal{O},f)$ where :

– $\mathcal{O} = \{g \in SU(2), g \neq \mathbf{1}\}$ is an open set in SU(2),

–f is the one–to–one mapping such that :

$$\forall\, U \in \mathcal{O} \qquad f(U) = \theta \quad ,$$

–$f(\mathcal{O})$ is the open 1-ball $]0,2\pi[$.

ii) **Lie algebras of linear groups :**

1–Let us first show that the Lie algebra of the linear group GL(n,R) is isomorphic with the n^2–dimensional Lie algebra over the real field of $n \times n$ real matrices, the bilinear mapping of this algebra being defined by means of the commutator (cf. Appendix B.1.2,i)).

The proof of Prop. C.73 is now closely followed. To the natural parametrization, considered above in i), of the matrices A in GL(n,R) by their matrix elements $A^i_j = f^i_j(A)$, there correspond n^2 tangent vectors T^j_i to the group at any element A defined by :

$$T^j_i(f^{i'}_{j'}) = \delta^{i'}_i\ \delta^j_{j'} \quad , \qquad \forall\, \phi \in F(A) \to T^j_i(\phi) = \frac{\partial \phi}{\partial f^i_j} \in R \quad ,$$

where $F(A)$ denotes the set of functions of the group into the real field, analytic at the point A ($f^i_j \in F(A)$). These n^2 tangent vectors form a basis of the tangent vector space at A, and any tangent vector T at A is uniquely defined by its components such that (Prop. C.73) :

$$T = T(f^i_j)\ T^j_i \quad .$$

Especially, for any left (resp. right) invariant infinitesimal transformation J on GL(n,R), J(A) is by definition a tangent vector to the group at A, and thus defined by its components with respect to the above basis :

$$J \longrightarrow J(A) = \{[J(A)](f^i_j)\}\ T^j_i \quad .$$

This allows us to define a mapping, which is clearly linear, of the Lie algebra $A(n,R)$ of $GL(n,R)$ into the vector space $M(n,R)$ over the real field of $n \times n$ real matrices, by :

$$\forall J \in A(n,R) \longrightarrow \tilde{J} = (\tilde{J}^i_j = [J(A)](f^i_j)) \in M(n,R) \quad .$$

The above mapping is injective since its kernel is reduced to $\{0\}$. Indeed if the matrix $\tilde{J}$ is equal to zero, then :

$$\forall i,j \qquad [J(A)](f^i_j) = 0 \Rightarrow J(A) = 0 \quad ,$$

and from the left (resp. right) invariant character of J :

$$\forall B \in GL(n,R) \qquad J(B) = dL_{BA^{-1}}[J(A)] = (\text{resp.}\, dR_{A^{-1}B}[J(A)]) = 0 \Rightarrow J = 0 \quad .$$

As both vector spaces $A(n,R)$ and $M(n,R)$ are n^2-dimensional, the mapping $J \to \tilde{J}$ is thus an isomorphism between these two vector spaces (clearly also between the tangent vector space to $GL(n,R)$ at any point A and $M(n,R)$, cf. Prop. C.81).

There remains to compute the image in the above mapping of the commutator $[J,J']$ of two left (resp. right) invariant transformations in terms of their respective images $\tilde{J}$ and $\tilde{J}'$. From the left invariant character of J one gets :

$$\forall B \in GL(n,R) \qquad [J(B)](f^i_j) = \{dL_{BA^{-1}}[J(A)]\}(f^i_j) = [J(A)](f^i_j \circ L_{BA^{-1}}) \quad .$$

The function $f^i_j \circ L_{BA^{-1}}$ of the group in R is such that :

$$\forall C \in GL(n,R) \xrightarrow{\ L_{BA^{-1}}\ } BA^{-1}C \xrightarrow{\ f^i_j\ } B^i_k (A^{-1})^k_\ell C^\ell_j = B^i_k (A^{-1})^k_\ell f^\ell_j (C) \quad ,$$

hence :

$$f^i_j \circ L_{BA^{-1}} = B^i_k (A^{-1})^k_\ell\ f^\ell_j \quad ,$$

and from the linear character of $J(A)$:

$$[J(B)](f^i_j) = B^i_k (A^{-1})^k_\ell [J(A)](f^\ell_j) = B^i_k (A^{-1})^k_\ell\ \tilde{J}^\ell_j \quad .$$

Using Def. C.58, one considers $[J(B)](f^i_j)$ as the value the function $J\, f^i_j$ assigns to any B in $GL(n,R)$. Since :

$$\forall B \in GL(n,R) \qquad B^i_k = f^i_k (B) \quad ,$$

this function reads :

$$J\, f^i_j = f^i_k (A^{-1})^k_\ell\ \tilde{J}^\ell_j \quad .$$

Now applying $J'(A)$ to this function yields :

$$[J'(A)](Jf^i_j) = [J'(A)](f^i_k)(A^{-1})^k_\ell \mathfrak{J}^\ell_j = \mathfrak{J}'^i_k(A^{-1})^k_\ell \mathfrak{J}^\ell_j \quad .$$

It follows that the matrix which corresponds to $[J,J']$ is given by :

$$[J,J'] \longrightarrow \mathfrak{J} A^{-1} \mathfrak{J}' - \mathfrak{J}' A^{-1} \mathfrak{J} \quad .$$

One easily checks that this bilinear mapping of $M(n,R) \times M(n,R)$ into $M(n,R)$ is anticommutative and fulfils the Jacobi identity and thus endows $M(n,R)$ with a Lie algebra structure, isomorphic to the Lie algebra of GL(n,R). A similar calculation for right invariant infinitesimal transformations yields :

$$[J,J'] \longrightarrow \mathfrak{J}' A^{-1} \mathfrak{J} - \mathfrak{J} A^{-1} \mathfrak{J}' \quad .$$

Taking the arbitrary point A in GL(n,R) as the unit element, one gets the usual Lie algebra of matrices where the bilinear mapping is defined by the commutator.

2–Let us now consider the subgroup O(n). The same steps as above can be used to find a Lie algebra of matrices, isomorphic to the Lie algebra of O(n). The only difference is that the n^2 matrix elements no longer define a local chart. Nevertheless in a neighbourhood of the unit element, i.e. for infinitesimal elements $\mathbb{1} + W$, the n(n-1)/2 matrix elements $W^i_j (1 \leqslant i < j \leqslant n)$ of the antisymmetrical matrix W are local parameters. Thus, in a neighbourhood of the unit element, the n(n-1)/2 functions f^i_j,

$$f^i_j(\mathbb{1} + W) = W^i_j \quad , \quad f^i_j = -f^j_i \quad ,$$

define a local chart. To this parametrization, there correspond n(n-1)/2 tangent vectors T^j_i to the group at the unit element. The isomorphism between the Lie algebra of O(n) and the Lie algebra of antisymmetrical matrices of $M(n,R)$ is defined as above by :

$$J \longrightarrow \mathfrak{J} = (\mathfrak{J}^i_j = [J(\mathbb{1})] \ (f^i_j) = \mathfrak{J}^j_i) \quad .$$

The above discussion can be repeated for the complex linear group GL(n,C) by splitting the real and complex parts of the matrices. The Lie algebra of $n \times n$ complex Hermitian and traceless matrices, where the bilinear mapping is defined by minus i times the commutator is studied in Sec. 4.3. This Lie algebra is isomorphic to the Lie algebra of the group SU(n).

APPENDIX D

THE BAKER–CAMPBELL–HAUSDORFF FORMULA

In this appendix the objects A,B,C,... we consider belong to a Lie algebra $\mathcal{A}$ over the complex field with the bilinear mapping of $\mathcal{A} \times \mathcal{A}$ into $\mathcal{A}$ defined by the commutator (Appendix B.1.2,i)). The adjoint operator Δ_A of A is defined by (Def. B.13) :

$$\forall A,B \in \mathcal{A} \qquad \Delta_A B = [A,B] \in \mathcal{A} \quad .$$

It is recalled that the subset $\Delta(\mathcal{A})$ of adjoint operators is a Lie algebra homomorphic to $\mathcal{A}$ (Prop. B.18).

Our aim is to establish formal expansions for several expressions. The convergence of these expansions will not be discussed here (cf. for instance Baker[3], Campbell[12], and Hausdorff[29]). For finite-dimensional matrices with bounded matrix elements we have already seen (proof of Theorem 4.2, Subsec. 4.1.2,i)) that the exponential series $\sum_{n=0}^{\infty} \frac{1}{n!} A^n$ is convergent, which implies in that case the convergence of the several expansions given below.

D.1. USEFUL FORMULAE

Prop. D.1. For all A and B in the Lie algebra $\mathcal{A}$ one formally has :

$$e^A B\, e^{-A} = e^{\Delta_A} B = B + [A,B] + \ldots + \frac{1}{n!} [\underbrace{A, \ldots, [A,B] \ldots]}_{n} + \ldots \quad .$$

Proof :

1–As in the time dependent perturbation theory, $B(t) = e^{tA}\, B\, e^{-tA}$ satisfies :

$$B(0) = B \quad , \quad \frac{d}{dt} B(t) = [A,B(t)] \quad ,$$

and therefore :

$$B(t) = B + \int_0^t d\tau\, [A,B(\tau)] \quad .$$

Solving this last equation by iteration, one obtains the announced formal expansion.

2–In terms of the adjoint operator Δ_A, the preceding proof takes a simple form. Indeed, one has :

$$\frac{d}{dt} B(t) = \Delta_A B(t) \quad ,$$

and with the condition B(0) = B the solution of this equation formally reads :

$$B(t) = e^{t\Delta_A} B \quad .$$

3–Another way to get the result consists in expanding both exponentials e^A and e^{-A} :

$$e^A B e^{-A} = \sum_{p,q=0}^{\infty} \frac{(-1)^q}{p!q!} A^p B A^q = \sum_{n=0}^{\infty} \frac{1}{n!} \sum_{p+q=n} (-1)^q \frac{n!}{p!q!} A^p B A^q \quad .$$

One then shows by induction that :

$$n \geqslant 1 \qquad \sum_{p+q=n} (-1)^q \frac{n!}{p!q!} A^p B A^q = \underbrace{[A, \dots ,[A,B]\dots]}_{n} \quad .$$

Remark : $\underbrace{[A, \dots ,[A,B]\dots]}_{n} = (-1)^n \underbrace{[\dots[B,A], \dots ,A]}_{n} \quad .$

Prop. D.2. Let A, in the Lie algebra $\mathcal{A}$, depend upon a real or complex variable t such that its first derivative with respect to t exists :

$$A = A(t) \qquad , \qquad A' = \frac{d}{dt} A(t) \quad .$$

A priori, A and A' are noncommuting quantities. One then formally has :

$$\frac{d}{dt} e^{A(t)} = \int_0^1 d\tau\, e^{\tau A(t)} A'(t)\, e^{(1-\tau)A(t)} = \int_0^1 d\tau\, e^{(1-\tau)A(t)} A'(t)\, e^{\tau A(t)}$$

$$= \frac{e^{\Delta_A} - 1}{\Delta_A} A' e^A = (A' + \frac{1}{2!}[A,A'] + \dots + \frac{1}{n!}\underbrace{[A, \dots ,[A,A']\dots]}_{n-1} + \dots) e^A$$

$$= e^A \frac{1 - e^{-\Delta_A}}{\Delta_A} A' = e^A (A' + \frac{1}{2!}[A',A] + \dots + \frac{1}{n!} \underbrace{[\dots[A',A], \dots ,A]}_{n-1} + \dots) \quad .$$

Proof :

1–Expanding both exponentials $e^{\tau A}$ and $e^{(1-\tau)A}$ and expressing the integral over τ by means of the first and second kind Euler functions, one obtains :

$$\int_0^1 d\tau e^{\tau A(t)} A'(t) e^{(1-\tau)A(t)} = \sum_{p,q=0}^{\infty} \frac{1}{p!q!} \int_0^1 d\tau \tau^p (1-\tau)^q A^p(t) A'(t) A^q(t) = \sum_{n=0}^{\infty} \frac{1}{(n+1)!} \sum_{p+q=n} A^p A' A^q \quad .$$

This last expansion is equal to the formal derivative of the exponential series, which ends the proof of the first expression given in Prop. D.2. The second one is obtained by changing τ into $1-\tau$.

2–From the above result and using Prop. D.1, one formally gets :

$$\frac{d}{dt} e^A = \int_0^1 d\tau\, e^{\tau A(t)} A'(t) e^{-\tau A(t)}\, e^{A(t)} = \int_0^1 d\tau e^{\tau \Delta_{A(t)}} A'(t)\, e^{A(t)} = \frac{e^{\Delta_A} - 1}{\Delta_A} A' e^A \quad .$$

Setting $z = \Delta_A$, the expansion :

$$\frac{e^z - 1}{z} = 1 + \frac{1}{2!} z + \dots + \frac{1}{n!} z^{n-1} + \dots$$

leads to the fourth expression of $\frac{d}{dt} e^{A(t)}$. The last two expressions given in Prop. D.2 are derived in a similar way from the second one.

3–One can give another proof of the above results if one assumes the existence of a B such that :

$$A' = [A,B] = \Delta_A B \quad .$$

This might not always be possible. However, if it is, the derivative of the exponential series formally reads, using Prop. D.1 :

$$\frac{d}{dt} e^A = A' + \frac{1}{2!}(AA'+A'A)+\ldots = [A,B] + \frac{1}{2!}[A^2,B]+\ldots = [e^A,B]$$

$$= (e^A B e^{-A} - B)e^A = (e^{\Delta_A} - \mathbf{1})B\, e^A = \frac{e^{\Delta_A} - \mathbf{1}}{\Delta_A} A' e^A \quad .$$

Prop. D.3. Let A, in the Lie algebra $\mathcal{A}$, depend upon a real or complex variable t such that its first derivative with respect to t exists. One then formally has :

$$\left(\frac{d}{dt}e^A\right)e^{-A} = \int_0^1 d\tau\, e^{\tau A(t)} A'(t)\, e^{-\tau A(t)} = \frac{e^{\Delta_A} - \mathbf{1}}{\Delta_A} A' = A' + \frac{1}{2!}[A,A'] + \ldots + \frac{1}{n!}\underbrace{[A,\ldots,[A,A']\ldots]}_{n-1} + \ldots ,$$

$$e^{-A}\frac{d}{dt}e^A = \int_0^1 d\tau\, e^{-\tau A(t)} A'(t)\, e^{\tau A(t)} = \frac{\mathbf{1} - e^{-\Delta_A}}{\Delta_A} A' = A' + \frac{1}{2!}[A',A] + \ldots + \frac{1}{n!}\underbrace{[\ldots[A',A],\ldots,A]}_{n-1} + \ldots$$

Proof. Direct consequence of Prop. D.2.

D.2. THE BAKER–CAMPBELL–HAUSDORFF FORMULA

D.2.1. DERIVATION OF FORMULAE

Prop. D.4. The Baker–Campbell–Hausdorff formula. For all A and B in the Lie algebra $\mathcal{A}$ one formally has :

$$e^A e^B = e^C \quad ,$$

where C belongs to $\mathcal{A}$, i.e. is given as a formal expansion in terms of repeated commutators of A and B. Several expressions of C can be derived, e.g. :

$$C = A + \int_0^1 dt\, f(e^{\Delta_A} e^{t\Delta_B})B = B + \int_0^1 dt\, g(e^{t\Delta_A} e^{\Delta_B})A$$

$$= \int_0^1 dt\, f(e^{t\Delta_A} e^{t\Delta_B})(e^{-t\Delta_B}A + B) = \int_0^1 dt\, g(e^{t\Delta_A} e^{t\Delta_B})(A + e^{t\Delta_A}B) \quad ,$$

where the functions f and g are given by their power series expansions :

$$f(z) = \frac{z \log z}{z-1} = 1 + \frac{1}{2}(z-1) + \ldots + \frac{(-1)^{n-1}}{n(n+1)}(z-1)^n + \ldots \quad ,$$

$$g(z) = \frac{\log z}{z-1} = 1 - \frac{1}{2}(z-1)+\ldots+\frac{(-1)^n}{n+1}(z-1)^n + \ldots \quad .$$

The first terms in the expansion of C read :

$$C = A+B+\frac{1}{2}[A,B] + \frac{1}{12}([A,[A,B]]+[B,[B,A]])+\ldots \quad .$$

Proof :

1–Let :

$$e^{C(t)} = e^{A} e^{tB} \quad ,$$

from Prop. D.2 one then has :

$$\frac{d}{dt} e^{C(t)} = e^{C(t)} B = e^{C(t)} \frac{1-e^{-\Delta_{C(t)}}}{\Delta_{C(t)}} C'(t) \quad .$$

Therefore, one obtains the formal differential equation :

$$C'(t) = \frac{\Delta_{C(t)}}{1-e^{-\Delta_{C(t)}}} B = \frac{e^{\Delta_{C(t)}} \Delta_{C(t)}}{e^{\Delta_{C(t)}}-1} B \quad .$$

Let us now compute $\Delta_{C(t)}$ in terms of Δ_A and Δ_B. From Prop. D.1 one has for all D in A :

$$e^{\Delta_{C(t)}} D = e^{C(t)} D\, e^{-C(t)} = e^{A}(e^{tB} D\, e^{-tB}) e^{-A} = e^{\Delta_A} e^{t\Delta_B} D \quad ,$$

hence, formally :

$$e^{\Delta_{C(t)}} = e^{\Delta_A} e^{t\Delta_B} \quad , \quad \Delta_{C(t)} = \log(e^{\Delta_A} e^{t\Delta_B}) \quad .$$

Finally, we integrate the differential equation for C'(t) from 0 to 1 with C(0)= A and C(1)= C , thereby obtaining the announced result :

$$C = A + \int_0^1 dt \; \frac{e^{\Delta_A} e^{t\Delta_B} \log(e^{\Delta_A} e^{t\Delta_B})}{e^{\Delta_A} e^{t\Delta_B} - 1} B \quad .$$

Now, setting $z = e^{\Delta_A} e^{t\Delta_B}$ in the expansion of the function f given in Prop. D.4 and expanding both exponentials, one gets the formal expression of C in terms of the only repeated commutators of A and B.

2–The three other expressions of C are derived in a similar way. For the second one, one sets :

$$e^{C(t)} = e^{tA} e^{B} \Rightarrow \frac{d}{dt} e^{C(t)} = A\, e^{C(t)} \quad ,$$

and for the last two (Prop. D.1) :

$$e^{C(t)} = e^{tA} e^{tB} \Rightarrow \frac{d}{dt} e^{C(t)} = e^{C(t)}(e^{-t\Delta_B} A+B) = (A+e^{t\Delta_A} B) e^{C(t)} \quad .$$

Prop. D.5. For all A and B in the Lie algebra A one formally has :

$$e^{A+B} = e^{A} e^{B} e^{C} \quad ,$$

where $\mathcal{C}$ belongs to $\mathcal{A}$, i.e. is given as a formal expansion in terms of repeated commutators of A and B. Several expressions of $\mathcal{C}$ can be derived, e.g. :

$$\mathcal{C} = -C + \int_0^1 dt \; f\left[e^{-\Delta_B} e^{-\Delta_A} e^{t(\Delta_A+\Delta_B)}\right] (A+B) \quad ,$$

where C and the function f are defined in Prop. D.4. The first terms in the expansion of $\mathcal{C}$ read :

$$\mathcal{C} = -\frac{1}{2}[A,B] + \frac{1}{6}[A,[A,B]] - \frac{1}{3}[B,[B,A]] + \ldots \quad .$$

Proof. One has :

$$e^{C} = e^{A} e^{B} \Rightarrow e^{\mathcal{C}} = e^{-C} e^{A+B} \quad ,$$

and applying the first expression given in Prop. D.4 one gets :

$$\mathcal{C} = -C + \int_0^1 dt \; f\left[e^{-\Delta_C} e^{t(\Delta_A+\Delta_B)}\right] (A+B) \quad .$$

Furthermore, it has been shown in the proof of Prop. D.4 that :

$$e^{-C} = e^{-B} e^{-A} \Rightarrow e^{-\Delta_C} = e^{-\Delta_B} e^{-\Delta_A} \quad ,$$

which yields the announced expression for $\mathcal{C}$.

D.2.2. SOME APPLICATIONS

i) The main consequences of the Baker–Campbell–Hausdorff formula in connection with group theory are discussed in Secs. 4.4 and 5.3. This formula gives a direct relationship between the local structure of a group and its Lie algebra.

ii) Two special cases :

1–A commutes with B :

$$[A,B] = 0 \quad \Rightarrow \quad e^{A} e^{B} = e^{A+B} \quad .$$

2–A and B commute with their commutators :

$$[A,[A,B]] = [B,[B,A]] = 0 \Rightarrow \begin{cases} e^{A} e^{B} = e^{A+B+\frac{1}{2}[A,B]} \\ e^{A+B} = e^{A} e^{B} e^{-\frac{1}{2}[A,B]} \end{cases} \quad .$$

This is for instance the case when A and B are linear combinations of the A_i, the commutators of which are multiples of the identity :

$$[A_i, A_j] = \alpha_{ij} \mathbf{1} \qquad \alpha_{ij} \in \mathbb{C} \quad .$$

Examples :

(1) The A_i are position or momentum operators :

$$[q_i,q_j] = [p_i,p_j] = 0 \quad , \quad [q_i,p_j] = i\hbar\, \delta_{ij}\, \mathbf{1} \quad .$$

(2) The A_i are creation or annihilation operators for Bosons :

$$[a_i^+,a_j^+] = [a_i,a_j] = 0 \quad , \quad [a_i,a_j^+] = \delta_{ij}\, \mathbf{1} \quad .$$

One should remark that if A and B are quadratic forms of objects A_i which fulfil the above commutation relations, $e^A e^B$ is then also the exponential of a quadratic form of the A_i. Indeed, the commutator of A and B is still a quadratic form of the A_i.

APPENDIX E

MATRIX REPRESENTATIVE OF FINITE AND INFINITESIMAL THREE-DIMENSIONAL ROTATIONS

Conventions :

1–Active point of view (Subsec. 1.2.1) of rotations R in a three-dimensional Euclidean vector space E_3.

2–With respect to a right-handed orthonormal basis a = (e_i) of E_3 (Subsec. 2.3.2.1,i)) :

$$v = v^i e_i \in E_3 \qquad R v = v^i R^j_i e_j \quad , \quad (R^j_i) \in SO(3) \quad . \tag{E-1}$$

When there is no ambiguity, R also denotes the matrix representative.

3–Column interpretation of the components v^i of v, which implies, Eq. (1–16) :

$$v^i \text{ ith row} \Rightarrow R^j_i \text{ jth row,} \qquad \text{ith column} \quad . \tag{E-2}$$

E.1. MATRIX REPRESENTATIVE OF THE ROTATION R(θ,$\hat{n}$)

E.1.1. MATRIX ELEMENTS IN TERMS OF θ AND $\hat{n}$

With respect to (e_i) the vector $\hat{n}$ reads :

$$\hat{n} = n^i e_i \quad , \qquad \sum_i (n^i)^2 = 1 \quad . \tag{E-3}$$

Several methods can be used to derive the matrix representative of the rotation R(θ ,$\hat{n}$) with respect to the basis (e_i). Three of them are now sketched :

1–Let $\hat{v} = v^i e_i$ be any unit vector orthogonal to $\hat{n}$; for instance W_1 or W_2 in Eq. (2–15). With respect to the right-handed orthonormal basis $\hat{v}, \hat{n} \times \hat{v}, \hat{n}$, the matrix representative of R(θ ,$\hat{n}$) reads, Eq. (2–28) :

$$R' = \begin{pmatrix} \cos\theta & -\sin\theta & 0 \\ \sin\theta & \cos\theta & 0 \\ 0 & 0 & 1 \end{pmatrix} \quad :$$

The matrix representative R with respect to the basis (e_i) is then obtained using Eqs. (1–13,14) :

$$R = t\, R'\, t^{-1} \qquad ,$$

where, according to Eq. (1–2), the columns of the matrix t are, respectively, defined by the components of $\hat{v}, \hat{n}\times v$, and $\hat{n}$ with respect to (e_i), i.e. :

$$t^j_1 = v^j \quad , \quad t^j_2 = \varepsilon_{jk\ell}\, n^k v^\ell \quad , \quad t^j_3 = n^j \quad .$$

2–Another way to get the result is to use Eq. (4–28) :

$$R(U)^k_j = \tfrac{1}{2}\,\mathrm{Tr}\,(\sigma_k\, U\, \sigma_j\, U^{-1}) \quad ,$$

where U in SU(2) is parametrized by θ and n^i as :

$$U = \cos\frac{\theta}{2}\,\mathbf{1} - i\sin\frac{\theta}{2}\, n^i\sigma_i \quad .$$

One is then led to compute the traces of the products of σ matrices ; this can be done using several times Eqs. (4–85,88, and 89a).

3–Any vector v can be decomposed into a vector parallel to the rotation axis and another one orthogonal to it, i.e. :

$$\forall v \in E_3 \qquad v = (\hat{n}.v)\hat{n} + [v-(\hat{n}.v)\hat{n}] \quad .$$

The first vector is invariant under $R(\theta,\hat{n})$ while the second one transforms according to Eq. (2–29). Therefore, one gets :

$$\forall v \in E_3 \qquad R(\theta,\hat{n})v = (\hat{n}.v)\hat{n} + \cos\theta[v-(\hat{n}.v)\hat{n}] + \sin\theta\; \hat{n}\times v$$

$$= \cos\theta\; v + (1-\cos\theta)(\hat{n}.v)\hat{n} + \sin\theta\; \hat{n}\times v \quad .$$

Finally, without taking care of the tensorial character of the indices, the matrix representative of $R(\theta,\hat{n})$ with respect to (e_i) reads :

$$R(\theta,\hat{n})^i_j = \cos\theta\,\delta^i_j + (1-\cos\theta)n^i n^j + \sin\theta\;\varepsilon_{ikj}\, n^k \quad . \tag{E-4}$$

For the special cases $\hat{n} = e_1, e_2$, and e_3, one obtains :

$$R(\theta,e_1) = \begin{pmatrix} 1 & 0 & 0 \\ 0 & \cos\theta & -\sin\theta \\ 0 & \sin\theta & \cos\theta \end{pmatrix} ,\; R(\theta,e_2) = \begin{pmatrix} \cos\theta & 0 & \sin\theta \\ 0 & 1 & 0 \\ -\sin\theta & 0 & \cos\theta \end{pmatrix} , R(\theta,e_3) = \begin{pmatrix} \cos\theta & -\sin\theta & 0 \\ \sin\theta & \cos\theta & 0 \\ 0 & 0 & 1 \end{pmatrix} . \tag{E-5}$$

E.1.2. ROTATION ANGLE AND ROTATION AXIS IN TERMS OF THE MATRIX ELEMENTS

i) **Rotation angle.** The rotation angle θ within $[0,\pi]$ is uniquely determined from Eq. (2–30) :

$$1+2\cos\theta = \mathrm{Tr}\; R = \sum_i R^i_i \quad . \tag{E-6}$$

ii) **Rotation axis.** The rotation axis, Eq. (2–23), is defined by the unitary vector $\hat{n}$ which is an eigenvector of R with the eigenvalue 1 :

$$R\,\hat{n} = \hat{n} \quad .$$

Since the matrix R is orthogonal, one gets from above :

$$\hat{n} = R^{-1}\,\hat{n} = R^{t}\,\hat{n} \quad \Rightarrow \quad (R-R^{t})\hat{n} = 0 \quad .$$

The nontrivial solution of this set of linear equations provides the non–normalized components of $\hat{n}$, i.e. :

$$n^{k} \propto \sum_{i,j} \varepsilon_{ijk}\, R^{i}_{j} \quad .$$

where there remains an overall indetermination of sign. Actually, the result can be obtained directly from Eq. (E–4) as follows :

$$\sum_{i,j} \varepsilon_{ij\ell}\, R(\theta,\hat{n})^{i}_{j} = \sin\theta \sum_{i,j} \varepsilon_{ij\ell}\, \varepsilon_{ikj}\, n^{k} = -\,2\sin\theta\, \delta_{\ell k}\, n^{k} \quad ,$$

and therefore, for θ nonzero :

$$n^{k} = -\,\frac{1}{2\sin\theta} \sum_{i,j} \varepsilon_{ijk}\, R^{i}_{j} \quad . \tag{E-7}$$

E.1.3. CHANGE OF COORDINATES UNDER INFINITESIMAL ROTATIONS

i) **Infinitesimal rotations.** For an infinitesimal angle ε the rotation $R(\varepsilon,\hat{n})$ takes the form given in Eq. (2–31) :

$$R(\varepsilon,\hat{n}) = \mathbf{1} + \varepsilon\,\hat{n}\,\times\; + \; 0(\varepsilon^{2}) \quad . \tag{E-8}$$

Its matrix representative with respect to (e_i) then reads :

$$R(\varepsilon,\hat{n})^{i}_{j} = \delta_{ij} + \varepsilon\,\varepsilon_{ikj}\, n^{k} + 0(\varepsilon^{2}) \quad . \tag{E-9}$$

The matrix representative of the infinitesimal rotations about the axes e_i can be expressed in terms of the matrices M_i defined in Eq. (4–73), i.e. in matrix notation :

$$R(\varepsilon,e_i) = \mathbf{1} + \varepsilon\, M_i + 0(\varepsilon^{2}) \quad , \quad (M_i)^{j}_{k} = \varepsilon_{ikj} \quad . \tag{E-10}$$

Let us now determine the velocity fields, Eq. (10–73), of the group of rotations $\mathcal{R} = (O,R)$ for a given fixed point O in the point space $E^{\bullet}_{3}$ (Subsec. 2.3.1,i)), at first for the Cartesian coordinates, Eq. (2–18), and next for the spherical polar ones, Eq. (2–20).

ii) **Velocity field for Cartesian coordinates.** Let us denote by $R(\varepsilon^{1}, \varepsilon^{2}, \varepsilon^{3})$ the product of three infinitesimal rotations of angle ε^{i} about the axis e_i. From Eq. (E–10) one gets up to first order in ε^{1}, ε^{2}, and ε^{3} :

$$x + dx = R(\varepsilon^{1},\varepsilon^{2},\varepsilon^{3})x = (\mathbf{1} + \varepsilon^{i}\, M_i)x \quad ,$$

$$dx = \varepsilon^{i}\, M_i\, x \Rightarrow dx^{j} = \varepsilon^{i}(M_i)^{j}_{k}\, x^{k} \quad , \tag{E-11}$$

and therefore :

$$dx^j = \varepsilon^i \mu_{Ci}^j(x) \qquad , \qquad \mu_{Ci}^j(x) = \varepsilon_{ikj}\, x^k \quad . \tag{E-12}$$

iii) **Velocity field for spherical polar coordinates.** Let us denote by x+dx the Cartesian coordinates which correspond to the spherical polar ones (r, θ+dθ, φ+dφ). One then has up to first order in dθ, dφ, and ε^i :

$$dx = r\begin{pmatrix}\cos\theta\cos\varphi\, d\theta - \sin\theta\sin\varphi\, d\varphi \\ \cos\theta\sin\varphi\, d\theta + \sin\theta\cos\varphi\, d\varphi \\ -\sin\theta\, d\theta\end{pmatrix} = \varepsilon^i M_i r\begin{pmatrix}\sin\theta\cos\varphi \\ \sin\theta\sin\varphi \\ \cos\theta\end{pmatrix} \quad .$$

Finally, one obtains :

$$d\theta = \varepsilon^i \mu_{Si}^1(\theta,\varphi) = \varepsilon^1(-\sin\varphi) + \varepsilon^2(\cos\varphi) + \varepsilon^3(0) \qquad , \tag{E-13a}$$

$$d\varphi = \varepsilon^i \mu_{Si}^2(\theta,\varphi) = \varepsilon^1(-\mathrm{cotg}\theta\cos\varphi) + \varepsilon^2(\mathrm{cotg}\theta\sin\varphi) + \varepsilon^3(1) \quad . \tag{E-13b}$$

E.2. MATRIX REPRESENTATIVE OF THE ROTATION R(φ,θ,ψ;a)

E.2.1. MATRIX ELEMENTS IN TERMS OF φ, θ, AND ψ

Let φ, θ, and ψ be the Euler angles defined with respect to the reference axes a as specified in Subsec. 2.3.2.2 (see Fig. 2.6). Using Eqs (2–42) and (E–5), a simple calculation yields the matrix representative of R(φ, θ, ψ ;a) with respect to the basis a = (e_i) :

$$(R(\varphi,\theta,\psi;a)^i_j) = \begin{pmatrix}\cos\varphi\cos\theta\cos\psi-\sin\varphi\sin\psi & -\cos\varphi\cos\theta\sin\psi-\sin\varphi\cos\psi & \cos\varphi\sin\theta \\ \sin\varphi\cos\theta\cos\psi+\cos\varphi\sin\psi & -\sin\varphi\cos\theta\sin\psi+\cos\varphi\cos\psi & \sin\varphi\sin\theta \\ -\sin\theta\cos\psi & \sin\theta\sin\psi & \cos\theta\end{pmatrix} . \tag{E-14}$$

E.2.2. ROTATION ANGLE AND ROTATION AXIS IN TERMS OF φ, θ, AND ψ

From Eqs. (E–6,7, and 14) the rotation angle α and the components n^i with respect to the orthonormal basis a=(e_i) of the unit vector n̂ such that :

$$R(\alpha,\hat{n}) = R(\varphi,\theta,\psi;a) \qquad , \tag{E-15a}$$

are determined by :

$$1+2\cos\alpha = (1+\cos\theta)\cos(\varphi+\psi)+\cos\theta \qquad , \tag{E-15b}$$

$$\begin{cases} 2\sin\alpha\, n^1 = -\sin\theta\,(\sin\varphi - \sin\psi) & , \\ 2\sin\alpha\, n^2 = \sin\theta\,(\cos\varphi + \cos\psi) & , \\ 2\sin\alpha\, n^3 = (1+\cos\theta)\,\sin(\varphi+\psi) & . \end{cases} \tag{E-15c}$$

E.2.3. VELOCITY FIELDS OF THE FIRST AND THE SECOND PARAMETER GROUPS

The parameter groups of any Lie group are defined in Subsec. 10.2.2.3,i), and the expressions of the corresponding velocity fields are given in Eqs. (10–93a,b).

i) **First parameter group.** The velocity field $\lambda_i(\varphi,\theta,\psi)$ of the first parameter group, which corresponds to left translations, can be obtained from :

$$R(\varphi+d\varphi,\ \theta+d\theta,\ \psi+d\psi;a) = R(\varepsilon^i)R(\varphi,\theta,\psi;a)\ ,\quad R(\varepsilon^i) = \prod_i R(\varepsilon^i,e_i)\ .$$

Indeed, using Eqs. (E–10,14), the identification up to first order in $d\varphi$, $d\theta$, $d\psi$, and ε^i of the matrix representatives of both sides of the above equation enables us to express $d\varphi$, $d\theta$, and $d\psi$ in terms of the ε^i. It is sufficient for that to choose three linearly independent relations, e.g. :

$$R(\varphi+d\varphi,\theta+d\theta,\psi+d\psi)=R(\varphi,\theta,\psi)+dR(\varphi,\theta,\psi)\ ,\quad \begin{cases} dR^2_3 = \cos\varphi\ \sin\theta\ d\varphi+\sin\varphi\ \cos\theta\ d\theta \\ dR^3_2 = \cos\theta\ \sin\psi\ d\theta+\sin\theta\ \cos\psi\ d\psi \\ dR^3_3 = -\sin\theta\ d\theta \end{cases} .$$

One then gets :

$$d\varphi = \varepsilon^i\lambda^1_i(\varphi,\theta,\psi) = \varepsilon^1(-\cos\varphi\ \mathrm{cotg}\theta) + \varepsilon^2(-\sin\varphi\ \mathrm{cotg}\theta) + \varepsilon^3(1)\ , \tag{E-16a}$$

$$d\theta = \varepsilon^i\lambda^2_i(\varphi,\theta,\psi) = \varepsilon^1(-\sin\varphi) + \varepsilon^2(\cos\varphi) + \varepsilon^3(0)\ , \tag{E-16b}$$

$$d\psi = \varepsilon^i\lambda^3_i(\varphi,\theta,\psi) = \varepsilon^1\left(\frac{\cos\varphi}{\sin\theta}\right) + \varepsilon^2\left(\frac{\sin\varphi}{\sin\theta}\right) + \varepsilon^3(0)\ . \tag{E-16c}$$

ii) **Second parameter group.** In the same way as above, the velocity field $\rho_i(\varphi,\theta,\psi)$ of the second parameter group, which corresponds to right translations, can be obtained from :

$$R(\varphi+d\varphi',\ \theta+d\theta',\ \psi+d\psi';a) = R(\varphi,\theta,\psi;a)\ R(\varepsilon^i)\ .$$

The result reads :

$$d\varphi' = \varepsilon^i\ \rho^1_i(\varphi,\theta,\psi) = \varepsilon^1\left(-\frac{\cos\psi}{\sin\theta}\right) + \varepsilon^2\left(\frac{\sin\psi}{\sin\theta}\right) + \varepsilon^3(0)\ , \tag{E-17a}$$

$$d\theta' = \varepsilon^i\ \rho^2_i(\varphi,\theta,\psi) = \varepsilon^1(\sin\psi) + \varepsilon^2(\cos\psi) + \varepsilon^3(0)\ , \tag{E-17b}$$

$$d\psi' = \varepsilon^i\ \rho^3_i(\varphi,\theta,\psi) = \varepsilon^1(\mathrm{cotg}\theta\cos\psi) + \varepsilon^2(-\mathrm{cotg}\theta\sin\psi) + \varepsilon^3(1)\ . \tag{E-17c}$$

One should note that, since SO(3) and SU(2) have isomorphic Lie algebras (Subsec. 4.3.4), the velocity fields of the first and the second parameter groups of SO(3) are, respectively, identical to the velocity fields of the first and the second parameter groups of SU(2). Thus, they can also be determined from the matrix relations in SU(2) :

$$U(\varphi+d\varphi,\ \theta+d\theta,\ \psi+d\psi) = [\mathbf{1}-i\,\frac{\varepsilon^i}{2}\,\sigma_i + 0(\varepsilon^2)]\ U(\varphi,\theta,\psi) \quad , \qquad \text{(E-18a)}$$

$$U(\varphi+d\varphi',\ \theta+d\theta',\ \psi+d\psi') = U(\varphi,\theta,\psi)\ [\mathbf{1}-i\,\frac{\varepsilon^i}{2}\,\sigma_i + 0(\varepsilon^2)] \quad , \qquad \text{(E-18b)}$$

where $U(\varphi, \theta, \psi)$ is given by Eq. (4–38).

Let us also remark that since one has, Eq. (2–46) :

$$R(\varphi,\theta,\psi;a)^{-1} = R(-\psi,-\theta,-\varphi;a) \quad , \qquad R(\varepsilon^i)^{-1} = R(-\varepsilon^i) \quad ,$$

the velocity fields of the first and the second parameter groups are related by :

$$\lambda^1_i(\varphi,\theta,\psi) = \rho^3_i(-\psi,-\theta,-\varphi)\,,\ \ \lambda^2_i(\varphi,\theta,\psi) = \rho^2_i(-\psi,-\theta,-\varphi)\,,\ \ \lambda^3_i(\varphi,\theta,\psi) = \rho^1_i(-\psi,-\theta,-\varphi) \quad , \qquad \text{(E-19)}$$

as it can be checked from the expressions derived above.

APPENDIX F

THE ROTATION MATRICES $\mathcal{D}^j$ AND d^j

F.1. THE MATRICES $\mathcal{D}^j$ AND d^j ACCORDING TO DIFFERENT AUTHORS

i) **The Euler angles** (φ, θ, ψ) we use are defined in Subsec. 2.3.2.2 with respect to a right-handed orthogonal reference frame. All the authors quoted below take the same definition with only sometimes a change of notation :

φ, θ, ψ	- Bohr and Mottelson[8], Appendix 1A, Fig. 1A-1
	- Fano and Racah[24], Sec. 5, Fig. 2
α, β, γ	- Edmonds[22], Sec. 1.3, Fig. 1.1
	- Messiah[39], Appendix C.10, Eq. (C.43) and Fig. C.3
	- Rose[54], Sec. 13, Eq. (4.7) and Fig. 3
	- Wigner[63], Sec. 10, Fig. 2.

ii) **The rotation matrices** $(\mathcal{D}^j_{m'm}(\varphi, \theta, \psi))$ adopted in these notes are defined by Eqs. (6–18,21) as a function of the three ordered variables φ, θ, and ψ. Let us now compare our definition with that of others. For all φ, θ, and ψ, one has :

$\mathcal{D}^j_{m'm}(\varphi, \theta, \psi)$	$= R^{(j)}_{m'm}(\varphi, \theta, \psi)$	- Messiah[39], Appendix C.10, Eq. (C.54)
	$= D^j_{m'm}(\varphi, \theta, \psi)$	- Rose[54], Eq. (4.12)
	$= \mathcal{D}^{j*}_{m'm}(\varphi, \theta, \psi)$	- Bohr and Mottelson[8], Eq. (1A-35)
	$= \mathcal{D}^{(j)}_{m'm}(-\varphi, -\theta, -\psi)$	- Edmonds[22], Eq. (4.1.12)
		- Fano and Racah[24], Eq. (D.2)
		- Wigner[63], Eq. (15.27).
$d^j_{m'm}(\theta)$	$= d^j_{m'm}(\theta)$	- Rose[54], Eq. (4.13)
		- Bohr and Mottelson[8], Eq. (1A-36)
	$= r^{(j)}_{m'm}(\theta)$	- Messiah[39], Appendices C.10,11,Eqs. (C.55,72)
	$= d^{(j)}_{m'm}(-\theta)$	- Edmonds[22], Eq. (4.1.15)
		- Wigner[63], Eq. (15.8)
	$= \delta^{(j)}_{m'm}(-\theta)$	- Fano and Racah[24], Eqs. (5.2), (D.3).

F.2. SPECIAL VALUES OF MATRICES d^j FOR $j = \frac{1}{2}$, 1, AND $\frac{3}{2}$

The $d^j(\theta)$ matrices are defined by Eq. (6–18). The adopted conventions are :

$$d^j_{m'm}(\theta) \quad \text{m'th row, mth column} \; ; \; \theta \in [0, \pi] \quad .$$

One recalls the symmetry properties :

$$d^j_{m'm}(\theta) = d^j_{mm'}(-\theta) = d^j_{-m-m'}(\theta) \quad .$$

TABLE F.1

The matrices $d^{\frac{1}{2}}(\theta)$, and $d^1(\theta)$.

m' \ m	$\frac{1}{2}$	$-\frac{1}{2}$
$\frac{1}{2}$	$\cos\frac{\theta}{2}$	$-\sin\frac{\theta}{2}$
$-\frac{1}{2}$	$\sin\frac{\theta}{2}$	$\cos\frac{\theta}{2}$

m' \ m	1	0	-1
1	$\frac{1+\cos\theta}{2}$	$-\frac{1}{\sqrt{2}}\sin\theta$	$\frac{1-\cos\theta}{2}$
0	$\frac{1}{\sqrt{2}}\sin\theta$	$\cos\theta$	$-\frac{1}{\sqrt{2}}\sin\theta$
-1	$\frac{1-\cos\theta}{2}$	$\frac{1}{\sqrt{2}}\sin\theta$	$\frac{1+\cos\theta}{2}$

TABLE F.2

The matrix $d^{\frac{3}{2}}(\theta)$.

m' \ m	$\frac{3}{2}$	$\frac{1}{2}$	$-\frac{1}{2}$	$-\frac{3}{2}$
$\frac{3}{2}$	$\cos^3\frac{\theta}{2}$	$-\sqrt{3}\cos^2\frac{\theta}{2}\sin\frac{\theta}{2}$	$\sqrt{3}\cos\frac{\theta}{2}\sin^2\frac{\theta}{2}$	$-\sin^3\frac{\theta}{2}$
$\frac{1}{2}$	$\sqrt{3}\cos^2\frac{\theta}{2}\sin\frac{\theta}{2}$	$\cos^3\frac{\theta}{2} - 2\cos\frac{\theta}{2}\sin^2\frac{\theta}{2}$	$\sin^3\frac{\theta}{2} - 2\cos^2\frac{\theta}{2}\sin\frac{\theta}{2}$	$\sqrt{3}\cos\frac{\theta}{2}\sin^2\frac{\theta}{2}$
$-\frac{1}{2}$	$\sqrt{3}\cos\frac{\theta}{2}\sin^2\frac{\theta}{2}$	$2\cos^2\frac{\theta}{2}\sin\frac{\theta}{2} - \sin^3\frac{\theta}{2}$	$\cos^3\frac{\theta}{2} - 2\cos\frac{\theta}{2}\sin^2\frac{\theta}{2}$	$-\sqrt{3}\cos^2\frac{\theta}{2}\sin\frac{\theta}{2}$
$-\frac{3}{2}$	$\sin^3\frac{\theta}{2}$	$\sqrt{3}\cos\frac{\theta}{2}\sin^2\frac{\theta}{2}$	$\sqrt{3}\cos^2\frac{\theta}{2}\sin\frac{\theta}{2}$	$\cos^3\frac{\theta}{2}$

F.3. RELATION WITH THE JACOBI POLYNOMIALS

F.3.1. DEFINITION AND SOME PROPERTIES OF THE JACOBI POLYNOMIALS

For more details cf. for instance Gradshteyn and Ryzhik[28] (Sec. 8.96) and Abramowitz and Stegun[2] (Chap. 22).

i) **Definition.** The Jacobi polynomials $P_n^{(\alpha,\beta)}$ are defined by the Rodrigues formula :

$$P_n^{(\alpha,\beta)}(x) = \frac{(-1)^n}{2^n n!}(1-x)^{-\alpha}(1+x)^{-\beta}\frac{d^n}{dx^n}[(1-x)^{\alpha+n}(1+x)^{\beta+n}] \tag{F-1}$$

$$= \frac{1}{2^n}\sum_{m=o}^{n}\binom{n+\alpha}{m}\binom{n+\beta}{n-m}(x-1)^{n-m}(x+1)^m \quad , \tag{F-2}$$

$$n = 0,1,\ldots \quad , \quad \alpha > -1 \quad , \quad \beta > -1 \quad .$$

ii) **Functional relation :**

$$P_n^{(\alpha,\beta)}(-x) = (-1)^n P_n^{(\beta,\alpha)}(x) \quad . \tag{F-3}$$

Many other relations can be found for instance in Ref. 28 loc. cit. .

iii) **Differential equation**. The Jacobi polynomials are the unique rational solution of the differential equation :

$$(1-x^2)\frac{d^2y}{dx^2} + [\beta-\alpha-(\alpha+\beta+2)x]\frac{dy}{dx} + n(n+\alpha+\beta+1)\,y = 0 \quad . \tag{F-4}$$

iv) **Generating function :**

$$\sum_{n=o}^{\infty} P_n^{(\alpha,\beta)}(x)t^n = 2^{\alpha+\beta} S^{-1}(1-t+S)^{-\alpha}(1+t+S)^{-\beta} \quad ,$$

$$S = \sqrt{1-2xt+t^2} \quad ; \quad |t| < 1 \quad . \tag{F-5}$$

v) **Orthogonality relation :**

$$\int_{-1}^{+1} dx\,(1-x)^{\alpha}(1+x)^{\beta}P_n^{(\alpha,\beta)}(x)\,P_m^{(\alpha,\beta)}(x) =$$

$$\frac{2^{\alpha+\beta+1}}{2n+\alpha+\beta+1}\,\frac{\Gamma(n+\alpha+1)\Gamma(n+\beta+1)}{\Gamma(n+1)\Gamma(n+\alpha+\beta+1)}\,\delta_{nm} \quad . \tag{F-6}$$

vi) **Connections with the hypergeometric series and the associated Legendre functions,** (Appendix H.1) :

$$P_n^{(\alpha,\beta)}(x) = \frac{(-1)^n\,\Gamma(n+1+\beta)}{n!\;\Gamma(1+\beta)}\,{}_2F_1(n+\alpha+\beta+1,-n;1+\beta;\tfrac{1+x}{2}) \quad , \tag{F-7}$$

$$P_{\ell-m}^{(m,m)}(x) = (-2)^m\frac{\ell!}{(\ell-m)!}(1-x^2)^{-\frac{1}{2}m}\,P_\ell^{-m}(x) \quad , \tag{F-8}$$

$$P_n^{(o,o)}(x) = P_n(x) \quad . \tag{F-9}$$

vii) **Special cases** (cf. Ref. 2 loc. cit., Table 22.1) :

$$P_o^{(\alpha,\beta)}(x) = 1 \quad , \tag{F-10a}$$

$$P_1^{(\alpha,\beta)}(x) = \frac{1}{2}\,[2(\alpha+1)+(\alpha+\beta+2)(x-1)] \quad , \tag{F-10b}$$

$$P_2^{(\alpha,\beta)}(x) = \frac{1}{8}\,[4(\alpha+1)(\alpha+2)+4(\alpha+\beta+3)(\alpha+2)(x-1)+(\alpha+\beta+3)(\alpha+\beta+4)(x-1)^2] \ . \tag{F-10c}$$

viii) Special values :

$$P_n^{(\alpha,\beta)}(1) = \binom{n+\alpha}{n} = \frac{\Gamma(n+\alpha+1)}{\Gamma(\alpha+1)\ \Gamma(n+1)} \quad . \tag{F-11}$$

F.3.2. RELATION BETWEEN $d^j_{mm'}$ AND THE JACOBI POLYNOMIALS

Comparison of Eqs. (6–18) and (F–2) leads us to the following relation :

$$d^j_{m'm}(\theta) = \left[\frac{(j+m')!\,(j-m')!}{(j+m)!\ (j-m)!}\right]^{\frac{1}{2}} \left(\cos\frac{\theta}{2}\right)^{m'+m} \left(-\sin\frac{\theta}{2}\right)^{m'-m} P^{(m'-m,m'+m)}_{j-m'}(\cos\theta)\,. \tag{F-12}$$

From the above equation and Eqs. (F–8) and (H–6), one finds :

$$d^\ell_{m\,0}(\theta) = (-1)^m \left[\frac{(\ell-m)!}{(\ell+m)!}\right]^{\frac{1}{2}} P^m_\ell\,(\cos\theta) = \left[\frac{(\ell+m)!}{(\ell-m)!}\right]^{\frac{1}{2}} P^{-m}_\ell(\cos\theta) \quad , \tag{F-13}$$

and therefore from Eq. (H–4) :

$$d^\ell_{00}(\theta) = P_\ell(\cos\theta) \quad . \tag{F-14}$$

APPENDIX G

CLEBSCH–GORDAN COEFFICIENTS, WIGNER'S 3j, 6j, AND 9j SYMBOLS

Notation : $\hat{j} = \sqrt{2j+1}$.

G.1. CLEBSCH–GORDAN COEFFICIENTS AND WIGNER'S 3j SYMBOLS

G.1.1. DIFFERENT NOTATIONS

The Clebsch–Gordan (C.G.) coefficients used in these notes are defined in Subsec. 6.2.3.1, and correspond to the widely used phase conventions discussed in Subsec. 6.2.3.4. They are related to the Wigner 3j symbols by :

$$\langle j_1 j_2 m_1 m_2 | j_3 m_3 \rangle = (-1)^{j_1 - j_2 + m_3} \hat{j}_3 \begin{pmatrix} j_1 & j_2 & j_3 \\ m_1 & m_2 & -m_3 \end{pmatrix} \quad . \qquad \text{(G-1)}$$

These C.G. coefficients are equal to :

$\langle j_1 j_2 m_1 m_2 \| j_3 m_3 \rangle$	- Messiah[39], Appendix C. , Eqs. (C.11,12,21, 22)
$\langle j_1 m_1 j_2 m_2 \| j_3 m_3 \rangle$	- Bohr and Mottelson[8], Eqs. (1A–4,6)
$(j_1 m_1 j_2 m_2 \| j_1 j_2 j_3 m_3)$	- Condon and Shortley[19], Chap. 3, Sec. 14
	- de Shalit and Talmi[56], Eq. (13.30)
	- Edmonds[22], Eqs. (3.5.1) and (3.6.11)
$(j_1 j_2 j_3 m_3 \| j_1 m_1, j_2 m_2)$	- Fano and Racah[24], Eqs. (7.5–7)
$r(j_3 m_3 \| m_1 m_2)$	
$C(j_1 j_2 j_3 ; m_1 m_2)$	- Rose[54], Eq. (4.24)
$C_{j_1 j_2}(j_3 m_3 ; m_1, m_2)$	- Blatt and Weisskopf[7], App. A, Sec. 5, Eq.(5.1)
$S^{(j_1 j_2)}_{j_3 m_1 m_2}$	- Wigner[63], footnote p.189 and Eq. (17.27).

G.1.2. SOME PROPERTIES

i) **Selection rules.** If one of the two conditions :

$$m_1 + m_2 = m_3 \qquad , \qquad |j_1 - j_2| \leqslant j_3 \leqslant j_1 + j_2$$

is not satisfied, $\langle j_1 j_2 m_1 m_2 | j_3 m_3 \rangle$ vanishes.

ii) **Reality.** The C.G. coefficients and the Wigner 3j symbols are real.

iii) Symmetry properties. The Wigner 3j symbol $\begin{pmatrix} j_1 & j_2 & j_3 \\ m_1 & m_2 & m_3 \end{pmatrix}$ is :

(1) invariant under a circular substitution of its three columns,

(2) multiplied by $(-1)^{j_1+j_2+j_3}$ under the interchange of any two of its columns,

(3) multiplied by $(-1)^{j_1+j_2+j_3}$ under the simultaneous change of sign of m_1, m_2, and m_3.

The above properties are called the usual symmetries of the Wigner 3j symbols. For other symmetry properties cf. Subsec. 6.2.3.3,iii) and Regge[51]. Consequences of the usual symmetries for the C.G. coefficients :

$$\langle j_1 j_2 m_1 m_2 | j_3 m_3\rangle = (-1)^{j_1+j_2-j_3} \langle j_2 j_1 m_2 m_1 | j_3 m_3\rangle \tag{G-2a}$$

$$= (-1)^{j_1+j_2-j_3} \langle j_1 j_2 -m_1 -m_2 | j_3 -m_3\rangle \tag{G-2b}$$

$$= (-1)^{j_1-j_3+m_2} \frac{\hat{j}_3}{\hat{j}_1} \langle j_3 j_2 m_3 -m_2 | j_1 m_1\rangle \tag{G-2c}$$

$$= (-1)^{j_2-j_3-m_1} \frac{\hat{j}_3}{\hat{j}_2} \langle j_1 j_3 -m_1 m_3 | j_2 m_2\rangle \quad . \tag{G-2d}$$

G.1.3. SPECIAL VALUES

i) Whenever at least one of the nine numbers

$$j_1 \pm m_1 \;,\; j_2 \pm m_2 \;,\; j_3 \pm m_3 \;,\; -j_1+j_2+j_3,\; j_1-j_2+j_3,\; j_1+j_2-j_3$$

vanishes, the sum over q in the general expression of a Wigner 3j symbol, Eq. (6–80), has only one term, e.g. :

1–The case $j_1 = m_1$, i.e. $\langle j_1 j_2 j_1 m_3 - j_1 | j_3 m_3\rangle$ given by Eq. (6–104).

2–The case $j_3 = j_1 + j_2$ where only q=0 contributes in Eq. (6–80), one then has :

$$\langle j_1 j_2 m_1 m_2 | j_1+j_2 m_1+m_2\rangle = \left\{\frac{(2j_1)!(2j_2)!(j_1+j_2+m_1+m_2)!(j_1+j_2-m_1-m_2)!}{[2(j_1+j_2)]!(j_1+m_1)!(j_1-m_1)!(j_2+m_2)!(j_2-m_2)!}\right\}^{\frac{1}{2}} . \tag{G-3}$$

3–Other usual cases :

$$\langle j\ 0\ m\ 0 | j\ m\rangle = \langle 0\ j\ 0\ m | j\ m\rangle = 1 \quad , \tag{G-4a}$$

$$\langle j\ j\ m\ -m | 0\ 0\rangle = \begin{pmatrix} j & j & 0 \\ m & -m & 0 \end{pmatrix} = \frac{(-1)^{j-m}}{\hat{j}} \quad , \qquad \langle j\ j\ -m\ m | 0\ 0\rangle = \begin{pmatrix} j & j & 0 \\ -m & m & 0 \end{pmatrix} = \frac{(-1)^{j+m}}{\hat{j}} \quad . \tag{G-4b}$$

ii) Tables of some special values :

$$\langle j_1 j_2 0 0 | j_3 0\rangle = \frac{(-1)^{j_1-j_2+\frac{1}{2}J}\, \hat{j}_3 (\frac{1}{2}J)!}{(\frac{1}{2}J-j_1)!(\frac{1}{2}J-j_2)!(\frac{1}{2}J-j_3)!} \left[\frac{(j_1+j_2-j_3)!(j_1+j_3-j_2)!(j_2+j_3-j_1)!}{(j_1+j_2+j_3+1)!}\right]^{\frac{1}{2}} . \tag{G-5}$$

for $J = j_1+j_2+j_3$ even, 0 otherwise.

TABLE G.1

Table of C.G. coefficients $\langle\lambda \frac{1}{2}\, m-\mu\ \mu | jm\rangle$.

$\lambda \backslash \mu$	$\frac{1}{2}$	$-\frac{1}{2}$
$j-\frac{1}{2}$	$\sqrt{\frac{j+m}{2j}}$	$\sqrt{\frac{j-m}{2j}}$
$j+\frac{1}{2}$	$-\sqrt{\frac{j+1-m}{2j+2}}$	$\sqrt{\frac{j+1+m}{2j+2}}$

Special cases :

$$\langle \tfrac{1}{2}, \tfrac{1}{2}, \tfrac{1}{2}, \tfrac{1}{2} | 1, 1\rangle = 1 \quad , \tag{G-6a}$$

$$\langle \tfrac{1}{2}, \tfrac{1}{2}, \tfrac{1}{2}, -\tfrac{1}{2} | 0, 0\rangle = \langle \tfrac{1}{2}, \tfrac{1}{2}, \tfrac{1}{2}, -\tfrac{1}{2} | 1, 0\rangle = \frac{1}{\sqrt{2}} \quad , \tag{G-6b}$$

$$\langle 1, \tfrac{1}{2}, 0, -\tfrac{1}{2} | \tfrac{1}{2}, -\tfrac{1}{2}\rangle = \frac{1}{\sqrt{3}} \quad , \tag{G-7a}$$

$$\langle 1, \tfrac{1}{2}, 1, -\tfrac{1}{2} | \tfrac{1}{2}, \tfrac{1}{2}\rangle = \sqrt{\frac{2}{3}} \quad . \tag{G-7b}$$

TABLE G.2

Table of C.G. coefficients $\langle\lambda\ 1m-\mu\mu \mid jm\rangle$.

$\lambda \backslash \mu$	1	0	-1
j-1	$\sqrt{\frac{(j+m-1)(j+m)}{(2j-1)2j}}$	$\sqrt{\frac{(j-m)(j+m)}{j(2j-1)}}$	$\sqrt{\frac{(j-m-1)(j-m)}{(2j-1)2j}}$
j	$-\sqrt{\frac{(j+m)(j-m+1)}{2j(j+1)}}$	$\frac{m}{\sqrt{j(j+1)}}$	$\sqrt{\frac{(j-m)(j+m+1)}{2j(j+1)}}$
j+1	$\sqrt{\frac{(j-m+1)(j-m+2)}{2(j+1)(2j+3)}}$	$-\sqrt{\frac{(j-m+1)(j+m+1)}{(j+1)(2j+3)}}$	$\sqrt{\frac{(j+m+2)(j+m+1)}{2(j+1)(2j+3)}}$

For the C.G. coefficients $\langle\lambda \frac{3}{2}m-\mu\mu | jm\rangle$ and $\langle\lambda\, 2m-\mu\ \mu | jm\rangle$ cf. for instance Condon and Shortley [19] (respectively, Tables 3^3 and 4^3).

G.2. SQUARE 6j SYMBOLS, WIGNER'S 6j SYMBOLS, AND COEFFICIENTS W OF RACAH

G.2.1. DEFINITIONS AND PROPERTIES

The definitions and the properties of the (square) 6j symbols are given in Subsec. 8.4.2.2. Some of the properties (the symmetry properties, the orthogonality relation, and the sum rules) are discussed in terms of graphical representations in Subsec. 9.1.3.2,iii) and 9.2.2,ii).

The square 6j symbol $\begin{bmatrix} j_1 & j_2 & j_3 \\ k_1 & k_2 & k_3 \end{bmatrix}$, the (Wigner) 6j symbol $\begin{Bmatrix} j_1 & j_2 & j_3 \\ k_1 & k_2 & k_3 \end{Bmatrix}$, and the coefficient $W(j_1 j_2 k_2 k_1; j_3 k_3)$ of Racah are related by :

$$\begin{bmatrix} j_1 & j_2 & j_3 \\ k_1 & k_2 & k_3 \end{bmatrix} = (-1)^{j_1+j_2+k_1+k_2} \hat{j}_3 \hat{k}_3 \begin{Bmatrix} j_1 & j_2 & j_3 \\ k_1 & k_2 & k_3 \end{Bmatrix}$$

$$= \hat{j}_3 \hat{k}_3 \, W(j_1 j_2 k_2 k_1; j_3 k_3) \quad . \tag{G-8}$$

G.2.2. SPECIAL VALUES, SQUARE 6j SYMBOLS WITH ZEROS

i) If any one of the twelve numbers given in Eq. (8–176) vanishes, the sum over ℓ in Eq. (8–175) reduces to a single one. For instance, if $k_1+k_2-j_3=0$, ℓ takes the value $j_1+j_2+j_3$, and one then obtains :

$$\begin{Bmatrix} j_1 & j_2 & j_3 \\ k_1 & j_3-k_1 & k_3 \end{Bmatrix} = (-1)^{j_1+j_2+j_3} \left[\frac{(-j_1+j_2+j_3)!(j_1-j_2+j_3)!(j_1+j_2+j_3+1)!}{(2j_3+1)!\,(j_1+j_2-j_3)!} \right.$$
$$\left. \times \frac{(2k_1)!(2j_3-2k_1)!(j_2-k_1+k_3)!(j_1-j_3+k_1+k_3)!}{(j_2+k_1-k_3)!(-j_2+k_1+k_3)!(j_2+k_1+k_3+1)!(j_1+j_3-k_1-k_3)!(-j_1+j_3-k_1+k_3)!(j_1+j_3-k_1+k_3+1)!} \right]^{\frac{1}{2}} . \tag{G-9}$$

Let us now give three useful cases of the above equation :

1– $k_1=j_3$, $k_2=0$, $k_3=j_1$:

$$\begin{bmatrix} j_1 & j_2 & j_3 \\ j_3 & 0 & j_1 \end{bmatrix} = (-1)^{j_1+j_2+j_3} \hat{j}_1 \hat{j}_3 \begin{Bmatrix} j_1 & j_2 & j_3 \\ j_3 & 0 & j_1 \end{Bmatrix} = \{j_1 j_2 j_3\} \quad . \tag{G-10}$$

As an example of the application of the tensorial formalism developed in Subsec. 8.3, let us recover the above result in an intrinsic way, i.e. from Eqs. (8–170a, 171a) :

$$T^{\{\{j_1 j_2\} j_3 j_3\} 0} = T^{\{j_1 \{j_2 j_3\} j_1\} 0} \Leftrightarrow {}^6R(j_1 j_2 j_3; j_3\alpha; j_1\beta)^{0\bullet\dot{0}} = \begin{bmatrix} j_1 & j_2 & j_3 \\ j_3 & 0 & j_1 \end{bmatrix} = \{j_1 j_2 j_3\} .$$

Indeed, one has :

$${}^6R(j_1 j_2 j_3; j_3\alpha; j_1\beta)^{0\bullet\dot{0}} = C^{j_3 \bullet j_3 \bullet\bullet 0} \, C^{j_1 \bullet j_2 \bullet \dot{j}_3} \, C^{\dot{j}_2 \ddot{j}_3 \ddot{j}_1} \, C^{\dot{j}_1 j_1 \bullet\bullet 0} \qquad \text{from Eq. (8-173a)}$$

$$= c_{j_1} c_{j_3} \frac{(-1)^{2j_1+2j_3}}{\hat{j}_1 \hat{j}_3} \, G^{j_3 \bullet j_3 \bullet\bullet} \, C^{j_1 \bullet j_2 \bullet \dot{j}_3} \, C^{\dot{j}_2 \ddot{j}_3 \ddot{j}_1} \, G^{\dot{j}_1 j_1 \bullet\bullet} \qquad \text{from Eq. (8-99)}$$

$$= c_{j_1} c_{j_3} \frac{(-1)^{2j_1+2j_3}}{\hat{j}_1 \hat{j}_3} \, C^{j_1 \bullet j_2 \bullet \dot{j}_3} \, C^{\dot{j}_2 j_3 \bullet \dot{j}_1} \qquad \text{from Eq. (8-105)}$$

$$= (-1)^{2j_3} \, W^{j_1 \bullet j_2 \bullet \dot{j}_3} \, W^{\dot{j}_2 j_3 \bullet \dot{j}_1} \qquad \text{from Eq. (8-86)}$$

$$= (-1)^{2j_3} \, W^{j_1 \bullet j_2 \bullet \dot{j}_3} \, W^{\dot{j}_1 \dot{j}_2 j_3 \bullet} \qquad \text{from Eq. (8-89a)}$$

$$= W^{j_1^\bullet j_2^\bullet j_3^\bullet} \; W^{j_1\bullet j_2\bullet j_3\bullet}$$ from Eqs.(8-84,102)

$$= \frac{1}{2j_3+1} \; G^{j_3^\bullet j_3 j_3\bullet} \qquad \{j_1 j_2 j_3\} = \{j_1 j_2 j_3\} \quad .$$ from Eq.(8-129a).

2- $k_1 = \frac{1}{2}$, $k_2 = j_3 - \frac{1}{2}$, $k_3 = j_2 - \frac{1}{2}$:

$$\begin{Bmatrix} j_1 & j_2 & j_3 \\ \frac{1}{2} & j_3-\frac{1}{2} & j_2-\frac{1}{2} \end{Bmatrix} = (-1)^{j_1+j_2+j_3} \left[\frac{(j_1+j_2+j_3+1)(j_2+j_3-j_1)}{2j_2(2j_2+1)2j_3(2j_3+1)}\right]^{\frac{1}{2}} \quad . \tag{G-11}$$

3- $k_1 = \frac{1}{2}$, $k_2 = j_3 - \frac{1}{2}$, $k_3 = j_2 + \frac{1}{2}$:

$$\begin{Bmatrix} j_1 & j_2 & j_3 \\ \frac{1}{2} & j_3-\frac{1}{2} & j_2+\frac{1}{2} \end{Bmatrix} = (-1)^{j_1+j_2+j_3} \left[\frac{(j_1+j_3-j_2)(j_1+j_2-j_3+1)}{(2j_2+1)(2j_2+2)2j_3(2j_3+1)}\right]^{\frac{1}{2}} \quad . \tag{G-12}$$

ii) One must take care in the following table that whenever any one of the selection rules for the couplings is not fulfilled, the considered square 6j symbol vanishes $(\propto\{j_1 j_2 J\})$.

TABLE G.3

Square 6j symbols with one zero.

$\begin{bmatrix} 0 & j_1 & j_2 \\ j_2 & J & J \end{bmatrix} = \begin{bmatrix} j_2 & j_1 & J \\ 0 & J & j_1 \end{bmatrix} = 1$	$\begin{bmatrix} j_1 & 0 & j_1 \\ j_2 & J & j_2 \end{bmatrix} = 1$
$\begin{bmatrix} j_1 & j_1 & 0 \\ j_2 & j_2 & J \end{bmatrix} = \begin{bmatrix} j_2 & j_1 & J \\ j_1 & j_2 & 0 \end{bmatrix} = (-1)^{j_1+j_2-J} \dfrac{\hat{J}}{\hat{j}_1\hat{j}_2}$	$\begin{bmatrix} j_1 & j_2 & J \\ J & 0 & j_1 \end{bmatrix} = 1$

G.3. SQUARE 9j AND WIGNER'S 9j SYMBOLS

G.3.1. DEFINITIONS AND PROPERTIES

The definitions and the properties of the (square) 9j symbols are given in Subsec. 8.4.2.2. Some of the properties (the symmetry properties, the orthogonality relation, the sum rules, and the expression in terms of 6j symbols) are discussed in terms of graphical representations in Subsec. 9.1.3.2,iv) and 9.2.2,ii). The square 9j symbols are equal to the functions χ of Hope and Jahne[32] and also to the coefficients of Coester and Jauch[18]. They are related to the Wigner 9j symbols by Eq. (8–172b) :

$$\begin{bmatrix} j_1 & j_2 & j_3 \\ k_1 & k_2 & k_3 \\ \ell_1 & \ell_2 & \ell_3 \end{bmatrix} = \hat{j}_3\hat{k}_3\hat{\ell}_1\hat{\ell}_2 \begin{Bmatrix} j_1 & j_2 & j_3 \\ k_1 & k_2 & k_3 \\ \ell_1 & \ell_2 & \ell_3 \end{Bmatrix} \quad . \tag{G-13}$$

G.3.2. SQUARE 9j SYMBOLS WITH ZEROS AND USEFUL FORMULAE

i) One must take care in the following tables that whenever any one of the selection rules for the couplings is not fulfilled, the considered square 9j symbol vanishes.

TABLE G.4

Square 9j symbols with one zero.

$$S = (-1)^{J+\mathcal{J}} \begin{Bmatrix} j_1 & j_2 & J \\ k_2 & k_1 & \mathcal{J} \end{Bmatrix}$$

$$\begin{bmatrix} 0 & \mathcal{J} & \mathcal{J} \\ J & k_2 & k_1 \\ J & j_2 & j_1 \end{bmatrix} = \begin{bmatrix} k_2 & J & k_1 \\ \mathcal{J} & 0 & \mathcal{J} \\ j_2 & J & j_1 \end{bmatrix} = (-1)^{j_2+k_1} \hat{j}_2 \hat{k}_1 S$$

$$\begin{bmatrix} J & k_2 & k_1 \\ 0 & \mathcal{J} & \mathcal{J} \\ J & j_2 & j_1 \end{bmatrix} = \begin{bmatrix} \mathcal{J} & 0 & \mathcal{J} \\ k_2 & J & k_1 \\ j_2 & J & j_1 \end{bmatrix} = (-1)^{j_1+k_2} \hat{j}_2 \hat{k}_1 S$$

$$\begin{bmatrix} \mathcal{J} & \mathcal{J} & 0 \\ j_1 & j_2 & J \\ k_1 & k_2 & J \end{bmatrix} = \begin{bmatrix} \mathcal{J} & j_1 & k_1 \\ \mathcal{J} & j_2 & k_2 \\ 0 & J & J \end{bmatrix} = (-1)^{j_2+k_1} \frac{\hat{k}_1 \hat{k}_2}{\hat{\mathcal{J}}} S$$

$$\begin{bmatrix} j_1 & j_2 & J \\ \mathcal{J} & \mathcal{J} & 0 \\ k_1 & k_2 & J \end{bmatrix} = \begin{bmatrix} j_1 & \mathcal{J} & k_1 \\ j_2 & \mathcal{J} & k_2 \\ J & 0 & J \end{bmatrix} = (-1)^{j_1+k_2} \frac{\hat{k}_1 \hat{k}_2}{\hat{\mathcal{J}}} S$$

$$\begin{bmatrix} j_1 & j_2 & J \\ k_1 & k_2 & J \\ \mathcal{J} & \mathcal{J} & 0 \end{bmatrix} = (-1)^{j_2+k_1} \hat{J} \hat{\mathcal{J}} S$$

TABLE G. 5

Square 9j symbols with two zeros.

$$\begin{bmatrix} 0 & j_1 & j_1 \\ j_2 & 0 & j_2 \\ j_2 & j_1 & J \end{bmatrix} = (-1)^{j_1+j_2-J}$$

$$\begin{bmatrix} j_1 & 0 & j_1 \\ J & j_2 & j_1 \\ j_2 & j_2 & 0 \end{bmatrix} = \begin{bmatrix} j_1 & J & j_2 \\ 0 & j_2 & j_2 \\ j_1 & j_1 & 0 \end{bmatrix} = 1$$

$$\begin{bmatrix} 0 & j_1 & j_1 \\ j_2 & J & j_1 \\ j_2 & j_2 & 0 \end{bmatrix} = (-1)^{j_1+j_2-J}$$

$$\begin{bmatrix} j_1 & j_1 & 0 \\ j_2 & 0 & j_2 \\ J & j_1 & j_2 \end{bmatrix} = \begin{bmatrix} j_1 & j_2 & J \\ j_1 & 0 & j_1 \\ 0 & j_2 & j_2 \end{bmatrix} = \frac{\hat{J}}{\hat{j}_1 \hat{j}_2}$$

$$\begin{bmatrix} 0 & j_1 & j_1 \\ j_2 & j_2 & 0 \\ j_2 & J & j_1 \end{bmatrix} = \begin{bmatrix} 0 & j_2 & j_2 \\ j_1 & j_2 & J \\ j_1 & 0 & j_1 \end{bmatrix} = \frac{\hat{J}}{\hat{j}_1 \hat{j}_2}$$

$$\begin{bmatrix} j_1 & j_1 & 0 \\ j_1 & J & j_2 \\ 0 & j_2 & j_2 \end{bmatrix} = \frac{1}{2j_1+1}$$

$$\begin{bmatrix} j_1 & 0 & j_1 \\ 0 & j_2 & j_2 \\ j_1 & j_2 & J \end{bmatrix} = 1$$

$$\begin{bmatrix} j_1 & j_1 & 0 \\ J & j_1 & j_2 \\ j_2 & 0 & j_2 \end{bmatrix} = \begin{bmatrix} j_1 & J & j_2 \\ j_1 & j_1 & 0 \\ 0 & j_2 & j_2 \end{bmatrix} = \frac{(-1)^{j_1+j_2-J}}{2j_1+1}$$

$$\begin{bmatrix} j_1 & 0 & j_1 \\ j_2 & j_2 & 0 \\ J & j_2 & j_1 \end{bmatrix} = \begin{bmatrix} j_1 & j_2 & J \\ 0 & j_2 & j_2 \\ j_1 & 0 & j_1 \end{bmatrix} = (-1)^{j_1+j_2-J} \frac{\hat{J}}{\hat{j}_1\hat{j}_2} \qquad \begin{bmatrix} J & j_1 & j_2 \\ j_2 & 0 & j_2 \\ j_1 & j_1 & 0 \end{bmatrix} = (-1)^{j_1+j_2-J}$$

$$\begin{bmatrix} j_1 & 0 & j_1 \\ j_1 & j_2 & J \\ 0 & j_2 & j_2 \end{bmatrix} = \begin{bmatrix} j_1 & j_1 & 0 \\ 0 & j_2 & j_2 \\ j_1 & J & j_2 \end{bmatrix} = (-1)^{j_1+j_2-J} \frac{\hat{J}}{\hat{j}_1\hat{j}_2} \qquad \begin{bmatrix} J & j_1 & j_2 \\ j_1 & j_1 & 0 \\ j_2 & 0 & j_2 \end{bmatrix} = \frac{1}{2j_1+1}$$

TABLE G. 6

Square 9j symbols with three zeros.

$$\begin{bmatrix} j_1 & j_2 & J \\ j_1 & j_2 & J \\ 0 & 0 & 0 \end{bmatrix} = \begin{bmatrix} j_1 & j_1 & 0 \\ j_2 & j_2 & 0 \\ J & J & 0 \end{bmatrix} = \frac{\hat{J}}{\hat{j}_1\hat{j}_2} \qquad \begin{bmatrix} j & j & 0 \\ j & 0 & j \\ 0 & j & j \end{bmatrix} = \frac{1}{2j+1}$$

$$\begin{bmatrix} j_1 & j_2 & J \\ 0 & 0 & 0 \\ j_1 & j_2 & J \end{bmatrix} = \begin{bmatrix} j_1 & 0 & j_1 \\ j_2 & 0 & j_2 \\ J & 0 & J \end{bmatrix} = \begin{bmatrix} 0 & 0 & 0 \\ j_1 & j_2 & J \\ j_1 & j_2 & J \end{bmatrix} = \begin{bmatrix} 0 & j_1 & j_1 \\ 0 & j_2 & j_2 \\ 0 & J & J \end{bmatrix} = 1 \qquad \begin{bmatrix} 0 & j & j \\ j & 0 & j \\ j & j & 0 \end{bmatrix} = 1$$

ii) Using the value of the 6j symbol $\begin{Bmatrix} \ell & s & j \\ s & \ell & 1 \end{Bmatrix}$ given by de Shalit and Talmi[56] (Appendix page 520), one gets :

$$\begin{bmatrix} \ell & s & j \\ 1 & 1 & 0 \\ \ell & s & j \end{bmatrix} = \frac{1}{2\sqrt{3}} \frac{\cdot \ell(\ell+1)+s(s+1)-j(j+1)}{\sqrt{\ell(\ell+1)s(s+1)}} \quad . \qquad \text{(G-14)}$$

Using the expression of $\begin{Bmatrix} \ell_1 & 1/2 & j_1 \\ j_2 & \ell & \ell_2 \end{Bmatrix} \langle \ell_1 \ell 0 0 | \ell_2 0 \rangle$ given by Bohr and Mottelson[9] (Eq. (3A–16a)), one has :

$$\begin{bmatrix} \ell_1 & \frac{1}{2} & j_1 \\ \ell & 0 & \ell \\ \ell_2 & \frac{1}{2} & j_2 \end{bmatrix} \cdot \begin{pmatrix} \ell_1 & \ell & \ell_2 \\ 0 & 0 & 0 \end{pmatrix} = (-1)^{\ell_1+j_1-\frac{1}{2}} \frac{\hat{j}_1}{\hat{\ell}_1} \begin{pmatrix} j_1 & j_2 & \ell \\ \frac{1}{2} & -\frac{1}{2} & 0 \end{pmatrix} \quad . \qquad \text{(G-15)}$$

APPENDIX H

ASSOCIATED LEGENDRE'S FUNCTIONS, LEGENDRE'S POLYNOMIALS, SPHERICAL HARMONICS, HARMONIC POLYNOMIALS, SOLID HARMONICS, MULTIPOLE EXPANSIONS, AND VECTOR SPHERICAL HARMONICS

H.1. ASSOCIATED LEGENDRE'S FUNCTIONS AND LEGENDRE'S POLYNOMIALS

For more details cf. for instance Gradshteyn and Ryzhik[28] (Secs. 8.7,8) or Abramowitz and Stegun[2] (Chap. 8).

i) **Definitions.** The Legendre polynomials P_ℓ are defined by the Rodrigues formula :

$$P_\ell(x) = \frac{1}{2^\ell \ell!} \frac{d^\ell}{dx^\ell} (x^2-1)^\ell \qquad ; \qquad \ell=0,1,\ldots \qquad , \tag{H-1}$$

from which we can obtain the series expression :

$$P_\ell(x) = \frac{1}{2^\ell} \sum_{n=0}^{\ell} (-1)^n \frac{(2\ell-2n)!}{n!(\ell-n)!(\ell-2n)!} x^{\ell-2n} \qquad . \tag{H-2}$$

$P_\ell(x)$ is a polynomial of degree ℓ, of parity $(-1)^\ell$, and which has ℓ zeros within $[-1,+1]$.

The associated Legendre functions P_ℓ^m are defined by :

$$P_\ell^m(x) = (1-x^2)^{\frac{1}{2}m} \frac{d^m}{dx^m} P_\ell(x) \qquad ; \qquad \ell=0,1,\ldots, \qquad m=\ell,\ell-1,\ldots,-\ell. \tag{H-3}$$

$P_\ell^m(x)$ is the product by $(1-x^2)^{\frac{1}{2}m}$ of a polynomial of degree $\ell-m$, of parity $(-1)^{\ell-m}$, and which has $\ell-m$ zeros within $[-1,+1]$. The Legendre polynomials are the associated Legendre functions for m=0, i.e. :

$$P_\ell(x) = P_\ell^0(x) \qquad . \tag{H-4}$$

Therefore, setting m=0 in the relations given below one gets the corresponding properties for the Legendre polynomials.

ii) **Functional relations :**

$$P_\ell^m(-x) = (-1)^{\ell-m} P_\ell^m(x) \qquad , \tag{H-5}$$

$$P_\ell^{-m}(x) = (-1)^m \frac{(\ell-m)!}{(\ell+m)!} P_\ell^m(x) \qquad , \tag{H-6}$$

$$(\ell-m+1) P_{\ell+1}^m(x) - (2\ell+1)x P_\ell^m(x) + (\ell+m) P_{\ell-1}^m(x) = 0 \qquad , \tag{H-7}$$

$$(1-x^2)\frac{d}{dx}P_\ell^m(x) = (\ell+1)x\,P_\ell^m(x) - (\ell-m+1)\,P_{\ell+1}^m(x) \tag{H-8}$$

$$= (\ell+m)\,P_{\ell-1}^m(x) - \ell x\,P_\ell^m(x) \quad . \tag{H-9}$$

Other relations can be found for instance in Edmonds[22] (Eqs. 2.5.21–24). A relation with the Jacobi polynomials is given in Eq. (F–8).

iii) Differential equation :

$$\left[(1-x^2)\frac{d^2}{dx^2} - 2x\frac{d}{dx} + \ell(\ell+1) - \frac{m^2}{1-x^2}\right] P_\ell^m(x) = 0 \quad . \tag{H-10}$$

iv) Generating function :

$$\sum_{\ell=m}^{\infty} P_\ell^m(x)\,t^\ell = (2m-1)!!\,(1-x^2)^{\frac{1}{2}m}\,t^m(1-2xt+t^2)^{-m-\frac{1}{2}} \quad ; \ |t|<1 \ . \tag{H-11}$$

v) Orthogonality relations :

$$\int_{-1}^{+1} dx\, P_k^m(x)\,P_\ell^m(x) = \frac{2(\ell+m)!}{(2\ell+1)(\ell-m)!}\,\delta_{k\ell} \quad , \tag{H-12}$$

$$\int_{-1}^{+1} \frac{dx}{1-x^2}\,P_\ell^m(x)\,P_\ell^n(x) = \frac{(\ell+m)!}{m(\ell-m)!}\,\delta_{mn} \quad . \tag{H-13}$$

vi) Connections with the hypergeometric series and the spherical harmonics :

$$P_\ell^m(x) = \frac{1}{\Gamma(1-m)}\left(\frac{x+1}{x-1}\right)^{\frac{1}{2}m} {}_2F_1(-\ell,\ell+1;1-m;\tfrac{1-x}{2}) \quad ; \quad |1-x|<2 \quad , \tag{H-14}$$

$$Y_{\ell m}(\theta,\varphi) = (-1)^m\left[\frac{(2\ell+1)(\ell-m)!}{4\pi(\ell+m)!}\right]^{\frac{1}{2}} P_\ell^m(\cos\theta)\,e^{im\varphi} \quad , \tag{H-15}$$

$$Y_{\ell o}(\theta,\varphi) = \left(\frac{2\ell+1}{4\pi}\right)^{\frac{1}{2}} P_\ell(\cos\theta) \quad . \tag{H-16}$$

vii) Special cases :

$$P_\ell^\ell(x) = (-1)^\ell\,(2\ell)!\,P_\ell^{-\ell}(x) = (2\ell-1)!!\,(1-x^2)^{\frac{1}{2}\ell} \quad . \tag{H-17}$$

The (associated) Legendre polynomials for $\ell=0,1,2$, and 3 are listed in Table H.1.

viii) Special values :

$$P_\ell(1) = 1 \quad , \qquad P_\ell(-1) = (-1)^\ell \quad , \tag{H-18a}$$

$$m \neq 0 \qquad P_\ell^m(1) = P_\ell^m(-1) = 0 \quad , \tag{H-18b}$$

$$P_\ell^m(0) = \begin{cases} (-1)^p\,\dfrac{(\ell+m)!}{2^\ell p!\,(p+m)!} & \text{if} \quad \ell-m=2p \\ 0 & \text{if} \quad \ell-m=2p+1 \end{cases} \quad . \tag{H-18c}$$

TABLE H.1

Associated Legendre polynomials.

ℓ	m	$P_\ell^m(x)$, $x = \cos\theta$
0	0	1
1	0	$x = \cos\theta$
1	1	$-(1-x^2)^{\frac{1}{2}} = -\sin\theta$
2	0	$\frac{1}{2}(3x^2-1) = \frac{1}{4}(3\cos 2\theta+1)$
2	1	$-3(1-x^2)^{\frac{1}{2}}x = -\frac{3}{2}\sin 2\theta$
2	2	$3(1-x^2) = \frac{3}{2}(1-\cos 2\theta)$
3	0	$\frac{1}{2}(5x^3-3x) = \frac{1}{8}(5\cos 3\theta+3\cos\theta)$
3	1	$-\frac{3}{2}(1-x^2)^{\frac{1}{2}}(5x^2-1) = -\frac{3}{8}(\sin\theta+5\sin 3\theta)$
3	2	$15(1-x^2)x = \frac{15}{4}(\cos\theta-\cos 3\theta)$
3	3	$-15(1-x^2)^{\frac{3}{2}} = -\frac{15}{4}(3\sin\theta-\sin 3\theta)$

H.2. SPHERICAL HARMONICS

i) **Definition.** The definition of the spherical harmonics $Y_{\ell m}(\theta,\varphi)$ is given by Eq. (10–239). It corresponds to the most commonly adopted convention. We also use the spherical harmonics $Y^{[\ell]}_{m}(\theta,\varphi)$ defined by Eq. (10–247), and which are the contravariant spherical components of a Hermitian tensor Y^{ℓ}. Let us now compare our notation with that of some others :

$Y_{\ell m}(\theta,\varphi) = (ic)^{\ell} Y^{[\ell]}_{m}(\theta,\varphi) = Y_{\ell m}(\theta,\varphi)$ - Bohr and Mottelson[8], Eq. (1A–42)
- Edmonds[22], Eq. (2.5.5)
- Messiah[39], Eq. (B.93)
- Rose[54], Eq. (III.20)
- Wigner[63], Eq. (15.3)

$= \Theta(\ell m)\Phi(m)$ - Condon and Shortley[19], Chap. 3, Sec. 10

$= (-i)^{\ell} y^{[\ell]}_{m}(\theta,\varphi)$ - Fano and Racah[24], Eq. (5.17).

ii) **Properties.** Some of them are given in Subsecs. 10.3.3.4,5, i.e. : the relation with the rotation matrix elements, the law of transformation under finite rotations, the closure and the orthogonality relations, the addition theorem, the coupling of two spherical harmonics, and the gradient formula. The expression of the spherical harmonics in terms of the associated Legendre functions is given in Eq. (H–15).

iii) **Functional relations.** Parity $(-1)^{\ell}$, i.e. under a space reflection $(\theta,\varphi)\to(\pi-\theta,\varphi+\pi)$ one has :

$$Y_{\ell m}(\pi-\theta,\varphi+\pi) = (-1)^{\ell}\, Y_{\ell m}(\theta,\varphi) \quad . \tag{H-19}$$

Complex conjugation :

$$Y^{*}_{\ell m}(\theta,\varphi) = (-1)^{m}\, Y_{\ell -m}(\theta,\varphi) \quad . \tag{H-20}$$

The action of the infinitesimal generators $L_{\pm}$, Eq. (10–232d), yields a recursion formula for spherical harmonics having the same ℓ :

$$L_{\pm} Y_{\ell m}(\theta,\varphi) = \sqrt{\ell(\ell+1)-m(m\pm 1)}\; Y_{\ell m\pm 1}(\theta,\varphi) = \sqrt{(\ell\mp m)(\ell+1\pm m)}\, Y_{\ell m\pm 1}(\theta,\varphi) \quad . \tag{H-21}$$

Iterating the above relation yields :

$$Y_{\ell m}(\theta,\varphi) = \left[\frac{(\ell\pm m)!}{(\ell\mp m)!\,(2\ell)!}\right]^{\frac{1}{2}} L_{\mp}^{\ell\mp m}\; Y_{\ell\pm\ell}(\theta,\varphi) \tag{H-22a}$$

$$= \left[\frac{(\ell-|m|)!}{(\ell+|m|)!}\right]^{\frac{1}{2}} L^{|m|}_{\mathrm{sgn}(m)}\; Y_{\ell o}(\theta,\varphi) \quad . \tag{H-22b}$$

On the other hand, from Eqs. (H–7,15) follows a recursion formula for spherical harmonics having the same m :

$$\left[\frac{(\ell+1+m)(\ell+1-m)}{(2\ell+1)(2\ell+3)}\right]^{\frac{1}{2}} Y_{\ell+1\, m}(\theta,\varphi) - \cos\theta\; Y_{\ell m}(\theta,\varphi) + \left[\frac{(\ell+m)(\ell-m)}{(2\ell+1)(2\ell-1)}\right]^{\frac{1}{2}} Y_{\ell-1\, m}(\theta,\varphi) = 0 \quad . \tag{H-23}$$

iv) **Special cases :**

$$Y_{\ell 0}(\theta,\varphi) = \left(\frac{2\ell+1}{4\pi}\right)^{\frac{1}{2}} P_{\ell}(\cos\theta) \quad , \tag{H-24a}$$

$$Y_{\ell\pm\ell}(\theta,\varphi) = (\mp 1)^{\ell} \left[\frac{2\ell+1}{4\pi(2\ell)!}\right]^{\frac{1}{2}} (2\ell-1)!!\, \sin^{\ell}\theta\; e^{\,i\pm\ell\varphi} \quad . \tag{H-24b}$$

The spherical harmonics with ℓ=0,1,2, and 3 are given in Table H.2.

v) **Special values.** From Eqs. (H–15,18a–c) follows :

$$Y_{\ell m}(0,\varphi) = \sqrt{\frac{2\ell+1}{4\pi}}\; \delta_{m0} \quad . \tag{H-25a}$$

$$Y_{\ell m}(\tfrac{\pi}{2},\varphi) = \begin{cases} (-1)^{m+p} \left[\dfrac{(2\ell+1)(\ell-m)!\,(\ell+m)!}{4\pi}\right]^{\frac{1}{2}} \dfrac{1}{2^{\ell}\; p!\,(p+m)!}\; e^{im\varphi} & \text{if } \ell-m=2p \\ 0 & \text{if } \ell-m=2p+1 \quad . \end{cases} \tag{H-25b}$$

TABLE H.2

Spherical harmonics $Y_{\ell m}$ and harmonic polynomials $\mathcal{Y}_{\ell m}$:

$x=r\sin\theta\cos\varphi$, $y=r\sin\theta\sin\varphi$, $z=r\cos\theta$, $x\pm iy=r\sin\theta\, e^{\pm i\varphi}$.

ℓ, m	$Y_{\ell m}(\theta,\varphi)$	$\mathcal{Y}_{\ell m}(x,y,z) = r^{\ell}\, Y_{\ell m}(\theta,\varphi)$
0 0	$\frac{1}{\sqrt{4\pi}}$	$\frac{1}{\sqrt{4\pi}}$
1 0	$\sqrt{\frac{3}{4\pi}}\cos\theta$	$\sqrt{\frac{3}{4\pi}}\, z$
1 ± 1	$\mp\sqrt{\frac{3}{8\pi}}\sin\theta\, e^{\pm i\varphi}$	$\mp\sqrt{\frac{3}{8\pi}}\,(x\pm iy)$
2 0	$\sqrt{\frac{5}{16\pi}}\,(2\cos^2\theta-\sin^2\theta)$	$\sqrt{\frac{5}{16\pi}}\,(-x^2-y^2+2z^2)$
2 ± 1	$\mp\sqrt{\frac{15}{8\pi}}\cos\theta\sin\theta\, e^{\pm i\varphi}$	$\mp\sqrt{\frac{15}{8\pi}}\,(x\pm iy)z$
2 ± 2	$\sqrt{\frac{15}{32\pi}}\sin^2\theta\, e^{\pm 2i\varphi}$	$\sqrt{\frac{15}{32\pi}}\,(x\pm iy)^2$
3 0	$\sqrt{\frac{7}{16\pi}}\,(2\cos^3\theta-3\cos\theta\sin^2\theta)$	$\sqrt{\frac{7}{16\pi}}\,(-3x^2-3y^2+2z^2)z$
3 ± 1	$\mp\sqrt{\frac{21}{64\pi}}\,(4\cos^2\theta\sin\theta-\sin^3\theta)\, e^{\pm i\varphi}$	$\mp\sqrt{\frac{21}{64\pi}}\,(x\pm iy)(-x^2-y^2+4z^2)$
3 ± 2	$\sqrt{\frac{105}{32\pi}}\cos\theta\sin^2\theta\, e^{\pm 2i\varphi}$	$\sqrt{\frac{105}{32\pi}}\,(x\pm iy)^2 z$
3 ± 3	$\mp\sqrt{\frac{35}{64\pi}}\sin^3\theta\, e^{\pm 3i\varphi}$	$\mp\sqrt{\frac{35}{64\pi}}\,(x\pm iy)^3$

H.3. HARMONIC POLYNOMIALS AND SOLID HARMONICS

i) Definition. A harmonic polynomial $H_\ell(x,y,z)$ is a homogeneous polynomial of degree ℓ in x,y, and z, and which satisfies the Laplace equation, i.e. :

$$\Delta H_\ell(x,y,z) = \left(\frac{\partial^2}{\partial x^2}+\frac{\partial^2}{\partial y^2}+\frac{\partial^2}{\partial z^2}\right) H_\ell(x,y,z) = 0 \quad . \tag{H-26}$$

The normalized harmonic polynomials or solid harmonics are defined by :

$$\mathcal{Y}_{\ell m}(x,y,z) \text{ or } \mathcal{Y}_{\ell m}(r,\theta,\varphi) = r^{\ell}\, Y_{\ell m}(\theta,\varphi) \quad , \tag{H-27}$$

where :

$$x = r\sin\theta\cos\varphi,\; y = r\sin\theta\sin\varphi,\; z = r\cos\theta \;,\; x\pm iy = r\sin\theta e^{\pm i\varphi} \;. \tag{H-28}$$

ii) Properties. The solid harmonics are harmonic polynomials. This, for instance, arises from the expression of the Laplacian operator :

$$\Delta f = \left(\frac{1}{r}\frac{d^2}{dr^2} r - \frac{L^2}{r^2}\right) f \qquad \text{if} \qquad f(r,\theta,\varphi)\Big|_{r=0} < \infty \quad , \tag{H-29}$$

where L^2 is the Casimir operator, Eq. (10–232e). Another way to obtain this result is to check that the $\mathcal{Y}_{\ell\pm\ell}$, which are multiples of $(x \pm iy)^\ell$, Eq. (H–24b), are harmonic polynomials. Then, since the infinitesimal generators $L_\pm$, Eq. (10–232d), commute with the scalar operator Δ , one derives from Eq. (H–22a) the $2\ell+1$ harmonic polynomials $\mathcal{Y}_{\ell m}$:

$$\mathcal{Y}_{\ell m}(x,y,z) = \left[\frac{(\ell\pm m)!}{(\ell\mp m)!(2\ell)!}\right]^{\frac{1}{2}} L_\mp^{\ell\mp m}\ \mathcal{Y}_{\ell\pm\ell}(x,y,z) \tag{H-30a}$$

$$= (\mp1)^\ell \left[\frac{(2\ell+1)(\ell\pm m)!}{4\pi(\ell\mp m)!}\right]^{\frac{1}{2}} \frac{1}{2^\ell \ell!} L_\mp^{\ell\mp m}(x\pm iy)^\ell \quad , \tag{H-30b}$$

with, Eq. (10–231b) :

$$L_\pm = \mp (x\pm iy)\frac{\partial}{\partial z} \pm 2z \frac{\partial}{\partial x\mp iy} \quad . \tag{H-30c}$$

Actually, these $2\ell+1$ linearly independent solid harmonics form a basis of the vector space over the complex field of harmonic polynomials of degree ℓ . This result can be established as follows. The dimension of the vector space over the complex field of homogeneous polynomials of degree ℓ in x,y, and z, is given by :

$$\sum_{p,q,r\geqslant0} \delta_{p+q+r,\ell} = \sum_{r=0} \ell+1-r = \frac{(\ell+1)(\ell+2)}{2} \quad . \tag{H-31}$$

A basis of this space is determined by the set of functions :

$$\{r^{\ell-\lambda}\ \mathcal{Y}_{\lambda m}\ ;\ \lambda=\ell,\ell-2,\dots,\ 1 \text{ or } 0 \quad , \quad m=\lambda,\lambda-1,\dots,-\lambda\} \quad .$$

Indeed, these functions are linearly independent, and furthermore their number is equal to the dimension of the space, i.e. :

$$\sum_{\lambda=\ell,\ell-2,\dots,1\text{ or }0} (2\lambda+1) = \frac{(\ell+1)(\ell+2)}{2} \quad . \tag{H-32}$$

Therefore, any harmonic polynomial H_ℓ can a priori be expressed as a linear combination of the previous basis functions, i.e. :

$$H_\ell = \sum_{\lambda,m} c_{\lambda m}\ r^{\ell-\lambda}\ \mathcal{Y}_{\lambda m} \quad , \qquad c_{\lambda m} \in C \quad .$$

Expressing $\Delta H_\ell = 0$ by means of Eq. (H–29), one finds :

$$\sum_{\lambda,m} c_{\lambda m} \frac{\ell(\ell+1)-\lambda(\lambda+1)}{r^2}\ r^{\ell-\lambda}\ \mathcal{Y}_{\lambda m} = 0 \quad .$$

Since the functions $r^{\ell-\lambda}\ \mathcal{Y}_{\lambda m}$ are linearly independent, one must have :

$$c_{\lambda m}[\ell(\ell+1)-\lambda(\lambda+1)] = 0 \qquad \lambda=\ell,\ell-2,\dots,1 \text{ or } 0 \quad ,$$

the only nontrivial solution of which being $\lambda=\ell$. One thereby obtains the announced result, i.e. any harmonic polynomial is a linear combination of the $2\ell+1$ linearly independent solid harmonics $\mathcal{Y}_{\ell m}$.

iii) **Application.** Let us establish that if :

$$OM = \alpha_1 OM_1 + \alpha_2 OM_2 \quad , \quad OM = (r,\theta,\varphi) \quad , \quad OM_i = (r_i,\theta_i,\varphi_i) \quad i=1,2 \quad ,$$

one then has the relation[40)] :

$$r^\ell Y_{\ell m}(\theta,\varphi) = \sqrt{4\pi} \sum_{\ell_1,\ell_2=0}^{\ell} \delta_{\ell_1+\ell_2,\ell}\left[\frac{(2\ell+1)!}{(2\ell_1+1)!(2\ell_2+1)!}\right]^{\frac{1}{2}} \langle\ell_1\ell_2 m_1 m_2|\ell m\rangle$$
$$\times (\alpha_1 r_1)^{\ell_1}(\alpha_2 r_2)^{\ell_2}\ Y_{\ell_1 m_1}(\theta_1,\varphi_1) Y_{\ell_2 m_2}(\theta_2,\varphi_2) \ . \qquad \text{(H-33)}$$

Since $\mathcal{Y}_{\ell m}(r,\theta,\varphi)$ is clearly a solution of the Laplace equations :

$$\Delta_1\ \mathcal{Y}_{\ell m}(r,\theta,\varphi) = \Delta_2\ \mathcal{Y}_{\ell m}(r,\theta,\varphi) = 0 \qquad ,$$

where Δ_i is the Laplacian operator with respect to the components of M_i, this function can be expressed in terms of products of solid harmonics $\mathcal{Y}_{\ell_1 m_1}(r_1,\theta_1,\varphi_1)$ and $\mathcal{Y}_{\ell_2 m_2}(r_2,\theta_2,\varphi_2)$. The homogeneous character of degree ℓ implies $\ell_1+\ell_2=\ell$ and the dependence upon α_1, and α_2 by $\alpha_1^{\ell_1}\ \alpha_2^{\ell_2}$. Furthermore, in order to preserve the behaviour under the rotation groups ℓ_1 and ℓ_2 must be coupled to ℓ. Finally, one obtains :

$$\mathcal{Y}_{\ell m}(r,\theta,\varphi) = \sum_{\ell_1,\ell_2=0}^{\ell} \delta_{\ell_1+\ell_2,\ell} C_{\ell\ell_1\ell_2}\alpha_1^{\ell_1}\alpha_2^{\ell_2}\langle\ell_1\ell_2 m_1 m_2|\ell m\rangle \mathcal{Y}_{\ell_1 m_1}(r_1,\theta_1,\varphi_1)\mathcal{Y}_{\ell_2 m_2}(r_2,\theta_2,\varphi_2) \ , \qquad \text{(H-34)}$$

where $C_{\ell\ell_1\ell_2}$ is a coefficient to be determined. This can be done by setting $\theta_i=\varphi_i=0$, and therefore $r=\alpha_1 r_1+\alpha_2 r_2$, in the above equation. From Eq. (H–25a) one then gets :

$$\sqrt{4\pi}\ \hat{\ell}(\alpha_1 r_1+\alpha_2 r_2)^\ell = \sum_{\ell_1,\ell_2=0}^{\ell} \delta_{\ell_1+\ell_2,\ell} C_{\ell\ell_1\ell_2}\langle\ell_1\ell_2 00|\ell 0\rangle(\alpha_1 r_1)^{\ell_1}\hat{\ell}_1(\alpha_2 r_2)^{\ell_2}\hat{\ell}_2 \ , \qquad \text{(H-35)}$$

and using the explicit expression for the C.G. coefficient, Eq. (G–3), one thereby finds the announced result.

One should note that the previous method can also be applied to the more general case where :

$$OM = \sum_{i=1}^{n} \alpha_i\ OM_i \qquad , \qquad \text{(H-36)}$$

which allows us to express $r^\ell\ Y_{\ell m}(\theta,\varphi)$ in terms of $\prod_{i=1}^{n}[(\alpha_i r_i)^{\ell_i}\ Y_{\ell_i m_i}(\theta_i\varphi_i)]$.

iv) Generating function. A generating function of $2\ell+1$ linearly independent harmonic polynomials $H_{\ell m}$ can be obtained using the following trick :

$$\Delta(OM.v)^{\ell} = \ell(\ell-1)(v.v)(OM.v)^{\ell-2} \quad , \qquad \text{(H-37)}$$

where the Laplacian operator acts on the coordinates of M, and v is any fixed vector. Then, choosing v as an isotropic vector, i.e. v.v=0, $(OM.v)^{\ell}$ is a harmonic polynomial of degree ℓ in x,y, and z. Hence, with for instance :

$$v = (-1+t^2, \ -i(1+t^2) \ , \ -2t) \quad ,$$

one obtains :

$$[-(x+iy) - 2zt + (x-iy)t^2]^{\ell} = t^{\ell} \sum_{m=-\ell}^{+\ell} H_{\ell m}(x,y,z)\ t^m \quad . \qquad \text{(H-38)}$$

H.4. MULTIPOLE EXPANSIONS

i) Expansion of $f(r_1,r_2,\cos\alpha)$. Since the Legendre polynomials form a complete set of orthogonal polynomials defined on $[-1,+1]$, any integrable function within this domain can be expressed as a linear combination of them, i.e. :

$$\forall x \in [-1,+1] \qquad f(x) = \sum_{\ell=0}^{\infty} (2\ell+1) f_{\ell}\ P_{\ell}(x) \quad . \qquad \text{(H-39)}$$

From the orthogonality relation, Eq. (H-12), the coefficients f_{ℓ} read :

$$f_{\ell} = \frac{1}{2} \int_{-1}^{+1} dx\ P_{\ell}(x)\ f(x) \quad . \qquad \text{(H-40)}$$

Let us apply the above result to any function f of two vectors OM_1 and OM_2, and which depends only on $\|OM_1\| = r_1$, $\|OM_2\| = r_2$, and on $x=\cos\alpha$, where α is the angle between OM_1 and OM_2, i.e. from Eq. (2-2) :

$$\cos\alpha = \frac{OM_1 . OM_2}{r_1 r_2} \quad .$$

Then, using the addition theorem, Eqs. (10-259), $P_{\ell}(\cos\alpha)$ can be expressed in terms of the spherical harmonics of the spherical polar coordinates θ_1, φ_1, and θ_2, φ_2 of, respectively, M_1 and M_2. One thereby finds :

$$f(r_1,r_2,\cos\alpha) = \sum_{\ell=0}^{\infty} (2\ell+1) f_{\ell}(r_1,r_2) P_{\ell}(\cos\alpha) \qquad \text{(H-41a)}$$

$$= 4\pi \sum_{\ell,m} f_{\ell}(r_1,r_2) Y^{*}_{\ell m}(\theta_1,\varphi_1) Y_{\ell m}(\theta_2,\varphi_2) = 4\pi \sum_{\ell} f_{\ell}(r_1,r_2)(-1)^{\ell}\ \hat{\ell}\{Y_{\ell}(\theta_1,\varphi_1) Y_{\ell}(\theta_2,\varphi_2)\}_{0} \qquad \text{(H-41b)}$$

$$= 4\pi \sum_{\ell,m} f_{\ell}(r_1,r_2) Y^{[\ell]*}_{m}(\theta_1,\varphi_1) Y^{[\ell]}_{m}(\theta_2,\varphi_2) = 4\pi c^{-2\ell} \sum_{\ell} f_{\ell}(r_1,r_2)\ \hat{\ell}\{Y^{\ell}(\theta_1,\varphi_1) Y^{\ell}(\theta_2,\varphi_2)\}^{0} \qquad \text{(H-41c)}$$

$$f_{\ell}(r_1,r_2) = \frac{1}{2} \int_{-1}^{+1} dx\ P_{\ell}(x)\ f(r_1,r_2,x) \quad . \qquad \text{(H-41d)}$$

The expressions of f_ℓ are given in Table H.3 for frequently encountered functions.

TABLE H.3

Multipole expansion of some functions. The notations are :
$\|OM_1\| = r_1$, $\|OM_2\| = r_2$, $\|M_1M_2\| = r$, $r_< = \min(r_1,r_2)$, $r_> = \max(r_1,r_2)$.
j_ℓ, y_ℓ, and $h^{(1)}_\ell$ denote as usual, the spherical Bessel functions of the first, the second, and the third kind (cf. Ref. 2 , Chap. 10). The second, the third, and the fifth expressions remain valid for all real or complex values of k. Green's functions of the operators Δ and $\Delta+k^2$ are :

$$\Delta(\frac{1}{r}) = 4\pi\delta_3(r) \ , \quad (\Delta+k^2)\frac{e^{ikr}}{r} = -4\pi\ \delta_3(r) \ , \quad (\Delta+k^2)\frac{\cos kr}{r} = -4\pi\ \delta_3(r) \ .$$

$f(r_1,r_2,\cos\alpha)$	$f_\ell(r_1,r_2)$
$\frac{1}{r}$	$\frac{1}{2\ell+1}\ \frac{r_<^\ell}{r_>^{\ell+1}}$
$\frac{e^{ikr}}{r}$	$ik\ j_\ell(kr_<)\ h^{(1)}_\ell(kr_>)$
$\frac{\cos kr}{r}$	$-k\ j_\ell(kr_<)y_\ell(kr_>)$
$e^{ir_1r_2\cos\alpha}$	$i^\ell\ j_\ell(r_1r_2)$
$e^{-k^2r^2}$	$(-i)^\ell j_\ell(i2k^2r_1r_2)\ e^{-k^2(r_1^2+r_2^2)}$

ii) **Expansion of f(θ , φ).** Since the spherical harmonics form a complete orthonormal basis of $L^2(S^2,d_2\hat{n})$, Eqs. (7–79a,b), any function f in this space reads :

$$f(\theta,\varphi) = \sum_{\ell,m} f_{\ell m}\, Y_{\ell m}(\theta,\varphi) \ , \qquad f_{\ell m} = \int_0^\pi \sin\theta\ d\theta \int_0^{2\pi} d\varphi\ Y^*_{\ell m}(\theta,\varphi) f(\theta,\varphi) \ . \qquad \text{(H-42)}$$

Let us apply the above result to the equation of a regular surface in spherical polar coordinates :

$$\|OM\| = r = f(\theta,\varphi) \qquad .$$

The symmetry properties if any of the surface will imply relations for the coefficients $f_{\ell m}$. Some of them are listed in Table H.4.

As an example, let us consider the case of an ellipsoid in the reference frame determined by its principal axes. This surface has all the symmetries listed in Table H.4, except perhaps the axial symmetry A_3. Therefore ℓ and m are even, and furthermore $f_{\ell m} = f_{\ell -m}$. In Cartesian coordinates the equation of the surface reads :

$$\sum_{i=1}^{3} \left(\frac{x^i}{a_i}\right)^2 = 1 \quad , \tag{H-43}$$

which implies :

$$\frac{1}{r^2} = f_{00}\, Y_{00}(\theta,\varphi) + f_{20}\, Y_{20}(\theta,\varphi) + f_{22}[Y_{22}(\theta,\varphi) + Y_{2-2}(\theta,\varphi)] \quad . \tag{H-44}$$

By identification one finds :

$$\frac{1}{r^2} = \frac{1}{3}\left(\frac{1}{a_1^2} + \frac{1}{a_2^2} + \frac{1}{a_3^2}\right) + \frac{2}{3}\sqrt{\frac{\pi}{5}}\left(-\frac{1}{a_1^2} - \frac{1}{a_2^2} + \frac{2}{a_3^2}\right) Y_{20}(\theta,\varphi)$$
$$+ \sqrt{\frac{2\pi}{15}}\left(\frac{1}{a_1^2} - \frac{1}{a_2^2}\right)\left[Y_{22}(\theta,\varphi) + Y_{2-2}(\theta,\varphi)\right] \quad . \tag{H-45}$$

As a special case, for an ellipsoid infinitely close to a sphere, i.e. with :

$$a_i = a(1+\varepsilon_i) \quad , \tag{H-46}$$

one obtains up to first order in ε_i :

$$r = a\left\{1 + \frac{1}{3}\ (\varepsilon_1+\varepsilon_2+\varepsilon_3) + \frac{2}{3}\sqrt{\frac{\pi}{5}}\ (-\varepsilon_1-\varepsilon_2+2\varepsilon_3)\ Y_{20}(\theta,\varphi)\right.$$
$$\left. + \sqrt{\frac{2\pi}{15}}\ (\varepsilon_1-\varepsilon_2)[Y_{22}(\theta,\varphi) + Y_{2-2}(\theta,\varphi)]\right\} \quad . \tag{H-47}$$

Actually, for such a study instead of characterizing the ellipsoid by the a_i it is more convenient to introduce the parameters a, β , and γ such that :

$$\frac{1}{a_i^2} = \frac{1}{a^2}\left[1 - \sqrt{\frac{5}{\pi}}\ \ \beta \cos\ (\gamma - \frac{2\pi}{3}\, i)\right] \quad , \tag{H-48}$$

where by permutations of the axes and change of sign of β the variations of γ can be restricted to $[0, \frac{\pi}{3}]$. The equation of the ellipsoid thereby takes the form :

$$\frac{1}{r^2} = \frac{1}{a^2}\left\{1 - 2\beta\cos\gamma\ Y_{20}(\theta,\varphi) - \sqrt{2}\beta\sin\gamma[Y_{22}(\theta,\varphi) + Y_{2-2}(\theta,\varphi)]\right\} \quad . \tag{H-49}$$

Then, $|\beta|$ characterizes the amplitude of the deviation from a sphere of radius a (which corresponds to $\beta=0$), while γ and the sign of β determine the shape of the ellipsoid. Indeed, as an example for $\beta < 0$, if $\gamma = 0$ the ellipsoid is prolate with the symmetry axis e_3 while it is oblate with the symmetry axis e_2 for $\gamma = \frac{\pi}{3}$. For small deformations, one has up to first order in β :

$$a_i = a\left[1 + \sqrt{\frac{5}{4\pi}}\ \ \beta \cos\ (\gamma - \frac{2\pi}{3}\, i)\right] \quad , \tag{H-50a}$$

$$r = a\left\{1 + \beta\cos\gamma\ Y_{20}(\theta,\varphi) + \frac{1}{\sqrt{2}}\,\beta\ \sin\gamma\ \ [Y_{22}(\theta,\varphi) + Y_{2-2}(\theta,\varphi)]\right\} \quad . \tag{H-50b}$$

Such ellipsoidal forms are encountered for instance in the study of molecule or nucleus shape (cf. the rigid rotator in Quantum Mechanics, Sec. 11.4).

TABLE H.4

We consider a right-handed orthonormal reference frame (O,e_i). P denotes the symmetry with respect to the origin or parity, A_3 is the axial symmetry around the axis e_3, $U_i=R(\pi,e_i)$ is the rotation of π around the axis e_i $(U_1=U_2U_3)$, and $P_{ij}=U_kP$ ($k\neq i$ and j) is the symmetry with respect to the plane determined by e_i and e_j. One reads in the second column the result of the symmetry operator on $Y_{\ell m}$, and in the third column the relation which $f_{\ell m}$ satisfies.

Symmetries	$Y_{\ell m}$	$f_{\ell m} =$
P	$(-1)^{\ell}\, Y_{\ell m}$	0 if ℓ odd
A_3	$e^{-im\theta}\, Y_{\ell m}\ ,\ \forall\theta$	0 if $m \neq 0$
U_1	$(-1)^{\ell}\, Y_{\ell -m}$	$(-1)^{\ell}\, f_{\ell -m}$
U_2	$(-1)^{\ell+m}\, Y_{\ell -m}$	$(-1)^{\ell+m}\, f_{\ell -m}$
U_3	$(-1)^{m}\, Y_{\ell m}$	0 if m odd
P_{12}	$(-1)^{\ell+m}\, Y_{\ell m}$	0 if $\ell+m$ odd
P_{13}	$(-1)^{m}\, Y_{\ell -m}$	$(-1)^{m}\, f_{\ell -m}$
P_{23}	$Y_{\ell -m}$	$f_{\ell -m}$

H.5. VECTOR SPHERICAL HARMONICS

The definition and the properties of the vector spherical harmonics $\mathbf{Y}_m^{\ell'[\ell]}$ are given in Subsec. 10.3.4.2,iii). Let us now compare our notation with that of some others :

$(ic)^{\ell+1}\,\mathbf{Y}_m^{\ell'[\ell]} = \Phi_{\ell';\ell m}$ - Bohr and Mottelson[8], Eqs. (1A-122,124)

$= \mathbf{Y}_{\ell\ell' m}$ - Edmonds[22], Eqs. (5.9.4,10)

$= \mathbf{T}_{\ell\ell' m}$ - Rose[54], Eqs. (5.44,57, and 60)

$= \mathbf{Y}_m^{\ell(\ell' 1)}$ - de Shalit and Talmi[56], Eqs. (16.17,18).

Hence, for all the authors quoted above one has :

$$c = i \quad .$$

REFERENCES

Let us insist that the following list of references is not intended to be, in any sense, an exhaustive bibliography about group theory and rotations. It quotes only those sources which we have ourselves found helpful and which we recommend to the reader for further information.

1. ABRAHAM, R., "Foundations of mechanics", W.A. Benjamin, Inc. New York, Amsterdam (1967).
2. ABRAMOWITZ, M. and STEGUN, I.A., "Handbook of mathematical functions", Dover Publications, Inc., New York (1970).
3. BAKER, H.F., Proc. London Math. Soc. (1) 34 (1902), 347 ; (1) 35 (1903), 333 ; (2) (1904), 293.
4. BARGMANN, V., Ann.of Math. 59 (1954), 1.
5. BARGMANN, V., Rev. of Mod. Phys. 34 (1962), 829.
6. BIEDENHARN, L.C. and VAN DAM, H., "Quantum theory of angular momentum", Academic Press New York and London (1965).
7. BLATT, J.M. and WEISSKOPF, V.F., "Theoretical nuclear physics", John Wiley and sons New York, Chapman and Hall, Limited, London (1952).
8. BOHR, A. and MOTTELSON, B.R., "Nuclear structure", Volume I, W.A. Benjamin, Inc. New York, Amsterdam (1969).
9. BOHR, A. and MOTTELSON, B.R., "Nuclear structure", Volume II, W.A. Benjamin, Inc. (1975).
10. BOURBAKI, N., "Eléments de mathématique, Fascicule XXVI, groupes et algèbres de Lie", Hermann Paris (1960).
11. BUDRIDE, S.D. and GUTMAN, A.M., Trudy Akademii Nauk Litovskoi SSR, B4 (1959), 11 (in Russian).
12. CAMPBELL, J.E., Proc. London Math.Soc. 28 (1897), 381.
13. CARTAN, E., "Sur la structure des groupes de transformations finis et continus", thèse, Paris (1894), "Oeuvres complètes", Gauthier–Villars, Editeur Paris (1952).
14. CARTAN, E., "La théorie des groupes finis et continus et la géométrie différentielle traitées par la méthode du repère mobile", Gauthier–Villars, Editeur Paris (1951).
15. CARTIER, P., private communication, I.H.E.S. France (1979).
16. CHEVALLEY, C., "Theory of Lie groups : I", Princeton University Press, Princeton (1946).
17. CHEVALLEY, C., "Théorie des groupes de Lie : Tome III, théorèmes généraux sur les algèbres de Lie", Hermann et Cie. Editeurs Paris (1955).
18. COESTER, F. and JAUCH, J.M., Helv. Phys. Acta. 26 (1953), 3.
19. CONDON, E.U. and SHORTLEY, G.H., "The theory of atomic spectra", Cambridge at the University Press (1977).
20. DANOS, M., Ann. Phys. 63 (1971), 319.

21. DYSON, F.J., J. Math. Phys. 3 (1962), 1199.

22. EDMONDS, A.R., "Angular momentum in quantum mechanics", Princeton University Press Rev. Ed. (1968).

23. EL–BAZ, E., "Traitement graphique de l'algèbre des moments angulaires", Masson et Cie, Editeurs Paris (1969).

24. FANO, U. and RACAH, G., "Irreducible tensorial sets", Academic Press Inc., Publishers New York (1959).

25. FROBENIUS, G., J. reine u. angew. Math. 84 (1878), 59 ; cf. DICKSON, L.E., "Linear algebras", Cambridge University Press, New York (1914) p.10.

26. FROBENIUS, G. and SCHUR, I., Proc. Berlin Acad. (1906), 186.

27. GOLDSTEIN, H., "Classical mechanics", Addison–Wesley Publishing Company, Inc. (1959).

28. GRADSHTEYN, I.S. and RYZHIK, I.M., "Table of integrals, series, and products", Academic Press, New York San Fransisco London (1965).

29. HAUSDORFF, F., Ber. Sachsischen Akad. Wiss. (Math. Phys. Klasse) Leipzig 58 (1906), 19.

30. HUREVITSCH, A., "Unitary representation in Hilbert space of compact topological group", Mat. Sb. 13 (55) (1943), 79–86.MR 6, 147.

31. JACOB, M. and WICK, G.C., Ann. of Phys. 7 (1959), 404.
JACOB, M., thesis, Paris (1961).

32. JAHN, H.A. and HOPE, J., Phys. Rev. 93 (1954), 318.

33. LEVINSON, I.B., Trudy Fiziko–tekhnicheskogo instituta Akademii Nauk Litovskoi SSR 2 (1957), 17–ibid 2 (1957), 31. Trudy Akademī Nauk Litovskoi SSR, B4 (1957), (in Russian).

34. LEVINSON, I.B. and CHIPLIS, I.V., Trudy Akademii Nauk Litovskoi SSR, B1 (1958), 3 (in Russian).

35. LICHNEROWICZ, A., "Théorie globale des connexions et groupes d'holonomie", Dunod Paris (1955).

36. LITTLEWOOD, D.E., "The theory of group characters", Oxford University Press, Oxford (1950).

37. MAHOUX, G., "Représentations physiques du groupe de Poincaré", cours à la Faculté d'Orsay (1972–1973), unpublished.

38. MEHTA, M.L., "Elements of matrix theory", Hindustan Publishing Corporation (I) Delhi–11 0007 (1977).

39. MESSIAH, A., "Quantum mechanics", Volumes I and II, North–Holland publishing company, Amsterdam (1962).

40. MOSHINSKY, M., Nucl. Phys. 13 (1959), 104.

41. MOUSSA, P. and STORA, R., lectures given at Herceg Novi summer school (1966), published in "Methods in subnuclear physics" Vol.II, edited by M. Nicolić, Gordon and Breach, pp 265–270.

42. MURNAGHAN, F.D., "The theory of group representations", Baltimore, the Johns Hopkins press (1938).

43. MURNAGHAN, F.D., "The unitary and rotation groups", Spartan books, Washington, D.C. (1962).

44. PONTRYAGIN, L.S., "Topological groups", Gordon and Breach, Science Publishers, Inc., Sec. Ed. (1966).

45. RACAH, G., "Group theory and spectroscopy", CERN 61–8. Reprint of lectures delivered at the Institute for Advanced Study, Princeton (1951).

46. RACAH, G., Phys. Rev. 62 (1942), 438.

47. RAYNAL, J., Nucl. Phys. A97 (1967), 572.

48. RAYNAL, J., Le Journal de Physique 31 (1970), 3.

49. RAYNAL, J., J. Math. Phys. 19 (1978), 467.

50. RAYNAL, J., J. Math. Phys. 20 (1979), 2398.

51. REGGE, T., Nuovo Cimento 10 (1958), 544.

52. REGGE, T., Nuovo Cimento 11 (1959), 116.

53. REGGE, T., Nuovo Cimento 14 (1959), 951.

54. ROSE, M.E., "Elementary theory of angular momentum", New York, John Wiley and sons, Inc., London, Chapman and Hall, Ltd. (1957).

55. SCHWINGER, J., "On angular momentum", U.S. Atomic Energy Commission, NYO–3071, (1952), reprinted in Ref. 6.

56. de SHALIT, A. and TALMI, I., "Nuclear shell theory", Academic Press, New York and London (1963).

57. SLATER, L.J., "Generalized hypergeometric functions", Cambridge at the University Press (1966).

58. VAN WINTER, C., Physica 20 (1954), 274.

59. VARSHALOVICH, D.A., MOSKALIEV, A.N., and KHERSONSKII, V.K., "Quantum theory of angular momenta", Leningrad (1975) (in russian).

60. VELTMAN, M., "Introduction to the theory of group representations", Ecole d'été de Gif sur Yvette France (1969), IN2P3 Orsay, unpublished.

61. VILENKIN, N Ja., "Fonctions speciales et théorie de la représentation des groupes", Dunod Paris (1969).

62. WEYL, H., "The theory of groups and quantum mechanics", London (1931).

63. WIGNER, E.P., "Group theory and its applications to the quantum mechanics of atomic spectra", Academic Press New York and London (1959).

64. YUTSIS, A.P., LEVINSON, I.B., and VANAGAS, V.V., "Mathematical apparatus of the theory of angular momentum", Israel program for scientific translations, Jerusalem (1962).

INDEX

This list is at the same time a subject index and an author index of the quoted references (in bold–faced type). When a word or an expression is defined it is underlined in the text of the book.

F

G

H

T

U

V

W